AF566947

Dislocation Modelling of Physical Systems

Proceedings of the International Conference

Gainesville, Florida, USA
June 22-27, 1980

Dislocation Modelling of Physical Systems

Proceedings of the International Conference

Gainesville, Florida, USA
June 22-27, 1980

Edited by
M.F. ASHBY
University of Cambridge, England
R. BULLOUGH
AERE Harwell, England
C.S. HARTLEY
University of Florida, USA
J.P. HIRTH
Ohio State University, USA

Pergamon Press
Oxford New York Toronto Sydney Paris Frankfurt

Pergamon Press Offices:

U.S.A.	Pergamon Press Inc., Maxwell House, Fairview Park, Elmsford, New York 10523, U.S.A.
U.K.	Pergamon Press Ltd., Headington Hill Hall, Oxford OX3 0BW, England
CANADA	Pergamon Press Canada Ltd., Suite 104, 150 Consumers Road, Willowdale, Ontario M2J 1P9, Canada
AUSTRALIA	Pergamon Press (Aust.) Pty. Ltd., P.O. Box 544, Potts Point, NSW 2011, Australia
FRANCE	Pergamon Press SARL, 24 rue des Ecoles, 75240 Paris, Cedex 05, France
FEDERAL REPUBLIC OF GERMANY	Pergamon Press GmbH, Hammerweg 6, Postfach 1305, 6242 Kronberg/Taunus, Federal Republic of Germany

British Library Cataloguing in Publication Data

Dislocation modelling of physical systems.
1. Dislocations in solids - Mathematical models - Congresses
I. Ashby, M F
531'.3825 QC191 80-41473

ISBN 0-08-026724-6

Printed in the United States of America

Foreword

Conferences devoted to limited topical areas provide convenient benchmarks that measure progress in the fields of study covered. By this standard, initial advances in understanding the relation of dislocations to physical properties occurred in the realm of solid state physics. In recent years the growing need for transferring fundamental concepts to engineering applications has stimulated much research in areas described generally as "modelling". This term refers to studies in which mathematical descriptions of atomic scale events lead to the interpretation of material parameters appearing in phenomenological equations describing the behavior of an ensemble of such events. Relating such material parameters - properties - to the nature of the governing events and their spatial and temporal distributions - structure - defines one of the most challenging and critical areas of research in materials science.

The immediate predecessor of this conference, held at the National Bureau of Standards in April, 1969, focussed attention on fundamental aspects of Dislocation Theory. While not ignoring the need for continuing advances in fundamental knowledge, the Organizing Committee of this Conference desired to emphasize the growing interest in relating dislocation theoretic concepts to engineering problems. This approach embraces a spectrum of disciplines and interest infrequently collected at a single conference. Accordingly, topic areas chosen for the program ranged from the fundamental, such as Properties of Single Dislocations, to the more applied, such as Fracture. Fracture and Dislocation-Point Defect Interactions illustrate classes of material behavior for which dislocation theory has provided a significant quantitative link between macroscopic and microscopic phenomena, while the relation of dislocation theoretic concepts to constitutive equations employed in descriptions of multiaxial deformation has proved more elusive. By focussing attention on the more and less successful attempts to enlist quantitative dislocation theory in the attack on problems which beset materials scientists, the Organizing Committee hopes to stimulate further research of this nature and to reveal critical areas for continuing fundamental study.

The Committee wishes to thank both the Division of Materials Research at the National Science Foundation for financial assistance with the Conference and the Metallurgy Program Director, Dr. R. J. Reynik, for his encouragement. Personnel of the University of Florida Division of Continuing Education rendered expert and timely administrative assistance not only with the local arrangements but also with the assembling of manuscripts for publication. The Graduate School and the College of Engineering of the University of Florida provided additional support necessary for the organizing of the Conference. Finally, the cooperation of the speakers and audience in conforming to an exhausting schedule of technical sessions and rigid deadlines for manuscript submission contributed most significantly to the success of the program.

For the Organizing Committee

C. S. Hartley
Stony Brook, New York
October, 1980

CONTENTS

RELATIONS BETWEEN DISLOCATION MECHANISMS AND FRACTURE TOUGHNESS

G. T. HAHN and W. F. FLANAGAN

Dept. of Mechanical and Materials Engineering
Vanderbilt University
Nashville, Tennessee

Abstract

This paper offers a review of dislocation concepts and models that enter into the analyses of the fracture toughness of crystalline materials. The parameters used to describe the toughness, i.e. K_{IC}, G_{IC}, J_{IC}, and δ_{IC} and their connections with the local stress and strain fields are introduced. The toughness of single phase crystals, and systems with brittle second phase particles in which cracks extend by cleavage are examined. The paper also deals with the toughness derived from the fibrous, void-nucleation-and-growth mechanism of crack extension for large (> 1μm), intermediate (0.01μm < d < 1μm), and small (d < 0.01μm) diameter particles Dislocation "unit" processes whose modeling enters into analyses of toughness are identified. These include: (i) the generation of dislocations at crack tips, particle-matrix interfaces, and voids; (ii) the movements of dislocations amid the "secondary" plastic zones of these entities, through small particles and within coarse slip bands; and (iii) the dislocation mechanisms of crack and void nucleation at interfaces and in heavily strained regions.

1. Introduction

Fracture toughness -- the resistance to crack extension -- is intimately connected with the amount and distribution of and the resistance to plastic deformation in the locale of the crack tip. For some facets of the problem, a macroscopic, continuum view of plastic deformation is appropriate. For others, analyses of the deformation in terms of dislocations is essential. In both cases, the continuum mechanics of crack-tip deformation provides the framework for analysis and the basic strain field descriptors K_{IC}, G_{IC}, J_{IC} and δ_{IC}, which are used to quantify this material property.

This paper examines contributions of dislocation concepts to the analyses of fracture toughness. The review is confined to the cleavage and fibrous, void nucleation and growth mode of crack extension under plane strain conditions; cyclic loading and environmental effects -- though important -- are not considered. The paper identifies dislocation mechanisms whose modeling enters into the analysis of toughness. These include: (i) the generation of dislocation at crack tips and prismatic glide loops at particle-matrix interfaces and voids, (ii) the movement of dislocations amid the "secondary" plastic zones of these entities, as well as through small particles and within coarse slip bands, and (iii) the dislocation mechanisms contributing to the nucleation of cracks and voids at interfaces and heavily strained regions. The void-nucleation-and-growth mechanisms attending large (d > 1μm), intermediate (0.01μm < d < 1μm) and small (d < 0.01μm) diameter particles are differentiated. The processes considered underly the toughness of single crystals, structural steels, and high strength precipitation and dispersion strenghened alloys. The paper examines existing treatments and identifies topics for further research.

2. Analysis of Fracture Toughness

The fracture toughness of a material is expressed by one of several descriptors of the crack tip strain field evaluated at the onset of crack extension. For linear elastic materials, G_{IC} and K_{IC}, the critical strain energy release rate and stress intensity parameter, are useful measures of fracture toughness (1). For non-linear elastic or elastic-plastic materials, whose stress-strain curve approximates power law hardening*, the critical value of the path independent integral, J_{IC}, or the critical crack opening displacement, δ_{IC}, are appropriate (2-5). These interrelated quantities** define the load and crack length at the onset of crack extension and the corresponding stresses and strains near the crack tip***. A more detailed discussion of these parameters is presented by Willis (6) in this volume. Their use in modeling crack extension is complicated by the fact that neither the K-field (Eqn. 4) nor the J-field (Eqn. 5) are reliable in the "process zone", the region ajacent to the crack tip where relevant dislocation and rupture processes take place. For elastic materials, evidence of large departures from the continuum solution within $\sim$10 atomic spacings of the crack tip is mentioned in the next section. Depatures from the J-field for plastic material are apparent within $\sim 2\delta_{IC}$ or (10μm - 1000μm) of the crack tip, and this is illustrated in Fig. 1 taken from the work of McMeeking (7) and Rice (8). Computer simulations of the lattice (9) and finite element models of the plastic zone (7) are able to evaluate stresses, strains, and displacements in the process zone, but more convenient generalizations of the numerical findings are needed.

3. Toughness of Single Phase Crystals

Energy conservation and Irwin's derivation of the work performed by an extending crack tip (1) provide the basis for equating the G_{IC}-value for a brittle material with the work of producing the crack surfaces. The Griffith relation (10):

$$G_{IC} = 2\gamma \qquad (6)$$

where $\gamma \simeq \frac{Eb}{30} \sim 1\ Jm^{-2}$ is the specific surface energy (11), is correct for an amorphous material when crack extension proceeds by the rupture of bonds directly ahead of the crack and no other work is done, but is approximate for crystalline cleavage****. An expression for the local

---

* $\frac{\varepsilon}{\varepsilon_o} = \left(\frac{\sigma}{\sigma_o}\right)^n$, where ε_o and σ_o are the reference stress and strain, and n is the hardening exponent; σ_o can also be viewed as the yield stress.

** $K_{IC} = \left[\frac{G_{IC}\,E}{1-\nu^2}\right]^{\frac{1}{2}}$ ......(1), where E is Young's modulus, ν Poisson's ratio, $K_{IC} \simeq \sigma\sqrt{\pi\alpha}$, where σ is the stress at failure and α is the crack length.

$J_{IC} = G_{IC}$ ......(2) and $\delta_{IC} = \frac{J_{IC}}{\sigma_o} D(n,\varepsilon_o)$......(3) where D = 0.78 for non hardening materials (5).

*** Continuum solutions of the stress fields are given by the following expression for linear elastic behavior:

$\sigma_{ij} = \left[\frac{G_{IC}\,E}{2\pi r(1-\nu^2)}\right]^{\frac{1}{2}} \tilde{f}_{ij}(\Theta)$ ......(4) and non linear elastic or elastic plastic behavior:

$\sigma_{ij} = \sigma_o \left[\frac{J_{IC}}{\sigma_o \varepsilon_o I_n r}\right]^{\frac{1}{n+1}} \tilde{f}_{ij}(\Theta, n)$ ...... (5) where r is the radial distance from the crack tip, f_{ij} and $\tilde{f}_{ij}$ are non singular functions of angular position, and I_n is an integration constant.

****Eqn. 6 is approximate for crystalline cleavage because the work of extending the crack a small increment on the atomic scale has a periodicity reflecting the regular spacing of the atomic bonds. As noted by Hsieh and Thomson (12), the <u>average</u> value of the work is exactly 2γ, but G_{IC} must be equated with the maximum value which is variously estimated to be 10%-400% larger than 2γ (9,12-14). A non crystallographic crack front or defects in the crystal can be expected to "smear" the highs and lows and bring G_{IC} closer to 2γ.

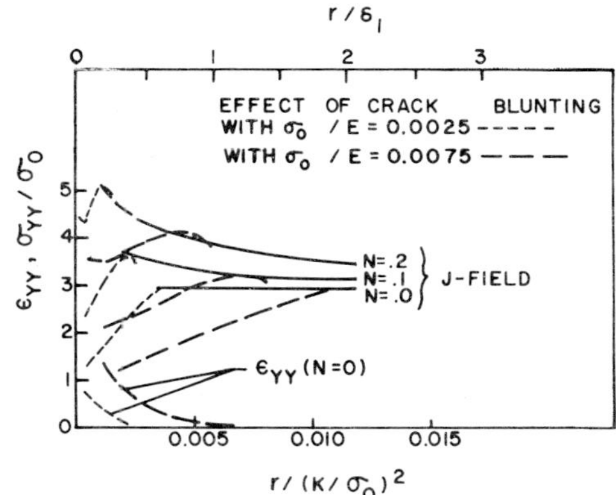

Fig. 1. Distribution of tensile stress (σ_{YY}) and tensile strain (ε_{YY}) in the plastic zone directly in front of the crack tip (θ=0) given by the J-field (Eqn.5) and numberical calculations by McMeeking (7,8,45).

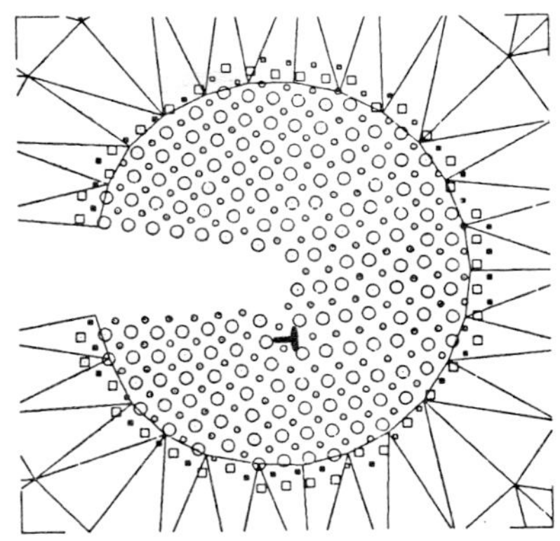

Fig. 2. Computer simulation of the atom positions at the tip of a (001) [010] crack in an α-iron lattice at a stress intensity level corresponding with G_I=4γ after Mullins and Dokainish (14). The atom positions are shown after 10^{-12} sec. at a simulated temperature of 400°K and reveal a dislocation just below the crack tip.

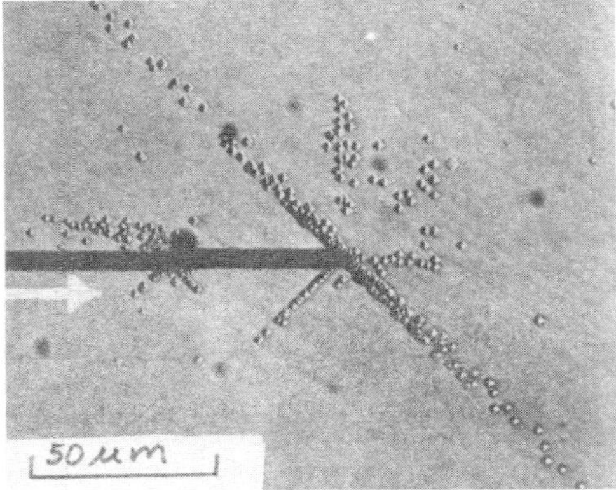

3a

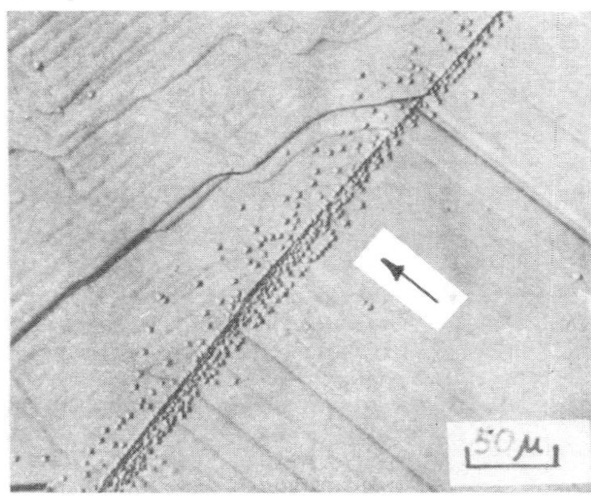

3b

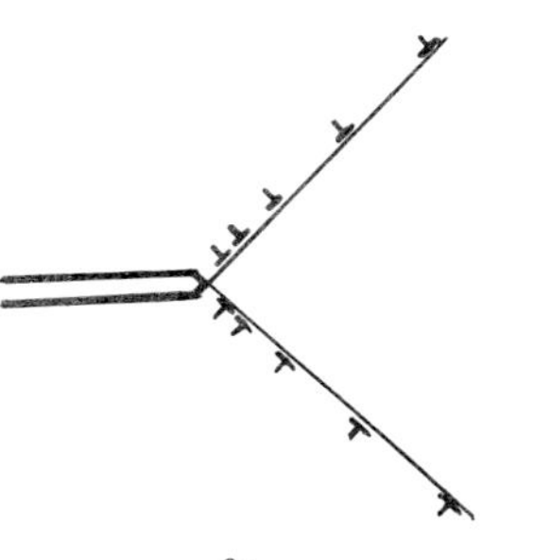

3c

Fig. 3. Dislocation arrays at the tip of a crack: (a) Dislocations (etch pits) on the surface normal to the fracture plane at the tip of an arrested crack in a LiF crystal after Gilman (16), (b) fracture surface of (a) after reinitiation, and (c) Bilby and Swinden (18) model. The line of etch pits in (b) corresponds with the position of the arrested crack front in (a). The absence of pits elsewhere indicates hat no dislocations are generated while the crack in propagating.

stress at the onset of crack extension is obtained by combining Eqn. (4) and (6):

$$\sigma_{ij} \text{ (at the onset of crack extension)} \simeq \left[\frac{E\gamma}{\pi r(1-\nu^2)}\right]^{\frac{1}{2}} f_{ij}\ (\Theta) \quad (7)$$

The very high stresses, $\sigma_{ij} \sim \frac{E}{10}$, obtained within a Burgers vector distance of the crack tip ($r < b$), show that the generation of dislocations prior to crack extension is a distant possibility. The fracture toughness then depends on n, the number of dislocations generated per repeat distance β:

$$G_{IC} \simeq 2\gamma + \frac{nQ}{\beta} \sim 0.4\ nEb \quad (8)$$

where Q is the line energy and $Q/\beta \sim 20\ Jm^{-2}$. A similar expression is obtained with Eqn. 3 by relating the crack opening displacement to the number of dislocations generated, $\delta_{IC} = nb$, and σ_o with the local stress for generation, i.e. $\frac{E}{10}$:

$$G_{IC} \simeq 1.5\ \delta_c \sigma_o \sim 0.2\ nEb \quad (9)$$

These expressions indicate a marked increase in toughness with the generation of even 1 or 2 dislocations at the crack tip.

Rice and Thompson (15) have produced a detailed analysis of the competition between dislocation generation and bond rupture at the crack tip. Their analysis is based on linear elastic descriptions of 3 forces that act on the dislocation and depend on its position. One is the force produced by the stress field of the crack (Eqn. 7), which is the driving force for generation. The other two are the image force and surface ledge force which dominate at small distances and oppose generation. Rice and Thomson postulate that generation can proceed when $\xi_c < \xi_o$, where ξ_c is the distance from the crack tip to the saddle point position, and ξ_o, is the dislocation core cut-off. The calculated values of ξ_c in Table 1 illustrate that the criterion accounts for the toughness of a large number of single phase materials at $0°K$. Rice and Thomson also formulate the energy barrier to the generation of a loops and estimate, U^o_{ACT}, the activation energy values (Table 1) that define $\dot{n}_g$, the rate of activation at finite temperatures: $\dot{n}_g = f(\exp U^o_{ACT}/RT)$. Their approach divides crystals into 2 categories:

(i) <u>Inherently tough crystals</u> like Ag, Al, and Ni, in which generation proceeds without activation, and

(ii) <u>Cleavage-prone crystals</u> like Fe, W, and LiF, which are equally tough at temperatures where $\dot{n}_g > \dot{n}_I$, where $\dot{n}_I$ is the value prescribed by the applied rate of loading**, but are brittle at lower temperatures.

---

* These include f_σ the force of the stress field of the crack, f_i, the image and f_1, the vector force

$$f_\sigma = \left[\frac{E\gamma b}{4\pi(1-\nu^2)\xi}\right]^{\frac{1}{2}} \sin\phi\ \cos\frac{\phi}{2}\ \cos\Psi \quad .....(10) \qquad f_i = -\frac{Eb(1-\nu\ \sin^2\Psi)}{8\pi(1-\nu^2)\xi} \quad ....(11)$$

$$f_1 = \frac{2}{\pi}\frac{\gamma\alpha\ \cos\Psi\ \sin\Psi}{\xi^2 + \alpha^2} \quad .....(12)$$

where $\xi = 4/b$, r is the distance to the crack tip, b is the Burgers vector, ϕ is the angle between the crack and the slip plane (corresponding to θ in Fig. 5), $b_e = b\cos\Psi$ is the edge component of the Burgers vector, $\alpha = e^{3/2}\ (\xi_o/2)$ and, ξ_o is the dislocation core cut-off.

** An estimate of $\dot{n}_I = \dot{\delta}_I/b$ which will keep the local stress from rising can be obtained by differentiating Eqn. 9:

$$\dot{n}_I \simeq \frac{n}{A}^{\frac{1}{2}}\ \dot{K}_I \quad .....(13)$$

where $\dot{K} = \dot{\sigma}\sqrt{\pi a}$ $\dot{\sigma}$ is the rate of stress application to the cracked body, n is the number of dislocations generated and $A \simeq \frac{0.04\ bE\sigma_e}{(1-\nu^2)}$ and σ_e is an effective value of stress reflecting the high stress to generate and a lower stress to move dislocations.

Table 1: EXAMPLES OF CRACK TIP DISLOCATION PARAMETERS CALCULATED BY RICE AND THOMSON (15)

Crystal	LOW TEMPERATURE TOUGHNESS	CORE CUT-OFF, ξ_o (units of b)		CRITICAL DISTANCE ξ_c (units of b)	LOOP ACTIVATION ENERGY, eV
Ag	TOUGH (G_{IC}>>2)	2	>	1.09	——
Al	" "	2	>	1.4	——
Ni	" "	2	>	1.7	——
Fe	BRITTLE(G_{IC}~2)	1.5	<	1.9	2.2
W	" "	1.5	<	4.0	329
LiF	" "	0.25	<	3.2	58
Zn	" "	1.5	<	4.3	107
Si	" "	0.25	<	2.2	111
C	" "	0.25	<	2.4	351

Although the Rice and Thompson analysis is conceptually satisfying, its results are suspect because the linear elastic continuum relations employed, e.g. Eqn. 7, are not reliable in the heavily strained region 1-2 atomic spacings from the crack tip or in the proximity of the free surface. As an alternative, a number of workers (9,12-14) have treated the crack tip region as a discrete array of atoms. These "computer simulations" involve crystalline arrays of ∿500 atoms that interact according to an atomic potential and are imbedded in and coupled with a model of the continuum. The recent results of Mullins and Dokainish (14) for an (001) (010) crack in a -iron lattice imbedded in a finite element mesh as reporduced in Fig. 2. The computations reveal large departures from the linear elastic continuum solution. Mullins and Dokaninish express the departure in terms of Δ, the ratio of the non linear to linear component of displacement. There calculations show that the <u>average</u> value of this ratio is Δ=0.25-0.35 for atoms in a 15° wedge extending 2nm (∿7 atomic spacings) from the crack tip and inclined at 45° to the crack plane. The large discrepancy between the experimental activation energy reported by St. John (22) for silicon: U°_{ACT} = 1.9 eV, and the theorectical value in Table 1, is a sign the departures have a large effect on the forces entering the Rice and Thomson analysis. Figure 2 illustrates a dislocation emitted spontaneously by a model that was subjected to a stress intensity 2x the Griffith K-level (i.e. G_I = 8.5 γ) for 10^{-12} sec. Its position implies a dislocation velocity of ∿500 ms^{-1} assuming it was nucleated at the start of the simulation. While it is not clear that the dislocation is necessarily stationary after 10-12 sec., Mullins and Dokainish suggest that the boundary of the model is not flexible enough with the current mesh to allow the dislocation to move farther. The inflexibility of boundaries and the essentially 2-dimensional nature of the model are problems common to the atomic simulation approach that stand in the way of refining the analysis of thermally activated generation.

4. Cleavage-Prone Crystals and the Transistion from Brittle to Tough Behavior

Repeated generation produces an array of dislocations on a slip plane emanating from the crack tip (16) as illustrated in Fig. 3b. Bilby, Cottrell and Swinden (17) and Bilby and Swinden (18) have calculated the dislocation positions in such an array. Their models, (Fig. 3c) which assure the stress to generate the dislocation at the crack tip is no greater than the stress to move the dislocation through the lattice, indicate a reverse pile-up-like arrangement with the dislocations more crowded close to the crack tip. The relatively high stresses for generation indicated by the Rice and Thomson analysis reduce the crowding near the tip and contribute to a "dislocation free" zone like the one discussed more fully by Chang and Ohr (19) in this volume.

The transistion from brittle to tough behavior of crystals occurs at the temperature where:

$$\dot{n}_g = \dot{n}_I \tag{14}$$

when loop generation is rate controlling. Alternatively, St. John's findings for silicon (22) indicate the transition temperature may also be determined by the kinetics of dislocation glide. This case is obtained when the resistance to movement dominates the term σ_e in Eqn. 13 and can be treated by inserting a velocity dependent friction stress in the Bilby Swinden model.

The redistribution of stress accompanying slip favors the activation of sources remote from the tip on parallel and/or orthogonol slip systems as indicated by the etch pit pattern in Fig. 4. In this way the generation of loops at a free surface is replaced by the operation of "easy" sources in the interior and the deformation pattern gradually approaches the continuum elastic-plastic solution. This means that the maximum normal stress, which may be $\sim E/10$ at the root of the crack when the first dislocation is generated, ultimately decreases to the level of $3\sigma_o$ to $5\sigma_o$* as indicated in Fig. 1, typically a $\sim$10- fold reduction. It is not clear whether the maximum normal stress necessarily decreases continuously with dislocation generation, or first increases, making bond rupture more likely. Edwards and Eyre (20) discuss this question in this volume. Since a decreasing normal stress precludes bond rupture, once the 2 rates $\dot{n}_g$ and $\dot{n}_I$ match, it implies a very abrupt transition. However, this rate is mitigated by the fact that the value of $\dot{n}_I$ needed to keep the local stress from rising (Eqn.13) increases with n for a constant rate of loading. Unless the loading rate diminishes with time, the local stress must rise to assist thermal activation to reach the increasing value of $\dot{n}_I$. This effect favors a gradual rather than an abrupt transition from brittle to ductile behavior as the temperature is increased. (see Figs. 5(a) and 5 (b).

Eqns. 13 and 14 illustrate that the temperature range of brittle-to-tough transition depends on the crack tip loading rate $\dot{K}$ (23). Since the effective loading rate of a rapidly advancing crack is very high**, dislocation generation will diminish or even cease when the crack begins to extend. This effect is illustrated in Fig. 3 for LiF. Burns and Webb (25) also show that energy dissipating movements of pre-existing loops cease ahead of a rapidly advancing crack in LiF. Tetelman and Robertson (26) assume $G_{ID} \simeq 2$; i.e. no dislocations are generated while a crack propagates and arrests in Fe-3Si, where G_{ID} is the toughness of a propagating rather than a stationary crack. However, these workers also devise an alternative method of evaluation G_{ID}, which, in retrospect, appears to be more reliable than their assumption***. This indicates that $G_{ID} \simeq 43\ Jm^{-2}$, and substatially larger than 2γ. River markings (steps in the cleavage plane) make up part of the difference, but the result suggests that some dislocations may have been generated by the moving crack. The analysis of this problem is discussed more fully by Vitek and Chell (27) in this volume.

5. Toughness of Cleavage-Prone Polycrystalline Materials with Brittle Particles

The studies of McMahon and Cohen (28) and others (29,30) provide direct evidence that the brittle fracture of steel proceeds by a mechanism involving a slip band blocked by a carbide as illustrated in Fig. 6. The fracture of the relatively brittle carbide particle subjects the adjoining ferrite grain to an advancing crack, which, for the reasons noted above, reduces the toughness of the ferrite. The cleavage stress σ_c -- the tensile stress needed for the crack to penetrate the ferrite -- has been analyzed by Smith (31) and by Almond, Timbers and Enbury (32) by treating the blocked slip band as a dislocation pile-up (32):

$$\sigma_c = \left[\frac{\sigma_s^2 L^2}{t^2} + \frac{4E\,G_{ID}}{\Pi(1-\nu^2)t}\right]^{\frac{1}{2}} - \frac{\sigma_s L}{t} \tag{16}$$

where $\sigma_s = \frac{kL^{\frac{1}{2}}}{2}$ is the effective pile-up stress, (33), $k \simeq 0.75\ MPam^{\frac{1}{2}}$, and L and t are the grain

* When the stress is averaged over distances large compared with the dislocation core.

** An elastic crack propagating with a velocity V produces a time variation of stress a distance r in front of the tip ($\Theta=0$) that is equivalent to a stress intensity rate $\dot{K}$ for a stationary crack (24): $\dot{K} \simeq \frac{(G_{ID}E)^{\frac{1}{2}}}{2r} V$......(15) Taking $G_{ID} = 2\gamma = Eb/15$, and $r = 10b$, $\dot{K} \sim 10^9 MPam^{\frac{1}{2}}/s$ for V = 1 ms, which is about 10^8 times faster than ordinary loading rates.

*** Tetelman and Robertson (26) derive the expression $G_{IC}/G_{ID} = (a/a_o)^2$, which relates the fracture toughness at the instant the crack begins to extend and just prior to arrest with the corresponding crack lengths for a crack propagating under the action of hydrogen pressure from a fixed amount of hydrogen in the cavity. The value of G_{IC} is deduced from r_o the extent of the plastic zone (18): $G_{IC} \simeq (5/E)\,\sigma_o^{\,2} r_o$, average values $a/a_o = 1.8$, and $r_o = 33\mu mm$ are deduced from Fig. 4.

Fig. 4. Dislocations (etch pits) produced by crack extention-crack arrest events in an iron-3% silicon single crystal. The driving force for the events is the pressure of hydrogen gas in the crack cavity introduced by cathodic charging. The crack propagation is accompanied by the expansion of a crack cavity, containing a fixed amount of hydrogen gas (since diffusion is slow relative to propagation) and, therefore, a decrease in the pressure which reduces the crack driving force, and causes the crack to arrest. Further diffusion and hydrogen gas pressure build up after the crack has stopped is responsible for reinitiation, and a succession of events; after Tetelman and Robertson (26).

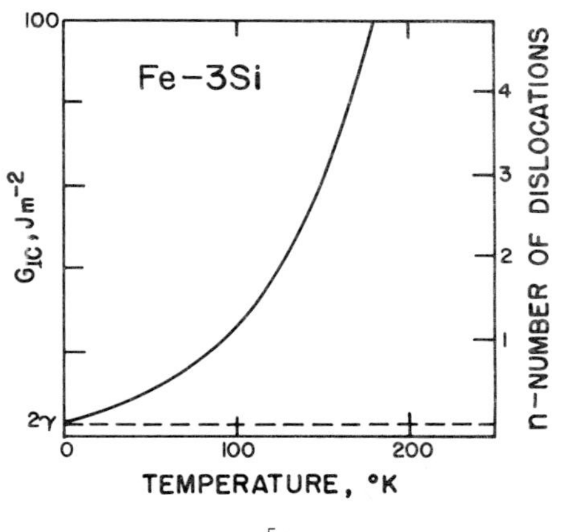

5a

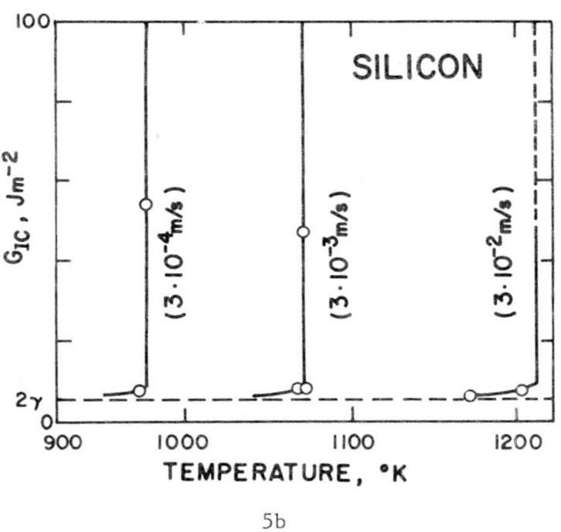

5b

Fig. 5. Examples of the fracture toughness-temperature transition of cleavage prone single crystals: (a) iron-3% silicon alloy crystals after Pilkington and Hull (21), and (b) pure silicon crystals at 3 loading rates after St. John (22).

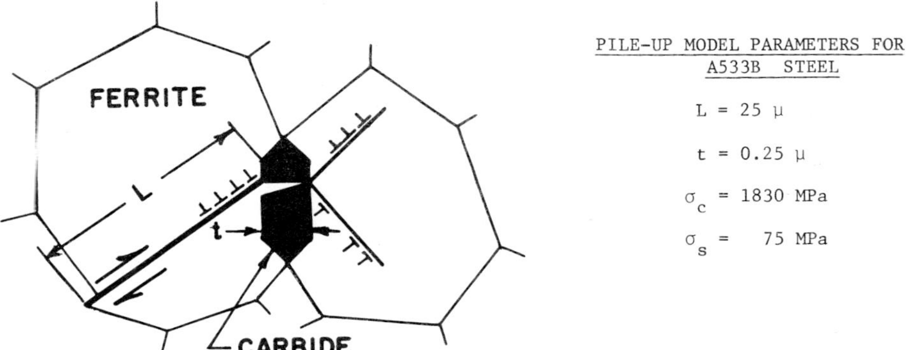

PILE-UP MODEL PARAMETERS FOR
A533B STEEL

$L = 25\ \mu$

$t = 0.25\ \mu$

$\sigma_c = 1830$ MPa

$\sigma_s = 75$ MPa

Fig. 6. Model of slip assisted, carbide induced cleavage initiation in steel (28-32). Typical values of the parameters L, t, and σ_c (see Section 5) are taken from the work of Ritchie, Server and Wullaert (36).

diameter (or length of the pile-up) and carbide thickness, respectively. The analysis assumes a static pile-up configuration (34); departures from static equilibrium are addressed by Korbel and Pawlek (35) in this volume. Typical values for A533B pressure vessel steel (36) in the Table in Fig. 6, together with Eqn. 16 define the toughness of the ferrite grain: $F_{ID} = 61\ Jm^{-2}$. The value is comparable to the one deduced for the advancing crack in Fe-3Si ($G_{ID} = 43\ Jm^{-2}$) and reflects the formation of ∿4 dislocations at the crack tip (Eqns. 8 and 9). The pile-up model offers a reasonable description of the effects of grain diameter and carbide size on σ_c (32) and provided inspiration for the development of the HSLA (high strength-low alloy) steels (37). However, it cannot be tested without an analysis of the crack-tip dislocation generation process that defines G_{ID}. One further problem is the relatively small number of carbide cracks that penetrate the ferrite, typically 1 in 10^4 - 10^5 ferrite grains (38). The statistics of the cracking have been examined by Cohen and Vukcevich (38) but the infrequency of penetration has not been explained in terms of the pile-up model.

Ritchie, Knott and Rice (39) have modeled the fracture toughness of steel, by inserting the Smith pile-up model in the heavily strained region ahead of the crack, as shown in Fig. 7. Their analysis assumed that the most heavily stressed carbide particles are located 1-2 grain diameters from the crack tip. Presumably, pile-ups acting on carbides nearer to the tip will be smaller than the grain diameter and ineffectual. They also assume that G_{ID} and σ_C are independent of temperature, with temperature and loading rate entering through the yield stress and its effect on the local tensile stress (see Fig. 7a). By locating the carbide cracking event several grain diameters from the main crack, the model can account for the relatively high fracture toughness values of steel: $K_{IC} \simeq 60\ MPam^{\frac{1}{2}}$ ($G_{IC} \simeq 16{,}000\ Jm^{-2}$) arising from the low toughness ferrite grains: $G_{ID} = 61\ Jm^{-2}$. However, the fact that a distance of 3-4 grain diameters must be assumed for good agreement (see Fig. 7b) has not been properly explained. One possibile reason is the infrequency of successful microcracking events, which renders cracking events closer to the crack tip statistically improbable. The model offers a reasonable accounting of the effects on toughness of loading rate (40) (see Fig. 7c) and irradiation hardening (36). It does not reproduce the relatively rapid increase of toughness at the transition temperature (see Fig. 7b). This is believed to be connected with the increase of G_{ID} with temperature which the analysis neglects.

6. Toughness of Ductile Materials with Large and Intermediate Size Particles

When the matrix phase resists cleavage, crack extension in systems with brittle particles can proceed by the void nucleation and growth process illustrated in Fig. 8a. Stresses arising from incompatabilities between an elastic particle and plastic matrix may be partially relaxed by slip in the matrix which produces a "secondary" plastic zone (see Fig. 9). Eventually, the stresses may produce fractures of the particle or its interface which serve as the void nuclei. The voids grow in volume under the action of hydrostatic stresses (41) via the formation and movement of prismatic loops in the matrix, (43) (see Fig. 9) and ultimately coalesce with the crack. The degree of stress relaxation, the vulnerability of the particle and the rate of growth are affected by the particle and void diameter. This is discussed in the following sections in terms of large ($d \gtrsim 1\mu m$), intermediate ($0.01\mu m \lesssim d \lesssim 1\mu m$) and small ($d \lesssim 0.01\mu m$) diameter particles. "Large" particles are usually extraneous inclusions, while "small" particles are usually precipitates or dispersoids intentionally added to improve strength.

6.1 Large Particles ($d \geq 1\mu m$)

Both Argon, Im and Safoglu (43) and Goods and Brown (44) agree that σ_{RR} the maximum radial stress, acting at the interface of relatively large elastic, spherical particles in a plastic matrix, i.e. $d \gtrsim 1\mu m$, is adequately described by "continuum" solutions:

$$\sigma_{RR} = \bar{\sigma} + \sigma_H \qquad (16)$$

where $\bar{\sigma}$ is the plastic flow stress and σ_H the hydrostatic component ($\sigma_H = 2.07\ \sigma_o$, in the region of maximum constraint ahead of a crack in a perfectly plastic material). The void nucleates when $\sigma_{RR} > \sigma_c$, where σ_c is the paritical stress for interface failure and the initial void diameter is comparable to the particle diameter. The formation and movement of prismatic loops about "large" voids is not viewed as difficult because of the relative proximity of potential sources. For this reason, void growth can also be treated with continuum plasticity in this case.

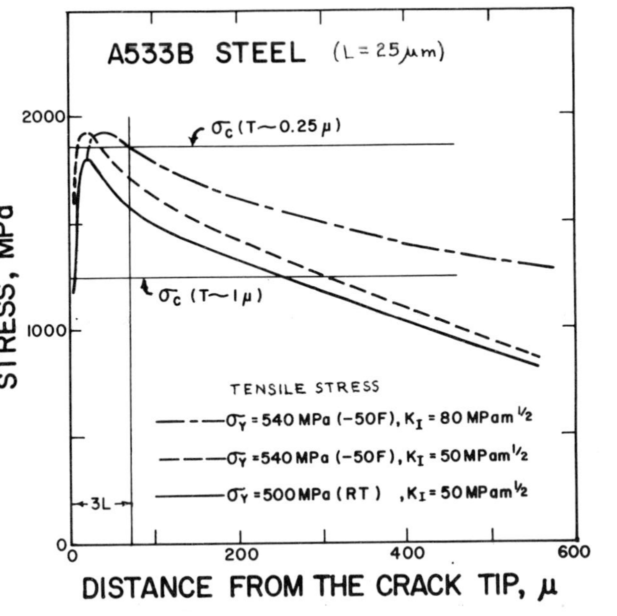

7a

Fig. 7. The Rice, Knott and Ritchie (39) model for cleavage crack extension in steel: (a) Estimated stress gradients in front of a crack in A533B steel based on Fig. 1 (b) corresponding fracture toughness predictions for A533B (39) and (c) fracture toughness (K_{ID}) predictions for dynamic loading of 1018 steel after Costin and Duffy (40).

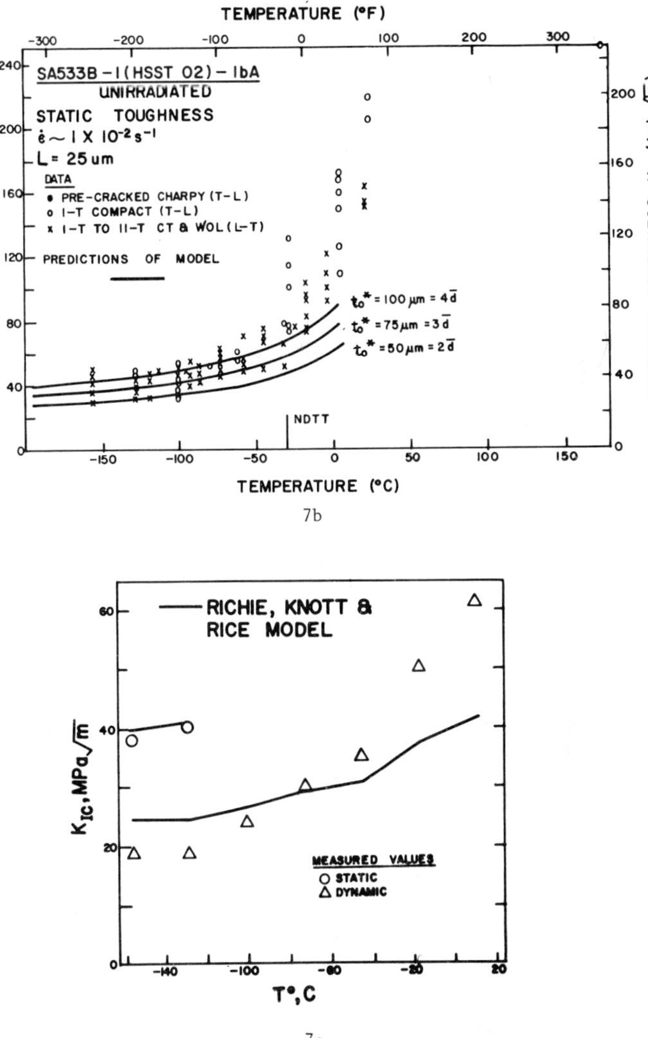

The main continuum parameters affecting the fracture toughness, in addition to the macroscopic elastic and plastic flow properties., are (i) X_o, the initial void spacing, (ii) x_o/d_o the initial void-spacing-to-diameter ratio, which is inversly related to the volume fraction of the voids (or particles)* and (iii) σ_c or ε_c, the critical stress or critical matrix strain for void nucleation. Rice and Johnson (45) have calculated the growth and coalescing of a spherical void in the strain field of a crack. Results defining the relation between relative toughness and the void spacing-to-diameter ratio, are shown in Fig. 8b and are in agreement with experiment. These calculations are for the case of a void that is present initially. The effect of a relatively high value of σ_c which delays void growth, is equivalent to reducing the original size of a void present initially, and this increases the toughness as shown in Fig. 8b.

Particles and voids in the path of a crack are subjected to an unusual stress-strain history, roughly equivalent to that experienced by a point, initially well to the right of the crack in Fig. 1, that approaches the tip. In the first stage the particle experiences small plastic strains and high radial stresses that increase to a maximum value**: σ_{RR} (max) $\simeq 4\sigma_o$. These conditions favor void nucleation but produce little or no growth. In the second stage, the surviving particle or void sees a falling stress and rapidly increasing plastic strain which promotes void growth. Increasing values of σ_c delay void nucleation in the first stage. However, this does not retard void growth as long as the void nucleates before the start of the second stage, i.e. when $\sigma_c < \sigma_{RR}$ (max) because the voids will be fully nucleated before the growth producing second stage begins. The strength of the particle/matrix interface must exceed σ_{RR} (max): $\sigma_c < 4\sigma_o$, before this will improve the toughness. In other words, systems with high volume fractions of "large", but well bonded particles can only display high levels of fracture toughness when the strength level is low, i.e. when $\sigma_o \lesssim \frac{E}{400}$, since $\sigma_c \simeq \frac{E}{100}$ (44,46).

6.2 Intermediate Size Particles (0.01μm < d < 1μm)

When details of the dislocation structure are on the scale of the secondary plastic zone size, then the continuum description of void nucleation is no longer appropriate and details of the dislocation structure must be considered. Ashby (42) first proposed that the local stresses are reduced by punching out symmetric arrays of prismatic loops (Fig. 9), making particles less vulnerable to decohesion. Eventually, the particle interface fails when the build-up of the back stress from these vacancy loops exceeds σ_c, the decohesion stress. Argon, Im and Safoglu (43) refine Ashby's approach by allowing the extent of the pile-up of vacancy loops to be determined by the friction stress and the strain incompatibility between the particle and the secondary zone. Their model defines ε_c, the plastic strain of the matrix for which the back stress just exceeds σ_c:

$$\varepsilon_c \simeq 0.83\,\varepsilon_o^{-1}\left[\frac{\sigma_c - \sigma_o - \sigma_H}{E}\right]^2 \tag{18}$$

where $\varepsilon_o = \frac{\sigma_o}{E}$, and σ_o is the yield stress of the polycrystalline material. Brown and Stobbs, (47) note that the size of the secondary plastic zone surrounding particles is dependent on the diameter of the particle and roughly independent of strain. Their analysis assumes forest hardening*** and concludes that the flow stress in the vicinity of the particle varies linearly with strain and inversely with its diameter. Goods and Brown (44) apply these results to estimate the critical strain****:

$$\varepsilon_c \simeq 5.7\,\frac{d}{b}\left[\frac{\sigma_c - \sigma_o - \sigma_H}{E}\right]^2 \tag{19}$$

---

* $f_v = (\pi/6)(X_o/d_o)^{-3}$, where f_v is the volume fraction.

** $\sigma_{RR}(\max) = \sigma + \sigma_H(\max)$, where $\sigma_H(\max) \simeq 2.07\,\sigma$ in front of the crack, and $\sigma \simeq 1.3\sigma_o$ to account for strain hardening.

*** $\sigma = \alpha\mu b\sqrt{\rho}$, where μ is the shear modulus, ρ is the dislocation density and $\alpha \simeq 0.15$

****Some confusion is inevitable since Goods and Brown employ the symbol σ_c to denote stresses arising from incompatibility, and σ_1 to denote the local stress for decohesion

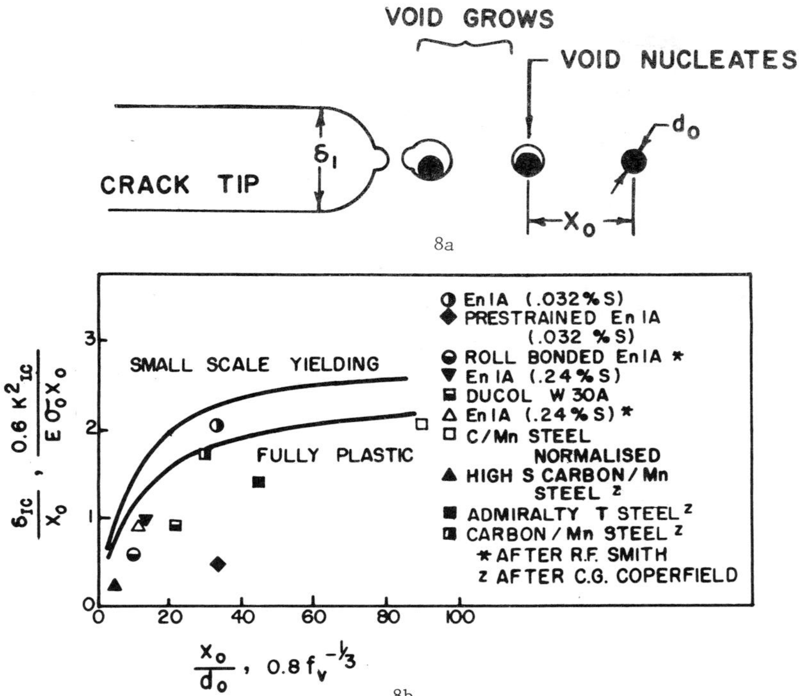

Fig. 8. Void nucleation and growth mode of crack extension: (a) schematic of mechanism and (b) comparison of calculations of the fracture toughness derived from continum models of hold growth in front of a crack by Rice and Johnson (45) with actual measurements after Green and Knott (63).

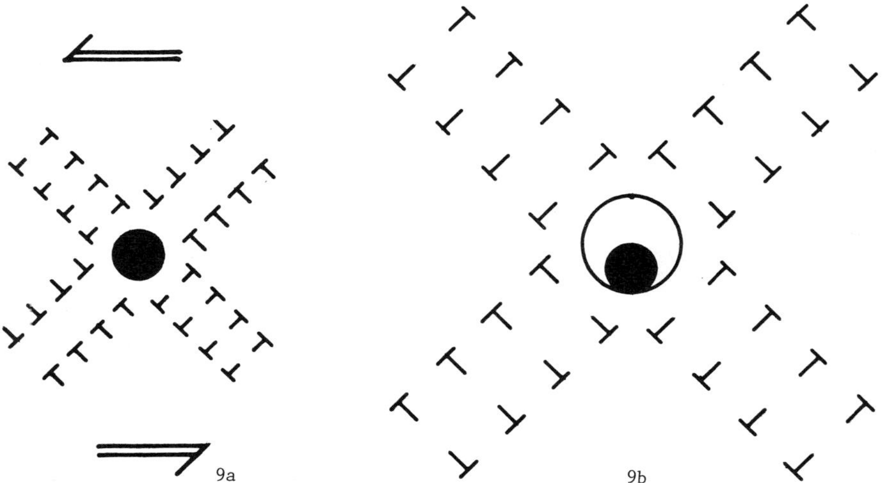

Fig. 9. Schematic representation of prismatic loops: (a) formed around a rigid elastic particle in a plastic alloy deformed matrix and (b) required for void growth. (42).

Since the flow stress and hyrdostatic stress contribute to σ_{RR} (Eqn. 17) and assist decohesion, σ_c in Eqns. 18 and 19 is reduced by $(\sigma_o + \sigma_H)$, as noted by Goods and Brown (43).

Figure 10 evaluates Eqns. 18 and 19 for the stress state in front of the crack: $\sigma_o + \sigma_H \simeq 3\sigma_o$. When $\varepsilon_c = \varepsilon_o$, void growth is not delayed because the voids nucleate during the first stage of the stress history, before they experience the second, growth producing stage. Large values of the critical strain, e.g. $\varepsilon_c > 0.1$, signify that nucleation does not occur during the first (rising stress) stage and therefore delays growth in the second (rising strain) stage, thereby enhancing the fracture toughness. The conclusions are entirely equivalent to those cited for "large particles", namely that high toughness levels are to be expected only at relatively low yield strength levels. Prismatic loops are likely to be more highly organized around smaller particles (d∿0.01μm) and the analysis of Argon, Im and Safoglu (43) should be valid. Plastic relaxation of the arrays is likely for larger particles, making the Goods and Brown (44) approach more reliable in the d = 0.1μm - 1μm range*. Figure 10 shows that the 2 analyses tend to agree when viewed in this light, and show that "intermediate" size particles are less resistant to void nucleation than "larger" ones.

The effect of particle diameter on void growth is clear. Once nucleated, the voids grow in the tensile direction until the bulk of the particle surface decoheres. Continued growth requires dislocation movements equivalent to the glide of prismatic interstitial loops away from the particles. This process is increasingly remote from other dislocations or easy sources as the void diameter decreases, with stress requirements for generating the loops likely to be in the range E/10 to E/100 for d = 0.01μm. This effect is equivalent to a hard region around the void and is expected to retard growth and increase toughness. Prismatic punching at "small" voids and its effect on the growth of "small" voids has not be treated explicitly.

7. Toughness of Ductile Materials with Small Brittle (d ≲ 0.01μ)

The nucleation of voids by small particles involves at least 4 separate considerations. First, and as noted above, there is less opportunity for relaxation within the secondary plastic zone as the particle diameter decreases. Secondly, prismatic punching becomes very difficult. This can preclude the formation of a secondary zone and prevent stress relaxation altogether. Brown and Woolhouse (48) consider a model for the formation of prismatic loops at the interface of growing precipitate which defines a critical strain, inversely related to particle diameter, for the production of stable loops:

$$\varepsilon_c \gtrsim \frac{b}{\sqrt{2}\pi(1-\nu)\,d}\left[\ln\left(\frac{4d}{b}\right) - 1 + \frac{3-2\nu}{4(1-\nu)}\right] \tag{20}$$

For b - 0.2 nm, and $\nu = 0.3$, this predicts that $\varepsilon_c = 0.03$ and $\varepsilon_c = 0.18$ for 0.01μm and 0.001μm diameter particles respectively. Since the elastic stresses associated with this level of strain are very high, it appears that "small" particles will either decohere or shear plastically before a secondary zone is formed. The third consideration involves the fact that the availability elastic strain energy (proportional to d^3) decreases more rapidly with decreasing particle size than the energy required for decohesion (which varies with d^2). Tanaka, Mori and Nakamura (49) have analyzed the decohesion of elastic particles in a plastic matrix in the absence of a secondary zone, on this basis. Their treatment shows that while the critical strain for the start of decohesion (derived from a stress criterion) is small, e.g. $\varepsilon_c \sim 0.01$ and independent of d, the critical strain for severing a fixed fraction of a particle's interface varies as $d^{-\frac{1}{2}}$. The fourth consideration is the void growth -- the volume expansion of the void -- is virtually impossible for "small" nuclei. This conclusion is supported by the Brown and Woolhouse (48) analysis, Eqn. 20, which shows that the generation of prismatic glide loops is difficult even with assistance of the misfit stresses, which are absent after decohesion.

These considerations for "small" particles suggest: (i) that if void nuclei forms during the first stage of the crack stress-history they will not grow in the second stage and (ii) that the particles may shear instead of decohering. Accordingly, the fracture toughness should be markedly improved in this regime of particle size. This is not the case. Fig. 11

---

* Argon, Im and Safoglu (43) do consider such plastic relaxation of the secondary dislocations by assuming power law hardening. Their plastic zone size still varies with strain and their analysis still predicts no dependence on particle diameter except for the case where the volume fraction of the particles is large enough to allow the impingement of the secondary plastic zones.

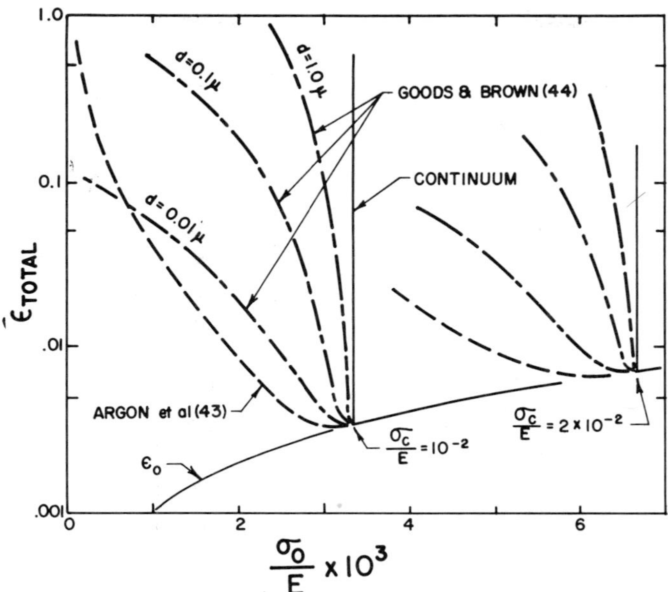

Fig. 10. Influence of yield stress level and particle diameter on the critical total (elastic) plus plastic) strain for void nucleation according to the Argon, Im and Safoglu (43) and Goods and Brown (44) dislocation models and on the continum solution for a non-strain hardening material (43).

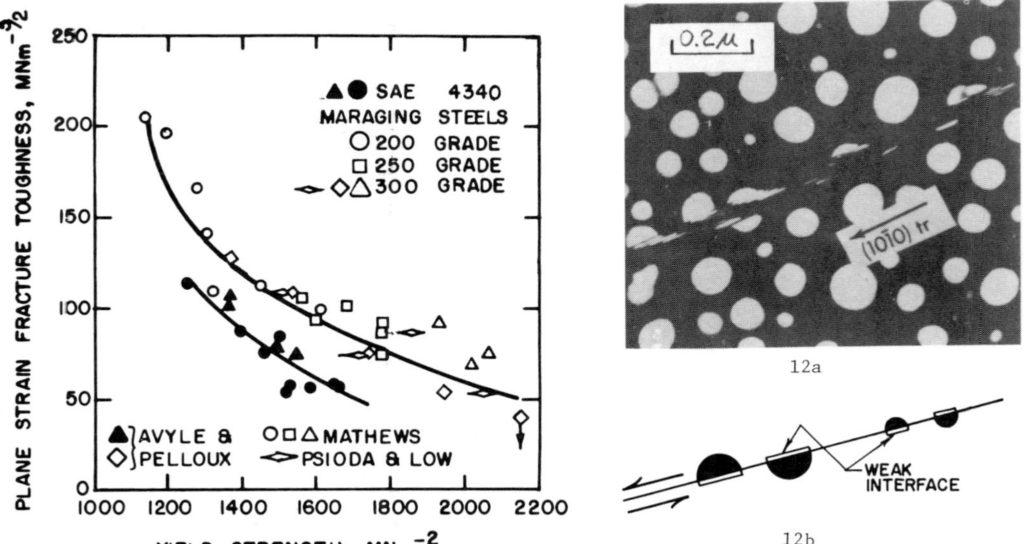

12a

12b

Fig. 11. Influence of the yield strength on fracture toughness of high strength alloys whose strength is inversely related to the size of "small" particles (50-52).

Fig. 12. Shearing of "small" particles: (a) coarse slip by shearing of α_2-particles in Ti-17% Al after Blackburn and Williams (64) and (b) potentially weak interfaces produced by shearing of "small" particles.

illustrates that the toughness decreases with increasing strength level and therefore with decreasing particle diameter, for systems that derive their strength from closely spaced precipitate particles. The Van Den Avyle and Pelloux (50) work demonstrates this loss of toughness even after larger inclusion particles have been removed. The reason for this is believed to be connected with instabilities arising from the combination of a "vertex-like effect in the description of yielding" of cross-slipping crystals, a problem recently treated by Asaro and Rice (53) and Asaro (54), together with reduced strain hardening. The reduced strain hardening may be a consequence of the reduction of the shear-resisting cross section of small shearing particles (Fig. 12a) along lines discussed by Hornbogen and Gahr (55), or be connected with local interface decohesion. A model analysed by Rau and Cook (56) shows that the stress concentrations produced by an array of dislocations cutting through a series of particles quickly leads to a nonhardening state. The net effect is the formation of coarse slip bands, reflecting enormous local shear strain concentrations, which are frequently observed in high strength precipitation hardened alloys (see Figs. 12a and 13). There is also direct and indirect evidence from a number of sources that cracks initiate and grow within the coarse slip bands. For example the orientation of cracks responsible for the fracture of unnotched bend coupons (57) and notched bars (58) and the zig-zag extension of cracks (see Fig. 14a) in high strength alloys, all correspond with the planes of maximum shear stress. A simple model for analysing the toughness associated with the zig-zag mode of crack extension, which relates $\sqrt{2}\delta_{IC}$, the crack tip displacement in the slip direction with $\lambda(\max)$ the maximum offset of the coarse slip bands, is illustrated in Fig. 14. The fracture toughness K_{IC} is related to δ_{IC} by way of Eqns. 1 and 3 can thus be expressed in terms of $\lambda(\max)$:

$$K_{IC} = \sqrt{2\sigma_o E\lambda(\max)} \tag{21}$$

Results in Table 2 illustrate this model is in rough accord with measurements. The higher toughness of the maraging steel (see Fig. 11) is associated with a tolerance for larger slip offsets. These are observed at the grain boundaries since the slip bands within the grains are noticeably finer than in 4340 steel (57).

Table 2: COMPARISON OF MEASURED SLIP BAND OFFSETS $\lambda(\max)$ WITH VALUES DERIVED FROM THE FRACTURE TOUGHNESS WITH EQN. 21. (57)

4340 STEEL				Ni-MARAGING STEEL		
σ_o, MP_a	K_{IC}, $MP_{am}^{1/2}$	$\lambda(\max)$, μm Measured	Eqn. 21	K_{IC}, $MP_{am}^{1/2}$	$\lambda(\max)$, μm Measured	Eqn. 21
1500	50	3-4	4	100		16
1750	30			75	5-10	8
2000				60		4

The details of the mechanism of crack initiation and linking up within the coarse slip band are not established. Voids may nucleate at some of the "weak" interfaces separating the sheared-off portions of the particle and matrix, (see Fig. 12b), and these may coalesce by way of the large shears within the bands as proposed by McClintock and Berg (59). Some evidence of shearing of the microvoids has been reported by Beachem and Yoder (60), but the dimples visible on the surface are not as small or elongated as one would expect from this sample view. Alteratively the voids may initiate at cell walls in the heavily deformed regions between the particles, along lines discussed by Gardner and Wilsdorf (61). It seems likely that dislocation mechanisms will be involved in the analysis of the slip band cracking process.

8. Summary and Conclusions

This review examined 4 distant modes of crack extension:

(a) Cleavage of single phase crystals
(b) Slip-induced cleavage in polycrystalline materials with brittle particles
(c) The fibrous mode of extension in ductile phases with "large" and "intermediate" size elastic particles, and

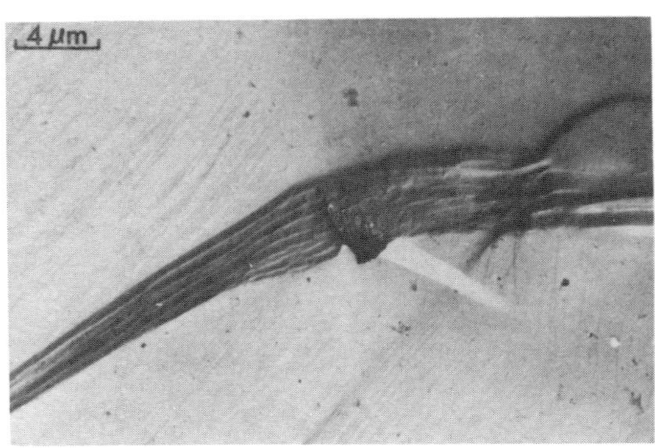

13a

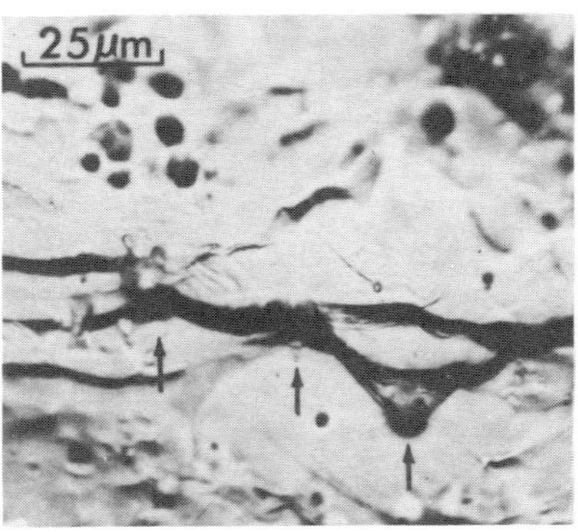

13b

Fig. 13. Examples of coarse slip bands observed on cellulose acetate replicas of the surface of (a) high purity 7000-type aluminum alloy and (b) 7075-T6 aluminum, both in the peak aged condition (57). The small tongue of replicating material in (a) is a sign of a small crack at the base of the slip band. Coarse slip bands in (b) are associated with "large" particles (see arrows).

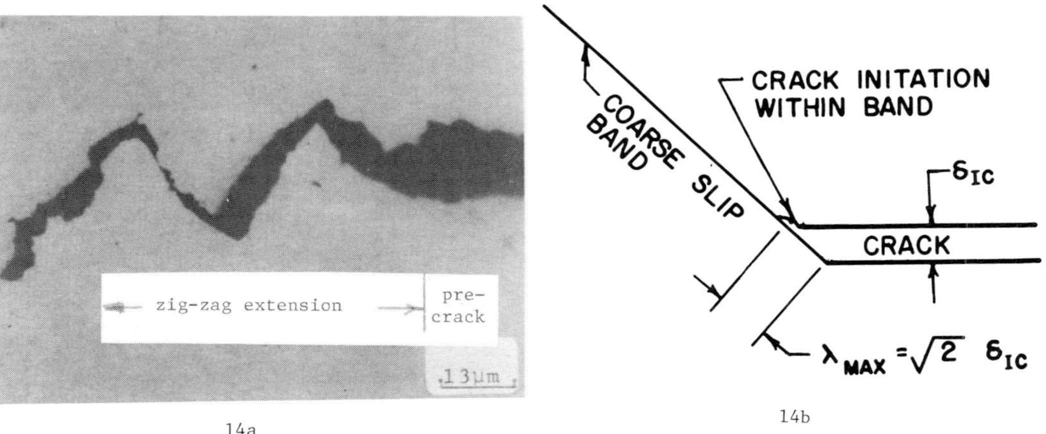

14a

14b

Fig. 14. Zig-zag mode of crack extension observed in high strength alloys: (a) example of zig-zag extension in an electroslag remelt heat of 4340 steel (σ_o = 1390 MPa, K_{IC} = 107 MPa $m^{\frac{1}{2}}$) after Van den Avyle and Pelloux (50) and (b) simple model of the onset of zig-zag extension.

(d) The zig-zag variant of (c) associated with "small" particles.

These are treated separately in this paper. However, crack extension in real materials usually involves several modes at once. For example, the brittle fracture of steel usually proceeds by cleavage together with a small fibrous component which makes a disproportionate contribution to the toughness (62). The fracture toughness of an aluminum alloy can reflect contributions from "large" and "intermediate" inclusions and "small" precipitate particles (57). For this reason, the expressions for toughness offered in this paper are viewed as limiting cases, and there remains a need for more sophisticated treatments that can anticipate both the amounts and, contributions of the different components. These analyses will, nevertheless, benefit from a better understanding of at least 11 dislocation mechanisms that facilitate analysis of the crack extension process:

(i) Thermally assisted generation of dislocation loops at the crack tip.
(ii) Dislocation multiplication and movement in the early stages of the formation of a crack-tip plastic zone.
(iii) Generation and movement of dislocations at the head of a rapidly advancing crack tip.
(iv) Formation of a dislocation pile-up within the plastic zone of a crack under static and dynamic conditions.
(v) Generation of prismatic dislocation loops at "small" and "intermediate" size elastic particles in a plastic matrix.
(vi) Movements of dislocations within the secondary zone of "intermediate" size particles.
(vii) Dislocation mechanisms of void growth for "intermediate" size particles.
(viii) Generation of prismatic loops at "small" and "intermediate" size voids.
(ix) Dislocation cutting of "small" particles.
(x) Dislocation movements and interactions within coarse slip bands.
(xi) Dislocation mechanisms of void or crack initiation and linking up in coarse slip bands.

Of these, the dislocation mechanisms contributing to the formation of coarse slip bands and their rupture are probably least well defined.

Acknowledgements

The authors wish to thank C. Moss for her contributions to the manuscript, and to acknowledge support from Vanderbilt University.

REFERENCES

1. G.R. Irwin, J. Appl. Mech, 24, 361 (1957)
2. J.R. Rice, J. of App. Mech., 35, 379 (1968)
3. J.R. Rice and G.F. Rosengren, J. Mech. Phys. Sol., 16, 1 (1968)
4. J.W. Hutchinson, J. Mech. Phy. of Sol., 16, 13 (1968
5. C.F. Shih, W.R. Andrews, M.D. German, R.H. Van Stone and V.P.D. Wilkinson, "Methodology for Plastic Fracture", Progress Report on EPRI Contract RP601-2, p. 3-1, July 31, 1978
6. J.R. Willis, (this volume)
7. R.M. McMeeking, J. Mech. Phys. Sol., 25, 357 (1977)
8. J.R. Rice, The Mechanics of Fracture, AMD, 19, 23 ASME 1977
9. P.C. Gehlen, G.T. Hahn, and M.F. Kanninen, Scr. Met. 6, 1087 (1972)
10. A.A. Griffith, Phil. Trans. Roy. Soc., London, Ser. A, 221, 163 (1920)
11. F.C. Frank and A.W. Stroh, Proc. Phys. Soc., 65B, 811 1952
12. C. Hseih and R. Thomson, J. Appl. Phys., 44, 2051 (1973)
13. J.E. Sinclair, Phil. Mag., 31, 647 1975
14. M. Mullins and M.A. Dokainish, (to be published)
15. J.R. Rice and R. Thomson, Phil. Mag., 29, 73 (1974)
16. J.J. Gilman, Trans. AIME, 209, 449 (1957)
17. B.A. Bilby, A.H. Cottrell, and K.H. Swinden, Proc. Roy. Soc., A, 272, 304 (1963)
18. B.A. Bilby and K.H. Swinden, Proc. Roy. Soc., A, 285, 22 (1965)
19 S.J. Chang and S.M. Ohr (this volume)
20. B.C. Edwards and B.L. Eyre (this volume)
21. R. Pikington and D. Hull, Conf. on Fracture Toughness, ISI 20, 5(1968)
22. C. St. John, Phil. Mag. 30, 1193 (1975)
23. A.K. Shoemaker and S.T. Rolfe, Eng. Fracture Mech. 2 (1971)
24. J.M.Krafft and G.R. Irwin, ASTM STP 381, 114 (1964)

25. S.J. Burns and W.W. Webb, J. Appl. Phys., 41, 2078 (1970)
26. A.S. Tefelman and W.D. Robertson, Acta Met., 11, 415 (1963)
27. V. Vitek and G.G. Chell (this volume)
28. C.J. McMahon and M. Cohen, Acta Met., 13, 591 (1965)
29. G. Dates, J. Iron Steel Inst., 206, 930 (1968)
30. M. Gell and P. Worthington, Acta Met., 14, 1265 (1966)
31. E. Smith, Physical Basis Yield and Fracture, Cont. Proc. Oxford, 36 (1966)
32. E.A. Almond, D.H. Timbres and J.D. Embury, Fracture 1969, 253, Ed. P.L. Pratt, Chapman and Hall, London 1969
33. N.J. Petch, J. Iron Steel Inst, 191, 25 (1953)
34. G.T. Hahn and A.R. Rosenfield, Acta Met., 16, 755 1968
35. A. Korbel and A. Pawelek (this volume)
36. R.O. Ritchie, W.L. Server and R.A. Wullaert, MIT Report FDL/R/79/1019, Feb 1979
37. F.B. Pickering, Micro Alloying 75, p.3
38. M. Cohen and M.R. Vukcevich
39. R.O. Ritchie, J.F. Knott and J.R. Rice, J. Mech. Phys. Solids, 21, 395 (1973)
40. L.S. Costin and J. Duffy, Trans. ASME, 101, 258 (1979)
41. F.A. McClintock, J. App. Mech 35, 363 (1968)
42. M.F. Ashby, Phil Mag., 14, 1157 (1966)
43. A.S. Argon, J. Im and R. Safoglu, Met. Trans 6A, 825 (1975)
44. S.H. Goods and L.M. Brown, Acta Met., 27, 1 (1979)
45. J.R. Rice and M.A. Johnson, Inelastic Behavior of Solids, Ed. M.F. Kanninen, McGraw-Hill, p. 641 (1970)
46. A.S. Argon, J.Im, Met. Trans., 6A, 839 (1975)
47. L.M. Brown and W.M. Stobbs, Phil. Mag., 34, 351 (1976)
48. L.M. Brown and G.R. Woolhouse, Phil. Mag., 21, 329 (1970)
49. K. Tanaka, T. Mori and T. Nakamura, Phil. Mag., 21, 267 (1970)
50. V.A. Van Den Avyle and R.M. Pelloux, ScD Thesis, MIT 9175
51. W.T. Mathews, U.S. Army Report AMMRC MC 73-6 (1973)
52. J.A. Psioda and J.R. Low, Jr., The Effect of Microstructure and Strength on the Fracture Toughness of an 18Ni, 300 Grade Maraging Steel, Carnegie Mellon Univ. Report to NASA on Graut NGR 39-087-003, Aug 1975
53. R.J. Asaro and J.R. Rice, J. Mech. Phys. Solids 25, 309 (9177)
54. R.J. Asaro, Acta Met., 27, 445 (1979)
55. E. Hornbogen and K.H. Zum Gahr, Metallography, 8, 101 (1975)
56. C.A. Rau, and T.S. Cook, Pratt and Whitney Aircraft Report PWA-4868, Nov. 1973
57. G.T. Hahn, C.R. Barns, and A.R. Rosenfield, Air Force Systems Command Report ARL TR 75-0194, June 1975.
58. V. Russo, A. Chakrabarti, and J. Spetnak, Met: Trans., 8A, 729 (1977)
59. F.A. McClintock, S.M. Kaplan and C.A. Berg, Jr., Int. J. Fracture Mech. 2, 614 (1966)
60. C.D. Beachem and G.R. Yoder, Met.Trans., 4, 1145 (1973)
61. R.N. Gardner and H.G.F. Wilsdorf, Met. Trans., 11A, 659 (1980)
62. G.T. Hahn, R.G. Hoagland, J. Lereim, A.J. Markworth and A.R. Rosenfield, ASTM STP 711 (1980)
63. G. Green and J. F. Knott, J. of Eng. Mat'ls and Tech, 98, 37 (1976)
64. M. J. Blackburn and J. C. Williams, Trans. ASM., 62, 398 (1969)

ON CRACK PROPAGATION ALONG CRYSTALLOGRAPHIC SLIP BANDS

D. A. Koss and K. S. Chan
Department of Metallurgical Engineering
Michigan Technological University
Houghton, MI 49931

Abstract

The influence of the crystallography of slip (fcc, hcp, and bcc) on crack propagation along planar slip bands is examined for two different mixed-mode crack configurations. If a crack propagates along a slip band and the Burger's vector of slip within that band is contained in the plane of the crack (i.e., coplanar slip), then non-coplanar slip is hindered and large normal stresses develop near the crack-tip. The magnitude of these normal stresses is largest for basal plane cracking in a hcp material. However, the calculations show that the magnitude of the plastic constraint factor is also large for an inclined crack in fcc and bcc alloys fracturing along crystallographic slip bands with the resulting normal stresses being comparable to a mode I crack in a polycrystalline material. Thus, cracking along crystallographic slip bands is characterized by a combination of intense coplanar slip and large normal stresses. Easy crack propagation across individual grains and a susceptibility to hydrogen embrittlement may result as a consequence.

Introduction

The propagation of a crack along a crystallographic slip band is a common occurrence in high strength alloys. Such crystallographic cracking is ductile in nature and occurs on an extensive scale in materials susceptible to planar, non-uniform slip, such as high strength Ni-, Al-, and Ti- base alloys. One of the features of fracture along crystallographic slip bands is that even though the crack path occurs along planes experiencing large shear stresses, the resulting fracture surface is usually characterized by a substantially brittle or cleavage-like appearance. This suggests that normal stresses as well as shear strains are important in this fracture process.

Koss & Chan (K & C) have very recently examined certain conditions for the propagation of a crack along a coplanar slip band (1). In their analysis of the crack-tip plastic zone, the plastic displacement is restricted to a shear parallel to the crack plane and such crack/slip interaction results in a relaxed "elastic-plastic" state of stress near the crack tip. Calculations by K & C indicate that once a crack with a coplanar slip band has formed, it becomes difficult to activate secondary slip with a Burger's vector inclined to the crack plane. As a result, large normal stresses can be generated near the crack-tip. The purpose of this paper is to explore certain implications of crack propagation along planar slip bands with regard to the crystallography of slip in fcc, hcp, and bcc alloys. Such cracking is always observed under combined tensile and shear loading, and thus the influence of crack orientation as regards the importance of mode II (in-plane shear) vs. mode III (anti-plane shear) mixed with the normal mode I opening will also be explored.

Crack Propagation Along Planar Slip Bands

The basis for understanding crack propagation along a coplanar slip band is contained in Fig. 1 and is discussed in more detail by K & C (1). The singularity in the elastic stress field near the tip of a mixed mode crack dictates that if r is the radial distance from the crack-tip, then $\sigma_{ij} \propto 1/\sqrt{r}$, where σ_{ij} is a stress component. Sufficiently close to the crack-tip, the stresses may be described by the tensor $\vec{\vec{\sigma}}$ which activates slip with a critical resolved shear stress τ_y, whose normal in $\vec{n}$ and with a Burger's vector $\vec{b}$ when:

$$\tau_y = (\frac{1}{b})\ \vec{b}\cdot\vec{\vec{\sigma}}\cdot\vec{n} \tag{1}$$

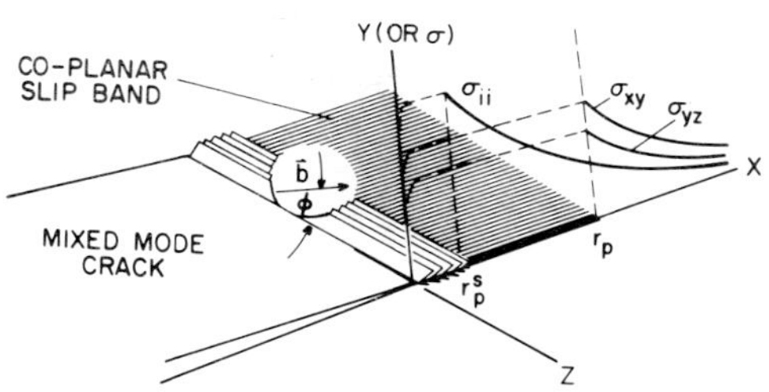

Fig. 1. A schematic illustration of the stress state ahead of a mixed mode crack with a coplanar slip band characterized by a slip vector $\vec{b}$ at an angle ϕ to the crack front. The coplanar slip extends to r_p while secondary slip is activated to r_p^s (after Koss and Chan, ref. 1).

Through the inverse square root dependence of the stress components on r, Eq. (1) can be used to estimate (ignoring plastic relaxation (2), crack-tip blunting, and work hardening) the plastic zone size r_p. For a material characterized by multiple, non-coplanar slip at the crack-tip (as is typical of a mode I crack), r_p indicates roughly the position at which all stress components are plastically relaxed. In contrast, if crack-tip plasticity is dominated by planar slip whose Burger's vector is also coplanar with the crack, then at r_p Eq. (1) is:

$$\tau_y = \sigma_{xy}^* \sin\phi + \sigma_{yz}^* \cos\phi, \tag{2}$$

and <u>only</u> the shear stresses σ_{xy} and σ_{yz} are plastically relaxed to values of σ_{xy}^* and σ_{yz}^*. At the same time, the normal stress components σ_{ii} continue to increase at $r < r_p$ in an elastic manner until a stress state is reached at $r = r_p^s$ at which slip with a non-coplanar Burger's vector is activated. This "relaxed" elastic-plastic stress state is schematically shown in Fig. 1. The stress tensor which activates non-coplanar slip in the plane of the crack thus is:

$$\vec{\vec{\sigma}} = \begin{bmatrix} \sigma_{xx} & \sigma_{xy}^* & 0 \\ \sigma_{xy}^* & \sigma_{yy} & \sigma_{yz}^* \\ 0 & \sigma_{yz}^* & \sigma_{zz} \end{bmatrix} \tag{3}$$

Eq. (3) shows that the increase in shear stress necessary to activate non-coplanar slip at $r = r_p^s$ according to Eq. (1) must arise from the <u>normal</u> stress components. In plane strain this stress state is obviously not conducive for the activation of secondary slip but is well suited for the development of large hydrostatic stress. Thus, once a crack begins to propagate along a coplanar slip band, it becomes difficult to activate non-coplanar slip which in turn generates large normal stresses near the crack-tip. This is shown schematically in Fig. 2. The combination of the large normal stresses and the intense coplanar slip should result in relatively easy crack propagation along the coplanar slip bands and in crack surfaces which, reflecting the large normal stresses, may have a cleavage-like or substantially brittle fracture appearance. This has been observed in a number of Al-, Ni-, and Ti- base alloys.

Application to Fcc, Hcp, and Bcc Alloys

Crack propagation along crystallographic slip bands is usually observed in single crystals, individual colonies or large grains of high strength alloys. As such, the crystallography of slip should have a significant influence on the ability of a given material to activate non-coplanar slip and therefore on the

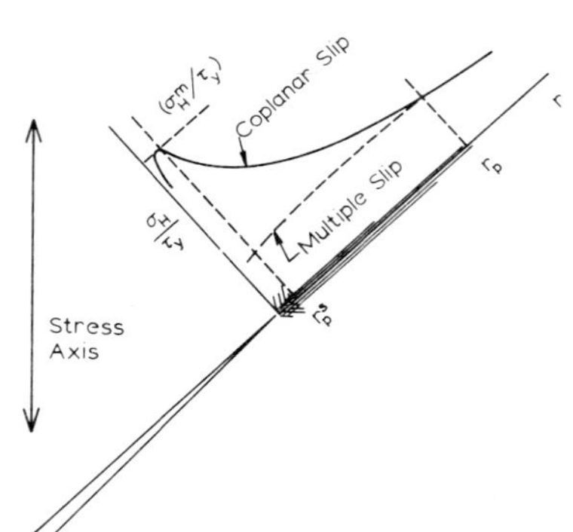

Fig. 2. The approximate distribution of the hydrostatic stress/yield stress ratio ahead of a mixed mode crack exhibiting either coplanar or multiple slip.

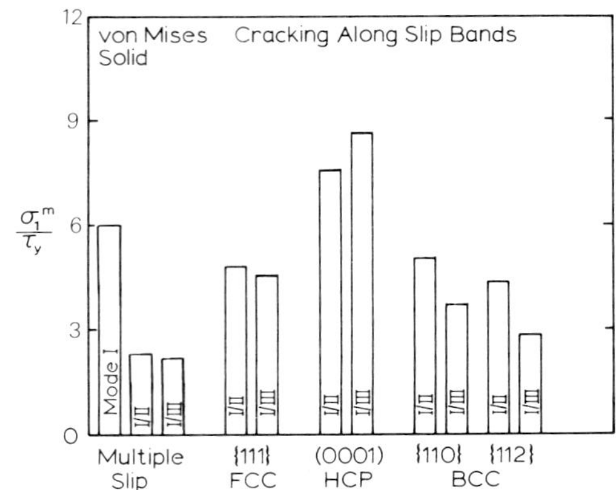

Fig. 3. A comparison of the approximate values of the maximum principal stress/yield stress ratios for cracks characterized either by multiple slip or coplanar slip in selected crystal systems. The crack geometries (i.e., mode I, I/II, or I/III) are defined in Table I.

Table I. A comparison of the magnitude of the approximate maximum normal stresses near the tip of cracks with differing orientations and with differing crack-tip slip behavior.

	Cracking/Multiple Slip			Crack Propagation Along Crystallographic Slip Bands							
Constraint	von Mises Solid*			Fcc Crystal		Hcp Crystal		Bcc Crystal Crack {110}-crack plane-{112}			
Factor+	Mode I	I/II	I/III	I/II	I/III	I/II	I/III	I/II	I/III	I/II	I/III
				(111) $[\bar{1}01]$**		(0001) $[11\bar{2}0]$**		(011) $[\bar{1}1\bar{1}]$**		(211) $[1\bar{1}\bar{1}]$**	
σ_H^m/τ_y	5.3	1.0	1.0	3.3	3.3	5.9	7.3	3.5	2.8	3.1	1.9
σ_1^m/τ_y	6.0 [∿6, continuum plasticity (3)]	2.3	2.1	4.7	4.2	7.6	8.6	5.0	3.7	4.5	2.8
σ_{yy}^m/τ_y	6.0	1.1	1.1	3.7	3.7	6.6	8.2	4.0	3.1	3.5	2.1
Secondary Deformation Systems:				$(\bar{1}11)[101]$	$(\bar{1}11)[110]$	twins $(10\bar{1}2)$ $[\bar{1}101]$	$(\bar{1}012)$ $[10\bar{1}1]$	$(12\bar{1})$ $[\bar{1}11]$	$(\bar{1}2\bar{1})$ $[111]$	$(21\bar{1})$ $[1\bar{1}1]$ $(2\bar{1}1)$ $[11\bar{1}]$	$(2\bar{1}\bar{1})$ $[111]$
Crack Orientations:	Mode I			Mode I/II (45°)		Mode I/III (45°)					

+ Values are calculated by assigning $r = r_p$ (von Mises) or $r = r_p^s$ (single crystal) in the following relations:
$\sigma_H^m = \frac{4}{9}\sigma\sqrt{\frac{a}{2r}}$, $\sigma_{yy}^m = \frac{\sigma}{2}\sqrt{\frac{a}{2r}}$, and σ_1^m is from the solution of: $\sigma^3 - \frac{8}{3}A\sigma^2 + (\frac{7}{3} - \frac{r_p^s}{r_p})A^2\sigma + (\frac{2}{3}\frac{r_p^s}{r_p} - \frac{2}{3})A^3 = 0$,
where $A = \frac{\sigma}{2}\sqrt{\frac{a}{2r_p^s}}$.

* Assuming $\sigma_y = 2\tau_y$

** Denotes y and x axes for mode I/II crack and y and z axes for mode I/III crack.

magnitude of normal stresses near the crack-tip. Crack propagation along crystallographic slip bands also occurs on planes inclined to the stress axis such that the mode I crack opening component is mixed with the mode II and/or mode III shear components.

Table I and Fig. 3 shows the influence of slip crystallography for fcc, hcp, and bcc deformation systems. These normal stresses have been calculated using Eqs. (1-3) and should be reasonable estimates; most of the details to such calculations are described in the appendix of the K & C paper.+ This procedure is based conceptually on Figs. 1 & 2 and as such, the normal stresses are determined by Irwin's elastic stress intensity factor at either $r = r_p$ (multiple slip) or $r = r_p^s$ (coplanar slip). While crude, this method none-the-less yields a value of σ_1^m for a mode I crack which agrees well with continuum plasticity estimates (3) as is shown in Table I. A principal difference in the two methods is that the present approach would place σ_1^m much nearer the crack-tip (3). Finally, it should be noted that these calculations assume isotropic elasticity, which is a very good assumption for materials with an anisotropy ratio of less than 2.

As shown in Table I and in Fig. 3, there are relatively large normal stresses near the tip of a mode I crack with multiple slip obeying a von Mises yield criterion at the crack-tip. However, there is a large decrease in normal stresses for a von Mises material when crack becomes inclined to the stress axis (for example, compare σ_H^m/σ_y or σ_1^m/σ_y for mode I vs. modes I/II or I/III). This is a principal reason why cracks initially inclined to the stress axis generally deflect into a mode I configuration upon subsequent propagation in a polycrystalline material. In contrast, crack propagation along inclined, coplanar, crystallographic slip bands is characterized by normal stresses which are much larger than inclined, "multiple" slip cracks. In fact, taken as a whole, σ_1^m, σ_H^m, and σ_{yy}^m as shown in Table I generally compared favorably in magnitude with that of a mode I crack. Thus it should not be a surprise that such cracks appear relatively content to continue propagating along crystallographic slip planes until microstructural features (i.e., grain boundaries, interphase boundaries, etc.) cause arrest or deflection or branching.

An interesting feature of the results in Table I and Fig. 3 is that, despite many secondary slip systems, large normal stresses can be generated by crystallographic slip band cracking in both fcc and bcc crystals. An implication of this behavior is that a cubic alloy which is otherwise immune to hydrogen embrittlement may become sensitive to hydrogen if heat treated so that planar slip and coplanar cracking occurs. This in fact has been observed in the case of age hardenable β Ti alloy tested in fatigue in a hydrogen environment (4).

As expected, very large normal stresses should be present near the tip of a crack propagating along a coplanar slip band in an hcp crystal. This is due to the difficulty of activating slip or twinning which causes an extension (or relaxation of the normal stresses) in the C direction, normal to the (0001). Twinning on the $\{10\bar{1}2\}$ $\langle\bar{1}101\rangle$ system is very common to most hcp alloys and should be activated near the crack-tip to relieve the normal stresses. Observations of Stage I, basal cracking in α and α-β Ti alloys show that $\{10\bar{1}2\}$ twins are indeed found near the tip of crystallographic cracks (5).

As shown in Table I, the influence of the crack orientation is such that the plastic constraint factors for mode I/II cracks are similar to those of a mode I/III crack for fcc and hcp material. This is surprising since there is not a larger difference since a mode II crack generates normal stresses while a mode III crack does not (assuming isotropic elasticity). All of the cases

---

+In the K & C paper (1), there is an error in the calculation of σ_1^m, which is used to demonstrate the model. The correct value, as indicated in Table I, should be $\sigma_1^m = 2.3\ \sigma_y$ for Fig. 2a and $\sigma_1^m = 2.0\ \sigma_y$ for Fig. 2b.

examined in Table I refer to cracks inclined to the stress axis at 45°. Calculations show that a more inclined crack (mode I component becomes smaller) has a smaller $r_p{}^s$ and therefore the normal stresses at the crack-tip exceed those in Fig. 3.

Acknowledgments

The authors wish to thank Professor R. Raj for helpful comments. The support of the Office of Naval Research through Contract No. N00014-76-C-0037 is gratefully acknowledged.

References

1. D. A. Koss and K. S. Chan, Acta. Met. (in print).
2. G. R. Irwin, Proc. 7th Sagamore Conf., p. IV-63 (1960).
3. J. R. Rice in Stress Corrosion Cracking and Hydrogen Embrittlement of Iron Base Alloys, p. 11, NACE, Houston, 1977.
4. J. R. Wilcox and D. A. Koss, to be published in the Proc. to the 3rd Int. Conf. on the Effect of Hydrogen on the Behavior of Metals, Jackson Hole, Wyoming (1980).
5. K. S. Chan and D. A. Koss, Mat'l. Sci. and Eng. 43, 177 (1980), also C. C. Wocjik and D. A. Koss, unpublished results.

A MODEL OF SHEAR CRACKS WITH DISLOCATION-FREE ZONES*

S. -J. Chang[+] and S. M. Ohr[++]

[+]Computer Sciences Division, Union Carbide Corporation, Nuclear Division, Oak Ridge, TN 37830
[++]Solid State Division, Oak Ridge National Laboratory, Oak Ridge, TN 37830

Electron microscopy observations of the distribution of dislocations in the plastic zone of cracks have shown that, in many instances, there is a zone near the crack tip which is free of dislocations. In order to treat this problem, an integral equation was formulated in terms of a distribution of dislocations. The distribution function of dislocations was obtained in a closed-form in terms of elliptic integrals of the first and third kinds. The condition of compatibility between the crack and the dislocation-free zone was obtained as a function of applied stress, similar to the condition derived in the BCS theory of fracture. Because of the presence of the dislocation-free zone the stress diverges at the crack tip, leading to a finite elastic stress intensity factor K. A simple analytical expression for K was obtained.

Introduction

In their dislocation theory of fracture, Bilby, Cottrell and Swinden (BCS) (1) considered a plastic zone consisting of an inverse pile-up of dislocations that was coplanar with the crack. The crack opening displacement (COD) was defined by the sum of the Burgers vectors of the dislocations that were introduced into the plastic zone at the crack tip. The theory has since been elaborated in various ways and has been widely applied to various aspects of fracture. Recently, direct observations have been made of the distribution of dislocations in the plastic zone of cracks during *in situ* deformation in the electron microscope (2,3). A hole was located at the center of a thin specimen which was subjected to tensile stress and failed in shear mode. The plastic zone was coplanar with the crack, the dislocations were in the form of an inverse pile-up and the number of dislocations observed in the plastic zone was consistent with the COD required to propagate the crack. These observations were generally in good agreement with the predictions of the BCS theory. The only exception was that in many instances there was a zone near the crack tip which was devoid of dislocations. The dislocations were observed to be emitted from the crack tip, passing through the dislocation-free zone and accumulated in the plastic zone. In the present paper, we have examined the equilibrium between the crack and the linear array of dislocations in the presence of a dislocation-free zone (DFZ) near the crack tip. A governing singular integral equation has been formulated to describe the regions of the crack, the DFZ and the plastic zone. The integral equation has been inverted to obtain the distribution function for the dislocations (4), and an expression for the elastic stress intensity factor derived. These results are discussed in terms of the model of elastic crack tip within the plastic zone proposed by Thomson (5), Weertman (6), and Rice and Thomson (9).

Theoretical Model

We have considered a centrally located crack of length 2c in an infinite isotropic elastic medium which is subjected to an externally applied shear stress $\sigma_{yz} = T$, causing the medium to deform in an antiplane mode. The cartesian coordinate system (x,y,z) is used to describe the crack

$$-c < x < c, \quad y = o$$

*Research sponsored by the Division of Materials Science, U. S. Department of Energy under contract W-7405-eng-26 with Union Carbide Corporation.

and two dislocation-free zones

$$-e < x < -c, \quad c \quad x < e, < y = o$$

which are located between the crack and the plastic zones. The discontinuity of the displacement across the crack plane (y = o) can be represented by a continuous distribution of screw dislocations lying in the plane y = o parallel to the z-axis. The resistance to the motion of these dislocations is zero in the crack but is equal to the friction stress in the plastically deformed regions. Because of the stress concentration established at the crack tip, the shear stress σ_{yz} is expected to be greater than the friction stress in the DFZ.

If we define $f(x')$ to be the dislocation distribution function, then the number of dislocations with Burgers vector b located in the interval dx' is $f(x')dx'$. The equilibrium condition that the resulting force on each dislocation is zero can be described by the dislocation pile-up equation

$$A \int_{-a}^{-e} + \int_{-c}^{c} + \int_{e}^{a} \frac{f(x')dx'}{x-x'} = -G(x) \tag{1}$$

where $A = \mu b/2\pi$, μ is the shear modulus and

$$\begin{aligned} G(x) &= T & -c < x < c \\ &= T - \sigma & -a < x < -e \text{ and } e < x < a. \end{aligned}$$

The method of treating singular integral equations of this type has been discussed extensively by Muskhelishvili (4). A well-known solution which has been obtained in the area of fracture is the BCS solution dealing with the distribution of dislocations in the plastic zone. The present problem is similar to the BCS problem except that a dislocation-free zone exists between the crack and the plastic zone. In obtaining the solution to the present problem, it is noted that the shear stress σ_{yz} is equal to the friction stress σ everywhere in the plastic zone and hence the distribution function $f(x)$ must be bounded in the plastic zone including its two end points. Under these boundary conditions, the solution to the integral equation is given by (4)

$$f(x) = -\frac{1}{A\pi^2}\left[\frac{R_1(x)}{R_2(x)}\right]^{1/2} \int_{-a}^{-e} + \int_{-c}^{c} + \int_{e}^{a} \left[\frac{R_2(x')}{R_1(x')}\right]^{1/2} \frac{G(x')dx'}{x'-x} \tag{2}$$

where $R_1(x) = (x^2-a^2)(x^2-e^2)$ and $R_2(x) = (x^2-c^2)$. The solution exists only if the condition

$$\int_{-a}^{-e} + \int_{-c}^{c} + \int_{e}^{a} \left[\frac{R_2(x)}{R_1(x)}\right]^{1/2} G(x)dx = o \tag{3}$$

is satisfied. This will be referred to as the DFZ condition. The integral in Eq. (2) is defined by its Cauchy value. If e is made to approach c so that the DFZ vanishes, then Eq. (2) is reduced to the BCS solution and Eq. (3) to the BCS condition.

Distribution Function of Dislocations and the DFZ Condition

When the appropriate expressions for $R_1(x)$, $R_2(x)$ and $G(x)$ are inserted in Eqs. (2) and (3), we have been able to show that the distribution function and the DFZ condition can be expressed in terms of complete and incomplete elliptic integrals. The incomplete elliptic integral of the first kind is defined by

$$F(\phi, k^2) = \int_0^{\phi} \frac{dx}{\sqrt{1 - k^2 \sin^2 x}}$$

and the third kind by

$$\Pi(\phi, n, k^2) = \int_0^{\phi} \frac{dx}{(1 - n \sin^2 x)\sqrt{1 - k^2 \sin^2 x}}$$

and the complete elliptic integrals are obtained by setting $\phi = \pi/2$. The expression for the distribution function f(x) is very lengthy and hence will not be presented here. The DFZ condition is simply given as

$$\frac{\pi T}{2\sigma} = \frac{e^2-c^2}{e\sqrt{a^2-c^2}} \; \Pi(\pi/2, \frac{a^2-e^2}{a^2-c^2}, k^2) \tag{4}$$

where

$$k^2 = \frac{(a^2-e^2)c^2}{(a^2-c^2)e^2} .$$

The DFZ condition presented here may be considered as a generalization of the BCS condition.

Figure 1 shows a plot of f(x) as a function of distance along the direction of crack propagation for four levels of applied stress. The size of the DFZ is chosen to be 2% of the crack length (e/c = 1.04). These curves are obtained by first assigning the values of a/c and e/c and then determining the value of T/σ from the DFZ condition, Eq. (4).

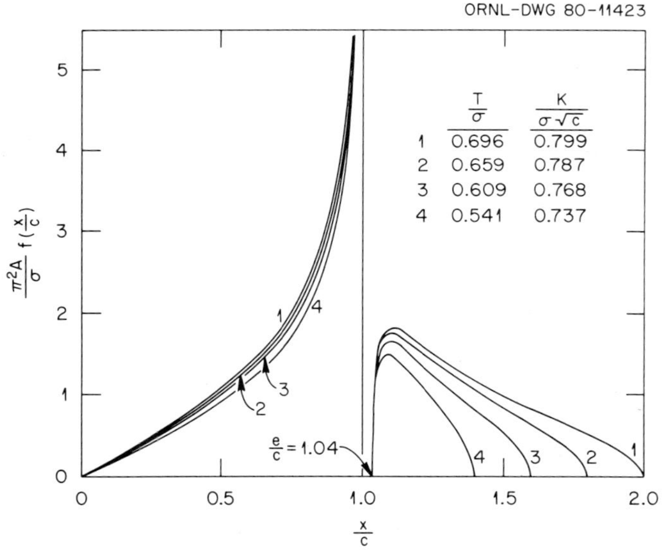

FIG. 1

Distribution function of dislocations f(x) obtained from the closed-form solution of Eq. (2) plotted as a function of distance along the crack at various levels of applied stress T.

The distribution function f(x) is obtained by substituting these values of a/c, e/c and T/σ into the solution of Eq. (2). The stress intensity factor $K/\sigma\sqrt{c}$ listed in Fig. 1 was obtained from the limiting value of f(x) $(c-x)^{1/2}$ as $x \to c$.

Figure 2 shows the displacement u(x/c) across the crack plane, which was obtained by integrating the distribution function f(x). It can be seen that the displacement is constant in the DFZ ($c < x < e$) and the magnitude there represents the value of the crack opening displacement (COD).

Elastic Stress Intensity Factor, COD and the Fracture Criteria

From the closed form solution of f(x), a simple expression for the stress intensity factor K is obtained from the limiting value of f(x) $(c-x)^{1/2}$ as $x \to c$. The result is

$$\frac{K}{\sigma\sqrt{c}} = \frac{2}{\sqrt{\pi}} \frac{(e^2-c^2)^{1/2}}{e} F(\pi/2, k^2) \quad , \tag{5}$$

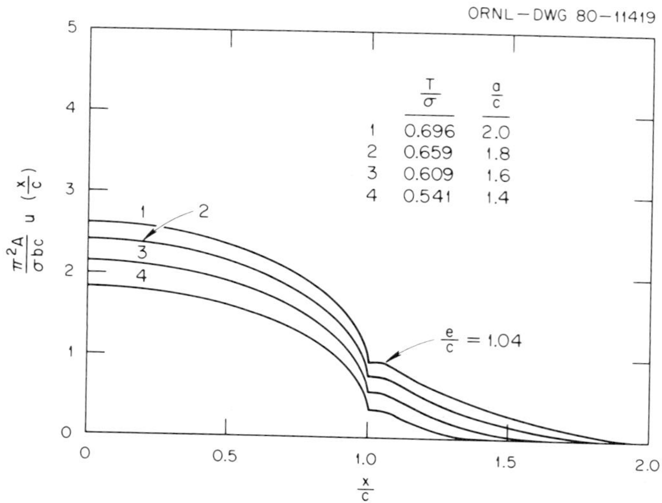

FIG. 2

Relative displacement u(x/c) plotted as a function of distance along the crack. The displacement is constant and is equal to the crack opening displacement in the dislocation-free zone ($1<x/c<1.04$).

where K is defined by $\sigma_{yz} = K(x-c)^{-1/2}/\sqrt{2\pi}$ for small x-c. The stress intensity factor K is zero for the BCS crack. This can be shown from Eq. (5) by noting that as $e \to c$, $(e^2-c^2)^{1/2}$ converges to zero much faster than the logarithmic divergence manifested by $F(\pi/2, k^2)$.

Figure 3 shows a plot of the stress intensity factor as a function of the applied stress for a range of values of a/c and e/c. It can be seen that for a specific value for K and for T there exist unique values for a and e. That is, when the values of K and T are specified it is not possible to have two different sizes of the DFZ. It is interesting to note that for small e and large a the stress intensity factor K is nearly independent of the plastic zone size and the applied stress, and hence depends solely on the size e of the DFZ. This is a remarkable result which can be used to determine experimentally the value of K by measuring the size of the DFZ, for instance, in the electron microscope. We have also made similar studies of the dependence of the COD on the applied stress, the plastic zone size and the DFZ.

In the present model, the energy release rate defined in terms of the J-integral should be the sum of the elastic energy release rate G ($= b\,K^2/4\pi A$) and the total force on the dislocations in the plastic zone given by COD x σ (7). G is often interpreted as the force acting on the locked dislocation at the crack tip (8). It is equal to the true surface energy density for brittle fracture. Since the critical condition for fracture is expressed as J_c = G + COD x σ, the fracture criterion can be defined in terms of the critical values of COD and K.

Recently, Thomson (4) and Weertman (5) proposed a model of fracture which includes an elastic core region next to the crack tip. They proposed an elastic stress-strain relationship for the regions of small strain and large strain greater than the ultimate theoretical strain ε_T, and plastic behavior for the region of intermediate strain. We have identified the elastic core region of their model with our dislocation-free zone at the crack tip.

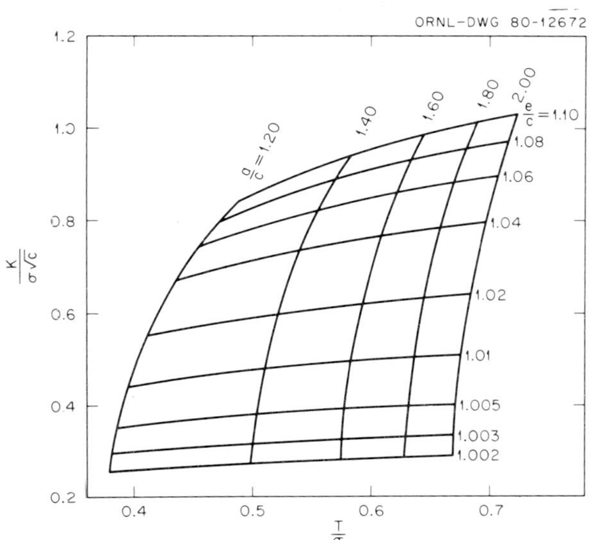

FIG. 3

Stress intensity factor K vs applied stress T for various sizes of the plastic zone a/c and the dislocation-free zone e/c.

In our model, the friction stress is constant throughout the plastic zone and the displacement is zero at x = a and increases to the maximum value equal to COD at x = e. The application of their model to our result would imply that, at the critical stage of ductile fracture, the value of COD is determined by the ultimate strain ε_T. If the total energy release reaches a critical value, then K, as well as e/c, should be at a critical value at fracture.

Rice and Thomson (9) proposed that the dislocations are generated at the crack tip when the elastic crack stress, represented by the stress intensity factor K, exceeds the image force on the dislocations exerted by the crack surface. We may, therefore, identify the critical value of K in our model with the minimum force capable of generating dislocations at the crack tip. The critical value of K determines the equilibrium configuration of the dislocations and the length of DFZ. The critical K can be measured directly from the length of the DFZ. A BCS crack, which may be considered as a crack with a vanishingly small DFZ, corresponds to a limiting case with a very small critical value of K.

Acknowledgement

The authors are indebted to F. W. Young, Jr. for his encouragement and to S. Kobayashi for many valuable discussions.

References

1. B. A. Bilby, A. H. Cottrell, and K. H. Swinden, Proc. R. Soc. A 272, 304 (1963).
2. S. M. Ohr and J. Narayan, Phil. Mag. A 41, 81 (1980).
3. S. Kobayashi and S. M. Ohr, Phil. Mag. A in press.
4. N. I. Muskhelishvili, "Singular Integral Equations," P. Noordhoff N. V., Groningen, 1953.
5. R. Thomson, J. Mater. Sci. 13, 128 (1978).
6. J. Weertman, Acta Met. 26, 1731 (1978).
7. S. -J. Chang, Int. J. Fracture, 16, R79 (1980).
8. B. A. Bilby and J. D. Eshelby, "Fracture," Vol. 1, H. Liebowitz ed., p. 99, Academic Press, New York (1968).
9. J. R. Rice and R. Thomson, Phil. Mag. 29, 73 (1974).

SOLUTE EFFECTS ON INTERGRANULAR FRACTURE

B.C. Edwards[+] and B.L. Eyre*
[+]Metallurgy Division, AERE Harwell, Oxon OX11 ORA, England
*Department of Metallurgy and Materials Science, University of Liverpool, PO Box 147, Liverpool L69 3BX, England

ABSTRACT

It is well-established that the segregation of solute elements, either individually or in combination, may promote intergranular embrittlement in a wide range of metal alloy systems. In this paper, the nature of the solute segregation and its influence on the macroscopic mechanical properties are reviewed followed by a discussion of the micromechanisms of the embrittlement.

A model for the grain boundary embrittlement by solute segregation is proposed which allows plastic flow ahead of a pre-existing crack prior to its eventual propagation in a brittle mode. The model considers the competition between the reduction in the crack tip tensile stress by plastic blunting and the modification of the stress field ahead of the crack tip by dislocation emission, which increases this stress. A reduction in the cohesive strength changes the balance in favour of brittle crack propagation. This is illustrated by recent quantitative measurements of the grain boundary fracture strength in molybdenum which provides direct evidence that oxygen segregated at boundaries results in small but significant reductions in the boundary strength thereby promoting intergranular fracture.

Recent results from studies of interatomic bonding at free surfaces provide a basis for understanding how impurity segregation may reduce the grain boundary cohesive strength, and these results are briefly reviewed.

1. Introduction

The common occurrence of ductile shear and transgranular cleavage failure mechanisms in polycrystalline metallic materials indicates that grain boundaries are not inherent preferred fracture paths. However, it is well-established that intergranular failure may occur over a wide range of temperatures and testing conditions in both b.c.c. and f.c.c materials indicating that grain boundaries are susceptible to embrittlement. Auger electron spectroscopy studies have demonstrated that intergranular failure in a wide range of alloy systems is associated with grain boundary segregation of solute atoms. The degradation of the properties of engineering alloys by such solute segregation is of major importance and examples include: (i) the reduced high temperature creep properties of ferritic and nickel based alloys; (ii) stress relief cracking of weldments during post weld heat treatment; (iii) the promotion of temper embrittlement in alloy steels; (iv) enhanced susceptability to environmental cracking processes including liquid metal embrittlement, stress corrosion cracking and hydrogen assisted cracking, (v) recrystallisation embrittlement of Group VI A refractory metals (Cr, Mo, W).

In an analysis of intergranular failure it is necessary to distinguish between crack propagation in situations where the solute segregation is produced by a prior thermal treatment and those in which the solute is able to segregate to the grain boundaries during the mechanical test, as for example in stress relief cracking and hydrogen assisted cracking. This paper addresses itself to an analysis of the effects of solute atoms on low temperature brittle fracture, with the solute grain boundary distribution being determined by a prior heat treatment at a higher temperature. In Section 2 of this paper, the parameters controlling grain boundary segregation are discussed, followed in Section 3 by a model of brittle fracture based on the competition between plastic blunting at a crack tip and the modification of the stress state ahead of the crack by dislocation emission. This model is illustrated, in Section 4, by recent quantitative measurements on the grain boundary fracture strength of molybdenum which provide direct evidence that the grain boundary segregation of oxygen results in small but significant reductions in the boundary strength thereby promoting brittle intergranular failure. In Section 5, the physical basis of how solute segregation influences the grain boundary cohesive strength is briefly reviewed with reference to recent surface physics results.

2. Solute Segregation

The description of interfacial segregation in an alloy system requires the construction of an isotherm which is a function of the grain boundary structure, the nature of the segregating specie(s), the bulk composition and the thermal history. The construction of such an isotherm, describing these variables and satisfying the boundary conditions, is thus extremely complex. However, Seah and Hondros (1) have shown that the solid state analogue of gas/surface adsorption can be used to describe segregation at solid/solid interfaces and it is now reasonably well-established that grain boundary segregation in binary alloys is generally of a Gibbsian (equilibrium) type. Grain boundary segregation can be correlated with decreases in grain boundary energy and this is considered to be the driving force for segregation. The segregation behaviour of multi-component systems (eg. low alloy steels) is more complex due to the possibility of non-equilibrium (transient) segregation being superimposed on the equilibrium-mode, eg. solute rejection from carbides (2-4). In addition, changes in the initial thermodynamic reference state may be produced due to matrix precipitation reactions occurring concomitantly with the segregation (4).

The thermodynamics of solute segregation in binary systems was formulated by McLean (5) who considered the grain boundaries as two dimensional phases having a higher solubility for the segregating species than the bulk crystal. The driving force for segregation was considered to be the change in free energy $\Delta G(seg)$, associated with the transfer of a solute atom, i, from a bulk lattice site B to a grain boundary site ϕ, and represented the excess lattice strains between the two sites. The segregation was shown to be described by the equation:

$$\frac{X^{\phi}}{1-X^{\phi}} \simeq X^{B} \exp\left(\frac{\Delta G_{seg}}{RT}\right)$$

where X^B is the concentration of the impurity in the bulk phase and X^{ϕ} its concentration in the grain boundary. The McLean treatment neglects any solute-solute interactions and the behaviour of multi-component systems cannot be explained using individual McLean isotherms. The segregation in ternary systems was treated by Guttmann (6,7) who showed that an attractive interaction between solutes, M, I, provided an additional driving force for segregation resulting in synergistic cosegregation. The segregation equations are of the same general form as that for the binary case except that the segregation energies are not constant but depend upon the solute (I)-solute (M) interaction and the intergranular concentration. In addition, a distinction was made between competitive and non-competitive segregation (8). In the case of competitive segregation, the equations are:

$$\frac{X_i^{\phi}}{1-X_I^{\phi} - X_M^{\phi}} \simeq X_i^{B} \exp\left(\frac{\Delta G_i(seg)}{RT}\right) \quad \text{for } i = I,M$$

$$\Delta G_I(seg) = \Delta G_I(ideal) + \alpha' X_M^{\phi} \; ; \quad \Delta G_M(seg) = \Delta G_M(ideal) + \alpha' X_I^{\phi}$$

where $\Delta G_I(ideal)$ and $\Delta G_M(ideal)$ are the segregation energies in the binary ideal solutions and

α' is the solute (M)-solute (I) interaction coefficient.

In the case of an attractive M-I interaction, $\alpha' > 0$, the segregation of one solute enhances the segregation of the other solute and may promote grain boundary segregation of non-surface active elements, eg. Ni in low alloy steels.

From a compilation of the experimental data of a number of binary systems, it was shown by Seah and Hondros (1) that the grain boundary enrichment factor (the ratio of the grain boundary solute content to the bulk content) was inversely proportional to the atomic solid solubility. AES studies have established in a wide range of binary systems (eg. Cu-Bi (9), Fe-S (1), Fe-Sn (1)) that impurity segregation is confined to within the first few atomic planes of the grain boundary. However, in low alloy steels where co-segregation of an impurity element with an alloy element occurs it has been shown in a number of different steels that the trace impurity is confined to within the first few atomic layers of the boundary, whereas the co-segregating alloy element has an enhanced concentration over $\sim$100 atomic planes from the grain boundary (4,10,11).

In summary, the degree of grain boundary segregation of solute atoms in a given solvent is inversely proportional to the solid solubility. In multi-component systems preferential chemical attractive interactions may exist between solutes which may enhance their respective segregation and induce the segregation of non-surface active elements.

3. The Criteria for Brittle Fracture

In this section a model for brittle fracture is presented in which we assume the prior existence of an atomically sharp crack. Possible mechanisms for the nucleation of this micro-crack include particle cracking, particle-matrix decohesion or dislocation pile ups. The propagation of this crack in a brittle manner involves bond breaking ahead of the crack, the driving force being the principal stress, σ_{11}, normal to the crack plane. In contrast, ductile fracture is defined as being plastic collapse ahead of the crack tip by a necking down process involving shear on inclined slip planes. In this case, the driving force is the resolved shear stress ahead of the crack.

The simplest case to consider is the brittle propagation of an atomically sharp crack in a linear elastic solid, ie. one in which there is no possibility of plastic deformation at the crack tip. Then the Griffith energy balance criterion (12) is both a necessary and sufficient condition for such propagation to occur, ie. the crack will propagate if the stored energy released at fracture is greater than the surface energy of the new surfaces. The energy release rate G is defined as the energy released per unit extension of the crack front per unit thickness. In the absence of plastic flow G is given by.

$$\left.\begin{aligned} G &= 2\gamma_s && \text{for cleavage} \\ &= 2\gamma'_s - \gamma_{gb} && \text{for intergranular fracture} \end{aligned}\right) \quad \ldots\ldots \quad (1)$$

where γ_s, γ'_s are the respective surface energies and γ_{gb} is the grain boundary energy.

The surface energy is defined as the energy/unit area required to break bonds at the crack tip. The surface energies do not necessarily correspond to the thermodynamic surface energies, even in the case of pure brittle cleavage, due to possible surface relaxation effects.

However, it is more realistic to consider an elastic-plastic solid and the problem is then to balance the critical conditions for a bond breaking against those for plastic deformation at the crack tip. This is the approach taken by a number of workers (13-16). Kelly et al (13) concluded that for brittle fracture to occur it is necessary to exceed the theoretical cohesive stress. The criteria for brittle fracture in the absence of crack blunting by thermal activation was defined as:

$$\frac{\sigma_{max}}{\tau_{max}} \geqslant \frac{\sigma_{th}}{\tau_{th}} \qquad ..... \qquad (2)$$

where σ_{max} is the maximum resolved tensile stress at the crack tip, τ_{max} is the maximum resolved shear stress at the crack tip, σ_{th} is the theoretical cleavage stress and τ_{th} is the theoretical shear stress. This criterion is based on the assumption that the stress to nucleate dislocations at the crack tip is the theoretical shear stress.

However, the attainment of the theoretical shear strength at the crack tip may not be a sufficient condition to blunt a crack. The shear stress is highly localised in the vicinity of the crack tip and this condition may not be sufficient to propagate the dislocations on slip planes extending away from the crack tip. This is analogous to the Griffith criterion in which the attainment of the theoretical cohesive strength at the crack tip is a necessary but not sufficient condition for brittle crack propagation. This was recognised by Rice and Thomson (4) who developed the concept of an activation energy U_{act} for the nucleation of a metastable dislocation loop at the crack tip. The activation energy U_{act} comprises the self energy of the dislocation loop, the energy to form a step at the crack tip as the dislocation loop is formed and the decrease in energy produced by the introduction of a dislocation into the crack tip stress field. If the activation energy is positive the work required for dislocation nucleation is greater than that for the production of two new surfaces and brittle fracture is produced; conversely, if the activation energy is negative the crack blunts by spontaneous dislocation nucleation and ductile fracture is produced. Alternatively, the critical conditions for brittle-ductile behaviour can be expressed in terms of the appropriate energy release rates G_{IC} and G_{crit}, where G_{IC} is the critical energy release rate for brittle fracture and G_{crit} is the critical energy release rate for dislocation nucleation. If $G_{crit}/G_{IC} > 1$, the load necessary for dislocation nucleation is greater than that for bond breaking and hence brittle fracture results. The converse applies for ductile behaviour. In the case of cleavage $G_{IC} = 2\gamma_s$. The above concept was extended to grain boundary fracture by Rice (15) and Mason (16).

The criteria used by Rice and co-workers (14-16) assume that if the condition for dislocation emission is achieved before that for bond breaking then the material will fail in a ductile manner. This assumes perfect elastic-plastic behaviour. We consider that this is too restrictive and is not consistent with experimental observations. More specifically, quite extensive plastic flow can occur ahead of a pre-existing crack prior to its eventual propagation in a brittle mode. This explains the wide variation in ernegy of fracture observed for example by Hull and co-workers (17,18) for cleavage fracture of tungsten and molybdenum crystals and by Kumar and Eyre (19) for intergranular fracture in molybdenum. It is therefore more realistic to consider brittle crack propagation from a pre-existing crack that has blunted as a result of plastic deformation. Using Inglis's solution (20) for stress and strain distributions around elliptical cracks Griffith (12) showed that the additional strain energy due to the presence of a crack is insensitive to crack tip shape. However, Inglis showed that the crack tip stress normal to the crack plane, σ_{11} depends on the crack tip radius ρ :-

$$\sigma_{11} = \sigma_{app}\ (1+2\sqrt{\frac{c}{\rho}}) \qquad \ldots\ldots \quad (3)$$

where σ_{app} is the applied stress and c is the half crack length for a centre crack in an infinite plate. Thus for the crack blunted by plastic flow the Griffith criterion remains a necessary condition but it may not be a sufficient condition for brittle fracture. As well as satisfying the energy release rate requirement, σ_{11} must also exceed the theoretical cohesion stress given by $(E\gamma/a_o)^{\frac{1}{2}}$. Thus following Gilman (21) and Westwood (22) a necessary condition for fracture is:

$$\sigma_{11} = \sigma_{app}\ (1+2\sqrt{\frac{c}{\rho}}) \geqslant \sqrt{\frac{E\gamma}{a_o}} \qquad \ldots\ldots \quad (4)$$

where a_o is the lattice parameter and γ is the surface energy of the new crack surface. When $\rho<<c$, the applied stress required to propagate a blunted crack is given by:

$$\sigma_{app} \geqslant \sqrt{\frac{E\gamma\rho}{4ca_o}} \qquad \ldots\ldots \quad (5)$$

It is important to note that σ_{app} increases as $\rho^{\frac{1}{2}}$.

Neumann (23) has shown how an initially sharp crack can blunt as a result of dislocation emission from the crack tip during the tensile loading part of a fatigue cycle. This is illustrated in Figure 1 in which the crack opening displacement (COD) and crack tip radius are indicated. Thus the crack tip radius, ρ, can be expressed in terms of the number of dislocations emitted:-

$$\rho = \frac{nb\sin\psi}{2} \qquad \ldots\ldots \quad (6)$$

where b is the dislocations Burgers vector and the angle ψ is defined in Figure 1. Sinclair (24) has calculated the number of dislocations emitted from a crack tip and their distributions on inclined slip planes. From such values the crack opening displacement can be calculated and for small scale yielding Sinclair has shown that an approximate treatment by Rice (25) is valid, ie:-

$$COD \simeq 2\rho = \frac{\pi(1-\nu)\sin^2\psi(1+\cos\psi)\sigma^2_{app}c}{8\mu\cos\psi(\sigma_o-\sigma_{app})} \qquad \ldots\ldots \quad (7)$$

where, $\sigma_o = \frac{\tau_o}{\sin\psi\cos\psi}$ is the uniaxial stress for flow on the inclined planes in the absence of a crack. Thus, in the general case of brittle fracture (either transgranular cleavage or inter-granular) it is necessary to consider how the conditions are achieved <u>following</u> plastic blunting of the crack tip. From equation (7) and assuming

$$\sigma_{11} = 2\sigma_{app}\sqrt{\frac{c}{\rho}} \quad \text{we obtain:-}$$

$$\sigma_{11} = 8\sqrt{\frac{\mu\cos\psi(\sigma_o-\sigma_{app})}{\pi(1-\nu)\sin^2\psi(1+\cos\psi)}} \qquad \ldots\ldots \quad (8)$$

If the stress σ_o for flow on the slip planes remains constant σ_{11} is reduced as the applied stress increases. The problem is therefore to understand how plastic deformation is made more difficult following plastic blunting of the crack tip, thus enabling σ_{11} to eventually exceed σ_{th}. We consider three mechanisms for increasing $(\sigma_o - \sigma_{app})$ within the crack tip plastic zone. One possible mechanism is that of work hardening along the slip bonds which results in an increase in σ_o. Then the criterion for achieving brittle fracture is that the rate of increase in the local crack tip tensile stress due to work hardening should exceed the rate of stress relaxation due to blunting. Failure to satisfy this criterion up to the point where $\sigma_{11} > \sigma_{th}$ will result in plastic collapse and ductile fracture. However, provided the criterion is satisfied it is possible for brittle fracture to occur after plastic deformation. Numerical calculations show that for small scale yielding, ie. $\sigma_{app} < 0.4\sigma_o$, which is the range over which the Rice equation holds, only a small increase in σ_o is required to maintain an increasing σ_{11}. The main problem with the work hardening explanation is in understanding the mechanism for increasing σ_o within the crack tip plastic zone. Initial spacings of obstacles such as precipitates, grain boundaries and grown in forest dislocations are likely to be large compared to the plastic zone size. However, obstacles may be created by the local plastic flow processes themselves, for example by the formation of non-conservative jogs on screw dislocations leading to the accumulation of debris from pinched off dipoles or point defect clusters.

Secondly, Sinclair (24) has shown that the edge dislocations emitted from the crack tip on inclined slip planes, as depicted schematically in Figure 2, exhibit a peak in their distribution close to the crack tip. A calculated distribution for a specific set of conditions as defined by ψ, σ_o, σ_{app} and b is shown in Figure 3. It can be seen that the dislocations exhibit a prominant peak in their distribution close to the crack tip which reflects the decrease in resolved shear stress with distance from the crack tip. Such a dislocation array will exert a back stress at the crack tip inhibiting further dislocation nucleation, thereby effectively increasing σ_o.

A third possibility to consider stems from the interactions between the elastic fields of the dislocations emitted from the crack tip, which will change the nature of the stress field ahead of the crack. Sinclair (24) has shown that the dislocations on the two slip bands (Figure 2) interact to increase the mean normal stress, $\bar{\sigma}_{11}$, within a crack tip radius ahead of the crack thus favouring brittle fracture over plastic flow. For the case of slip bands inclined at 60° to the crack plane the maximum value of the normal stress, σ_{11}(max), on the x axis, ie. in the crack plane, can be expressed (26) approximately as

$$\sigma_{11}(\max) = 5.0\mu(\bar{h})^{\frac{3}{4}} \qquad \ldots\ldots \qquad (9)$$

where $\bar{h} = \frac{\rho}{L\sin\psi}$ is the mean dislocation density on the slip bands and L is the slip band length. Again the criterion for brittle fracture to occur is that the increase in σ_{11}(max) due to dislocation interactions must be sufficient for it to exceed σ_{th}. During the early stages of plastic blunting when $\frac{L}{\rho}$ is small the dislocation density can achieve high values leading in turn to high values for σ_{11}(max), ie. $\gtrsim 0.1\mu$. This suggests that the dislocation interactions can lead to σ_{th} being exceeded close to the crack tip thus enabling brittle fracture to occur following plastic deformation. A further result from Sinclair's analysis (26) is that σ_{11}(max) is proportional to $(\sigma_o - \sigma_{app})^{\frac{3}{4}}$ which demonstrates a stronger dependence of σ_{11}(max) on σ_o when dislocation interactions are considered, compared to the elastic solution given in equation (8).

The important general conclusion to emphasise is that during the early stages of plastic deformation at a crack tip the crack tip normal stress, σ_{11}, relaxes slowly. It is conceivable that this will be more than off-set by one, or a combination of the above processes, namely, work hardening, an increased activation energy for dislocation nucleation and elastic interaction between dislocations on inclined slip planes.

Considering now the specific case of grain boundary fracture and how the criterion for brittle fracture along the boundaries might be influenced by solute segregation, this requires that the surface energy is replaced by a net energy term in which the gain in interfacial energy is accounted for, viz. $(2\gamma_s - \gamma_{gb})$.

Then a necessary condition for brittle intergranular propagation following plastic blunting is from (5)

$$\sigma_{app} \geqslant \sqrt{\frac{E(2\gamma_s' - \gamma_{gb})\rho}{4ca_o}} \quad \ldots\ldots \quad (10)$$

Thus segregated solute atoms at the grain boundaries may enhance the embrittlement susceptibility by either inhibiting plastic blunting of the crack and/or reducing the net surface energy. The former possibility implies an increase in the activation energy for dislocation emission from the crack tip, ie. an increase in γ_{step} in the terminology of Mason (16). γ_{step} is the work per unit area of new surface to form the "step" left behind at the crack tip after the nucleation of the dislocation loop. However, for reasons discussed in Section 5 it is expected that the segregated solute will reduce γ_{step} thus making dislocation nucleation easier. Moreover, solute atoms on the grain boundary will not have a significant effect on the flow stress, σ_{δ} opposing dislocation propagation on slip planes inclined to the boundary. Alternatively, plastic relaxation processes may occur on boundary planes inclined to and ahead of the crack tip, for example, at triple points in which case it is conceivable that the segregated solute stress will inhibit the relaxation process.

With regard to the second possibility involving a reduction in $(2\gamma_s - \gamma_{gb})$ this would result from a reduction in the cohesive strength across the boundary. We consider this to be a general consequence of segregation as illustrated by the experimental results obtained from molybdenum which are discussed in Section 4. A physical model to account for such reductions is presented in Section 5.

4. Grain Boundary Fracture in Molybdenum

A systematic study has recently been conducted by Kumar and Eyre (19) on the recrystallisation embrittlement of molybdenum. The embrittlement was shown by Auger Electron Spectroscopy to be associated with oxygen segregation to the grain boundaries and a series of mechanical tensile tests were conducted on "bamboo" specimens containing an identical type of grain boundary traversing the cross section. The variation of the total work of fracture as a function of grain boundary oxygen concentration is presented in Figure 4 for both unnotched and pre-cracked specimens. Transmission electron microscopy studies on the unnotched specimens with high levels of grain boundary oxygen ($X_o^\phi = 0.47$) showed a virtual absence of dislocations adjacent to the fracture surface, consistent with a purely elastic fracture. The large increase in the work of fracture for the unnotched samples with decreasing oxygen segregation was attributed to the influence of plastic flow prior to unstable brittle intergranular fracture.

In the case of the pre-cracked specimens exhibiting grain boundary segregation of oxygen, it is reasonable to assume a purely elastic failure. From the measured fracture stress and crack dimension the term $(2\gamma_s'-\gamma_{gb})$ was shown to increase by 60% for a decrease in the grain boundary oxygen level, X_o^ϕ from 0.47 to 0.11. This corresponds to an increase of 28% in the elastic fracture stress.

The crack tip radius of curvature ρ and σ_{11} may be calculated using equations (7) and (8), the results are presented in Table 1. It is interesting to note that σ_{11} increases by 22% as X_o^ϕ decreases from 0.47 to 0.11 which is in good agreement with the 28% increase predicted from surface energy calculations. It is interesting to note that in the paper by Kumar and Eyre (19) values of ρ and σ_{11} were obtained using the crack opening displacement (COD) expression given by Burdekin and Stone (27). The values of ρ obtained are larger than those obtained using equation (7) and this is consistent with the analysis of Sinclair (24) who showed that the Bilby-Cottrell-Swinden solution (23) (which is equivalent to that of Burdekin and Stone (27)) overestimates the value for the COD.

The above results demonstrate a general trend for the crack tip to blunt as X_o^ϕ decreases and the variation of σ_{11} with X_o^ϕ does not reflect the large increase in the measured fracture stress. The calculated value for σ_{11} was $\sim E/10$ for both unnotched and pre-cracked specimens which indicates that crack propagation occurs when the theoretical fracture stress is exceeded at the crack tip. These experimental observations support the model for brittle fracture outlined above and demonstrate that substantial plastic flow may proceed brittle crack propagation. A decrease in the solute concentration at the grain boundaries only has a relatively small effect on increasing the cohesive strength of the bonds at the boundary but this increase may be sufficient to allow dislocation emission from the crack tip thereby promoting blunting. It is this change in crack tip geometry which results in the subsequent large increase in the work of fracture. A delicate balance exists at the crack tip between initiation of plastic flow (via dislocation emission) and bond breaking and this determines the subsequent macroscopic fracture properties.

5. The Physical Basis for Intergranular Failure

The physical basis for the reduction in the cohesive strength of grain boundaries by the segregation of impurities is not well understood. Current grain boundary theories evoke either mechanical hard sphere models of neutral atoms (29,30) or solid state bulk concepts (31-33). Both these approaches are inadequate for explaining the changes in grain boundary cohesion.

The hard sphere model considers the reduction in grain boundary strength to be primarily caused by a size effect and neglects any effects of chemical interactions. In low alloy steels co-segregation of alloy-trace impurity elements occurs and a chemical interaction between the species has been determined using Mössbauer spectroscopy (34) and by the analysis of the fine structure of AES spectra obtained from intergranular facets (35,36). These observations confirm Guttman's theory (7) of chemically induced segregation and imply that an understanding of the reduction in grain boundary cohesion requires a consideration of the electronic structure of the segregated species and its effect on the surrounding metallic matrix.

The application of bulk solid state theories developed for infinite regular periodic crystals whilst explaining bulk properties (37) cannot be applied to grain boundaries which represent a strong perturbation of crystal periodicity (38,39). A more realistic approach is to consider a grain boundary to be analogous with a free surface, ie. a strong perturbation of solid state periodicity and symmetry with electron states perpendicular to the surface considered localised. Model calculations of free surface bonds are only in a preliminary stage (38) but do present a qualitative description of the problem as described by Losch (39) who attempted to describe bulk interfacial chemical reactions using concepts of free surface chemisorption studies. Both these phenomena have chemical reactions of an "adsorbate" with a "substrate" in common.

The major impurities of concern in iron and iron base alloys are shown in Table 2. In the case of pure iron the deleterious impurities are S, Se and Te whereas in the case of alloy steels the major impurities are P, Sn and Sb which co-segregate with a major alloying element (Mn, Ni or Cr). In the case of pure nickel sulphur has been shown to be a major deleterious impurity promoting intergranular failure (40). It is important to note that the above impurities have a similar electronic configuration in their outer unfilled shells (np^x) which implies that they have a strong tendency to form covalent bonds of a localised character.

Extensive surface physics studies have been conducted on the chemisorption of S on Ni (41) and Losch (39) used these results to illustrate his model of grain boundary cohesion. Analogous situations are considered to exist for the chemisorption of S on Fe and the effects of "alloy-impurity" complexes on the Fe matrix in alloy steels. In the case of S chemisorption on Ni it was shown that the formation of covalent S (3p) - Ni (4s) bonds is accompanied by a disruption of the neighbouring Ni (4s) - Ni (4s) metallic bonds. In Losch's grain boundary model (39) it is considered that charge transfer occurs from the metal atom (M) to the impurity atom (I) resulting in covalent (M-I) bonds which are localised in character (ie. next nearest neighbours do not contribute significantly) and the disruption of neighbouring metallic bonds. It is considered that the bond strength is reduced in the immediate neighbourhood of the segregated layer with fracture occurring preferentially along the grain boundaries by the breaking of the neighbouring M-M bonds (rather than the M-I bonds). The fracture path is considered to run parallel to the grain boundary impurity layer but may cross it due to irregularities in the boundary (Figure 5).

Similar arguments of charge transfer reducing the cohesive strength may be applied to the phenomena of hydrogen assisted cracking (42). Photoelectron studies (43) of H chemisorption on

Ni free surfaces has demonstrated a Ni (4s) - H (1s) interaction. The formation of the localised covalent Ni (4s) - H (1s) bond results in charge transfer from Ni to H and a disruption in the 4s-bonds of adjacent metal atoms. The 4s-bond disruption results in a decrease of the metallic cohesion between the H bonded Ni atom and its metallic neighbours. Chemisorption studies (43) of H on Fe also suggest the formation of Fe (4s) - H (1s) interaction.

Hydrogen is most probably dissolved in the bulk as a proton and preferentially segregates to lattice discontinuities (ie. dislocations or grain boundaries) with the H forming covalent (M-H) bonds in the vicinity of dislocations or other lattice defects. An interstitial H atom attracted by an edge dislocation is shown in Figure 6. The formation of such covalent bonds disrupts the neighbouring matrix bonds which lose their non-localised, non-directional nature, thus enhancing the ease of ductile, cleavage or intergranular fracture. The effect of covalent bond formation on the free energy separation diagram and the cohesive strength of the surrounding metallic matrix is shown schematically in Figure 7.

6. Conclusions

(1) The degree of grain boundary segregation of solute elements is inversely proportional to their solid solubility. In multi-component systems preferential chemical interactions may exist between solutes which enhance their respective segregation characteristics. Solute impurity segregation has been shown to be confined to within the first few atomic layers of the grain boundary.

(2) A model for brittle fracture has been proposed which considers plastic blunting of the crack tip prior to brittle crack propagation. Pre-requisites for brittle failure are the Griffith criterion and that the theoretical bond strength σ_{th} be achieved in the crack tip region. Plastic blunting of the crack results in a reduction in σ_{11} but this stress may be subsequently increased by work hardening on the slip planes and/or by a modification of the stress state ahead of the crack produced by dislocation interactions. The grain boundary segregation of solute reduces the cohesive strength of the boundary, thereby favouring brittle fracture.

(3) Experimental results from molybdenum have shown a reduction in σ_{th} with increasing levels of oxygen at the grain boundaries. Crack blunting prior to brittle fracture was observed for low oxygen segregation levels which produced large increases in the work of fracture. These observations are consistent with the above model.

(4) The physical basis for the reduction in σ_{th} with solute segregation may be explained by recent surface physics results which indicate the formation of covalent bonds between the solute(s) and matrix which modify the adjacent metallic bonds with a concomitant decrease in σ_{th}.

Acknowledgements

The authors would like to thank Drs J Sinclair and R Bullough for their helpful discussions and constructive criticisms during the preparation of this paper. Particular thanks are extended to Dr J Sinclair for allowing us access to his unpublished results.

References

1. M.P. Seah and E.D. Hondros, Proc. R. Soc. A335, 191 (1973)
2. E.B. Kula and A.A. Anctil, J. Mat., ASTM 4, 817 (1969)
3. J.R. Rellick and C.J. McMahon Jr., Met. Trans. 5, 2439 (1974)
4. B.C. Edwards, B.L. Eyre and G. Gage, Acta Met. 28, 335 (1980)
5. D. McLean, Grain Boundaries in Metals, Oxford Univ. Press (1957)
6. M. Guttmann, Surface Sci. 53, 213 (1975)
7. M. Guttmann, Met. Trans. 8A, 1383 (1977)
8. M. Guttmann and D. McLean, "Interfacial Segregation", Eds. W.C. Johnson, J.M. Blakely, p.261, ASM (1979)
9. B.D. Powell and H. Mykura, Acta Met. 21, 1151 (1973)
10. H.L. Marcus, L.H. Hackett Jr., and P.W. Palmburg, p.90, Temper Embrittlement of Alloy Steels, STP, 499
11. A. Joshi and D.F. Stein, ibid, p.59
12. A.A. Griffith, Phil. Trans. Roy. Soc., A221, 163 (1921)
13. A. Kelly, W.R. Tyson and A.H. Cottrell, Phil. Mag. 15, 567 (1967)
14. J.R. Rice and R. Thomson, Phil. Mag. 29, 73 (1974)
15. J.R. Rice, "Effect of Hydrogen on the Behaviour of Materials", Eds. A.W. Thompson and I.M. Bernstein, p.455, American Institute of Mining Engineers (1976)
16. D.D. Mason, Phil. Mag., 39, 455 (1979)
17. P. Beardmore and D. Hull, J. Inst. Metals, 94, 14 (1966)
18. J. Cordwell and D. Hull, Phil. Mag., 19, 951 (1969)
19. A. Kumar and B.L. Eyre, Proc. Roy. Soc. London A370, 431 (1980)
20. C.E. Inglis, Proc. Int. Naval Architect, L.V. 1, 219 (1913)
21. J.J. Gilman, "Plasticity", Pergamon Press, New York (1960)
22. A.R.C. Westwood and M.H. Kandar, Phil. Mag., 8, 787 (1963)

23. P. Neumann, Acta Met., 22, 1155 (1974)

24. J. Sinclair, Nucl. Metall., 20, 388 (1976)

25. J.R. Rice, J. Mech. Phys. Solids, 22, 17 (1974)

26. J. Sinclair, Private Communication

27. F.M. Burdekin and D.E.W. Stone, J. Strain Analysis, 1, 145 (1966)

28. B.A. Bilby, A.H. Cottrell and K.H. Swinden, Proc. Roy. Soc. London A279, 304 (1963)

29. E.D. Hondros and D. McLean, Phil. Mag., 29, 771 (1974)

30. M.P. Seah, Proc. Roy. Soc. London A349, 535 (1976)

31. A.R. Troiano, Trans. ASM, 52, 54 (1960)

32. E.S. Machlin, Scripta Met., 12, 111 (1978)

33. L. Marchut and C.J. McMahon Jr., "Electron and Positron Spectroscopics of Materials", Ed. O. Buck, Academic Press, New York (1978)

34. B.C. Edwards, B.L. Eyre and T.E. Cranshaw, Nature 269, 47 (1977)

35. W. Losch and I. Andreoni, Scripta Met., 12, 277 (1978)

36. W. Losch, Acta Met., 27 567 (1979)

37. D.G. Pettifor, CALPHAD 1, 305 (1977)

38. K.H. Johnson, C.Y. Yang, D. Vvedensky, R.P. Messmer and D.R. Salahab, "Interfacial Segregation", Eds. W.C. Johnson and J.M. Blakely, ASM (1979)

39. W. Losch, Acta Met., 27, 1885 (1979)

40. C.L. White, Ph.D. Thesis, Michigan Technical University (1974)

41. S.P. Walch and W.A. Goddard III, Surface Sci., 72, 645 (1978)

42. W. Losch, Scripta Met., 13, 661 (1979)

43. J.E. Demnth, Surface Sci., 65, 369 (1977)

44. K.Y. Yu, W.E. Spicer, I. Lindau, P. Pianetta and S.F. Lin, Surface Sci., 57, 157 (1976)

Table 1

Values of σ_{11}, σ_f, ρ and $(2\gamma_s - \gamma_{gb})$ as a Function of Grain Boundary Oxygen Content in Molybdenum

Material Condition	Grain Boundary Oxygen level X_o^ϕ	Fracture Stress σ_f, MNm^{-2}	$(2\gamma_s - \gamma_{gb})$ Jm^{-2}	From Kumar & Eyre (19) ρ m	From Kumar & Eyre (19) σ_{11} GNm^{-2}	Calculated using eqns. (7),(8) ρ m	Calculated using eqns. (7),(8) σ_{11} GNm^{-2}
Pre-cracked specimens	0.47	30.8	2.03	1.70×10^{-9}	32.2	1.07×10^{-9}	40.2
	0.14	48.6	2.83	1.89×10^{-9}	39.4	1.22×10^{-9}	49.0
	0.11	47.7	3.32	1.76×10^{-9}	39.4	1.13×10^{-9}	49.0
Unnotched specimens	0.17	350			38.0		
	0.05	940			33.5		

Table 2

The Outer Electronic Configuration of the Principal Embrittling Elements

Period n	IIIB (ns^2, np^1)	IVB (ns^2, np^2)	VB (ns^2, np^3)	VIB (ns^2, np^4)
2	B	C	N	O
3		Si	P	S
4		Ge	As	Se
5		Sn	Sb	Te
6		Pb	Bi	
	enhance grain boundary strength?	alloy steels		Pure Fe

Group and Spectral Notation

Unfilled np^x shells have strong tendency to form covalent bonds of a localised character.

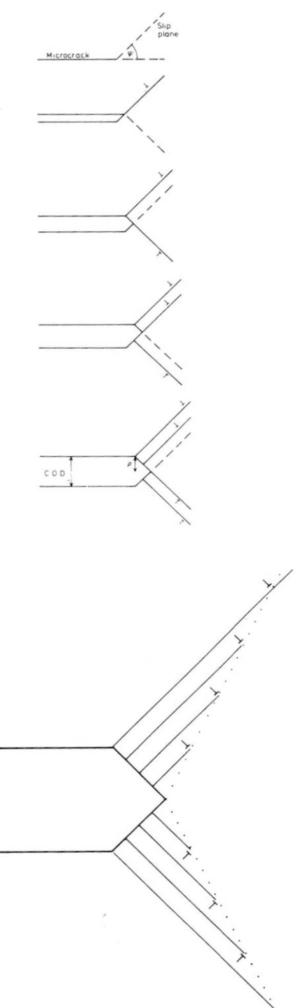

Figure 1 Schematic representation of crack blunting of an initial atomically sharp crack, by dislocation emission from the crack tip along slip planes inclined at angles $\pm\psi$ through the current position of the crack tip.

Figure 2 The dislocation distribution around a blunted crack used in the calculations of Sinclair (24). The dislocations lying along the dotted arcs are treated as a continous distribution. Yield is assumed to occur continuously and be symmetric on both sides of the crack plane.

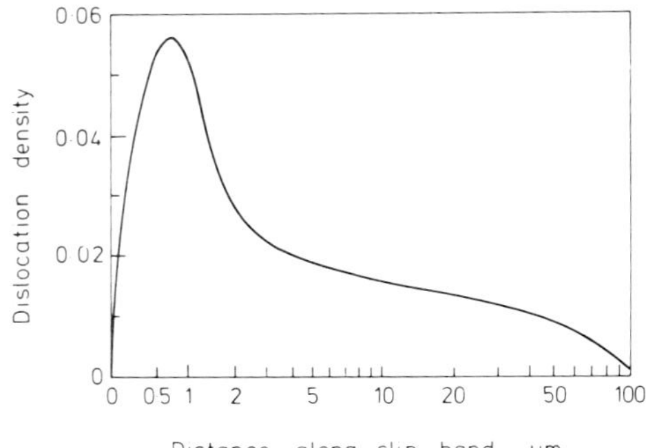

Figure 3 The calculated dislocation density, h, along a slip plane inclined at angle $\psi \pm 60^{\circ}$ to the crack plane. Calculations made for a crack half length c = 100µm; slip band length L = 100µm; crack tip facet width = 1µm; applied stress σ_{app} = 1400MNm^{-2}; flow stress τ_{o} = 1120MNm^{-2}. The figure is plotted on a non-linear scale in order to show detail at both small and large distances from the crack tip (after Sinclair (24)).

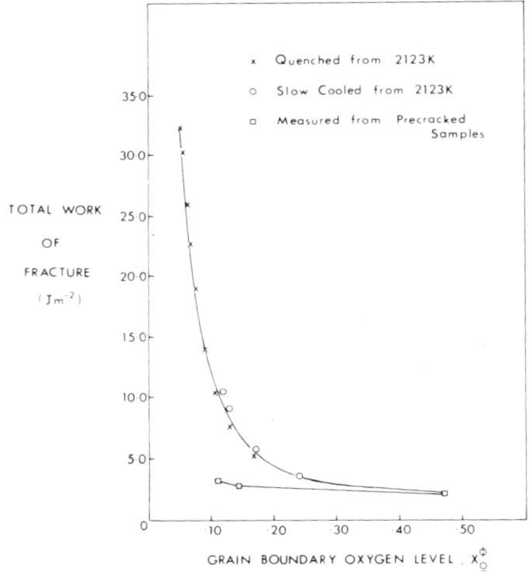

Figure 4 A plot of the total work of fracture as a function of grain boundary oxygen content, X_o^{ϕ}, in pre-cracked and unnotched "bamboo" tensile specimens of molybdenum (19).

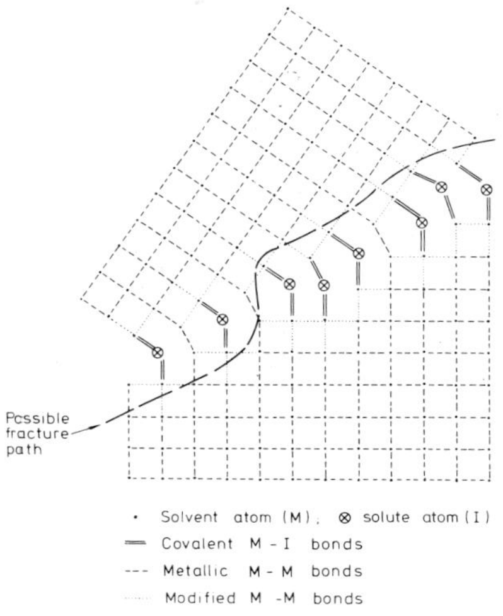

Figure 5 The formation of covalent metal (M) - impurity (I) bonds at grain boundaries modify the adjacent metallic bonding (M-M) resulting in a reduced bond strength. A possible grain boundary fracture path is shown produced by the breaking of the modified M-M bonds. The fracture path is parallel to the impurity layer but may cross it due to irregularities in the boundary (after Losch (39)).

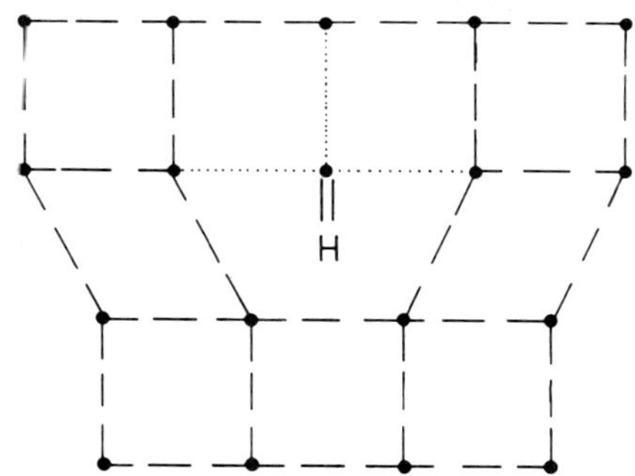

Figure 6 The interaction of hydrogen with an edge dislocation producing covalent M-H bonds and a disruption of the neighbouring metallic M-M bonds (after Losch (42)).

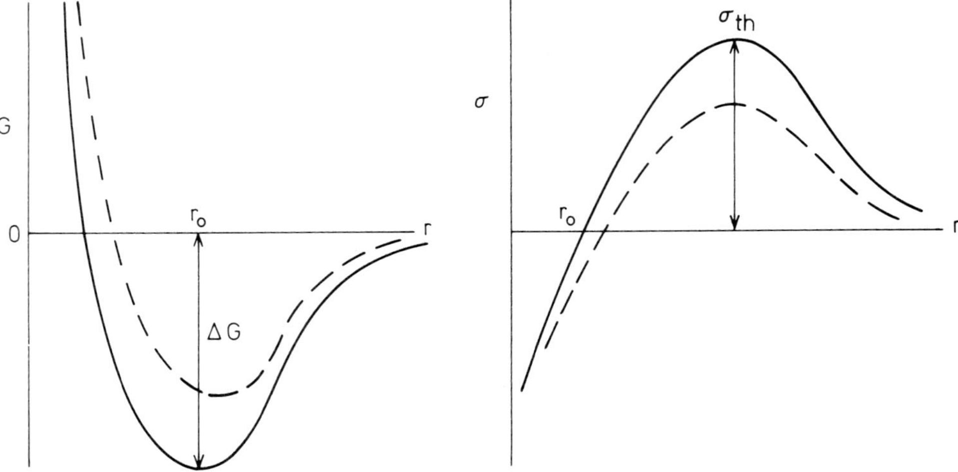

Figure 7 The effect of covalent bond formation on the free energy separation diagram (G-r) and the cohesive strength, σ, of the metallic bonds adjacent to the segregated layer. (——— metallic M-M bonds; ------ modified M-M bonds).

"DISLOCATION CELL STRUCTURES AND A MECHANISM FOR EROSION OF DUCTILE MATERIALS"

Gordon A. Sargent*, M. K. Keshavan and H. Conrad*
Department of Metallurgical Engineering and Materials Science
the University of Kentucky, Lexington, Kentucky 40506

ABSTRACT

Surface damage produced by the erosion of an annealed Stellite alloy has been studied by the techniques of transmission electron microscopy (TEM) and x-ray diffraction (XRD). Specimens of the alloy were eroded by gas carried angular alumina particles of two sizes (mean sizes 10μm and 30μm), at velocities of 28 and 68 m/s for angles of impingement between 10° and 90°.

The depth of the surface damaged layer was measured as a function of the above experimental conditions, using an x-ray diffraction line broadening analysis technique, and was found to increase with increased particle velocity and angle of impingement.

From a detailed analysis of the x-ray diffraction line profiles it was possible to determine the dislocation cell size and cell size distribution and to estimate the dislocation density within the damaged layer. From these measurements mean dislocation cell sizes of about 70Å were obtained corresponding to dislocation densities of about 10^{12} cm^{-2}.

Dislocation structures were also observed by the technique of transmission electron microscopy. It was found that the cell sizes, cell size distributions and dislocation densities determined from this technique were in good agreement with those calculated from the x-ray line broadening analysis.

INTRODUCTION

The phenomenon of erosion has been the subject of many investigations over the past ten years (1-10). The emphasis of this work, however, has been primarily to develop empirical relationships by which the erosion behavior of a specific material could be predicted as functions of particle velocity and angle of impingement on the surface. The importance of a materials structural condition and the relationship between plastic flow processes and erosion have only been given consideration in a few studies (4-6).

To date very little work has been done to characterize the structural damage produced at and below the surface of the material subjected to erosion. Wright and Edington (11-12) and Ives and Ruff (13-14) have, however, studied topographic and subsurface structural features associated with erosion damage using the techniques of SEM and TEM.

In this paper a measure of the structural damage caused by erosion, has been obtained using the x-ray line broadening analysis. Data such as depth of damage, dislocation cell sizes and densities were determined by this technique and these results were compared with similar data obtained by more direct microstructural studies using TEM.

*Also associated with the Institute for Mining and Minerals Research Kentucky Center for Energy Research Laboratory, Iron Works Pike, P. O. Box 13015, Lexington, Kentucky 40583.

EXPERIMENTAL TECHNIQUE

The material used for this study was annealed Stellite 6B alloy (obtained form the Stellite Division of Cabot Corporation). Specimens for the erosion tests were in the form of 2.5 cm squares by 0.2 cm thickness. The loss of material due to erosion was determined by weighing the specimens both prior to and after a test. The erosion was carried out in a sand blaster type of erosion tester. Angular alumina particles of 240 and 400 grit (mean size 30 and 10μ respectively) were used for the experiments. The particles were accelerated to velocities Of 28 and 68 m/s by pressurized air. The particle velocity was measured by the two-disc method of Ruff and Ives (15).

The SEM and x-ray diffraction studies were carried out on the 2.5 cm square specimens which were used for the weight loss measurements. Thinner slices were cut from the specimens for the TEM studies, and small discs 3mm in diameter were punched out to fit the TEM specimen holder. The eroded side of the disc was protected by laquer and the specimen was thinned from the opposite side by jet electropolishing.

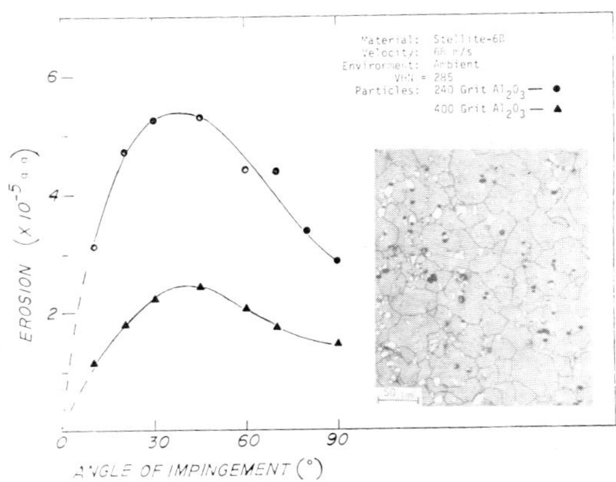

Figure 1. Effect of angle of impingement on the erosion of annealed Stellite-6B (microstructure shown in insert) using 240 and 400 grit alumina particles.

RESULTS AND DISCUSSIONS

The initial optical microstructure and the effects of particle size and angle of impingement on the erosion of annealed Stellite 6B alloy are shown in Figure 1. It can be seen that increasing the particle size increases the erosion at all angles and that a maximum in the erosion occurs at an angle of about 35° for both particle sizes.

After erosion the topography of the surface was observed by SEM. Figure 2(a) shows that at small angles (10°) the particles produce long tracks on the surface, which may be related to a cutting or micro-machining mechanism of material removal. At 90°, Figure 2(b), the surface damage is more characteristic of that resulting from continuous battering, with material being extruded up and around each individual particle.

In a previous study by the present authors (16) x-ray diffraction line broadening measurements were used as a technique to characterize the subsurface damage caused by erosion under a variety of experimental conditions. It was found that after erosion there is typically a decrease in intensity and an increase in the breadth of the diffraction profiles. Figures 3 (a), (b), (c), show the line broadening obtained after erosion with particles of different sizes, different velocities, and for different angles of impact.

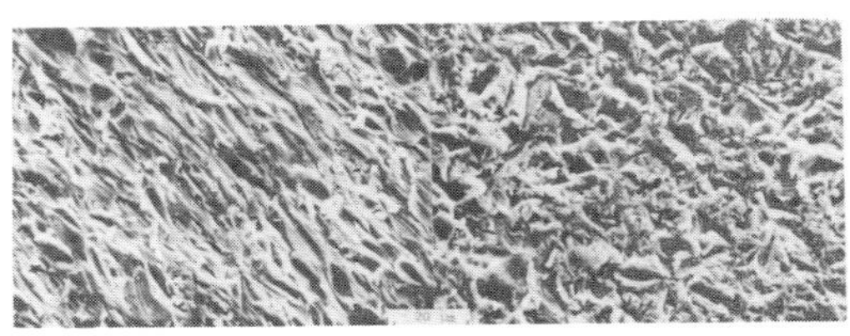

Figure 2. Scanning electron micrographs of eroded Stellite-6B surfaces using 240 grit angular alumina particles at 68 m/s for (a) 10° and (b) 90° angle of impingements.

ELECTROPOLISH AND ERODE-(111) — × - with 240 grit Al_2O_3 particles
ELECTROPOLISH & ERODED-(111) — △ - with 400 grit Al_2O_3 particles
ELECTROPOLISHED STELLITE-(111) — □

INTENSITY (ARB.UNITS) ×10³
DIFFRACTION ANGLE (TWO THETA)

(a)

ERODED STELLITE-6B AT 28 M/S — △
ERODED STELLITE-6B AT 68 M/S — ×
ELECTROPOLISHED STELLITE-6B-(111) — □

INTENSITY (ARB.UNITS) ×10³
DIFFRACTION ANGLE (TWO THETA)

(b)

ERODED STELLITE-(111) — × α = 10°
ERODED STELLITE-(111) — △ α = 90°
ANNEALED STELLITE-(111) — □

INTENSITY (ARB.UNITS) ×10³
DIFFRACTION ANGLE (TWO THETA)

(c)

ERODE AND ELECTROPOLISH-(111) — △ - depth removed = 4um
ELECTROPOLISH & ERODED-(111) — ×
ELECTROPOLISHED STELLITE-(111) — □

INTENSITY (ARB.UNITS) ×10³
DIFFRACTION ANGLE (TWO THETA)

(d)

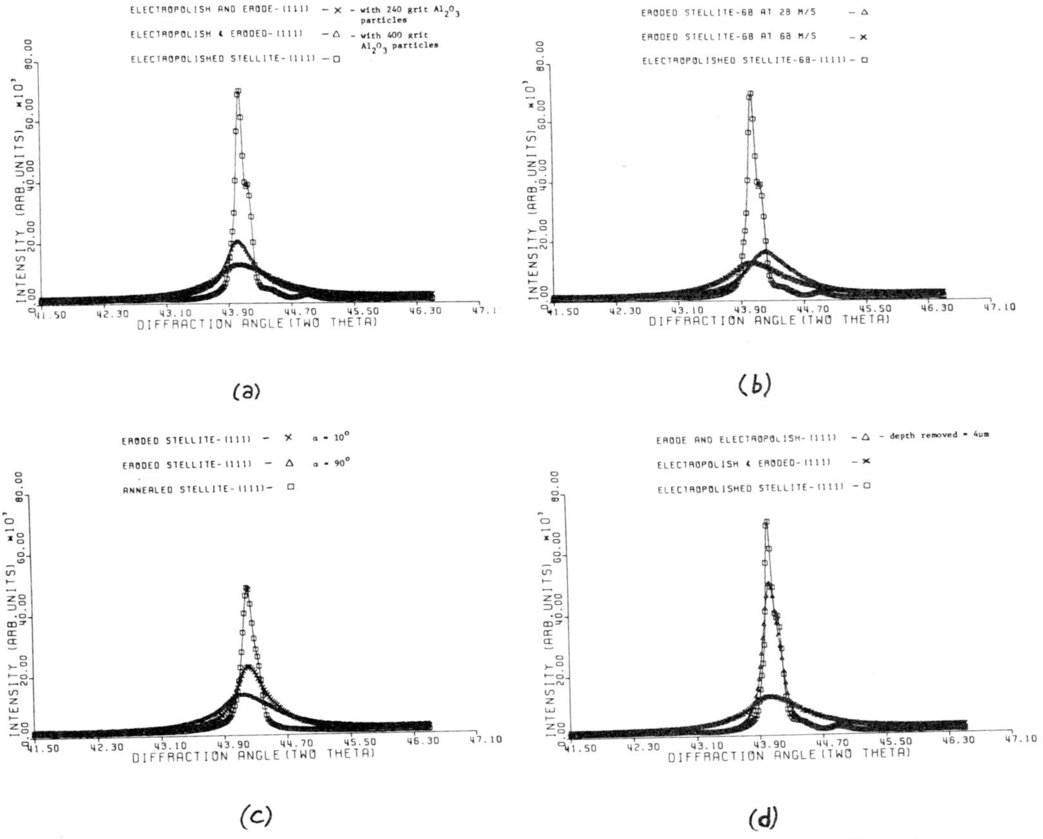

Figure 3. (111) x-ray diffraction profiles of the initial annealed Stellite-6B surface a) following erosion with 240 and 400 grit alumina particles at 68 m/s and 90° impact b) effect of velocity c) effect of angle d) effect of removal of surface material by electropolishing.

To ensure that such line broadening was due totally to strain and not due to increased surface roughness caused by the erosion, both the annealed and the eroded specimens were given a stress-relieving treatment at 700°C for 30 minutes. After this procedure it was found that the diffraction profiles were identical and hence surface roughness could be disregarded as contributing to the line profile broadening.

The depth of the damaged layer due to particle impingement under the various experimental conditions of velocity, angle of impingement and particle size was also determined. In each case the diffraction profiles of the annealed and eroded samples were obtained as shown in Figure 3(a), (b), (c). The eroded surfaces were then electropolished to various depths and the corresponding diffraction profiles were again measured as shown in Figure 3(d). Figure 4 shows the variation below the surface, as a function of erosion conditions, of the half-breadth of the diffraction profile B_D corrected for instrumental broadening as given by

$$B_D = \sqrt{B^2\,(hk\ell)_{eroded} - B^2\,(hk\ell)_{annealed}}$$

It is of interest to note that the damage layer extends to about 16μm below the surface for specimens eroded by 240 grit (30μm) alumina particles at 68 m/s and 90° angle of impingement. There is little change in maxim depth of the damaged layer upon reducing the velocity to 28 m/s. However, the depth of the damaged layer does decrease slightly with lower angles of impingement.

From an analysis of the diffraction profiles using either the Warren-Averbach (17) (double-order) or the Ganesan-De Angelis (18) (single-order) methods it is possible to determine the

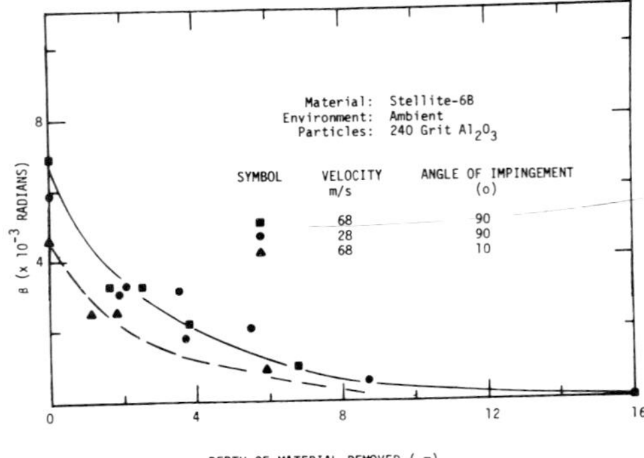

Figure 4. Variation of half-breadth (β) of the diffraction profiles as a function of depth of material removed.

strain in the material due to the deformation caused by the erosion and to estimate the dislocation cell size and size distribution function. Similar data can be also obtained by TEM analyses although in this case cell sizes are more difficult to measure. Figure 5 shows a TEM of the structure of Stellite-6B prior to, and after erosion. The cell size distribution obtained from a large number of such micrographs of the surface region of eroded specimens are shown in Figure 6, together with the size distributions obtained for both the single-order and double-order x-ray diffraction line broadening analyses. Cell sizes measured were in the range 20 to 200Å, for both the x-ray and TEM studies, however, the mean cell size from the line broadening studies was about 70Å which was slightly less than the mean value of 110Å measured by the TEM work.

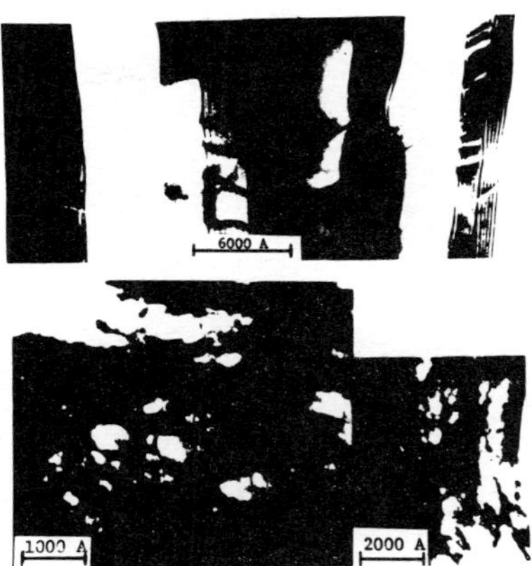

Figure 5. Transmission electron micrographs of Stellite-6B before and after erosion.

Dislocation densities were calculated using the relationship

$$\rho \simeq \frac{1}{(\text{Dislocation cell size})^2}$$

This yields dislocation densities in the range $\rho \simeq 2.8 \times 10^{11}$ to $11 \times 10^{12}/\text{cm}^2$ with an average value of 2×10^{12} as calculated from the x-ray line broadening studies and 8×10^{11} calculated from TEM. In view of the high dislocation densities obtained it is understandable why particle size and velocity do not have a significant effect on the degree of damage.

The cellular dislocation structure and the depth of the damaged layer produced during the erosion of this ductile alloy is very similar in appearance to that observed in some forms of wear. It seems reasonable to propose, therefore, that material loss in both cases may occur by the same mechanism. Suh (19) has proposed a delamination theory of wear in which it is suggested that dislocation tangles at cell boundaries are locations for the formation of voids at which cracks may be nucleated. Subsequent propagation of the cracks along the cell boundaries results in separation of material from the heavily damaged surface region.

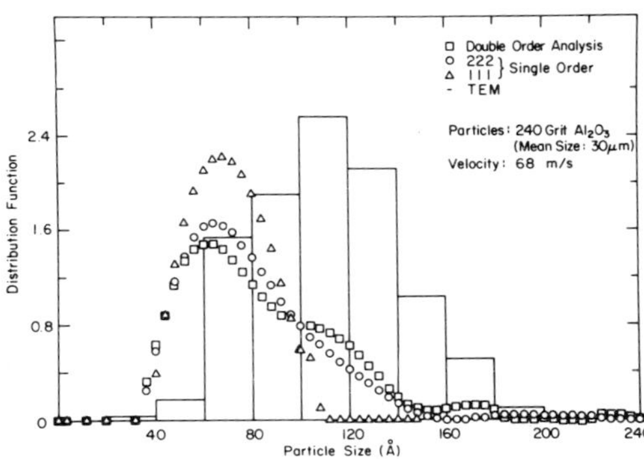

Figure 6. Distribution function of dislocation cell sizes calculated from x-ray diffraction profiles compared to those measured by TEM.

REFERENCES

1. J. G. A. Bitter, "Wear", 6, 5 and 169, (1963).
2. I. Finne, J. Wolak and Y. Kabil, Journal of Metals, 2, 682, (1967).
3. W. Head and M. E. Harr, Wear, 15, 1, (1970).
4. W. H. Jennings, W. J. Head and C. R. Manning, Jr., Wear, 40, 93, (1976).
5. J. H. Neilson and A. Gilchrist, Wear, 11, 111, (1968).
6. G. L. Sheldon, J. Basic Eng. (Trans. ASME Ser. D), 92, 619, (1970).
7. C. E. Smeltzer, M. E. Gulden and W. A. Compton, J. Basic Eng. (Trans. ASME Ser. D), 92, 639, (1970).
8. G. L. Sheldon and A. Kanhere, Wear, 21, 195, (1972).
9. G. P. Tilly, Wear, 23, 87, (1963).
10. G. P. Tilly and W. Sage, Wear, 16, 447, (1970).
11. J. W. Edington and I. G. Wright, Wear, 48, 131, (1978).
12. J. W. Edington and I. G. Wright, Wear, 48, 145, (1978).
13. L. K. Ives and A. W. Ruff, Wear, 46, 149, (1978).
14. L. K. Ives and A. W. Ruff, ASTM STP 664, W. F. Adler, Ed., American Society for Testing and Materials, 5, (1979).
15. A. W. Ruff and L. K. Ives, Wear, 35, 195, (1975).
16. M. K. Keshavan, G. A. Sargent, H. Conrad, to be published in the Proceedings of the 12th Annual IMS-ASM Meeting, Tamiment, PA, July 8-11, (1979).
17. B. E. Warren and B. L. Averbach, J. Appl. Phys. 21, 595, (1950).
18. P. Ganesan, H. K. Kuo, A. Saavedra and R. J. De Angelis, J. of Catalysis, 52, 310, (1978).
19. N. P. Suh, Wear, 25, 111, (1973).

DISLOCATIONS AND THE FATIGUE STRENGTH OF METALS

L. M. Brown
Cavendish Laboratory, Madingley Road,
Cambridge CB3 OHE

Abstract

Recent work has shown that cyclic straining induces well-defined dislocation structures in ductile crystals. The structures seem to result from the cross-slip of screw dislocations and from the mutual trapping of edge dislocations. A brief review is given of recent experimental work on these structures, and an attempt is made to explain them in terms of dislocations. There is a pattern of long-range internal stress associated with the structures, and the significance of this for crack nucleation is stressed. It seems possible that the fatigue endurance limit of metals is the stress level above which screw dislocations on different slip planes cross-slip and annihilate each other rather than trap each other without annihilation.

§ 1. Introduction

Recent experimental work on the structure of persistent slip bands and the initiation of cracks by cyclic straining has greatly clarified the understanding of the fatigue strength of metals. The subject has been almost entirely observational, with minimal intervention by dislocation theory. It seems fair to say that this state of affairs will continue for some time, because the theories so far proposed are obviously incomplete and deficient. Nevertheless, it is important to try to model the micromechanical processes which ultimately produce one of the crucial limitations to the use of ductile metals for components subjected to alternating stress and strain.

In this article, two quite different uses are made of dislocation theory. In § 3, the interactions between real, observable crystal dislocations are analysed to try to understand cyclic hardening, and the observed saturation of the hysteresis loops (the phenomenon of saturation is one in which, after a certain number of cycles of prescribed plastic strain, the material enters a nearly unchanging state and further hysteresis loops remain the same shape until fracture commences). On the other hand, in § 4, the macroscopic internal stresses associated with the saturated state are modelled with the aid of fictitious dislocations. The main purpose of these dislocations is to solve rather complicated problems in elasticity, and to perform certain averages of the stress fields of the real dislocations. It is important at the outset that the reader should distinguish the two types of model.

In § 2, a brief review is presented of the current state of knowledge. Space does not permit an ample review, but it is hoped that by consulting the references the interested reader can fill out the picture. It is most unfortunate that the scope of the article cannot be enlarged to permit anything other than the discussion of single-crystal behaviour; it must just be taken on trust that the ideas apply in slightly modified form to the far more important (but more complicated) case of polycrystals.

§ 2. The Current State of Knowledge

Fatigue is a weakness suffered by ductile crystals. Brittle materials undergo brittle fracture, and cannot display the progressive nucleation and propagation - in a seemingly brittle manner - of a crack which develops only small amounts with each cycle.

After large numbers of cycles at constant plastic strain amplitude, the dislocation arrangement in ductile metals becomes a 'two-phase' arrangement, with a rather ill-defined matrix structure, consisting of clumps of dipoles with clear vein-like channels between them; embedded in the 'matrix' is a phase of well-ordered wall structure, which consists of periodic walls of dislocation dipoles with a definite spacing and orientation - they are perpendicular to the direction of slip in crystals oriented for single slip. The wall-structure phase carries all the plastic strain, and produces characteristic extrusions and intrusions at the surface of the crystal. These surface features were observed many years ago and called 'persistent slip bands'. It has been known, again for many years, that the persistent slip bands play a key role in the development of fatigue cracks on polished and otherwise crack-free surfaces.

These features are shown schematically in fig. 1. The wall spacing, labelled d, is typically a few micrometres, but it is very temperature-dependent. In copper, d = 1.4 μm at room temperature, but decreases to 0.45 μm at liquid helium temperature. The walls maintain their spacing right across the crystal, which may contain 10^4 or so such walls. The width of the walls, denoted by w, is typically about one-tenth of the spacing. The height of the walls (or the width of the persistent slip band) denoted h_b is very variable, but seldom seems to be less than about a wall spacing; it can be of macroscopic dimensions.

The variability of the wall height results from the fact that the volume fraction of persistent slip bands depends linearly on the imposed plastic strain amplitude (2). This result can be understood if the band carries a definite and limited plastic strain amplitude, so that the crystal adjusts the volume fraction of band to accommodate the external strain. Again in copper, at room temperature, the strain in the band $\varepsilon_b \simeq 0.75\%$. The saturation stress, that is, the maximum stress level achieved in the hysteresis loop at saturation, is also a characteristic of the wall structure because in saturation only the wall structure is plastically active. In copper this stress is 27 MPa, and very temperature dependent: the saturation stress is approximately inversely proportional to the wall spacing. Now, as has been pointed out by a variety of writers (2,3,4,5), the saturation stress (determined in a test at constant plastic strain amplitude) plays a singular role: if, in a stress controlled test, the stress level exceeds the saturation stress, the crystal fills with persistent slip and ultimately fractures; if the stress level is less than the saturation stress, no persistent slip bands form and the crystal can safely undergo very large numbers of cycles. The saturation stress is the same as the endurance limit of the crystal. This behaviour is neatly described by the two-phase model (2) in which the wall structure is regarded as a condensed phase and the matrix structure as an uncondensed phase: the endurance limit plays the role of the vapour pressure of the condensed phase. There is also a critical plastic strain amplitude below which persistent slip is not observed, and this is interpreted as the plastic strain which the matrix will bear; its value in copper is about .006%.

The characteristic dislocation structure of the persistent slip is produced in a wide variety of crystal structures. In addition to face centred cubic, similar structures have been observed in Mg (6), silicon-iron (7), AgCℓ (8), and in germanium at high temperatures (9). The phenomenon is clearly a dislocation phenomenon, not crucially affected by detailed crystallography. The structure can also be found in such materials after work-hardening (10) or in the presence of precipitation or dispersion hardening (11). In circumstances where the saturation stress is less than the flow stress due to other hardening mechanisms, the nucleation of persistent slip leads to fatigue softening and the cyclic straining has a particularly disastrous effect on strength properties.

The detailed dislocation arrangements in the wall structure have been studied by a number of authors. The walls consist of dislocation dipoles, all derived entirely from primary Burgers vectors. There is no need to consider slip on secondary systems to model the persistent slip phenomenon. Densities and size distributions of edge dislocation dipoles and of screw dislocations have been determined by electron microscopy (12,13,14) and the results have been confirmed by X-ray techniques (15). There is little point in reproducing the detailed figures here, except to point out that in the walls the dislocation density can be extremely high, of the order of $5 \times 10^{15} m^{-2}$.

A striking feature of the dislocation dipoles is that the narrower dipoles tend to be of vacancy type (16). This observation is relevant to copper, but the production of vacancies by cyclic straining is probably a general phenomenon: for example, vacancy loops nucleate and grow in Mg at room temperature (6) and this is most likely the result of the break-up of very fine dislocation dipoles of vacancy type.

One very striking observation which must be mentioned in any brief review is that of Neumann (17), who observed marked instabilities during cyclic hardening. In Neumann's experiments, the cycling is controlled by slowly increasing the maximum stress level in each cycle. The resulting strain amplitude shows marked oscillations, rising to a maximum and then falling back to a very low value. This continues as the stress is increased to a level corresponding to the saturation stress, when the hysteresis loop becomes stable. These experiments demonstrate the delicacy of the balance of forces which control the dislocation structures.

Figure 2 shows a hitherto unpublished electron micrograph of wall structure in Ni which has been cyclically hardened at constant plastic strain amplitude to saturation. It shows all the features outlined above. It is important to realise that the dark patches in such pictures are unresolved clumps of dislocation dipoles: individual dipoles can be revealed only by techniques of weak-beam microscopy which allow higher resolution. The interested reader is referred to the references quoted above: it is impossible in a brief space to give much feel for the structures of their mechanical behaviour. The reviews by Mughrabi (5) and by Mughrabi et al.(18) are particularly useful.

§ 3.1 Possible Dislocation processes in Cyclic Hardening and Saturation

All recent authors are agreed that the main reason for cyclic hardening is the accumulation of edge dislocation dipoles caused by the irreversible motion of screw dislocations, the irreversibility resulting from cross-slip. The problem is to translate this concensus into a useful theory. The treatment given here suffers serious shortcomings, as the reader will see, but at the present tentative stage of theoretical development it seems better to give an account of current ideas than to remain silent.

An absolutely essential idea is that of the passing stress required to make dislocations of equal and opposite Burgers vector pass each other. If the separation of glide planes is S, then two screw dislocations will pass each other at a stress

$$\sigma_S = \frac{\mu b}{4\pi S} \qquad 3.1a$$

and two edge dislocations at a stress

$$\sigma_E = \frac{\mu b}{8\pi(1-\nu)S} \qquad 3.1b$$

(μ is the elastic shear modulus, b is the Burgers vector, and ν is Poisson's ratio.)

It is an important fact that for both edge and screw dislocations the passing stress in independent of its sense, whatever the initial configuration of the dislocation. The passing stress can be viewed as the stress required to break up a dislocation dipole, so that equations 3.1a and 3.1b define critical heights of dipole for a given stress level: wider dipoles break up, but narrower ones remain intact.

It is widely agreed that the stress required to force screw dislocations between the walls is also an important flow stress for the structure at all stages of cyclic hardening. The main evidence at saturation is reviewed in the appendix to this paper, where it is shown empirically that

$$\sigma_{sat} \approx (2 \pm .3)\,\frac{\mu b}{(1-\nu)d} \qquad 3.2$$

This equation says that the saturation stress is about twice the Orowan bowing stress for a screw dislocation anchored at both ends by walls of spacing d. Earlier indications by Kwadjo and Brown (6) that the saturation stress may be even higher than this are incorrect, the error arising from a failure to use a sufficiently sophisticated formula for the bowing stress. On the other hand, corrections introduced by Basinski et al. (1) indicate that the saturation stress may exceed the Orowan stress by only a factor of 1.3 rather than a factor of 2. All authors agree that the saturation stress exceeds the bowing stress.

Figure 3 indicates in a schematic way a possible process for cyclic hardening. In fig. 3a, screw dislocations push to and fro between the walls, suffering little resistance except (i) the dipoles they drag and (ii) possible internal stress caused by the fact that the walls are only partially penetrable by the edge dislocations making up the dipoles. It seems fairly clear that in the hardening stage there is an internal stress contribution of appreciable magnitude (19,20). At the edges of the walls, however (regions marked s in the figure), there must be dipole exchange, because such dipoles have to be fine enough to withstand the applied stress together with the local internal stress. As the stress levels rise, more dipoles are laid down, decreasing the length of the bowing screw dislocations and producing a differential strength within the walls: the wall centres are weak, but their surfaces strong (fig. 3b). At the centre of the accumulated dipoles, the dipole spacing will be that more appropriate to the stress level at which the dipole was formed. However, the dipole cannot break up, because the wall is not wide enough to allow long range dislocation motion. However, after the stress has approximately doubled, the centre of the wall gives way and dipole accumulation starts at a higher stress level (fig. 3c). This schematic diagram is consistent with the ideas of Kuhlmann-Wilsdorf and Laird (21) and with recent observations by Winter (22). It is not intended that the 'wall' widths in fig. 3 should be absolutely identical, but the clear areas between the walls should be of uniform width, as indeed Winter finds.

The refinement process sketched here allows strain bursts of the type observed by Neumann (17). To see this, it is necessary to estimate the mean free path of the screw dislocations. At a given stress level, screws on slip planes separated by less than S (equation 3.1a) will trap each other and form a screw dipole which can contribute no further to the strain. There is thus a capture cross-section for screw dislocations; if the mean free path is Λ and the screw dislocation density is ρ_s, we have

$$\rho_s \Lambda S = 1 \qquad 3.3$$

(No factor of two arises because, as in the kinetic theory of gases, we imagine all but one of the dislocations to be stationary and shrunk to a point; the moving one has cross-section 2S per unit length. However, only half the dislocations (assumed random in sign) are available for annihilation.)

However, if the movements of the edge dislocations within the walls contribute a negligible strain, the homogeneous plastic strain available per quarter cycle is

$$\varepsilon_p = \tfrac{1}{2}\rho_s b\Lambda = \frac{b}{2S} = \frac{2\pi\sigma}{\mu} \qquad 3.4$$

(The factor $\frac{1}{2}$ arises because the mean free path refers to movement over the whole half cycle.)

These equations show that there will be strain bursts, for when the centres of the walls collapse the area swept on each cycle by the screw dislocations will increase dramatically - by a factor of two - and then as the obstacles build up, the swept area declines. The magnitude of each strain burst will be successively greater, as the mean free path increases with overall stress level. We have approximate relationships between wall spacings, stress-amplitudes and strain-amplitudes at the nth and (n+1)th burst

$$\frac{\varepsilon_n}{\varepsilon_{n+1}} = \frac{\sigma_n}{\sigma_{n+1}} = \frac{d_{n+1}}{d_n} = 2 \qquad 3.5$$

This behaviour is qualitatively that observed by Neumann (17), but the observed ratios of successive stress- and strain-amplitudes are about 1.2 rather than 2. Smaller ratios will result from any co-operative effect within the collapsing walls which can produce premature collapse; also, smaller ratios can result from structural inhomogeneity in which only a fraction of the walls contribute to each burst. Strain bursts can only be observed under stress-control, obviously. Co-operative collapse of dipole arrays has been discussed previously by Neumann (23), but his picture is two-dimensional and rather different from that envisaged here.

The question now arises, what is it that prevents an infinite sequence of bursts? Why does the cyclic hardening, and its accompanying refinement of structure, come to an end with the production of persistent slip with a well-defined stress- and strain-amplitude?

§ 3.2 A Tentative Theory of the Saturation

An answer to the question just posed, is that screw dislocations travelling on planes separated by less than a critical value of S (S_c, say) cross-slip and annihilate each other. If the stress level is increased above the stress corresponding to the critical spacing, no further strain is available without unlocking edge dislocations from the walls - a difficult process. Thus under stress control, at the critical stress level corresponding to the last strain burst, the crystal fills with persistent slip (wall structure) and then must enter another mode of deformation, less dependent upon screw dislocations. If we imagine a test carried on under strain control, sufficient persistent slip is created to accommodate the strain, and no more - the strain available in the band is limited by the critical mean free path.

Thus equation 3.4 gives us a relationship between the saturation stress and the plastic strain amplitude in the persistent slip band

$$\varepsilon_b = \frac{2\pi\,\sigma_{sat}}{\mu} \qquad 3.6$$

This equation is approximately borne out by the data: in copper, $\varepsilon_b \sim .75\%$ and $2\pi\sigma_{sat}/\mu \sim .4\%$ so the equality holds to within a factor of two. A further test of the ideas is possible by another relationship (equations 3.2, 3.6):

$$\varepsilon_b = \frac{4\pi}{(1-\nu)}\,\frac{b}{d}$$

This enables us to relate the plastic strain amplitude in the band with the microscopically observed wall spacing; we find $4\pi b/(1-\nu)d \simeq .5\%$ and again the agreement for copper is satisfactory. One final test can be made: it is possible to estimate S_c, the critical separation of screw dislocations. We find, again for copper,

$$S_c = \frac{b}{2\varepsilon_b} \simeq 200\ \text{Å}$$

Mughrabi's work with neutron pinned dislocations (5) enabled him to measure the smallest screw dislocation dipole in copper at room temperature: it is 500 Å wide, about a factor of two larger than our estimate.

The theory presented here is closely related to ideas put forward by Mughrabi (5,18). According to Mughrabi, the screw dislocations travel in groups of three or four, each with the same sign. However, the main idea that the annihilation of screw dislocations by cross-slip is an essential feature of dislocation movement within the persistent slip band is common to his theory and to the one presented here. One satisfactory feature of such theories is that they build in a strong temperature dependence from the start: S_c is expected to depend strongly on temperature, and since S_c determines the scale of the wall structure, the striking temperature dependence which is a feature of the observations can be accounted for. The ideas of Kuhlmann-Wilsdorf and Laird (21) do not have this feature.

Unfortunately, the theory is deficient in many respects. There is, throughout, the constantly recurring factor of two discrepancies between theory and experiment. One trouble seems to be that the bowing screws encounter some resistance in addition to the Orowan stress: the factor of two in equation 3.2 is not understood. (The treatment by Basinski et al.(1) makes this only a factor of 1.3, but still the discrepancy exists.) Another fault in the theory is that it does not indicate how to calculate the complete hysteresis loop - for this, the behaviour of the edge dislocations must be modelled. In spite of these deficiencies, the theory makes one very definite prediction: namely, that the plastic strain amplitude in a persistent slip band is proportional to the saturation stress (equation 3.5). This prediction has not yet been tested. If it is borne out, the theory must be along the right lines, but if different behaviour is observed, it seems that the whole basis of the theory must be reconsidered. For the time being, it seems reasonable to assume that the fatigue endurance limit of metals is controlled by the cross-slip of screw dislocations, and that the value of the endurance stress is that stress which can just separate two screw dislocations which are on the point of annihilation.

Some further commentary on the theory may be helpful. As it stands, the theory predicts a square hysteresis loop: screw dislocations emerge from the walls, and move to annihilation at constant stress in one half-cycle. Such a model requires the easy nucleation of screw dislocations from the walls, but because the stress required to separate an edge dislocation dipole is independent of the sense of the stress, it is not possible to have a dipole which is stable at the end of one cycle (positive stress, say) which can be broken up at the beginning of the next cycle (negative stress, say). On the other hand, the emergence of screw dislocations from walls is a common observation, and one must assume that configurations of tripoles and quadrupoles which are stable under one sign of stress break up under another. Some theoretical work by Neumann (23) indicates that this can be the case.

There are several possible sources of extra hardening to explain the discrepancy between the bowing stress and the saturation stress. One is the internal stress produced by edge dislocations laid down at the boundaries of impenetrable walls. On this view, the walls act almost as undeformable particles in dispersion hardening; they will introduce a 'work-hardening' which is proportional to the volume fraction of wall, accompanied by a large Bauschinger effect. This may well be the case in the early stages of cyclic hardening (19,20), but cannot be important at saturation because the amount of work-hardening in the hysteresis loop is very small: the loops are, indeed, very nearly square. Another and more probable source of extra resistance is the stress required for one dislocation to overcome neighbouring screw dislocations while it is itself bowing between the walls. This seems to be a very difficult situation to analyse, requiring an understanding of the co-operative effects of many simultaneously bowing and interacting dislocations. Another possibility for extra hardening is simply debris left between the walls, for which evidence can be cited (24).

The discrepancies of factors of two may not be very significant considering the approximate treatment of statistical processes which enables one to write equations 3.3 and 3.4. An interesting consequence of the incomplete nature of the theory is that it makes no prediction of the density of screw dislocations, but only of the product of the dislocation density and the mean free path. In a simple and tentative theory, this failure may be an advantage, because one can concentrate on the essential ideas and forget the secondary complications. However, any complete theory must enable the dislocation densities to be calculated and compared with experiment.

§ 3.3 Summary

The mechanism of cyclic hardening is mainly the accumulation of edge dislocation dipoles caused by the cross-slip of screw dislocations. This leads to the idea that the bowing stress for screw dislocations to pass between clumps of dipoles must be equal to the stress required to break up the weakest of the dipoles laid down as the dislocation bows; this in turn implies that previously deposited dipoles are weaker than currently deposited ones. One thus visualises a structure in which older parts are inherently unstable, and break up when the stress rises sufficiently. The process comes to an end when screw dislocation dipoles of sufficient stability to withstand the applied stress cannot survive: they cross-slip and mutually annihilate. This picture leads in a natural way to the explanation of characteristic stress and strain amplitudes associated with persistent slip bands. The theory makes one definite prediction, not so far tested: the strain amplitude within a band should be a strongly decreasing function of temperature, following exactly the saturation stress. This prediction means that for a given plastic strain amplitude, the volume fraction of persistent slip band should decrease with decreasing temperature. Until this behaviour has been checked, it seems foolish to elaborate and improve the theory.

§ 4.1 Internal Stresses associated with Persistent Slip Bands

The fact that the dipoles in the walls of the persistent slip band are mostly of vacancy type indicates that the band is accompanied by macroscopic internal stresses.

There are several ways of understanding the resulting stress distribution. The model of the band drawn in fig. 1 shows each wall of vacancy dipoles replaced by a single fictitious dipole at the ends of the wall. This replacement is a way of modelling the internal stress which ignores the fluctuating stresses associated with the individual dislocations of each dipole, but emphasises the long-range stresses associated with the array. Provided the action of the stress field is considered at points much further from the array than the width of the persistent slip band, that is, much further away than h_b, such a replacement will be valid. Observations and calculations suggest that the typical fictitious Burgers vector is not much different in magnitude from a single real Burgers vector: the fictitious Burgers vector is equal to the real Burgers vector multiplied by the ratio of the summed heights of the real dipoles to the height of the imaginary dipole, h_b. Such an argument shows clearly that the material in the persistent slip band is subject to a tension along the line AA'; this tension is compensated by compression outside the band.

It is important to recognise that this state of stress is produced by slip alone. It results from a preference for vacancy dipoles over interstitial dipoles, which may arise either from difficulties with producing extrinsic stacking fault in close-packed materials, or from the elastic energy penalty incurred by interstitial defects compared with vacancy defects. Motion of screw dislocations is able to convert interstitial dipoles to vacancy dipoles (4) and indeed to transfer the 'half-plane' of a vacancy dipole to the surface of the persistent slip band and so to produce an extrusion: see fig. 4. This motion is of the sort envisaged long ago by Mott (25) and is the fundamental reason for the surface features caused by persistent slip.

Another way of understanding the stress distribution is shown in fig. 5. As an aid to clarity, the band has been drawn at right angles to the tensile axis. First, fig. 5a, cut the band from the matrix; then mark out those regions which will be bounded by vacancy dipoles. Thirdly, fig. 5c, remove the 'half-planes' of the dipoles and allow the dipoles to collapse: as a consequence, the band will get shorter as shown. To make the band fit the matrix, tractions must be applied (fig. 5d); then the band may be welded back into place and the tractions released, fig. 5e. Finally, the material which formed the 'half-planes' may be glued to the surfaces to form the extrusions of the persistent slip band. This style of analysis shows directly that the band is in tension, and it shows that the height of the extrusions is related directly to the strain in the band.

The internal stresses described here have several important properties. Apart from the stresses at the very surface of the band (fig. 5e) the stress field is a simple tension parallel to the primary Burgers vector. It therefore exerts no force upon primary dislocations, and can build up in the course of cycling without any effect upon the distribution of such dislocations. However, if the stress level exceeds that for flow on secondary systems, secondary slip will occur and it will occur on critical and conjugate slip planes; observation of such slip late in a fatigue test can be regarded as experimental evidence for the build-up of stresses (16). One can also estimate the magnitude of the tensile strains involved, because the corresponding tensile stress must be less than that which will cause flow on secondary systems; the maximum magnitude of the elastic tensile strain is about 5×10^{-4}.

Another feature of the stress field is that it provides a driving force for decohesion of the matrix from the persistent slip band. The elastic strain energy in the band is released when a crack runs along the boundary of the band. This type of cracking has recently been studied by Kendall (26) and it plays a key role in the strength of adhesive joints. The simplest familiar example of such cracking is the peeling of paint from a wall: the crack between the paint and the wall runs parallel to the direction of the residual stress in the paint, the residual stress caused usually by shrinkage of the paint when it dries. The point is that the crack runs parallel to the slip band, and this is the distinguishing feature of Stage I fatigue cracks. (A good, short introduction to the subject of fatigue fracture can be found in the book by McClintock and Argon (28)).

It is obviously worth getting an improved understanding of such stress fields, and the next three sections are devoted to this purpose.

§ 4.2 Using Fictitious Dislocations to Model the Stresses due to Persistent Slip

The drawing of the persistent slip band shown in fig. 1 suggests a simple way of calculating stress distributions in the neighbourhood of the band. Head (27) presents complete solutions for edge dislocations in a half-space. By integrating his expressions it is possible to obtain the stresses in the persistent slip band in the neighbourhood of one surface of the band. The procedure is as follows:

Head gives expressions for the stresses due to a single dislocation of Burgers vector b making an arbitrary angle ϕ with the x-axis, the dislocation being distant a from a traction-free surface which coincides with the y axis (fig. 6 shows the geometry). Head's expressions are of the form

$$\sigma_{ij}/D = f_{ij}(x,y',a,\phi) \qquad 4.1$$

Here, σ_{ij} stands for the σ_{xx}, σ_{yy}, σ_{xy} components of the stress, and

$$D = \mu b/2\pi(1-\nu) \quad ,$$

the assumption being, as usual, isotropic elasticity. Equation 4.1 states in abbreviated form combined results for a dislocation with $\phi = 0$ (Burgers vector perpendicular to the surface) and $\phi = \pi/2$ (Burgers vector parallel to the surface).

To adapt Head's expressions for our purposes, we let b dt be the Burgers vector in the interval dt distant t from the surface, measured along the slip plane (see fig. 6). The quantity D thus becomes

$$D = \mu b'/2\pi(1-\nu)d \qquad 4.2$$

where b' is the fictitious Burgers vector and d is the wall spacing: the quantity b'/d is of course the tensile strain in an infinite persistent slip band.

Now the co-ordinates can be transformed as follows:

$$\left.\begin{aligned} y' &= y + t\sin\phi \\ a &= t\cos\phi \end{aligned}\right) \qquad 4.3$$

The required stresses are now calculable as integrals

$$\sigma_{ij}/D = \int_{t=0}^{\infty} f_{ij}(x,y,\phi,t)\,dt \qquad 4.4$$

The integrals involved are all elementary, although there is much scope for algebraic error. In the sections to follow, the results are merely quoted and, although they have all been checked in various ways, it is difficult to be certain that every part of the algebraic expressions is correct.

§ 4.3 Surface Stress near a Persistent Slip Band

An important part of the solution is the stress in the surface, which can have only one component, σ_{yy}. For a single dislocation at (a,0), we find

$$f_{yy} = \frac{8a^2y'\cos\phi - 8ay'^2\sin\phi}{(a^2 + y'^2)^2} \qquad 4.5$$

and for an array

$$\sigma_{yy}/D = \int_{t=0}^{t_{max}} \frac{-4t^3\sin2\phi + 8yt^2\cos\phi(1+\sin^2\phi) - 4y^2t\sin2\phi}{(t^2 - 2yt\sin\phi + y^2)^2}\,dt \qquad 4.6$$

It is clear that the leading term of the integral in equation 4.6 will be logarithmic, so that there is a logarithmic singularity in the stress. Thus attention must be paid to the limits of integration: t_{max} would otherwise be infinity. The integral can be evaluated and the result is

$$\sigma_{yy}/D = -2\sin2\phi\,\ell n\left[1 - \frac{t_{max}\sin\phi}{y} + \frac{t^2_{max}}{y^2}\right] + 4\left[\phi + \frac{\pi}{2}\,\mathrm{sgn}(y)\right]\cos2\phi - 3\sin4\phi + \tan^2\phi(2\sin2\phi - \sin4\phi) \qquad 4.7$$

Equation 4.7 contains terms which may not be quite correct: they indicate a finite value of the stress at infinite distance from the dislocation array. However, it is probable that this problem arises from Head's original solution, which, in treating a half-plane, ignores corrections from more distant surfaces which must exist. The problems do not appear in the solution for a full persistent slip band, which consists of two equal and opposite layers of edge dislocations, separated by an interplanar spacing h. This solution is

$$\sigma_{yy}/D = -2\sin2\phi\,\ell n\left\{\frac{1 - \dfrac{t_{max}\sin\phi}{y} + \dfrac{t^2_{max}}{y^2}}{1 - \dfrac{t_{max}\sin\phi}{y+h\sec\phi} + \dfrac{t^2_{max}}{(y+h\sec\phi)^2}}\right\} + 2\pi\cos2\phi[\mathrm{sgn}(y) - \mathrm{sgn}(y+h\sec\phi)] \qquad 4.8$$

There seems little doubt that both terms in this expression are correct. When $\phi = 0$, the surface stress becomes simply a 'top-hat' function, zero outside the band and compressive within it. This behaviour can be rationalised with reference to figs. 5d and 5e. When tractions are applied to the band, it will suffer a transverse shrinkage because of Poisson's ratio. On welding, and release of the tractions, the interior of the band is fully constrained and cannot undergo relaxation, whereas that part of it near the surfaces relaxes somewhat and experiences transverse expansion, which, because it is constrained by the matrix, produces a compressive surface stress component σ_{yy}.

It is of great interest that when the angle $\phi \neq 0$, the surface stress shows logarithmic singularities at the lines of intersection between the bounding planes of the persistent slip band and the surface. If the band is in tension, so that the half-planes of the edge dislocations are external to the band, there is an infinite tensile stress where the band meets the surface in an obtuse angle, and an infinite compressive stress where it meets the surface in an acute angle. This behaviour can easily be rationalised by thinking about the stresses associated with the dislocation arrays: where a half-plane points into the surface, compression results; and where a half-plane points into the interior, tension results. This situation is illustrated in fig. 7. Typically, t_{max} will be equal to the diameter of a test specimen, say 1 cm, and the width of a persistent slip band will vary from a few microns to a few millimetres. Under these circumstances, the argument of the logarithm in equation 4.8 is dominated by the quadratic terms, and the stress changes from tension to compression midway between the planes defining the persistent slip band. The infinities in stress are of course not physically realised, and the continuum model will become unreliable at distances closer than the wall spacing to the lines of intersection between the free surface and the dislocation arrays. Even so, very large stresses may be achieved; in many instances close to the theoretical fracture stress.

This analysis helps to explain a result by Neumann (29), namely that fatigue cracks originate on the edge of a persistent slip band. Observation indicates that the edge where the cracks start is the edge at which the slip steps make an obtuse angle with the surface. If the band is in tension, as has so far been assumed, the crack should start at the opposite edge. However, when the band is first formed it will contain a lower concentration of vacancy dipoles than the matrix, so it will be in compression and at this early stage the crack will nucleate under the concentrated stresses. As the cyclic straining persists, the band stresses should reverse, and the crack will propagate as a Stage I fatigue crack when the stresses are high enough.

§ 4.4 Stresses at the Tips of Stage I Fatigue Cracks

The methods of the previous sections may be taken further to study the distribution of stress around idealised cracks propagating in Stage I of fatigue.

The first point to clarify is the energy release rate associated with the crack propagation. If in the band of fig. 1 the tensile strain is ε_t, then the energy per unit volume stored in the band is $\frac{1}{2}E\,\varepsilon_t^2$. This estimate is based on the assumption that the band is thinner than the matrix so far as strains parallel to AA' are concerned. If the band breaks free from the matrix, by the propagation of cracks down the interface between the band and the matrix, all of the elastic energy in the band is released; the energy per unit area is $\frac{1}{2}E\,\varepsilon_t^2 h_b$. Thus it becomes energetically favourable to propagate such cracks if

$$\varepsilon_t \geq \left(\frac{8\gamma_{eff}}{Eh_b}\right) \qquad 4.9$$

where γ_{eff} is the surface energy per unit area of the material: four surfaces are exposed when the cracks run. If γ_{eff} is taken to be a typical surface energy of 2 Jm^{-2}, and if the width of the band h_b is taken to be 1 mm, it becomes energetically favourable to propagate such cracks in copper if ε_t exceeds 3×10^{-4}. This value is of the same order of magnitude as the estimated strains in the persistent slip band. It is important to note two features of this calculation: firstly, this type of fracture condition (eqn.4.9) is quite different from the more familiar Griffith condition; in particular, the condition does not depend upon the length of the crack, but upon a dimension perpendicular to the crack, which is the dimension over which the elastic energy is released. Secondly, such cracks can propagate in metals only if no plastic blunting occurs at the tip of the crack. If the usual plasticity occurs, γ_{eff} is raised by orders of magnitude from the value used here, and crack propagation cannot be driven by internal stress alone.

It is therefore important to calculate the stress at the tip of a sharp crack. Considerable progress can be made as follows:

Head's equations can be used to find expressions for the complete stress field of an edge dislocation array running perpendicular to a free surface: the geometry is shown in fig. 9. If the array Ⓐ along OX lies perpendicular to a traction-free surface - YOY, the resulting stresses are

$$\sigma_{xx}/D = -\tan^{-1}\frac{x}{y}\left[4 + \frac{3x^2}{y^2}\right] + \frac{x(2y^2 - 3x^2)}{y(x^2 + y^2)} + \frac{3\pi x^2}{2y^2}\,\mathrm{sgn}(y)$$

$$\sigma_{yy}/D = 4\,\frac{\pi}{2}\left\{\mathrm{sgn}(y) - \tan^{-1}\frac{x}{y}\right\} - \frac{4x^3 y}{(x^2 + y^2)^2} \qquad 4.10$$

$$\sigma_{xy}/D = -\,2\,x^2\,\frac{2x^2 + 3y^2}{(x^2 + y^2)^2}$$

The stresses given here agree with those given earlier in equation 4.7, with $\phi = 0$, but show more completely how the different components behave near the surface. Far from the origin, σ_{xx} possesses a step discontinuity across the x-axis, but near the origin, σ_{xx} is zero and σ_{yy} possesses a step discontinuity across the x-axis. Loosely speaking, the free surface has the effect of transferring the discontinuity from the σ_{xx}-component to the σ_{yy}-component. The behaviour of equation 4.10 is not quite correct, because in the limit of very large x, the shear component σ_{xy} does not go to zero; this problem is once again probably due to the fact that Head's solution ignores the effect of other distant surfaces: certainly a distant stress-free surface parallel to the y-axis will have the effect of removing this stress.

A sharp crack can be modelled by insisting upon a traction-free surface which runs from - X to 0 along the x-axis; this is shown as the heavy line in fig. 9. This can be achieved using the stresses of equation 4.10 in the following way: construct arrays Ⓑ and Ⓒ as shown. Array Ⓑ runs along OY with -XOX traction-free, and array Ⓒ runs along -YO with -XOX traction-free. The arrays Ⓑ and Ⓒ can be used to eliminate the σ_{yy}-components along the y-axis, while at the same time they add no tractions to the x-axis. The arrays Ⓑ and Ⓒ are fictional 'image' type dislocations which generate the required stress-field. The stresses due to Ⓑ are found simply from equation 4.10 by replacing x by y, and y by -x; the stresses due to Ⓒ are found by replacing x by -y, and y by x. The correct solution is found to be: in region I, the sum of the stresses due to array Ⓐ and array Ⓑ, Ⓐ being twice the strength of Ⓑ; in region II, the stress is that due to array Ⓑ alone; in region III, the stress is that due to aray Ⓒ alone; and in region IV, the stress is the sum of those due to array Ⓒ and array Ⓐ, Ⓐ being twice the strength of Ⓒ. The resulting stress-field is too cumbersome to reproduce here, but it can be verified that all stress components are continuous and differentiable everywhere except for the appropriate discontinuity along the positive x-axis, and that the negative x-axis is traction-free. The stresses thus represent the field due to an infinitely sharp Kendall crack* running parallel to the uniform continuous array of edge dislocations represented by array Ⓐ. The only difficulty with this solution is the behaviour of σ_{yy} at large x, and this problem probably results from the absence of the distant free surfaces; it seems unlikely that this can affect stresses near the origin. Now the main point of the analysis is as follows: the stresses are everywhere finite, and in magnitude never exceed $2\pi D$: that is, they will not be greater than the flow stress. This behaviour is in contrast with that of the sharp Griffith crack, where the stresses at the tip of the crack become infinite, and vary as the reciprocal square-root of distance from the tip. In the case of the Griffith crack, if it is energetically favourable to propagate it, then bonds at the tip are stressed to very high stresses: the Griffith condition for propagation is both necessary and sufficient. In the case of the Kendall crack, the necessary condition that the energy be lowered is not

---

*It is necessary to find a name for this type of crack and, because of Kendall's pioneering work, it seems appropriate to use his name.

sufficient to guarantee the breaking of bonds at the crack tip. Such cracks must spread by the aid of local fluctuating stresses, which can be provided by the stress-fields of the real dislocations which make up the dipoles of the wall structure, or by chemical attack from the environment, which can provide bond-breaking reactions. It is a well-known feature of the nucleation and early propagation of fatigue cracks that the environment plays an important role.

§ 4.5 Summary of Results

This section has been written on the assumption that persistent slip bands are associated with macroscopic internal stresses: that is to say, stresses with appreciable value when averaged over dimensions ranging from hundreds of microns to centimetres. If the band is in tension, the axis of the tensile strain being parallel to the primary Burgers vector, then the intersection of the band with the free surface can produce large tensile surface stresses, apparently capable of opening a short notch or crack. The theory makes definite predictions about where the cracks should be found, and it seems to be in agreement with observation. When the cracks propagate, however, the large tensile stresses disappear, and in the limit when the crack can be modelled as a sharp crack in an infinite medium, the stresses near the tip of the crack are everywhere finite and the crack can propagate in a brittle manner provided some stress is available either from local dislocation stresses or from chemical agents. Such behaviour is qualitatively in accord with what is known of Stage I fatigue cracks. It is important to realise that the concepts of fracture mechanics, which apply to plastically blunted cracks and in which the crack length plays a crucial role, do not apply at all to the cracks discussed here.

5. CONCLUSIONS

It seems that the beginning of a consistent picture of the fatigue strength of metals is emerging. As a result of much careful and elegant experimental work, it is clear that the cross-slip of screw dislocations plays a crucial role in the mechanism of crack nucleation. The endurance limit seems to be simply interpreted as that stress which is just capable of separating two screw dislocations of opposite sign which are on the point of cross-slipping to annihilate each other. This idea can be taken as a working hypothesis while further experiments are performed, particularly on the temperature dependence of the strain amplitude in a persistent slip band. It seems also fairly clear that persistent slip bands are accompanied by internal stresses of a characteristic pattern: they are 'fibre-loaded' in the direction of the slip vector. Such stresses do not affect the motion of the dislocations within the band, but they promote decohesion of the band and surrounding matrix (just as fibre-loading promotes fibre debonding). Many features of the nucleation and early propagation of Stage I fatigue cracks find an explanation in these terms.

The ultimate origin of the 'fibre-loading' is the non-linear interaction between dislocations which, either because of stacking-fault structure or because of non-linear elasticity, causes more vacancy dipoles to be generated than interstitial dipoles. Thus fatigue cracks may be said to be caused by the interactions between dislocations which first localise the slip to form persistent slip bands, and then which bias the dislocation population to favour vacancy dipoles so as to build macroscopic internal stress.

Much exciting work remains to be done to see how reliable this picture is, and to fill in many of the rather important details. It is of great interest to be able to predict from first principles the shape of the hysteresis loops produced by persistent slip: such loops are universal curves, depending only on the micromechanics of primary dislocations. From an experimental point of view, too little is known about multiple slip, about the stability of the wall structure on annealing, and about the magnitude of the internal stress and how it builds up with cycling. Such information will lead to a quantitative theory of fatigue damage which is bound to have a considerable impact on the science of materials.

APPENDIX

Effective Elastic Constants, the Wall Spacing and the Saturation Stress

Throughout this paper an isotropic approximation to the anisotropic elastic constants has been used. The effective shear modulus has been chosen to give the energy per unit length of a screw dislocation correctly, and Poisson's ratio has been chosen to give the energy per unit length of an edge dislocation correctly. The elastic constants (at room temperature) and energy factors were calculated for Cu and Ni from the paper by Foreman (30) and for Mg from the paper by Chou and Eshelby (31). The resulting values are:

	Cu	Ni	Mg
μ/GPa	42.0	73.2	16.6
ν	.44	.39	.29
b/nm	.256	.249	.321

The stress to move a screw dislocation between walls of spacing d is then

$$\sigma = \frac{\mu \; b \; \ell n \frac{R}{b}}{(1-\nu) d \; 2\pi} \qquad \text{A.1}$$

The outer cut-off R which enters the line tension should probably be taken to be a wall width, R ≃ 500 b. This assumption makes the value of the logarithm about 6.2, so that it cancels the 2π in equation A.1. Current data are displayed in the table.

TABLE

Material	σ_{sat}/MPa	Homologous Temp. of Observation T/T_m	Wall spacing d/μm	$\frac{\sigma_{sat}(1-\nu)d}{\mu b}$
Cu[†]	25.2	.22	1.3 ± .08	1.7
	48.4	.06	0.7 ± .05	1.7
	73.5	.003	0.45 ± .03	1.7
Mg[†]	2	.33	8	2.1
	7.4	.08	2	1.9
Ni[††]	55	.17	1.3	2.4

[†] Basinski et al. (1). The difference between this treatment and that offered by Basinski et al. is as follows: they estimate that part of the bowing stress which is temperature-independent, and they take the bowing length to be the wall spacing less the wall width. These corrections approximately cancel each other. They then take the factor R/b in equation A.1 to be 5000 b, rather than 500 b, thus making the logarithm 8.5 rather than 6.2. This use of the wall spacing rather than the wall width is clearly incorrect (32), and results in values of the numbers in the final column closer to 1.4 than to 2.

[††] Kwadjo and Brown (6).

ACKNOWLEDGMENTS

It is a pleasure to thank Dr. A. T. Winter for many discussions and comments; also Dr. E. H. Yoffe who independently obtained some of the results of § 4. I am grateful to Professor P. Neumann for supplying fig. 7.

References

1. Z. S. Basinski, A. S. Korbel and S. J. Basinski, Acta Met. 28, 191 (1980).
2. A. T. Winter, Phil. Mag. 30, 719 (1974)
3. C. Laird, Mater. Sci. and Engng. 22, 231 (1976)
4. L. M. Brown, Metal Science 11, 315 (1977).
5. H. Mughrabi, Proc. 5th Int. Conf. on Strength of Metals and Alloys, ed. P. Haasen, V. Gerold and G. Kostorz, p 1615. Pergamon Press, (1979)
6. R. Kwadjo and L. M. Brown, Acta Met. 26, 1117 (1976
7. H. Mori, M. Tokuwame and T. Miyazaki, Phil. Mag. 40, 409 (1979)
8. S. L. Ogin and L. M. Brown, this conference.
9. W. R. Scoble and S. Weissmann, Crystal Lattice Defects 4, 123.
10. A. T. Winter, Phil. Mag. 31, 411 (1975).
11. J. D. Atkinson, L. M. Brown, R. Kwadjo, W. M. Stobbs, A. T. Winter and P. J. Woods, 3rd Int. Conf. on Strength of Metals and Alloys, Cambridge 1973, Metals Society.
12. J. C. Grosskreutz and H. Mughrabi, Constitutive Equations in Plasticity (ed. A. S. Argon) p 251. M.I.T. Press (1975)
13. J. G. Antonopoulos and A. T. Winter, Phil. Mag. 33, 87 (1976).
14. U. Essmann and H. Mughrabi, Phil. Mag. A40, 731 (1979)
15. M. Wilkens, K. Herz and H. Mughrabi, Z. für Metall. 71, 376 (1980)
16. J. G. Antonopoulos, L. M. Brown and A. T. Winter, Phil. Mag. 34, 549 (1976)
17. P. Neumann, Z. für Metall. 59, 927 (1968).
18. H. Mughrabi, F. Ackermann and K. Herz in Fatigue Mechanisms, ASTM-NBS-NSF Symposium, ed. J. T. Fong, A.S.T.M. (1979)
19. D. Kuhlmann-Wilsdorf, Mater. Sci. Engng. 39, 231 (1979).
20. K. V. Rasmussen and O. B. Pedersen, Proc. 5th Int. Conf. on Strength of Metals and alloys, ed. P. Haasen, V. Gerold and G. Kostorz, Pergamon Press
21. D. Kuhlmann-Wilsdorf and Laird, C., Mat. Sci. and Engng. 27, 137 (1977)
22. A. T. Winter, Phil. Mag. A37, 457 (1978).
23. P. Neumann in Constitutive Equations in Plasticity, Ed. A.S. Argon, M.I.T. Press (1975).
24 J. Picqueras, J. C. Grosskreutz and W. Frank, Phys. Stat. Sol.(a) 11, 567 (1972).
25. N. F. Mott, Acta Met. 6, 195 (1958).
26. K. Kendall, Phil. Mag. 36, 507 (1977).
27. A. K. Head, Phil. Mag. 44, 92 (1953).
28. F. A. McClintock and A. S. Argon, Mechanical Behaviour of Materials, Ch.18 Addison-Wesley (1966).
29. P. Neumann, Z. Metall. 58, 780 (1967).
30. A. J. E. Foreman, Phil. Mag. 15, 1011 (1967).
31. Y. T. Chou and J. D. Eshelby, J. Mech. Phys. Solids 10, 27 (1962).
32. D. J. Bacon, D. M. Barnett and R. O. Scattergood, Progress in Materials Science 23, 51 (1979). See particularly p 212.

FIG.1
Schematic diagram of persistent slip band

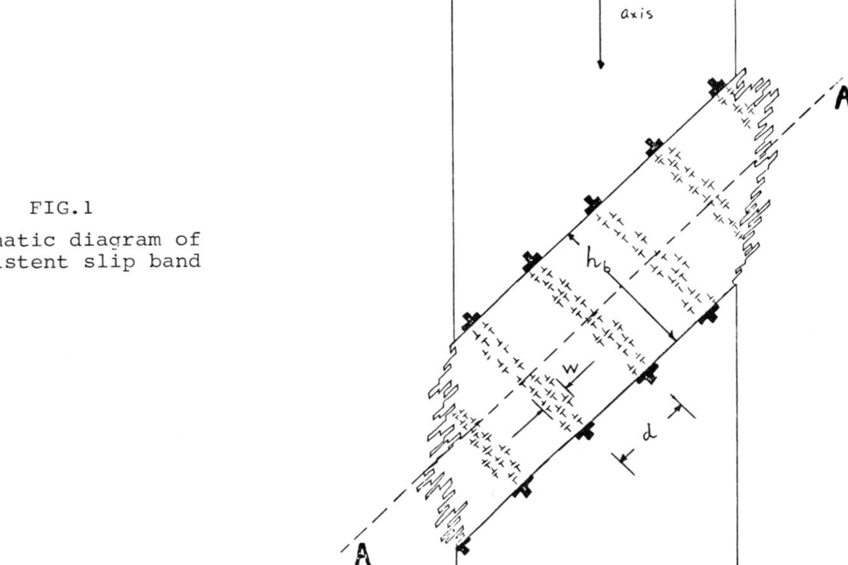

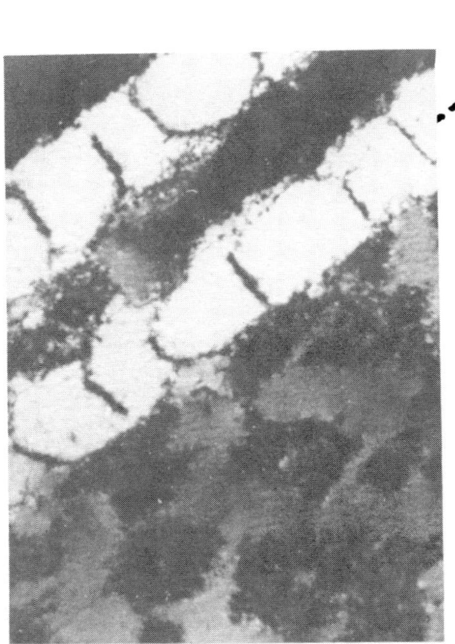

FIG.2
Wall structure in nickel. The line AA' corresponds to that shown in Fig.1; the tensile axis is vertical. Picture courtesy Dr. P. L. Gai.

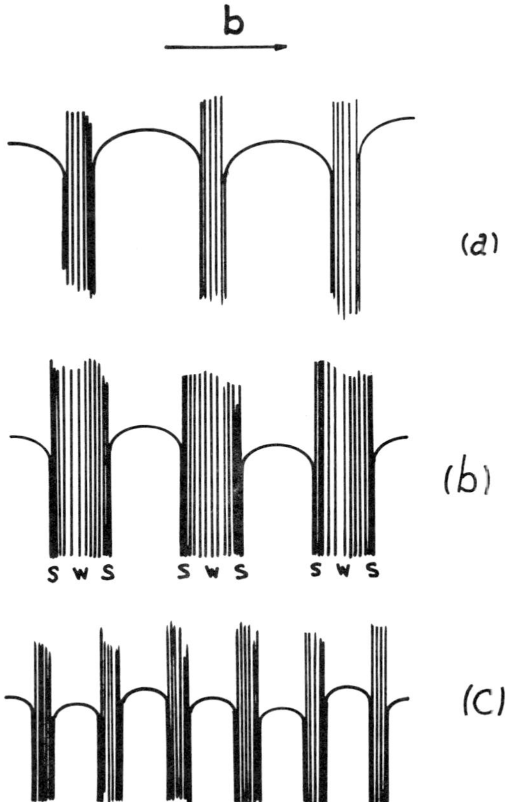

FIG.3

Proposed sequence of structural refinement in cyclic hardening. Screw dislocations bow between dipole clumps, whose spacing decreases as the stress increases. The slip plane coincides with the paper.

FIG.4

The Mott mechanism. The screw dislocation SS' glides on a prismatic surface, leaving a vacancy dipole in the crystal and a step on the surface. The line AA' corresponds to Fig.1: the tensile axis is vertical.

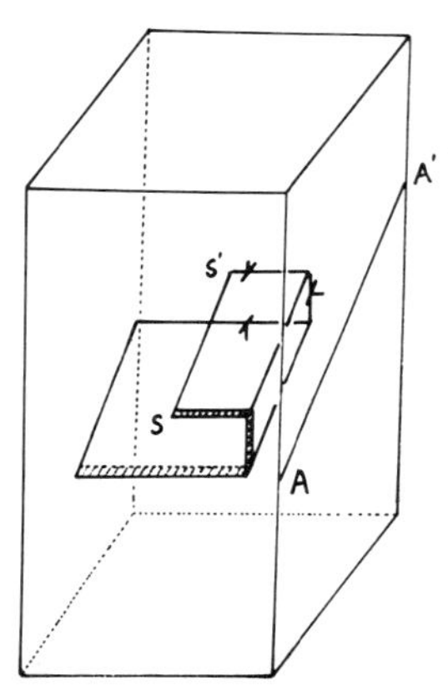

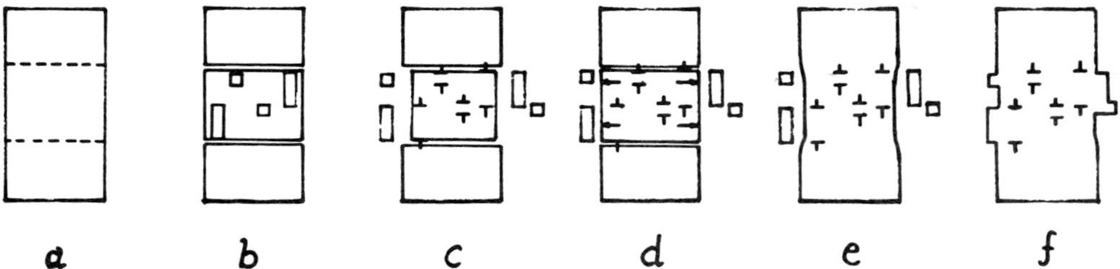

FIG.5

A persistent slip band according to Eshelby's style of continuum mechanics. (a) The band is marked out; (b) the band is cut from the matrix and half-planes of dipoles are marked; (c) the dipoles are allowed to collapse and the material from the half-planes is stored; (d) tractions are applied to make the band fit the matrix; (e) the band is welded again to the matrix; and (f) the material from the half-planes is returned to the matrix to make extrusions.

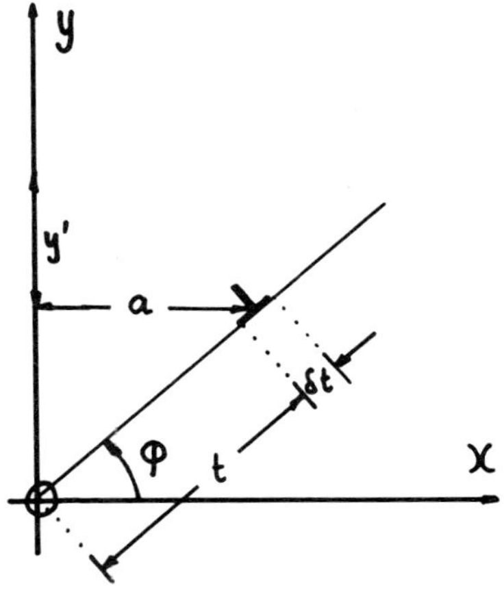

FIG.6

Coordinates for a dislocation near a free surface which is perpendicular to the x-axis.

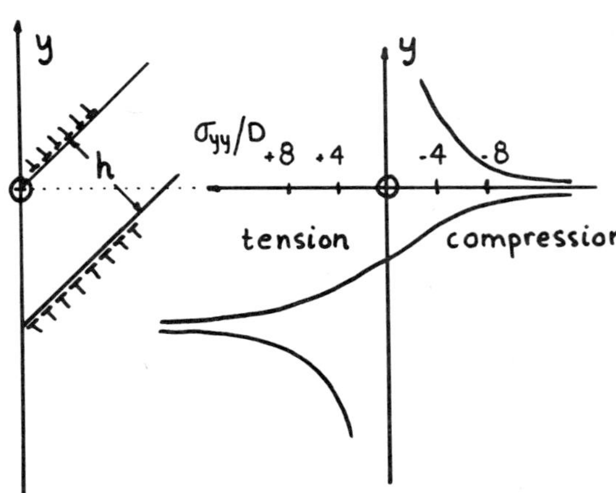

FIG.7

Surface stress at the intersection of a persistent slip band with the surface. The stress is large and tensile where the band makes an obtuse angle with the surface, if the dislocations have the sign shown.

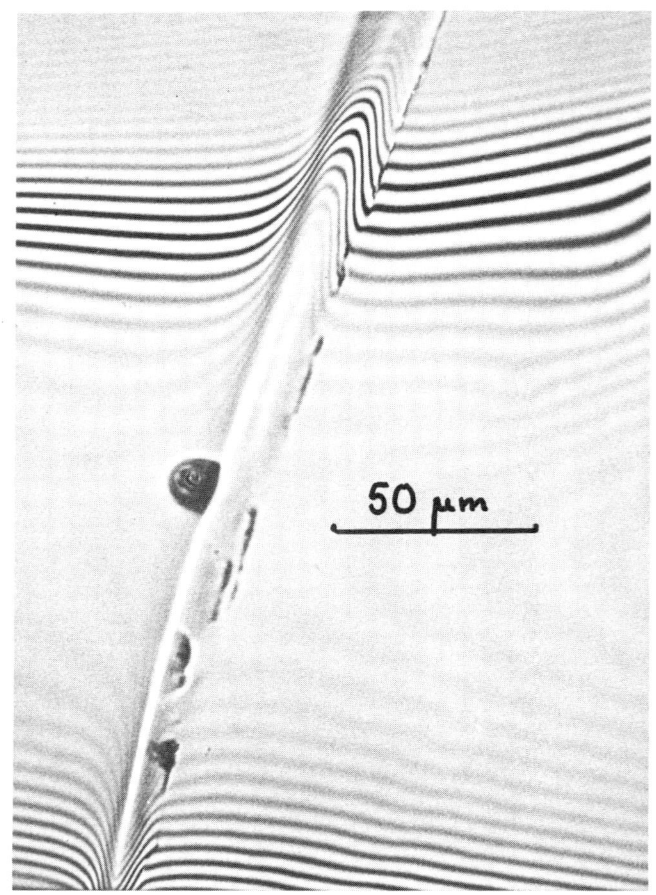

FIG.8

Surface features associated with persistent slip in copper. Surface relief is indicated by optical interference. Cracks form on one edge of the extrusion associated with the band, and experiment shows that it is the edge at which the slip steps make an obtuse angle with the surface. Picture courtesy of Prof. P. Neumann.

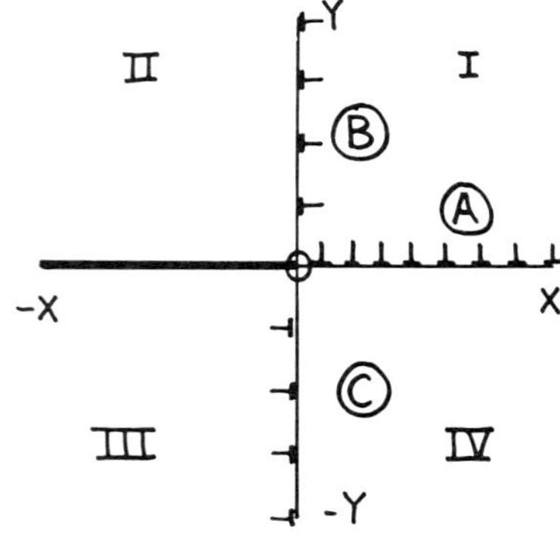

FIG.9

The synthesis of the stress field around a crack running along the negative x-axis, by superposition of semi-infinite dislocation arrays.

A DISLOCATION MODEL FOR FACETED MODE OF FATIGUE CRACK GROWTH

K. Sadananda
Material Science and Technology Division
Naval Research Laboratory
Washington, D. C. 20375

Abstract

Accumulated evidence indicates that fatigue crack growth in planar slip materials occurs by a crystallographically well-defined faceted mode with facets along either slip planes or cleavage planes. It is argued here that the generally accepted plastic blunting model can not explain the faceted growth and a dislocation pile-up model is proposed for the growth along slip planes or cleavage planes. Crack growth according to this model occurs in discrete increments whenever the energy to pump additional dislocations onto the pile-up near the crack tip is more than the energy to extend the crack. The model provides a rationale for several experimental observations including the role of grain size and microstructure on fatigue crack growth.

Introduction

It has been known (1) that crack propagation under fatigue occurs first by Stage I followed by Stage II, where Stage I growth is generally limited to one or two grains from the specimen surface and occurs along slip bands. Thus crack growth in Stage I follows a definite crystallographic orientation. On the other hand, Stage II growth is generally non-crystallographic and usually occurs along a plane of maximum normal stress. This growth is frequently associated with the formation of striations on the fracture surface.

There is an increasing experimental evidence that in high strength materials (2-16), particularly those that have a propensity for planar slip, fatigue crack growth occurs by a faceted mode with facets showing the features characteristic of a cleavage mode such as river lines. Unlike in the normal Stage II mode, the facets have a definite crystallographic orientation. A faceted mode of crack growth was observed even in relatively ductile materials (17) that have low stacking fault energy and thus restricted cross slip. In these materials, a faceted mode of crack growth was observed at low ΔK (stress intensity range) values close to the threshold ΔK, before it changed to a ductile striation mode at high ΔK values.

Identification of the plane of the facets with the slip plane of the material in some cases led to the belief that the faceted mode of crack growth in these materials is essentially Stage I mode of crack growth which extends to a large part of the fatigue fracture surface (8,9). In some other materials (11,12), facets were along the cleavage planes indicating that fatigue crack growth in these materials occurs by progressive cleavage. We have observed (18) faceted growth along the $\{100\}$ cleavage plane in Udimet 700 even at temperatures as high as 850°C in both air and vacuum. Meyn (10,11), on the other hand, has shown that in vacuum crack growth in an Al-alloy occurs either along the $\{100\}$ or $\{111\}$ plane but in air it occurs only along the $\{100\}$ plane. The choice of the plane could also depend on the symmetry of the slip systems with respect to stress axis as shown by Richards (4). That the faceted mode of crack growth is not necessarily just an extension of Stage I mode of crack growth is clear since it occurs on cleavage planes as well. Also this mode of crack growth differs from the normally observed Stage II growth because of the presence of river lines characteristic of the cleavage process and the absence of commonly observed ductile striations. In the following, the micromechanics of the faceted mode of crack growth is examined using a dislocation approach. A model is proposed to account for the faceted mode of crack growth along either cleavage planes or slip planes. The model is used to provide a rationale for several experimental observations.

Fatigue Crack Growth Model

Laird (19) proposed several years ago a fatigue crack growth model generally known as the plastic blunting which could explain the formation of striations on the fracture surface. The crack growth by this process occurs most efficiently if the material ahead of the crack tip deforms homogeneously. Also the net forward movement of the crack tip by the plastic blunting process depends on the net irreversible plastic strain accumulated in each cycle and not so much on the nature of the distribution of dislocations in the plastic zone. Because of this reason crack growth by this process is not very sensitive to microstructure.

The faceted mode of crack growth is different from the plastic blunting process and to see this difference one can speculate how the growth by the process would occur in an extreme case where the slip is restricted to one slip plane on either side of the crack plane as shown in Fig. 1a. Slip during the tensile half of the cycle would cause extended slip steps at the crack tip and during the compressive part would reverse part of the dislocations back leaving a net residual plastic strain. Continuation of the above process would result in only crack tip blunting and not crack growth, Fig. 1b. On the other hand, crack growth may still occur by inducing a cleavage process at the crack tip. The dislocation pile-ups that form at the crack tip induce normal stresses on the crack plane which augment the crack tip stress fields until it becomes energetically more favorable to cleave the material ahead of the crack tip than to pump more dislocations onto the slip plane as shown schematically in Fig. 1c. Crack growth occurs in discrete increments with each increment accomplished by the local stress build up due to the plastic flow during cycling, Fig. 1d.

For conditions under which slip becomes more asymmetric with respect to the crack plane, cleavage occurs more preferentially along the slip plane and the schematic illustration of this process is shown in Fig. 2. Unlike in the case of Fig. 1, stress fields of the dislocations do not aid in the cleavage process since the normal component of their stress field vanishes on the slip plane. However, decrease of the strain energy of the whole system is accomplished with the nucleation of a crack which absorbs most of the crystal dislocations in its wake. Thus a mixed mode of crack (tensile and shear) could form in the presence of the normal and shear stress fields of the main crack with additional energy provided by the strain energy of the slip dislocations. The growth again occurs by discontinuous increments with each increment accomplished by the strain energy build up by plastic flow during cycling.

A quantitative analysis of the above processes is made using the concepts from dislocation theory and assuming a cyclic stress-strain relation for the material ahead of the crack tip as given by

$$\Delta \sigma = B (\Delta \epsilon)^{\beta} \tag{1}$$

where $\Delta \sigma$ and $\Delta \epsilon$ are stress and strain respectively, and B and β are constants for a given material and test temperature. While the detailed formulation of the problem is presented elsewhere (20), we shall discuss here only the main results of the analysis and their implication in terms of fatigue crack growth behavior of planar slip materials.

Crack Growth Rates

The analysis of the energetics of the crack growth processes in Figs. 1 and 2 indicates that the rate of fatigue crack growth by the faceted mode is given by

$$\frac{da}{dN} = \frac{p\,F}{\pi(1-\nu^2)\ \sin\theta\ f_1(\theta)} \left\{ \frac{\Delta K}{B} \sqrt{\frac{2}{D}} \right\}^{1/\beta} \sqrt{\frac{a\ D}{2}} \tag{2}$$

where D is the dislocation pile-up length and a is the crack length and for a case where $D << a$, the above equation can be approximated as

$$\frac{da}{dN} \ \alpha \ G \left(\frac{\Delta K}{B}\right)^{1/2\beta} \left(\frac{1}{D}\right)^{\frac{1-4\beta}{4\beta}} \tag{3}$$

where

$$G = \frac{p\ F}{\pi(1-\nu^2)\sin\theta\ f_1(\theta)\ \sqrt{2}} \left(\frac{2}{\pi}\right)^{1/4\beta} \tag{4}$$

For the faceted growth along a cleavage plane, $p = 1/2$ and

$$F = \frac{8\ f_1^2\ (\theta)}{\left[\dfrac{8\ \gamma_T\ \mu}{K^2\ (1-\nu)} - \left\{6 \sin\theta\ \cos\ \theta/2\ f_1(\theta) + 1\right\}^2\right]} \tag{5}$$

while for the growth along a slip plane, p = 1 and

$$F = \frac{8 \; f_1^2 \; (\theta)}{\left[\dfrac{8 \; \gamma_T \mu}{K^2(1-\nu)} \quad - \quad f_2^2 \; (\theta) \right]} \tag{6}$$

In the above equations B and β are the same as those defined in eq. 1, θ_3 is the angle of the slip plane with respect to the plane of the main crack, $f_1(\theta) = \cos^2 \theta/2$ and $f_2(\theta) = \cos^3 \theta/2$, ν is the Poisson's ratio, K is the stress intensity factor and ΔK is the stress intensity range, γ_T is the surface energy, and μ is the shear modulus.

Discussion

Plane of the Facet

Since the faceted mode of crack growth is observed either along slip planes or along cleavage planes, it is instructive to examine what factors determine the plane for the faceted growth in a given material. If crack growth can occur on either plane then it will preferentially occur along a plane that provides the larger growth rate since in terms of the external loading mechanism this requires lower energy for the same increment in crack length. The ratio of crack growth rates by the faceted mode along a slip plane and a cleavage plane for a given ΔK and D values is given by

$$\frac{(da/dN)_s}{(da/dN)_c} \approx \frac{F_s}{F_c} = \frac{8 \; \gamma_{Tc} \quad - \; K^2(1-\nu) \left\{ 1+3 \sin^2 \theta_s/2 \quad \cos^2 \theta_s/2 \right\}}{8 \; \gamma_{Ts} \quad - \; K^2(1-\nu) \quad \cos^6 \theta_c/2} \tag{7}$$

where subscripts s and c are used to distinguish the terms corresponding to slip and cleavage planes, respectively. The above equation shows that if the surface energies γ_{Tc} and γ_{Ts} are nearly the same then the ratio of crack growth rates depends on the orientation of the slip plane and the crack plane. For a symmetric case, $\theta_c \approx 0$, and $\theta_s > 0$ the ratio becomes less than 1, that is, the crack grows preferentially along a cleavage plane in agreement with experimental results (4). With the increasing asymmetry of slip, θ_c increases and the tendency for the crack to propagate along the slip plane increases. If surface energies are not the same the above equation shows that the ratio of crack growth rates is inversely related to the ratio of surface energies of the two planes. Since in most cases, the surface energy on a cleavage plane is lower, then the crack preferentially grows along a cleavage plane.

Effect of Grain Size

The analysis shows that crack growth along either cleavage or slip planes depends on the parameter D which is the length of the pile-up. This parameter is sensitive to the microstructure since presence of insurmountable obstacles on the slip plane would limit the extent of slip. For a relatively homogeneous material, the grain boundaries may provide natural barriers for the propagation of slip and the dislocations may begin to pile-up near the boundary before the slip can propagate in the next grain. Crack growth rate, according to this argument, will be sensitive to grain size. In contrast to the faceted mode, crack growth by the plastic blunting process (19) as analysed by Tomkins (21) is relatively insensitive to changes in microstructure as supported by several experimental observations.

The effect of grain size on fatigue crack growth has received much attention (22-30) in the past and the evidence appears to be contradictory since some authors show an absence of a grain size effect (29) and others show a large effect (30). These results can be rationalized on the basis of the present analysis. In particular, most of the grain size effects were observed in the materials or in the low ΔK region where crack growth occurs by the faceted mode. Analysis of the crack growth data of several Ti-alloys (20) indicates that the growth rates for a given ΔK value varies exponentially with grain size as shown in eq. 3. For more complex alloys the dislocation obstacle spacing may be less than the grain size and for such cases the crack growth rate depends on the mean free path for dislocations as indicated recently by Chakrabortty and Starke (31).

Equation 3 further indicates that the Paris exponent m in $da/dN = n \, \Delta K^m$ is equal to $1/2 \, \beta$. Landgraf (32) has shown that β values for several alloys fall in the range of 0.05 to 0.2. If these alloys deform by planar slip, then eq. 3 implies that their Paris exponents fall in the range of 10 to 2.5 if crack growth occurs by the faceted mode. Since β generally decreases with increasing planarity of slip, the Paris exponent increases in agreement with several experimental results.

The analysis thus explains how the faceted mode of fatigue growth along a slip plane or cleavage plane occurs in planar slip materials and what the important variables are that determine the rate of crack growth in such materials.

Acknowledgements

It is a pleasure to acknowledge Drs. P. Shahinian, G. R. Yoder, and D. A. Meyn of the Naval Research Laboratory for many helpful discussions. The present research effort is supported by the Office of Naval Research.

References

1. P. J. E. Forsyth, Proceedings of the Symposium on Crack Propagation, Cranfield, The College of Aeronautics, 1, 76 (1962).
2. R. P. Wei, Trans. ASM, 60, 279 (1967).
3. K. R. L. Thompson and J. V. Craig, Met. Trans., 1, 1047 (1970).
4. C. E. Richards, Acta Met., 19, 583 (1971).
5. G. C. Garrett and J. F. Knot, Acta Met., 23, 841 (1975).
6. R. N. Wright and A. S. Argon, Trans. Am. Inst. Mech. Engrs., 1, 3065 (1970).
7. J. C. Chesnutt, C. G. Rhodes and J. C. Williams, in "Fractography-Microscopic Cracking Processes," C. D. Beachem and W. R. Warke, eds., ASTM STP 600, 99 (1976).
8. M. Gell and G. R. Leverant, Acta Met., 16, 553 (1968).
9. D. J. Duquette, M. Gell and J. W. Piteo, Met. Trans., 1, 3107 (1970).
10. D. A. Meyn, Trans. ASM, 61, 52 (1968).
11. D. A. Meyn, Met. Trans., 2, 853 (1971).
12. G. R. Yoder, L. A. Cooley and T. W. Crooker, Met. Trans. A, 8, 1737 (1977).
13. D. Eylon, Met. Trans. A, 10A, 311 (1979).
14. A. Yuen, S. W. Hopkins, G. R. Leverant and C. A. Rau, Met. Trans., 5, 1833 (1974).
15. H. D. Williams and G. C. Smith, Phil. Mag., 13, 835 (1965).
16. J. L. Robinson and C. J. Beevers, Met. Sci. J., 7, 135 (1973).
17. R. W. Hertzberg and W. J. Mills, in "Fractography-Microcracking Processes," C. D. Beachem and W. R. Warke, eds., ASTM-STP 600, 220 (1976).
18. K. Sadananda and P. Shahinian, Met. Trans., Submitted for Publication (1980).
19. C. Laird, in Fatigue Crack Propagation, ASTM-STP 415, 131 (1967).
20. K. Sadananda, Met. Trans., Submitted for publication (1980).
21. B. Tomkins, Phil. Mag., 18, 1041 (1968).
22. M. F. Carlson and R. O. Ritchie, Scripta Met. 11,1113 (1977)
23. J. Masounave and J. P. Bailon, Scripta Met., 10, 165 (1976)
24. A. W. Thompson, Eng. Frac. Mech., 7, 61 (1975)
25. W. W. Gerberich and N. R. Moody, in "Fatigue Mechanisms", ASTM STP 675, 292 (1979).
26. A. J. McEvily, Jr., and R. C. Boettner, Acta Met., 11, 725 (1963).
27. A. W. Thompson and R. J. Bucci, Met. Trans., 4, 1173 (1973).
28. J. L. Robinson and C. J. Beevers, Metal Science Journal, 7, 153 (1973).
29. A. W. Thompson, J. D. Frandsen and J. C. Williams, Metal Science, 9, 46 (1975).
30. G. R. Yoder, L. A. Cooley and T. W. Crooker, J. of Engrg. Mat. and Tech., Trans. ASME, 101, 86 (1979).
31. S. B. Chakrabortty and E. A. Starke, Jr., Met. Trans. A, 10 A, 1901 (1979).
32. R. W. Landgraf in "Achievement of High Fatigue Resistance in Metals and Alloys", ASTM-STP 467, 3 (1970).

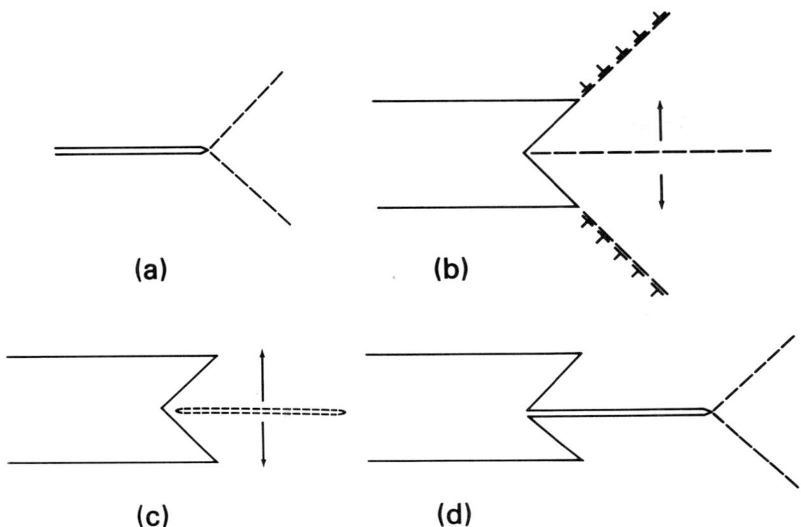

FIG 1.

A model for the faceted mode of fatigue crack growth on a cleavage plane, (a) initial crack, (b) cumulative plastic strain on restricted slip planes in planar slip materials, (c) formation of a crack due to normal stresses and (d) crack length increment with subsequent slip at the crack tip during further cycling.

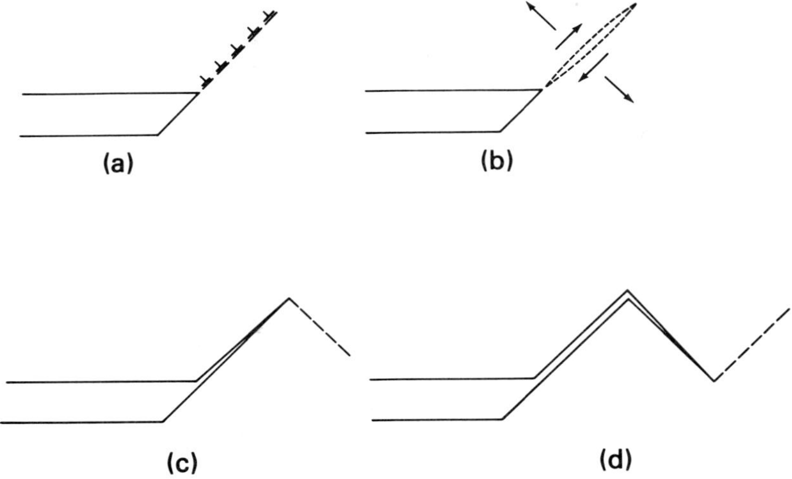

FIG 2.

A model for the faceted mode of fatigue crack growth along a slip plane. (a) initial crack with extended slip on a particular plane, (b) formation of a crack on the slip plane, (c) crack extension with subsequent slip along another favorably oriented plane and (d) continuous faceted crack growth along slip planes.

IN SITU OBSERVATION OF DISLOCATION BEHAVIOUR IN FATIGUE PRCCESS BY HVEM

A.Yamamoto and T.Imura

Department of Metallurgy, Faculty of Engineering, Nagoya University, Furo-cho, Chikusa-ku, Nagoya, 464, Japan

Abstract

Progressive formation of dislocation substructures during fatigue deformation has been investigated by means of HVEM. The main results are : 1) In the transition stage of the fatigue hardening curve, irregular wall-like structures were formed firstly in the core region of the specimen. These substructures change to wall structures in the following way. (1) The width of the walls become regular (2) The walls line up at regular intervals (3) The walls straighten (4) The walls align perpendicular to the primary Burgers vector ; 2) In the transition stage, the fatigue hardening in the core of a specimen precedes that in the surface region ; 3) It was found by in situ experiments that mobile dislocations were observed mostly in the wall structures, and in vein structures those were observed very infrequently.

the wall structure can be seen clearly. The change occurs gradually in the order shown in Fig. 8-1 ∿ 8-4 at the transition stage.

One example of the motion of individual dislocations in the wall structure observed by in situ fatigue test is shown in Fig.9. The specimen used was pre-fatigued for 4,000 cycles at the same strain amplitude. Walls and veins are seen in Fig.9. When the tensile stress was applied to the specimen, a mobile segment bowed out from the wall (Fig.9 B), and then slipped to the next wall (cf. Fig.9 C). Such a motion of dislocations was observed only in the wall structures and not observed in the vein structures. Detailed observations of the dislocation configurations within the veins and the walls were performed by weak-beam method, which are shown in Fig.10. A number of mobile dislocations are observed in the vicinity of the walls, although those are not observed in the veins (cf. Fig.10).

Conclusion

1) In the transition stage of the fatigue hardening curve, irregular wall-like structures were formed firstly in the core region of the specimen. These structures change to wall structures in the following way. (1) The width of the walls become regular. (2) The walls line up at regular intervals. (3) The walls straighten. (4) The walls align perpendicular to the primary Burgers vector.
2) In the transition stage, the fatigue hardening in the core of a specimen precedes that in the surface region.
3) It was found by in situ experiments, that mobile dislocations were observed mostly in the wall structures, and in vein structures those were observed very infrequently.

References

1. H. Mughrabi, F. Ackermann and K. Herz, Symposium on fatigue mechanism, Kansas, 1978
2. A. Winter, Phil. Mag. 30, 719, (1974)
3. J. Finney and C.Laird, ibid, 31, 339, (1975)
4. A.Yamamoto, H.Saka and T.Imura, Kristal und Technik, 14, 1275, (1979)
5. A.Yamamoto, C.Morita, T.Tono, S.Saimoto, S.Saka and T.Imura, Pro. Fifth Int. Con. on High Voltage Electron Microscopy, Kyoto, p.133 (1977)

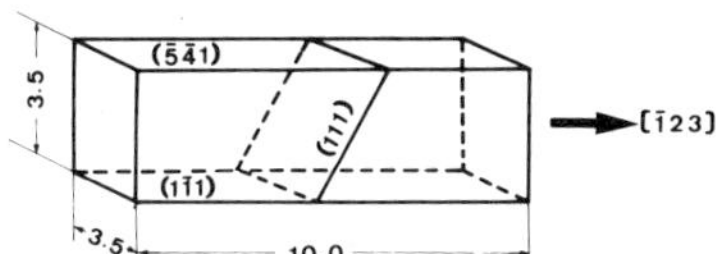

FIG.1
Geometry and the size of the fatigue specimen of Cu- 2 at % Al alloy

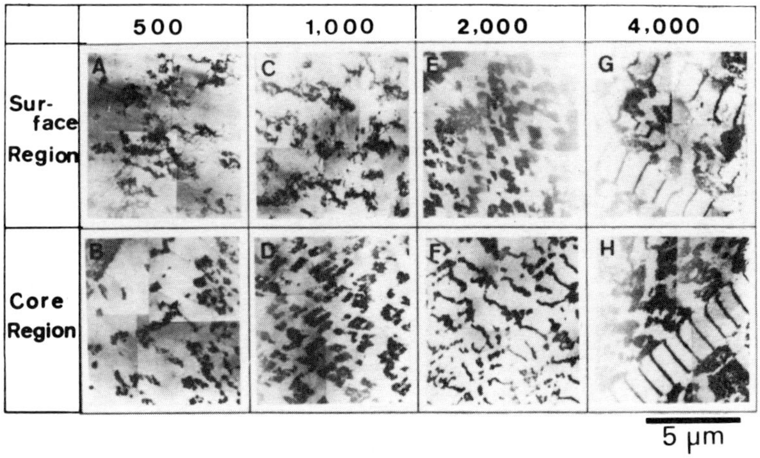

FIG.2
Dislocation substructures formed in the surface and the core region of the specimens at various stages of fatigue deformation
A, B and C: Bundle structures
D and E: The vein structures
F: Irregular wall-like structures and the vein structures
G and H: The wall structures and the vein structures

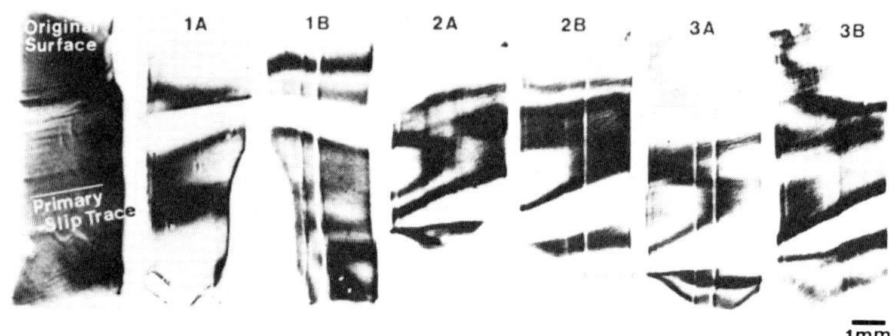

FIG.3
X-ray topographs of the specimens sliced into components after pre-fatigue deformation

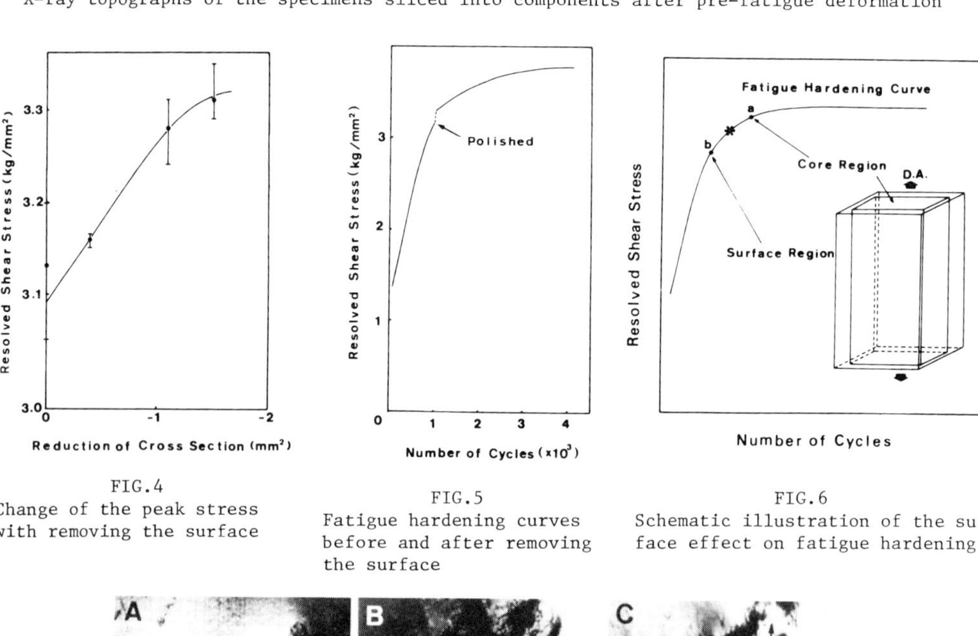

FIG.4
Change of the peak stress with removing the surface

FIG.5
Fatigue hardening curves before and after removing the surface

FIG.6
Schematic illustration of the surface effect on fatigue hardening

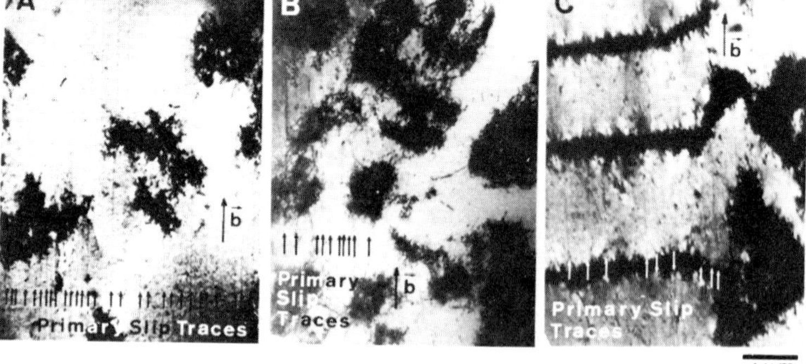

FIG.7
Slip traces formed in the in situ cyclic deformation. A:Homogeneous slip, B and C :Slip is localized

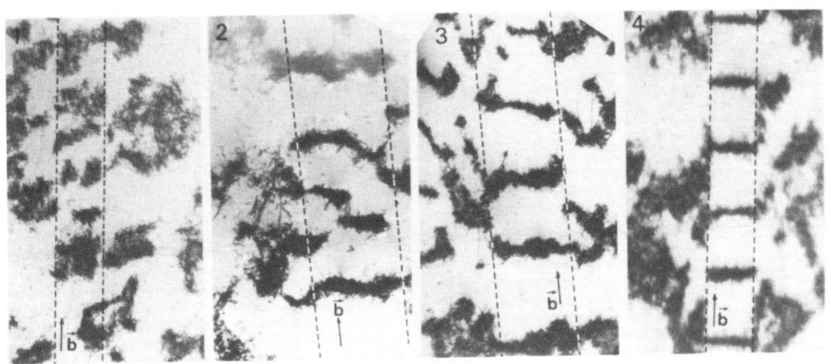

FIG.8
Change of the dislocation substructure into the wall

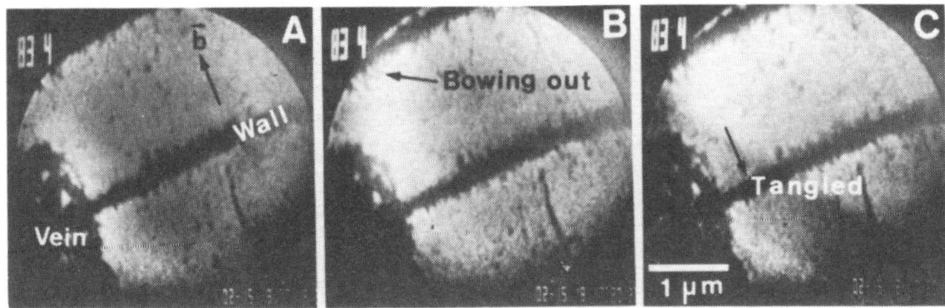

FIG.9
Motion of dislocation in the wall structure

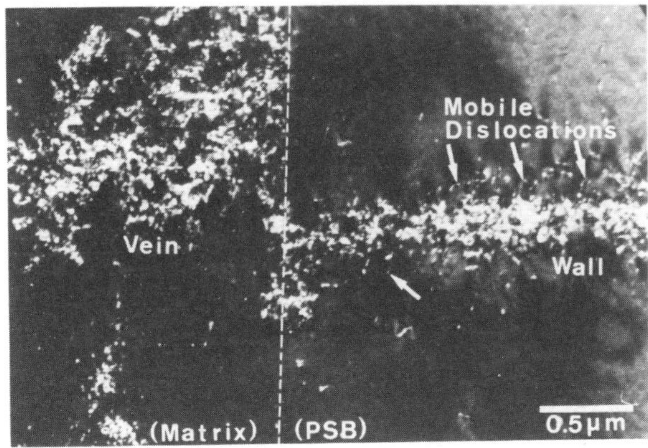

FIG.10
Detailed observation of the vein and the wall structure by weak-beam method

ENGINEERING FRACTURE CRITERIA

J.R. Willis
School of Mathematics
University of Bath, Bath, U.K.

Abstract

It is explained how engineering fracture criteria are developed by seeking parameters that characterize those features of the macroscopic stress field around a crack that determine events occurring in the fracture process zone. In many cases, a single parameter can be identified and crack extension must then take place when the parameter attains a critical value, which can be determined by experiment, without reference to the detail of the fracture process. In other cases, however, convincing arguments cannot be made for the selection of a single parameter: consideration of microscopic mechanisms then cannot reasonably be avoided.

1. Introduction

According to any of the usual constitutive models (elastic, plastic, viscoelastic, etc.) of a material, the presence of a crack intensifies the stress or strain near its tip and it is this intensification that induces the structural changes that result in crack extension. These generally occur in a small region, adjacent to the crack tip, which may be termed the fracture process zone. Events taking place in this region naturally perturb the 'macroscopic' stress and strain fields predicted by disregarding its existence. The modelling of the fracture process zone and its interaction with the macroscopic fields is required for the understanding of fracture and forms one of the areas of interest for this Conference. The interest of an engineer, however, centres upon the identification of particular features (which, ideally, should be easy both to compute and to measure) of the macroscopic field, which play a decisive role in determining events in the fracture process zone and hence govern crack extension. The success of engineering fracture mechanics follows, in fact, from the possibility of describing the key macroscopic features in terms of a single parameter in a wide variety of situations. Then, whatever the model of the fracture process zone, events within it are correlated with this one parameter, crack extension is bound to be associated with its attaining a critical value and the need to understand the fracture process is avoided: all that is needed is a series of engineering tests to determine the critical value of the parameter. When such conditions prevail, microscopic modelling of the fracture process zone should in principle provide an estimate of the critical value of the macroscopic fracture characterizing parameter and assist thereby in the development of fracture-resistant materials. It is perhaps of greater importance, however, to determine when a one-parameter characterization is possible and, when it is not, to determine what new features would have to be incorporated into a macroscopic criterion for crack extension.

This article will describe the theory by which one-parameter characterizations are justified, while emphasising limitations whose resolution requires detailed modelling. The section that follows contains a brief account of the form of the stress fields around cracks in elastic and elastic-plastic bodies and an even briefer account of the methods by which they may be calculated. Subsequent sections then describe the main parameters that are employed to characterize fracture, together with comments upon their possible range of applicability.

2. Stresses around Cracks

2.1 Elastic Fields

Consider a linearly elastic body, such as the one sketched in Fig. 1, containing a crack which occupies a part of the plane $x_1 < 0$, $x_2 = 0$ $-\infty < x_3 < \infty$. If the body is subjected to loading so that the resulting stress and strain fields are independent of the coordinate x_3 then, it can be shown that, in the vicinity of the edge $x_1 = 0$, $x_2 = 0$ of the crack, the stress components σ_{ij} have the form

$$\sigma_{ij} \sim (2\pi r)^{-\frac{1}{2}} \{K_I \tilde{\sigma}^{I}_{ij}(\theta) + K_{II} \tilde{\sigma}^{II}_{ij}(\theta) + K_{III} \tilde{\sigma}^{III}_{ij}(\theta)\} + T_{ij} + \ldots , \tag{1}$$

relative to the polar coordinates (r,θ) based on the crack tip shown in Fig. 1. The functions $\tilde{\sigma}_{ij}(\theta)$ are universal functions, first discussed explicitly by Williams [1]. They have symmetries that correspond to symmetric plane-strain loading (Mode I), antisymmetric plane-strain loading (Mode II) and anti-plane shear loading (Mode III). Their exact form is not important for the present discussion but it may be noted that $\tilde{\sigma}^{I}_{22}(0) = \tilde{\sigma}^{II}_{12}(0) = \tilde{\sigma}^{III}_{23}(0) = 1$. The stress intensity factors K_I, K_{II}, K_{III} are constants that have to be found by solving the stress analysis problem. The constant tensor T_{ij} is likewise obtained from the full solution except that, to satisfy conditions on the crack faces, only the components T_{11}, T_{13}, T_{33} may differ from zero. At points sufficiently close to the crack tip, the terms of order $r^{-\frac{1}{2}}$ dominate the right side of (1) which, in fact, is a series whose next term will be of order $r^{\frac{1}{2}}$, somewhat similar to the first term but with different θ-variations also discussed in [1]. The asymptotic form (1) applies also in the vicinity of any smooth crack edge having the x_3-axis as tangent at the origin. Stress intensity factors K_I, K_{II}, K_{III} have been found for a great variety of configurations and procedures for their appoximate determination in other configurations have been developed; see, for example [2].

2.2 Elastic-Plastic Fields

(a) Near-tip fields

Calculations analogous to those that produced the functions $\tilde{\sigma}_{ij}(\theta)$ in (1) have been carried out by Hutchinson [3] and Rice and Rosengren [4] for a power-law hardening material with constitutive relation

$$e_{ij}/e_o = 3\alpha(\sigma_e/\sigma_o)^{n-1} s_{ij}/\sigma_o, \tag{2}$$

where σ_o is the initial tensile yield stress (or another suitable reference stress), $e_o = \sigma_o/E$, s_{ij} represents deviatoric stress and

$$\sigma_e^2 = 3 s_{ij} s_{ij}/2. \tag{3}$$

The singularity analysis requires only that (2) should apply at sufficiently high stresses; the resulting singular fields (usually termed HRR fields) have the form

$$\begin{aligned} \sigma_{ij} &= \sigma_o K_\sigma r^{-1/(n+1)} \tilde{\sigma}_{ij}(\theta,n), \\ e_{ij} &= \alpha e_o K_\sigma^n r^{-n/(n+1)} \tilde{e}_{ij}(\theta,n), \end{aligned} \tag{4}$$

the plastic intensity factor K_σ being undetermined. The universal asymptotic forms (4) were derived in [3,4] for the case of symmetric (mode I) loading. Similar forms could be found for modes II and III but the general mixed-mode case could not be obtained from these by superposition, as in (1).

The fields (4) are not directly applicable in the limit of perfect plasticity ($n \to \infty$): it can be shown that the stress given by the first of (4) as $n \to \infty$ agrees with that obtained from the slip-line pattern shown in Fig. 2, but the associated strain field, which has an r^{-1} singularity, is not unique, since other θ-variations are also possible, depending upon the global solution to the problem.

DISLOCATION MODELING

Although the fields (4) were derived from the "deformation theory" law (2) they do, in fact, also conform to "flow theory" because each element is subject, as K_σ increases, to proportional loading. The situation is different, however, at the tip of a moving crack, since material unloads as the crack tip passes, as depicted in Fig. 3. A singularity analysis has been performed by Amazigo and Hutchinson [5] for a linearly hardening material. The elastic unloading reduces the strain singularity from order $r^{-\frac{1}{2}}$ for the static crack (n=1 in equations (4)) to order r^{-s}, with $0 < s < \frac{1}{2}$ for the moving crack. The exponent s tends to zero with diminishing hardening; in the non-hardening limit, Rice and Sorensen [6] obtained a $\log r^{-1}$ singularity from an approximate analysis based upon the slip-line field of Fig. 2.

(b) Global solutions

The complete solution to an elastic-plastic problem generally requires a large computation; see, for example, Sumpter and Turner [7], McMeeking and Parks [8]. The only models simple enough to permit analytical study are that of Dugdale [9] and Bilby, Cottrell and Swinden [10], which confines yielding to a thin strip ahead of the crack, and the still simpler "superdislocation" model of Atkinson and Kanninen [11]. For a crack of length 2a in an infinite body subject to remote tension σ_∞, the model of Dugdale, Bilby, Cottrell and Swinden predicts a plastic zone of length s, given by

$$s/a = \sec(\pi\sigma_\infty/2\sigma_o) - 1 \tag{5}$$

and a relative displacement δ_t of the crack faces at the tip (which is consistent with the r^{-1} strain singularity obtained from solving the full elastic-perfectly plastic equations) given by

$$\delta_t/a = 8e_o\pi^{-1} \log[\sec(\pi\sigma_\infty/2\sigma_o)]. \tag{6}$$

In (5) and (6), σ_o represents the stress across the strip of yielding, roughly equal to the tensile yield stress, and $e_o = \sigma_o/E$.

Relatively simple calculations are also generated by "fully plastic" problems, in which a constitutive equation such as (2) is taken to hold even at small strains where, strictly, linear elasticity is operative; the error in the global features of the solution will be unimportant if loads are such that large plastic strains are generated. The particular case of non-hardening rigid-plastic materials is particularly simple, as slip-line theory can be employed; see McClintock [12]. An example of a more general "fully plastic" analysis is contained in Goldman and Hutchinson [13].

Finally, a systematic asymptotic approach to elastic-plastic fracture problems has been developed by Edmunds and Willis [14,15,16]. They formulated elastic-plastic problems within the framework of matched asymptotic expansions [17] to obtain complementary series for the fields close to and remote from the crack tip. Interaction between the regions near to and remote from the crack tip was allowed for iteratively and led, ultimately, to expressions for the field at any point in the form of series whose coefficients could be determined from elastic solutions alone. The functions in the series, at least near to the crack tip, were defined by a sequence of standard boundary value problems for a semi-infinite crack in an infinite body. They have to be found numerically but are universal functions, depending upon material properties and not on specimen geometry. The method breaks down as limit load is approached but it does provide systematically a set of easily-determined parameters that characterize the near-tip field, even when plasticity is fairly extensive.

3. Fracture Characterizing Parameters

3.1 Linear Elastic Fracture Mechanics

When the stress in a cracked body is calculated according to linear elasticity theory, it will be approximated, up to any chosen degree of accuracy, by the singular term in (1), within some radius R of the crack tip, independently of the magnitude of the loading. The radius R has, of course to be small relative to relevant dimensions of the body, such as crack length. Plastic relaxation in fact will occur near the crack tip and a rough estimate of the extent of yielding is obtained by finding the value of r at which the elastic singular stress field satisfies the yield criterion. Thus, for mode I loading, yield will occur over a region whose characteristic dimension is of order $(K_I/\sigma_o)^2$, where σ_o represents yield stress in simple tension. "Small scale yielding" occurs when the load level is such that

$$(K_I/\sigma_o)^2 << R. \tag{7}$$

When the condition (7) is met, a "boundary layer" formulation of the full elastic-plastic problem is possible for, as depicted in Fig. 4, the plasticity will influence the stress field only slightly at a distance of order R or greater from the crack tip. It suffices, therefore, to consider an elastic-plastic problem for a semi-infinite crack in an infinite body, which is subjected just to the leading term of (1) at points remote from the crack tip (inclusion of T_{ij} is a refinement that will be discussed later). This canonical elastic-plastic problem is independent of specimen geometry and has been studied by a number of workers; the most accurate solution, which accounts for finite strains at the crack tip, is probably that of McMeeking [18]. Calculations including allowance for a fracture process zone, also with dimensions much smaller than R, could, of course, also be performed.

Suppose, now, that the crack is actually subjected to mode I loading. In that case, the foregoing discussion demonstrates that whatever is occurring at the crack tip is driven by K_I alone and hence that crack extension must begin to occur when K_I reaches a critical value K_{Ic}. Independently of the fracture mechanism, therefore, it is necessary only to perform a set of experiments to establish the value of K_I at which crack extension begins. The asymptotic form (1) contains the implicit assumption of plane strain (so that K_I is either constant or varies slowly along the crack edge), so that test specimens must be sufficiently thick. Also, the specimen must be large enough for the inequality (7) to be satisfied up to the critical value of K_I. The usual requirement (which is less stringent than (7)) is that

$$a \geq 2.5(K_{Ic}/\sigma_o)^2, \tag{8}$$

where a denotes the smaller of specimen thickness, crack length or remaining ligament length: extensive tests have shown this to suffice in practice.

For mixed-mode loading, the small-scale yielding argument demonstrates that crack extension can be influenced only by K_I, K_{II} and K_{III}. When symmetry is absent, however, the crack is likely to deviate from a straight path and the determination of the correct critical combination of K_I, K_{II} and K_{III}, together with prediction of the crack path, cannot proceed without some microscopic modelling. Calculations of the local elastic stress intensity factors at the tip of a crack with a very short kink have been made by several workers (see, for instance, [19]), and small-scale yielding zones have been found under mixed loading by Shih [20]. It may be mentioned, too, that Cotterell and Rice [21] studied the stability of a slightly kinked crack, subject to symmetric loading defined by K_I and T_{11} in equation (1) and concluded, on the assumption that the crack chooses its path so as to remain locally in mode I, that the straight path was stable for $T_{11} < 0$ and unstable for $T_{11} > 0$. Thus, even within linear fracture mechanics, knowledge of K_{Ic} is adequate only for a certain (but rather important) class of problems.

3.2 Non-linear Fracture Mechanics

In view of the unresolved issues that mixed-mode loading presents even for linear fracture mechanics, it is hardly surprising that progress at load levels for which the inequality (7) is violated is limited to symmetric loading. Thus, the present objective of non-linear fracture mechanics may be taken as the identification of a single parameter that takes over the role of K_I at higher levels of symmetric loading.

(a) The J-integral

For an elastic body (possibly admitting non-linear behaviour), an energy function $W(e_{ij})$ is defined as the energy per unit volume at strain e_{ij} and the stress σ_{ij} is its gradient, so that

$$\sigma_{ij} = \frac{\partial W}{\partial e_{ij}} \tag{9}$$

and

$$W(e_{ij}) = \int_o^{e_{ij}} \sigma_{k\ell}\, de_{k\ell}\ , \tag{10}$$

the integral being independent of the path from 0 to e_{ij} in strain space. It can be shown [22] that, if a crack in a non-linearly elastic body extends by an amount δa, then the energy that flows into the crack edge is $J\delta a$ per unit length, where J is defined by the contour integral

$$J = \oint_{\Gamma} (W dx_2 - \sigma_{ij} n_j \partial u_i / \partial x_1 ds) \tag{11}$$

and Γ is a contour enclosing the crack tip. It is easy to verify, too, that the integral (11) is independent of the choice of the path Γ so that Γ could, in particular, be taken either around the boundary of the specimen or around a small circle enclosing the crack tip.

If, now, the load level is such that significant non-linearity occurs only within a very small neighbourhood of the crack tip and this is surrounded by a region in which the $r^{-\frac{1}{2}}$ singularity of equation (1) is dominant (giving small scale yielding), evaluation of J from a contour in the latter region gives

$$J = K_I^2/E' , \tag{12}$$

where $E' = E/(1-\nu^2)$ for plane strain conditions, ν denoting Poisson's ratio. Thus, fracture under conditions of small scale yielding implies $J = J_c = K_{Ic}^2/E'$.

Now consider any load level. The crack tip will be subject to some singular behaviour, different from that described by (1) because of the non-linearity but still, because J can be evaluated on a path lying within the region dominated by the singular field, the value of J will measure its intensity. Substituting the HRR singular fields (4) into (11), for example, produces

$$J = \alpha \sigma_o e_o K_\sigma^{n+1} I_n , \tag{13}$$

where the numerical factor I_n has been tabulated by Hutchinson [3]. Thus, the value of J determines the value of the plastic intensity K_σ. By an argument similar to that given for the linear singular field (1) in small scale yielding, if the relevant non-linear singular field (such as (4)) is dominant within a distance R_1 of the crack tip, this will be perturbed by the fracture process zone, whose characteristic dimension is δ, say. If δ is comparable to R_1, then the singular field is destroyed and a full analysis is necessary. If, however, $\delta << R_1$, the process zone will be surrounded by the singular field, which itself is characterized by its intensity K_σ or, equivalently, by J, and crack extension is bound to be initiated when J reaches a critical value, J_c. Furthermore, this critical value J_c must correspond to K_{Ic} through equation (12), because small scale yielding provides a particular realization in which the inequalities $\delta << R_1 << R$ happen to be satisfied. It is, of course, conceivable that δ could be of the order of R_1 while the small scale yielding requirement (7) remained satisfied. In such a case, while linear elastic fracture mechanics could be phrased in terms of J through (12), extensions beyond the regime of (7) would probably not be possible.

Now consider elastic-plastic bodies: the stress at any point depends not only upon the current value of strain but also upon its history and a unique energy function $W(e_{ij})$ cannot be defined. It is possible, nevertheless, to define W by equation (10), in which the integral is evaluated for the actual strain history for the material point in question, and then J can be defined by (11). However, the J so defined is neither path-independent, nor can it be interpreted as an "energy release rate", unless the loading of the body happens to produce a deformation that would also have been produced in a suitably chosen non-linearly elastic material, with a constitutive relation such as (2), for instance. An example in which this occurs exactly is in the mode III loading of an elastic-perfectly plastic body. Proceeding under the assumption that J is both approximately path-independent and can be interpreted as an energy release rate, Begley and Landes [23] successfully demonstrated the practical use of a J-criterion for evaluating fracture toughness from specimens that violated the small-scale yielding restriction (8), and J is now the most commonly used fracture characterizing parameter in situations that involve extensive yielding.

Some theoretical support for the use of J in elastic-plastic fracture has been provided by finite-element studies, but limitations have also been identified. Finite element calculations have generally confirmed the approximate path-independence of J [7] but there is evidence [24] (conflicting, to some extent, with the finding in [7]) that J becomes increasingly path-dependent as plastic flow develops in a centre-cracked plate. Also, it has been demonstrated [8] that a singular HRR field exists at the tip of a crack in a body conforming to incremental plasticity with power-law hardening, over a region of extent R_1 that depends both upon hardening exponent and upon specimen geometry. This work showed that, while R_1 was generally large enough for the HRR field to surround a plausible process zone for compact tension and bend specimens, R_1 became unacceptably small in the intermediate yielding range for the centre cracked plate.

An additional theoretical restriction in the use of J should be noted. Although a "deformation theory" description may be acceptable before the crack extends, it cannot possibly apply after extension, since the elastic unloading that occurs as the crack extends is inherently strain-history dependent. Therefore, although in small-scale yielding, even crack extension might be expected to correlate with K_I and, similarly, J, the applicability of J in the non-linear regime should, strictly, be confined just to the initiation of crack extension. As will be discussed later, however, definition of initiation is not entirely straightforward (modelling is required again) and there are actually circumstances in which J can be employed to correlate small amounts of crack growth.

(b) The measurement of J

In their original experiments, Begley and Landes [23] determined J by use of the formula

$$J = -\frac{1}{b}\frac{\partial U}{\partial a}, \tag{14}$$

where b denotes specimen thickness, a is crack length, U represents the work done by the applied loads up to a given deflection and the derivative is taken at constant deflection. This entailed loading specimens containing cracks of lengths a and a + δa and then differencing the values of U: crack extension in a single specimen could not be allowed because the deformation theory approximation would then not have applied. More recently, however, methods have been developed for obtaining J from a single load-displacement record. The most comprehensive account is that of Ernst et. al. [25]; here, just the simplest (and most commonly used) formula will be given. First, it follows by dimensional analysis that, for any particular specimen,

$$J = \frac{\eta U}{b(w-a)}, \tag{15}$$

where w denotes specimen width and η is some dimensionless function of the deformation. If, however, the deformation is concentrated near the ligament ahead of the crack, as it is, for example, in a deeply-cracked bend specimen, the load-displacement record itself will depend only upon (w-a) and η can be determined, using (14). For a bend specimen, to which is applied a remote rotation θ, stress can depend upon x only in the combination x/(w-a) and hence the couple $M(\theta)$ required to produce the rotation θ, and the work $U(\theta)$, are both proportional to $(w-a)^2$. Equation (14) then implies that $\eta=2$.

In practice, therefore, most determinations of J in fact amount to determinations of U or, if a formula more complicated than (15) is employed, some other property of the load-deflection curve, which is defined unambiguously even during crack extension.

(c) The crack opening displacement

The strip-yielding model of Dugdale, Bilby, Cottrell and Swinden predicts a discontinuity δ_t in displacement at the tip of a crack, given explicitly by equation (6) for a crack in a large plate subject to tensile loading. This discontinuity is undoubtedly a near-tip feature which plausibly characterizes the condition of strain near the crack tip and it therefore has appeal as a fracture characterizing parameter: the fracture criterion

$$\delta_t = \delta_c \tag{16}$$

has been used extensively ever since it was proposed by Wells [26]. A practical difficulty (not usually regarded as serious) is the requirement of measuring δ_c accurately; this is because real crack tips never have the regular appearance implied by theoretical models. There is also a theoretical difficulty in that, although non-hardening materials allow a finite crack opening displacement, hardening materials give crack-tip fields such as (4), with zero displacement at the crack tip. Therefore, some working definition of δ_t needs to be adopted based, inevitably, upon measuring the separation of the crack faces at some point behind the crack tip.

A question now arises regarding the equivalence or otherwise of a J- or a δ_t- criterion for fracture. For the model of Dugdale, Bilby, Cottrell and Swinden, the answer is immediate for the J-integral can be evaluated on a contour Γ just surrounding the yielded strip to give

$$J = \sigma_o \delta_t . \tag{17}$$

Also, for any model of plastic behaviour, under small scale yielding conditions, both J and δ_t must depend upon K_I and so one is a function of the other. Dimensional considerations then give

$$J = m\sigma_o \delta_t , \tag{18}$$

where m is a constant. Finite element studies in small-scale yielding (see, for example, [18]) usually give $1 < m < 3$. Under more general conditions, however, the simple relationship (18) need not hold (that is, m might become a function of both the load level and specimen geometry). This was shown explicitly in Mode III by Edmunds and Willis [14] who gave, in the non-hardening limit, rigorous series representations for J and δ_t that conformed to (18) only to lowest order in the loading parameter (that is, in the small-scale yielding limit). Whether or not (18) is true in hardening materials depends upon the definition that is adopted for δ_t. If δ_t is measured within the region of dominance of the singular HRR field, then (18) is bound to be true. If, however, δ_t is measured on the crack faces at the edge of the plastic zone, it is not [7,15].

Agreement or otherwise of the δ_t-criterion with the J-criterion is not, of course, an argument for or against its validity. It could be that δ_t, in representing a near-tip feature, might be a useful fracture-characterizing parameter even when J is not; the matter is not definitively settled at present.

4. Crack Growth

(a) Stable growth

Even at very low levels of loading, plastic yielding is associated with very large strains in the vicinity of the crack tip and some small increase in crack surface area. Exact quantification is difficult but, conventionally, the crack extension Δa when the crack opening displacement is δ_t is taken as $\delta_t/2$. If this estimate (which, roughly speaking, assumes that the blunted crack tip has a semicircular shape) is combined with the estimate (17) for δ_t as a function of J, it follows that

$$\Delta a = J/2\sigma_o . \tag{19}$$

This defines the "blunting line" on Fig. 5. Extension described by (19) is not regarded as "real" fracture: this is taken to commence when Δa begins to increase more rapidly than given by (19), through the appearance of some distinctive mechanism (for example cleavage or growth and coalescence of voids), taking place at the blunted crack edge. In performing a test to evaluate J_c, therefore, J is deduced from the load-displacement record by a formula such as (15), and plotted against crack extension, Δa, which is monitored separately. Usually , the experiment does not accurately reproduce the blunting line but a well-defined J versus Δa plot is obtainable for varying amounts of "real" crack extension; this is also shown in Fig. 5. The critical value J_c of J is now inferred by extrapolating the measured curve back to its intersection with the blunting line. It is important not to extrapolate back to Δa = 0 because, in many tests, the measured J versus Δa curve has large slope and an appreciably lower estimate of J_c would result [27].

In view of the universally observed tendency for J to increase with Δa during tests on tough materials, Paris et al. [28] have proposed that the J-resistance curve (that is, the plot of J versus Δa) should be regarded as a material property. This requires, of course, that the curve should be independent of specimen size and shape and implies, too, that a sensible universal definition of J can be made: measured values of J are obtained from formulae like (15) which, when crack extension has taken place, amount to definitions that vary from specimen to specimen. Before considering this question further, however, it may be remarked that, independently of the notion of J, some resistance to crack extension is to be expected from the singularity analyses that have been performed for extending cracks in plane strain conditions. They predict (see Section 2.2(a)) crack tip singularities that are less severe than those applying to stationary cracks and so some increase in loading is likely to be required to generate whatever critical conditions over the process zone that are needed for fracture to continue. Another source of resistance to extension is important in relatively thin specimens. This is associated with the large amount of plastic work that is absorbed in tearing the material in the shear lip zones that form at the intersections of the crack edge with the free faces of the specimen. The stress fields are fully three-dimensional and have not so far been analyzed. Clearly, however, if a plane-strain resistance curve is sought, their effect has to be minimized by selection of sufficiently thick specimens (thicker, perhaps, than are required just for the determination of J_c) and also by side grooving; see, for example, [29].

Reverting now to the possible use of J as a parameter to characterize resistance, Hutchinson and Paris [30] have proposed a set of limitations that should ensure that J can be sensibly defined: it amounts, in fact, to restricting the amount of growth to a large "process zone", outside which the field can essentially be described by deformation theory. The situation envisaged in [30] is depicted in Fig. 6. When the crack has extended an amount Δa, unloading occurs over a region whose size is of order Δa and there is a rather larger region in which loading is non-proportional and deformation theory inadequate. However, outside of this region, there is supposed to be a region that has been essentially unperturbed by the crack extension and is governed (in the case of power-law hardening) by the HRR singular field, which itself is characterized by J. The restrictions implied by all of this were studied approximately, by considering what happens when an HRR field is shifted an amount da while J itself increases by dJ. From the second of equations (4), together with (13), the change de_{ij} in strain is then given as

$$de_{ij} = k_n J^{(n/n+1)} r^{-(n/n+1)} \left\{ \frac{n}{n+1} \frac{dJ}{J} \tilde{e}_{ij}(\theta) + \frac{da}{r} \tilde{\beta}_{ij}(\theta) \right\}, \tag{20}$$

where the term $\tilde{\beta}_{ij}(\theta)$ curves from differentiating e_{ij} in the direction of the crack (k_n is the constant $(\alpha\sigma_0 e_0 I_n)^{-n/(n+1)}$). The first term in (20) is of the original J-controlled form (4) and dominates, provided that

$$\frac{\delta a}{r} << \frac{\delta J}{J} . \tag{21}$$

This motivates defining a length scale D by the relation

$$\frac{1}{D} = \frac{1}{J}\frac{dJ}{da} . \tag{22}$$

The HRR field then applies within an annular region for which

$$D << r < R_1 , \tag{23}$$

where R_1 defines the distance from the crack tip within which deformation theory predicts that the HRR field is dominant. It is to be expected that R_1 is some fraction of a typical specimen dimension d (though this fraction may be small: see Section 3.2(a), so that a necessary condition for the possibility of J-controlled crack growth is that

$$\omega = \frac{d}{J}\frac{dJ}{da} >> 1. \tag{24}$$

The preceding discussion suggests that J is to be regarded as a useful parameter for calibrating test data rather than a fundamental parameter of universal relevance. In contrast, the crack tip displacement (or, rather, the shape of the crack opening near the crack tip) has the virtue of being unambiguously defined and (at least in principle) measurable, even when the restriction (24) is violated. The possibility of describing resistance in terms of the shape of the crack opening (with this idealized as a wedge) has been explored by Shih et. al. [31]. For non-hardening materials, Rice and Sorensen [6] studied the shape of the crack tip in more detail and obtained the equation

$$\dot{\delta} = \alpha\dot{J}/\sigma_0 + \beta(\dot{a}/\sigma_0 E) \log[R/a-x] \tag{25}$$

for the displacement δ at position x, the crack tip being located at x = a. The parameters α and R are not given by the singularity analysis that was employed. The meaning of J is unambiguous in the case of contained yielding, for then it can be defined by (11) with the contour Γ in the elastic region. Assuming now that the crack profile remains invariant during extension, so that δ is a function of r = a-x, and regarding a as the time-like variable in (25), this equation can be integrated; the result is given by Hermann and Rice [32] in the form

$$\delta = (\beta r\sigma_o/E)\log(\rho/r), \tag{26}$$

where

$$\rho = eR\exp[(\alpha/\beta)(E/\sigma_o^2)\,dJ/da]. \tag{27}$$

The assumption of constant profile then led to the extension criterion ρ = constant. This would provide a differential equation for J, if the dependence of R upon J were known. Strictly, it is not but, by plausible choice of the functional form and of the parameter α, Hermann and Rice were able to correlate their theory with measured resistance curves. At high loads, incidentally, this correlation also required a plausible definition of J.

The important point to make here is that, in this growth situation in which J theory is not rigorously applicable, some success is being achieved from consideration of the shape of the crack tip which is characterized, in effect, by the single parameter ρ. Thus, apparently, there is even some hope of characterizing growth without study of the fracture mechanism, except that this whole approach relies upon the assumption that the same local mechanism of fracture is always operative. The topic is not yet at a definitive state of development.

(b) Fatigue

Finally, very brief mention will be made of the important topic of fatigue crack growth. The simplest engineering approach was initiated by Paris, Gomez and Anderson [33] who proposed that, under conditions of small scale yielding, since crack tip conditions are governed by the stress intensity factor K_I, it might be appropriate to correlate fatigue crack growth rates (during Stage II of growth, when the crack extends essentially in the 'opening' mode, mode I) with ΔK, the difference between the greatest and smallest values of K_I. If no features of the material structure are admitted, dimensional considerations impose the relation

$$da/dN = \alpha(\Delta K/\sigma_o)^2 \tag{28}$$

for the growth-rate per cycle da/dN, where α is a constant. The fact that this simple relationship is not usually observed (see, for example, [34]) is a clear demonstration that some characteristic length, associated with the microstructure, must be introduced.

Growth-rates proportional to $(\Delta K)^n$, with n different from 2, are frequently obtained from fatigue tests. Clearly, however, ΔK cannot always be the only relevant parameter, since catastrophic failure will occur in a single cycle if the maximum K_{max} of K_I in the cycle is sufficiently large. A common method of incorporating this effect is to include some dependence upon the ratio K_{max}/K_{min}; this, however, is an ad hoc procedure at present. An effect more recently noted is that, during the unloading part of any cycle, the crack faces may come into contact, particularly at asperities, and so modify the stresses at the crack tip. This was recognised from an experimental study by Elber [35,36] and has subsequently been supported by finite element calculations [37] and a simple explicit analysis of the strip yield model by Budiansky and Hutchinson [38]. Budiansky and Hutchinson demonstrated that, under cyclic loading during which the crack extends only a small fraction of the plastic zone size in each cycle, the crack tip displacement at maximum load was reduced from the value obtained in monotonic loading, both directly because the material at the crack faces had been subjected to stretching, and indirectly because the faces came into contact at a value of K_I larger than K_{min}, the lowest value attained in the cycle. It was found that the crack tip displacement could be expressed in the form of equation (28), but with ΔK representing the difference between K_{max} and K_I at contact of the crack faces. Detailed predictions of growth rates would require a model more realistic than the strip-yielding model employed, however.

Finally, there does not even seem to be experimental agreement on the features of fatigue crack growth at high loads or (to some extent equivalently) of the growth of cracks whose lengths are not much greater than the extent of the zone of cyclic plastic deformation. Although complete calculations in the cyclic loading case are not available, the analyses of Edmunds and Willis [14,15,16] on monotonic loading demonstrate that, at intermediate load levels, specimen geometry will influence strains within the plastic zone, initially through the terms T_{ij} in equation (1) and, at higher load levels still, through other parameters that can be obtained from elastic solutions. If some unique field analogous to the HRR fields (4) is always set up at the crack tip, then effects of T_{ij} and similar parameters will be insignificant so long as the crack tip field dominates over a length scale greater than that applying to the process zone. Cyclic zones are typically smaller than those induced by monotonic loading, however, roughly because the effective yield stress is doubled [39], so this condition for dominance may well not be met. In that case, growth rates should be influenced by T_{ij}. Experiments by Miller [40] display such an effect but those of Liu et al. [41] do not: further experimentation and theoretical modelling, performed with the object of identifying the limits of validity of simple "ΔK" theory, are desirable.

References

1. M.L. Williams, J. Appl. Mech. 24, 109 (1957).
2. H. Tada, P.C. Paris and G.R. Irwin, The Stress Analysis of Cracks Handbook, Del Research Corporation, Hellertown, Pennsylvania (1973).
3. J.W. Hutchinson, J. Mech. Phys. Solids 16, 13 (1968).
4. J.R. Rice and G.F. Rosengren, J. Mech. Phys. Solids 16, 1 (1968).
5. J.C. Amazigo and J.W. Hutchinson, J. Mech. Phys. Solids 25, 81 (1977).
6. J.R. Rice and E.P. Sorensen, J. Mech. Phys. Solids 26, 163 (1978).
7. J.D.G. Sumpter and C.E. Turner, J. Mech. Eng. Sci. 18, 97 (1976).
8. R.M. McMeeking and D.M. Parks, Elastic-Plastic Fracture, ASTM STP 668, p.175, American Society for Testing and Materials, Philadelphia (1979).
9. D.S. Dugdale, J. Mech. Phys. Solids 8, 100 (1960).
10. B. Bilby, A.H. Cottrell and K.H. Swinden, Proc. Roy. Soc. A279, 1 (1964).
11. C. Atkinson and M.F. Kanninen, Int. J. Fract. 13, 151 (1977).
12. F.A. McClintock, Fracture: An Advanced Treatise, Vol. 3, p.47, Academic Press, New York (1971).
13. N.L. Goldman and J.W. Hutchinson, Int. J. Solids Struct. 11, 575 (1975).
14. T.M. Edmunds and J.R. Willis, J. Mech. Phys. Solids 24, 205 (1976).
15. T.M. Edmunds and J.R. Willis, J. Mech. Phys. Solids 24, 225 (1976).
16. T.M. Edmunds and J.R. Willis, J. Mech. Phys. Solids 25, 423 (1977).
17. M.D. Van Dyke, Perturbation Methods in Fluid Mechanics (annotated edition), The Parabolic Press, Stanford, Ca. (1975).
18. R.M. McMeeking, J. Mech. Phys. Solids 25, 357 (1977).
19. K.K. Lo, J. Appl. Mech. 45, 797 (1978).
20. C.F. Shih, Mechanics of Crack Growth, ASTM STP 590, p3, American Society for Testing and Materials, Philadelphia (1976).
21. B. Cotterell and J.R. Rice, Int. J. Fract. 16, (1980).
22. J.R. Rice, Fracture: An Advanced Treatise, Vol. 2, p.191, Academic Press, New York (1968).
23. J.A. Begley and J.D. Landes, Fracture Toughness, ASTM STP 514, p.1, American Society for Testing and Materials, Philadelphia (1972).
24. M.E. Karabin, Jr. and J.L. Swedlow, Fracture Mechanics, ASTM STP 677, p.600, American Society for Testing and Materials, Philadelphia (1979).
25. H. Ernst, P.C. Paris, M. Rossow and J.W. Hutchinson, Fracture Mechanics, ASTM STP 677, p.581, American Society for Testing and Materials, Philadelphia (1979).
26. A.A. Wells, Symposium on Crack Propagation, College of Aeronautics, Cranfield, Paper B4 (1961).
27. J.D. Landes and J.A. Begley, Post-Yield Fracture Mechanics, p.211, Applied Science Publishers, Barking, Essex (1979).
28. P.C. Paris, H. Tada, A. Zahoor and H. Ernst, Elastic-Plastic Fracture, ASTM STP 668, p5, American Society for Testing and Materials, Philadelphia (1979).
29. J.P. Gudas, J.A. Joyce and D.A. Davis, Fracture Mechanics, ASTM STP 677, p.474, American Society for Testing and Materials, Philadelphia (1979).
30. J.W. Hutchinson and P.C. Paris, Elastic-Plastic Fracture, ASTM STP 668, p.37, American Society for Testing and Materials, Philadelphia (1979).
31. C.F. Shih, H.G. de Lorenzi and W.R. Andrews, Elastic-Plastic Fracture, ASTM STP 668, p.65, American Society for Testing and Materials, Philadelphia (1979).
32. L. Hermann and J.R. Rice, Brown University Report no. E(11-1)3084/76.
33. P.C. Paris, M.P. Gomez and W.E. Anderson, Trends in Engineering Vol. 13 no. 1, University of Washington (1961).
34. L.P. Pook, Developments in Fracture Mechanics - 1, p.183, Applied Science Publishers, Barking, Essex (1979).
35. W. Elber, Engineering Fracture Mechanics 2, 37 (1970).
36. W. Elber, Damage Tolerance in Aircraft Structures, ASTM STP 486, p.230, American Society for Testing and Materials, Philadelphia (1971).
37. J.C. Newman, Jr., Mechanics of Crack Growth, ASTM STP 590, p.281 (1976).

38. B. Budiansky and J.W. Hutchinson, J. Appl. Mech. 45, 267 (1978).
39. J.R. Rice, Fatigue Crack Propagation, ASTM STP 415, p.247, American Society for Testing and Materials, Philadelphia (1967).
40. K.J. Miller, Metal Science 11, 432 (1977).
41. A.F. Liu, J.E. Allison, D.F. Dittmer and J.R. Yamane, Fracture Mechanics, ASTM STP 677, p.5, American Society for Testing and Materials, Philadelphia (1979).

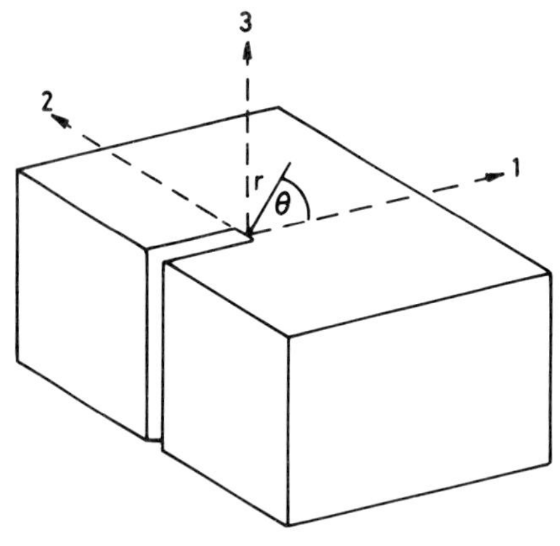

Fig. 1.

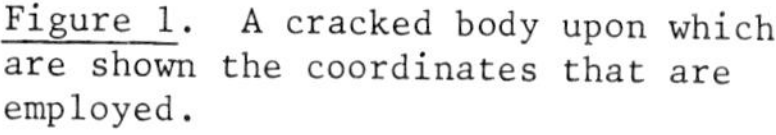

Figure 1. A cracked body upon which are shown the coordinates that are employed.

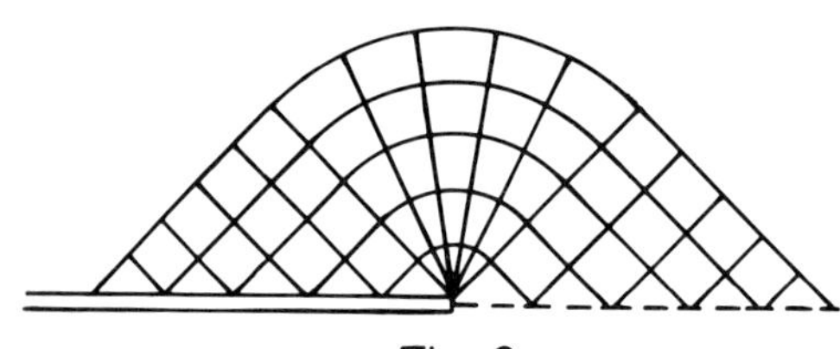

Fig. 2.

Figure 2. The slip-line field at the tip of a crack in a non-hardening plastic specimen.

Fig. 3.

Figure 3. Schematic diagram of the plastic deformation field at the tip of a growing crack.

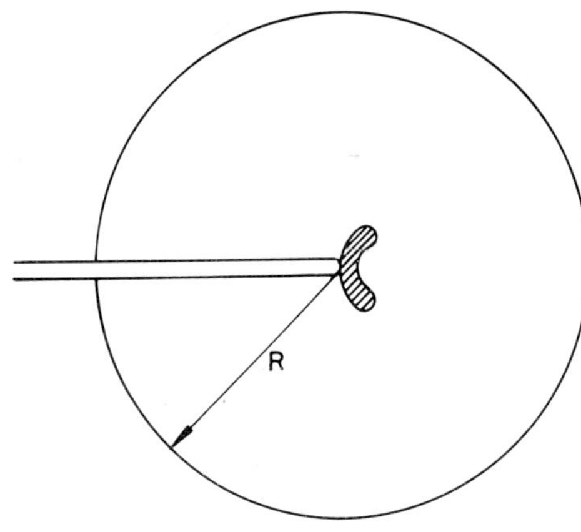

Fig. 4.

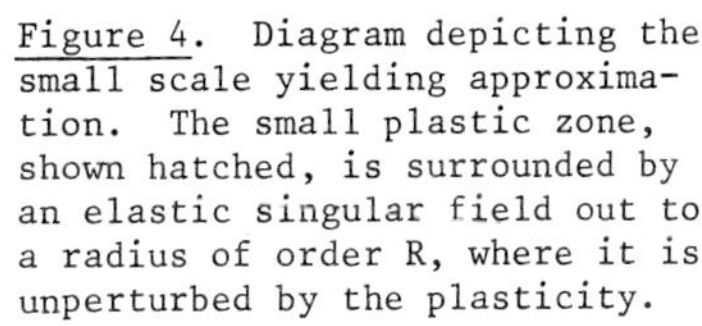

Figure 4. Diagram depicting the small scale yielding approximation. The small plastic zone, shown hatched, is surrounded by an elastic singular field out to a radius of order R, where it is unperturbed by the plasticity.

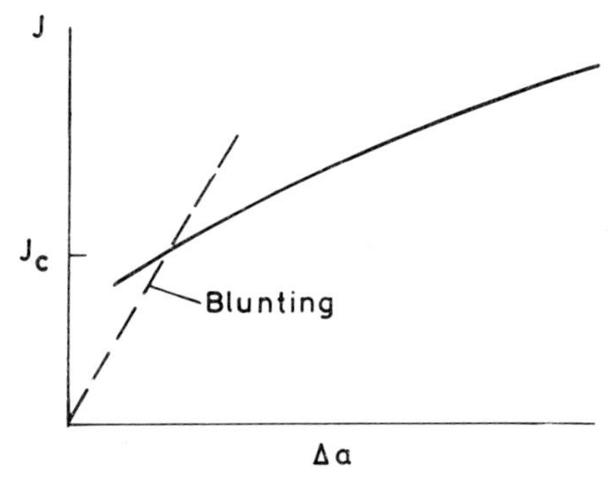

Fig. 5.

Figure 5. Schematic diagram of a J versus Δa plot. The solid curve is extrapolated back to the blunting line defined by equation (19) to define J_c.

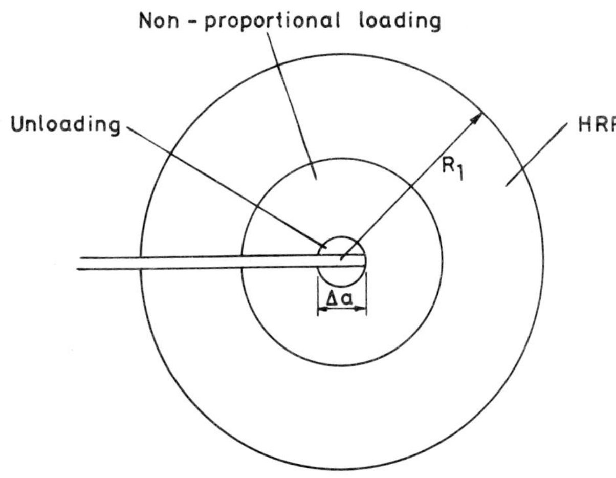

Fig. 6.

Figure 6. Schematic diagram depicting the possibility of J-controlled growth. The "deformation theory" singular HRR field applies out to a radius R_1, supposed much greater than Δa and the intermediate radius within which non-proportionality is significant.

DISLOCATION MODELS IN FRACTURE

V. Vitek[*] and G. G. Chell[**]

[*]Department of Materials Science and Engineering and the
Laboratory for Research on the Structure of Matter
University of Pennsylvania
Philadelphia, PA 19104 U.S.A.

[**]Materials Division, Central Electricity Research Laboratories
Kelvin Avenue, Leatherhead, Surrey, U.K.

ABSTRACT

Dislocation theory usually deals either with dislocations in a contiuum or with crystal dislocations. These two alternatives also appear in applications of the dislocation theory to fracture. The purely continuum theory of dislocations has been used principally in stress analysis and in fracture mechanics whilst lattice dislocations have been considered in studies of the micromechanisms of cracking.

In this paper we first review the use of dislocations in stress analysis and linear elastic-fracture mechanics. In particular, we discuss in more detail, calculations of the stress intensity factors for complicated crack configurations. The dislocation models used in elastic-plastic fracture mechanics are then summarized and their impact on the interpretation of the CTOD and critical J_c, discussed. Specifically, the contribution of the dislocation interpretation towards understanding the phenomenon of warm pre-stressing is shown. Applications of dislocation models in fatigue and creep fracture are also presented.

The mechanisms of microcrack nucleation and propagation based on pile-up models are first discussed in the part dealing with microscopic models of brittle fracture. A new model of microcrack propagation in which the bond breaking and dislocation emission at the microcrack tip are regarded as concomitant processes is then presented.

1. Introduction

The concept of the dislocation was first introduced by Volterra (1) and Somigliana (2, 3) to describe such sources of internal stresses in a continuum which are connected with a closure failure in displacements. It was realized only much later (4-6) that in crystals there are defects causing closure failures, i.e., crystal dislocations, and that their motion and multiplication are responsible for plastic deformation. The concepts of crystal and continuum dislocations are analogous. The main difference is that in the continuum any closure failure (Burgers vector) is allowed while in the crystal it is determined by the crystal structure. On the other hand, the long-range elastic stresses and strains of both continuum and lattice dislocations are the same and the plastic strain in a volume of a material is related to the Burgers vector content in this volume. The plastic deformation in a continuum may thus be described by a continuous distribution of dislocations. This leads to a clear separation of plastic strains and the elastic strains which arise due to the plastic deformation, the latter being now the elastic strains of the corresponding dislocations. Hence the theory of the continuous distribution of dislocations in a continuum, originally developed by Kondo (7), Kröner (8, 9), and Bilby et al. (10, 11) may serve as a basis for the development of an alternative theory of macroscopic plasticity. On the other hand when studying micromechanisms of any phenomena involving plastic deformation, we have to take into account the motion, configurations and distributions of lattice dislocations.

These two applications of dislocation theory have both been used in the theory of fracture. The dislocation description of macroscopic plasticity has found numerous applications in studies relating to crack tip plasticity, i.e., in elastic-plastic fracture mechanics. The best known model in this field is that originally proposed by Bilby, Cottrell and Swinden (12) which has been further developed by a large number of authors. Furthermore, the concept of dislocations in a continuum has been employed in an even more formal way by using the solutions for the stress field of a dislocation as a Green's function in the stress analysis of crack configurations. In this case the cracks are modelled by distributions of "crack" dislocations which have to be determined so as to satisfy the required boundary conditions. These dislocation distributions are usually continuous pile-ups and the first analysis of this type was made in references (13-15).

Microscopic models of the mechanisms of cracking which use the dislocation theory are principally those of microcrack nucleation and subsequent propagation. The former case usually involves the investigation of the formation of pile-ups of lattice dislocations which serve as local stress concentrators causing crack nucleation, as originally suggested by Zener (16). In the latter case the emission of lattice dislocations from the microcrack tip is studied in order to elucidate when the microcrack starts to propagate and when propagation is prevented and only blunting of the microcrack tip occurs (e.g. ref. 17, 18).

In this paper we review the above mentioned applications of dislocation theory to fracture mechanics and the micromechanisms of cracking. Several cases studied by the present authors will be presented in more detail. First the use of dislocations in stress analysis is shown with the emphasis on methods of determining stress intensity factors for various configurations of cracks. The application of the dislocation models in elastic-plastic fracture mechanics (EPFM) is then presented. The physical meaning of both the critical crack tip opening displacement (CTOD) and J-integral (J_c), and the contribution of dislocation models towards their interpretation will be discussed. A specific example is an interpretation of the phenomenon of warm pre-stressing. Applications of the dislocation models to fatigue and creep fracture will also be briefly outlined. In the part of the paper which deals with micromechanisms the dislocation models of crack nucleation, in particular the pile-up models, will be discussed. A model of microcrack propagation which is concomitant with dislocation emission from its tip will be presented and it will be shown how a relationship between the plastic work, γ_p, and the ideal work to fracture, γ, can be established.

2. Linear Elastic Fracture Mechanics (LEFM)

Fracture mechanics is concerned with macroscopically sized crack-like defects and the effects these have on the load bearing capacity of structures. In LEFM the fracture parameter of interest is the stress intesity factor, K, which determines the magnitude of the stress singularity at the tip of a loaded sharp crack in an elastic material (19). This singularity goes ar $r^{-\frac{1}{2}}$, where r is the distance from the tip. It is postulated that fracture ensues when the value of K reaches a critical value, K_c, called the fracture toughness. For engineering purposes it is regarded as a fundamental material parameter and it has been measured for many structural materials. On the other hand the goal of the studies of micromechanisms of fracture is to determine the dependence of K_c on microstructural metallurgical parameters.

It is not feasible in this article to give a comprehensive survey of the contribution of dislocation theory to fracture mechanics. Of necessity some details must be omitted, particularly concerning the mathematical equivalence of dislocation and continuum mechanics theory. However, a review of this aspect and the application of dislocations to the theory of cracks up to 1968 was given by Bilby and Eshelby (20).

In contrast to crystal dislocations which can occur in unstressed material, crack dislocations occur because of external stressing. The extra plane of material associated with crystal dislocations becomes a plane of removed material when discussing crack dislocations. A crack is formed by the imposition of barriers which confine the dislocations to the region occupied by the crack, the tips of which are coincident with the barriers. Because crack dislocations are formal rather than real, edge dislocations can climb as easily as they glide; they are free to move and experience no resistance to their motion. Furthermore since in the framework of linear isotropic elasticity parallel dislocations with perpendicular Burgers vectors do not interact, a planar crack may be described by three independent distributions of dislocations, two consisting of edge and one of screw dislocations. These give rise to the three independent modes of deformation. Mode I deformation corresponds to the application of a tensile stress σ_{22} which produces relative displacements ϕ_2 between the crack surfaces, where the plane of the crack is in x_1-x_3 plane corresponding to $x_2 = 0$ and the crack is infinitely long in the x_3 direction with the crack tips at $x_1 = \pm a$. Subscripts 1, 2 and 3 signify x_1, x_2 and x_3 respectively. The stress intensity factor for this mode is labelled K_1, and the crack is simulated by freely climbing edge dislocations with Burger vectors in the x_2 direction. Modes 2 and 3 correspond to applied shear stress σ_{12} and σ_{32}, displacements ϕ_1 and ϕ_3 and stress intensities K_2 and K_3, respectively. They are simulated by freely gliding edge and screw dislocations with Burgers vectors in the x_1 and x_3 directions, respectively.

3. Stress Analysis

When solving any two dimensional elasticity problem the stress field of an infinite dislocation can be used as the Green's function. For cracks infinitely long in the x_3-direction the solution can be obtained by finding a dislocation distribution inside the cracks such that the tractions arising from their stress fields cancel the tractions due to the external load at the crack surfaces, rendering them traction free. In general this leads to the following system of three coupled integral equations for the densities, ρ_k, of dislocations with Burgers vectors pointing in three orthogonal directions (k = 1, 2, 3):

$$\int \rho_k (\underset{\sim}{r}') \, \sigma_{ij}^{D,k} (\underset{\sim}{r}, \underset{\sim}{r}') \, n_j(\underset{\sim}{r}') \, d\underset{\sim}{r}' + T_i(\underset{\sim}{r}) = 0 \quad (i = 1, 2, 3) \tag{1}$$

where Einstein summation is assumed over the repeating indices. Both $\underset{\sim}{r}$ and $\underset{\sim}{r}'$ are position vectors of the points on the cracks. $\sigma_{ij}^{D,k}$ (r, r') is the ij component of the stress tensor at the point $\underset{\sim}{r}$ of a dislocation with the unit Burgers vector in the direction k, positioned at the point $\underset{\sim}{r}'$. $n_j(r')$ is the j component of the unit vector normal to the crack at the point $\underset{\sim}{r}'$ and and T_i are the tractions due to the external loading. Equations (1) are Cauchy type singular integral equations which in some special cases may be solved analytically (e.g. 21) and in general by standard numerical methods (e.g. 22).

The solutions of eq. (1) represent pile-ups of continuously distributed dislocations. The dislocation densities near the crack tip always converge towards (23)

$$\rho_k (\underset{\sim}{r}) = \alpha \frac{K_k}{\sqrt{2\pi |\underset{\sim}{a} - \underset{\sim}{r}|}} \tag{2}$$

where $\underset{\sim}{a}$ is the position vector of the crack tip and $\alpha = 2(1-\nu)/G$ for k = 1, 2 and $\alpha = 2/G$ for k = 3; G is the shear modulus and ν Poisson ratio. Thus when ρ_k are known the stress intensity factors can be found using eq. (2).

Analytical solutions can be obtained for planar crack configurations in the isotropic case. Taking σ_{ij}^{D} as the stress field of a dislocation in an infinite isotropic body, equations (1) are no longer coupled and read

$$\frac{1}{\pi\alpha} \int \frac{\rho\ (x')}{x - x'} \, dx' + \sigma(x) = 0 \tag{3}$$

where σ is the applied shear or tensile stress and ρ the corresponding dislocation density. The solutions can be found using the theory of analytical functions and Cauchy integrals (21, 24). The solutions for a number of planar pile-ups described by eq. (3) were given by Head and Louat (15). For example the cases of a single crack of length 2a and of a pair of cracks in a uniformly stressed infinite body have been solved in ref. (15) for any mode of loading. In the former case the dislocation density is

$$\rho\ (x) = \frac{\sigma\alpha}{\pi} \quad \frac{x}{\sqrt{a^2 - x^2}} \quad .$$

Clearly as $x \to a$ $\rho(x)$ is given by eq. (2) with $K = \sigma\sqrt{\pi a}$. A number of other problems for constant external load were treated by Smith (e.g. 25). In general, of course, nonuniform loading can be handled using this method and analytical or numerical solutions can be obtained e.g. (25).

The dislocation models of cracks become particularly useful when solving problems with a complex crack geometry. An example are branched and bent cracks which in the framework of plane strain and constant tensile loading have been treated by Vitek (26) and Lo (27). In ref. (26) a more general case of slit-like cracks emanating from a cylindrical hole of elliptical cross section with semi axes a and b (Fig. 1), was treated. Only the branches were modelled by dislocation distributions but not the elliptical hole and/or the main crack in the case of $b \to 0$. This means that $\sigma_{ij}^{D,k}$ (Green's function) in eq. (1) was written as a sum of the stress

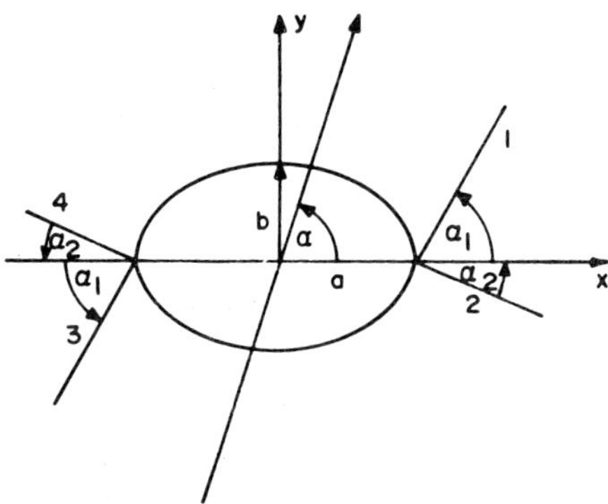

Fig. 1: Schematic picture of a cylindrical hole of elliptical cross-section with four slit-like crack branches (marked 1-4).

field of a unit dislocation in an infinite isotropic body and of an image stress field which ensures zero tractions on the surface of the hole*. In ref. (27) any slit-like main crack was considered and thus the branches or bends as well as the main crack were modelled by dislocation distributions. Therefore $\sigma_{ij}^{D,k}$ in eq. (1) was simply taken as the stress field of a dislocation in an infinite isotropic body. Detailed calculations of the stress intensity factors for branched and bent cracks have been made in refs. (26) and (27). In both studies very good agreement was found with more elaborate calculations which employed a conformal mapping technique (32-34) but not with the approximate calculations in (35).

Stress analyses which employ continuous distributions of dislocations have been used in a number of other cases. In ref. (36) penny-shaped cracks were modelled in terms of a distribution of concentric dislocation loops. The free surfaces and internal holes in a homogeneous medium have been represented by dislocations distributed at the surfaces so as to ensure zero tractions (37). This method of surface dislocations was then applied in the stress analysis of cracks emanating from free surfaces (38) and in the studies of the interaction between cracks and holes (39). Recently, dislocation representation of cracks has also been used in dynamic fracture. In this case the fundamental solution $\sigma_{ij}^{D,k}$ is that of a dislocation of a unit Burgers vector in the direction k, moving with a velocity $\underset{\sim}{v}$. The cracks are again represented by dislocation distributions ρ_k but these are now functions of $\underset{\sim}{v}$ rather than of the position $\underset{\sim}{r}$. Since again zero tractions are required on crack surfaces, the integral equations determining ρ_k, are analogous to eq. (1) with $\underset{\sim}{r}'$ replaced by $\underset{\sim}{v}$; furthermore $\sigma_{ij}^{D,k}$

*The image stress field was derived in ref. (28) for the case of an edge dislocation (plane strain) and a cylindrical hole of any elliptical cross section in an isotropic material. For the anisotropic case and a slit (i.e. b = 0 in Fig. 1) it was also derived in references (29-31). For the case of screw dislocations (anti plane strain) the image stress field can easily be evaluated using the method of analytical functions.

and T_i are explicitly dependent on time (40). This method has been used for the stress analysis and evaluation of dynamic stress intensity factors for dynamically loaded fast moving cracks in an infinite body (40) and also for cracks moving with constant velocity perpendicular to the boundary of a linear elastic half space (41).

In general, stress analyses carried out with the help of dislocation distributions are straightforward and although analytical solutions can be found only in a few special cases, accurate numerical solutions can be obtained relatively easily. Hence, in many cases this approach is superior to other stress analysis methods in particular to the commonly used finite element methods.

4. Elastic-Plastic Fracture Mechanics (EPFM)

4.1 Crack Tip Opening Displacement

The application of LEFM to fracture problems is limited to those circumstances where the effects of crack tip plasticity are small, e.g., low stress failures. Here the elastic fields which are characterized by K provide the boundary conditions for the development of non-linear deformation (42, 43). The breakdown of LEFM is heralded by unrealistic predictions of fracture behavior, in particular in the presence of appreciable yielding the fracture load is overestimated.

The crack tip opening displacement (CTOD) is the easiest EPFM parameter to understand. It measures the plastic displacement in the region of the crack tip where intense straining and blunting occurs. It is particularly useful when discussing stable ductile cracking, fatigue and creep crack growth because it reflects the damage accumulated during these processes.

One of the simplest ways of allowing for plastic deformation at a crack tip is the Bilby, Cottrell and Swinden (BCS) model (12). Here the plastic region is restricted to the plane of the crack and is simulated by a continuous distribution of dislocations whose motion is opposed by a friction stress σ_1 (Fig. 2). The equation of equilibrium giving the dislocation distribution, ρ, is similar to equation (3) and for a single embeded crack becomes

$$\int_{-a-s}^{a+s} \rho\,(x')\,\sigma_D\,(x',x)\,dx' + S(x) = 0 \qquad (4)$$

where a is the half length of the crack, s the extent of the plastic zone and $S(x) = \sigma(x)$ for $x<a$, $S(x) = \sigma(x) - \sigma_1$ for $a<x<a+s$. The dislocations representing the plastic zone smear out the elastic stress singularity so that the induced stresses are finite. The condition for this is that $\rho(x) = 0$ at the end of the plastic zone. This condition determines the size of the zone. The BCS model may be solved in two ways; either both the crack and plastic zone are simulated by dislocations or only the plastic zone is. In the former case $\sigma_D = 1/\pi\alpha(x-x')$. In the latter case $\rho\,(x) = 0$ for $-a<x<a$ and $\sigma_D = 1/\pi\alpha(x-x') + \sigma_D^{im}$, where σ_D^{im} is an image stress field of a unit dislocation which ensures zero tractions on the crack surface (see also section 3).

Of particular importance is the CTOD, $\phi(a)$, which is given by

$$\phi\,(a) = \int_{a}^{a+s} \rho\,(x)\,dx \qquad (5)$$

For the case of a crack of length 2a in an infinite body subjected to a uniform tensile stress σ (12)

$$\phi\,(a) = \frac{2\alpha\sigma_1 a}{\pi}\,\ln\sec\left(\frac{\pi}{2}\,\frac{\sigma}{\sigma_1}\right) \qquad (6)$$

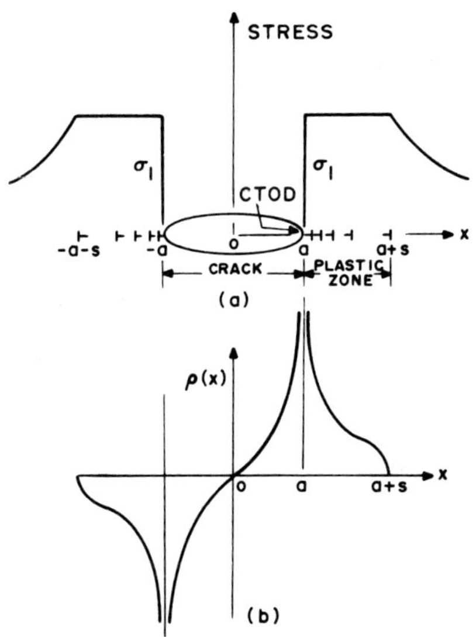

Fig. 2: Schematic representation of BCS model for mode I deformation. (a) Plastic zone dislocations and stress field. (b) Dislocation density in crack and plastic zone.

It has been postulated (44) that fracture occurs when $\phi(a)$ attains a critical value ϕ_c. It follows e.g. from eq. (6) that for $\sigma \ll \sigma_1$, $\phi = \alpha K^2/4\sigma_1$, and thus in the small scale yielding limit this criterion and the criteria based on a critical K_c are equivalent. However, the criterion of the critical CTOD is assumed to apply even when extensive yielding occurs.

The BCS model has been solved for a number of crack configurations (45, 46), and for a disc shaped crack using a continuous distribution of circular dislocation loops (47). The model has also been modified to take account of work-hardening by making the friction stress σ_1 linearly dependent on the plastic displacement (48), and, approximately, by assuming that σ_1 depends on the level of the applied stress (49). Besides sharp cracks blunt notches of finite root radius have also been treated in anti-plane strain (50) and plane strain (51). The problem of cracks emanating from blunt notches has also been studied (52). Further developments of the dislocation model have included yielidng on planes inclined to the plane of the crack. In anti-plane strain this was considered by Lardner (53), and in plane strain when the yielding planes are 45° to the crack by Bilby and Swinden (48). The problem of yielding at an arbitrary angle was treated by Vitek (54) and Riedel (55). In some cases large discrete dislocations, rather than continuous distributions have been used to represent the plastic zone (30, 56).

In all the foregoing works the CTOD can be calculated analytically or numerically and fracture prediced when $\phi(a) = \phi_c$. However the CTOD is not the only EPFM parameter commonly used.

4.2 The J-Integral

For a non-linear elastic continuum Eshelby (57) has derived an expression for the total force on all singularities and inhomogeneities enclosed within a surface. When all the singularities and inhomogeneities are infinitely long in the x_3-direction the component of force in the x_1-direction is given by (20,57).

$$F_1 \equiv J = \int_{\Sigma} (W dx_2 - \sigma_{ij}\, u_{i,1}\, d\Sigma_j) \tag{7}$$

where J is the so-called J-integral of fracture mechanics. The integration is carried out along the closed circuit Σ and $W = \int \sigma_{ij} de_{ij}$ is the elastic energy density. σ_{ij} and e_{ij} are the stress and strain tensors and u_i the displacement vector; the normal notation of differentation is used. In the case of a single dislocation equation (7) reduces to the Peach-Koehler formula. Equation (7) was independently derived in a fracture mechanics context by Rice (58) as the potential energy release rate per unit length (J) for an increment of crack extension. In a linear elastic medium if Σ encloses a sharp crack tip, it can be shown that

$$J = \mathcal{G} = \alpha K^2/4 \tag{8}$$

where $\mathcal{G}$ is the strain energy release rate (19), the crack extension force of LEFM appropriate to the particular mode of crack loading (58). Furthermore, in the BCS model if Σ encloses the crack and all its associated plasticity then

$$J = \sigma_1 \ \phi(a) \tag{9}$$

Equation (9) is simply obtained as follows (59): In equilibrium the total force on a single dislocation in the yielded region must be $\sigma_1 b$ to balance the resistance $-\sigma_1 b$ due to the friction stress σ_1, so that for n dislocations $F_1 \equiv J = n\sigma_1 b = \sigma_1 \phi(a)$ since $nb = \phi$ (a) (b is the Burgers vector).[1]

Equation (8) suggests the failure criterion $J = J_c$ which in the small scale yielding limit is again equivalent to $K = K_c$. From eq.(9) it also follows that in the framework of the BCS model this criterion is equivalent to $\phi(a) = \phi_c$. However, for yielding on inclined planes the simple equality of (9) is no longer valid and Vitek (54) has shown that J and $\phi(a)$ do not provide equivalent descriptions of fracture.

Equation (7) is applicable to a set of discrete dislocations. In the case of a continuous distribution of dislocations and a contour Σ which cuts through the plastic zone the elastic displacement field is not uniquely defined and eq. (7) should be modified to read (59, 60)

$$Q_1 \equiv J_e = \int_\Sigma (\overline{W} dx_2 - \sigma_{ij} \beta^e_{i1} \, d\Sigma_j) \tag{10}$$

where the distortion tensor $\beta_{ij} = \beta^e_{ij} + \beta^p_{ij} = u_{i,j}$,

$$\overline{W} = \int \sigma_{ij} d\beta^e_{ij}$$

and superscripts e and p refer to elastic and plastic components respectively. The integral J_e is applicable to incremental plasticity, including cases where unloading is performed (60), and where the friction stress and applied load vary with temperature (61). If the contour Σ encloses the crack and plastic zone then $F_1 = Q_1$ ($J = J_e$) and the integrals are path independent (57 58). If Σ cuts through the plastic zone then both F_1 (for discrete dislocations) and Q_1 (for a continuous distribution) are path dependent, reflecting the omission of the force on all the dislocations left outside of Σ. For the BCS model this was discussed in detail in ref. (62). Clearly if Σ is shrunk onto the crack tip $F_1 = Q_1 = 0$, showing that there is no force on the crack tip itself (59).

In elastic-plastic finite element calculations the integral J is often determined using (7) with ε_{ij} and u_i replaced by their total (elastic plus plastic) values (63, 64.) For a contour that cuts through the plastic zone the integral defined in this way has no physical meaning, whereas $Q_1 = J_e$ still retains its significance as the total force in the x_1-direction on all the dislocations enclosed by Σ (60, 64, 65).

The relative merits and the regions of applicability of EPFM parameters, CTOD, J and J_e have been discussed by Vitek and Chell (65). They concluded that J_e is the most suitable EPFM for describing the onset of fast fracture (cleavage) while the CTOD is more suitable time dependent growth (stable ductile tearing). In the former case the failure criterion, $J_e = \alpha K_c^2/4$, states that fracture will occur when the total force on a crack and its associated plasticity (dislocations) attains a critical value. (This implicitly assumes that J_e characterizes the stress field at the tip of the crack and hence may be related to the micromechanism of the fracture process). Since J_e is path dependent if Σ cuts through the plastic zone, and since there are circumstances when dislocations may be immobilized (e.g. when the force on them is less than the friction force opposing motion), Chell and Vitek (66) proposed a more general failure condition. This is that failure will occur when J_e attains the critical value equal to $\alpha K_c^2/4$ provided J_e is evaluated around a contour that encloses only mobile dislocations. This condition, and the path dependence of J_e, is vital to the explanation of the phenomenon of warm prestressing that will be discussed below.

4.3 Application of J_e to warm prestressing.

The fracture resistance of ferritic steels failing by cleavage at a temperature T_2 can be increased by initially loading a cracked structure to near failure at temperature $T_1 > T_2$, partially or fully unloading, and then cooling to T_2 under constant load (Fig. 3). This phenomenon is called warm prestressing (WPS) and has recently been explained and quantified in terms of J_e (61, 66). The representation and interpretation of plasticity in terms of dislocations played a crucial part in the explanation of this phenomenon.

The relevant stages in the WPS process are illustrated in Fig. 4 in terms of the BCS model. The warm temperature loading at T_1 produces plastic zone s_1 which develops under a friction (yield) stress $\sigma_1(T_1)$. In equilibrium the stress in s_1 equals $\sigma_1(T_1)$. At this stage J_e can be evaluated using any contour, Σ_1 for example, that remains outside s_1 (Fig. 4a). On unloading some dislocations are forced back into the crack and a region of compressive yielding s_2, is formed where the stress is $-\sigma_1(T_1)$ (see, for example, ref 69). The stress outside of s_2 is greater than $-\sigma_1(T_1)$ but less than $\sigma_1(T_1)$ so that the dislocations in the region $a+s_2 \leq x \leq a+s_1$ become immobilized (Fig. 4b). The failure parameter J_e for this state must be evaluated so as to include only the force on mobile dislocations, i.e., along a contour such as Σ_2. On cooling

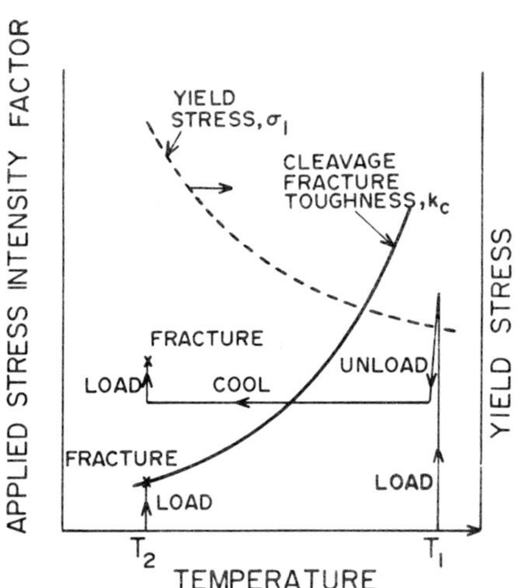

Fig. 3 The enhancement of the fracture resistance of ferritic steels by warm prestressing.

to T_2 the friction stress increases from $\sigma_1(T_1)$ to $\sigma_1(T_2)$ with the result that all the dislocations generated by the loading sequences at T_1 now become immobilized. The contour Σ_2' for calculating J_e must now be shrunk onto the crack tip and $J_e = 0$ (Fig. 4c). This is the case even if no unloading occurs at T_1. Hence in this state fracture cannot occur at T_2 even though in terms of the stress intensity factor applied at T_1 the fracture toughness at T_2 is exceeded (Fig. 3). For fracture additional loading is required at T_1 and this gives rise to new dislocation emission from the crack tip and the formation of a tensile plastic zone, s_3, where the stress level is $\sigma_1(T_2)$ (61). In the region $a \leq x \leq a+s_3$ the dislocations are mobile and J_e, evaluated around a contour such as Σ_3, is non-zero (Fig. 4d) and hence fracture will occur when the criterion $J_e = \alpha K_c^2(T_2)/4$ is attained. The foregoing theory of WPS has been verified

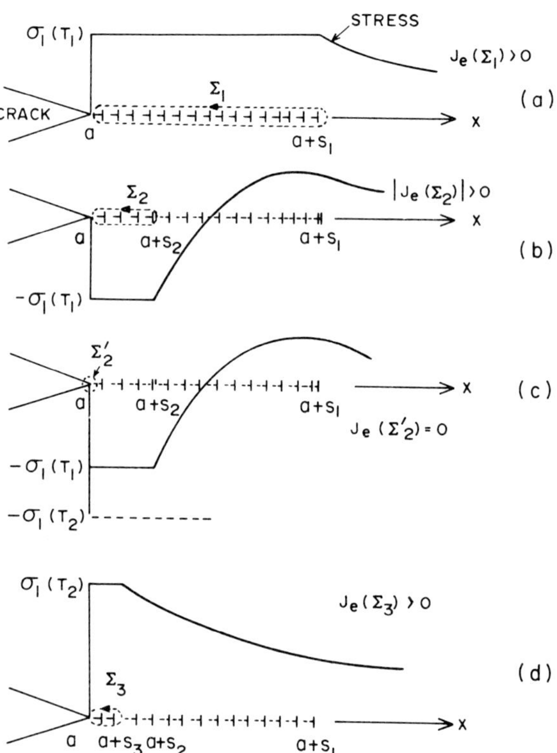

Fig. 4: Representation of warm prestressing in terms of BCS model. Mobile dislocations are marked by ⊣ and immobile ones by -⊣ . (a) Initial loading at T_1. (b) Unloading at T_1. (c) Cooling under load from T_1' to T_2. (d) Loading to failure at T_2.

experimentally for a wide range of WPS cases (61, 66). This validation is shown in Fig. 5 in the form of the experimentally measured stress intensity factors at failure plotted against the predicted values. Considering that the analysis was made using average values for the fracture toughness, which typically have a scatter of $\pm$ 15 percent, the agreement between theory and experiment is excellent. This success demonstrates the important role dislocations can play in the modelling and physical interpretation of fracture behaviour.

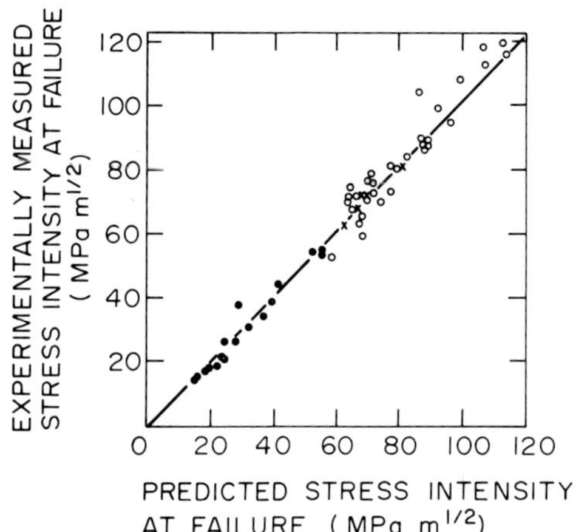

Fig. 5: Experimental validation of warm prestressing theory (ref. (61)).

5. Fatigue and Creep

The BCS model has also been applied to describing time dependent sub-critical crack growth such as fatigue and creep. In fatigue the damage parameter is identified as the modulus of the change in plastic displacement that occurs during cyclic loading (67-69). The increments of plastic displacement are accumulated for each load cycle and the crack assumed to advance when the sum of these and the plastic displacement due to static loading reaches a critical value ϕ_c. The method has been applied to disc-shaped cracks (70) and used to produce a generalized theory of fatigue crack propagation (71, 72). More recently the retardation in fatigue crack growth rates due to overload cycles has been explained using the dislocation model (73).

The initiation and growth of cracks under creep conditions have been studied by Vitek (74, 75) and Riedel (76) using dislocations to describe creep damage. The strain ahead of the crack tip was simulated by a time dependent continuous distribution of dislocations that satisfies the condition

$$\frac{\partial \rho(x,t)}{\partial t} = -h \frac{\partial \dot{\varepsilon}(x,t)}{\partial x}$$

where $\rho(x,t)$ is the dislocation density at a point x ahead of the crack at time t, and h is a constant gauge length representing the width of the plastic zone. The strain rate $\dot{\varepsilon}(x,t)$ was assumed to be related to the local tensile stress $\sigma(x,t)$ through the equation

$$\dot{\varepsilon}(x,t) = A\ (\sigma(x,t)/G)^m$$

where A is a constant and m an integer. The solution of this problem leads to a time dependent BCS model in which the stress ahead of a crack at any given time is almost constant in a region (which can be identifed with the plastic zone) ahead of the crack and decreases with time. Crack propagation is assumed to occur when the CTOD resulting from the dislocations reaches a critical value.

A dislocation description has also been employed in studies of diffusion controlled intergranular creep cavitation and crack growth (77-80). Here the back stress produced by the deposition of atoms ahead of the crack can be identified with that due to a continuous distribution of dislocations (81) whose density satisfies the equation.

$$\rho(x,t) = -\frac{\partial \ell(x,t)}{\partial x}$$

where $\ell(x,t)$ is the width of the deposited layer of atoms. However, in this case the dislocation distribution is not only formal and may represent a distribution of discrete grain boundary dislocations.

6. Micromechanisms of Brittle Fracture

6.1 Models of Microcrack Nucleation and Extension

By comparing the fracture stress in tension with the yield stress in compression Low (82) showed that in fine grain materials yielding preceded fracture and in coarse grain materials yielding and fracture were coincident. This conclusion, which is generally valid for metallic materials, suggests that local yielding is necessary to produce a nucleus for brittle fracture. Zener (16) was the first to propose that if yielding can occur dislocation pile-ups may form at obstacles and the strong stress concentration near their tips may cause microcrack nucleation.

The first dislocation model based on this idea was developed by Stroh (83, 84) who took as an obstacle a grain boundary and assumed that the microcrack develops either in the grain neighboring that with the pile up or along the grain boundary. In the former case cleavage and in the latter case intergranular fracture ensues. This model is analogous to that of Hall and Petch (85, 86) for slip propagation through a polycrystal. The pile up has length d, which can be identified with the grain size, and it is formed under the effect of an applied shear stress σ_s. There is an intrinsic friction stress σ_i so that the dislocations move under an effective stress $\sigma_e = \sigma_s - \sigma_i$. The dislocations forming the pile-up are individual lattice dislocations but to find their distribution they can be approximated by a continuous distribution of dislocations, which is then determined by eq. (3) with $\sigma(x)$ replaced by σ_e. This means that the pile-up is equivalent to a crack of length d loaded in mode II by the stress σ_e. The maximum tensile stress ahead of such a pile up is

$$\sigma_{tp} = \sigma_e \left(\frac{4d}{3r}\right)^{\frac{1}{2}} \tag{11}$$

where r is the distance from the tip of the pile-up. A microcrack nucleated ahead of the pile-up will propagate if the Griffith condition

$$\frac{k^2 (1 - \nu)}{2G} = 2\gamma_0 \tag{12}$$

is satisfied; k is the stress intensity factor at the tip of the microcrack and γ_0 is the work required for forming unit area of crack surface. Stroh (83) neglected the nonuniformity of the stress ahead of the pile up and the fact that the pile-up will change as the microcrack forms because dislocations can run into it. He assumed that the microcrack is loaded by σ_{tp} (eq. 11) evaluated at $r = a/2$ where a is the length of the microcrack. Eq. (12) gives then the following condition for fracture:

$$\sigma_e = \left(\frac{3\, G\, \gamma_o}{\pi(1-\nu)d}\right)^{\frac{1}{2}} \tag{13}$$

Thus it is the effective shear stress which has to reach a critical value before fracture can start. However, as first emphasized by Orowan (87) and since then proved by many authors (for reviews see e.g. 88, 89), it is the tensile stress which controls brittle fracture. Furthermore, σ_e is only slightly temperature dependent and, therefore, fracture should be temperature insensitive if controlled by this whilst the opposite is true. This model was further modified (90, 91) by taking into account the nonuniform stress ahead of the pile-ups and the pile-up relaxation which takes place as the microcrack is formed. (In these papers the wedge shaped microcrack was taken to be parallel with the pile-up and thus the pile-up loaded the microcrack only in mode II). However, the condition for fracture obtained in these studies was within a small numerical factor the same as that given by eq. (13).

An alternate microscopic theory of brittle fracture was proposed by Cottrell (92) who suggested that <001> dislocations can serve as microcrack nuclei in body-centered-cubic metals. These dislocations can be formed at the intersections of two <111> slip systems and will lie on {100} planes. In this theory the tensile stress was considered implicitly and a criterion for the critical tensile stress was, indeed, obtained. However, the model does not discriminate properly between local and general yielding conditions and, furthermore, can only be applied to {100} cleavage while, for example, intergranular brittle fracture is a common phenomenon.

The effect of the tensile stress can be introduced into the Stroh's model if we assume that the microcrack is loaded not only by the pile-up stress but also by a local external tensile stress σ_t. The Griffith condition (12) yields then

$$\sigma_t^2 \frac{3a}{8d} + \sigma_e^2 \left[1 + \frac{\sigma_t}{\sigma_e}\left(\frac{3a}{2d}\right)^{\frac{1}{2}}\right] \geqq \frac{3\gamma_o G}{\pi(1-\nu)d} \qquad (14)$$

Assuming the same value of γ_o for both microcrack nucleation and propagation then if inequality (14) is satisfied for $a = 0$ it is also satisfied for any $a > o$. This clearly leads again to eq. (13) and no dependence on σ_t is found. However, if a sudden increase of γ_o occurs after the microcrack of the length c_o was formed, eq. (14) has to be satisfied for $a = c_o$ and the higher value of γ_o, and this gives a condition for the critical value of σ_t. This is, indeed, the case if fracture initiates on second phase particles (e.g. carbides) as first suggested by McMahon and Cohen (93). In this case the nucleation is relatively easy i.e. γ_o is small at this stage. However, as the crack enters the metal matrix (either along a cleavage plane or intergranularly) γ_o rapidly increases because fracturing the metal is in general more difficult. The minimum crack length c_o can then be identified with the size of the corresponding inclusion. This model is illustrated in Fig. 6 and was developed in more detail by Smith (94) who also considered the nonuniformity of the stress ahead of the pile-up and the relaxation of the wedge shaped microcrack parallel with the pile up. (Thus only mode II loading by the pile-up was considered). The condition for the critical tensile stress derived in (94) is to within small numerical factors identical with eq. (14).

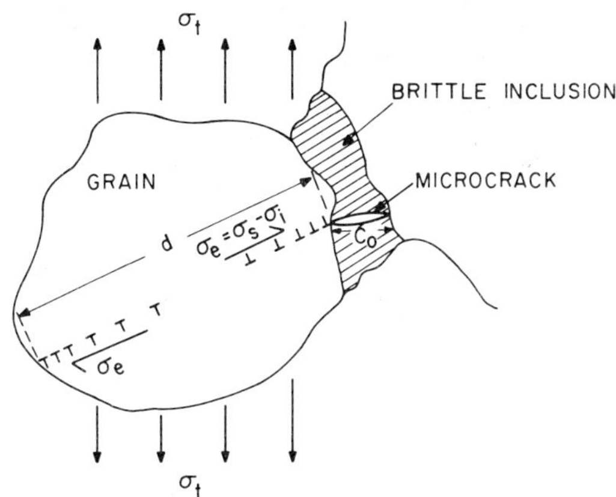

Fig. 6: A schematic picture of a microcrack formed in a brittle inclusion at a grain boundary with the help of the pile-up of dislocations inside the grain.

This condition for microcrack extension has frequently been used as a basis for the micromechanistic description of brittle fracture. An example is the theory of the temperature dependence of the fracture toughness in mild steels developed in ref. (95). Here propagation of a loaded macrocrack ensues when a microcrack nucleates, propagates and links up with it under the influence of the tensile stress resulting from the macrocrack. In this case σ_t in eq. (14) is identified with this tensile stress. In general, it has been found that this condition leads to physically reasonable answers if γ_o is taken about one order of magnitude larger than the corresponding cohesive (surface) energy (94-97). This apparently reflects the fact that plastic work is produced during the microcrack propagation in metallic materials. A dislocation model which relates the cohesive energy and the plastic work will be discussed in the following section.

6.2 Relation Between Plastic Work and Ideal Work to Fracture

The importance of plastic work in fracture of metallic materials has been recognized for many years. Orowan (87) was the first to suggest that $2\gamma_o$ in eq (12) should be replaced by $2\gamma + \gamma_p$ where 2γ is the ideal work to fracture, i.e. the cohesive energy, and γ_p is the plastic work. Both γ and γ_p are constant material parameters and in general $\gamma_p >> \gamma$. With this postulate it is not possible, however, to rationalize the embrittlement which occurs in alloys when certain impurity elements segregate to grain boundaries (see e.g. 98, 99). In all cases the segregation is limited to a very narrow region along the grain boundaries and the plastic properties of the material are not influenced. Hence, only γ can be expected to be significantly changed by segregation while γ_p is not directly affected. This dilemma can be resolved by recognizing that γ_p must be a function of γ rather than an independent material parameter (100). (A similar idea was apparently first mentioned by Rice (101) in the context of macrocrack propagation. Here the plastic work is dissipated throughout the yielded region associated with the microcrack and is thus several orders of magnitude greater than the plastic work generated by microcrack propagation). Implicit in this approach is the assumption that bond breaking and dislocation emission at the microcrack tip are concomitant processes. This leads to more general questions concerning the physical basis for the brittle versus ductile behaviour of materials.

In dealing with this fundamental problem of material behaviour a number of investigators assumed that bond breaking and dislocation emission are mutually exclusive processes. For example, a microscopic criterion for brittle fracture has been based on comparison of the ideal fracture stress with the yield stress on appropriate slip planes in refs. (102, 103). A detailed model of dislocation nucleation at the crack tip was developed by Rice and Thomson (17) who showed that in some materials emission will occur spontaneously while in others there may be a barrier for dislocation nucleation. The distinction between the two cases depends, however, rather sensitively on the treatment of the dislocation core. It was suggested then that brittle fracture takes place if a dislocation emission is prevented at the instant that the energy release rate connected with the microcrack propagation becomes equal to the ideal work of fracture, i.e., when $k = (4\gamma G/(1-\nu))^{\frac{1}{2}}$. In this case a very small k would be required for microcrack extension and in order to reconcile this with the large critical stress intensity (fracture toughness) of the macrocrack the screening of the microcrack by the plastic zone of the macrocrack was porposed (104).

However, there is ample experimental evidence that dislocation emission occurs during crack propagation. For example, it has been found by X-ray diffraction (105, 106), SEM channeling patterns (107) and by TEM (108) that there is a high dislocation density adjacent to fracture surfaces. Furthermore, it was found that in the case of intergranular fracture this dislocation density decreases with increasing concentration of embrittling impurities suggesting thus again the existence of a relationship between γ_p and γ. A model of the simultaneous dislocation emission and bond breaking at a microcrack tip has recently been developed in refs. (18) and (109). The rate of dislocation emission was assumed to be proportional to the velocity of dislocations, v, which depends on the local stress, τ, according to the power law

$$v = v_o\ (\tau/G)^n \tag{15}$$

where v_o and n are empirical parameters. This form has frequently been used to represent experimentally measured dislocation velocities and plastic strain rates (e.g. 110, 111). An analogous problem has been solved for time dependent plastic flow near a crack tip during power law creep (74, 76, 112) (see also Section 5). These calculations were made for plastic zones coplanar with the crack, whilst dislocations are actually emitted on planes inclined to the crack plane, but for small zone sizes the former approximation is sufficient. The results gave the dependence of the stress, τ, ahead of the microcrack, the total Burgers vector of dislocations emitted, ϕ, and the extent of the microcrack plastic zone, s, on time, k and material parameters G, ν, v_o and n (109, 112). For any given value of k the plastic work w_p can then be calculated as a function of time and of the above mentioned material parameters.

The bond breaking ahead of the microcrack tip was assumed to be driven by the time dependent stress, τ, ahead of the microcrack which results from the microcrack stress field and stress of emitted dislocations. This stress decreases with time as more dislocations are emitted. Stretching of only the bond between the atom pair nearest to the crack tip was considered and a harmonic restoring force, which vanishes for a certain atom-pair separation δ, was assumed. The strength of the spring, κ, between the two atoms is then related to the cohesive energy γ by the relation $\kappa = 4\gamma b^2/\delta^2$ where b^2 is the area per atom.

The time t_B needed for breaking the bond can be calculated as a function of k and material parameters and using the results of the calculations of dislocation emission we can calculate for a given value of k the plastic work produced during the time t_B i.e. the dependence $w_p(k)$ is obtained. A sufficient thermodynamical condition for fracture to take place is that the energy of the system decreases as the microcrack extends; this means

$$E = -\frac{k^2}{2G}(1-\nu) + 2\gamma + w_p(k) < 0 \qquad (16)$$

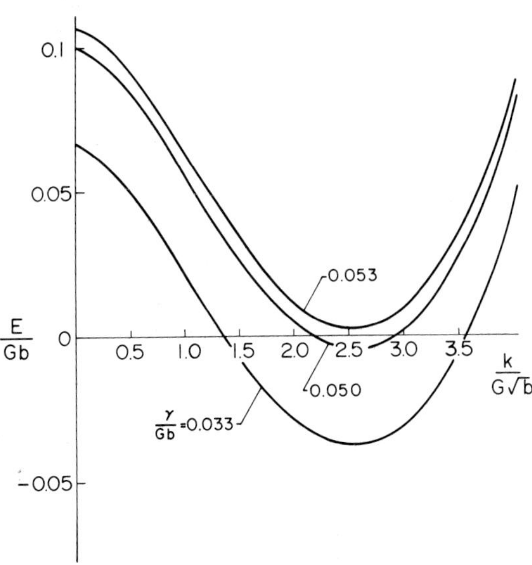

Fig. 7: Dependence of E on k for three different values of γ and $\phi_o = 8.37\times10^5$ cms^{-1}. $\phi_o = v_o \cdot bg$ where g is a dimensionless constant relating rate of dislocation emission and their velocity (109) and b the dislocation Burgers vector.

The dependence of E on k for several values of γ is shown in Fig. 7. Two possibilities now exist: (i) The condition (16) is never satisfied, (ii) it is satisfied for $k_G<k<k_M$ where k_G and k_M are the lower and upper values of k for which E = 0. In the former case w_p increases too steeply for the condition (16) to be satisfied and such a material is inherently ductile. In the latter case fracture occurs for $k = k_G$. The plastic work at fracture can be calculated

as $\gamma_p = w_p(k_G)$ and it is a function of γ and other material parameters. A typical dependence on γ is shown in Fig. 8.

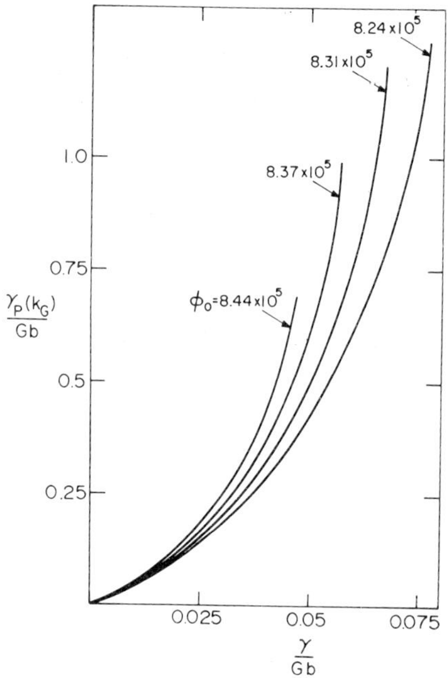

Fig. 8: Dependence of γ_p on γ for four different values of ϕ_o.

The model explains at least qualitatively the basic reasons why some materials are brittle and others ductile. It is applicable to both cleavage and intergranular fracture. In the case of cleavage, however, γ has a fixed value and γ_p is, therefore, also fixed. On the other hand in the case of intergranular fracture γ varies with the amount of segregation. Hence, a material ductile in a pure state may become brittle when alloyed and segregation occurs. This corresponds to a transition from the above mentioned case (i) to case (ii). An example is the embrittlement of copper by bismuth (113). At the same time the model can explain the sharp decrease in fracture toughness often observed when relatively small changes in segregation take place (e.g. 98, 99). The reason is the strong dependence of γ_p on γ so that a relatively small change in γ may cause a large change in γ_p.

The calculations in (18) and (109) were carried out for a range of values of n, $v_o(\phi_o)$ and δ and it was found that only when $1<n<2$ is the plastic work at fracture about an order of magnitude higher than γ; for $n>2$ it is close to γ. The case of $1<n<2$ corresponds to dislocations which move at a speed close to the speed of sound (110, 111). Since the experiments (94-97) suggest that γ_p at fracture is about an order of magnitude larger than γ it means that dislocations which are being emitted during the microcrack propagation move at very high speeds. This is certainly a reasonable conclusion but it also means that the quasistatic treatment outlined here is not fully applicable and a full dynamic study has to be performed in order to develop the theory into a quantitative state.

7. Conclusions

The applications of dislocation models to the mechanics of macrocrack propagation and microcrack nucleation and propagation have been discussed. In the former case by simulating cracks and their associated plastic zones by formal dislocations two advantages may be obtained over the standard continuum mechanical theories of fracture mechanics. First, it is sometimes mathematically and in terms of calculation, numerically easier to obtain the required solution (e.g. determining stress intensity factors). Second, the use of dislocations can provide extra physical insight into the importance and interpretation of failure parameters and fracture criteria (e.g. the case of warm prestressing and J_e).

On the micromechanistic level dislocation theory provides a tool for more fundamental description of microcracking particularly in cases where slip processes play a dominant role. Here the dislocations considered are real crystal dislocations unlike in the case of macrocrack where they provide a formal description of displacement discontinuities in a continuum solid. Relatively simple models of dislocation pile-ups at grain boundaries and other obstacles lead to the identification of the microstructural parameters (e.g. grain sizes, brittle inclusion sizes) and mechanical quantities (e.g. shear stress, tensile stress, plastic work) which control the fracture processes.

Thus dislocation theories have played important roles in the development, interpretation and prediction of fracture phenomena. There are still many cases where they may continue to provide valuable insight. An example is the problem of macrocrack growth under conditions where fatigue and creep interact and correspondingly on a micromechanistic level, the development of microcracks under time dependent creep conditions. Plenty of research is still required in order to establish links between the micromechanisms of cracking and macroscopic fracture parameters, such as fracture toughness. It seems not at all unlikely that dislocation theories which have been so successful in describing the two extremes of macro- and micro-cracking will also play an important role in this intermediate area.

Acknowledgements

One of the authors (V. Vitek) would like to acknowledge the support of the National Science Foundation, MRL Program, under Grant No. DMR79-23647. This work was performed in part at Central Electricity Research Laboratories and is published by permission of the Central Electricity Generating Board (U.K.).

References

1. V. Volterra, Ann. Sci. Ec. Norm. Sup., Paris 24, 40 (1907).
2. C. Somigliana, Atti Accad. Naz. Lincei Rc. 23, 463 (1914).
3. C. Somigliana, Atti Accad. Naz. Lincei Rc. 24, 655 (1915).
4. E. Orowan, Z. Phys. 89, 634 (1934).
5. M. Polanyi, Z. Phys. 89, 660 (1934).
6. G. I. Taylor, Proc. Roy. Soc. London A145, 362 (1934).
7. K. Kondo, Proc. Second Jap. Congress Appl. Math., p. 41 (1952).
8. E. Kröner, Z. Phys. 142, 463 (1955).
9. E. Kröner, Kontinuumstheorie der Versetzungen und Eigenspannungen, Ergeb. Angew. Math. 5, 1 (1958).
10. B. A. Bilby, R. Bullough and E. Smith, Proc. Roy. Soc. A234, 263 (1955).
11. B. A. Bilby, Prog. Solid Mech. 1, 329 (1960).
12. B. A. Bilby, A. H. Cottrell and K. Swinden, Proc. Roy. Soc. A272, 304 (1963).
13. J. D. Eshelby, Phil. Mag. 40, 903 (1949).
14. G. Leibfried, Z. Phys. 130, 214 (1951).
15. A. K. Head and N. Louat, Australian J. Phys. 8, 1 (1955).
16. C. Zener, Fracturing of Metals (New York: AMS), p. 3 (1948).
17. J. R. Rice and R. Thomson, Phil. Mag. 15, 567 (1974).
18. M. L. Jokl, Jun Kameda, C. J. McMahon, Jr. and V. Vitek, Proc. Conf. Micromechanisms of Crack Extension, Metal Science, to be published (1980).
19. G. R. Irwin, Handbuch der Physik 6, 551 (1958).
20. B. A. Bilby and J. D. Eshelby, Fracture, Volume 1, (H. Liebowitz ed.), Academic Press: New York, 99 (1968).
21. N. J. Muskhelishvili, Singular Integral Equations, translated by J. R. M. Radock, Noordhof-Laden (1953).
22. F. Erdogan, G. D. Gupta and T. S. Cook, Mechanics of Fracture, ed. G. C. Sih, Noordhof-Laden, p. 368 (1973).

23. E. Smith, Int. J. Fracture Mech. 1, 204 (1965).
24. R. W. Lardner, Mathematical Theory of Dislocations and Fracture, University of Toronto Press (1974).
25. E. Smith, Proc. Roy. Soc. A305, 387 (1968).
26. V. Vitek, Int. J. Fracture 13, 481 (1977).
27. K. K. Lo, J. Appl. Mech. 45, 797 (1978).
28. V. Vitek, J. Mech. Phys. Solids 24, 67 (1976).
29. C. Atkinson, Int. J. Fracture Mech. 2, 567 (1966).
30. C. Atkinson, and D. L. Clements, Acta Metall. 21, 35 (1973).
31. R. J. Asaro, J. Phys. F: Metal Physics 5, 2249 (1975).
32. S. N. Chatterjee, Int. J. Sol. and Structures 11, 521 (1975).
33. H. Kitagawa, R. Yuuki and T. Ohira, Eng. Fract. Mech. 7, 515 (1975).
34. H. Kitagawa and R. Yuuki, Int. Conf. Fracture, ed. D. M. R. Taplin, University of Waterloo Press, Vol. 3, p. 201 (1977).
35. M. A. Hussain, S. L. Pu and J. Underwood, Assoc. Nat. Symp. Fract. Mech. II, ASTM 560, 2 (1973).
36. M. J. Marcinkowski and R. W. Armstrong, J. Appl. Phys. 43, 2548 (1972).
37. M. J. Marcinkowski, J. Appl. Phys. 46, 496 (1975).
38. M. J. Marcinkowski and E. S. P. Das, Int. J. Fracture 10, 181 (1974).
39. K. Jagannadham and M. J. Marcinkowski, Int. J. Fracture 14, 155 (1978).
40. L. B. Freund, Mechanics Today, ed. Nemat-Nasser, Pergamon Press, Vol. 3, p. 55 (1976).
41. P. Burgers and L. B. Freund, to be published (1980).
42. J. R. Rice and G. F. Rosengren, J. Mech. Phys. Solids 16, 1 (1968).
43. J. W. Hutchinson, J. Mech. Phys. Solids 16, 13 (1968).
44. A. H. Cottrell, I. S. I. Spec. Rep. No. 69, 281 (1961).
45. B. A. Bilby, A. H. Cottrell, E. Smith and K. H. Swinden, Proc. Roy. Soc. A279, 1 (1964).
46. E. Smith, Int. Conf. on Fracture, Sendai, Japan, Chapter A, 73 (1965).
47. L. M. Keer and T. Mura, Int. Conf. on Fracture, Sendai, Japan, Chapter A, 49 (1965).
48. B. A. Bilby and K. H. Swinden, Proc. Roy. Soc. A285, 22 (1965).
49. G. Chell, Int. J. Fracture, 10, 128 (1974).
50. E. Smith, Proc. Roy. Soc. A299, 455 (1967).
51. V. Vitek, J. Mech. Phys. Solids 24, 277 (1976).
52. G. G. Chell and V. Vitek, Eng. Fract. Mech. 11, 251 (1979).
53. R. W. Lardner, Int. J. Fracture Mech. 4, 299 (1968).
54. V. Vitek, J. Mech. Phys. Solids 24, 263 (1976).
55. H. Riedel, J. Mech. Phys. Solids 24, 277 (1976).
56. C. Atkinson and T. R. Kay, Acta Metall. 19, 679 (1971).
57. J. D. Eshelby, Solid State Physics 3, 79 (1956).
58. J. R. Rice, Fracture, Volume 2 (H. Liebowitz ed.), Academic Press, New York, 192 (1968).
59. B. A. Bilby, 3rd Int. Conf. Fracture, Volume XI, Verein Deutscher Eisenhüttenleute, Düsseldorf, PL1 (1973).
60. H. Miyamoto and K. Kageyama, Proc. 1st. Int. Conf. on Numerical Methods in Fracture Mechanics, Swansea, (eds. A. R. Luxmoie and D. R. J. Owen), Swansea University, 479 (1978).
61. G. G. Chell, J.R. Haigh and V. Vitek, Int. J. Fract. (to be published) (1980).
62. G. G. Chell and P. T. Heald, Int. J. Fract. 11, 349 (1975).
63. J. R. Rice, Int. J. Fract. 11, 352 (1975).
64. G. G. Chell and P. T. Heald, Int. J. Fract. 12, 172 (1976).
65. V. Vitek and G. G. Chell, Mat. Sci. Eng. 27, 209 (1977).
66. G. G. Chell and V. Vitek, Int. J. Fract. 13, 882 (1977).
67. J. Weertman, Int. J. Fract. Mech. 2, 460 (1966).
68. J. Weertman, Int. J. Fract. Mech. 5, 13 (1969).
69. B. A. Bilby and P. T. Heald, Proc. Roy. Soc. A305, 429 (1968).
70. P. T. Heald, Int. J. Fract. Mech. 4, 365 (1968).
71. L. N. McCartney and B. Gale, Proc. Roy. Soc. A322, 223 (1971).
72. L. N. McCartney and B. Gale, Proc. Roy. Soc. A333, 337 (1973).
73. L. N. McCartney, Int. J. Fract. 14, 213 (1978).
74. V. Vitek, Int. J. Fract. 13, 39 (1977).
75. V. Vitek, Proc. Int. Conf. Fracture Mechanics and Technology, Hong-Kong (eds. G. C. Sih and C. L. Chow) Sijthoff and Noordhoff, p. 161 (1977).
76. H. Riedel, Mat. Sci. Eng. 30, 187 (1977).
77. T. J. Chuang, Thesis, Brown University, Providence, R. I. (1974).
78. V. Vitek, Acta Metall. 26, 1345 (1978).

79. T. J. Chuang, K. I. Kagawa, J. R. Rice and L. B. Sills, Acta Metall. 27, 265 (1979).
80. V. Vitek, Proc. Int. Conf. Micromechanisms of Crack Extension, Metal Science, to be published (1980).
81. R. Raj, Metall. Trans A6, 1499, (1975).
82. J. R. Low, Symposium on Relation of Properties to Microstructure, AMS, p. 163 (1954).
83. A. N. Stroh, Proc. Roy. Soc. London A223, 548 (1955).
84. A. N. Stroh, Adv. Phys. 6, 418 (1957).
85. E. O. Hall, Proc. Phys. Soc. 64B, 747 (1951).
86. N. J. Petch, J. Iron Steel Inst. 173, 25 (1953).
87. E. Orowan, Trans. Inst. Eng. Shipbuilders Scot. 89, 165 (1945).
88. A. H. Cottrell, Fracture, Swampscott Conf., eds. B. L. Averbach et al. (New York: John Wiley), p. 20 (1959).
89. J. F. Knott, Fundamentals of Fracture Mechanics, Butterworths: London (1973).
90. E. Smith, Acta Metall. 14, 985, 991 (1966).
91. E. Smith and J. T. Barnby, Metal Sci. J. 1, 56 (1967).
92. A. H. Cottrell, Trans. AIMMPE 212, 192 (1958).
93. C. J. McMahon, Jr. and M. Cohen, Acta Metall. 13, 591 (1965).
94. E. Smith, Proc. Conf. Phys. Basis of Yield and Fracture, Inst. Phys. and Phys. Soc. (London), Oxford, p. 36 (1966).
95. R. O. Ritchie, J. F. Knott and J. R. Rice, J. Mech. Phys. Solids 21, 395 (1973).
96. J. F. Knott, Proc. ICF4, Volume 1, University of Waterloo Press, p. 61 (1977).
97. D. A. Curry and J. F. Knott, Metal Sci. 12, 341 (1979).
98. C. J. McMahon, Jr., Temper Embrittlement in Steels, ASTM STP 407, p. 127 (1968).
99. C. J. McMahon, Jr., Mat. Sci. Eng. 25, 233 (1976).
100. C. J. McMahon, Jr., and V. Vitek, Acta Metall. 27, 507 (1979).
101. J. R. Rice, Proc. First Int. Conf. Fracture, Sendai. (eds. T. Jokobori, T. Kawasaki and J. L. Swedlow), p. 309 (1966).
102. A. Kelly, Strong Solids, Oxford University Press: London (1966).
103. A. Kelly, W. Tyson and A. H. Cottrell, Phil. Mag. 15, 567 (1967).
104. R. Thomson, J. Mater. Sci. 13, 128 (1978).
105. D. K. Felbeck and E. Orowan, Welding J. Res. Suppl. 34, 570 (1955).
106. Jun Kameda and C. J. McMahon, Jr., Metal Trans. 11A, 91 (1980).
107. D. E. Newbury, B. W. Christ and D. C. Joy, Metal. Trans. 5A, 1505 (1974).
108. A. Kumar and B. Eyre, to be published (1980).
109. M. L. Jokl, V. Vitek and C. J. McMahon, Jr., to be published (1980).
110. W. G. Johnston and J. J. Gilman, J. Appl. Phys. 30, 129 (1959).
111. D. F. Stein and J. R. Low, J. Appl. Phys. 31, 362 (1960).
112. D. J. F. Ewing, Int. J. Fract. 14, 101 (1978).
113. A. Joshi and D. F. Stein, J. Inst. Metals 99, 178 (1971).

VARIATIONAL FORMULATIONS IN DEFECT MECHANICS: CRACKS AND DISLOCATIONS

Alicia Golebiewska Herrmann

Division of Applied Mechanics
Stanford University, Stanford, CA

Abstract

The forces on cracks and dislocations are discussed, first for the static case, then including dynamic effects. It turns out that the formal procedure used here works in the same way for both dislocations and cracks, giving e.g. the dynamic generalization of the Peach-Koehler force in case of continuous distribution of dislocations.

The general variational formalism, however, cannot be given in a unified way, due to incompatibilities of fields describing dislocations. For the latter, the Lagrangian is given in terms of fields of velocities, distortions and their derivatives and source functions, such as dislocation density and flux tensors.

In consideration of bodies with defects, cracks and dislocations are often mentioned together, as examples of such defects. Both are connected with some irregularities in energy, the presence of either of them is connected with "material" forces. The difference which is easily recognized is the presence of self-stresses (or internal or residual stresses) in case of dislocations, and absence of the latter for cracks.

The aim of the present contribution is to examine somewhat closer the differences and similarities of forces acting on cracks and dislocations in the framework of variational principles.

Because in fracture mechanics the force on a tip of a crack is often called (or related to) an energy release rate, let us start our examination with statics, for which the total energy is represented by a strain energy density W .

For the "perfect" medium, without cracks or dislocations, the energy is a function of strain $\varepsilon_{ij}(X_k)$, where X_k denote the material or Lagrangian coordinates. Thus

$$W = W[\varepsilon_{ij}(X_k)] \tag{1}$$

and

$$\frac{\partial W}{\partial X_i} = \frac{\partial W}{\partial \varepsilon_{jk}} \frac{\partial \varepsilon_{jk}}{\partial X_i} \tag{2}$$

Further

$$\frac{\partial W}{\partial \varepsilon_{jk}} = \sigma_{jk} \tag{3}$$

where σ_{ji} is a symmetric stress tensor, and the equation of static equilibrium is

$$\sigma_{jk,j} = 0 \tag{4}$$

in the absence of body forces. Let u_i denote the components of the displacement vector. From (2) we obtain

$$\frac{\partial}{\partial X_k}[W\delta_{ki} - \sigma_{jk}u_{j,i}] = 0 \tag{5}$$

Globally, after integration of both sides of (5) and using the Gauss theorem, we have

$$\int_S (Wn_i - T_j u_{j,i})\, dS = 0 \tag{6}$$

Comparing (6) with formula (1b) of [1] we can deduce that the force J_k can be referred to as "incompatibility" in the strain energy due to the presence of a defect (e.g. a crack); formally this force can be derived, if we assume that the elastic energy of a body with a crack can be given by

$$W = W[\varepsilon_{ij}(X_k), X_i] \tag{1a}$$

instead of (1), i.e. the energy can be an explicit function of $\vec{X}$ which is expressed by a discontinuity of the energy across the crack face (even in the case of a homogeneously applied stress field [2]) which in turn is due to the presence of traction-free surfaces of the crack inside the body. Taking the same steps for (1a) as for (1), we obtain

$$\left(\frac{\partial W}{\partial X_i}\right)_{exp} = \frac{\partial}{\partial X_k}(W\delta_{ki} - \sigma_{jk}u_{j,i}) \tag{5a}$$

instead of (5). Usually, the derivative of the potential with respect to a coordinate is referred to as a force and this what we found is actually the density of a material force (because the derivatives in (5a) are taken with respect to material coordinates) acting on a crack. Actually, similarity of the material force and well-known body forces is even stronger: absence of body forces is expressed as

$$\sigma_{ik,k} = 0 \tag{4}$$

while absence of material forces as

$$b_{ik,k} = 0 \tag{5}$$

where

$$b_{ik} = W\delta_{ik} - \sigma_{jk}u_{j,i} \tag{7}$$

can be called a material momentum.

Repeating the same analysis for dislocations, we will use an incompatible distortion field β_{ij}, such that

$$\varepsilon_{k\ell m}\beta_{im,\ell} = \alpha_{ik} \tag{8}$$

where α_{ij} denotes a dislocation density tensor. We assume that energy is a functional of β_{ij}, and a stress tensor

$$\sigma_{ij} = \frac{\partial W}{\partial \beta_{ij}} \tag{9}$$

satisfies the static equilibrium equation

$$\sigma_{ij,j} = 0$$

Proceeding as before, we obtain (with (8))

$$\frac{\partial W}{\partial X_i} = \sigma_{jk}\beta_{jk,i} = \sigma_{jk}(\beta_{ji,k} - \varepsilon_{kip}\alpha_{jp})$$

and thus

$$\frac{\partial}{\partial X_k}(W\delta_{ik} - \sigma_{jk}\beta_{ji}) = -\varepsilon_{kip}\sigma_{jk}\alpha_{jp} \tag{10}$$

If we ascribe the nonvanishing of the left-hand side to the field-force, we have an explicit form of that force in terms of the dislocation density

$$f_i = -\varepsilon_{kip}\sigma_{jk}\alpha_{jp} \tag{11}$$

For a single straight dislocation, we can find a force per unit of length, which, up to the sign, coincides with Peach-Koehler force [c.f. [3]]:

$$F_i = \varepsilon_{ik\ell}\sigma_{j\ell}b_j\lambda_k$$

where b_i denotes the Burgers vector, and λ_k the unit tangent to the dislocation line.

The dynamic generalization of the forces given by (5a) and (11) is straightforward. We have only to remember that the quantity to deal with now is not the total energy, but rather the Lagrangian $\mathscr{L}$ for the system, which for many mechanical cases can be represented as the difference of kinetic and potential energy. For the case of a crack we have

$$-\left(\frac{\partial \mathscr{L}}{\partial X_i}\right)_{exp} = \frac{d}{dt}\left(\frac{\partial \mathscr{L}}{\partial v_j} u_{j,i}\right) + \frac{\partial}{\partial X_k}(-\mathscr{L}\delta_{ik} - \sigma_{jk}u_{j,i}) \tag{12}$$

where $v_i = \frac{dx_i}{dt} = \left.\frac{\partial u_i}{\partial t}\right|_{\vec{X} = const}$ denotes a velocity of a material particle $\vec{X}$. This expression replaces (5a). With the notation

$$b_i = \frac{\partial \mathscr{L}}{\partial v_j} u_{j,i} \tag{13}$$

from (12) we have

$$-\left(\frac{\partial \mathscr{L}}{\partial X_i}\right)_{exp} = \frac{db_i}{dt} + \frac{\partial b_{ij}}{\partial X_j} \tag{14}$$

Thus for the dynamic case, the force is represented as a four-dimensional divergence of b_i, b_{ij}, which can be called the material momentum vector and tensor, respectively. They are related to those introduced in [4] or [5].

In considering dislocations, we have to add the equations expressing incompatibility in space and time derivatives:

$$\dot{\beta}_{ik} - v_{i,k} = J_{ik} \tag{15}$$

where J_{ik} is the dislocation flux tensor. Equations of equilibrium become

$$\rho \dot{v}_i - \sigma_{ij,j} = 0 \tag{16}$$

Then, using (8), (15) and (16), we derive

$$\frac{db_i}{dt} + \frac{\partial}{\partial X_k} b_{ik} = f_i + g_i \tag{10a}$$

where

$$b_i = \frac{\partial \mathscr{L}}{\partial v_j} \beta_{ji} \quad , \quad b_{ik} = -\mathscr{L}\delta_{ik} - \sigma_{jk}\beta_{ji}$$

$$g_i = \rho v_j J_{ji}$$

and f_i is given by (11). If again we interpret, by comparison to (12), the right-hand side of the last relation as a force, we arrive at a dynamic analog of the Peach-Koehler force for a continuous distribution of moving dislocations.

Dynamic generalizations bring out another noteworthy aspect of cracks and dislocations. In the static case the only independent variable was X_i . In dynamics, $\mathscr{L}$ can depend in general on time t . Differentiating the Lagrangian with respect to time, we obtain (using (16)):

$$\frac{d}{dt}\left(\frac{\partial \mathscr{L}}{\partial v_i} v_i - \mathscr{L}\right) + \frac{\partial}{\partial X_k}(-\sigma_{jk}v_j) = -\left(\frac{\partial \mathscr{L}}{\partial t}\right)_{exp} \tag{17}$$

The quantities appearing on the left-hand side have a well-known physical meaning:

$$e = \frac{\partial \mathscr{L}}{\partial v_i} v_i - \mathscr{L}$$

is the energy density and

$$e_k = -\sigma_{jk}v_j$$

is the energy flux. The explicit dependence of $\mathscr{L}$ on t implies nonconservation of energy and is connected with dissipation or energy flow. These, in turn, in many processes, are related with the work done by or on the system. Let p denote the power, i.e. the rate of work per unit of time. Eq. (17) can then be written as

$$p = \frac{d}{dt} e + \frac{\partial}{\partial X_k} e_k \tag{18}$$

For dislocations we can give the explicit value of p . Making use of (15), (16) and (8)

$$p = \dot{e} + e_{k,k} = v_j(\rho \dot{v}_j - \sigma_{jk,k}) + \frac{\partial \mathscr{L}}{\partial v_i}\dot{v}_i - \sigma_{jk}v_{j,k} - \frac{d\mathscr{L}}{dt} = \sigma_{jk}J_{jk}$$

For a single dislocation this can be rewritten as

$$p = \sigma_{ij}\varepsilon_{j\ell k}V_\ell \alpha_{ik} = -f_i V_i$$

where f_i is given by (11) and V_i is the dislocation velocity. The interpretation of this seems straightforward. To move a dislocation with a velocity $\vec{V}$ we have to deliver a power equal to the product of that velocity and the force exerted on the dislocation and thus constraining it.

The most general case considered could be a crack in a dislocated medium. Mathematically this could be described by a Lagrangian which depends explicitely on X_i and is subjected to additional constraints: equation (8) for statics; equations (8) and (15) for dynamics. The density of the force acting on a crack is

$$j_i = b_{ij,j} - f_i \tag{18}$$

in static case, and

$$j_i = \dot{b}_i + b_{ij,j} - f_i - g_i \tag{19}$$

for dynamics, where b_i, b_{ij} and f_i, g_i were defined before. We leave these results in the form of densities rather than trying to integrate them, because path-independence of integrals obtained by integrating (18) or (19) is rather questionable, and could possibly be proved only in some special cases. The usefulness of the expressions for the force densities is not clear at this time. But since by the same procedure we were able to derive the Peach-Koehler force on a dislocation and also the force on a crack tip, it might be expected that these force densities will also prove to have a physical significance.

The use of the Lagrangian for the derivation of forces suggests that all results obtained until now can be derived also from a unified variational principle. Actually, we were assuming that all results for the cracks can be "translated" into those for dislocations by formal replacement of $\dot{u}_i$ by v_i and $u_{j,i}$ by β_{ji} in the Lagrangian.

We can verify that if $\mathscr{L}$. is of the form

$$\mathscr{L} = \rho v_i v_i - W(\varepsilon_{ij}, X_k)$$

we obtain, from a variational principle, the equations of motion and the equations of balance of material momentum, which represent forces on a crack. In addition we also obtain equations representing balance of energy (c.f. (17)). This procedure is discussed in [6].

However, extension of the general variational formalism for dislocations is not straightforward. As it is well known, (8) and (15) express incompatibility of the basic field used in continuum mechanics, namely the displacement vector. Thus, instead of dealing with 3 independent fields u_i we have to deal with 12: v_i, β_{ij}. In addition, derivatives of u_i become fields themselves, and in order to obtain Euler equations for them, the Lagrangian has to depend now on their derivatives also. The proposed action integral for the medium with dislocations has the form

$$\begin{aligned} S &= \int dt \int d^3X\, \mathscr{L}(X_i, t; v_i, \beta_{ij} \dot{v}_i, \dot{\beta}_{ij}, v_{i,k}, \beta_{ij,k}) \\ &= \int dt \int d^3X \Big[v_i c_{ik\ell m} J_{\ell m,k} + \frac{1}{2} \rho \dot{v}_i \dot{v}_i - \frac{1}{2} c_{ijk\ell} v_{i,j} v_{k,\ell} \\ &\quad + \rho\beta_{ij} (\dot{J}_{ij} - c_{ik\ell m} \varepsilon_{mjs} \alpha_{\ell s,k}) + \frac{1}{2} \rho \dot{\beta}_{ij} \dot{\beta}_{ij} - \frac{1}{2} c_{ijk\ell} \beta_{jp,\ell} \beta_{ip,s} \Big] \end{aligned} \tag{20}$$

The Euler equations following from the principle of stationary action for the fields v_i, β_{ij} are:

$$\rho\ddot{v}_i - c_{ik\ell m}v_{\ell,mk} = c_{ik\ell m}J_{\ell m,k} \tag{21}$$

$$\rho\ddot{\beta}_{ij} - c_{ik\ell m}\beta_{ij,mk} = \rho\dot{J}_{ij} - c_{ik\ell m}\varepsilon_{mjs}\alpha_{\ell s,k} \tag{22}$$

These equations can be obtained directly from (8), (15) and (16). It is to be noted that the operator acting on fields v_i and β_{ij} in equations (21) and (22) has the same form, and this brings the dislocation theory close to electromagnetism where equations for electromagnetic potentials are given in terms of the same operator, namely the d'Alembertian. This analogy was discussed in [7].

The presence of α_{ik}, J_{ik} and their derivatives represents the explicit dependence of $\mathscr{L}$ on t and X_i and in the framework of the Lagrangian formalism this expresses the existence of forces in dislocated media and flow of energy.

We would like to underline that nonconservation of energy and existence of forces on dislocations are independent in this framework (compare (10a) and (19)). Only in the static case is the energy release rate identical to the force. In the general, dynamic case, existence of forces is connected with explicit dependence of $\mathscr{L}$ on X_i, while non- or conservation of energy is connected with explicit dependence on time. Energy flux appears in the relation expressing balance of energy and not balance of material momentum.

Equations (21) and (22), obtained from (20) are equivalent to those known usually as dislocation equations; however, it was possible to obtain the general force field due to the presence of dislocations (10a) in a much simpler way, assuming dependence of $\mathscr{L}$ on v_i, β_{ij} only.

The explanation of these two features is rather simple. As it is well known [8], dislocations and disclinations change basically the geometry of the material space. We could have kept u_i as the basic field; however, we would have to use the covariant derivatives (instead of partial) which do not commute.

This approach will be discussed in a separate paper. The idea of the role of curvature and Cartan torsion in the theory of dislocations and disclinations have been worked out in many papers, e.g. [9], [10], [11]. The content of a paper under preparation, however, is to place these geometric notions into the framework of variational formalism.

Acknowledgment

This research was supported in part by AFOSR F49620-79-C-0217 and Electric Power Research Grant RP-609-1 to Stanford University. The author is most grateful to Professor D.M. Barnett of Stanford University for many valuable discussions with regard to this paper.

References

1. B. Budiansky and J.R. Rice, Journal of Applied Mechanics 40, 201 (1973)
2. A. Golebiewska Herrmann and G. Herrmann, "On Energy-Release Rates for a Plane Crack" to be published.
3. J.P. Hirth and J. Lothe, "Theory of Dislocations", McGraw-Hill Book Company (1968).
4. J.D. Eshelby, Solid State Physics - Advances in Research and Applications, p. 79, Academic Press Inc., New York (1956).
5. A. Golebiewska Herrmann, "On Conservation Laws of Continuum Mechanics", International Journal of Solids and Structures (in print).
6. A. Golebiewska Herrmann, "Material Momentum Tensor and Path-Independent Integrals of Fracture Mechanics" in preparation.
7. A. Golebiewska-Lasota, International Journal of Engineering Science 17, 329 (1979).
8. I.A. Kunin, "Methods of Tensor Analysis in the Theory of Dislocations", Israel Program for Scientific Translations, Jerusalem (1968).
9. E. Kröner, Kontinuumstheorie der Versetzungen und Eigenspannungen, Springer-Verlag, Berlin (1958).
10. M. Zorawski, Theorie Mathematique des Dislocations, Dunod, Paris (1967).
11. H. Günther, Phys. Stat. Sol. 49, 551 (1972).

THE INTERACTION ENERGY BETWEEN INTERSTITIAL ATOMS AND DISLOCATIONS AND ITS RELEVANCE TO IRRADIATION DAMAGE PROCESSES

R. Bullough, D.W. Wells*, J.R. Willis† and M.H. Wood
Theoretical Physics Division
A.E.R.E. Harwell, Oxfordshire, U.K.
*Harwell Student Assistant; now at
Clare College, Cambridge, U.K.
†School of Mathematics
University of Bath, Bath, U.K.

Abstract

To assess the suitability of alloy materials for the structural components of fast fission or fusion reactors, it is necessary to understand the various processes that influence the evolution of the microstructure of such materials when subjected to irradiation. In particular the vacancies and interstitials created during the irradiation will interact with existing dislocation lines and, under appropriate conditions can be lost at such sinks. To estimate the quantitative nature of such point defect loss it is necessary to calculate the point defect-dislocation interaction energies accurately. In this paper we describe such calculations for both straight dislocations and dislocation loops in both a stress-free body and in a body subjected to an external uniaxial tensile stress. The relatively large distortion field associated with the interstitial compared to the vacancy ensures that only the dislocation-interstitial interactions need be considered. The stress-free interaction is dominated by the first order size effect and is considered to be the fundamental cause of such important phenomena as void swelling. On the other hand when a stress is present, important stress induced inhomogeneity interactions can prevail that will distinguish between dislocations of differing orientations relative to the applied stress. This distinction leads to stress induced differential loss of interstitials within the dislocation distribution and is probably the fundamental cause of irradiation creep.

The interaction energies for the two dislocation morphologies will be used to calculate the appropriate sink strengths required for the quantitative rate theory analysis of the microstructure evolution. Various predictions, particularly relating to irradiation creep will be presented and discussed.

1. Introduction

Rate theory has been developed and used by numerous authors to study the evolution of the microstructure in crystalline materials subjected to irradiation damage [1-5]. In such theoretical models the real spatially varying microstructure is replaced by a homogeneous lossy continuum in which the various sink types, in the microstructure, are given appropriate sink strengths. These sink strengths have been evaluated using either a self-consistent embedding procedure [1,6,7], or by an appropriate cellular method [5,8]. Though the former approach also yields the higher order correction terms to such strengths [6,7] when the overall sink density is high, it is only directly applicable when the sink is neutral and does not have a long range field interaction with the intrinsic point defects [7]; on the other hand the cellular method, though suffering from an inability to deal with such high sink density interactive effects [7], does enable the strengths of sinks with long range fields to be estimated [5,7]. The edge dislocation line is such a sink and the determination of its strength as a function of the line morphology is an important ingredient of the rate theories of irradiation creep and swelling.

Although various estimates of dislocation sink strengths have been published their accuracy is open to doubt because of various approximations introduced to facilitate the calculations. For example, in order to integrate analytically the relevant cellular continuity equations for the point defect concentrations either the field is simplified [9] or the displacement damage source term is replaced by a (questionably) equivalent cellular boundary condition [5]; furthermore, the actual dislocation line morphology is often replaced by a geometrically simpler morphology [1, 10]. The loss of accuracy due to any of these simplifications is very difficult to estimate and can, we believe, be unacceptably large for certain morphologies.

In this paper we present a number of sink strength calculations for both straight edge dislocations and edge dislocation loops that introduce no such simplifications. The cellular continuity equations are solved numerically to a sufficiently high accuracy to enable reliable sink strengths to be calculated. The general explicit interaction fields for straight edge dislocations and circular edge dislocation loops in a body subjected to uniaxial tension are presented in section 2 and the cellular models appropriate for the evaluation of the sink strengths of such dislocations are described in section 3. Sections 4 and 5 describe numerical results obtained for the straight dislocations and loops respectively and finally in section 6 the overall relevance of the results to irradiation creep and other aspects of the microstructure evolution is discussed.

2. The Interaction Fields of Dislocations

In this section the result obtained by Bullough and Willis [11] for the interaction energy between a general applied stress and a misfitting inhomogeneity is used to calculate the required interaction energies between dislocations and the self-interstitials in a body subjected to an external stress. The straight and circular edge dislocation only are considered and the respective coordinate systems used for these two sink types are shown in figures 1 and 2. The straight edge dislocation is located along the x_2-axis and has the Burgers vector (0,0 b) in the (x_1, x_2, x_3) Cartesian coordinates. The circular edge dislocation (figure 2) lies in the (x_1, x_2) plane with its centre at the coordinate origin and its Burgers vector is again (0,0,b); for this morphology it is also convenient to use the associated (r,θ,z) cylindrical polar coordinates. The cell features illustrated in these figures are referred to in section 3.

Following Bullough and Willis [11] the self-interstitial is represented by a small misfitting inhomogeneity of volume V with an associated relaxation volume strain e^o. Defect modulus measurements and atomic simulation calculations of self-interstitials indicate that such interstitials are best represented by an inclusion with a shear modulus much less than the corresponding matrix modulus and a bulk modulus not differing much from that of the matrix; in the present analysis we adopt, for simplicity, the extreme value of zero shear modulus for the inclusion and no difference in bulk modulus between the inclusion and the matrix. Only the interaction with the self-interstitial is considered; the interaction with a vacancy is much smaller and is ignored in the present discussion [12,13]. Finally, again for simplicity of presentation, we set Poisson's Ratio ν = 1/3 in all the various expressions.

With these simplifications the interaction energy between the general applied strain field e^A_{ij} and the interstitial* may be written [11]

$$\phi = -V\mu \left\{ \frac{15}{8}\, {}'e^A_{ij}\, {}'e^A_{ij} + 4e^A e^o \right\}, \tag{1}$$

where μ is the shear modulus, $e = e_{ii}$, and ${}'e_{ij} = e_{ij} - 1/3\, e\delta_{ij}$ is the deviatoric part of e_{ij}. The applied strain is now identified with the sum of the internal dislocation field e_{ij} at the interstitial and the strain field due to the external stress e^E_{ij}.

2.1 The Straight Edge Dislocation

The dislocation lies along the x_2 axis, as shown in figure 1, and with the interstitial located at the point (x_1, x_3) the relevant dislocation strain field $e^D_{ij}(x_1, x_3)$ has the diagonal components [14]

*Henceforth the term interstitial will always refer to 'self' interstitial.

FIG.1a.

FIG.1b.

FIG.2a.

FIG.2b.

FIG.2c.

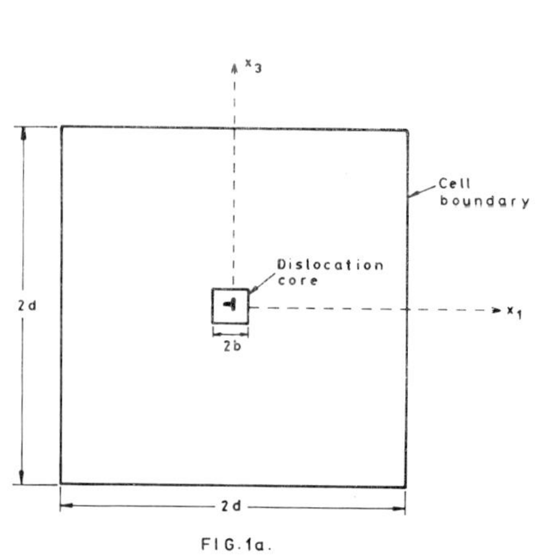

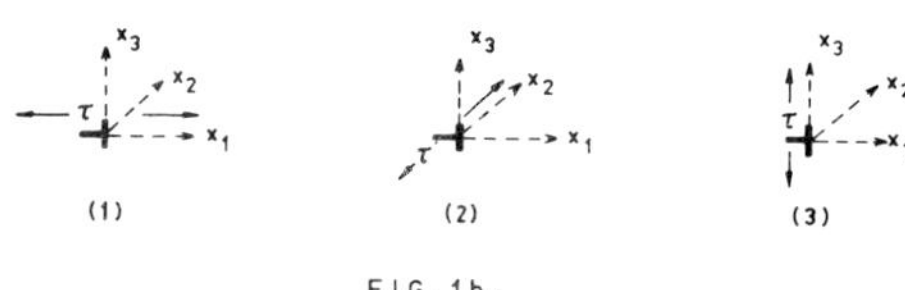

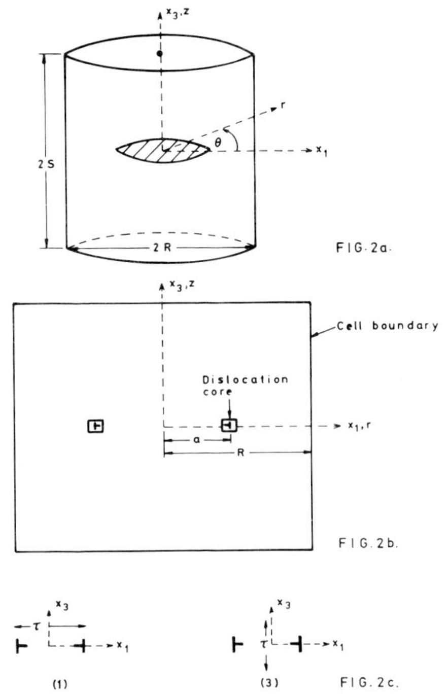

FIG 1a

The rectilinear cell and relevant coordinates for the straight dislocation. The core is an ideal sink and there is zero flow across the cell boundary.

FIG 1b

The three orientations of the tensile stress τ relative to the straight edge dislocation line lying along the x_2 axis. The edge dislocation has the Burgers vector (0,0,b) in the (x_1,x_2,x_3) coordinates shown.

FIG 2a

The complete cylindrical cell with an interstitial circular edge dislocation loop at its centre. The hatched area indicates the extra circular lattice plane.

FIG 2b

A cross-section of the cylindrical cell containing the x_3 axis. The section through the loop (of radius a) is indicated by the core sections. The core is an ideal sink and there is zero flow across the cell boundary.

FIG 2c

The two orientations of the tensile stress τ relative to the circular edge dislocation loop lying in the $x_3 = 0$ plane. The Burgers vector of the loop is (0,0,b) and its radius is a.

$$\left.\begin{aligned} e^D_{11} &= \frac{b}{8\pi}\frac{x_1}{r^4}[x_1^2 - 5x_3^2] \\ e^D_{22} &= 0 \\ e^D_{33} &= \frac{b}{8\pi}\frac{x_1}{r^4}[x_1^2 + 7x_3^2] \\ e^D &= \frac{b}{4\pi}\frac{x_1}{r^2} \end{aligned}\right\} \quad (2)$$

where $r^2 = x_1^2 + x_3^2$.

If τ is the uniaxial external tensile stress with its axis parallel to one of the x_k directions, as shown in figure 1b, then the three associated strain fields e^{Ek}_{ij} (k = 1, 2 or 3) have the diagonal components, for orientation (1), τ parallel to x_1:

$$e^{E1}_{11} = 3\tau/8\mu,\ e^{E1}_{22} = e^{E1}_{33} = -\tau/8\mu$$

for orientation (2), τ parallel to x_2:

$$e^{E2}_{11} = e^{E2}_{33} = -\tau/8\mu,\ e^{E2}_{22} = 3\tau/8\mu \quad (3)$$

for orientation (3), τ parallel to x_3:

$$e^{E3}_{11} = e^{E3}_{22} = -\tau/8\mu,\ e^{E3}_{33} = 3\tau/8\mu.$$

Substitution of the applied strain fields

$$e^{Ak}_{ij} = e^{D}_{ij} + e^{Ek}_{ij} \qquad k = 1,\ 2 \text{ and } 3 \quad (4)$$

into (1) yields the interaction energies ϕ^k for the three orientations of the external stress:

$$\left.\begin{aligned} \phi^1 &= \phi_o[1 + \frac{5}{64e^o}(\frac{\tau}{\mu})[x_1^2 - 17x_3^2]/r^2] \\ \phi^2 &= \phi_o[1 - \frac{5}{32e^o}(\frac{\tau}{\mu})] \\ \phi^3 &= \phi_o[1 + \frac{5}{64e^o}(\frac{\tau}{\mu})[x_1^2 + 19x_3^2]/r^2] \end{aligned}\right\} \quad (5)$$

where

$$\phi_o = -\frac{\mu V b e^o}{\pi}\frac{x_1}{r^2} \quad (6)$$

is the stress free first order or size effect interaction. In the derivation of (5) from (1), the higher order terms of $O(r^{-2})$ and the constant terms have been omitted. The higher order terms are always negligible in the applications considered in the present work.

The forms and relative magnitudes of the interaction energies (5) and (6) are illustrated by the contour maps in figures 3 and 4. The contours in figure 3 are the equipotentials of ϕ_o in the (x_1,x_3) plane about the edge dislocation as depicted in figure 1. The integers (±n) identified with the contours refer to interaction energies of ± 2.5n x 10^{-2} eV; these energy magnitudes follow from (6) with typical values for the physical parameters: V = 1.2 x 10^{-29} m^3, μ = 20GPa, e^o = 1 and we note kT = 5 x 10^{-2} eV at 550K so that, for example, the n = ±2 contours correspond to an interaction energy magnitude close to kT at a typical irradiation creep temperature. The square core region of size 2b is also indicated, where all the elastic interaction energies are cut off; b, the magnitude of the Burgers vector is also used as the convenient unit of length. The contours in figure 4a, b and c are respectively the equipotentials of $\phi^1 - \phi_o$, $\phi^2 - \phi_o$ and $\phi^3 - \phi_o$ from (5) for the straight edge dislocation subjected to a uniaxial tensile stress τ parallel to the respective x_1, x_2 and x_3 directions. With the same physical parameter values as used for the contours of ϕ_o in figure 3 plus a stress τ = 100 MPa, the integers (±n) identified with the contours refer to energies of $\pm 5n \times 10^{-5}$ eV. These three contour maps illustrate the considerable variation of the purely stress induced component of the interaction energies with the orientation of the stress axis. Thus it is immediately clear that the interactions are greatest in figure 4c when the stress is orthogonal to the extra plane of the edge dislocation and least in figure 4b when the stress is applied along the dislocation line. It is precisely because the stress orientation in figure 4c causes the largest interaction that such dislocations will preferentially attract more interstitials during irradiation and cause the material to creep by their preferential climb [13,15]. The evaluation of the sink strengths of such straight edge dislocations, then enables the quantitative effect of such small variations in the dislocation-interstitial interactions on processes such as irradiation creep to be calculated and will be discussed in the next section. In this connection we note the absolute magnitude of the induced energies in figure 4 are approximately 1/500th of the magnitude of the first order size effect interaction energies in figure 3 at corresponding positions around the edge dislocation; the stress induced interaction is thus a relatively small perturbation on the size effect interaction.

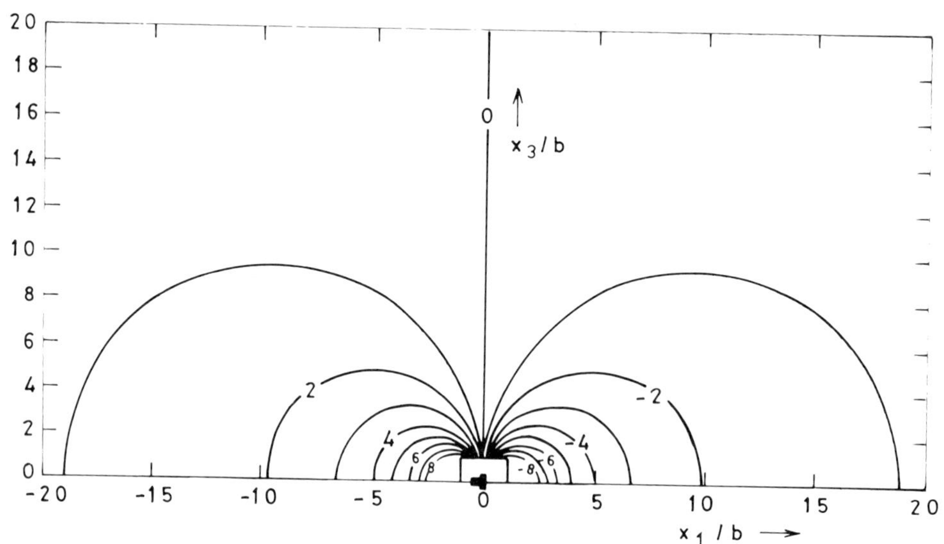

FIG 3

The equipotential contours of ϕ_o (equation (6)) in the upper half of the (x_1,x_3) plane around an edge dislocation lying along the x_2-axis. The contour intervals are in units of 2.5×10^{-2} eV and the energies have been cut off at the core dimension shown; the contours are symmetric about the x_1 axis. (μ = 20GPa, V = 1.2×10^{-29} m^3, e^o = 1)

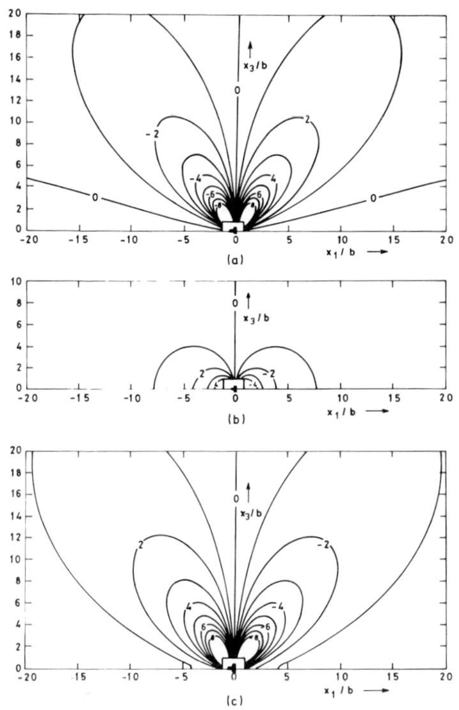

FIG 4

The equipotential contours of $\phi^1 - \phi_o$, $\phi^2 - \phi_o$ and $\phi^3 - \phi_o$ in 4a, 4b and 4c respectively. These expressions, given by (5), define the stress induced components of the total interaction energies when the orientation of the stress is varied relative to the edge dislocation: ϕ^k corresponds to the stress parallel to the x_k axis. The contour intervals are in units of 5×10^{-5} eV and the energies have been cut-off at the core dimension shown; the contours are symmetric about the x_1 axis. (μ = 20 GPa, V = 1.2×10^{-29} m^3, $e^o = 1$, τ = 100 MPa)

2.2 The Circular Edge Dislocation Loop

The axial symmetry of the loop about the x_3 axis, as shown in figure 2, ensures that the required diagonal components of its strain field e^D_{ij} have the general form at the arbitrary point $\underline{x} = (x_1, x_2, x_3) = (r \cos\theta, r \sin\theta, z)$:

$$\left.\begin{aligned} e_{11}^{D}(\underline{x}) &= [x_1^2\, e_{rr}^{D}(r,z) + x_2^2\, e_{\theta\theta}^{D}(r,z)]/r^2 \\ e_{22}^{D}(\underline{x}) &= [x_2^2\, e_{rr}^{D}(r,z) + x_1^2\, e_{\theta\theta}^{D}(r,z)]/r^2 \\ e_{33}^{D}(\underline{x}) &= e_{zz}^{D}(r,z) \\ e^{D}(\underline{x}) &= e_{rr}^{D} + e_{\theta\theta}^{D} + e_{zz}^{D} = e^{D}(r,z) \end{aligned}\right\} \tag{7}$$

In these relations $e_{rr}^{D}(r,z)$, $e_{\theta\theta}^{D}(r,z)$ and $e_{zz}^{D}(r,z)$ are the principal cylindrical components of the loop strain field evaluated in the $\theta = 0$ plane ($x_2 = 0$).

The explicit forms for these cylindrical strain components can be deduced from the general integral expression [16] for the distortion field of an arbitrary dislocation loop and it can be shown that:

$$\left.\begin{aligned} e_{rr}^{D}(r,z) &= \frac{3b}{16\pi}\left[\frac{2}{3} I_{11} - J_{1331}\right] \\ e_{\theta\theta}^{D}(r,z) &= \frac{3b}{16\pi}\left[\frac{2}{3} I_{22} - J_{2332}\right] \\ e_{zz}^{D}(r,z) &= -\frac{3b}{16\pi}\left[\frac{4}{3}(I_{11} + I_{22}) - (J_{1331} + J_{2332})\right] \\ e^{D} &= -\frac{b}{8\pi}\left[(I_{11} + I_{22})\right]. \end{aligned}\right\} \tag{8}$$

$I_{ij}(r,z)$ and $J_{ijk\ell}(r,z)$ are the integrals

$$\left.\begin{aligned} I_{ij}(r,z) &= \int_0^{2\pi} \left[\left(\frac{1}{P}\right)_{,i}\right]_{\theta=0} x_j'\, d\theta' \\ J_{ijk\ell}(r,z) &= \int_0^{2\pi} \left[P_{,kij}\right]_{\theta=0} x_\ell'\, d\theta' \end{aligned}\right\} \tag{9}$$

where

$$P = [r^2 + z^2 + a^2 - 2ar\cos(\theta-\theta')]^{\frac{1}{2}}.$$

and $(x_1', x_2', x_3') = (a\cos\theta', a\sin\theta', 0)$; the comma notation in the integrals refers to differentiation with respect to the cylindrical coordinates r, θ, z: thus $P_{,1} = \frac{\partial P}{\partial r}$, $P_{,2} = \frac{\partial P}{\partial \theta}$, $P_{,3} = \frac{\partial P}{\partial z}$, etc. All the integrals in (8) can be expressed explicitly in terms of the complete elliptic integrals

$$E(k) = \int_0^{\frac{\pi}{2}} (1 - k^2 \sin^2 x)^{\frac{1}{2}}\, dx, \quad K(k) = \int_0^{\frac{\pi}{2}} (1 - k^2 \sin^2 x)^{-\frac{1}{2}}\, dx \tag{10}$$

where

$$k^2 = 4ar/[(r+a)^2 + z^2], \quad k'^2 = 1 - k^2$$

and subsequent results will be presented in terms of these quantities.

The axis of the applied tensile stress τ will be considered either perpendicular to the plane of the loop, parallel to the x_3 direction, or in the plane of the loop, parallel to the x_1 direction; these two orientations generate the respective internal strains e_{ij}^{E3} and e_{ij}^{E1}

given by (3) and are depicted in figure 2c. The in-plane direction parallel to the x_1 axis has been chosen without loss of generality and we present below the interaction energy for this particular direction; subsequently for utilitarian reasons we replace this non-axially symmetric energy by its average over the azimuthal angle θ.

Substitution of the applied strain fields

$$e_{ij}^{Ak} = e_{ij}^{D} + e_{ij}^{Ek} \qquad k = 3 \text{ and } 1 \tag{11}$$

from (7), (8) and (3) into (1) yields the interaction energies ψ^3 and ψ^1 for the two orientations of the applied stress:

$$\left.\begin{aligned} \psi^3 &= \psi_o + \psi_\tau^3 \\ \psi^1 &= \psi_o + \psi_\tau^1 \end{aligned}\right\} \tag{12}$$

where the stress free interaction energy ψ_o for the loop has the axisymmetric form,

$$\psi_o(r,z) = \frac{\mu V b e^o}{2\pi}(I_{11} + I_{22}), \tag{13}$$

with

$$I_{11} + I_{22} = \frac{2}{[(r+a)^2 + z^2]^{\frac{1}{2}}}\left\{K(k) + \left(\frac{a^2-r^2-z^2}{(r-a)^2 + z^2}\right)E(k)\right\} \tag{14}$$

and the elliptic integrals are defined in (10). The stress induced parts, ψ_τ^3 and ψ_τ^1, in (12) are similarly:

$$\psi_\tau^3(r,z) = \frac{5Vb\tau}{128\pi}\left[10(I_{11} + I_{22}) - 9(J_{1331} + J_{2332})\right] \tag{15}$$

and

$$\psi_\tau^1(r,\theta,z) = -\frac{5Vb\tau}{128\pi}\Big[2(3\cos^2\theta + 1)I_{11} + 2(3\sin^2\theta + 1)\, I_{22} - 9\cos^2\theta\, J_{1331} - 9\sin^2\theta\, J_{2332}\Big]. \tag{16}$$

The axisymmetric nature of ψ_τ^3 is evident from (15) with

$$J_{1331} + J_{2332} = I_{11} + I_{22} + \frac{4az^2}{[(r+a)^2 + z^2]^{5/2}}\left\{\frac{2}{k^2k'^4}\left[r(k^4-k^2+1) + ak^2(k^2-2)\right]E(k) + \frac{1}{k^2k'^2}\left[r(k^2-2) + ak^2\right]K(k)\right\} \tag{17}$$

where the elliptic integrals are given by (10). ψ_τ^1, on the other hand, is not axisymmetric since the τ-axis is particular (the x_1 axis). It is convenient to define an axially symmetric mean interaction energy by;

$$\bar{\psi}_\tau^1(r,z) = \frac{1}{2\pi}\int_o^{2\pi}\psi_\tau^1(r,\theta,z)d\theta \tag{18}$$

when τ is in the plane of the loop; from (16) this yields

$$\bar{\psi}_\tau^1 = -\frac{5Vb\tau}{256\pi}\left[10(I_{11} + I_{22}) - 9(J_{1331} + J_{2332})\right] \tag{19}$$

where $(I_{11} + I_{22})$ and $(J_{1331} + J_{2332})$ are given by (14) and (17) and we note from (15) and (19)

$$\overline{\psi}_{\tau}^{1} (r,z) = -\frac{1}{2} \psi_{\tau}^{3} (r,z).$$

The contours in figure 5 are the equipotentials of ψ_o, defined by (13), in the (r,z) plane section through the loop centre; the loop is interstitial in character of radius a = 10b and the integers (± n) identified with the contours refer to interaction energies of $\pm 4.5 \times 10^{-2} n$ eV corresponding to μ = 20 GPa, V = 1.2×10^{-29} m^3 and e^o = 1 for the interstitial in steel. Again, as for the straight dislocation contours, the energies have been cut off at the core distance b from the dislocation and all lengths are in units of this distance.

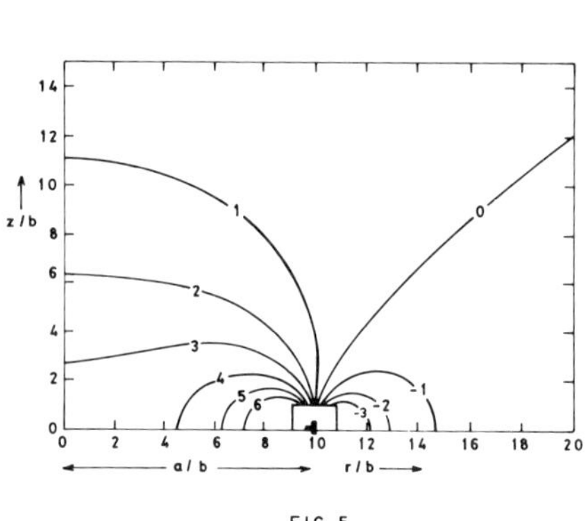

FIG. 5.

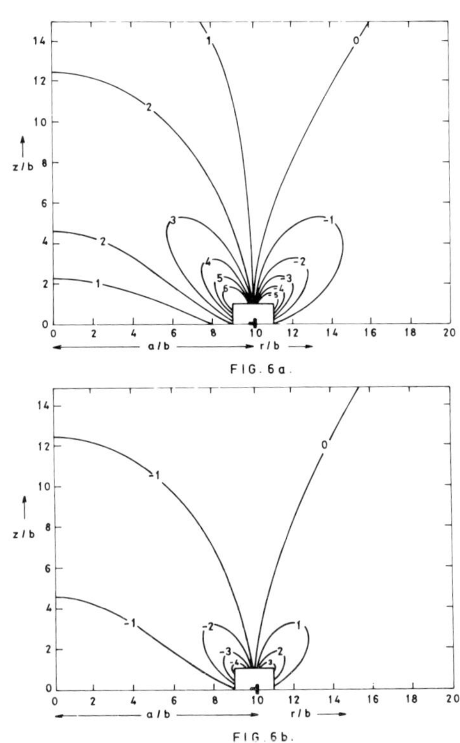

FIG. 6a.

FIG. 6b.

FIG 5

The equipotential contours of ψ_o (equation (13)) in the upper quartile of the (r,z) plane perpendicular to the loop plane. a, the loop radius, is 10b and all lengths are in units of b, the magnitude of the Burgers vector. The loop is interstitial in character; the contour intervals are in units of 4.5×10^{-2} eV and the energies have been cut off at the core dimension shown. The contours are axially symmetric about the z axis and there is reflection symmetry in the z = 0 plane. (μ = 20 GPa, V = 1.2×10^{-29} m^3, e^o = 1)

FIG 6

The equipotential contours of $\psi^3 - \psi_o = \psi_\tau^3$ and $\overline{\psi}^1 - \psi_o = \overline{\psi}_\tau^1$ in 6a and 6b respectively. These expressions given by (15) and (19), define the variation of the stress induced components of the total interaction energies between an interstitial edge dislocation loop of radius a and an interstitial with orientation of the applied tensile stress τ; ψ_τ^3 applies when the stress is aligned normal to the plane of the loop (parallel to the z-axis (x_3)) and $\overline{\psi}_\tau^1$ is the average induced interaction when the stress lies in the plane of the loop. The contour intervals are in units of 2.5×10^{-3}eV and the core cut off is shown. The contours are axially symmetric about the z-axis and there is reflection symmetry in the z=0 plane (μ=20GPa, V=1.2×10^{-29} m^3, e^o=1, τ = 100MPa)

The contours in figure 6a and b are respectively the equipotentials of $\psi^3 - \psi_o = \psi^3_\tau$ and $\bar{\psi}^1 - \psi_o = \bar{\psi}^1_\tau$ from (15) and (19) for the interstitial edge dislocation loop subjected to an axial tensile stress or a tensile stress in the plane of the loop respectively. With the same physical parameter values as used for the contours of ψ_o in figure 5 plus a stress of τ = 100MPa, the integers (± n) identified with the contours refer to energies of ± 2.5 x 10^{-3} n eV. From these two contour maps it is immediately evident that the stress induced interaction energy is largest when the applied tensile stress is normal to the plane of the loop, as is clear, of course, from (20). Dislocation loops orthogonal to such applied stresses will thus attract more interstitials than loops in other less favourable orientations and thereby cause the material to creep by their preferential climb. We note that the absolute magnitude of these induced energies in figure 6 are approximately 1/20th of the magnitude of the first order size effect interaction energies in figure 5, at corresponding positions adjacent to the loop. This is a much larger ratio than was found for the straight dislocation (1/500th) and suggests the loop morphology may be a more significant contributor to the irradiation creep strain than the straight line morphology. This point will be investigated and discussed later when we evaluate the appropriate sink strengths and assess the magnitude of the consequent irradiation creep strains, etc.

Finally, to conclude this section, we show in figure 7 two equipotential contour maps of ψ^1_τ (r,θ,z) in the two orthogonal (r,z) planes θ = 0 (x_2 = 0) and θ = 90° (x_1 = 0); in these maps the energies identified by the contour integers are identical to those in figure 6. The variations between these two contour maps provide the basis for the validity of adopting the average interaction energy $\bar{\psi}^1$ (figure 6b) to represent the interaction when the stress axis lies in the plane of the loop.

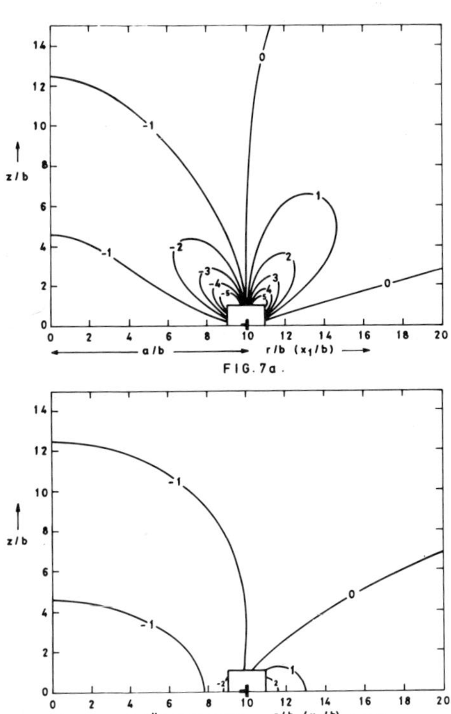

FIG 7

Equipotential contours of the induced interaction energy $\psi^1 - \psi_o = \psi^1_\tau$ (r,θ,z) for the two orthogonal (r,z) sections: θ = 0(a), θ = 90°(b), to illustrate the basis of the mean interaction energy $\bar{\psi}^1$ in figure 6b. The contour intervals are identical to those defined in figure 6, but there is now only reflection symmetry about the z and x_1 or x_2 axes in the particular (r,z) plane, but no axial symmetry about the z axis as in the figures 5 and 6.

3. Method of Dislocation Sink Strength Calculation

The so-called "single sink cellular method" [5,7,17] has been used in the present work to obtain an estimate of the sink strengths of the various dislocations under zero or differing orientations of the external stress. In this method the dislocation is located at the centre of a suitable cell. The average interstitial concentration within the cell is then obtained by solving the diffusion plus drift steady state continuity equation over the cell with an interstitial source term representing the damage production rate and zero flow boundary condition at the cell boundary; the dislocation itself is treated as an ideal sink for interstitials. The dimension of the cell is related to the volume density of this sink type in the actual body. From the average concentration and the magnitude of the source term it is then possible to define a sink strength for the dislocation. Before presenting the mathematical procedure in more detail it should be emphasized that the absence of at least one other sink type within the cell means that the single sink model cannot be a complete model of a real evolving system. The presence of at least two sink types is necessary for a net accumulation of vacancies or interstitials to occur at a sink. When the cell contains only one sink, the approach should be regarded simply as a method for obtaining an estimate of the sink strength of a sink type, that happens to be particularly convenient when the sink has a long range interaction with the interstitials; it cannot, for example, possibly provide a measure of the sink interactive effects that will prevail at high sink densities and even when the sink density is low its failure to really represent the physical system means there is no rigorous proof that the resulting sink strength is accurate [7]. However, when the interaction field is zero we do know that the present single sink method, with one sink per cell, leads to sink strengths that agree exactly with the lowest order results from the corresponding consistent embedding procedure [6,7], where all the existing sink types are included. We then feel some confidence that the present method can also be used when such interactive fields are present to yield at least reliable lowest order sink strengths. Needless to say the corresponding embedding procedure with interactive fields present would lead to a much more complex computational analysis than required in the present model but complete justification of the single sink cellular method must await such calculations [7]. In the meantime, and on the basis of past experience, we presume that our present method does yield acceptable measures of the lowest order dislocation sink strengths.

3.1 The Straight Edge Dislocation

The cell is an infinite cylinder with a square cross-section of side length 2d, as shown in figure 1 and the edge dislocation lies along the axis of the cylinder (x_2-axis). If interstitials are created uniformly within the cell at a fractional rate K and then migrate under the influence of an interaction energy $\phi(x_1,x_3)$ with the dislocation, the steady state fractional concentration $c(x_1,x_3)$ of interstitials satisfies the field equation:

$$\nabla^2 c + \frac{1}{kT}\nabla c.\nabla\phi + \frac{c}{kT}\nabla^2\phi = -\frac{K}{D} \tag{21}$$

where k is Boltzmann's constant, T is the absolute temperature and D is the diffusion constant*.

At the cell boundaries the interstitial flow is zero: that is

$$\frac{\partial c}{\partial x_1} + \frac{c}{kT}\frac{\partial\phi}{\partial x_1} = 0 \quad \text{at } x_1 = \pm d \tag{22}$$

and

$$\frac{\partial c}{\partial x_3} + \frac{c}{kT}\frac{\partial\phi}{\partial x_3} = 0 \quad \text{at } x_3 = \pm d. \tag{23}$$

---

*The discussion will, in general, refer specifically to interstitial point defects since only they have a significant interaction with dislocation. For vacancies we can reasonably set $\phi \equiv 0$. The subscripts i and v will be attached to c, D, etc., only when the simultaneous presence of vacancies and interstitials is required. Bulk recombination has been omitted in (21) since we know from previous experience that explicit sink strength modification due to bulk recombination is generally unimportant [8,18,19,20]; however if it were to be included we would need to replace (21) by a pair of simultaneous field equations for c_i and c_v, with their respective ϕ_i and $\phi_v(\equiv 0)$, coupled by the non-linear bulk recombination loss term ($\alpha\ c_i\ c_v$). Such calculations have been made by Bullough, Eyre and Perrin [17] but with simplified interaction fields.

The primitive cell is actually the upper half of the complete cell with reflection symmetry across $x_3 = 0$. We note here that the interaction energies ϕ such as given explicitly by (5) and (6), refer to an isolated edge dislocation in an infinite medium; their gradients do not vanish at the cell boundaries as they should in a periodic lattice of dislocations. The model again thus departs from a strict representation of the situation prevailing in a real body and it is necessary to include the explicit field gradients of ϕ in (22) and (23) to ensure there is no leakage, due to the non-zero field gradient terms, from the cell. In reality this use of the infinite body interaction field is almost unavoidable when the two-dimensional nature of the field in included. Thus it would be necessary to specify the relative orientations of all neighbouring parallel edge dislocations to specify the complete ϕ with vanishing gradients at the cell boundary. Such a specification would not be sensible, since, in any case, the basic notion of all the dislocations being straight and parallel is somewhat unphysical to say the least. We therefore prefer to retain the relative simplicity of the infinite body interaction fields and adopt the zero flow boundary conditions (22) and (23) to compensate for the lack of dislocation-dislocation field interactions. This problem is much less severe for the dislocation loop morphology (see section 3.2) since interactive fields from a loop (ψ) falls off much more rapidly with distance than does ϕ from a straight dislocation.

The final boundary condition is

$$c = 0 \quad \text{at} \quad x_1, x_3 = \pm\, b; \tag{24}$$

that is at the dislocation core, as indicated in figure 1a. The dislocation is thus treated as an ideal sink for interstitials.

The partial differential equation (21) is then solved, subject to the boundary conditions (22), (23) and (24) by a discretization procedure in which the mesh dimension is always taken as b, the magnitude of the Burgers vector and core 'radius'. In the actual numerical work b is taken as the unit of length and c is replaced by the 'non-dimensional' concentration c' where

$$c' = (D/Kb^2)c. \tag{25}$$

The resulting set of simultaneous linear equations has been solved accurately, using the Harwell Subroutine MA28A, for a range of cell dimension parameter d which can be conveniently identified with the line dislocation density ρ by

$$\rho = (4d^2)^{-1}. \tag{26}$$

The interaction energies used are either the stress free $\phi=\phi_o$, given by (6) or one of the complete stress induced energies $\phi = \phi^i$ (i=1, 2 or 3), given by (5). The stress free calculations require a value for the ratio

$$P = \mu V e^o/kT \tag{27}$$

and the stress induced calculations require P and a value for the external stress ratio

$$Q = \frac{1}{e^o}\,(\tau/\mu). \tag{28}$$

The mean concentration $\bar{c}'$ over the cell is then obtained by quadrature and the required sink strength

$$k^2 \equiv Z\rho, \tag{29}$$

where Z will be referred to as the bias parameter, then follows by equating the generation rate K to the loss rate to such a dislocation sink in a lossy continuum:

$$K = Dk^2\bar{c} = DZ\rho\bar{c} \tag{30}$$

The final expression for Z, in terms of $\bar{c}'$, then follows from (25), (26) and (30):

$$Z = (2d')^2/\bar{c}' \tag{31}$$

where $d' = d/b$. In the stress free situation (equation (6)) $Z \to Z^o$ and in the three orientations with the stress present (equation (5)) $Z \to Z^1$, Z^2 or Z^3 with the subscripts i or v added if needed; Z_v^o thus refers to the bias parameter with no interaction field present ($\phi \equiv 0$ in (21)) and may be obtained by the same numerical procedure as used to get Z_i^o. The vacancy bias parameters with a stress present are, where required, taken identical to Z_v^o. Various results for the straight dislocation are presented in section 4.

3.2 The Edge Dislocation Loop

The cell adopted for the loop is the finite circular cylinder, shown in figure 2, of length 2S and radius R. The loop lies, with its centre at the coordinate origin, in the $(x_1 x_2)$ or (r,θ) plane and the axial symmetry of the relevant interaction potentials, ψ_o, ψ^3 and ψ^1 ensures the two dimensional (r,z) nature of the required interstitial concentration. The required steady state concentration c again satisfies an equation of the form (21) but, where now, the two dimensional differential operators refer to the cylindrical (r,z) coordinates, defined in the cross-section of the cell in figure 2b, and with $\phi(x_1,x_3)$ replaced by $\psi(r,z)$. At the cell boundaries, the zero flow conditions, equivalent to (22) and (23) are

$$\frac{\partial c}{\partial r} + \frac{c}{kT}\frac{\partial \psi}{\partial r} = 0 \quad \text{at } r = R \tag{32}$$

and

$$\frac{\partial c}{\partial z} + \frac{c}{kT}\frac{\partial \psi}{\partial z} = 0 \quad \text{at } z = \pm S \tag{33}$$

Although a periodic array of such circular cylindrical cells cannot fill space, we minimize any error in geometrical space filling by adjusting the cell volume $2\pi R^2 S$ to be consistent with the specified dislocation loop density N, such that

$$N = (2\pi R^2 S)^{-1}. \tag{34}$$

In addition we always set S = R so that the upper quartile of the (r,z) region $0 \leqslant r \leqslant R$, $0 \leqslant z \leqslant S$ is square; the symmetry of the variables is such that this region is primitive with axial symmetry along r = 0 and reflection symmetry across z = 0. The boundary conditions (32) and (33) again explicitly include the interaction field ψ to preclude any spurious leakage at the boundaries. The error in using this infinite body ψ field is small unless the radius of the loop a exceeds about R/2 and for this reason we set this as an upper bound on a in a given cell size.

The final boundary condition is

$$c = 0 \quad \text{at} \quad r = a \pm b,\ z = \pm b \tag{35}$$

that is on the square sectioned toroidal core as indicated in figure 2b. The partial differential equation (21) is again solved by a discretization procedure using a uniform mesh of size b over the upper quadrant of the (r,z) cell domain in figure 2b. As for the straight dislocation it is convenient to replace c by the scaled concentration c' defined by (25) and the equation has been solved accurately for a wide range of cell dimensions defined by the parameter R which can then be identified with the dislocation loop volume concentration N by (34), with S=R:

$$N = (2\pi R^3)^{-1}. \tag{36}$$

For each such cell size R the loop radius has been varied up to a = R/2 which enables the total line dislocation density ρ, in the form of loops, to be written

$$\rho = 2\pi a N = a/R^3. \tag{37}$$

Such calculations have been made for either the stress free situation $\psi = \psi_o$, given by (13) or for the situation when the uniaxial stress τ is present and the stress induced interaction energies ψ^3 or ψ^1, given by (15) and (19) respectively, are relevant. Values for the parameters P and Q given by (27) and (28) respectively are again required to specify these interaction energies. The relatively complex nature of ψ^3 and ψ^1 compared to ϕ^1 for the straight dislocation means great care has to be taken in the evaluations of the various derivatives of these functions required in (21). To guarantee the correctness of our results we first obtained $\nabla\phi$, $\nabla^2\phi$ and $\nabla\psi$, $\nabla^2\psi$ analytically, the latter pair entirely in terms of the complete elliptic integrals E, K and then checked the validity of these results by numerically differentiating ϕ and ψ on a very much finer grid than that used for solving (21). Having established the accuracy of our analytic expressions for the required derivatives, these could then be speedily evaluated at the nodal points of our discretization mesh. In the absence of the external stress the relevant interaction energies ϕ_o and ψ_o are, of course, harmonic functions and so the Laplacian term in (21) is not present and only the first derivatives of ϕ_o and ψ_o are required. However, unfortunately, the stress induced interaction energies are not harmonic functions and then the Laplacian in (21) is required. The required bias parameter Z for the loop can then be deduced from

$$Z = K/D\rho\bar{c} \tag{38}$$

which, from (25) and (37), can be expressed in terms of the mean scaled concentration $\bar{c}'$, R and a:

$$Z = R'^3/a'\bar{c}' \tag{39}$$

where $R' = R/b$ and $a' = a/b$. For the stress free situation (equation (13)) $Z \to Z^o$ and in the two orientations with the stress τ present (equations (15) and (19) respectively), $Z \to Z^3$ and Z^1 with the subscripts i or v added if needed. Z^o_v refers to the situation when $\psi \equiv 0$ and is taken as the bias parameter for the interaction of vacancies with the loops.

Various results for the dislocation loop are presented in section 5.

4. Results for Straight Edge Dislocations

The field equation (21) has been solved numerically for a range of values of the parameters d', P and Q defined in section 3.1. Such solutions are displayed in figure 8 for the stress free situation (Q = 0) and d' = 30. The integers n on the isoconcentration contours refer to values of $c'(x_1,x_3) = 100n$ where c' is related to c, the actual fractional concentration, by (25). Figure 8a is the zero field solution $P \equiv 0$, 8b and 8c are obtained for P = 60 and 120 respectively. The corresponding mean scaled concentrations $\bar{c}'$ and the resulting bias parameters are presented in table 1; two additional P values are also included in the table. The contours in figure 8 show a marked sensitivity to the presence of the interaction energy ϕ. When $\phi = 0 (P \equiv 0)$ the equiconcentration contours are, as expected, concentric circles away from the immediate vicinity of the square core boundary and away from the outer rectilinear boundary. The effect of increasing ϕ is to increase the size of a partially denuded zone displaced to the compressive side of the edge dislocation. This is precisely analogous to the completely denuded zone that appears in the pure drift solution and which in that case is a genuine characteristic [12]. In the absence of the field ϕ, the equation (21) can be solved analyticaly over a cylindrical cell of size R and core radius r = b with the zero radial flow boundary condition corresponding to (22) and (23) at R and the ideal sink condition (24) at b. This yields the expression for the bias parameter Z:

$$Z = 2\pi/[\ell n\, R' - \frac{3}{4} + R'^{-2} - R'^{-4}/4] \tag{40}$$

where $R' = R/b$. If we identify R' with $2d'/\sqrt{\pi}$ to ensure the cylindrical cell has a cross-sectional area equal to that of the rectilinear cell we find from (38), for d' = 30, Z = 2.27 which is reasonably close to the corresponding computed zero field result (2.47); the remaining discrepancy arises because simple area scaling cannot completely allow for the difference in shape of the two cells. The rectilinear results (table 1) are to be preferred since the cells do actually fill space and a square shaped core is somewhat more physical than the circular shape. Finally we refer to the alternative method for calculating sink strengths in which the source term K in (21) is removed and the zero flow condition on the cell boundary is replaced by a constant concentration. Although we have misgivings about the physical sense of this method it is briefly described in the appendix for the general case when the straight dislocation is under external stress; the existence of an analytic solution is the main justification for including the method. In relation to the present zero stress situation we can extract the corresponding bias parameter for the constant concentration method from the result (A8) in the appendix. We find

$$Z = 2\pi/\ell n\, R' \tag{41}$$

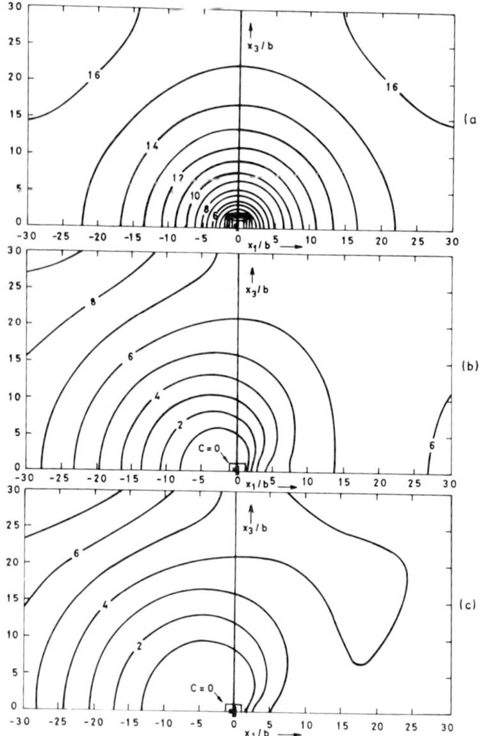

FIG. 8

The equiconcentration contours representing the solutions of (21) for the straight edge dislocation along the x_2-axis. The cell size is defined by d' = d/b = 30, the dislocation is stress free and the parameter P takes the value: P=0 (8a), P=60 (8b), P=120 (8c). The integers n on the contours refers to c' = 100n where c' is the scaled concentration defined by (25). Results for these and additional values of P are summarized in table 1 which also presents the various mean concentrations and the corresponding bias parameters.

which clearly agrees with (40) as R' becomes large. At d' = 30 (41) yields Z = 1.78 which is significantly at variance with (40) or the accurate result in table 1 and confirms the care with which results using the constant outer boundary condition must be viewed.

When the tensile stress is present ($Q \neq 0$) the various Z^i values can be used to deduce a stress induced preferred absorption (SIPA) irradiation creep rate. Such a rate can then be compared with relevant experimental data and with previous and less rigorous derivations. Thus, if we consider the only significant point defect sinks in the microstructure to be long straight edge dislocations with a total line density of ρ then we can write

$$\rho^1 = \rho^2 = \rho^3 = \frac{1}{3}\rho \tag{42}$$

where the superscripts identify the three orientations of the edge dislocations relative to the tensile stress axis. When the temperature is low enough to preclude vacancy emission from the dislocations (with a consequent thermal creep contribution) the simple rate theory analysis of SIPA [13] yields the creep rate $d\varepsilon/dt$ in the x_3 direction:

$$\frac{d\varepsilon}{d\Delta} = [Z_i^3 - \frac{1}{2}(Z_i^1+Z_i^2)]/[Z_i^1+Z_i^2+Z_i^3] \tag{43}$$

where Δ = Kt and the dislocation density is assumed large enough for losses by mutual recombination to be negligible. Any preferential losses of the vacancies have been neglected and the i subscripts in (43) have been retained for clarity only. The creep strain ε in (43) is a measure of the shape change and includes a component of anisotropic volume change; the strict deviatoric strain is two thirds of ε.

The Z^i bias parameters have thus been calculated for a number of cell sizes corresponding to different dislocation densities and a range of P and Q values. These straight dislocation results are summarised in table 2 and illustrated in figure 9. This figure shows the dependence of the creep rate on the shear modulus μ where we have expressed the former in the traditional form $\tau^{-1} d\varepsilon/d\Delta (\mathrm{TPa})^{-1}$; the linear dependence on τ of our computed results, implied by this form of presentation, has been thoroughly checked by varying τ over a wide range of values. The particular curves in figure 9 are for d' = 30, 20 and 10 and have been computed with the stress τ = 100 MPa. The parameter P, from (27), has the value

$$P = 1.58\mu \text{ } (\mu \text{ in GPa units})$$

which corresponds to $V=1.2\ 10^{-29}$, $e^o=1$ and $kT = 7.58\ 10^{-21}$ J($T \sim 550^oK$). The dashed curves in figure 9 refer to the analytic bias parameters derived in the appendix (equation (A21)) by the constant boundary concentration method.

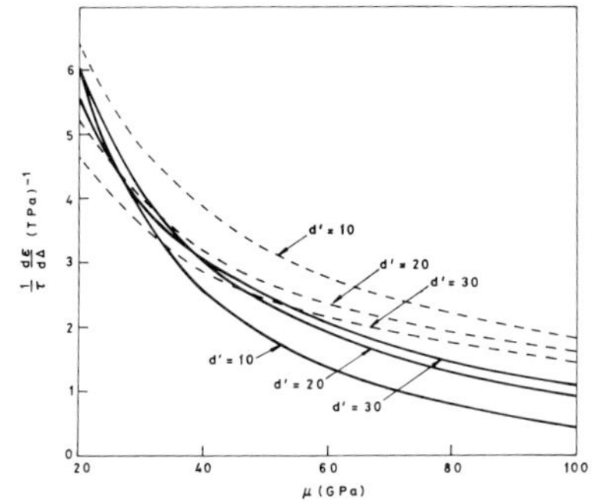

Various points can be made concerning the computed results in figure 9. The curves show relatively weak dependence on dislocation density, they indicate a creep rate inversely proportional to μ and can be quite well represented by the simple formula.

$$\frac{1}{\tau}\frac{d\varepsilon}{d\Delta} = \frac{0.12}{\mu e^o} \tag{44}$$

which has precisely the form suggested by Bullough and Hayns [13] on the basis of a much cruder calculation; the magnitude is lower than that obtained by Bullough and Hayns by a factor between 3 and 4. Nichols [21] has already drawn attention to the fact that the Bullough-Hayns estimate is probably too high and we agree with his conclusion*. We do not, however, agree that this reduction in the expected SIPA creep rate precludes SIPA from being the dominant mechanism of irradiation creep. The spread of the data and the uncertainty of the physical and irradiation parameters make such judgements based on absolute magnitudes extremely uncertain. We feel strongly that the linear form of the theoretical creep variations with τ and the predicted relative independence on dislocation density coupled with the fact that sensible values of the physical parameters defining P and Q can still be chosen to encompass the data, lend considerable support for SIPA as the most likely dominant source of such creep. We shall comment on this further when we present analogous results for the dislocation loops under stress in the next section.

FIG. 9

FIG. 9

The variation of irradiation creep rate with shear modulus for the straight edge dislocation network: —— obtained using the cellular method for various cell sizes d' = d/b; ---- derived from the analytic solution using the constant concentration method described in the Appendix (equation (A21)). The range of P consequent upon varying μ is given in table 2 together with the explicit bias parameter values.

*This does not, however, mean we fully understand his calculation [21] upon which he based the criticism of our previous work [13].

5. Results for Edge Dislocation Loops

The field equation (21) for the concentration within the cylindrical cell has been solved numerically for a wide range of values of the parameters P, Q, defined by equations (27) and (28), the cell size $R' = R/b$ and loop radius $a' = a/b$. The stress free results (Q = 0) for the dislocation loop morphology can be used to answer two somewhat controversial questions: (1) how does the relative bias parameter ($Z_i - Z_v$) vary with loop radius and (2) does a vacancy edge dislocation loop have the same bias for interstitials as the similar interstitial loop. Examples of concentration contours around such loops are shown in figures 10, 11 and 12 for a cell size $R' = 20$ and loop radius $a' = 6$. The integers n on the contours in these three figures refer to values of $c'(r,z) = 20n$ where c' is related to the actual fractional

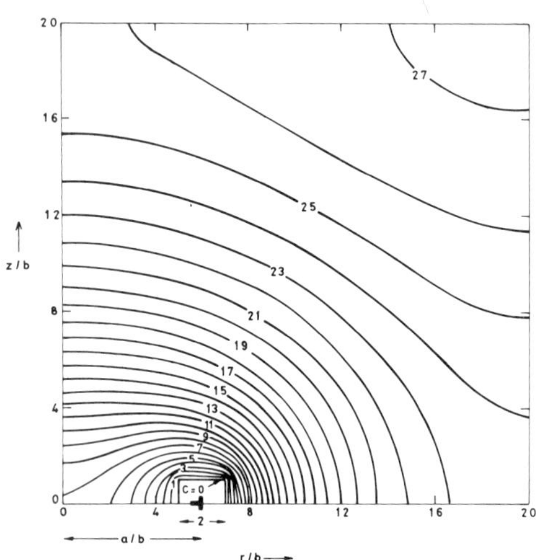

FIG 10

Equi-concentration contours around an interstitial loop of radius 6b with zero interaction energy ψ_o. The numbers n = 1...27 on the contours define the scaled concentration values $c' = 20n$. The mean "concentration" in the cell is $\bar{c}' = 471.086$ and the consequent bias parameter $Z_v = 2.8303$ from (39).

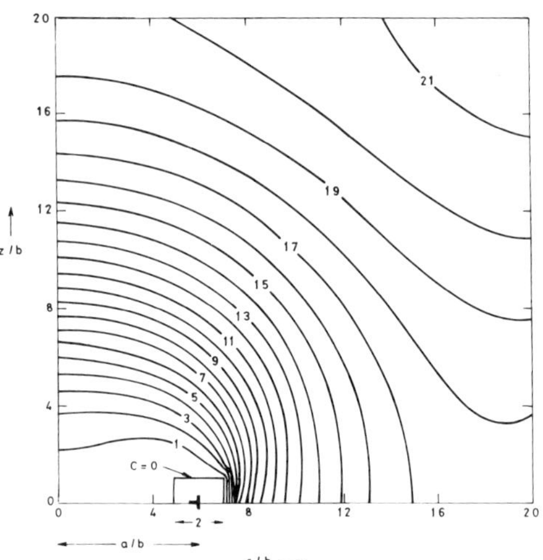

FIG 11

Equi-concentration contours around an interstitial loop of radius 6b with an interaction energy ψ_o (equation 13) defined by P = 20 present. The numbers n = 1...21 on the contours define the scaled concentration values $c' = 20n$. The mean "concentration" in the cell is $\bar{c}' = 351.0024$ and the consequent bias parameter $Z_i = 3.7986$ from (39).

concentration c by (25). Figure 10 is for zero interaction energy (P=0) and can be regarded as relevant for estimating the vacancy bias parameter Z_v. From the average concentration $\bar{c}' = 471.086$ we obtain from (39), $Z_v = 2.8303$, for this loop size and loop density. The contours in figure 11 correspond to an interaction energy of strength defined by P = 20 for the same loop and cell size as in figure 10. In figure 12 the sign of the interaction energy has been reversed (P = -20) and thus corresponds to a vacancy rather than interstitial loop interaction with interstitials. Although the concentration contours look very different in figures 11 and 12, the mean concentrations and hence bias parameters are not very different being $\bar{c}' = 351.0024$, $Z_i = 3.7986$ and $\bar{c}' = 345.328$, $Z_i = 3.8611$ respectively. This small discrepancy between the bias parameters of the same sized vacancy and interstitial loop has arisen because of the finiteness of the cell and it can easily be proved that the bias parameters are identical

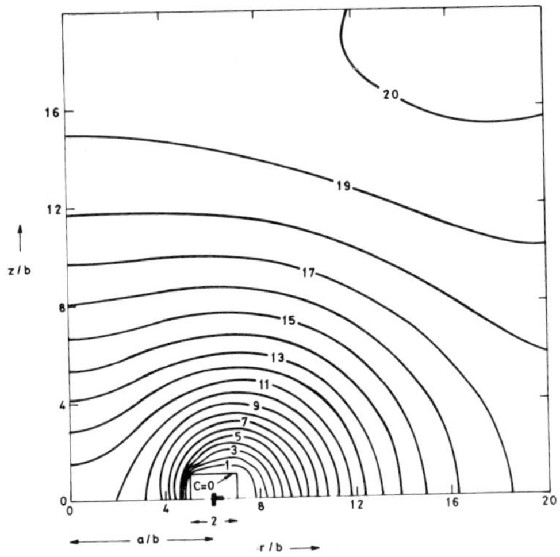

FIG 12

Equi-concentration contours around a vacancy loop of radius 6b with an interaction energy ψ_o (equation (13)) defined by P = -20 present. The numbers n = 1...20 on the contours define the scaled concentration values c' = 20n. The mean "concentration" in the cell is $\bar{c}'$ = 345.328 and the consequent bias parameter Z_i = 3.8611 from (39).

when loop-loop interactive effects are negligible, that is, when the interaction energy is negligible on the cell boundary (see the appendix); thus in the present study we find the discrepancy becomes negligible as the loop radius is decreased for a given cell size.

The computed variations of the bias parameters Z_i and Z_v and (Z_i-Z_v) with loop radius for the cell size R' = 20 are shown in figure 13 and tabulated in table 3. Although the variations of the individual bias parameters are not monotonic functions of a the relative bias does show a definite monotonic increase as a increases. Note, to avoid spurious boundary effects we never permit the loop radius to exceed half the cell dimension R/2. Precisely analogous results are shown in figures 14 and 15 for R' = 40 and 60 respectively. All the results in figures 13-15 have been obtained with the relatively low interaction energy defined by P = 20. For interest and final confirmation of the important conclusion that the relative bias of such loops always increases with the loop radius, we show in figure 16 the corresponding results for the stronger interaction energy (P = 100) for R' = 40. The Z_v curve in figure 16 has been taken for P = 20 in case it may be argued that the vacancy has a relatively weak finite long range size effect interaction with the loop. In all other cases Z_v has been deduced with P = 0.

When a tensile stress τ is present (Q ≠ 0) a Z^i can be deduced for the stress axis orthogonal to the loop plane (i = 3) and when the stress axis lies in the plane (i = 1), as indicated in figure 3c. Corresponding to the straight edge dislocation result (43) an irradiation creep rate dε/dt can be expressed in terms of these two Z^i values when only such edge dislocation loops are present and bulk recombination is negligible:

$$\frac{d\varepsilon}{d\Delta} = (Z_i^3 - Z_i^1)/(Z_i^3 + 2Z_i^1) \tag{45}$$

The Z^3 and Z^1 bias parameters* have thus been calculated for a number of cell sizes and a range of P and Q values. It was first checked that both Z^3 and Z^1 are precisely linear functions of the stress τ for a given value of P and μ. Such a variation is shown in figure 17 for P = 60, μ = 60 GPa, a = 8b and a cell size R = 20b. With this linear dependence on stress confirmed we can then usefully present the computed variation of the creep rate from (45), in the form τ^{-1} dε/dΔ, with the shear modulus μ. These results are presented in figures 18, 19 and 20 for the respective cell sizes R' = 20, 40 and 60; the curves in each figure are for a number of loop radii a' ranging from 2 to 29 but not exceeding half the respective cell radius. The immediate feature notable in these figures is their qualitative similarity to the curves for the straight dislocations in figure 9. They have been obtained with the same strength of the interaction energy as used for the straight dislocations; that is

$$P = 1.58\mu \quad (\mu \text{ in GPa units}).$$

*The i subscript is dropped unless we wish to distinguish the vacancy. Vacancy absorption contributions to irradiation creep are neglected.

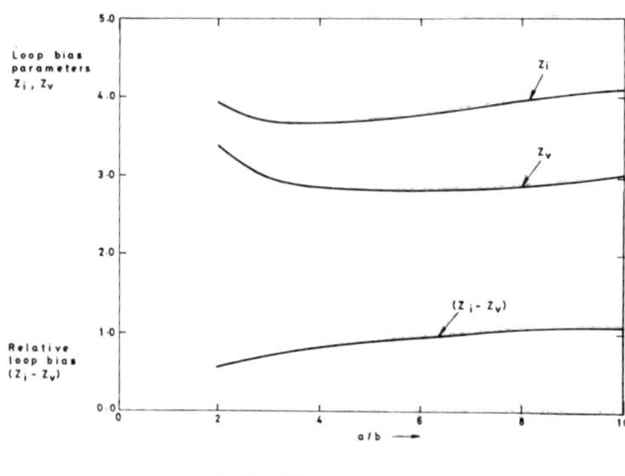

FIG 13

The variation of the bias parameters and the relative bias for dislocation loops, with loop radius. The interaction energy ψ_o for Z_i is defined by P = 20 with zero for Z_v. The cell size is R' = 20.

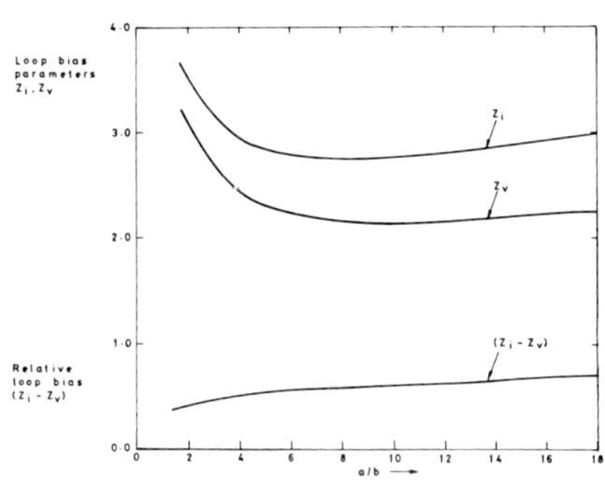

FIG 14

As for figure 13 with the larger cell size R' = 40.

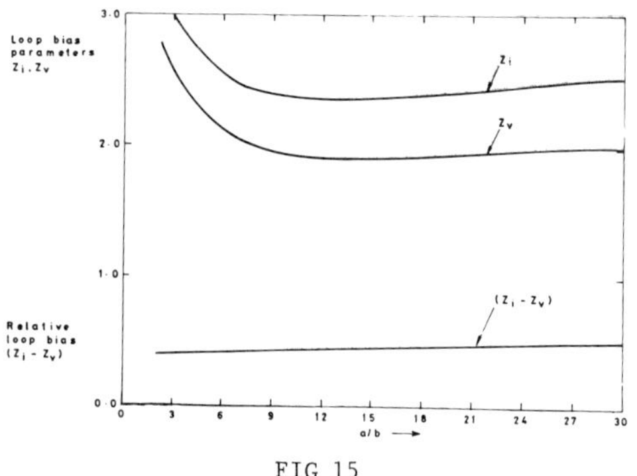

FIG 15

As for figure 13 with the larger cell size R' = 60

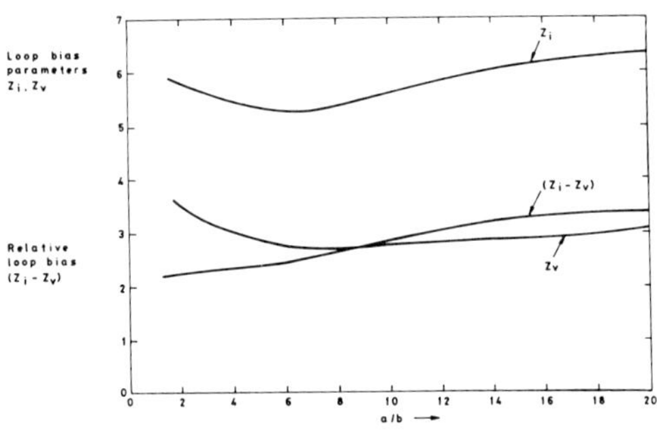

FIG 16

The variation of the bias parameters and relative bias for dislocation loops, with loop radius. The interaction energy ψ_o for Z_i is defined by P = 100 and for Z_v by P = 20. The cell size is R' = 40.

In fact the predicted creep rates for the loops are somewhat greater than those deduced from the straight dislocation model and help to substantiate the significance of the SIPA mechanism of irradiation creep. Of further note in these figures is the decreasing variability of the creep rate with loop size as the loop density decreases (cell size increases). The spread of the curves in figure 20 (R' = 60: N = 3 x 10^{22} m^{-3} with b $\sim$ 0.3 nm) is really quite minimal.

6. Discussion

The single sink cellular method has been used in the present work to obtain numerical estimates of the bias parameters for straight and circular edge dislocations. The complete stress induced interaction energies for both these dislocation morphologies have been calculated and used in the sink strength calculations. These interaction energies are, however, appropriate for isolated dislocations in an infinite body and we are using them in a finite cell. When

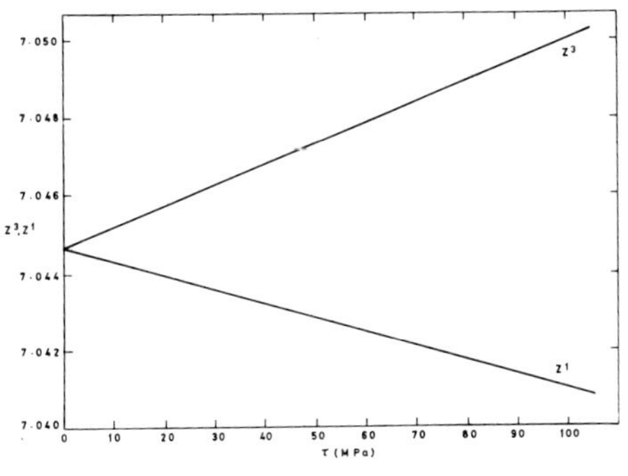

FIG 17

Illustrating the precise linearity of the bias parameters Z^3, Z^1 with the applied stress τ for a loop radius of $a' = 8$ and cell size $R' = 20$. ψ^3 and $\bar{\psi}^1$ are defined by $P = 60$ and $\mu = 60$ GPa.

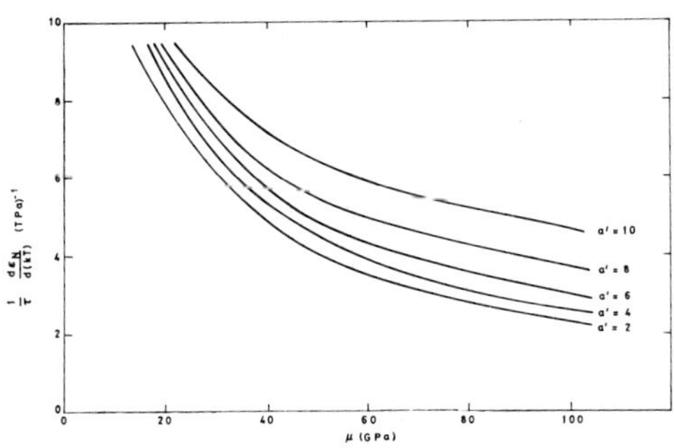

FIG 18

The variation of irradiation creep rate with shear modulus for a range of edge dislocation loop sizes. The interaction energy parameter $P = 1.58\mu$ (μ in GPa units), the curves have been obtained from (45) with $\tau = 100$ MPa. The cell size is $R' = 20$.

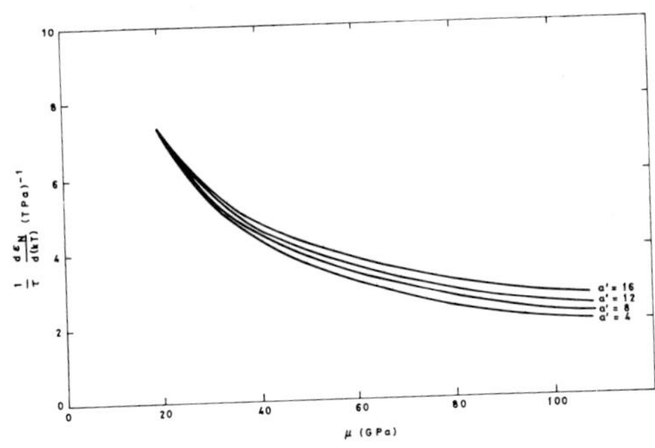

FIG 19

As for figure 18 with the larger cell size $R' = 40$

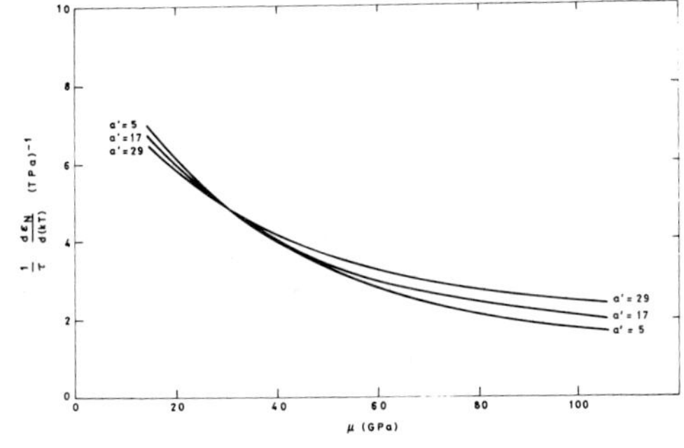

FIG 20

As for figure 18 with the larger cell size $R' = 60$

the complete angular form of these energies is included, as in the present work, it is clearly very difficult to allow sensibly for the vanishing of these interactions at the cell boundary. If one replaces the interaction energy ψ_o for the unstressed straight dislocation by a simple radial function, as first suggested by Bullough and Newman [12] in their studies of strain ageing, then it is possible to adjust analytically such a radial function to have zero gradient at the cell boundary and such a procedure has been explored by Heald and Miller [9] in the sink strength context. However, the error involved in such a procedure is uncertain and the lack of importance of the angular form of the interaction energy when motion is by pure drift alone with no diffusion flow [12], may not be a reliable guide. In fact, perhaps the most encouraging feature of the present work is the beautiful consistency between the dislocation loop results and the straight dislocation results. In particular the closeness of the predicted creep rates shown in figures 9 and 20 helps to substantiate the use of the infinite body interaction energies for the straight dislocation; we know the situation for the loop is much better since the stress field of a loop falls off with distance from the loop much faster than the corresponding stress field of the straight dislocation. Furthermore, the consistency is not just on the dominant size effect interaction but survives the relatively small induced perturbation interactions arising from the external stress and whose explicit presence is exposed by calculating the SIPA creep rates.

The stress free bias parameter for the loop was found to be an increasing function of the loop radius. This result is at variance with several published calculations [22-24] and we can only assume that their use of an effective capture radius in place of the actual dislocation core has somehow buried the essential result. In fact we believe such an increasing relative bias is obviously correct. Thus, let us consider the interaction free Z_v variation with the loop radius, such as indicated in figure 13. Then at a given distance from a loop the dilatation stress field must fall as the loop radius decreases; now apart from the 'geometrical' content of Z_i, which is present in Z_v, the excess of Z_i over Z_v for a certain loop radius is due to the presence of that stress field. When the loop radius decreases Z_i <u>must</u> move closer to Z_v since the stress field is decreasing (due to the cancellation of the dislocation stress field across the loop - it eventually becomes a single interstitial!). It follows that (Z_i-Z_v), the relative bias parameter, must decrease with decreasing loop radius in agreement with our calculations. In addition, of course, we believe, this result is essential to explain the observed <u>growth</u> of small vacancy loops in the presence of network dislocations or larger interstitial loops in certain metals such as zirconium [25] and molybdenum [26] when genuine neutral sinks such as voids are not present. This relative bias variation with loop size, due to the interaction energy, is complementary to the analogous variation obtained by Bullough, Hayns and Woo [10] arising from the interactions between the loops and existing network dislocations. Fortunately, both effects indicate an increasing relative bias with increasing loop radius.

As far as we are aware our analytic expression for the stress induced interaction energy for an edge dislocation loop has not been previously derived. It is essential to obtain this energy accurately particularly since it is not a harmonic function and thus the Laplacian term in (21) must be accurately computed. The maintenance of such accuracy is greatly facilitated by evaluating the interaction terms at the mesh nodes from the analytic expressions.

Although the present calculations indicate a SIPA creep rate that is a factor of about three below the previous estimate by Bullough and Hayns [13] we nevertheless consider that the sensible form of the predicted creep rate and the relative uncertainty of the physical and irradiation parameters together enable even the present reduced estimate to encompass much of the available irradiation creep data. We therefore remain convinced that the SIPA mechanism makes an important contribution to the irradiation creep process.

References

1. A.D. Brailsford and R. Bullough, J. Nucl. Mat. 44, 121 (1972).
2. S.D. Harkness and Che-Yu Li, Metall. Trans. 2, 1457 (1971).
3. A.D. Brailsford and R. Bullough, Vacancies 76, Proc. of Conf. on 'Point Defect Behaviour and Diffusional Processes', Univ. Bristol, 13-16 Sept. 1976. (Edit. by R.E. Smallman and and J.E. Harris: Metals Soc.) p.108.
4. R. Bullough and M.R. Hayns, Proc. 2nd Int. Symp. Cont. Models of Discrete Systems, Mont Gabriel, Canada, June 1977.
5. P.T. Heald and M.V. Speight, Acta. Met. 23, 1389 (1975).
6. A.D. Brailsford, R. Bullough and M.R.Hayns, J. Nucl. Mat. 60, 246 (1976).
7. A.D. Brailsford and R. Bullough, to be published.
8. R. Bullough, M.R. Hayns and M.H. Wood, Proc. Int. Conf. on 'Fundamental Mechanisms of Irradiation Creep and Growth', Chalk River, Canada 8-10 May 1979, to be published in J. Nucl. Mat.
9. P.T. Heald and K.M. Miller, CEGB Report RD/B/N3798, (1976).
10. R. Bullough, M.R. Hayns and C.H. Woo, J. Nucl. Mat. 84, 93 (1979).
11. R. Bullough and J.R. Willis, Phil. Mag. 31, 855 (1975).
12. R. Bullough and R.C. Newman, Rep. on Prog. in Physics 33, 101 (1970)
13. R. Bullough and M.R. Hayns, J. Nucl. Mat. 57, 348 (1975).
14. J.P. Hirth and J. Lothe 'Theory of Dislocations' McGraw-Hill (1968).
15. P.T. Heald and M.V. Speight, Phil. Mag. 29, 1075 (1974).
16. T. Mura, Phil. Mag. 8, 843 (1963).
17. R. Bullough, B.L. Eyre and R.C. Perrin, Nuc. Appl. Tech. 9, 346 (1970).

18. A.D. Brailsford, J.R. Matthews and R. Bullough, J. Nucl. Mat. 79, 1 (1979).
19. M.R. Hayns, J. Nucl. Mat. 79, 323 (1979).
20. R. Bullough and M.H. Wood, to be published (Harwell Res. Rep. TP.820, 1979).
21. F.A. Nichols, J. Nucl. Mat. 84, 207 (1979).
22. W.G. Wolfer, Proc. of Int. Conf. on 'Fundamental Aspects of Radiation Damage in Metals', Vol II, p.812, Gatlinburg, USA Oct. 6-10, 1975. CONF-751006-P2.
23. M.H. Yoo and W.H. Butler, Phys. Stat. Sol. (b) 77 (1) 181 (1976).
24. W.A. Coghlan, in "Computer Simulation for Materials Applications", p.166, NBS, Gaithersburg, Md., USA 1976 (see also review by L.K. Mansur, ORNL Workshop, June 1976, p.19).
25. D.O. Northwood, Atomic Energy Review 15(4), 548 (1977).
26. J.H. Evans, Nature 278, 728 (1979).
27. I.G. Margvelashvili and Z.K. Saralidze Fiz. Iverd. Telu. 15, 2665 (1973). (English translation: Sov. Phys: Sol. State 15, 1774 (1974)).

Appendix

The Constant Boundary Concentration Method: Analytic Stress Induced Bias Parameters for the Straight Dislocation

In this method we replace the spatially homogeneous source term K in (21) by the constant concentration boundary condition $c = c(R)$ on the boundary of a cylindrical region, radius R, surrounding the dislocation. The field equation for $c(r,\theta)$ within the cylindrical region of radius R is thus

$$\nabla^2 c + \nabla c . \nabla(\chi) + c\nabla^2(\chi) = 0 \tag{A1}$$

with the boundary conditions

$$c = 0 \quad \text{at} \quad r = b \tag{A2}$$

$$c = c(R) \quad \text{at} \quad r = R \tag{A3}$$

In (A1) χ is identified with the interaction energies ϕ^i (equation (5)) for the three orientations of the external tensile stress τ by

$$\chi^i = \phi^i/kT. \tag{A4}$$

It is convenient to write χ^i in the form, from (5) and (A4);

$$\chi^i = \chi_o[1 + \varepsilon^i \cos^2 \theta + \delta^i \sin^2 \theta] \tag{A5}$$

where

$$\chi_o = - L \frac{\cos \theta}{r} , \tag{A6}$$

$L = Pb/\pi$ (P is given by equation (27)), and

$$\begin{aligned} \varepsilon^1 &= \frac{5}{64} \left(\frac{\tau}{\mu}\right) \frac{1}{e^o} , \quad \delta^1 = - \frac{5}{64} \left(\frac{\tau}{\mu}\right) \frac{17}{e^o} \\ \varepsilon^2 &= \delta^2 = - \frac{5}{32} \left(\frac{\tau}{\mu}\right) \frac{1}{e^o} \\ \varepsilon^3 &= \frac{5}{64} \left(\frac{\tau}{\mu}\right) \frac{1}{e^o} , \quad \delta^3 = \frac{5}{64} \left(\frac{\tau}{\mu}\right) \frac{19}{e^o} \end{aligned} \tag{A7}$$

When R is large, the sink strength of the dislocation is equal to the flux into the dislocation from this source at $r = R$ or, equivalently, the flux across the outer boundary $r = R$. The required three bias parameters Z^i corresponding to the three interaction energy functions χ^i are given by

$$Z^i = \frac{R}{c(R)} \int_o^{2\pi} \frac{\partial c^i(R,\theta)}{\partial r} \, d\theta \tag{A8}$$

where the c^i ($i=1$, 2 or 3) are the corresponding solutions of (A1) with χ replaced by χ^i. It is obviously convenient to replace c^i by the normalised concentration $\bar{c}^i = c^i/c(R)$ and this is treated as the new variable with obvious changes to (A1) and (A3).

To solve (A1) for $\bar{c}^i$ we follow Margvelashvili and Saralidze[27] by introducing the transformation

$$\bar{c}^i(r,\theta) = g^i(r,\theta)e^{-\frac{1}{2}\chi^i(r,\theta)} \tag{A9}$$

in terms of which the problem becomes

$$\nabla^2 g^i - \frac{1}{4}(\nabla\chi^i)^2 g^i + \frac{1}{2}(\nabla^2\chi^i)g^i = 0 \tag{A10}$$

with

$$g^i(r_o,\theta) = 0 \tag{A11}$$

$$g^i(R,\theta) = e^{\frac{1}{2}\chi^i(R,\theta)} \tag{A12}$$

In fact, if R is large, $\chi^i(R,\theta) \sim 0$ and the outer boundary condition (A12) is replaced by

$$g^i(R,\theta) = 1. \tag{A13}$$

Now for the potential (A5), if only linear terms in ε^i and δ^i are retained, equation (A10) becomes

$$\nabla^2 g^i - \frac{L^2}{4r^4}\left\{1 + \frac{3\varepsilon^i+\delta^i}{2} + \left(\frac{\varepsilon^i-\delta^i}{2}\right)[2\cos 4\theta - \cos 2\theta]\right\} g^i + \frac{L(\varepsilon^i-\delta^i)}{4r^3}\cos 3\theta\, g^i = 0. \tag{A14}$$

With the variable change

$$r = 1/x \tag{A15}$$

equation (A14) gives

$$\frac{\partial^2 g^i}{\partial x^2} + \frac{1}{x}\frac{\partial g^i}{\partial x} + \frac{1}{x^2}\frac{\partial^2 g^i}{\partial\theta^2} - \frac{L^2}{4}\left\{1 + \frac{3\varepsilon^i+\delta^i}{2} + \left(\frac{\varepsilon^i-\delta^i}{2}\right)[2\cos 4\theta - \cos 2\theta]\right\} g^i + \frac{L(\varepsilon^i-\delta^i)}{4x}\cos 3\theta\, g^i = 0 \tag{A16}$$

which may be solved by iteration. The zero order radial solution, obtained by disregarding terms of order ε^i, δ^i, is

$$g_o^i(r) = \frac{K_o(\frac{L}{2r})I_o(\frac{L}{2b}) - K_o(\frac{L}{2b})I_o(\frac{L}{2r})}{K_o(\frac{L}{2R})I_o(\frac{L}{2b}) - K_o(\frac{L}{2b})I_o(\frac{L}{2R})} \tag{A17}$$

Since $L/b >> 1$, (A17) can be simplified to

$$g_o^i(r) = K_o(\frac{L}{2r})/K_o(\frac{L}{2R}) \tag{A18}$$

Now we notice from (A16) that the complete radial solution $g^i(r)$ of (A16) to 1st order in ε^i, δ^i is precisely (A18) with L replaced by L^i; that is

$$g^i(r) = K_o(\frac{L^i}{2r})/K_o(\frac{L^i}{2R}) \tag{A19}$$

where

$$L^i = L[1 + \frac{3\varepsilon^i+\delta^i}{4}] \tag{A20}$$

Thus, from (A8), (A9) and (A19) the required bias parameters are

$$Z^i = \frac{\pi L^i}{R} K_1\left(\frac{L^i}{2R}\right)/K_o\left(\frac{L^i}{2R}\right). \tag{A21}$$

The angular terms in $\bar{c}^i$ do not contribute to the integral in (A8) and χ^i is set to zero on $r = R$ in (A9). The Z^i values given by (A21) have been used to calculate the dashed curves for the creep rate variation with shear modulus in figure 9. If $L/R < 1$ (which, incidentally, is not always true with the interaction energy parameters used in the present work), the Bessel functions in (A21) can be replaced by their small argument expansions and then the resulting Z^i can be expanded again to 1st order in ε^i, δ^i to yield the explicit result:

$$Z^i \sim \frac{2\pi}{\ell n(4R/L)-\gamma}\left\{1 + \frac{(3\varepsilon^i+\delta^i)}{4(\ell n(4R/L)-\gamma}\right\} \tag{A22}$$

where $\gamma = .5772..$ is Euler's constant. When ε^i and δ^i are zero (A22) is close in value to the corresponding stress free result given by Heald and Speight [5].

On the equality of the Bias Parameters for Vacancy and Interstitial Loops

The analogous constant concentration method for the stress free dislocation loop can be easily adopted to demonstrate the invariance of the bias parameters to the sign of the Burgers vector of the loop, from which it follows that a vacancy or interstitial edge dislocation loop (of the same size and strength) should have equal bias parameters. Thus consider a single loop in a cell V, at whose boundary ∂V_c the concentration of interstitials is maintained at a fixed value c_o. The cell V is bounded internally by the surface ∂V_L of the dislocation core at which the ideal sink boundary condition $c = 0$ is assumed. The problem is thus defined by the equations analogous to (A1) to (A3):

$$\nabla^2 c + \nabla c.\nabla\chi = 0 \qquad x \in V \tag{A23}$$

$$c = 0 \qquad x \in \partial V_L \tag{A24}$$

$$c = c_o \qquad x \in \partial V_c \tag{A25}$$

where χ is (1/kT) times the energy of interaction between an interstitial and the stress field of the dislocation. For the circular edge dislocation loop, for example, $\chi = \psi_o/kT$ where ψ_o is given by (13). For an isotropic medium, χ is harmonic being proportional to the dilatation field of the dislocation; this has already been assumed in equation (A23).

We now make the general transformation of Margvelashvili and Saralidze, as in (A9) for the straight dislocation;

$$c = g\, e^{-\chi/2}. \tag{A26}$$

In terms of g, which is now an arbitrary three dimensional function, equations (A23), (A24) and (A25) become

$$\nabla^2 g - \tfrac{1}{4}(\nabla\chi)^2 g = 0 \quad , \quad x \in V, \tag{A27}$$

$$g = 0 \qquad x \in \partial V_L \tag{A28}$$

$$g = c_o e^{\chi/2} \qquad x \in \partial V_c \tag{A29}$$

which are the generalizations of (A10), (A11) and (A12).

The objective is to solve (A23), (A24) and (A25) or equivalently (A27), (A28) and (A29), to find both the flux into the loop and the mean concentration $\bar{c}$. The required sink strength is then defined as their ratio. The flux density is given by

$$-D(\nabla c + c\nabla\chi) \equiv -D(\nabla g + \tfrac{1}{2}g\nabla\chi)e^{-\chi/2} \tag{A30}$$

where D is the diffusion coefficient, and the sink strength in general plainly depends upon the sign of χ. However, if V is so large that $\chi \sim 0$ near to ∂V_c, the boundary condition (A29) approximates to

$$g \sim c_o, \quad x \in \partial V_c \tag{31}$$

and g then depends upon χ only through the term $(\nabla\chi)^2$ in (A27). In this case therefore, g is independent of the sign of χ. Also if the flux term (A30) is evaluated at ∂V_c the total flux emerges as,

$$F = -D \int_{\partial V_c} \nabla g.n \, dS \tag{A32}$$

again since $\chi \sim 0$ on ∂V_c, while the dominant contribution to c comes from points remote from the dislocation, so that $\bar{c} \sim c_o$. Hence, asymptotically, in the limit of large V, the sink strength is independent of the sign of χ. Note, however, that the simple boundary condition (A28) on the dislocation core, is a crucial requirement.

The result can also be established for the single sink cellular model used in the main body of this paper. However, the presence of the homogeneous source term in equation (21), which replaces the constant concentration condition at the cell boundary, introduces rather formidable mathematical problems; for example, a simple transformation analogous to (A26) does not appear to exist. The result has, however, been established using a lengthy and somewhat complex asymptotic matching technique which would seem inappropriate to present in the present paper. Suffice it to emphasize that our present numerical results, particularly for small loops in large cells are completely consistent with the conclusion that the bias parameters for similar vacancy and interstitial loops are equal.

TABLE 1

Mean concentration $\bar{c}'$ and bias parameter Z for the stress free straight edge dislocation; see figure 8 for P = 0, 60 and 120 concentration distributions. The cell size is given by d' =d/b=30 (corresponding to a dislocation density of $\rho \sim 3\text{x}10^{15}$ m/m^3 for b $\sim$ 0.3nm)

$P=\mu Ve^o/kT$	$\bar{c}'=(D/Kb^2)\bar{c}$	$Z=(2d')^2/\bar{c}'$	Z_i-Z_v	$(Z_i-Z_v)/Z_i$
0	1452.277	2.479	-	-
30	854.988	4.2106	1.7317	0.4113
60	585.897	6.1444	3.6655	0.5965
90	458.662	7.8489	5.3700	0.6842
120	381.9652	9.4249	6.9461	0.7370

TABLE 2

Results* for the Straight Edge Dislocations Under a Tensile Stress τ = 100MPa.

μ (GPa)	P	d' = d/b	Orientation of τ parallel to x_1 axis		x_2 axis		x_3 axis	
			$\bar{c}'$	Z^1	$\bar{c}'$	Z^2	$\bar{c}'$	Z^3
20	31.6	10	50.009	7.9985	49.998	8.0004	49.914	8.0138
		20	305.11	1.3110	305.01	1.3114	304.50	1.3136
		30	854.89	1.0528	854.63	1.0531	853.32	1.0547
40	63.2	10	31.709	12.615	31.710	12.614	31.685	12.624
		20	200.85	1.9915	200.83	1.9917	200.66	1.9934
		30	580.30	1.5509	580.22	1.5511	579.74	1.5524
60	94.8	10	23.784	16.818	23.786	16.816	23.775	16.824
		20	154.46	2.5896	154.46	2.5897	154.38	2.5911
		30	452.69	1.9881	452.65	1.9883	452.40	1.9894
80	126.4	10	19.214	20.818	19.216	20.815	19.211	20.822
		20	127.07	3.1480	127.07	3.1479	127.02	3.1491
		30	376.28	2.3919	376.26	2.3919	376.11	2.3929
100	158.0	10	16.211	24.674	16.214	24.671	16.210	24.676
		20	108.67	3.6809	108.67	3.6807	108.64	3.6818
		30	324.57	2.7729	324.56	2.7730	324.46	2.7738

TABLE 3

Variation of $\bar{c}'$, Z_i and Z_v with Loop Radius a' = a/b for cell size R' = R/b = 20. Illustrated in figure 13.

a'	P = 20		P = 0		$(Z_i - Z_v)$
	$\bar{c}'$	Z_i	$\bar{c}'$	Z_v	
2	1010.65	3.958	1190.62	3.360	0.598
3	718.03	3.714	892.21	2.989	0.725
4	543.99	3.677	700.85	2.854	0.823
5	430.63	3.715	567.80	2.818	0.897
6	351.00	3.799	471.08	2.830	0.969
7	293.45	3.895	397.96	2.872	1.023
8	250.75	3.988	342.19	2.922	1.066
9	218.92	4.060	298.06	2.982	1.078
10	195.61	4.089	264.08	3.029	1.060

*The Z_i values used to construct figure 9 have been obtained to more significant figures than presented in this table.

ON THE SIMULTANEOUS FORMATION OF INTERSTITIAL - AND VACANCY-TYPE LOOPS DURING IRRADIATION

A. Si-Ahmed and W.G. Wolfer

Fusion Engineering Program, Nuclear Engineering Department
University of Wisconsin, Madison, WI 53706, U.S.A.

Abstract

During the irradiation of metals at temperatures above Stage III, dislocation loops are formed. Both interstitial - and vacancy-type loops are observed. Their simultaneous formation is due to non-linear elasticity effects on the strain fields of loops. This difference is only a necessary condition for the nucleation of vacancy-type loops. Other conditions, such as temperature, dose rate, and the average interstitial capture efficiency of the entire sink structure, must also be met. These conditions are investigated, and it is found that vacancy loops can nucleate at low tempratures and when the capture efficiency of the entire sink structure exceeds a critical value. With continuing irradiation, both interstitial - and vacancy loops grow to large radii, and the capture efficiency drops below the critical value. At this point, further vacancy-loop nucleation is terminated.

Introduction

Radiation-induced changes of the dislocation structure in metals at temperatures above Stage III are driven by the preferential absorption of interstitials at edge dislocations. This preferential capture is the result of a drift diffusion of the point defects in the stress field of dislocations. According to previous calculations (1) the drift-induced currents of point defects to both interstitial - and vacancy-type loops are identical. It has, therefore, been always assumed that vacancy-type loops could not nucleate and grow simultaneously with interstitial-type loops. Nevertheless, there are numerous observations [references cited in (2)] attesting to the simultaneous growth of both types of loops.

To explain the apparent contraction between the earlier theoretical predictions and these observations, two proposals have been advanced recently (2,3). Our explanation (2), to be discussed further in section 2, is based on the non-linear elasticity contribution to the strain field of small dislocation loops. In contrast, Bullough et al. (3) suggest that the loop bias increases with increasing loop radius. Therefore, interstitial-type loops larger than vacancy-type loops absorb an excess of interstitials, whereas the latter receive an excess of vacancies. According to our analysis, the loop bias is a decreasing function of the loop radius, but the interstitial-type loops have a larger bias than the vacancy-type loops of the same radius.

A difference in preferential absorption at both loop types is only a necessary, but not a sufficient condition for vacancy-type loop formation. Other conditions must be met also for vacancy-type loops to nucleate. It is the purpose of this paper to investigate these other conditions. They arise from the fact that a significant nucleation barrier exists for vacancy-type loops, whereas practically none exists for interstitial-type loops (4).

Capture Efficiency of Dislocation Loops

The current of point defects to a circular dislocation loop of radius R can be written as

$$J=Z(R)[4\pi^2R/\ln(8R/a)]D(C-C^{\ell}(R)) \qquad (1)$$

where Z(R) is the capture efficiency, which accounts for the drift of the point defects in the strain field of the dislocation. The remaining factors in Eq. (1) represent the point-defect current in the absecne of the drift as given by Seeger and Goesele (5). Here, a is the dislocation pipe radius, D the duffusion coefficient, C the average point defect concentration, and

$$C^{\ell}(R)=C^{eq}\exp\{\Delta W(R)/kT\} \tag{2}$$

is the point defect concentration in local thermodynamic equilibrium with the dislocation loop. C^{eq} is the corresponding concentration in the ideal crystal, and $\Delta W(R)$ is the change in the energy of the loop when one point defect is absorbed.

The capture efficiency can be represented (2) by

$$Z(R)=\ell n(8R/a)/\ell n(8R/c) \tag{3}$$

where the capture radius c is greater than the pipe radius a. Both radii were determined by matching the expression (3) to results of the infinitesimal loop (1) for small R and to the results for the straight edge dislocation (6). Two sets of values, a_i = 8 b_o, c_i = 20 b_o, and a_v = 2.7 b_o, c_v = 3.5 b_o, are thereby obtained for the capture efficiencies Z_i and Z_v for interstitials and vacancies, respectively. All these radii are proportional to the Burgers vector b_o. The latter is a measure of the magnitude of the strain field of the dislocation loop in the linear elasticity theory.

As shown recently (2), nonlinear elastic effects can be accounted for if b_o is replaced by an apparent Burgers vector b(R). Its value can be determined from the volume change δV_ℓ produced by the creation of dislocation loops. The volume of point defects contained in a loop is given by

$$V_\ell=\pm\pi R^2 b_o+\delta V_\ell=\pm\pi R^2 b(R) \quad , \tag{4}$$

where the plus (minus) sign holds for an interstitial-type (vacancy-type) loop. The volume change δV_ℓ can be computed according to the Zener expression as given by Seeger and Haasen (8), if the pressure derivatives of the elastic constants are known. Using the derivatives for nickel from the recent compilation of Guinan and Steinberg (8), the apparent Burgers vector can be computed from Eq. (4). The results are shown in Figure 1. The ratio $b(R)/b_o$ is also equal to the ratio V_ℓ/V_ℓ^o, where $V_\ell^o=\pm\pi R^2 b_o$ is the volume of atoms inserted or extracted when creating the dislocation loop. In other words, V_ℓ/V_ℓ^o is the volume per point defect contained in the loop divided by the atomic volume Ω. As seen from Fig. 1, the volume per point defect deviates from the atomic volume as the loop radius decreases. When extrapolated to one point defect, it is found that the volume relaxation due to an interstitial is equal to 1.81 Ω, and the volume relaxation due to a vacancy is -0.18 Ω. Both values are in good agreement with measured values of the relaxation volumes for interstitials and vacancies (9).

The apparent Burgers vector b(R) can now be substituted for b_o in the capture radii c_i and c_v to reflect the second-order elasticity effects on the loop strain field and on the capture efficiency ratio $Z_i(R)/Z_v(R)$. Since $b(R) > b_o$ for interstitial-type loops and $b(R) < b_o$ for vacancy-type loops, the two loops acquire different capture efficiencies. This difference decreases with increasing loop radius, as shown in Figure 2.

Vacancy Loop Nucleation

The formulation of a nucleation theory for vacancy-type loops is identical to the void nucleation theory as developed by Katz, Wiedersich (10) and Russell (11), except for the substitution of the point defect fluxes as given now by Eq. (1). Based on our previous formulation for void nucleation (12), the nucleation barrier $\Delta G(R)$ for vacancy-type loops depends on two important contributions. $\Delta G(R)$ increases as the net bias

$$Z_i(R)/Z_v(R)-\langle Z_i/Z_v\rangle$$

increases. Here, the capture efficiency ratio $\langle Z_i/Z_v\rangle$ is averaged over all the sinks in a manner defined previously (12). The second contribution to $\Delta G(R)$ arises from the difference in the thermal vacancy concentration

$$C_v^{\ell}(R)-\langle C_v\rangle$$

where $\langle C_v\rangle$ is the average vacancy concentration in thermal equilibrium with the sinks. Again, this average was defined previously (12).

In order to obtain the average vacancy and interstitial concentrations C_i and C_v the usual rate equations are solved for a given Frenkel pair production rate and a given recombination volume. The latter was computed with a recently derived expression (13) which shows excellent agreement at low temperature with measured values by Nakagawa et al. (14). At

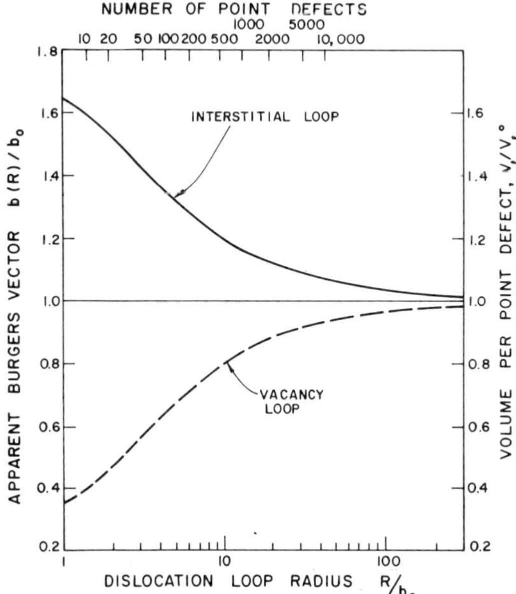

Figure 1 Apparent Burgers vector or relaxation volume per point defects contained in loops of given radius R. b_0 is the Burgers vector for the undeformed crystal.

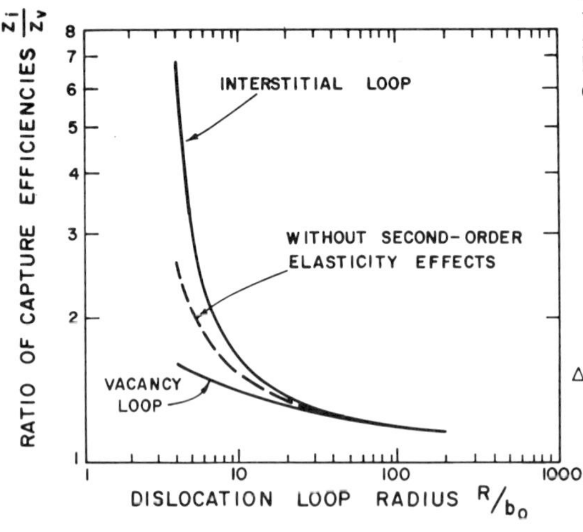

Figure 2 (above) Ratio of the capture efficiencies for interstitials, Z_i, and vacancies, Z_v, as a function of the loop radius

Figure 3 (right) Nucleation barrier for vacancy-type loops. Dose rate is typical for ion-bombardment experiments.

temperatures of 200 and 400°C, the recombination volumes are 34Ω and 27Ω, respectively. It was further assumed that 10% of all displacement events contribute to freely migrating point defects.

Using the above parameters as well as those of Ref. 12, $\Delta G(R)$ can be computed. Typical results are shown in Fig. 3 for the nucleation barrier of vacancy-type loops. The inserted table gives also the steady-state nucleation rates. It is seen that the nucleation barrier increases with decreasing average capture ratio $\langle Z_i/Z_v \rangle$ with a concomitant sharp drop in the nucleation rate. This is most dramatically demonstrated by the results of Figure 4 which shows that vacancy-type loops can only nucleate in a material whose sinks are highly biased for preferential interstitial absorption. This condition is only met when small dislocation loops constitute the dominant sink. As no nucleation barrier exists for interstitial-type loops (4), they can readily and abundantly form. Therefore, they produce a highly biased sink structure which favors also the nucleation of vacancy-type loops. However, as the loops grow to larger sizes, the average capture ratio $\langle Z_i/Z_v \rangle$ drops continously until it approaches a value of 1.2 typical of a dislocation network. It is this decrease of $\langle Z_i/Z_v \rangle$ which eventually terminates the nucleation of vacancy-type loops, rather than the increase of the sink density. To demonstrate this more clearly, Table 1 gives the steady-state nucleation rates for two sink densities. Clearly, the sink density has a negligible effect on the nucleation rate of vacancy-type loops.

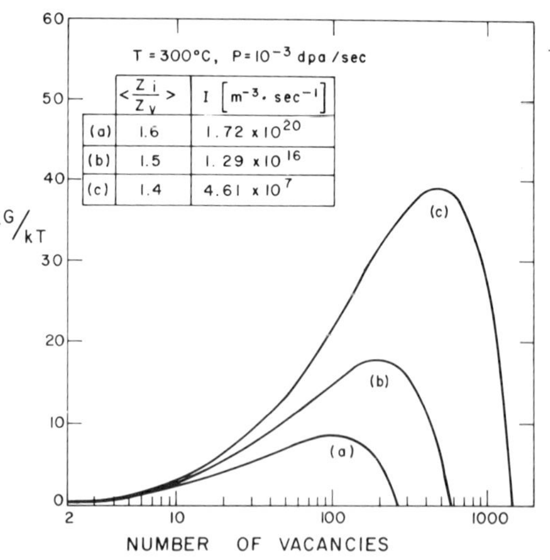

	$\langle Z_i/Z_v \rangle$	$I\ [m^{-3} \cdot sec^{-1}]$
(a)	1.6	1.72×10^{20}
(b)	1.5	1.29×10^{16}
(c)	1.4	4.61×10^{7}

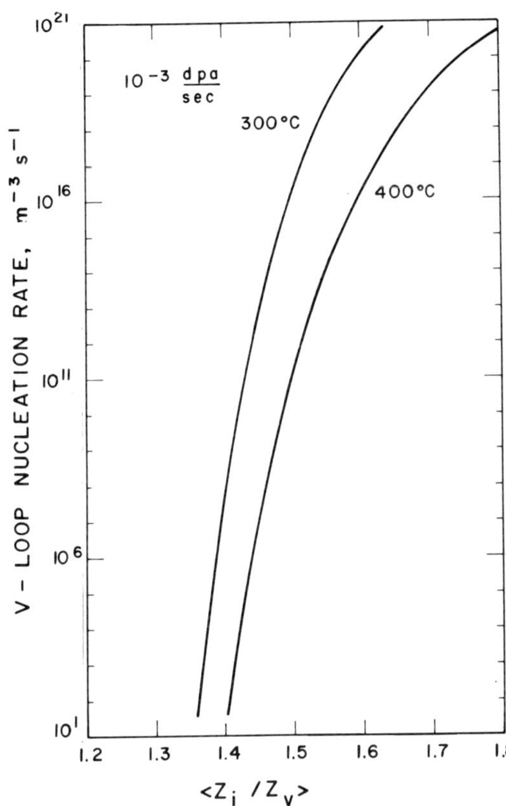

Figure 4 Nucleation rate for vacancy-type loops as a function of the average capture ratio.

A further condition for the nucleation of vacancy-type lopps is that the re-emission rate of vacancies from the loops be small in comparison with the capture rate of vacancies. Consequently, their nucleation is only possible at low temperatures. The temperature limit depends, however, on the Frenkel pair production rate, as can be clearly seen from the results in Fig. 5. For ion bombardment conditions with a typical production rate of 10^{-3} dpa/s, the upper temperature limit is about 420°C. In contrast, for breeder reactor irradiations with 10^{-6} dpa/s, the temperature limit is at about 320^{0}C.

Discussion

The conditions for the nucleation of vacancy-type loops are that their interstitial capture efficiency be lower than that of interstitial-type loops; that the temperature of irradiation be lower than a critical value which depends on the dose rate; and that the average interstitial capture efficiency for the total sink structure be larger than about 1.45. In order to satisfy thelast condition, the sink structure must be dominated by small dislocation loops.

Since there exist no significant barrier for interstitial loop nucleation (4) they form immediately after the irradiation begins. In fact, calculations by Hayns (15) and by Ghoniem and Cho (16) have shown, that interstitial loop densities of the order of 10^{20} loops/m^3 are reached within less than a second during fast reactor irradiation in materials with an initial line dislocation density of 10^{12} m^{-2}. For this value of the loop density, the average radius need only be of the order of 3 b_0 to obtain a loop sink strength equal to the dislocation sink strength. In this case, the interstitial capture averaged over all sinks is large, and vacancy-loop nucleation can proceed. If the initial line dislocation density is 10^{13} m^{-2}, the interstitial loop density reaches a lower value (15,16) around 10^{19} loops/m^3. In order for the loops to become a sink of comparable strength, they must now grow to a radius of about 30 b_0. As seen from Fig. 2, however, their capture efficiency is about 1.3 at that radius. The average interstitial capture efficiency is then 1.25, which is too low for vacancy-type loop nucleation to occur. Therefore, we estimate that the initial dislocation line density can be no higher than 5×10^{12} m^{-2} for vacancy-type loops to form.

TABLE 1. Effect of Sink Density on Vacancy-Type Loop Nucleation ($m^{-3}s^{-1}$) at T = 400^{0}C and for 10^{-3} dpa/s

Average Capture Ratio	Sink Density, m/m^3 5×10^{12}	10^{14}
1.4	4.36×10^{-1}	4.23×10^{-1}
1.5	1.65×10^{11}	1.59×10^{11}
1.6	1.52×10^{16}	1.48×10^{16}

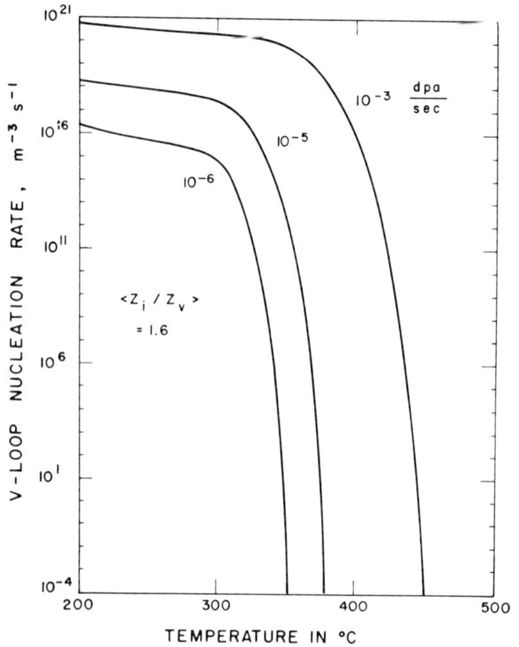

Figure 5 Nucleation rate for vacancy-type loops as a function of temperature and dose rate.

The nucleation rates given above are for the steady-state which is reached after the incubation time. For a dose rate of 10^{-6} dpa/s and an irradiation temperature of 300°C, the incubation time for vacancy-type loop nucleation was found to be about 12 hours, and it decreases with increasing temperature and increasing dose rate. This incubation time is much longer than for the interstitial-type loop formation, but in general very short compared to the total irradiation time.

A well documented case of the simultaneous formation of both loop types is the one for zirconium alloys (17). Here, both loop types form on the prism planes but not on the basal planes. The observed loop densities are between 10^{22} and 5×10^{22} m^{-3}. The dislocation density, even in cold-worked materials, is 10^{12} m^{-2} or lower. Therefore, the conditions for vacancy loop nucleation are certainly met. Considering these loop densities and the observed (17) loop diameters (30 to 90 nm), the ratio of the loop radius to the loop separation distance is only of the order of 0.1. Therefore, we conclude that proximity effects on the capture efficiencies are indeed small and of secondary importance compared to the effect produced by nonlinear elasticity.

Acknowledgements

This research was supported by the Division of Basic Energy Sciences, U.S. Department of Energy, under contract ER-78-S-02-4861 with the University of Wisconsin.

References

1. W.G. Wolfer and M. Ashkin, J. Appl. Phys. 46, 547; 46, 4108 (1975).
2. W.G. Wolfer and A. Si-Ahmed, Phys. Letters 76A, 341 (1980).
3. R. Bullough, M.R. Hayns, and C.H. Woo, Harwell Report TP.783, Feb. 1979.
4. F.A. Garner, W.G. Wolfer, and H.R. Brager; in Effects of Radiation on Structural Materials, 9th Internat. Symp., ASTM STP 683, 1979, p. 160.
5. A. Seeger and U. Goesele, Phys. Letters 61A, 423 (1977).
6. W.G. Wolfer and M. Ashkin, J. Appl. Phys. 47, 791 (1976).
7. A. Seeger and P. Haasen, Phil. Mag. 3, 470 (1958).
8. M.W. Guinan and D.J. Steinberg, J. Phys. Chem. Solids 35, 1501 (1974).
9. P. Ehrhart, J. Nucl. Materials 69 & 70, 200 (1978).
10. J.L. Katz and H. Wiedersich, J. Chem. Phys. 55, 1414 (1971).
11. K.C. Russell, Acta Met. 19, 753 (1971).
12. W.G. Wolfer and M.H. Yoo, in Rad. Effects and Tritium Technology for Fusion Reactors, Proc. of the Internat. Conf., Eds. J.S. Watson, F.W. Wiffen, CONF-750989, 1976, Vol. II, p. 458.
13. W.G. Wolfer and A. Si-Ahmed, to be published.
14. M. Nakagawa, K. Boening, P. Rosner, and G. Vogel, Phys. Rev. B16, 5285 (1977).
15. M.R. Hayns, J. Nucl. Materials 56, 267 (1975).
16. N.M. Ghoniem and D.D. Cho, Phys. Stat. Sol. (a) 54, 171 (1979).
17. D.O. Northwood, R.W. Gilbert, L.E. Bahen, P.M. Kelly, R.G. Blake, A. Jostens, P.K. Madden, D. Faulkner, W. Bell, and R.B. Adamson, J. Nucl. Materials 79, 379 (1979).

The Effect of Dislocation-Point Defect Interactions on Rate Processes in Metals

by F. A. Nichols

At steady state during irradiation, since vacancies and interstitials are produced in equal numbers, they must disappear in equal numbers. A net difference in atomic fluxes for a particular type of sink can occur only if there is at least one other type of sink present and then only if one sink preferentially absorbs either vacancies or interstitials. Of course, if one type of sink receives an excess of interstitials, it follows that a second type of sink must receive an excess of vacancies. (Recombination, which we shall ignore here (1,2), always removes equal numbers.) Thus, the magnitude of the differential flux must depend on the relative sink strengths of at least two types and, in particular, there can be no differential flux if only one type of sink is present. However, most estimates of these differential fluxes have been based upon mathematical models which contain only one type of sink, for which there can be no differential fluxes. This difficulty has been circumvented in basically two different ways.

One approach was introduced by Bullough, Eyre and Perrin (3) who analyzed one cylindrical cell around a long edge dislocation, across whose outer boundary a condition of zero flux was imposed for both vacancies and interstitials. Steady state was invoked and thus exactly equal fluxes of vacancies and interstitials into the dislocation core were necessitated. However, rather than deal with these loss rates directly, these authors calculated the concentrations of vacancies and interstitials at the outer surface. These concentrations were then utilized in a separate, independent calculation of the growth rate of voids and so the lack of conservation was not detected. We thus feel that no quantitative reliability can be placed on the estimate (4) derived from this model of a 2% mismatch in vacancy and interstitial fluxes into the dislocation. More recently, an analogous approach has been used to analyze a dislocation loop (5). The different interaction field assumed between the point defects and the sink is the only basic difference except in this case the authors calculate the average defect concentration in the cell (as originally suggested by Wiedersich (6)) and they apply this to a totally independent, reaction-rate-theory model to calculate rate processes.

A second type of approach, used by Heald (7) and Wolfer (8), is to use again a single cell around a dislocation but to ignore defect production and approximate its effect by imposing a fixed concentration of defects on the outer cell boundary. The imbalance here then arises completely from (implicitly assumed) imaginary sources external to the cell. Another attempt at assessing flux imbalances utilized an unsupported ad hoc assumption and solved no diffusion problem at all (9). Because of the uncertainties and inconsistencies of the available models, we have developed an approach which we feel represents the first internally consistent analysis of the partitioning of vacancies and interstitials between various sinks with which they interact differently (10,11).

The basic concept is illustrated schematically in Fig. 1 which depicts diffusional cells for dislocations, cavities and grain boundaries.

The interaction energy between a point defect and a dislocation is taken from the result of Bullough and Willis (12), who employed the "misfitting inhomogeneity" model of Eshelby (13), which is essentially of the form

$$E_\alpha = \frac{A}{r}\left[e_\alpha + \sigma\,(B\Delta\kappa_\alpha \pm C\Delta\mu_\alpha)\right]\begin{Bmatrix}\sin\theta\\ \cos\theta\end{Bmatrix} \tag{1}$$

where A, B, and C are constants for a given material and temperature, the subscript α signifies either vacancy or interstitial, e is the relaxation strain of the defect, σ is an applied tensile stress, $\Delta\kappa$ is the relative difference between effective bulk modulus of the defect and that of the matrix, $\Delta\mu$ is the same measure with respect to shear modulus and $\sin\theta$ goes with the positive sign whereas $\cos\theta$ goes with the negative sign of the last term. This is for the case of two edge dislocations, one with Burgers vector parallel to the axis of σ and the other normal to it. It is clear that the angular dependence of E introduces considerable mathematical difficulty, but there is appreciable precedent for ignoring it or "averaging" it out in some manner.

Bullough and Newman (14) dealt with the strain-aging problem in the drift approximation, i.e. ignoring concentration gradients, and used the first term of Eq. 1 with $\sin\theta \equiv -1$, thus obtaining a purely radial field with a magnitude equal to the maximum attractive value of the angular field. Their functional form was identical to that of Cottrell and Bilby (15) but their rate was ~1.87 times as large.

Friedel (16) treated the same problem, again in the drift approximation, and for a radial field with magnitude equal to the maximum attractive value obtained a rate which exceeded the Cottrell-Bilby result by the factor $(27/2\pi^2)^{1/2} = 1.17$.

Schroeder (17) surmises that for fields of the form $\sin\theta/r$, the flow is determined by the attractive regions only and cites a previous analytical approximation by Schroeder and Dettmann (18) and numerical results by Profant and Wollenberger (19), both dealing with dipole-dipole interactions, as agreeing quite well with this expectation. He states that the full angular solution decreases the flow rate, allowing for both concentration and interaction energy gradients, by about 0.7 compared with a radial interaction with magnitude equal to the maximum attractive value. The actual ratio of currents given by Profant and Wollenberger for the radial field vs the full angular field is 1.59.

With the above as background, we have replaced Eq. (1) with a radial field but, since in the results quoted using the maximum amplitude consistently overestimated the rates, we have instead used its average value over the attractive region (20), i.e. we have used for the dislocation field

$$E_{\alpha D} \doteq -\frac{2A}{\pi r}\left|e_\alpha + \sigma\,(B\Delta\kappa_\alpha \pm C\Delta\mu_\alpha)\right| \qquad (2)$$

for all of our calculations. For interaction between a point defect and spherical cavity, we have used the radial field calculated by Wolfer and Ashkin (21).

Since all cells are assumed to have radially symmetric fields, simple conservation yields

$$\nabla\cdot J = \frac{1}{r}\frac{d}{dr}(rJ) = \dot{a} \qquad (3)$$

for all cells where J is the defect flux and $\dot{a}$ is the production rate.

We also allow for the effect of the interaction fields on the defect diffusivity and obtain for each defect type

$$J = -D\exp-\left(\frac{E^S}{kT}\right)\nabla\left[c\exp\left(\frac{E^G}{kT}\right)\right] \qquad (4)$$

where D is the unperturbed diffusivity, E^S and E^G are the interaction energies in the saddle-point and ground states respectively and c is concentration.

In any particular problem, Eq. (3) is solved for J in the appropriate geometry and equated to Eq. (4). Integrating over the cell inner and outer boundaries thus introduces the quantities $c\exp(E^G/kT)$ evaluated at these two positions for each defect in each cell. We then specify the values at the sink surfaces and require that the values on the cell boundaries all be equal. For the final boundary condition, we require that the algebraic sum of all currents at the cell boundaries be zero, i.e. whatever defects leave one cell must enter some other cell or cells.

(a) SIPA Creep Model

For this problem we employ two cylindrical cells, as in Fig. 1 (a), with the same inner (core) radius r_o and outer radius R, with R set by the dislocation density. The resulting creep rate, displays several features common to all such problems.

1. The thermal contribution is an additive component directly dependent on point defect diffusivity and the difference in concentrations at the different sinks. Its magnitude is affected by the sum of integrals over the two cells involving the interaction energies in the saddle point only.

2. The additive, radiation-induced component is directly proportional to the production rate, totally independent of defect diffusivities and the concentrations at the different sinks and is determined by differences in equivalent integrals over the two cells involving again only the interaction in the saddle point. It is totally independent of the interaction in the ground state if it differs from that in the saddle point.

These basic results were shown in a more approximate model by Schilling (22) and in a more generalized treatment by Scattergood et al (23).

Examination of Eq. (2) shows that the sign of the third term provides the only difference in E for the two cells, and this produces the net flux. It has been demonstrated theoretically (24) and experimentally (25) that the dumbbell interstitial is extremely weak to certain shear modes and so to obtain quantitative results we have, with Bullough and Hayns (9), taken $\Delta\mu_I$ = 1, $\Delta\mu_V$ = 0 to apply at the saddle point and the entire predicted radiation effect is then due to the interstitials. Choosing other values typical of stainless steels we produced (26) the results shown in Fig. 2. The calculated difference in radiation-induced fluxes to the two types of dislocation was ∿0.07%. Both the thermal and radiation components are linearly dependent on applied stress but the radiation component (proportional to $\dot{a}$) has only a very weak temperature dependence. These results are qualitatively compatible with experimental trends but the magnitude of the predicted creep rate is about five times smaller than the Bullough-Hayns prediction and hence is at least about five times smaller and more likely 20 to 25 times smaller than experimental results when allowance is made for close-pair recombination of defects within a high-energy cascade. Thus, we have argued that this is probably not a dominant mechanism for radiation creep.

(b) Void Swelling Model

For this problem we employed (10,11) all three types of cells shown in Fig. 1. Only the larger first term of Eq. (2) is now significant because the required asymmetry in defect interactions is supplied by the much weaker interactions with the cavities and the zero interactions assumed for the grain boundaries. The same separation into thermal and radiation-induced components occurs, with the former depending on defect diffusivity and concentrations at the various sinks and the latter being totally independent of these. Again, only interactions in the saddle-point produce a net radiation effect. Available estimates for defect relaxation strain apply to the ground state and are subject to considerable uncertainty. Thus, in the present numerical evaluations we have not attempted to make such a distinction. This is at least partially validated by the small values of activation volume for both vacancies and interstitials which have been reported for several metals (11). For results typical of stainless steels, we have employed (27) e_I = 1.4 and e_V = - 0.23, with other parameters as before, to produce the results shown in Fig. 3 for a grain diameter of 20μm and in Fig. 4 for various grain sizes, both for high dislocation and cavity sink strengths. Fig. 3 also shows the effects of gas pressure in the cavities and externally applied hydrostatic pressures, where local equilibrium has been invoked at all sink surfaces. Three production rates, $\dot{a} = 10^{-6}$, 10^{-4} and 10^{-2} dpa s^{-1}, are shown. Introducing the bias factor B, we can write for the void swelling rate

$$\dot{\delta} = \dot{a}\ S_C(0)\ S_D(0)\ B \tag{5}$$

where $S_C(0)$ and $S_D(0)$ are the sink strengths of cavities and dislocations, respectively, i.e. the relative fractional volumes from which they drain defects as calculated internally in the model. There are wide variations in B with different sink strengths but for comparable cavity and dislocation sink strengths (hence rather large swelling rates) B ∿0.6 obtains

Some recent results of Scattergood (28), displayed in Fig. 5, address the question of the effect of replacing the angularly dependent interaction energy of Eq. (1) with the radial form of Eq. (2). Scattergood utilized one dislocation cell of the form in Fig. 1 (a) but made the outer radius a second sink by assuming the concentration there to be zero. His numerical results show, as expected from our discussion, that a radial interaction field with magnitude equal to the maximum attractive value significantly overestimates the bias effect employing both concentration and E^G gradients. They also show that there is no fixed ratio between these results, but for values typical of interstitials, a radial field using the average of the attractive region still overestimates the bias by ∿2.5 for Scattergood's geometry. If we assume this same factor applies to our void swelling model discussed above, then the value of B ∿0.6 should possibly be reduced to ∿0.24. This value would then be quite consistent with experimentally deduced bias factors of 0.06 (29) and 0.08 (30) if we assume that ∿20 to 30% of the defects escape a typical cascade (an effect ignored in the experimental comparisons). This would be in excellent agreement with independent estimates from basic radiation damage studies (31-33).

(c) Thermal Rates

Finally, we mention the predicted effects of the interaction fields on thermal rates such as the Nabarro creep rate in the high-temperature region of Fig. 1. For extreme ranges of values of e_V, we have determined (34) that rates may be altered up to a factor of ∿10 and that apparent activation energies can differ from the diffusion values by as much as ∿5%. Such results may be of interest in basic defect studies but may go undetected in many kinetic studies.

References

1. M. R. Hayns, J. Nucl. Mater. 79 (1979) 323.
2. R. Bullough and M. H. Wood, "Numerical Calculations of the Effect of Recombination on Sink Strengths in the Rate Theory of Void Swelling, Irradiation Creep and Growth", TP. 820, AERE, Harwell (1979), to be published.
3. R. Bullough, B. L. Eyre and R. C. Perrin, Nucl. Appl. Technol. 9 (1970) 346.
4. A. D. Brailsford and R. Bullough, J. Nucl. Mater. 44 (1972) 121.
5. R. Bullough, AERE Harwell (1979), to be published in J. Nucl. Mat.
6. H. Wiedersich, Radiation Effects 12 (1972) 111.
7. P. T. Heald and M. V. Speight, Acta Met. 23 (1975) 1389.
8. W. G. Wolfer and M. Ashkin, J. Appl. Phys. 47 (1976) 791.
9. R. Bullough and M. R. Hayns, J. Nucl. Mater. 57 (1975) 348.
10. F. A. Nichols, "Bias Factors for Radiation Creep, Growth and Swelling", to be published in J. Nucl. Mater.
11. F. A. Nichols and Y. Y. Liu, "An Internally Consistent Model for Radiation-Induced Void Swelling", to be published by ASTM.
12. R. Bullough and J. R. Willis, Phil. Mag. 31 (1975) 855.
13. J. D. Eshelby, in Vacancies '76, The Metals Society, London (1977), p. 3.
14. R. Bullough and R. C. Newman, Rep. Prog. Phys. 33 (1970) 101.
15. A. H. Cottrell and B. A. Bilby, Proc. Phys. Soc. 62 (1949) 49.
16. J. Friedel, Dislocations, Pergamon Press, New York (1964), p. 405.
17. K. Schroeder, in Vacancies '76, The Metals Society, London (1976), p. 51.
18. K. Schroeder and K. Dettmann, Z. Physik B22 (1975) 350.
19. M. Profant and H. Wollenberger, Phys. Stat. Solidi (b) 71 (1975) 515.
20. P. T. Heald and R. Bullough, in Vacancies '76, The Metals Society, London(1977), p. 134.
21. W. G. Wolfer and M. Ashkin, J. Appl. Phys. 46 (1975) 547.
22. W. Schilling, J. Nucl. Mater. 69/70 (1978) 465.
23. R. O. Scattergood, F. V. Nolfi and A. P. L. Turner, Scripta Met. 12 (1978) 939.
24. P. H. Dederichs, C. Lehmann, H. R. Schober, A. Scholz and R. Zeller, J. Nucl. Mater. 69/70 (1978) 176.

25. L. E. Rehn, K.-H. Robrock and V. Spirić, in Fundamental Aspects of Radiation Damage in Metals, Oak Ridge Report CONF-751006-P1 Vol. 1 (1975), p. 234.
26. F. A. Nichols, J. Nucl. Mater. 84 (1979) 207.
27. P. T. Heald, in Vacancies '76, The Metals Society, London (1977), p. 11.
28. R. O. Scattergood, Argonne National Lab. (1980), unpublished research.
29. M. H. Yoo and J. O. Stiegler, Phil. Mag. 36 (1977) 1305.
30. R. Bullough, B. L. Eyre and K. Krishan, Proc. Roy. Soc. London A346 (1975) 81.
31. T. C. Reiley and P. Jung, in Radiation Effects in Breeder Reactor Structural Materials, M. L. Bleiberg and J. W. Bennett Editors, the Metallurgical Society, AIME, New York (1977) 295.
32. R. S. Averback, R. Benedek and K. L. Merkle, J. Nucl. Mater. 69/70 (1978) 786.
33. U. Theis and H. Wollenberger, J. Nucl. Mater. 88 (1980) 121.
34. Y. Y. Liu and F. A. Nichols, unpublished research.

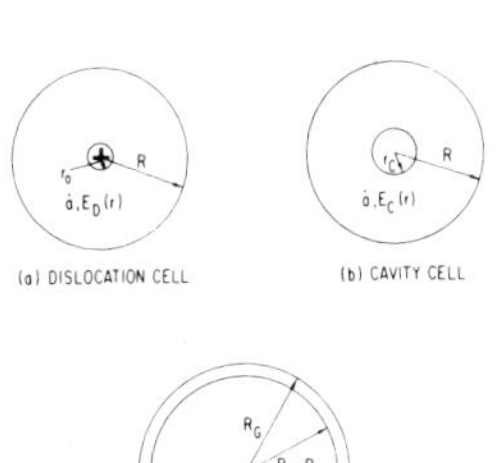

Fig. 1. Schematic of Three Diffusion Cells.

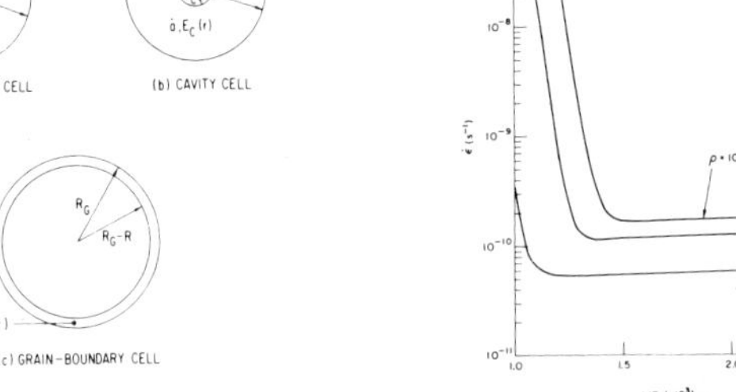

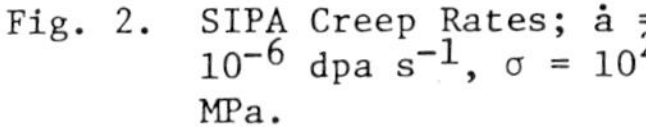

Fig. 2. SIPA Creep Rates; $\dot{a} = 10^{-6}$ dpa s^{-1}, $\sigma = 10^2$ MPa.

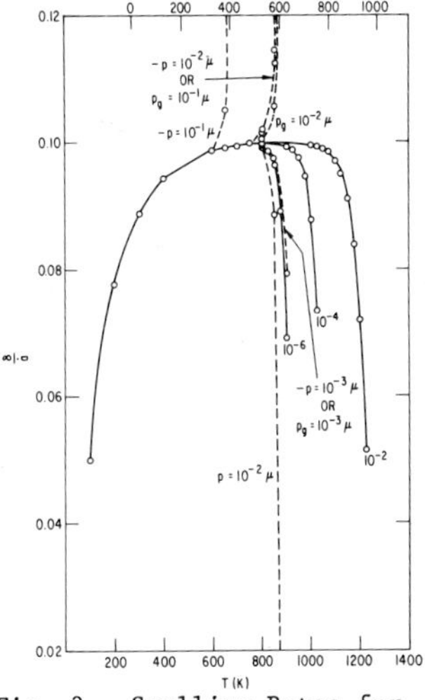

Fig. 3. Swelling Rates for Large-Grained Material, High Sink Density.

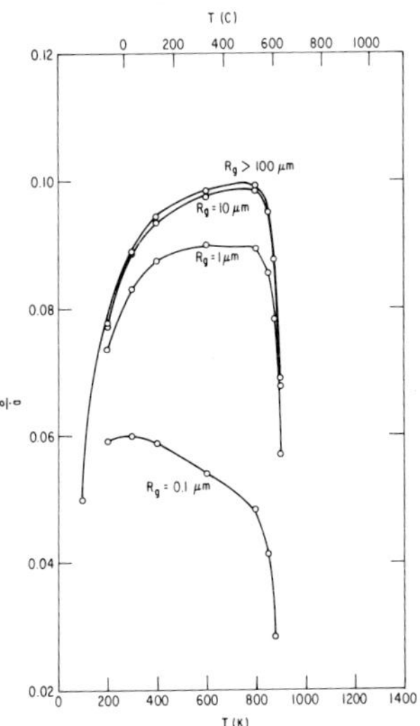

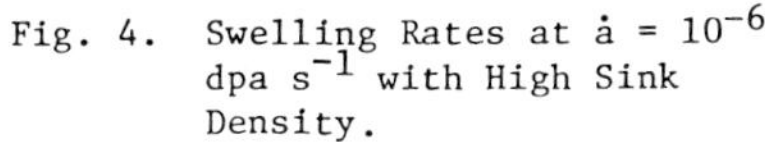

Fig. 4. Swelling Rates at $\dot{a} = 10^{-6}$ dpa s^{-1} with High Sink Density.

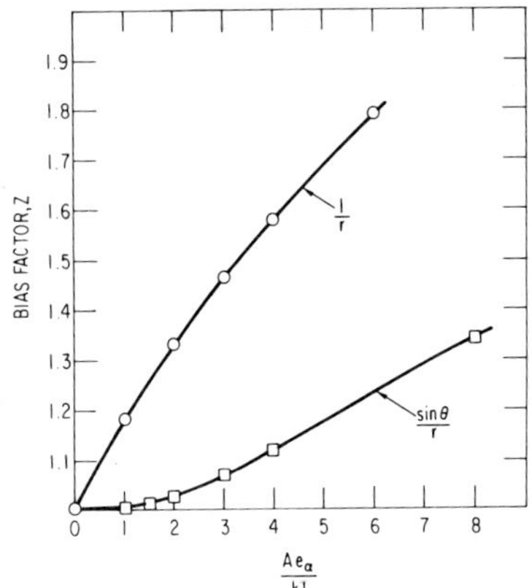

Fig. 5. $Z \equiv I_D/I_D$ $(E_D = 0)$ for Dislocation with Cylindrical, Non-Interacting Sink.

RADIUS DEPENDENCE OF THE SINK STRENGTH OF A DISLOCATION LOOP*

W. A. Coghlan and M. H. Yoo

Metals and Ceramics Division
Oak Ridge National Laboratory
Oak Ridge, Tennessee 37830

Abstract

The sink strength of a dislocation loop has been calculated using a numerical method. We have shown that the resulting loop radius dependence is consistent with the analytical results based on an electrostatic analog. In addition, the results are consistent with recent calculations made for very small loops. We have given an expression that describes the sink strength as a function of loop radius, defect relaxation volume, and the materials elastic constants. This expression enables us to characterize the reaction rate constants in modeling studies of loop growth and other applications of reaction rate theory.

1. Introduction

Diffusion of point defects under the influence of a force field is an important physical process involved in phase transformation, radiation damage, and inelastic deformation of materials. When the force field originates from the interaction of a point defect with another defect or defect cluster such as dislocation loop, the role of such interaction in the drift flow of point defects becomes particularly important in the overall diffusion kinetics of cluster evolution.

Dislocation loops are usually the first observable defect clusters in materials with the supersaturated concentrations of point defects. Existing radiation damage models inadequately treat the growth kinetics of these loops because the defect interaction with these loops is complicated and the resulting diffusion problem does not yield an analytic solution. In the previous solutions to this problem (1-4), an approximation was made by treating the dislocation loop as an infinitesimal loop which is valid only for very small loops. This approximation was obviated by taking full account of the finite toroidal feature of a dislocation loop in solving the diffusion equations for point defect migration (5,6). One essential conclusion emerging from these previous works (1-6) is that the capture efficiency for the loop decreases as the loop size increases. However, more recent results by other investigators (7,8) are said to be in contradiction to this conclusion, as discussed by Bullough et al. (9).

The purpose of this paper is to reassess the physical sources of the dependence of the capture efficiency, or the sink strength, of a prismatic dislocation loop on its loop size. In addition, we believe that much of the controversy regarding differences in solutions centers on differences in definitions of the terms involved. We will therefore make a clear statement of how we treat this problem and will make comparisons with solutions found from other points of view.

*Research sponsored by the Division of Materials Sciences, U.S. Department of Energy under contract W-7405-eng-26 with the Union Carbide Corporation.

2. Diffusion Equation and an Analytic Solution

In the continuum theory of lattice defects, the flux of point defects of concentration $C(\underset{\sim}{r},t)$ under the influence of an interaction force field $-\nabla E$ is given to a first approximation by*

$$\underset{\sim}{J} = -D(\nabla C + \frac{C}{kT}\nabla E) , \quad (1)$$

where D is the diffusion coefficient of point defects, k the Boltzmann constant, and T the absolute temperature. The diffusion equation follows from the continuity condition $\nabla\cdot\underset{\sim}{J} = -\partial C/\partial t$ as

$$\frac{\partial C}{\partial t} = D\nabla\cdot(\nabla C + \frac{C}{kT}\nabla E) , \quad (2)$$

where the diffusion coefficient is assumed to be isotropic and independent of concentration and position.

In the theory of diffusion-controlled reaction kinetics, the drift effect due to the interaction potential can be taken into account by choosing an appropriate reaction volume. The size of this volume will change depending on the strength of the interaction with the defect. In general calculating the size of this volume is difficult, but for one case the calculation is relatively easy. When dislocation loops are randomly distributed and are small compared to the radius of their sphere of influence, $R_s = (3/4\pi N_\ell)^{1/3}$, where N_ℓ is the loop number density, an analytic solution based on an electrostatic analogy (11) can be used to give the net defect current to a dislocation loop as

$$I = -\int_S \underset{\sim}{J}\cdot d\underset{\sim}{S} = 4\pi b r^* D(C_2 - C_1) , \quad (3)$$

where b is the Burgers vector of the dislocation loop, C_1 and C_2 are the defect concentrations at the inner and outer boundaries, respectively, and the capture radius, $r_c^* = \pi r_\ell/\ell n\,(8r_\ell/r_c)$, is the radius of an equivalent spherical reaction volume. The r_ℓ and r_c are the radii of the dislocation loop and the core respectively in units of b. As the loops become larger the diffusion fields begin to interact. We have chosen the formalism provided by Brailsford et al. (12) to include this effect. In this case the above capture radius increases by a factor F^2 where $F = (1 + \pi N_\ell r^{*3}b^3)^{1/2} + (\pi N_\ell r^{*3}b^3)^{1/2}$. The resulting defect strengths per unit length dislocation line based on this solution are shown in Fig. 1 for several loop densities for $r_c = 1$.

There are two points to be made concerning Fig. 1 that are relevant to future discussions. First, for a single loop, the capacitance solution shows that the sink strength per unit length of edge dislocation increases for decreasing loop size. Second, that adding the correction due to overlapping diffusion fields results in an increase in sink strength as the loops get larger. Further increasing the densities to unreasonably high values would shift the minimum sink strength to smaller loop sizes. When the elastic interaction energy E, between a point defect and a prismatic dislocation loop is included in Eq. (2); the solution can be found only by a numerical integration method, which is discussed in the ensuing section. Once the numerical solution is obtained, the net current of point defects can be obtained from Eqs. (1) and (3), from which a new larger capture radius can be evaluated. We will show that this increase in capture radius over that calculated in the absence of an interaction potential, Δr^*, decreases with decreasing loop size.

---

*See, for instance, Appendix B of Ref. 10 for a derivation of Eq. (1) by considering the forward and backward current densities based on the change in potential energy at equilibrium sites.

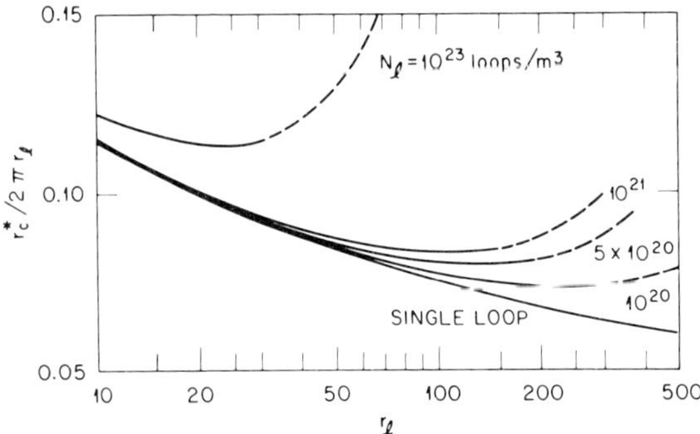

FIG. 1

The loop radius dependence of the sink strength of a dislocation loop per unit length edge dislocation based on the capacitance solution analog. The curves are dashed in Fig. 1, 4, 5, and 6 for loop radii greater than one-half of the radius of the sphere of influence, and the curves do not extend beyond this radius.

3. A Numerical Solution

Equation (2) was solved numerically for a dislocation loop enclosed by a cylinder for an interaction energy based on the first-order size effect and the inhomogeneity effect. We found that in the absence of an applied stress this second contribution to the interaction energy causes very little change in the capture radius. The first-order size effect depends on the gradient of the hydrostatic component of the loop stress field, which were evaluated in terms of tabulated elliptic integrals based on the work of Bastecka and Kroupa (13). The defect concentration is maintained at the thermal equilibrium value at the dislocation core and at a constant value on the surface of the cylinder.

The numerical method used is described in detail in Ref. (5) and is a straightforward relaxation procedure where the defect concentration field is allowed to relax according to an expanded and transformed version of Eq. (2). The reasons for the variable transformation and the cylindrical expansion are also discussed in Ref. (5). After a steady state solution has been found, the defect current, I, and resulting capture radius, r^*, are found using Eq. (3). Finally, a value of r_∞^*, the capture radius of a single loop in an infinite medium, is found using a spherical analog, $r_\infty^* = r^* r_2/(r_2 + r_\infty^*)$, where r_2 is the radius of a sphere having the same outside dimensions as the cylinder. For modeling purposes we will use reaction volumes having radii, r_∞^*, to replace dislocation loops and will use the formalism of Brailsford et al. (12) to include the effect of overlapping diffusion fields.

Before we show the results of the calculation, there is one significant problem to be dealt with; namely, the choice of the inner dislocation core cut-off radius r_c. For the case discussed above excluding the interaction potential, we chose 1b for r_c which would logically be the radius where a defect loses its identity and becomes part of the dislocation core. For the case involving the interaction, the choice is not as simple, however. The fundamental diffusion equation, Eq. (1), is derived for small values of ∇E. In one dimension this derivation includes a second order expansion of an exponential and requires that $\eta < 1$, where $\eta = b|\nabla E|/kT$ The use of this equation in the regions where $\eta > 1$ can result in an underestimate of the diffusion flux and in a smaller sink strength for the loop. Fortunately, there is a solution to this problem. In Fig. 2, we have plotted r_∞^* as a function of η. Constant values of η result in nearly perfect tori about the dislocation loop. At the top of Fig. 2 we have given typical values for r_c for interstitials and vacancies. The important point is that r_∞^* is independent of η for $\eta \gtrsim 0.5$. We therefore chose a value of $\eta = 1$ to determine the core radius for all of our subsequent calculations. Figure 3 shows values of r^* as a function of the loop radius r_ℓ. The solid lines are the result of a useful empirical equation for the capture radius,

$$r_\infty^* = 0.58\, r_\ell^{P(r_\ell,\xi)} \quad \text{for} \quad \begin{matrix} 10 \leqslant r_\ell \leqslant 500 \\ 0 \leqslant \xi \leqslant 700 \end{matrix} \tag{4}$$

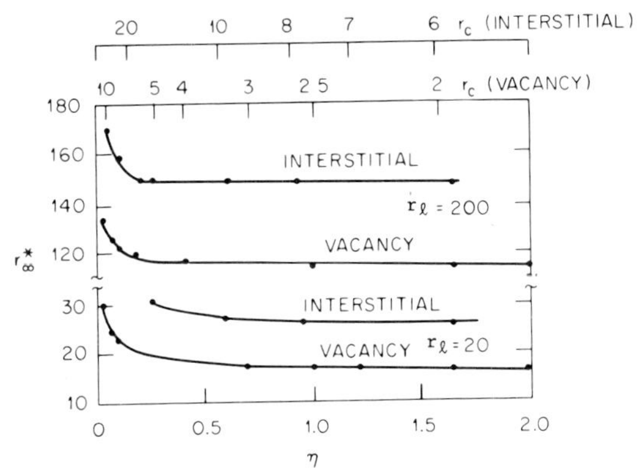

FIG. 2

The capture radius of a single loop in an infinite medium as a function of core radius.

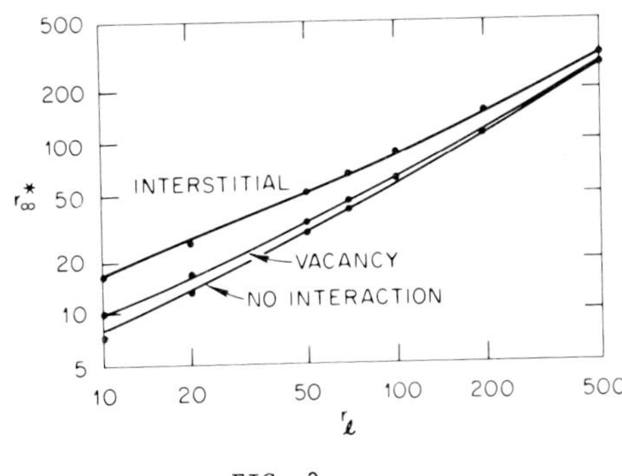

FIG. 3

The capture radius of a single loop based on a numerical calculation for interstitials, vacancies, and a defect having no interaction versus loop radius. The relaxation volumes used for these calculations are 1.8 Ω and −0.2 Ω for interstitials and vacancies, respectively.

where

$$P = 0.9852 + 2.779/r_\ell^{1.25} + 0.05492\ (\xi/r_\ell)^{0.75} - 0.0023\ (\xi/r_\ell)^{1.5}$$

$$\xi = \frac{\mu\Omega\,|\Delta V/\Omega|}{\gamma kT}$$

$$\gamma = \frac{3(1-\nu)}{1+\nu}$$

and μ is the shear modulus, Ω the atomic volume, $\Delta V/\Omega$ the defect relaxation volume, and ν Poisson's ratio.

The constants in the equation were calculated using a least square fit to 70 calculated points in log r^* vs log r_ℓ space. Further discussions involving the calculations will be based on this expression.

4. Discussion

As we mentioned in the introduction, there are several important physical processes that involve knowing the sink strength of a dislocation loop. The one which interests us most is radiation damage. Under irradiation, dislocations in a metal form as loops, which grow, unfault, and become network dislocations. The difference in the capture radii of a loop for interstitials and vacancies is the driving force for this process and also provide the defect bias that leads to void growth. Figure 4 is a plot of this difference based on Eq. (4) for a single loop and for a reasonable loop density of 5×10^{20} loops/m^3.

In the absence of an interaction potential there would be no difference; but as the figure shows, the calculated difference represents a sizeable fraction of the loop radius. It is also important to observe that the over-lap of the diffusion fields enhances the difference between the two defects. This is not surprising since the drift contribution to the flux has a longer range for interstitials.

Figure 4 also illustrates that the difference in the capture radii of vacancies and interstitials, resulting from differences in the interaction potentials, decreases with decreasing

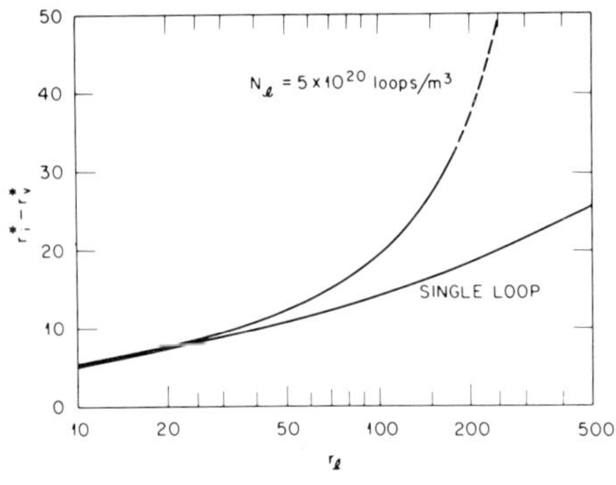

FIG. 4

The difference in captive radii for dislocation loops as a function of loop size.

loop size. This decrease simply reflects the fact that the ranges of the stress field of a loop decreases with loop size. If the calculation could be continued to much smaller loop sizes, we would expect that the sink strength difference might disappear. Our calculation does not permit these calculations on very small loop sizes.

Figure 5 is a plot that allows the numerical calculation to be compared with analytical results in Fig. 1. It is important to notice that including the interaction term in the solution does not change the form of the solution very much. It simply increases the magnitude of the defect sink strength per unit length of edge dislocation.

The last figure, Fig. 6 is a plot of the difference in sink strength per unit length edge dislocation versus loop radius. Its purpose is to illustrate the minimum value that occurs for a given loop density and also to provide a comparison with the values calculated by Bullough et al.[9]. The relative loop bias given in Ref. (9) is proportional to the value plotted in Fig. 6. The product of 4π and the values in Fig. 6 illustrate the agreement between the two calculations. For a 20b radius loop our values are about 30% higher for a loop concentration of 10^{23} loops/m^3. For our calculations we used a relaxation volume of 1.8 Ω which is higher than that used in Ref. (9) where we estimate that a value of 1.0 Ω was used. In light of the differences in the calculations we believe the agreement is good.

In summary, we have shown the results of a calculation of the sink strength of a dislocation loop as a function of loop size. We have shown that the results are in agreement with the diffusion based analytic solution for a torus in an infinite medium. Also, the adoption of the effective medium approach makes it possible to use the solution in realistic radiation damage situations. Finally, the results for small loop sizes are in general agreement with calculations made by others for observed concentrations of dislocation loops.

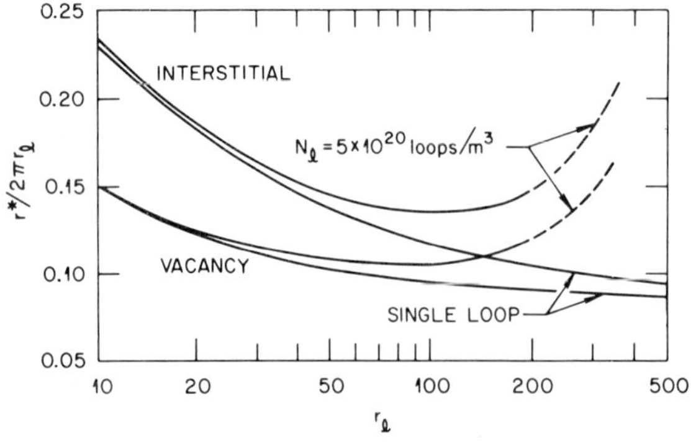

FIG. 5

A summary of the defect strength of a loop per unit length edge dislocation.

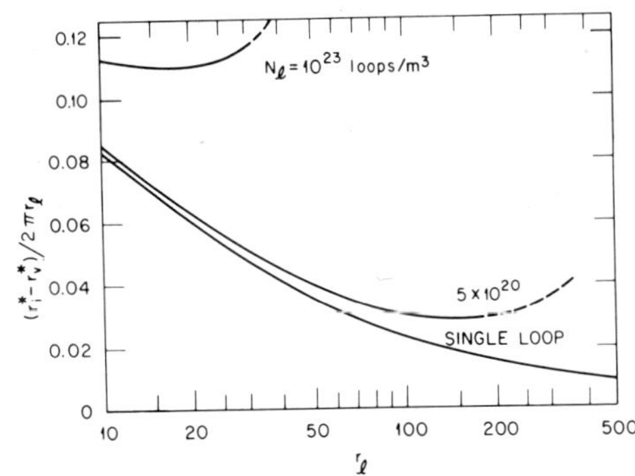

FIG. 6

A plot of the difference in defect strength per unit length edge dislocation versus loop radius.

REFERENCES

1. W. G. Wolfer and M. Ashkin, J. Appl. Phys. 46, 547 (1975).
2. M. H. Yoo, W. H. Butler, and L. K. Mansur, Proc. Int. Conf. Fundamental Aspects of Radiation Damage in Metals, eds., M. T. Robinson and F. W. Young, Jr., Gatlinburg, Tenn., October 1-3, 1975 (CONF-75006-P2), p. 804.
3. W. G. Wolfer, Proc. Int. Conf. Fundamental Aspects of Radiation Damage in Metals, eds., M. T. Robinson and F. W. Young, Jr., Gatlinburg, Tenn., October 1-3, 1975 (CONF-75006-P2), p. 812.
4. M. H. Yoo and W. H. Butler, Phys. Stat. Sol. B77, 181 (1976).
5. W. A. Coghlan, Proc. Int. Conf. Computer Simulation for Materials Application, eds., R. J. Arsenault, J. L. Beeler, and J. A. Simmons, Gaithersburg, MD, April 19-21, 1976, p. 166.
6. W. A. Coghlan and M. H. Yoo, Trans. Am. Nucl. Soc. 27, 330 (1977).
7. R. Bullough, M. R. Hayns, and C. H. Woo, J. Nucl. Mater. 84, 93 (1979).
8. C. H. Woo, W. S. Liu, and M. S. Wuschke, Whiteshell Nucl. Res. Establ. Report (AECL-6441), December, 1979.
9. R. Bullough, D. W. Wells, J. R. Willis, and M. H. Wood, in this issue.
10. M. H. Yoo and L. K. Mansur, J. Nucl. Mater. 62, 282 (1976).
11. A. Seeger and V. Gosele, Phys. Letters, 61A, 423 (1977).
12. A. D. Brailsford, R. Bullough, and M. R. Hayns, J. Nucl. Mater. 60, 246 (1976).
13. J. Bastecka and F. Kroupa, Czech. J. Phys. B14, 443 (1964).

A FRANK LOOP UNFAULTING MECHANISM IN fcc METALS DURING NEUTRON IRRADIATION

D. S. Gelles
Hanford Engineering Development Laboratory
Richland, Washington 99352

Abstract

The unfaulting mechanism whereby sessile Frank dislocation loops evolve into a complex tangle of glissile dislocations during irradiation of face centered cubic metals is not well understood. It is generally presumed that such loops grow by absorption of point defects until interactions develop which provide sufficient impetus for nucleation of an unfaulting event. The loops then become glissile, interact and form a dislocation network. The present study identifies an alternate mechanism which has been observed to occur in an austenitic precipitation-strengthened commercial alloy irradiated in the EBR-II fast reactor. The mechanism requires an interaction between the sessile $\frac{a}{3}\langle 111\rangle$ Frank loop and a moving glissile $\frac{a}{2}\langle\bar{1}\bar{1}0\rangle$ perfect dislocation. An unfaulting $\frac{a}{6}\langle\bar{1}\bar{1}2\rangle$ dislocation is created which, as it moves, eliminates the Frank loop and leaves only a perfect dislocation with the original $\frac{a}{2}\langle\bar{1}\bar{1}0\rangle$ Burgers vector. The process is demonstrated by an example illustrated with transmission electron micrographs. The impact of this alternate mechanism on the development of the dislocation microstructure in a metal undergoing irradiation creep is discussed in relation to relevant irradiation creep models.

Introduction

Irradiation creep is an important behavioral response which must be considered in the development of structural materials for fast breeder reactors. Irradiation creep occurs as a result of enhanced concentrations of point defects produced by bombardment with highly energetic neutrons. Increases in point defect concentrations lead to increased dislocation climb rates which in turn generally result in enhanced creep rates. Two approaches can be taken to quantify irradiation creep response. Creep tests can be performed in-reactor to obtain data for specific alloy conditions under specific test conditions (1) or, alternatively, irradiation creep response can be modeled mathematically and predictions can be made for various inputs, as test condition changes, to the model. The former approach is limited by the fact that every possible test condition cannot be tested and the latter approach is limited by inaccurate modeling due to oversimplification of the model and due to misinformation regarding controlling mechanisms. Ideally, a marriage of the two approaches is required to best predict irradiation creep response for any given application. The present work provides insight into an important aspect of the evolution of dislocation structure in an irradiation environment: the transition from a Frank loop dominated microstructure to one involving perfect dislocation networks. The impact of this process on irradiation creep response is then discussed.

Autocatalytic Frank Loop Unfaulting

It is generally presumed that the Frank loop unfaulting process can occur in an autocatalytic manner (2-4). Once a given loop grows sufficiently large, it may intersect another microstructural feature which can generate sufficient localized stresses to nucleate an unfaulting dislocation loop of the type $\frac{a}{6}\langle 11\bar{2}\rangle$. This Shockley partial can then glide in the plane of the Frank loop and react with the Frank partial according to the equation

$$\frac{a}{6}[11\bar{2}] + \frac{a}{3}[111] = \frac{a}{2}[110] \ , \qquad (1)$$

thereby transforming the Frank loop into a perfect prismatic dislocation loop.

Perfect Dislocation/Frank Loop Unfaulting

An alternate mechanism exists, however, which allows Frank loop unfaulting. It has been demonstrated to apply in the case of Frank loop unfaulting in quenched pure aluminum deformed at room temperature (5,6) and, as will be demonstrated, it is found to apply in a precipitation strengthened commercial alloy during irradiation creep. It differs from the above in that nucleation of an $\frac{a}{6}\langle 11\bar{2}\rangle$ dislocation loop is not required.

This unfaulting mechanism is initiated by the interaction of a sessile $\frac{a}{3}\langle 111\rangle$ Frank loop with a glissile $\frac{a}{2}\langle\bar{1}\bar{1}0\rangle$ perfect dislocation. A reaction results by which an $\frac{a}{6}\langle\bar{1}\bar{1}2\rangle$ dislocation is created according to the equation

$$\frac{a}{2}[\bar{1}\bar{1}0] + \frac{a}{3}[111] = \frac{a}{6}[\bar{1}\bar{1}2] \ . \qquad (2)$$

The Shockley partial created can sweep across the Frank loop, remove the stacking fault and react with the opposite side of the Frank loop according to the equation

$$\frac{a}{6}[\bar{1}\bar{1}2] + \frac{a}{3}[\bar{1}\bar{1}\bar{1}] = \frac{a}{2}[\bar{1}\bar{1}0] \ . \qquad (3)$$

The process diagramed in Figure 1 shows the interaction of a Frank loop in the plane of the figure and a perfect dislocation moving on some other plane and intersecting the loop. Thus, the Frank loop can be annihilated, the only remnant being a coil in the $\frac{a}{2}[\bar{1}\bar{1}0]$ dislocation approximately on the Frank loop {111} plane. Thus, the unfaulting product of a perfect dislocation/Frank loop interaction immediately becomes part of the perfect dislocation network but the autocatalytic unfaulting product consists of an individual prismatic dislocation loop.

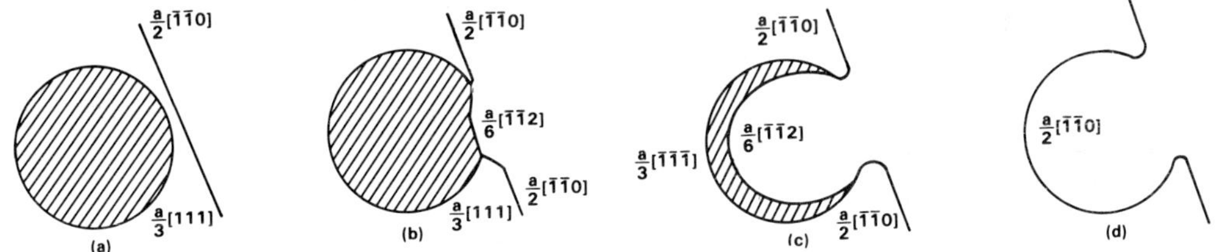

FIG. 1
The Frank loop unfaulting mechanism which requires the interaction of a perfect dislocation with the Frank loop.

Experimental Demonstration

A pressurized tube specimen of solution treated Nimonic PE16 (specimen AV75) was irradiated to a fluence of 2 x 10^{22} n/cm^2 (E >0.1 MeV) at 545°C under a hoop stress of 167 MPa (1). Nimonic PE16 is a commercial gamma prime [$Ni_3(Al,Ti)$] precipitation strengthened austenitic superalloy. The specimen was examined by transmission electron microscopy and found to contain a partially unfaulted dislocation structure. In many regions of the specimen, unfaulting had not occurred and only Frank loops were present. In other regions, however, Frank loops had been almost completely replaced by a perfect dislocation network. In the transition regions, many examples of crescent-shaped stacking fault features were found, and several of these were observed to be connected to the perfect dislocation network.

An example is given in Figure 2 which shows such a transition region in a (112) foil imaged using $11\bar{1}$ dislocation contrast so that [1] stacking fault fringes for three of the four sets of Frank loops can be seen (the fourth appearing edge-on) and [2] half of the perfect dislocations are in contrast. Locations in the micrograph where perfect dislocation/Frank loop interactions were confirmed by stereo examination are indicated by arrows. Analysis of the $\frac{a}{3}\langle 111\rangle$ and $\frac{a}{2}\langle 110\rangle$ Burgers vectors involved in such interactions demonstrated that the reaction of equation (2) was responsible for the unfaulting. Several examples can also be found where large dislocation coils have developed in the perfect dislocation network identifying locations where Frank loops have recently been annihilated.

Figure 3, a low magnification montage of a whole grain, demonstrates the localized effects of the unfaulting behavior. In this example, the specimen is in an (011) foil orientation, again imaged using $11\bar{1}$ dislocation contrast so that two sets of Frank loops are steeply inclined and appear edge-on, and two sets are more circular in shape. Several large regions, generally elliptical in shape, can be found which are unfaulted, one of which is so labeled. Sources of

FIG. 2
Dislocation structure in a specimen of solution treated Nimonic PE16 after irradiation to 2.0 x 10^{22} n/cm^2 (E >0.1 MeV) at 545°C under a hoop stress of 167 MPa showing a region undergoing Frank loop unfaulting; arrows identify locations where perfect dislocation/Frank loop interactions are confirmed by stereoscopic examination.

unfaulting response can be identified in this figure as being either grain boundaries or large blocky precipitates (which have punched-out matrix dislocations as a result of quenching strains).

Therefore, it is found that in the case of Nimonic PE16 irradiated under stress at 545°C, Frank loop unfaulting occurs as a result of interactions with perfect dislocations. Perfect dislocations are present even in highly annealed material, produced by quenching strains around large blocky precipitates or associated with grain boundaries. In fact, the dislocation generation capability of grain boundaries during cold working (7) appears to apply for irradiation creep response as well. However, the motion of these perfect dislocations will reflect an externally applied stress, for those dislocations oriented for optimum motion under stress will move most rapidly. This is demonstrated in Figure 3 by the noncircular unfaulted regions. As the irradiation proceeds, the area swept out by the perfect dislocations becomes larger and larger until almost the entire Frank loop population has been annihilated.

The same mechanism for Frank loop unfaulting likely applies in the case of AISI 316, a solid solution strengthened austenitic stainless steel. Brager et al (8) showed that when perfect dislocations and Frank loops were able to coexist, the Frank loop size was governed by the mean spacing between dislocations. Thus, loop unfaulting occurred as a result of interactions with the perfect dislocation network and not by an autocatalytic process.

Consequences

The evolution of the dislocation network during irradiation creep has generally been envisioned as follows. Interstitial atoms preferentially coalesce to form Frank loops. Anisotropy in the loop populations develops in proportion to the deviatoric normal stress on the given plane (8-10). The loops grow initially primarily as a result of increased interstitial capture efficiencies of

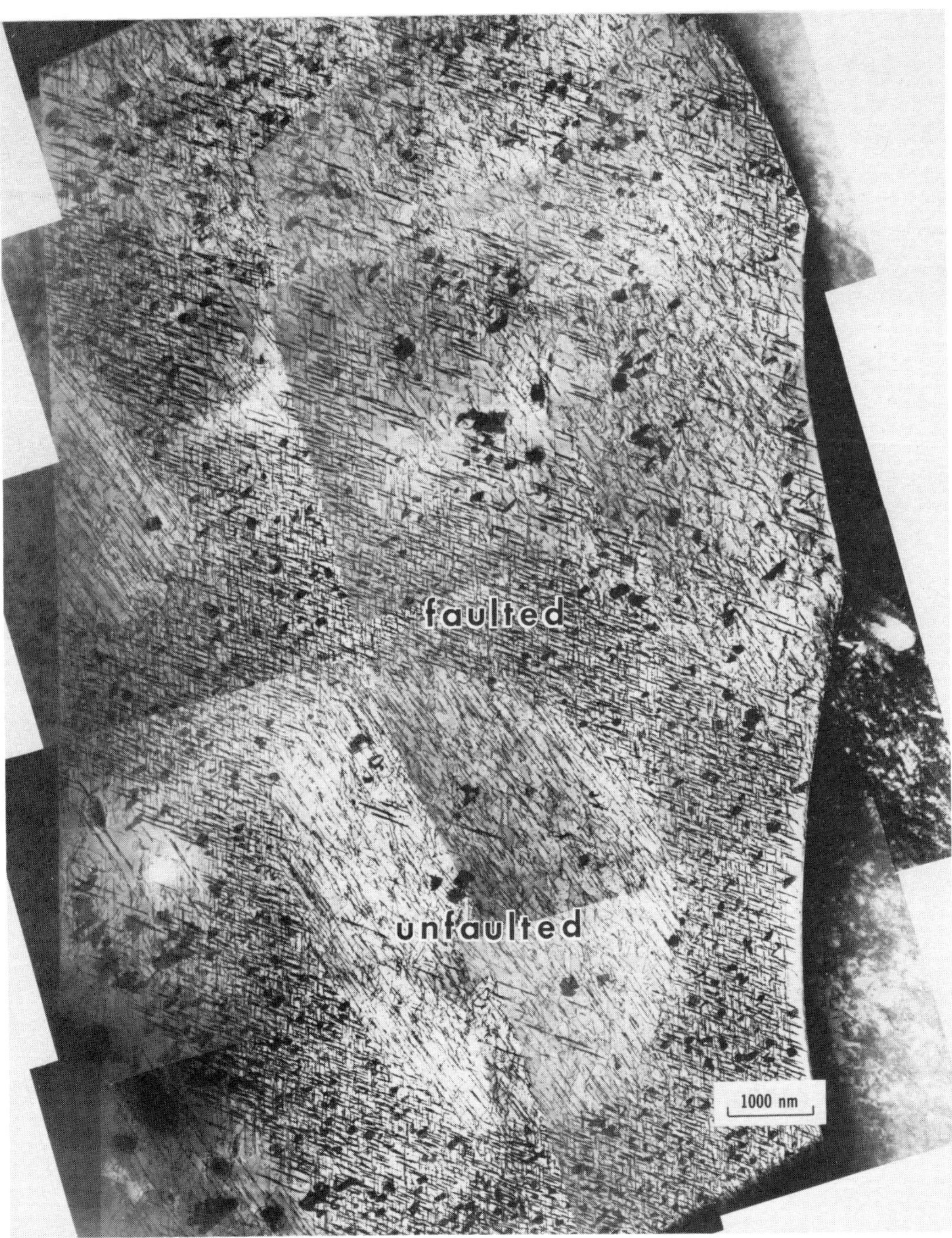

FIG. 3
Montage showing variations in the dislocation microstructure of a single grain in a specimen of Nimonic PE16 ST after irradiation to 2.0 x 10^{22} n/cm^2 (E >0.1 MeV) at 545°C under a hoop stress of 167 MPa.

the small loops (9). The loops continue to grow in response to the irradiation induced point defect fluxes until they unfault in some autocatalytic manner, thereby forming individual prismatic loops. The prismatic loop populations are presumed to retain the anisotropy observed in the Frank loop populations. Thus, the irradiation creep rate continues unaffected. The prismatic loops grow into a dislocation network and a steady state microstructure finally develops.

The present results demonstrate that an autocatalytic unfaulting process does not control development of a perfect dislocation network in Nimonic PE16 at a temperature of 545°C. Also, there is strong evidence that a steady state dislocation network develops in a similar manner in AISI 316 both in the solution treated and cold worked conditions. As perfect dislocation interactions appear to control the Frank loop unfaulting process in both Nimonic PE16 and AISI 316 and as these two materials constitute the major source of irradiation creep data, irradiation creep modeling must allow for the consequences of this unfaulting process.

A major consequence of a perfect dislocation/Frank loop interaction mechanism for controlling loop unfaulting is that anisotropy in the Frank loop populations generated by irradiation creep will not necessarily be retained after unfaulting. A given Frank loop can be unfaulted by any one of three perfect dislocation Burgers vectors out of the six possible. As the Frank loop contains opposing Burgers vectors on opposite sides of the loop, the sign of the unfaulting perfect dislocation is unimportant: the reaction will either take place on one side of the Frank loop or the other. Thus, a given Frank loop can be unfaulted by half of the perfect dislocation Burgers vectors available. Furthermore, it is possible to annihilate all Frank loops in a given area using perfect dislocations having only two of the six possible Burgers vectors because each perfect dislocation can unfault two sets of Frank loops (or half the population). Whereas it has been previously assumed that the distribution of perfect dislocation Burgers vectors must be fairly uniform, at most reflecting the anisotropy produced in Frank loop populations by irradiation creep, it is now apparent that much larger variations in the distribution of Burgers vectors representing perfect dislocations may be produced. This could lead to much larger macroscopic shape changes during irradiation creep.

The present work also indicates that the perfect dislocation sources which effect the Frank loop unfaulting reaction during irradiation creep are one and the same as those for the low temperature plastic deformation case. Grain boundary ledges and precipitate particle interfaces emit the perfect dislocations which primarily control dislocation network evolution. It is the perfect dislocation network which is responsible for steady state irradiation creep at high fluences.

However, a major difference exists between Frank loop unfaulting during cold working (5,6) as opposed to during irradiation creep. During irradiation creep, climb plays a significant role in dislocation mobility whereas during cold working only glide is important. Washburn and coworkers have demonstrated that unfaulting of a Frank loop can occur by an alternative interaction mechanism, one involving a perfect dislocation capable of gliding on the plane of the Frank loop. (The product of the reaction at the loop perimeter is a perfect dislocation with a Burgers vector different than that of the unfaulting dislocation.) Evidence for this alternative reaction has not been identified in irradiation creep specimens. The reason is thought to be due to the enhanced climb mobility available for dislocation motion during irradiation creep. Much stronger interaction forces are available for the unfaulting mechanism defined by equation (2) and this reaction therefore dominates the unfaulting response during irradiation creep.

References

1. M. M. Paxton, B. A. Chin, E. R. Gilbert and R. E. Nygren, J. Nucl. Mat., 80, 144 (1979).
2. F. R. N. Nabarro, Theory of Crystal Dislocations, Oxford University Press, London (1967), p. 361.
3. J. P. Hirth and J. Lothe, Theory of Dislocations, McGraw-Hill, New York (1968), p. 305.
4. V. P. Swart and S. Kritzinger, Eighth International Congress on Electron Microscopy, Canberra, Vol. 1 (1974), p. 456.
5. G. Saada and J. Washburn, J. Phys. Soc. of Japan, 18, Sup. #1 (1963), p. 43.
6. J. L. Strudel and J. Washburn, Phil. Mag., 9, 491 (1964).
7. L. E. Murr and E. Venkatesh, Metallography, 11, 61 (1978).
8. H. R. Brager, F. A. Garner and G. L. Guthrie, J. Nucl. Mat., 66, 301 (1977).
9. F. A. Garner, W. G. Wolfer and H. R. Brager, Effects of Radiation on Structural Materials, J. A. Sprague and D. Kramer (Eds), American Society for Testing and Materials, STP 683, Philadelphia, Pennsylvania (1979), p. 160.
10. D. S. Gelles, F. A. Garner and H. R. Brager, "Frank Loop Formation in Irradiated Metals in Response to Applied and Internal Stresses," Proceedings, ASTM 10th International Symposium on Effects of Radiation on Materials, Savannah, Georgia (June 1980).

ON THE KINETICS OF THE LOCKING OF A SCREW DISLOCATION DUE TO REORDERING OF AN INTERSTITIAL SOLUTE IN A BCC METAL

R. E. Reed-Hill
Department of Materials Science and Engineering
University of Florida
Gainesville, Florida 32611

Abstract

Some recent studies of the kinetics of the return of the yield point, due to the reordering of interstitial atoms in a BCC metal, have yielded relaxation times significantly smaller than the corresponding Snoek relaxation times. Calculations of the kinetics of the locking of a screw dislocation in a BCC metal have been made. These show that for distances from the dislocation less than A/kT reordering can occur significantly faster than predicted by Snoek kinetics. An Evans and Douthwaite calculation of the curve giving the force, required to pull a screw dislocation from an equilibrium atmosphere, versus the distance was also made. This gave a maximum force at approximately 0.5A/kT. These results suggest that the observed magnitude of the reordering yield point is due largely to the reordering of the interstitial atoms very close to the dislocation.

Delobelle and Oytana (1) have presented evidence that strongly suggests that in the Ta-O, Nb-O and Fe-C-N systems the return of the yield point due to the reordering of interstitial atoms in the stress fields of the dislocations follows the simple exponential law

$$\Delta\sigma = \Delta\sigma_{MAX}\,[1-\exp(-t/\tau_R)]$$

where t is the time, τ_R is a relaxation time, $\Delta\sigma$ is the difference in stress between the upper yield point and the flow stress at the end of the prestrain and $\Delta\sigma_{MAX}$ is the value of $\Delta\sigma$ due to reordering at infinite time.

Recent experimental determinations (2) of the relaxation time, τ_R, using vanadium containing oxygen, based on the time for $\Delta\sigma$ to reach 0.63 $\Delta\sigma_{MAX}$, gave values of τ_R equal to about a half or less than a half of the corresponding relaxation time for the Snoek effect in vanadium due to oxygen. Delobelle and Oytana have also reported that in the systems which they studied the strain-aging relaxation times associated with the reordering of interstitial atoms were significantly shorter than those measured for the Snoek effect using internal friction methods.

A possible cause for this disagreement between the two relaxation times may be as follows. Internal friction devices such as the torsion pendulum are usually operated so that the damping is strain-rate independent. This normally requires that the strain level in the specimen during a test be held below about 10^{-5}. On the other hand, the yield point due to the reordering of the interstitial atoms around a dislocation is probably largely due to the reordering close to the dislocation. Here reordering occurs under strain (or stress) levels several orders of magnitude greater than those in a torsion pendulum.

This problem has been attacked by computing the interaction energy between a screw dislocation and the interstitial atoms surrounding it as a function of the time available for reordering. In these calculations, following Schoeck and Seeger (3), and Cochardt, Schoeck and Wiedersich (4), it was assumed that the jump rate from one of the three possible <100> interstitial sites is controlled by an activation energy $q_i = q-u_i$ where q is the activation energy for interstitial diffusion and the u_i are the interaction energies

$$u_i = A/r \cos[\phi-(i-1)2\pi/3] \quad i = 1,2,3$$

between the screw dislocation and the interstitial atom and where A is the interaction parameter, r is the radial distance from the dislocation line and ϕ is the angle between r and the [01$\bar{1}$] direction.

If the total line interaction energy U_o is considered as a function of the time as well as of the coordinates r and ϕ of an atom with respect to the dislocation line then

$$U_o(t) = \int_o^{2\pi} \int_{r_1}^{r_2} \int_o^t \Sigma c_i(t,r,\phi)\, u_i(r,\phi)\, dr d\phi dt \quad c = 1,2,3$$

where the c_i are concentrations expressed in numbers of atoms per unit volume.

In order to evaluate the above integral, the dependence of the c_i on time must first be determined. This can be done by numerically integrating with respect to time the time derivatives of the concentrations. Expressions for the $\dot{c}_i$ may be obtained by considering the differences between the fluxes of interstitial atoms into and out of each type of site. Thus

$$\dot{c}_1 = c_2(1/\tau_{21}) + c_3(1/\tau_{31}) - c_1(1/\tau_{12}) - c_1(1/\tau_{13})$$

$$\dot{c}_2 = c_1(1/\tau_{12}) + c_3(1/\tau_{32}) - c_2(1/\tau_{21}) - c_2(1/\tau_{23})$$

$$\dot{c}_3 = -[\dot{c}_1 + \dot{c}_2]$$

where, in general, $1/\tau_{ij}$ is the interstitial atom jump rate from an i to a j site. If the u_i are zero, the interstitial atom jump rate is simply $1/\tau = 1/\tau_o \exp -q/kT$ where τ_o is a constant equal to $24a^2/D_o$ (a is the BCC metal lattice parameter and D_o is the interstitial diffusion coefficient frequency factor) and q is the activation energy for interstitial diffusion.

In the vicinity of a screw dislocation the u_i are finite and the jump rates are

$$1/\tau_{12} = 1/\tau_{13} = 1/2\tau_o \exp -(q-u_1)/kT = 1/2\tau \exp u_1/kT$$

$$1/\tau_{21} = 1/\tau_{23} = 1/2\tau_o \exp -(q-u_2)/kT = 1/2\tau \exp u_2/kT$$

$$1/\tau_{32} = 1/\tau_{31} = 1/2\tau_o \exp -(q-u_3)/kT = 1/2\tau \exp u_3/kT$$

Substituting these jump rate expressions into those for the $\dot{c}_i$ gives

$$\dot{c}_1 = c_o/2\tau\ [\exp u_3/kT - (\exp u_3/kT + 2 \exp u_1/kT)c_1/c_o + (\exp u_2/kT - \exp u_3/kT)c_2/c_o]$$

$$\dot{c}_2 = c_o/2\tau\ [\exp u_3/kT + (\exp u_1/kT - \exp u_3/kT)c_1/c_o - (\exp u_3/kT + 2 \exp u_2/kT)c_2/c_o]$$

$$\dot{c}_3 = -[\dot{c}_1 + \dot{c}_2]$$

To determine if the rate of locking due to the reordering of the interstitial atoms is dependent on the value of r the function

$$u_o(r,t) = \int_o^{2\pi} \int_{.03\tau_{R\ SN}}^t \Sigma c_i(t,r,\phi)\, u_i(r,\phi)\, d\phi dt$$

was evaluated by numerical integration with respect to both the time t and the angle ϕ at each of a number of arbitrarily selected values of r. The $u_o(r,t)$ thus obtained allowed the interaction energy, summed around the dislocation at a fixed value of r, to be followed as a function of time. The time interval from 0.03 to $3.0\tau_{R\ SN}$ was surveyed using time increments of $0.03\tau_{R\ SN}$. In the integration with respect to ϕ a 15° increment was used.

The results are shown in Fig. 1 where curves showing the variation of the interaction energy with respect to time are plotted for 5 values of r between 0.25R and 1.00R, where R = A/kT. Each of these curves was normalized by dividing the u_o data by the relevant value of u_{oMAX}. A sixth curve is also shown in Fig. 1 that corresponds to

$$u_{SN} = u_{oMAX}[1-\exp-(t/\tau_{R\ SN})]$$

It should be noted that at $r = A/kT$ the interaction energies u_i become equal to the thermal energy kT. This distance in iron at 300 K is 44.44 Å or 44.44 x 10^{-10}m or about 9b assuming $A = 1.84 \times 10^{-20}$ dyn-cm^2 (1.84×10^{-29} J-m).[3]

Figure 2 shows a plot from r = 0.25R to 5.0R of the ratio of the computed relaxation time τ_R to the Snoek relaxation time $\tau_{R\ SN}$. The computed relaxation times represent the respective times required for u_o/u_{oMAX} to reach 0.63. The data in Figs. 1 and 2 clearly show that for distances less than $r = A/kT$ the reordering kinetics are faster than the Snoek kinetics. In fact, as shown in Fig. 2, the apparent relaxation time is only about $1/3\tau_{R\ SN}$ at r = 0.25R (4.5b). On the other hand, at r = 5R (about 90b) the computed relaxation time has risen sufficiently so that it is within about a half a percent of the Snoek relaxation time. At distances greater than this the relaxation kinetics can be assumed to closely approximate the Snoek kinetics. This implies that the reordering very close to the dislocation line probably in large measure controls the magnitude of the yield point return in a kinetics study when one takes account of the observed yield point return kinetics that are more rapid than the Snoek kinetics.

If this is true, it may also be possible that under equilibrium conditions (with maximum ordering) the major contribution to the observed value of the yield point return may come from the reordering of the interstitial atoms close to the dislocation.

To test this conclusion, calculations similar to those of Evans and Douthwaite were made. These involved computing the interaction energy between a screw dislocation and its equilibrium atmosphere as the dislocation is pulled from the center of its atmosphere assuming that the displacement occurs too rapidly for reordering of the interstitial atoms to occur. There are five parameters in this calculation. These are r_1 and ϕ_1 giving the coordinates of a point in the crystal relative to the center of the atmosphere. These coordinates define the equilibrium concentrations $\bar{c}_1$, $\bar{c}_2$ and $\bar{c}_3$ of the interstitial atoms. The coordinates r_2 and ϕ_2 then locate the given point in the crystal in terms of the displaced position of the dislocation. The displacement distance L is the fifth parameter. The interaction energy was computed by numerically integrating with respect to r_2 and ϕ_2 from r_2 = 0.225R to 6.98R using increments of 0.1125R for r_2. It was determined that extending the upper limit beyond 6.98R had a negligible effect on the variation of the interaction energy in the area of interest, i.e., below a maximum displacement distance L of about 1.125R. The angle ϕ was varied from 0 to 180° in increments of 7.5° taking advantage of the two-fold symmetry of the problem. Values of the interaction energy were computed corresponding to dislocation displacements from L = 1/32R to 35/32R in steps of 1/32R. Figure 3 shows the interaction energy U_L thus calculated as a function of the displacement distance L.

The derivative of U_L with respect to L is proportional to the force required to displace the dislocation. The variation of this quantity with respect to the displacement distance is shown in Fig. 4. Note that it reaches a maximum at approximately 0.50R (22.2 Å in Fe at 300 K) where the force to remove the dislocation from its atmosphere should thus be at its maximum. This result is also consistent with the assumption that the locking of the screw dislocation is largely associated with the reordering of the interstitial atoms close to the dislocation.

Acknowledgments

The financial support for this work was furnished by the Department of Energy, Contract No. DE-AS05-76ER03262. The assistance of S. C. Park and S. Krishnamurthy with some of the calculations is also acknowledged.

References

1. P. Delobelle and C. Oytana, Scripta Met., 12, 765 (1978).
2. L. P. Beckerman, Ph.D. Dissertation, University of Florida, Gainesville, Florida (1980).
3. G. Schoeck and A. Seeger, Acta Met., 7, 469 (1959).
4. A. W. Cochardt, G. Schoeck and H. Wiedersich, Acta Met., 3, 533 (1955).
5. J. T. Evans and R. M. Douthwaite, Acta Met., 21, 49 (1973).

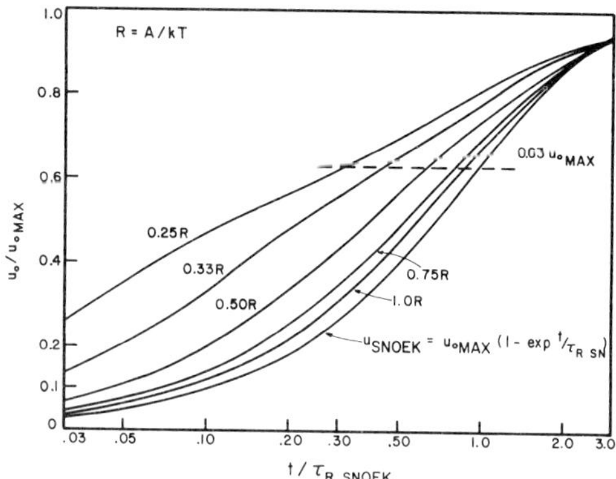

Fig. 1. The variation with time of u_o/u_{oMAX} for several distances from the dislocation line between 0.25R and 1.0 R.

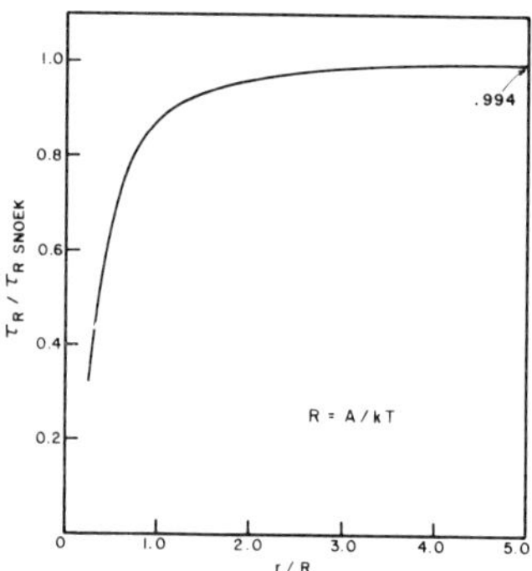

Fig. 2. The dependence of the ratio $\tau_R/\tau_{R\,SN}$ on the distance from the dislocation line.

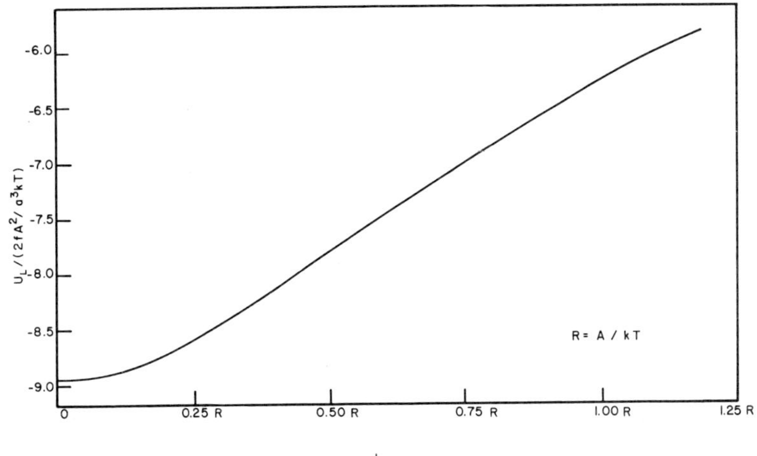

Fig. 3. The total interaction energy U_L as a function of the displacement distance when a screw dislocation is pulled from its equilibrium atmosphere. Note f is the interstitial concentration in atom fraction.

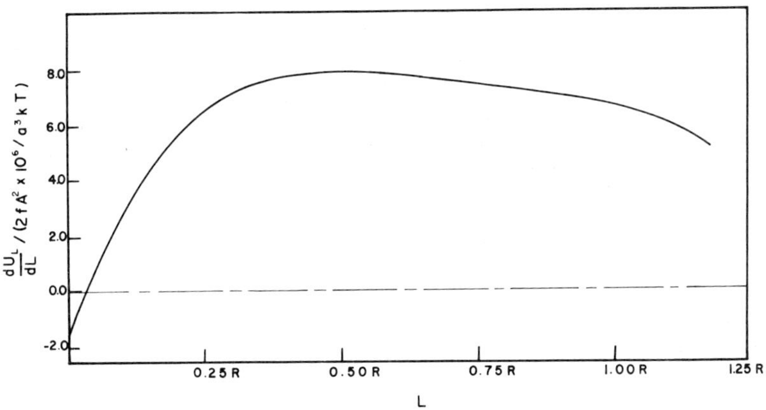

Fig. 4. The variation dU_L/d_L, which is proportional to the force required to pull a screw dislocation from its equilibrium atmosphere, as a function of the displacement distance. Note f is the atom fraction of the interstitial.

A MECHANISM RATIONALIZING STRAIN AGING KINETICS IN NICKEL 200

W. R. Cribb
Brush Wellman Inc.
South River Road
Elmore, Ohio 43416

R. E. Reed-Hill
Dept. of Materials Sci. & Engr.
University of Florida
Gainesville, Florida 32611

Abstract

A model is presented which can explain the second stage of static strain aging in Nickel 200. Unlike strain aging phenomena in many bcc metals which are rationalized by the Cottrell-Bilby model ($t^{2/3}$ kinetics), Nickel 200 exhibits either a logarithmic or $t^{1/7}$ lower yield stress return.(1) The development of a yield return is quantitatively rationalized by assuming that immediately vacancies created during deformation become trapped by interstitial carbon atoms.(2) The carbon-vacancy pairs thus formed act as strain dipoles and quickly give rise to an initial yield point return (Stage I). Stage II, which terminates before peak aging, is governed by stress-assisted diffusion of vacancies toward dislocations with trapping at interstitial carbon atoms.

Introduction

This paper utilizes the concept first set forth by Rose and Glover (3) to develop a quantitative relationship between the lower yield return and the stress-assisted vacancy migration of carbon-vacancy pairs to dislocation. Prior publications (1,2,4) describe the ordering of carbon vacancy pairs for very short aging times and provide more background regarding the justification for the concept as it applies to Nickel 200.

Experimental Procedure and Results

Nickel 200 specimens containing 0.18% Mn and 0.10% C were processed in a fashion described elsewhere(1). Aging after 5 pct tensile strain at 273K was conducted at 373, 408, 428, 448 or 473K for up to approximately 10^6s. The behavior of $\Delta\sigma$, the increase in lower yield stress above the prestrain flow stress, is shown in Fig. 1. A logarithmic plot, Fig. 2, demonstrates that the lower yield stress return can be adequately described using a logarithmic or $t^{1/7}$ power law equation. To assist in establishing a mechanism for the static strain aging before peak aging, the apparent activation energy for the lower yield return was calculated using the times required to achieve stress increases of 15 and 20 MPa. On this basis, the activation energy for the yield point return in Nickel 200 was determined to be approximately 25,200 cal/mole.

Discussion

Methodology

It is assumed that in Stage II of strain-aging (i.e., before the peak in $\Delta\sigma$), very few vacancies are lost by reaching the dislocation core or by moving to other sinks since trapping at carbon atoms dominates. As free vacancies migrate toward the dislocation from a distance, it is also assumed that a saturated Snoek atmosphere is formed near the dislocation with a fraction, f, of the carbon atoms paired to vacancies. To relate the increase in $\Delta\sigma$ to time, a "saturation radius", R_s, bounding the carbon-vacancy pair atmosphere is defined that grows with time. This is schematically illustrated in Fig. 3. This limits the carbon-vacancy pair concentration to fc_c within the radius R_s where c_c is the atomic fraction of carbon atoms.

The relationship of the "saturation radius" to the aging kinetics may be more clearly visualized as follows. At the beginning of the second stage of aging (after initial Snoek ordering) the concentration of the most favored carbon-vacancy pair orientation, as a result of continued vacancy migration, becomes constant, i.e., fc_c, for distances closer to the dislocation than R_s. The value of f must be much less than unity since all carbon atoms cannot be paired to vacancies because of equilibrium considerations. As vacancies continue to accumulate within the atmosphere, R_s increases with time. The growth of the saturated Snoek atmosphere causes the energy of the dislocation and its carbon-vacancy pair atmosphere to decrease with time. By relating the energy decrease to the increased force required to free a dislocation from its atmosphere, the kinetic dependence of $\Delta\sigma$, the yield stress increase after aging, is calculated. The activation energy associated with this process is that for long range vacancy migration in the presence of carbon impurity since the diffusion process, in conjunction with trapping at carbon atoms, controls the supply of carbon-vacancy pairs to the saturated Snoek atmosphere.

Quantitative Model Of The Static Strain Aging In Nickel 200

From the Cottrell-Bilby model and assuming that the vacancy interaction energy with a dislocation behaves as Ar^{-2} (modulus interaction), one may deduce the number of vacancies that have arrived at a dislocation in time t as

$$n_v(t) = n_v^o \pi b \left(\frac{8AD}{kT}\right)^{1/2} t^{1/2} = \frac{C_v}{\Omega} \pi b \left(\frac{8AD}{kT}\right)^{1/2} t^{1/2} \tag{1}$$

where n_v^o is the average initial vacancy concentration per unit volume, A the effective interaction constant of a vacancy, D the apparent diffusivity of a vacancy in the presence of carbon impurity atoms, t the time in seconds, C_v the atomic fraction of vacancies, and Ω the atomic volume of the solvent.

Assuming that saturation of carbon atoms occurs within a radius r of the dislocation and that the growth of the "saturation radius" is described reasonably well by a two-dimensional growth model, then:

$$\frac{dn_v}{dt} = fn_c \left(2\pi rb\frac{dr}{dt}\right) \tag{2}$$

where dn_v/dt is the accumulation rate of vacancies moving into a cylinder of radius r and depth b between r and r + dr in time dt. The concentration of vacancies in this cylinder is assumed to be fn_c where n_c is the number of carbon atoms per cm^3 and is assumed to be unchanging and independent of position.

Integration of Eq. 2 results in $n_v(t) = f\pi R_s^2 bn_c = f\pi R_s^2 bc_c/\Omega$ (3)
where c_c is the atom fraction of carbon.

Equation 3 describes the number of carbon-vacancy pairs accumulated within r at time t. Each available vacancy is assumed to be attached to a carbon atom. Equating Eq. 1 to Eq. 3 and rearranging gives the growth of the ordered Snoek atmosphere:

$$R_s = \left(\frac{c_v}{fc_c}\right)^{1/2} \left(\frac{8AD}{kT}\right)^{1/4} t^{1/4} \tag{4}$$

Thus, the atmosphere grows approximately as $t^{1/4}$.

Equation 1 is, strictly speaking, the accumulation of vacancies near the dislocation with time. The meaning of Eq. 3, however, is the number of vacancies that have accumulated within R_s; that is, the vacancies are not arriving at the core of the dislocation as Eq. 1 assumes but are accumulating as R_s grows. Thus, the exact meaning of Eq. 1 is relaxed somewhat to allow for the estimation of the growth of a Snoek atmosphere.

It has been shown (4) that the decrease in energy associated with Snoek ordering of carbon-vacancy pairs around a dislocation after a time span approximately equal to that required for one reorientation event is approximately

$$U_o = \frac{8\pi n_o A^2}{3kT} \ln \frac{R_o}{r_o} \tag{5}$$

where n_o is the average volume concentration of the ordering species (dipoles) and r_o is approximately the dislocation core radius, 1b. R_o is the upper limit of integration on U and is $\rho^{-1/2}/2$, i.e., one-half of the mean dislocation spacing.

In order to estimate the kinetics of pinning for the present case, it is assumed that only those carbon-vacancy pairs within the "saturation radius" contribute to the decrease in energy. This is reasonable since the assumed interaction energy between a dislocation and a carbon-vacancy pair decreases as r^{-1}. Thus, the upper limit of the integration is approximately R_s rather than one-half the mean dislocation spacing and Eq. 5 becomes

$$U_o = \frac{8\pi fc_c A^2}{3\Omega kT} \ln \frac{R_s}{r_o} \tag{6}$$

where $r_o \simeq b$ and n_o becomes fc_c/Ω.

An estimate of the increased applied tensile stress necessary to free the dislocation from its Snoek atmosphere is approximately given by $F_{max} \simeq \frac{b}{m} \Delta\sigma$ (6,7).

Thus, $$\Delta\sigma \simeq \frac{mkT}{2bA} U_o \qquad (7)$$

where m is an orientation factor between 2 and 2.75. Using the result of Eqs. 6 and 7, one deduces that

$$\Delta\sigma = \frac{4m\pi f c_c A^2}{3\Omega b} \ln \frac{R_s}{r_o} \qquad (8)$$

Thus, the increase in $\Delta\sigma$ is related to the size of the ordered atmosphere. Substituting the time dependent relation for R_s, Eq. 4 into Eq. 8 results in

$$\Delta\sigma = \frac{m\pi f c_c A}{3\Omega b} \ln t + \frac{m\pi f c_c A}{3\Omega b} \left(-\frac{Qm}{RT}\right) + \frac{4m\pi f c_c A}{3\Omega b} \ln \left(\frac{C_v}{f c_c}\right)^{1/2} \left(\frac{8AD_o}{kT}\right)^{1/4} \left(\frac{1}{r_o}\right) \qquad (9)$$

This model neglects (a) migration of carbon atoms, and (b) movement of vacancies within the "saturation radius", R_s, to the dislocation where annihilation might be assumed to occur by elimination of jogs or rapid "pipe" diffusion to other sinks.

Equation 9 is the increase in flow stress with time caused by the stress enhanced diffusion of vacancies toward the dislocation to form a growing saturated and ordered carbon-vacancy pair atmsophere. Accurately speaking, Eq. 9 is the upper yield point increase which is caused by breakaway of dislocations from their atmospheres. As an approximation, however, the lower yield stress increment, $\Delta\sigma$, should follow the time dependent behavior of the upper yield point much in the same way as in bcc metals which obey Cottrell-Bilby kinetics (8).

The time dependence of $\Delta\sigma$ is reflected only in the first term of Eq. 9, i.e., ln t. Furthermore, the first term also predicts that the slopes of $\Delta\sigma$ versus ln t curves should be independent of temperature. This feature is reasonably well exhibited in the $\Delta\sigma$ versus log t of Figure 1. The temperature dependence of static strain aging from this model lies in the second and third terms of Eq. 9. The predominant temperature dependence is reflected in the second term. Thus, the model predicts (for times less than those to achieve maximum hardening) straight $\Delta\sigma$-ln t aging curves which are parallel and are simply displaced in time depending upon the size of Q_m and T.

As an example, the 428 K aging curve of Fig. 1 was chosen to demonstrate the model given by Eq. 9. The values assumed for the various parameters to obtain a reasonable fit to the data are listed below:

$m = 2.5$	$D_o = 1\ cm^2/s$
$\Omega = a^3/4 = b^3 \sqrt{2}$	$Q_m = 25{,}200$ cal/mole
$b = 2.49 \times 10^{-8}$ cm	$C_v = 1.2 \times 10^{-9}$
$c_c = 5 \times 10^{-3}$	$r_o = b$
$A = 0.2$ eV-b $= 4.98 \times 10^{-9}$ eV-cm	$f = 0.065$

These values when substituted into Eq. 9 give

$$\Delta\sigma(MPa) = 2.414 \ln t - 3.042 \times 10^4\ T^{-1} - 2.414 \ln T + 88.679 \qquad (10)$$

Table I lists the values for the slopes and intercepts (at t=1 second) of the $\Delta\sigma$ versus ln t curves calculated from Eq. 10 and those determined from the experimental data. In addition, Fig. 4 illustrates the comparsion between the model and the experimental data. The agreement is reasonable. Note that the principal unknown which could not be estimated was f, the maximum fraction of carbon atoms which can be occupied by vacancies. The value assumed for f is 0.065 and would imply that the solubility limit of carbon-vacancy pairs near a dislocation, that is, the saturation limit, is considerably less than the total number of carbon atoms. This is a reasonable result in view of the comments made by Cochardt et al. (9) relating to the saturation concentration of carbon in iron near a dislocation. They estimate a maximum carbon concentration of 0.07 atom fraction near edge dislocations and about twice as much near screw dislocations. In addition a variation in f could account for the difference between the upper and lower yield points since the model presented relates to the upper yield point.

The value for the average free vacancy concentration, c_v^f, of 10^{-9} atom fraction is in fair agreement with the 7×10^{-8} value obtained by assuming a binding energy of approximately 0.3 eV between carbon atoms and vacancies. Thus, the model as derived can rationalize the aging behavior for times less than the time to achieve a maximum in $\Delta\sigma$.

Acknowledgments

The support of the Energy Research and Development Agency under Contract AT-(40-1)-3262 is gratefully acknowledged. Helpful discussion with Dr. C. S. Hartley is also appreciated.

References

1. W. R. Cribb and R. E. Reed-Hill, Met. Trans. A, 8A, 71 (1977).
2. W. R. Cribb and R. E. Reed-Hill, Met. Trans. A, 9A, 887 (1978).
3. K. S. B. Rose and S. G. Glover, Acta Met., 14, 1505 (1966).
4. W. R. Cribb, PhD Dissertation, University of Florida, 1975.
5. F. R. N. Nabarro, Theory of Crystal Dislocations, Oxford University Press, 409 (1967).
6. G. Schoeck, Phys. Rev., 102, 1458 (1956).
7. G. Schoeck and A. Seeger, Acta Met., 7, 469 (1959).
8. J. D. Baird, Met. Rev., 5, 1 (1971).
9. A. W. Cochardt, G. Schoeck and H. Wiedersich, Acta Met., 3, 532 (1955).

TABLE I

AGING CURVE PARAMETERS

	Raw Data		Equation 10	
Temperature (K)	Slope (MPa/ln s)	Intercept (t = 1s)	Slope (MPa/ln s)	Intercept (t = 1s)
373	2.09	-4.92	2.41	-7.17
408	2.65	-3.36	2.41	-0.39
428	2.60	1.62	2.41	2.98
448	2.32	6.69	2.41	6.04
473	-	-	2.41	9.50

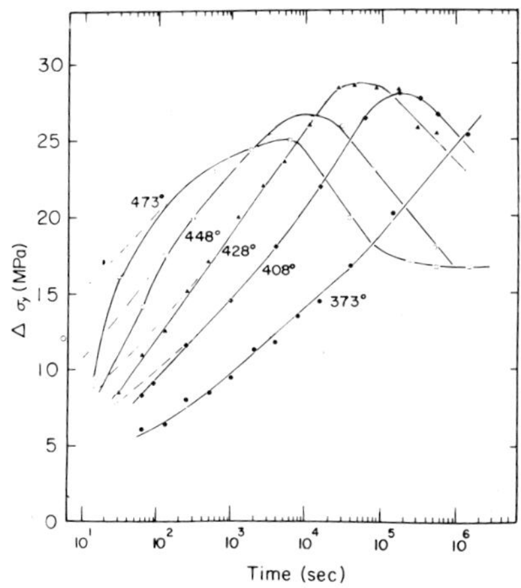

FIG. 1
The variation of $\Delta\sigma$ with log t

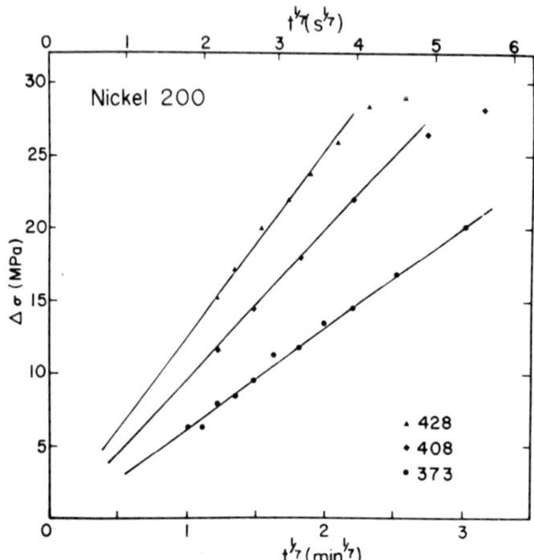

FIG. 2
Plots of $\Delta\sigma$ vs t $^{1/7}$ for the three lower aging temperatures.

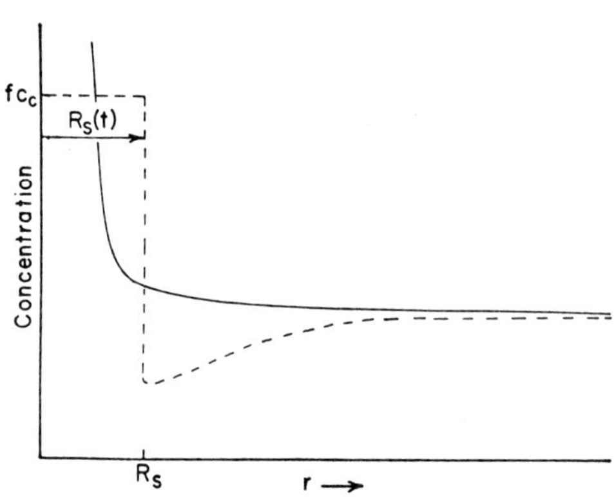

FIG. 3
Schematic illustration of the growth of a saturated carbon-vacancy atmosphere. R_s is time dependent and the concentration within R_s is assumed to be a fraction, f, of the carbon centration.

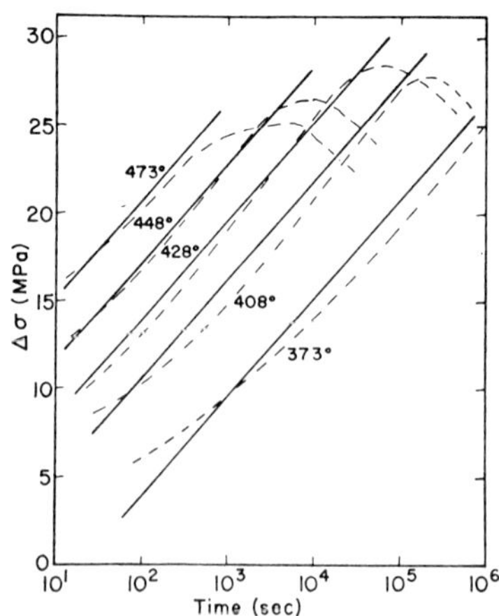

FIG. 4
The aging curves obtained from model (Eq. 9); dashed lines are the experimental data.

DISLOCATION KINETICS AT NOT-SO-CONSTANT STRUCTURE*

U. F. Kocks and H. Mecking**

Materials Science Division, Argonne National Laboratory, Argonne, IL 60439

Abstract

Departures from the idealization of dislocation kinetics "at constant structure" are discussed. They stem from changes in the mobile dislocation density, changes in the structure sampled by dislocations as they percolate through the slip plane, changes in dynamic pile-up configurations, changes in the density and arrangement of dislocations through dynamic recovery, and changes in the solute concentration along dislocations through aging. All of these adjustments in internal parameters lead to transients in macroscopic behavior after an abrupt change in external conditions. The type, strength, and length of observed transients can give an indication as to which structure change may be responsible for it. This provides the basis for a judgment of what is the most appropriate back-extrapolation procedure in the evaluation of experiments.

Another effect of changing structure is that even "instantaneously" evaluated macroscopic properties may depend on the current rate of change of structure. It is suggested that this effect may play an important role in the interpretation of both the flow-stress plateau at intermediate temperatures and the high-temperature drop in flow stress.

Introduction

Twenty-five years ago, Adams and Cottrell [1] published a series of experiments that culminated the rapidly developing understanding of the time concerning the dependence of the flow stress on temperature and strain rate. They introduced an experimental technique in which the temperature is abruptly changed during a stress strain test, and the flow stresses before and after the change are compared. This contrasts with a comparison of continuous stress strain curves taken at different temperatures. The important concept that makes this more cumbersome method of abrupt changes necessary was probably an extension to low temperatures of the recovery/glide theory of creep [2], in which it was realized that the evolution of the dislocation structure during straining could follow kinetics different from that controlling glide at a given structure. Adams and Cottrell in effect postulated that structure variations were eliminated by comparing the flow stresses immediately before and after a change in conditions, so that glide kinetics could be investigated in this way.

The same experiments also confirmed for copper the previous observation in aluminum[3] that transients in the stress strain behavior sometimes follow upon such a change in conditions. In particular, Stokes and Cottrell identified the phenomenon of work softening: a transient

* Work supported by the U. S. Department of Energy.

** Permanent address: Technical University, Hamburg-Harburg, W. Germany.

negative rate of strain hardening after an increase in temperature. It was interpreted as a partial collapse of the dislocation structure under the combined action of straining and thermal activation. In the meantime, an equivalent transient with opposite sign has been found, under some conditons, upon a decrease in temperature (or an increase in strain rate)[4,5]. Other types of transients were also discovered [4,6], which were associated with adjustments in the mobile dislocation distribution [4] or density [6], in analogy to the initial (slow) yield drop and anomalous primary creep in some ceramic materials [7,8,9]. Finally, a particularly obvious transient has long been known that is a consequence of solute atom mobility in strain aging.

All these transients have in common that some element of the structure is assumed to change while they are in progress. Indeed, whenever the flow stress changes while the strain rate and temperature remain the same, the mechanical state of the system must change in some way. Often, as in the case of the transients mentioned above, the changes in state come to end after an little while: the structure variable that was changing reaches an equilibrium value compatible with the external conditions and with the other, slowly-changing internal parameters.

Under these circumstances, it is not self-evident how an abrupt-change experiment should be evaluated: should the new flow stress be measured as instantaneously as possible, so that all aspects of the structure are as little changed as possible; or should one extrapolate through the transient, so that the extra structural element, which is changing here, comes to its new equilibrium under the changed external conditions? This question has, on occasion, assumed the flavor of a philsophical battle. We believe that the answer should in fact be different under different circumstances -- and that, in general, something important can be gained by not ignoring the existence of the transient by either one of these methods.

The predecessor of this conference (or should one say: the ancestor?) saw two papers that dealt specifically with the complexities of this problem. In one [10], the existence of a number of different characteristic times in dislocation kinetics was emphasized. (Actually, different strain increments are usually more appropriate.) In another paper [11], it was proposed that the rate of change of structure be incorporated into the kinetic equation for glide: it may be finite even "at constant structure" (in the sense of a Taylor expansion). In that case, even "instantaneous" flow stress changes reflect evolution kinetics, as will be discussed in the last section.

The plan of this article is as follows. First, the "simple theory" of flow stress changes at constant structure will be summarized. Then, the phenomenological form of transients as they are observed after abrupt changes in both stress/strain and creep tests is reviewed, and it is emphasized that these transients cannot be assumed to be absent even in continuous tests (such as stress relaxation), where they are not easily identified. The next two sections deal with adjustments in mobile dislocation density and with the percolation of dislocations through regions of varying glide resistance. While these processes lead to significant transients only under rather restricted conditions, adjustments in dynamic pile-up distributions are shown, in the following section, to be able to explain a commonly observed transient. The next two sections deal with the two most important transients: those produced by dislocation rearrangement during dynamic recovery and with solute redistribution during dynamic strain-aging, both of which also exert a significant influence on kinetic behavior during continuous tests. As a result, the notion of a "flow stress at constant structure" is applicable in the classical sense only to deformation at relatively low temperatures.

Dislocation Kinetics

The theory of the flow stress at constant structure was reviewed (and supplemented) in great detail by Kocks, Argon, and Ashby in 1975 [12]. For the usual case of jerky glide, the strain rate $\dot{\varepsilon}$ is an average of a small "unit strain" ε_o achieved after each waiting time t_w:

$$\dot{\varepsilon} = \varepsilon_o/t_w \tag{1}$$

The temperature dependence can be described by an Arrhenius term:

$$t_w^{-1} = \nu_G \exp\left[-\Delta G(\tau)/kT\right] \tag{2}$$

Here, ν_G is an appropriate attempt frequency. Equations (1) and (2) together give the strain rate as a function of temperature and of stress, which enters through the activation energy ΔG. If the temperature and strain rate are prescribed, the stress is called the <u>flow stress</u> τ under these conditions.

The Arrhenius description is useful only if the preexponential factor $\dot{\varepsilon}_o \equiv \varepsilon_o \nu_G$ is constant, or at least depends on temperature and stress less strongly than the exponential, or in a well-known way. There are two general mechanistic expressions for ε_o , depending on whether the mobile dislocations can be regarded as virtually straight or whether glide is essentially two-dimensional. In the first case,

$$\varepsilon_o = b\ \rho_m\ L \tag{3}$$

where b is the Burgers vector, ρ_m the length of "mobile" (waiting plus running) dislocations per unit volume, and L the distance travelled after a successful activation event. The jerky motion associated with widely spaced thermal-activation events can, in this case, be reasonably described by an average velocity $\bar{v}$, which is L/t_w:

$$\dot{\varepsilon} = b\ \rho_m\ \bar{v} \tag{4}$$

In most cases, on the other hand, glide is more appropriately represented by a number N_w of dislocation <u>segments</u> per unit volume, waiting to be activated and sweeping out an <u>area</u> a after a successful release. Then,

$$\varepsilon_o = b\ N_w\ a \tag{5}$$

and the meaning of a "velocity" becomes tenuous.

The density of mobile dislocations or mobile segments has played a major role in previous discussions of transient phenomena [4,8,9,10]. More important structure changes, however, enter through the activation energy, not through the preexponential factor.

The dependence of ΔG on stress and on structure are intimately linked. This may seem at first surprising, inasmuch as the activation energy relates to a <u>local</u> event, which depends only on the local properties of the obstacles and the local force applied to them. But since the macroscopically measured quantity is a <u>stress</u>, which depends not only on the necessary local <u>force</u> but also on the obstacle <u>spacing</u>, ΔG depends on a structure parameter when it depends on stress. The most appropriate measure of this structure dependent material property is [12] the mechanical threshold $\hat{\tau}$: the stress at which plastic flow can occur without the benefit of thermal activation, i.e., when $\Delta G = 0$. It depends on both the strength and the spacing of the obstacles. Thus, the activation energy can be best described as a function of $\tau/\hat{\tau}$: the stress (or: flow stress at a given temperature and strain rate), divided by the mechanical threshold.

Frequently, the flow stress is assumed to have an additive athermal contribution τ_o (on which more later). Then one can write

$$\Delta G = \mu b^3 \cdot g\left(\frac{\tau - \tau_o}{\hat{\tau} - \tau_o}\right) \tag{6}$$

Here, ΔG has also been normalized by the shear modulus μ times b^3. The free-enthalpy nature of ΔG can be explicitly taken account of by allowing a temperature dependence proportional to that of μb^3. These are special assumptions, but often sufficiently close for a description of dislocation interaction processes.[12]

The end result of such a treatment of thermally activated plastic flow is the prediction of the diagram shown in Fig. 1, or at least of the right combination of variables on its axes. One can show that the normalization of the flow stress by μ and of the temperature (times the Boltzmann constant k) by μb^3 must go hand-in-hand. Furthermore, temperature and strain rate must appear in the combination $(kT/\mu b^3)\cdot\ln(\dot{\varepsilon}_o/\dot{\varepsilon})$, which is the value of the function g in eq.(6). The curves shown in Fig. 1 thus display the dependence of g on τ which was left open up to now. Its general characteristics are as follows. The flow stress must have a maximum value ($\hat{\tau}$) at T = 0 K, which is a structure-dependent material property, and it may have a finite minimum value (τ_o) up to high temperatures. Where τ reaches τ_o, $\Delta G = F_o$, the total interaction energy of dislocation and obstacle. In Fig. 1, F_o was chosen to have two widely distinct values: about 0.5 eV and 2.5 eV. (The latter would typically put the point $\tau = \tau_o$ above the melting point.) Beyond these <u>scale</u> parameters, the <u>shape</u> of the curves which, in principle, reflects the obstacle profile, is insensitive to details.[12]

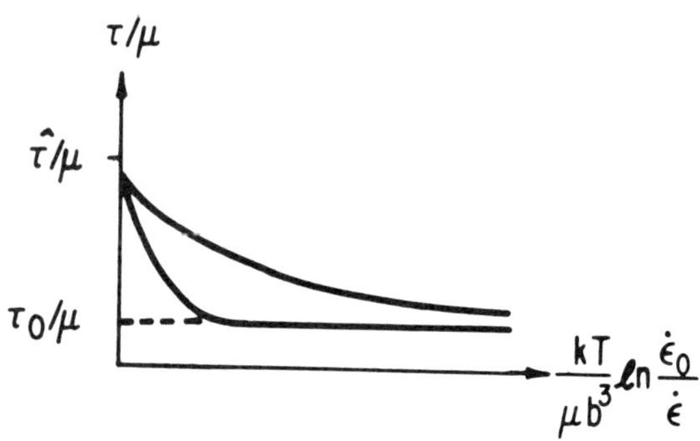

Fig. 1 - Theoretical prediction: the combined thermal-activation parameter g lowers the flow stress (over shear modulus) from the mechanical threshold $\hat{\tau}$.

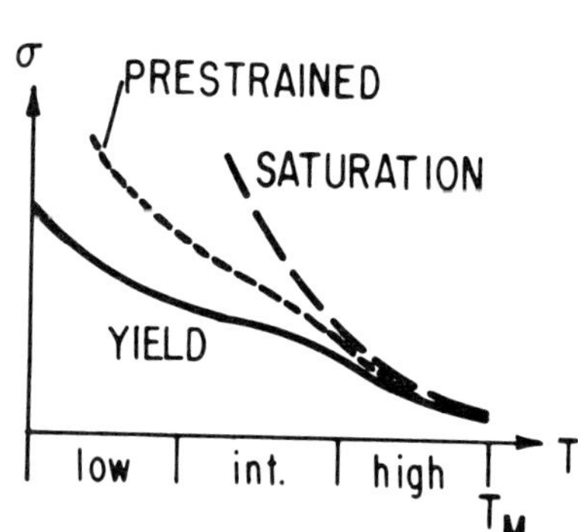

Fig. 2 - Typical experimental flow-stress vs temperature relation in three temperature regimes.*

Typical experimental diagrams of flow stress versus temperature have only a limited similarity to these predictions. Figure 2 shows a qualitative example. The agreement is very good for the yield stress at low temperatures. At intermediate temperatures, there is often the semblance of a plateau (especially if one has divided by μ), which could be related to an "athermal" flow stress contribution τ_o, or to any mechanism involving an especially high value of the total activation energy F_o. In any case, it is not well explained to many investigators' satisfaction. At high temperatures, there is a drop in yield stress that cannot be explained on the basis of glide kinetics at constant structure. We shall discuss both regimes more fully below.

The flow stress in the above discussion has been identified with the yield stress, and the "constant structure", therefore, with that obtained from the previous thermo-mechanical history. It may be physically due to a large variety of mechanisms. A particular contribution of interest here is that due to dislocation/dislocation interactions, since it is the main cause of flow stress changes during deformation. Treating the total dislocation density ρ as the principal structure parameter, the structure dependence of the mechanical threshold can be written as

$$\hat{\tau} = \hat{\tau}_f + \hat{\alpha}\ \mu\ b\ \sqrt{\rho} \qquad (7)$$

All contributions other than the dislocation/dislocation interaction component have here been lumped into a "friction stress" $\hat{\tau}_f$, which was assumed to be linearly additive. This superposition rule is reasonable [13] and observed [14] in a number of cases, such as a lattice resistance or solution hardening — but in just those cases, it is not reasonable to assume τ_f to be athermal. Thus, the flow stress at finite temperature and strain rate has two thermally activated contributions, not one as in eq.(6):

$$\tau = \tau_f(\dot{\varepsilon},T) + \alpha(\dot{\varepsilon},T)\ \mu\ b\ \sqrt{\rho} \qquad (8)$$

The proportionality constant $\hat{\alpha}$, which was introduced in eq.(7), depends on the strength of the dislocation/dislocation interaction, which was, in eq.(8), assumed to be effectively lowered to $\alpha(\dot{\varepsilon},T)$ by thermal activation.[12]

The dependence of the "flow stress at constant structure" on temperature and strain rate now follows from differentiating eq.(8). For strain-rate changes, one obtains

$$\beta \equiv \left.\frac{\partial\tau}{\partial\ln\dot{\varepsilon}}\right|_{T,\rho} = \frac{\partial\tau_f}{\partial\ln\dot{\varepsilon}} + (\tau - \tau_f)\cdot\left.\frac{\partial\ln\alpha}{\partial\ln\dot{\varepsilon}}\right|_T \qquad (9)$$

* The symbol σ was used for the experimental flow stress in this diagram because, at high temperatures, it may no longer reflect a plane glide resistance as τ in eq.(6).

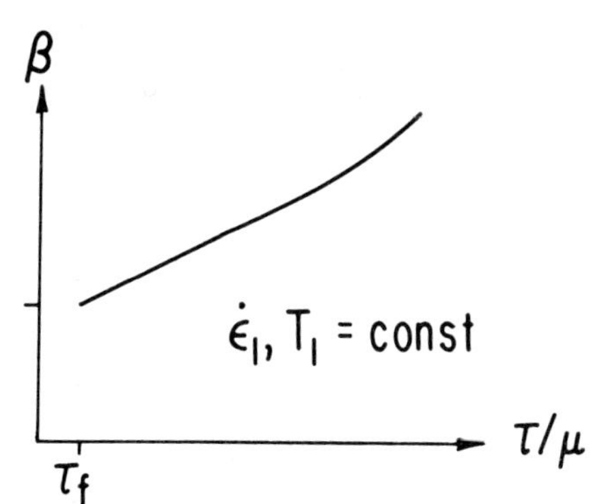

Fig. 3 - Rate sensitivity of the flow stress versus the flow stress, both as they vary with prestrain. The initial value characterizes the rate sensitivity of the friction stress τ_f.

Equation (9) is particularly useful, because it permits an experimental separation of the various components [13] by testing the strain-rate sensitivity as a function of prestrain, and plotting it against the flow stress at that prestrain. (The standard temperature T_1 and strain rate $\dot{\varepsilon}_1$, from which the excursions are being made, are kept constant.) Figure 3 shows such a diagram. When the data do not extrapolate through the origin, τ_f is thermally activated in a way different from the dislocation contribution (more so in the case of Fig. 3). When the data plot as a straight line, the rate sensitivity of the dislocation flow stress, $\partial \ln\alpha/\partial \ln\dot{\varepsilon}$, is strain independent; this is generally true in "stage II" strain hardening, but is increasingly violated as the strain-hardening rate gets lower.[15]

Strain hardening is expressed in terms of the dislocation density evolution by

$$\theta \equiv \left.\frac{\partial \tau}{\partial \varepsilon}\right|_{\dot{\varepsilon},T} = \left.\frac{\partial \tau}{\partial \rho}\right|_{\dot{\varepsilon},T} \cdot \frac{d\rho}{d\varepsilon} \qquad (10)$$

It can be decomposed into an athermal hardening rate θ_h (of which stage II is a special case [15]), diminished by a dynamic recovery rate θ_r:

$$\theta = \theta_h - \theta_r(\tau,\dot{\varepsilon},T) \qquad (11)$$

The strain-rate and temperature dependence of evolution, which appears in the dynamic-recovery rate, is physically independent of that of the flow stress: it is presumably based on thermally activated rearrangement of previously stored dislocations, as will be discussed in detail later.

Figure 4(a) shows typical stress strain curves for two different temperatures and/or strain rates. They illustrate the stronger rate sensitivity at large strains. They are also meant to indicate that the rate of strain hardening, θ, tends to zero at large strains. If it does so at a finite stress, this stress may be called a "saturation stress", and the behavior is then equivalent to steady-state creep. The evolution toward steady state in creep is displayed in Fig. 4(b): the strain rate continuously decreases, because of structural hardening.

The result of abrupt changes in external conditons after some prestrain is also demonstrated in Fig. 4: it gives a (so-called "reversible") flow stress change smaller than the difference between the flow stresses reached at the same strain by independent continuous paths (a); equivalently (b), the strain rate after a stress change in a creep test overshoots that which would have been reached at the same strain in a continuous test under the new stress.

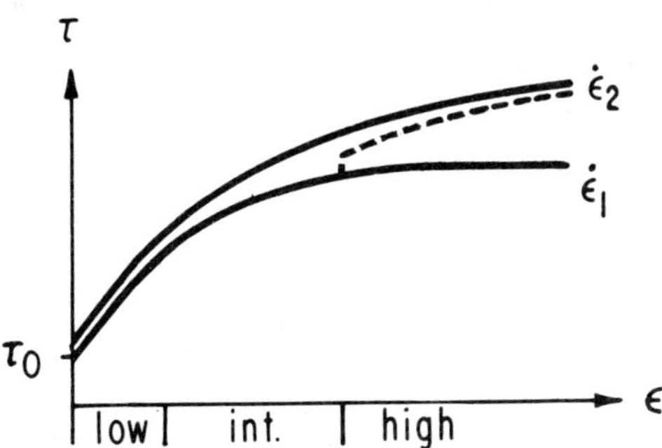

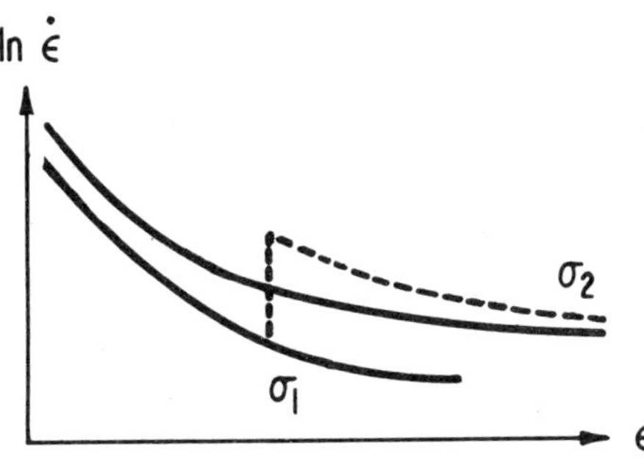

Fig. 4 - Typical stress strain curves (a) and creep curves (b), for two sets of external conditions, and for an abrupt change between them. In the "low-strain" regime, hardening is athermal and substantial; in the "high-strain" regime, steady state is approached.

If such changes after one particular history were undertaken to many different strain rates or, more effectively, many different temperatures, one could plot the "flow stress at constant structure" in a diagram like Fig. 2, as it was in fact done there: dotted. Finally, if a "saturation stress" exists, one can plot its dependence on temperature in the same diagram: it was done dashed. An important point to be noted from a comparison of the three curves in Fig. 2 is that the influence of "high-temperature" processes appears to make itself felt at lower and lower temperatures the higher the prestrain. We believe that the correlation is actually with the low strain-hardening rate (rather than the high strain), and that this is evidence for the pervasive influence of dynamic recovery.

The abrupt-change tests in Fig. 4 were highly idealized in two respects: they do not show any transients; and the entire curves after the change are identical to those obtained in continuous tests, with only a shift in strain. This means that the flow stress is here a unique state variable. In fact, only this comparison at constant stress made it possible to state that there are "no transients" shown in Fig. 4. The curves plotted may actually look like "transients" -- but only in the same sense as the entire strain-hardening curve is a "transient" on the way to flow stress saturation; or in the same sense that pre-steady-state creep is called "transient" creep. In reality, everything that is not in steady state may be labelled transient, and every transient is evidence for one state parameter changing. In this paper, we mean to deal with transients other than that due to continuous hardening, in either stress/strain or creep tests.

Before discussing any of the observed transients in detail and with some physical interpretation, we shall now summarize in a phenomenological way the various types of transient that actually occur.

Transients and Back-Extrapolations

There are, from their appearance, two types of transients: overshoots and gradual approaches. From a physical point of view, however, a more appropriate distinction is between two types which, conforming to a nomenclature common in creep, we shall label "normal" and "inverted".

Figure 5 shows these forms of transients for both stress strain tests (upper row) and creep tests (lower row). While the shapes drawn are meant to be realistic (more so than ignoring the transients entirely), they are still idealized in the sense that the sharp kinks and cusps would tend to be smoothed out in reality.

Figure 5(a) illustrates the response in the absence of transients. An abrupt increase in strain rate leads to an abrupt increase in flow stress, and unaltered hardening behavior thereafter. Similarly, an abrupt increase in stress in a creep test leads to an abrupt increase in strain rate, followed by deceleration as usual. If the imposed change is reversed, the response is reversed.

Figure 5(b) shows normal transients. In the creep test (lower part), the creep rate immediately after an abrupt change in stress is higher, as before, but it now decelerates rapidly. The analogous situation in stress strain tests is shown above. Here, the material hardens more rapidly after the change, which is physically the same as decelerating more rapidly in the creep test. Note that the same transient has a "gradual" appearance in a stress strain diagram, and "overshoot" character in creep tests (as plotted). Upon reversal of the imposed changes after some strain increment, the corresponding transients are reversed. Note that their nature of being "gradual" or "overshooting" is the same, but the sign of the curvature has now changed. Thus, neither a curvature criterion [16] nor a distinction between gradual and overshooting character provides a reasonable classification scheme. For these reasons, we discriminate between "normal" and "inverted" transients.

After a certain amount of strain, the rate of deceleration in a creep test, and the rate of hardening in a stress strain test, approach that observed in continuous tests at the same stress and strain rate. If one let the test go on longer than shown, without imposing a reverse change, one could, with increasing confidence, back-extrapolate the new continuous curve to the point of change. This back-extrapolation is shown dotted, and the corresponding jump is marked by 'ℓ' (for long transient), while 'i' marks the "instantaneous" change. We shall see below that the long extrapolation, while common[5], is probably not meaningful.

Figure 5(c) demonstrates the nature of inverted transients. In the creep test, the transient is now gradual: upon an increase in stress, the creep rate increase immediately, and then increases some more. This is also called an "anomalous creep transient", or "accelerating creep". (Note that it is decelerating upon a decrease in stress.) In Fig. 5(c), we have superposed on this behavior the normal long-term deceleration, so that the strain/time curves would actually be sigmoidal. The corresponding state of affairs in tensile tests is the yield drop shown in the upper part of the figure. It is not a sharp yield drop as after aging, but a more gradual one. The back-extrapolation methods are again self-evident provided straining has gone on sufficiently long after the change. The resulting "equilibrium" change is labelled 's' (for short transient).

Finally, Fig. 5(d) shows a combination of a short inverted transient with a long normal one. This is actually the most generally observed behavior above intermediate strains and temperatures. It is seen that back-extrapolation methods now become more problematical, especially since even the continuous curves are curved significantly on the scale of the long transient.

The quantitative desription of a transient requires the specification of at least two numbers: its strength and its length. The latter seems to be better given in terms of strain rather than time (when the temperature is not too high); it is marked as $\Delta\varepsilon$ in the figures.

Normal transients are typically observed when dynamic recovery is prevalent: at intermediate (and possibly high) temperatures, and at low temperatures at larger strains. They also occur after recovery. The stress change during the transients is of the order of 10% of the flow stress, and the length of the transient is usually a few percent strain.

Inverted transients have been observed in metals after strain-rate changes at low temperatures, when special care has been taken to make the testing machine very responsive [4,17]. The instantaneous stress jump tends to be about double as large as the ('s') back-extrapolated one, and the strain involved is less than one percent. Much more pronounced "transients" of similar shape are commonly observed in a number of ceramic materials [7,9,18], but they occur upon initial loading only, when the dislocation density is small. Certain solid solutions [19] form an intermediate case.

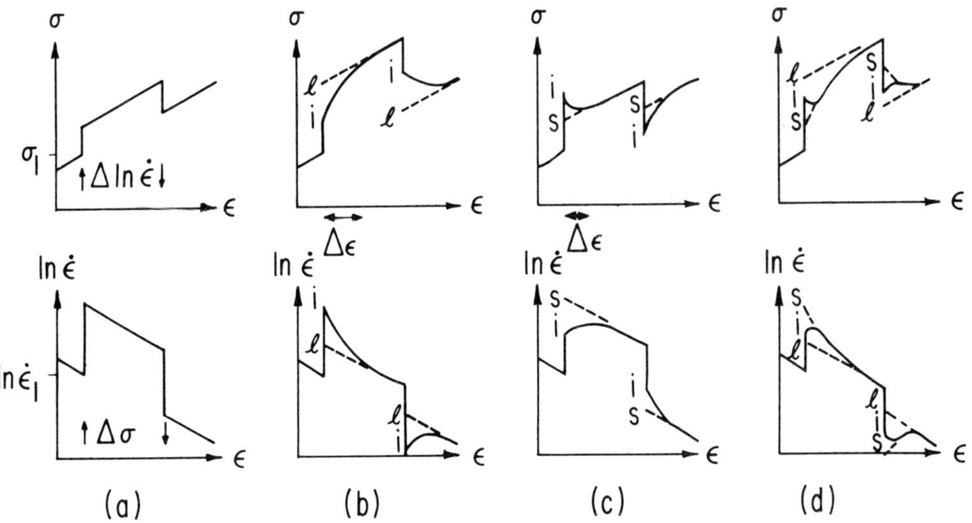

Fig. 5 - Response to abrupt changes in strain-rate in a stress/strain test (upper row), and to abrupt changes in stress in a creep test (lower row). (a) No transients. (b) "Normal" transients: they have "gradual" character in stress strain tests, "overshoot" character in the creep curves; the sign of the curvature depends on the sign of the imposed change; the "long" back-extrapolation ('ℓ') is later shown to be inappropriate. (c) "Inverted" transients and "short" back-extrapolation ('s'). (d) A combination of both types of transient in succession.

The transients of interest here relate to abrupt changes. They could be identified as transients only because the behavior settled down to slower evolution after the rapid changes. In continuous tests, the same structural changes may occur but do not evidence themselves by any noticeable feature of the observed curves. For example, the equilibrium that is approached by the structural variable responsible for a particular transient may well depend on the current flow stress (or on the average dislocation density); in that case, it will change slowly and steadily as the test goes on. The continuous curve certainly does not represent a change in only the major structure parameter, but the "equilibrium" or "steady-state" values of all internal parameters as they continually adjust themselves.

A particular test for which this consideration is relevant is stress relaxation. For while all changes are here completely continuous, both the stress and the strain rate vary rapidly and thus all internal parameters may have rapidly changing equilibrium values. These may not be actually reached if it takes much strain to do so (but they would be approached efficiently if the transient is time-activated). The point to be emphasized is that different tests measure different combinations of structural evolution if there is more than one parameter.

Dislocation Multiplication

A common qualitative explanation for the occurrence of an inverted transient after an abrupt change in stress or strain-rate is based on eq.(4). It is argued that, while the dislocation velocity (and even its average $\bar{v}$) can change instantaneously, the dislocation density (and even its mobile component ρ_m) cannot. For example, after an abrupt increase in stress, $\bar{v}$ increases immediately, and ρ_m increases slowly to a new steady-state value determined by the new conditions; thus, the strain rate would increase first abruptly and then by an additional amount in a gradual manner. Equivalently, an abrupt change in the prescribed strain rate would demand an immediate increase in stress to accelerate the then mobile dislocations, but it could subsequently decrease as the mobile density increases.

If this interpretation were correct, the most appropriate evaluation procedure would be to determine the stress (or strain-rate) jump as instantaneously as possible, if one wants to measure the relation between stress and dislocation propagation rate.[4] A more quantitative analysis to be given below will show that this mechanism is unlikely to be applicable to strain hardened metals. First, however, it is useful to consider the conditions under which it is an important mechanism: namely, materials with a low initial dislocation density and a high friction stress [8,9,18].

When the initial dislocation density is low, the distinction between total and mobile densities may be dropped without loss of generality. (In fact, we only need assume that ρ_m is proportional to ρ.) Also, the dislocations tend to be long and fairly straight. In this case, differentiation of eq.(4) gives the change of the creep rate with strain at constant stress. It may be written as

$$\left.\frac{\partial \ln\dot{\varepsilon}}{\partial \varepsilon}\right|_{\tau} = \left.\frac{\partial \ln\rho}{\partial \varepsilon}\right|_{\tau} \cdot \left\{1 + \left.\frac{\partial \ln\bar{v}}{\partial \ln\rho}\right|_{\tau}\right\} \qquad (\rho_m \propto \rho) \qquad (12)$$

Here, it was assumed that the dislocation velocity is a function of stress and dislocation density. In particular, $\bar{v}$ is usually set proportional to some power of an "effective stress", such as the friction stress in eq.(8) (with $\dot{\varepsilon}$ replaced by $\bar{v}$). In that case, the last term in eq.(12) becomes

$$\left.\frac{\partial \ln\bar{v}}{\partial \ln\rho}\right|_{\tau} = - \left.\frac{\partial \ln\bar{v}}{\partial \ln\tau}\right|_{\rho} \cdot \left.\frac{\partial \ln\tau}{\partial \ln\rho}\right|_{\bar{v}} = - m_i \cdot \frac{\tau-\tau_f}{\tau} \cdot \frac{1}{2} \qquad (13)$$

Here, the "instantaneous" relative stress dependence of the strain rate was labelled m_i.

Inserting eq.(13) into eq.(12) gives for the creep acceleration:

$$\left.\frac{\partial \ln\dot{\varepsilon}}{\partial \varepsilon}\right|_{\tau} = \left.\frac{\partial \ln\rho}{\partial \varepsilon}\right|_{\tau} \cdot \left\{1 - \frac{m_i}{2}(1 - \tau_f/\tau)\right\} \qquad (\rho_m \propto \rho) \qquad (14)$$

Since the dislocation density presumably increases monotonically with strain, it is the sign of the term in braces that determines whether creep is accelerating or decelerating -- and, equivalently, whether the slope of the stress strain curve is negative or positive. The conditions for accelerating creep, and thus an "inverted transient" are easily seen. When $m_i < 2$, the right-hand side of eq.(14) is always positive. For larger values of m_i, this is true only when the friction stress is a substantial portion of the flow stress; however, as deformation progresses, τ_f becomes a smaller and smaller fraction of τ, and thus straining must eventually decelerate when $m_i > 2$. (Actually, under the same conditions, m_i itself would also tend to grow, even if it was near one initially, since it was defined, in eq.(13), in terms of the total flow stress, not the "effective stress".)

The result of this analysis is that one would expect inverted primary creep and a (slow) yield drop in materials with a very high rate sensitivity (such as diamond-structure single crystals[9]) and/or a large friction stress (such as solution hardened alloys and bcc materials at low temperature), but never in strongly strain-hardened materials -- so long as $\rho_m \propto \rho$.

The conditions discussed up to now were "transient" in only a very limited sense. For the assumption that the mobile and total dislocation densities were identical, or at least proportional to each other, amounted to assuming a single structure parameter to be controlling. This single parameter had two opposing effects, which took their turn at dominating the macroscopic properties, giving rise to nonmonotonic behavior. An abrupt <u>change</u> in conditions will, under these circumstances, <u>not</u> lead to further transient behavior: the continuous curve, of whatever slope, will be rejoined "instantaneously".

A true transient due to an adjustment of the mobile dislocation density would have to be based on ρ_m being an independent parameter. For this case, we rederive eq.(14) from eq.(4), keeping ρ and ρ_m separate:

$$\left.\frac{\partial \ln \dot{\varepsilon}}{\partial \varepsilon}\right|_\tau = \left.\frac{\partial \ln \rho_m}{\partial \varepsilon}\right|_\tau - \frac{m_i}{2} \cdot (1 - \tau_f/\tau) \cdot \left.\frac{\partial \ln \rho}{\partial \varepsilon}\right|_\tau \tag{15}$$

Actually, the only difference between eqs.(14) and (15) is an additive term $\partial \ln(\rho_m/\rho)/\partial\varepsilon$ in eq.(15): a relative change in the ratio of mobile to total dislocation density. Obviously, this term can be large only when the ratio is small -- and then it would, by the very same mechanism, increase and approach one rapidly. Thus, we see again that an inverted transient due to mobile dislocation adjustments can be of any importance only so long as ρ_m is small.

Finally, let us consider what possibilities could reasonably be postulated for any independent adjustment of the mobile dislocation density. This is commonly done on the basis of population dynamics [6,12]. Using the symbol $d\rho_m^+$ for an <u>increase</u> in mobile dislocation density (by generation, about which we make no further assumptions), the net change can be expressed as

$$d\rho_m = d\rho_m^+ - \rho_m \frac{dy}{\Lambda} \tag{16}$$

The rate of trapping is assumed to be proportional to the current number of mobile dislocations and to the distance dy they have travelled, so that it can be described by a mean free path Λ. The product $\rho_m dy$ is proportional to the strain increment, so that one can write:

$$\frac{d\rho_m}{d\varepsilon} = \frac{d\rho_m^+}{d\varepsilon} - \frac{1}{b\Lambda} \tag{17}$$

Now if there are dislocations other than mobile dislocations, they should certainly influence the trapping probability, i.e., the mean free path. Thus, the last term in eq.(17) is a function of ρ, not ρ_m. In fact, it is not clear how ρ_m can enter the right-hand side of eq.(17) at all, so that its steady-state value can be determined by setting the left-hand side equal to zero. If ρ_m does enter, its steady-state value will be a function of the total density. It is not unreasonable to assume that then again $\rho_m \propto \rho$, in <u>steady state</u>. If this is the case, the steady-state value of the mobile density does not change upon an abrupt change in external conditions, since the total density does not, and thus there will be no reason for any true transient; the behavior described by eq.(14) applies.

For any explanation of the observed inverted transient upon abrupt changes in strain hardened materials on the basis of mobile dislocation adjustments, it is therefore necessary that ρ_m depend on the <u>external</u> conditions: <u>applied</u> stress or strain rate. On the basis of dimensional arguments, it is often assumed that $\rho_m \propto \sigma^2$. As is schematically shown in Fig. 6(a), this would produce a negligible stress jump when $m > 2$. Inverted transients in strain-hardened metals, on the other hand, are typically of the form and scale shown in Fig. 6(c).

The only remaining possibility for explaining a substantial transient on the basis of mobile-dislocation adjustments is that the mobile dislocation density be a strong function of stress -- about as strong as the dislocation mobility, or even proportional to it[6]. Under such circumstances, eq.(4) is obviously an inappropriate way to describe plastic deformation: the waiting time in eqs.(1) and (2) should then be related to dislocation generation [6,10], and ε_o is proportional to a new quantity, the source density[12]. The important kinetic process is that of generation, which is measured by the back-extrapolated jump.

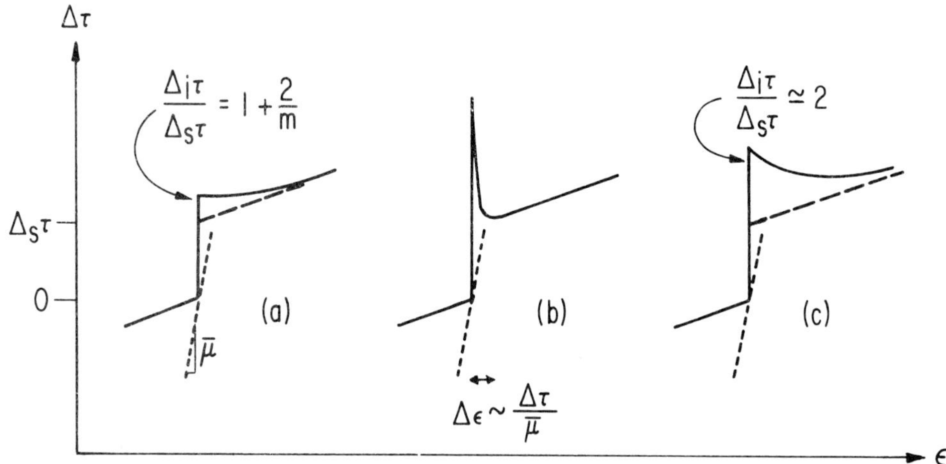

Fig. 6 - Schematic form of inverted transient for some of the discussed mechanisms: (a) mobile-dislocation adjustment in the absence of a friction stress; (b) dislocation percolation through soft and hard regions; (c) dynamic-pile-up adjustment when the soft regions have a low, athermal glide resistance. The solid lines are plotted against plastic strain; the dotted lines indicate the scale: $\bar{\mu}$ is the combined effective modulus of specimen and machine.

Dislocation Percolation

The "structure" relevant to the plastic behavior of a material is that encountered by the dislocations as they move through various regions of the crystal. It is not usually the same everywhere; and while it is true that, during macroscopic flow, many dislocations sample many regions, they do not do so at random. In fact, there is a definite and obvious correlation: flow will always begin in the softest regions, and only when these become exhausted or hardened will it spread to harder areas. It is likely, in general, that the rate sensitivities of the two regions are also different; then, the macroscopically measured rate sensitivity changes during loading, and a transient may be observed upon abrupt changes in loading conditions.

Figure 7 illustrates two possibilities, which are realized when two different stresses are applied to the region for which the line glide resistance diagram[12] can be schematically represented as shown. Under the higher applied stress τ, dislocation kinetics in the soft central region is limited only by phonon drag; most of the time is spent by the mobile dislocations waiting for thermal activation at the stronger obstacles. Now if the applied stress were raised "instantaneously" by a small amount, the few dislocations actually in transit in the soft regions could respond immediately by increasing their velocity, but eventual long-range slip would be controlled by the change in waiting time at the strong obstacles. Since phonon drag is much more rate sensitive than thermal activation, the "instantaneous" rate sensitivity measured macroscopically would be larger: the transient would be of the inverted type.

At the lower of the two applied stresses τ shown in Fig. 7, dislocation motion is thermally activated in both regions. While the activation energy, and thus the waiting time, is obviously smaller at each individual obstacle in the soft regions, the activation area may be larger (as shown), and thus the rate sensitivity smaller. Now, the "instantaneous" response, which is still due to the soft regions, is less rate sensitive than the steady-state response: the transient is "normal". This case would be expected, for example, as a consequence of statistical variations in the spacing of a single set of obstacles[20]: the soft regions then correspond to regions of larger-than-average spacing, and thus larger activation area.

The crucial question in this type of transient is the strain scale. It is dictated by the size of the soft regions. One would generally expect the variations in glide resistance to be on a scale small compared to the specimen dimensions, and more or less equi-axed. Then, the very softest regions, which will deform first, are entirely surrounded by hard regions, which still deform elastically.[21] The total plastic strain in the soft regions can, at this stage, be only of the order of the elastic strains ("elastically constrained plastic flow"). As the stress is raised, the soft regions begin to touch, and dislocations "percolate" from one soft region to the next. The critical condition for long-range slip is passage of the dislocations through the "critical gates".[20,21] Since, at this point, the end of the transient has been reached, the length of this "percolation transient" is limited to the order of elastic strains. This was schematically shown in Fig. 6(b).

In normal tensile testers, the maximum loading rate is also limited by the elastic response (of both specimen and machine). For this reason, it should not be possible to observe percolation transients in typical stress strain tests. Under very good servo-control, however, they might be seen: when the effective stiffness is infinite, and the switching time is small. Similarly, one should be able to observe percolation transients in creep tests at low strain rates, when the time required for effective changes in conditions is associated with a strain that is small compared to the elastic strain increment. Finally, percolation transients may exert a disturbing influence on stress relaxation tests, where the total strain is of the order of the elastic strain and where, because of the continuous nature of the test, transient behavior is not immediately obvious.

When percolation transients are in fact observed, but the interest is in macroscopic (long-range) plastic flow, the proper evaluation procedure is to back-extrapolate through the transient. On the other hand, the rate sensitivity observed in small-strain tests such as internal friction or shock loading is more likely to relate to the soft regions; for comparisons with tensile tests, it is then important to ascertain whether the corresponding small-strain effect was resolved under the conditions of the test.

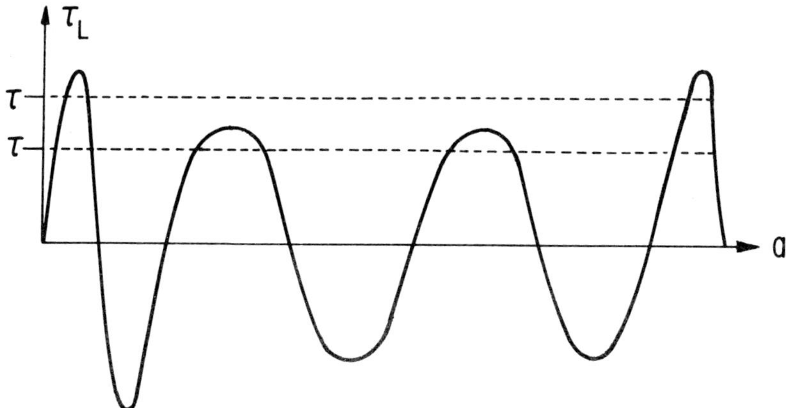

Fig. 7 - Schematic diagram of line glide resistance τ_L versus area swept, a, for two regions, "soft" and "hard" (or "critical gates"). Two levels of applied stress τ.

Dynamic Pile-Ups

If the difference in the glide resistance of the soft and hard regions, τ_1 and τ_2, respectively, is sufficiently large, as in the example shown in Fig. 8, dislocations will no longer travel as single entities: dynamic pile-ups will form, which have the effect of smoothing out the glide resistance variations over the slip plane. The average number p of dislocation in such a pile-up depends on the geometrical distribution of soft and hard regions, and on the difference $\tau_2-\tau_1$. It will, in general, change with temperature and strain rate if τ_1 and/or τ_2 do. After an abrupt change in conditions, p must therefore readjust; but since this requires a finite amount of strain, a transient will result. The type of transient that should be observed depends, as in the case of percolation, on whether it is the soft region or the hard region that is the more rate sensitive. There are, however, major differences in scale as well as in detail between the transients due to the percolation of individual dislocations and those due to the reajustment of dynamic pile-ups.

For a quantitative treatment, we will use the well-known formalism developed for large pile-ups; but we expect the results to be applicable, at least in a semi-quantitative way, whenever at least one dislocation segment waits at each critical gate.[12] The two governing relations are that for the stress at the head of the pile-up (which is set equal to the flow stress τ_2 there, for a dynamic pile-up),

$$\tau_2 - \tau_1 = p\cdot(\tau - \tau_1) \tag{18}$$

and that for the equilibrium (or steady-state) number of dislocations in the pile-up,

$$p = \frac{L}{b}\,\frac{(\tau-\tau_1)}{\mu} \tag{19}$$

where L is the spatial separation of the hard regions in the slip plane (Fig. 8). Eliminating τ from eqs.(18) and (19) gives, in steady state,

$$p = \sqrt{\frac{L}{b}\,\frac{(\tau_2-\tau_1)}{\mu}} \tag{20}$$

and eliminating p,

$$\tau = \tau_1 + \sqrt{\frac{\mu b}{L}\,(\tau_2-\tau_1)} \tag{21}$$

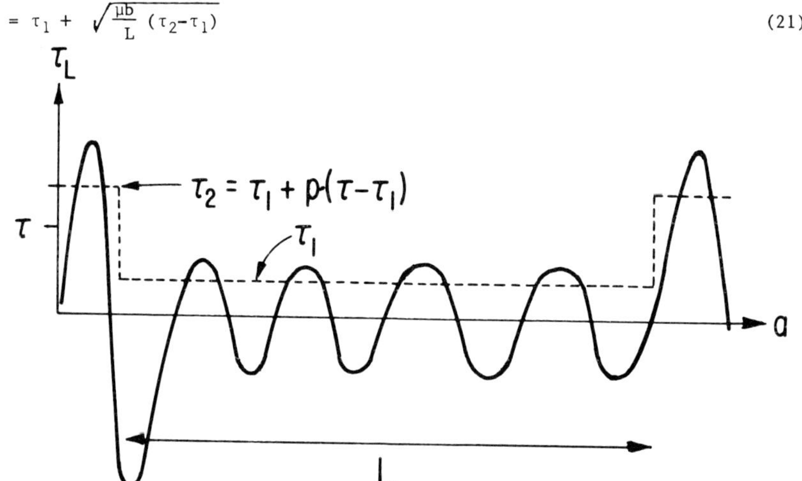

Fig. 8 - As Fig. 7, but with a greater difference between the glide resistance of the soft and the hard regions. This leads to dynamic pile-ups of length L, and two local effective stresses, τ_1 and τ_2.

Upon a sudden change in the imposed stress or strain rate, the adjustment of the pile-ups occurs in two successive steps. First, the dislocations waiting in front of each hard obstacle will move into equilibrium positions corresponding to the new stress. This process is controlled by the rate sensitivity of the soft regions. As in the case of the percolation of individual dislocations, this transient behavior involves only very small strains, and will usually be lost in the loading transient when strain-rate changes are prescribed. The process is nevertheless important in an indirect way: it builds up the stress at the head of the pile-up from $\tau_2+\Delta\tau$ (the change in applied stress) to the new value of τ_2 according to eq.(18), with unchanged p. At the end of this transient, and thus for practical purposes "instantaneously", the flow stress is controlled by the new waiting time at the hard obstacles, but the old value of the pile-up strength. Thus, the "instantaneous flow stress change" is obtained by differentiating eq.(18) at p=constant (and, as usual, T=constant):

$$\frac{\tau}{m_i} \equiv \left.\frac{\partial\tau}{\partial\ln\dot{\varepsilon}}\right|_p = \frac{\partial\tau_1}{\partial\ln\dot{\varepsilon}} + \frac{1}{p}\cdot\left(\frac{\partial\tau_2}{\partial\ln\dot{\varepsilon}} - \frac{\partial\tau_1}{\partial\ln\dot{\varepsilon}}\right) \tag{22}$$

The next step is the adjustment of the number of dislocations per pile-up to its new equilibrium value given by eq.(19). How much strain is involved in this transient depends very much on the specific model. It can be quite significant, as in the following example. If the distance between sources of new dislocations is large compared to that between strong obstacle configurations ("critical gates"), then many pile-ups in series have to be successively supplied with new dislocations, and the distances travelled by these dislocations are large compared to the size of the soft regions. Conversely, when the pile-up strength must be decreased, the requisite number of dislocations must travel from each pile-up to the surface (or other sink) before the new equilibrium situation is reached.

The stress level at the end of this longer transient (which we will nevertheless label "short") is described by eq.(21). Its differentiation gives

$$\frac{\tau}{m_s} \equiv \left.\frac{\partial\tau}{\partial\ln\dot{\varepsilon}}\right|_{s.s.} = \frac{\partial\tau_1}{\partial\ln\dot{\varepsilon}} + \frac{1}{2p}\cdot\left(\frac{\partial\tau_2}{\partial\ln\dot{\varepsilon}} - \frac{\partial\tau_1}{\partial\ln\dot{\varepsilon}}\right) \tag{23}$$

Equations (22) and (23) are very similar. It is seen that the type of transient to be observed depends on the sign of the difference between the two rate sensitivities. When τ_2 is more rate sensitive than τ_1 (or τ_1 is zero), the initial stress jump is larger than the back-extrapolated one, and the transient is inverted; when the soft regions are more rate sensitive, normal transients should occur. (This situation is just the reverse of that in the small-strain percolation transient!)

Of specific interest are the two limiting cases when one or the other of the two rate sensitivities can be neglected. If the strong obstacles are athermal, the ratio of the two measurable stress jumps in the normal transient is $m_s/m_i=1-1/2p$ (neglecting a second-order term in 1/p). On the other hand, if the glide resistance in the soft regions is athermal, the back-extrapolated flow stress jump is precisely one-half of the "instantaneous" one: $m_s/m_i=2$. This was shown in Fig. 6(c) (neglecting the very short transient of Fig. 6(b)). It is a particularly interesting case, since it closely corresponds to the behavior observed in strain-hardened pure FCC metals.[4,6,22] Inasmuch as both the mobile-dislocation multiplication mechanism and the percolation transient failed to give a plausible explanation of this effect, we are tempted to conclude that the observations can be interpreted as giving evidence for the existence of dynamic pile-ups. Note that this conclusion is entirely independent of the number of dislocations in the pile-ups (we would not expect it to be much larger than 1), or any other details of the glide resistance diagram; it is essentially a consequence of the square-root in eq.(21). That the soft region must be athermal for this explanation to be applicable corresponds to what would be expected on the basis of a statistical variation in the spacing of strong obstacles.

If this interpretation is correct, it becomes a matter of convenience whether the experiments are evaluated by back-extrapolation or as "instantaneouly" as possible, since both measurements are proportional to each other.[22] Note, however, that when the soft regions are not strictly athermal, the strict proportionality disappears, and the back-extrapolated measurement survives as that of relevance for long-range slip. Simultaneous measurement of both stress jumps could yield additional information on the distribution of obstacles.

Dislocation Rearrangement

In Figure 5(b), we pointed out the existence of "normal" transients in both creep and stress strain curves. Figure 9 displays the latter in more detail, for "up" changes as well as "down" changes, each by two different amounts. (By "up" is meant up in stress, and up in strain rate or down in temperature.) Curve A illustrates the well-known phenomenon of (transient) "work-softening".[3] Curve B shows that a transient exists just as well, even though not as drastic, when the "down" change was not sufficient to produce actual softening, i.e. a negative slope of the stress strain curve. Finally, curves C and D demonstrate that the same type of transient exists after "up" changes.

These transients are observed with increasing strength as deformation proceeds. In fact, they are not observed in the steep, athermal hardening region that appears, in many cases, in the initial part of the stress strain curve. For this reason, they have been associated [23,24,15] with dynamic recovery, in a more general sense than was proposed by Stokes and Cottrell [3]. We shall now give an outline of a physical mechanism that could be responsible for these observations. Subsequently, we will address the question of what influence these processes may have on the instantaneous flow stress changes.

The evolution of the dislocation structure with strain can be described by two independent processes, which give rise (eq.(10)) to the two components of strain hardening (eq.(11)). The first is an athermal storage process, which is determined by the geometry and statistics of the distribution of obstacles in the slip plane.[20] It is in the nature of these storage processes that "forward" internal stresses develop, which provide a driving force for subsequent rearrangement of the stored dislocations into lower-energy configurations.[25,26] Such rearrangement processes may take place simultaneously with the ongoing deformation, giving rise to the second, negative contribution to strain hardening.

The rate of rearrangement will be commensurate with the current strain rate only for those dislocation segments that are under a sufficiently high driving force. One may assume [25] that the athermal process would, by itself, generate a spectrum of internal driving forces and resistances; in other words, a spectrum in the degree of metastability of the stored dislocation segments. At the upper end of this spectrum, dislocations become ineffective as obstacles to slip at the same rate as they are statistically stored; thus, no net storage will in fact occur above this critical level of the forward internal stress. It has been proposed [25] that the

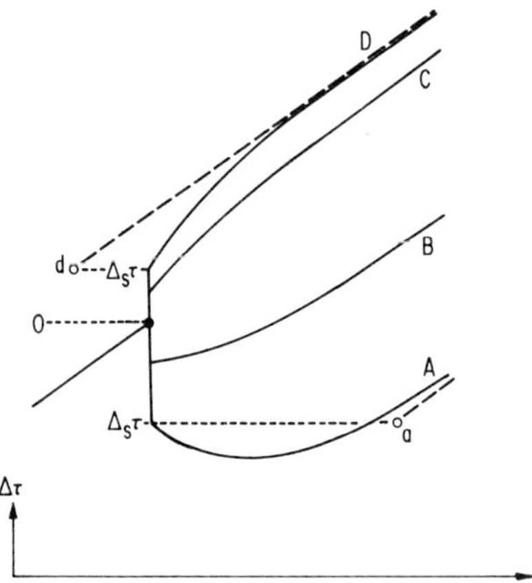

Fig. 9 - Long transients observed after an abrupt change in strain rate or temperature, in the dynamic-recovery regime. Curve A is classical "work softening"[3]. Initial short transients, if present, are nor shown.

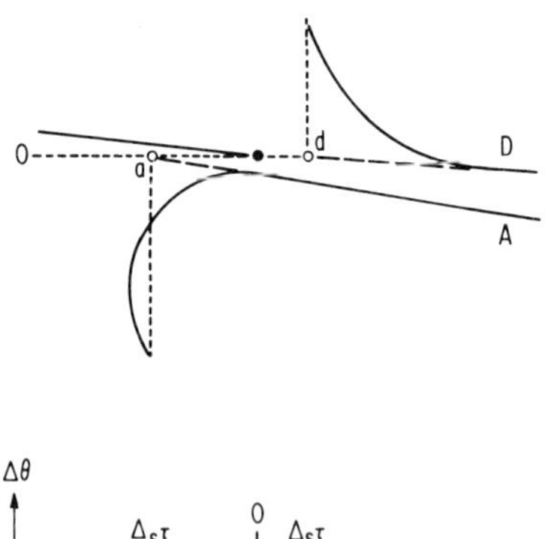

Fig. 10 - Interpretation of Fig. 9 as hardening transients. The behavior rejoins the continuous curve after the transient. Back-extrapolation to the new flow stress defines the points a and d for a one-parameter approximation.

"saturation stress τ_s"[27] can be used as a measure of this critical internal stress, inasmuch as in the steady-state limit, no net storage occurs at all.

When the strain rate or temperature is abruptly changed, the critical value $\tau_s(\dot{\varepsilon},T)$ for a balance between storage and recovery rates changes immediately, but the edge in the internal-stress spectrum can only change as a consequence of the first rearrangement processes. In the case of "down" changes, there will be an initial rapid recovery of the segments left in the part of the spectrum between the old and the new level of τ_s. This gives rise to a transient low or even negative strain-hardening rate (curves B and A in Fig. 9). Conversely, after an "up" change, the part of the spectrum above the old "edge" and below the new τ_s must first be replenished; this occurs essentially at the athermal storage rate, and thus the initial work-hardening rate is often close to θ_h (as in curve D).

The transients shown in Fig. 9 are therefore transients in the hardening rate, more so than in the flow stress. This will be better appreciated by plotting θ vs τ in the neighborhood of the change, as is done in Fig. 10, in a qualitative way, for the curves A and D in Fig. 9. In a θ vs τ diagram, as opposed to a stress vs strain plot, the curve eventually rejoins the extrapolation of the continuous curve from before the change. The amount of strain that is accumulated before this situation has been reached is determined by the integral over $d\tau/\theta$ in the transient. As can be easily seen, this strain is smaller in the case of curve D, and larger in the case of curve A, than what would have been accumulated along the continuous curve. It is for this reason that the curves after the change are, in Fig. 9, displaced to the left and right, respectively, from the extrapolation of the continuous curve. It is also for this reason that the back-extrapolation of the new curve to the strain before the change has lost all meaning ($\Delta_\ell\tau$ in Fig. 5(c)). The obvious back-extrapolation through the hardening transient in Fig. 10, i.e. to the points a and d, in fact corresponds, in Fig. 9, to a back-extrapolation of the new "equilibrium" curve to the flow stress after the change, i.e. again to the points a and d, respectively. This would appear to be the only physically reasonable evaluation of the data when one wishes to use a single-parameter description, as is so often helpful[28], and thus must ignore the "long transient". This corresponds to the idealization shown in Fig. 4.

Finally, we point out that the rearrangement of dislocations during dynamic recovery may well be associated with some strain [11,23,24,29,15], in particular in view of the polarization of the work-hardened structure [25,26,30]. In connection with the transients discussed above, this "rearrangement strain rate" has recently been used to explain the pronounced asymmetry of "instantaneous" flow stress changes upon "up" and "down" changes that is evident in Fig. 9.[15]

The more important effect of the rearrangement strain rate is that, at high temperatures, it may be responsible for most of the deformation.[31,13] Then, flow kinetics is no longer governed by thermally activated glide (eqs.(1) through (8)), but by thermally activated recovery (of the static or dynamic kind). This view could explain the yield-stress drop at high temperatures, as well as its correlation with an essential absence of strain hardening in this regime. It may also be applicable to the long-time regime of stress-relaxation curves.

When this mechanism is operative, the structure is never truly "constant": it changes at a rate given by the external and structure conditions. Of course, this rate of change can be evaluated at a particular, current value of the structure parameter, and in this sense "instantaneous" changes in properties are still "isostructural".

Dynamic Strain-Aging

The changes in structure we have discussed until now all related to readjustments in the dislocation density or arrangement; the metallurgical structure was assumed constant. During deformation at higher temperatures, changes in the number and size of precipitates ("aging") may occur simultaneously, along with many other kinetic processes. These will not be discussed in the present paper. Changes in the distribution of solute atoms, however, will now be considered. The very fact that solutes interact with dislocations, which is what makes them one kind of "obstacles" to glide, also causes them to seek low-energy positions near dislocations rather than at random places in the lattice. The effect of this tendency, provided the solute mobility is sufficiently high, is an increase with time of the effective obstacle concentration along the dislocations (or the effective interaction strength). When this happens simultaneously with the deformation, it is called "dynamic strain-aging".

The flow stress is then a function of time and temperature even if thermal activation took no part in the overcoming of the obstacles by the dislocations. It can therefore be described by a dependence of the mechanical threshold on aging time t_a and temperature T:

$$\tau_a = s_a(\dot{\varepsilon},T) \cdot \hat{\tau}_a(t_a,T;c,\rho) \tag{24}$$

Here we have affixed a subscript 'a' to various quantities to indicate that this is the contribution that depends on aging, without committing ourselves to any particular mechanism, or to how this contribution is to be superposed with others. For reasons to be discussed below, we have indicated that $\hat{\tau}_a$ depends on both the solute concentration c and the dislocation density.

The aging process may occur simultaneously with the straining; then it is called "dynamic strain-aging". The accumulation of solutes on the dislocations is now widely considered to take place during the waiting time of the dislocations during their jerky glide [32-35], and not as the formation of a dragging cloud during their motion. Then aging time t_a and waiting time t_w are identical, and the connection to strain rate follows from eq.(1):

$$\mathrm{d}\ln t_a = \mathrm{d}\ln t_w = \mathrm{d}\ln\varepsilon_o - \mathrm{d}\ln\dot{\varepsilon} \tag{25}$$

Thus, for ε_o=constant,

$$\left.\frac{\partial\ln\tau_a}{\partial\ln\dot{\varepsilon}}\right|_{T,\rho} = \left.\frac{\partial\ln\tau_a}{\partial\ln\dot{\varepsilon}}\right|_{T,\hat{\tau}} - \left.\frac{\partial\ln\hat{\tau}_a}{\partial\ln t_a}\right|_{T,\rho} \tag{26}$$

When dynamic strain-aging takes place, it makes a negative contribution to the rate sensitivity.

Figure 11 demonstrates the superposition between the two opposing tendencies. The left-most curve represents thermally activated obstacle cutting at constant structure, i.e., at a fixed mechanical threshold $\hat{\tau}$. The dependence of $\hat{\tau}_a$ on an aging time t_a is shown by the dashed curve on the right: at zero time, $\hat{\tau}_a$ is unaffected, at infinite time it saturates (to a value assummed to be twice $\hat{\tau}_a(0)$ in the figure) . The other two thin, solid lines then describe thermally activated flow at two higher (but constant) values of $\hat{\tau}_a$; at the points 1 and 2, the aging time is equal to the waiting time. The heavy, solid curve is the locus of all points for which $t_w = t_a$. It exhibits a negative rate sensitivity for strain rates above the point M. (The regime of negative rate sensitivity is also limited at the high-strain-rate end, for other reasons such as phonon drag.)

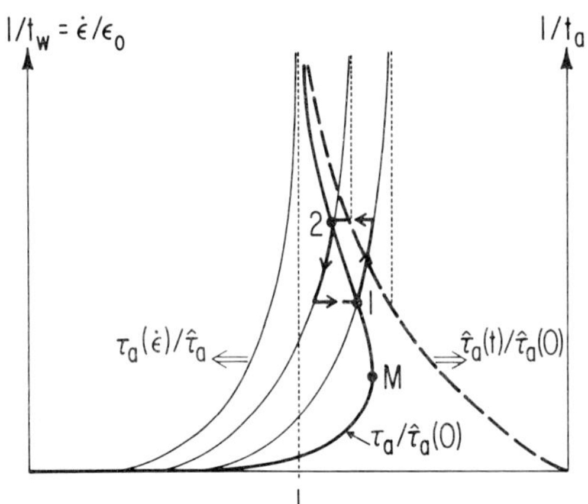

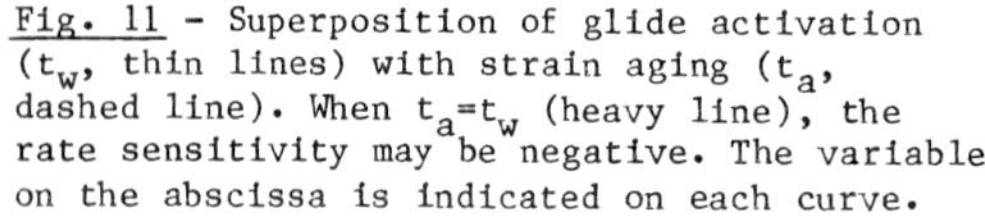
Fig. 11 - Superposition of glide activation (t_w, thin lines) with strain aging (t_a, dashed line). When t_a=t_w (heavy line), the rate sensitivity may be negative. The variable on the abscissa is indicated on each curve.

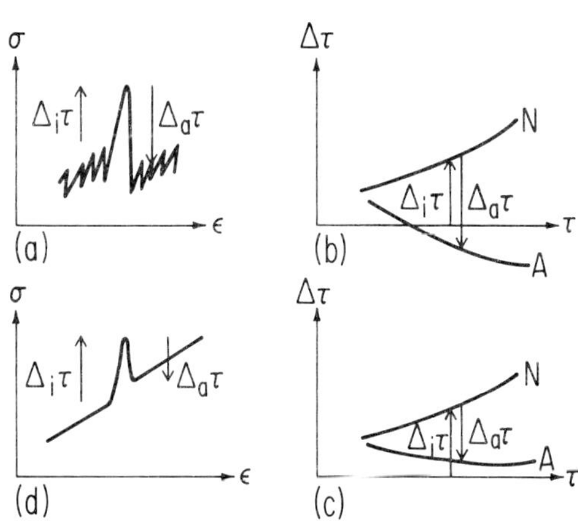

Fig. 12 - The negative rate sensitivity during jerky flow appears after a stress peak (a). It is due to the strain-dependent part (b). Even when the rate sensitivity remains positive (c) and flow is smooth (d), there may be aging.

Let us now examine what the negative rate sensitivity during dynamic strain-aging really means. It results entirely from the assumption that aging is allowed to proceed during the waiting times and affects the current flow stress. Immediately before and after an instantaneous change, however, the concentration along the dislocation, and thus $\hat{\tau}_a$, must be constant, and the second term is missing from eq.(26): the rate sensitivity is positive. In Fig. 11, an abrupt increase in strain rate from point 1 will therefore first lead to an increase in stress, and only when aging has reached its "steady state" will the stress have decreased to point 2. One should thus observe a transient of a "length" t_w, or equivalently ε_o. This is in accord with typical experiments, as shown schematically in Fig. 12(a).[36,35,37]

Figure 12(a) also indicates the macroscopically jerky flow during dynamic strain-aging. It is a consequence of the fact that, when the rate sensitivity is negative, spontaneous increases in strain rate tend to occur.[38,28] Indeed, if the stress were prescribed and continuously raised, points 1 and 2 in Fig. 11 could never be reached: the strain rate would become unstable at point M. In a tensile test, a minimum strain rate is prescribed by the boundary conditions (higher values being possible through localization and/or jerkiness of flow [28]). This condition along with strain hardening makes one reach points such as 1 and 2 -- if only as trigger points for instabilities.

Dynamic strain-aging is known to be strain dependent. This fact provides valuable clues for an understanding of the detailed mechanism, and it is relevant here. Figure 12(b) displays the rate sensitivity as a function of the flow stress reached during prestraining, for a "normal" material (curve N, as in Fig. 3), and under conditions of dynamic strain-aging (curve A).[35] Only in the regime where the rate sensitivity is negative is jerky flow observed. The divergence of the curves at small strains suggests, however, that aging occurs even then, it just is not strong enough to make the total (or net) rate sensitivity negative, according to eq. (26).

The rate sensitivity plotted as curve A in Fig. 12(b) was that referring to a measurement after the transient shown in Fig. 12(a), i.e., after aging has reached its steady state. We now postulate that the instantaneous change $\Delta_i\tau$ shown in Fig. 12(a) represents what is shown as $\Delta_i\tau$ in Fig. 12(b): the jump that would be observed in the absence of aging. Then, aging is responsible for the subsequent (negative) change $\Delta_a\tau$, in both figures.

Figure 12(c) illustrates an intermediate situation: in curve A, the strain-dependent contribution to the rate sensitivity is negative over some range, but it never dominates. Mulford and Kocks proposed [35,39] that this still be considered as part of dynamic strain-aging and due to solute mobility. In analogy to the interpretation given in Fig. 12(a), one should ct then expect a transient in this case also — only that now $\Delta_a\tau < \Delta_i\tau$: Fig. 12(d). (Note that flow is smooth under these conditions.) Such transients may not be observable under normal testing conditions since the strain involved should be small, as in percolation transients.

The "Flow Stress at Constant Structure"

In the last example, we have seen that solute mobility could be responsible for a low (even though positive) rate sensitivity. Usually, however, when a positive rate sensitivity is observed, there is no obvious need to depart from the classical interpretation of a flow stress measured at constant structure. A low rate sensitivity, such as in the "plateau" on a τ vs T diagram (Fig. 2), is then frequently assumed to be due to an "internal stress" contribution. If solute mobility plays a role, it could instead be due to aging, which compensates the intrinsic rate sensitivity.

A straight-forward way to distinguish these two cases is to investigate the dependence of the rate sensitivity on prestrain. If, as in curve A, Fig. 12(c), the slope of the data is negative, then one component of the rate sensitivity (namely, the strain dependent part) is negative -- and any negative component can only be due to an influence of aging. Even when this slope is positive, however, as in curve N, it could be the result of a decrease due to aging from a more positive slope. Indeed, these slopes are generally smaller in solution hardened alloys than in the corresponding pure or precipitation hardened metals.[14] While this could also be interpreted on the basis of a change in stacking fault energy [14], an interpretation in terms of dislocation aging is reasonable.

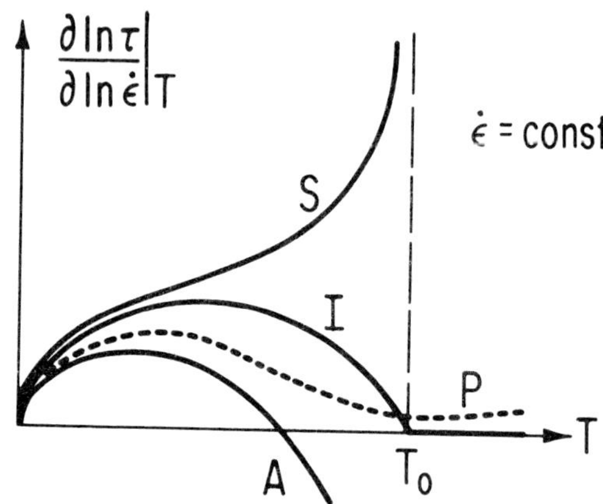

Fig. 13 - The rate sensitivity at constant strain rate, for Short-range obstacles, an Internal stress, Aging, and the Plateau.

A further tool [12,39] for studying flow stress mechanisms is a plot of rate sensitivity (at constant base strain rate) versus temperature. Figure 13 illustrates some common types of behavior. In general, there must be a sharp increase near T = 0 K. Dislocation penetration of short-range obstacles (curve S) leads to another sharp increase near the temperature T_o, where the flow stress approaches zero. If there is an additive internal-stress component (curve I), the rate sensitivity must vanish above T_o. When the increase in solute mobility with temperature leads to effective aging of the dislocations (curve A), the rate sensitivity will become negative. In a detailed study of nickel cobalt alloys [39], it was found that the yield behavior always approximately followed curve S, but that the strain dependent part of the rate sensitivity showed evidence of aging. In the highly-alloyed material, this component did become negative as in curve A (even though the total rate sensitivitry was positive and there was no jerky flow). At lower concentrations, however, the data followed a curve such as P in Fig. 13: it does not prove that aging took place, but it certainly makes this the easiest interpretation.

In summary, the assumption that the flow stress as measured in the plateau region is one "at constant structure" is tenuous; rather, it is likely to reflect some degree of simultaneous adjustment in the degree of interaction between solute atoms and dislocations, even when this does not manifest itself in jerky flow.

At even higher temperatures, the assumption that the flow stress reflects the deformation kinetics "at constant structure" is also likely to be violated. Here, it is the strain rate associated with the ongoing dislocation rearrangement that must be taken account of, and may be responsible for the drop in flow stress. This interpretation gains support from the observation (Fig. 2) that the flow stress vs temperature relation in the high-temperature regime is similar to that for the saturation stress at all temperatures: dynamic-recovery kinetics should be responsible for both.

It thus emerges that the classical separation of constant-structure kinetics and evolution kinetics is strictly applicable only at low temperatures[29] and low strains[15]. When either the strain-hardening rate or the strain-rate sensitivity attain low values, the current rate-of-change of structure may significantly influence "instantaneous" properties. A detailed study of transients after abrupt changes should help identify the dominant mechanisms.

Summary

In addition to the continual structure changes that cause strain hardening, other more dynamical changes in dislocation configuration and interaction accompany straining. In continuous tests, they each achieve an internal equilibrium or steady state under the given external and overall structure conditions; but after imposed abrupt changes, they lead to various forms of transient behavior.

Transients can be classified into two types, normal and inverted, and can be further described by their strength (such as the amount of flow stress overshoot in an inverted transient) and their length (in time or strain).

The most frequently discussed mechanism for transient behavior, namely an adjustment in mobile dislocation density, is shown to be of consequence only when the total dislocation density is low, and the rate sensitivity is very large, or the friction stress is high. In strain-hardened materials, the strength of the transient should be negligible.

A transient based on the partition of dislocations between those actually in transit through "soft" regions and those waiting at "hard" obstacles to percolate through the structure is also generally negligible — in this case because its length is so small: the total transient strain must be of the order of the elastic strains.

A new explanation is offered for the inverted transients observed, under some conditions, in metals after strain-rate or temperature changes. It is based on the adjustment of dynamic pile-ups (even if they consist, in the average, of but a few dislocations), which are built up to smooth out the differences between the soft and the hard regions of the specimen. Since the obstacle spacing in the soft regions is expected to be relatively large, the glide resistance is here essentially athermal; then the ratio between the stress jump at constant pile-up strength and that in equilibrium is exactly two.

A gradual change in structure due to the rearrangement of previously stored dislocations into lower-energy configurations during dynamic recovery is the chief mechanism responsible for "normal" transients. Inasmuch as they are transients in strain-hardening rather than in glide kinetics, the appropriate back-extrapolation has to be undertaken in a strain-hardening vs stress, not in a stress vs strain diagram. The rearrangement process (since it is polarized) should also be associated with a macroscopic strain rate, which may serve to explain the drop in flow stress at high temperatures and at very low strain rates.

Among metallurgical changes, only the aging of dislocations by mobile solute atoms is discussed. It causes a negative partial rate sensitivity, which is observed only after an initial positive, inverted transient. This transient, as well as a detailed study of the strain dependence of the phenomenon, may be helpful in an investigation of the possibility that dislocation aging is also responsible for the plateau in the flow-stress temperature relation for solution hardened alloys.

In view of these various dynamical structure changes, a determination of dislocation kinetics at truly "constant structure" becomes less and less feasible as the temperature is raised and steady-state deformation conditions are approached. In the same regime, any description in terms of a single structure parameter, desirable as it may be, becomes increasingly problematical. Under these conditions, it is in fact a fortunate circumstance when transients are observed: then, ongoing changes in an additional structure parameter are obvious, and one has the option to decide how to treat them. The general observation must be made, however, that if one does decide to ignore any additional parameters in the interest of simplicity, the most "instantaneous" determination is the least meaningful; one must attempt an appropriate back-extrapolation.[40]

References

1. M. A. Adams and A. H. Cottrell, Phil. Mag. 46, 1187 (1955)
2. A. H. Cottrell and V. Aytekin, J. Inst. Met. 77, 389 (1950)
3. R. J. Stokes and A. H. Cottrell, Acta Met. 2, 341 (1954)
4. Z. S. Basinski, Phil.Mag. 4, 393 (1959)
5. U. F. Kocks, H. S. Chen, D. A. Rigney, R. J. Schaefer, in "Work Hardening", J. P. Hirth and J. Weertman, eds. (Gordon and Breach), p.151 (1968)
6. H. Mecking and K. Lücke , Scripta Met 4, 427 (1970)
7. W. G. Johnston and J. J. Gilman, J. Appl. Phys. 30, 129 (1959)
8. W. G. Johnston, J. Appl. Phys. 33, 2716 (1962)
9. H. Alexander and P. Haasen, Solid State Physics 22, 27 (1968)
10. W. deRosset and A. V. Granato, in "Fundamental Aspects of Dislocation Theory", J. A. Simmons, R. deWit and R. Bullough, eds. (NBS Spec. Pub. 317), p.1099 (1971)
11. U. F. Kocks, in "Fundamental Aspects of Dislocation Theory", J. A. Simmons, R. deWit and R. Bullough, eds. (NBS Spec. Pub. 317), p.1077 (1971)
12. U. F. Kocks, A. S. Argon, M. F. Ashby, Prog. Mater. Sci. 19 (1975)
13. U. F. Kocks, in "Strength of Metals and Alloys", P. Haasen, V. Gerold, and G. Kostorz, eds. (Pergamon), 1661 (1980)
14. R. A. Mulford, Acta Met. 27, 1115 (1979)
15. H. Mecking and U. F. Kocks, Acta Met., in press
16. J. E. Bird, A. K. Mukherjee, J. E. Dorn, in "Quantitative Relation between Properties and Microstructure", D. G. Brandon and A. Rosen, eds. (Israel University Press), p.255 (1969)
17. H. Mecking and K. Lücke, Mater. Sci. Eng. 1, 349 (1967)
18. D. Griggs, J. Geophys. Res. 79, 1653 (1974)
19. O. D. Sherby and P. M. Burke, Prog. Mater. Sci. 13, 323 (1968)
20. U. F. Kocks, Phil. Mag. 13, 541 (1966)
21. U. F. Kocks, in "Constitutive Equations in Plasticity", A.S.Argon, ed.(MIT Press), 81 (1975)
22. C. Y. Cheng and U. F. Kocks, 2nd Int'l. Conf. Strength Metals and Alloys (ASM), 199 (1970)
23. H. Mecking and K. Lücke, Acta Met. 17, 279 (1969)
24. K. Lücke and H. Mecking, in "Inhomogeneities in Plastic Deformation", R. E. Reed-Hill, ed. (Amer. Soc. Metals), p.223 (1973)
25. U. F. Kocks and H. Mecking, in "Strength of Metals and Alloys", P. Haasen, V. Gerold, G. Kostorz, eds. (Pergamon), p. 345 (1979)
26. U. F. Kocks, T. Hasegawa, R. O. Scattergood, Scripta Met. 14, 449 (1980)
27. U. F. Kocks, J. Eng. Mater. Tech. (ASME H) 98, 76 (1976)
28. U. F. Kocks, Prog. Mater. Sci., in press
29. T. H. Alden, Phil. Mag. 25, 785 (1972)
30. T. Hasegawa, U. F. Kocks, and T. Yakou, in preparation
31. T. H. Alden, Met. Trans. 6A, 1597 (1975)
32. A. W. Sleeswyk, Acta Met. 6, 598 (1958)
33. P. G. McCormick, Acta Met. 20, 351 (1972)
34. A. van den Beukel, phys. stat. sol. (a) 30, 197 (1975)
35. R. A. Mulford and U. F. Kocks, Acta Met. 27, 1125 (1979)
36. S. H. van den Brink, A. van den Beukel, P. G. McCormick, phys. stat. sol. (a) 30, 469 (1975)
37. P. Wycliffe, U. F. Kocks, and J. D. Embury, Scripta Met., in press.
38. P. Penning, Acta Met. 20, 1169 (1972)
39. R. A. Mulford and U. F. Kocks, Scripta Met. 13, 729 (1979)
40. J. M. Makin, Phil. Mag. 3, 309 (1958)

HIGH VOLTAGE ELECTRON MICROSCOPY AND MODELS OF DEFORMATION

J.L. Martin*
Laboratoire d'Optique Electronique du CNRS, BP 4347, 31055 Toulouse Cedex, France

Abstract

HVEM allows in situ experiments to be performed in which dynamic properties of the substructure can be directly observed with a good resolution. The limitations of this technique are evaluated and three examples of plastic behaviour are described, for which the main features of the corresponding deformation model could be determined from such observations: creep of aluminium at intermediate temperatures, heterogeneous deformation of FCC alloys, creep under irradiation of 316 steel at 350°C.

A reliable dislocation model describing a given plastic behaviour, should take into account the data of the corresponding macroscopic deformation tests, together with conventional electron microscope observations of the deformed specimen and when possible, observation of dislocation properties under load, in a high voltage electron microscope (HVEM). The latter experiments provide unique information about dislocation motion and interactions with a good resolution.

They really started in the early seventies, when various straining stages were built, allowing different in situ deformation conditions (1), about ten years later than the first encouraging 100 kV experiments (2). The increase of penetration at high voltages (several microns in light elements) allows one to observe dislocation properties similar to those in the bulk specimen in several cases and recent articles review fruitful experiments in this field (3 to 7).

The aim of the present paper, after summarizing the limitations of such observations, is to report some experiments carried out in Toulouse, on the microstructure properties under load of some FCC metals and alloys, mostly above room temperature in an attempt to improve the corresponding deformation models.

1. Limitations and Advantages of In Situ Experiments in the HVEM

Limitations are mainly due to radiation damage and surface effects. A suitable foil thickness should be larger than the characteristic dimension of the microstructure and in addition, the obstacle to dislocation motion should be strong enough for moving defects to resist surface forces. Strong metals or alloys should then be well adapted, if not too brittle in thin foils. To increase the observable thickness, higher voltages can be used, but this parameter saturates, and when the threshold voltage is reached, radiation damage occurs which may alter or impede dislocation motion. As a rule, light elements have the highest penetration depth at their threshold voltage, despite the low values of this latter parameter. If working conditions are necessary above the threshold, intermediate or high temperatures should be preferred to enhance the recombination of point defects at sinks. The evaluation of these limitations, together with the definition of safe experimental conditions have been attempted (8)(9), but available data on

---

*Now at Laboratoire de génie atomique, EPFL, 33, av. de Cour, CH-1007 Lausanne, Switzerland

surface effects and radiation damage are quite insufficient to determine accurately quantitative limits. The best validity test is to compare the in situ observations to other conventional data obtained under similar deforming conditions and check whether they agree or not.

In spite of these limitations, deformation parameters can be quantitatively estimated, namely local stresses (through the dislocation curvatures), strains (through the number of moving dislocations and the corresponding swept areas), strain rates, mobile dislocation densities, waiting times of defects at obstacles. In addition, the achievable resolution often allows one to identify the geometry of dislocation-obstacle interactions. As a rule, the understanding of the strain rate requires a careful identification, sampling and summing up of all the observed individual events.

The examples illustrated below are respectively connected to creep properties of aluminium at intermediate temperatures, heterogeneous deformation of face centered cubic alloys and creep under irradiation of 316 steel.

2. Creep of Aluminium at Intermediate Temperatures in Stage II

The metal investigated was 99.3% pure, of a 100 μm average grain size and had been creep tested between 0.3 and 0.5 T_M under stresses of 30 to 80 M Pa. The test was interrupted in the so-called "constant strain rate stage" (stage II) when the subgrain structure was present. Microsamples were then made with the same tensile axis and further deformed in the HVEM under similar conditions. The results of macroscopic creep tests can be found in (10) while the in situ observations are reported in (11) and are summarized below.

The microstructure consists of subgrains with an average diameter of a few microns which contain a 3 dimensional network and are bounded by well defined subboundaries with a 100 Å mesh size. A careful inspection of the subboundary geometry through weak beam experiments indicates that the Frank criterion is fulfilled within the accuracy of the method, i.e. the long range stress field, if any, is of the order of the applied stress. In addition, the dislocation segments in the network lie in their corresponding slip planes. Most of the subboundaries can be classified into symmetrical tilt walls, walls of mixed character with two Burgers vectors at 90° or 120° respectively, the latter kind being the most numerous. This is an indication that during creep, only two (or three) of the six possible sources are preferentially activated. During deformation, several phenomenae can be observed (12):

- Subboundary migration: This confirms Warrington's optical observations (13). It occurs by glide of the individual dislocations on 111 planes but also 110 and 100. The triple points at adjacent subboundaries are mobile and therefore, the subboundary bulging under stress, if any, seems to be too small to account for the anelastic creep which is macroscopically observed during partial unloading of the sample. Migration accounts for only about 1/10 of the total deformation and is responsible for the isotropic shape of the subgrains and their size changes during load modification. Occasionally subboundary destruction is also observed (14), but this phenomenon occurs rather seldom.
- Movement of individual dislocations: Dislocation sources are observed inside the subgrains, and are of the single ended type, the milling arm being pinned at impurity clusters, at junction segments of two dislocations after reaction, at superjogs of a sufficient length. An important point is that there is less than one source in each subgrain, so that the emitted loops have to cross subboundaries. They cut through the 3 dimensional network rather easily with several cross slip events at intersections, probably of the Washburn type (15), with a flying time through the cell which is negligible with respect to the waiting time at the subboundaries.
- Dislocation-subboundary interactions: Passing through subboundaries is a difficult process, since junctions have to recombine in a small size network. A mechanism has been imagined (11), involving cross slip, with a high critical stress (but lower than the corresponding Orowan stress) and an activation volume proportional to the mesh size: it can be interpreted as a dynamic recovery process and accounts for the wavy slip traces observed in the vicinity of subboundaries and the zig-zag shapes of the defects, observed just after escaping.

The mechanism controlling the creep rate of Aluminium at intermediate temperatures, in the constant strain rate range, appears to be the thermally activated crossing of subboundaries by individual defects, which occurs through a cross slip process.

3. Heterogeneous Glide in Face-Centered Cubic Alloys

When the solute concentration increases, macroscopic deformation tests indicate a flow stress dependence on temperature and strain rate which cannot be interpreted easily because of several concurrent phenomenae, namely dislocation splitting, dislocation solute interactions, glide localization. Two different types of alloys have been investigated: Ni-30 wt % Cr and Cu-x at % Ge (with x = 2 and 6) following conventional macroscopic tests (16)(17).

- Ni-30 wt % Cr has a rather low stacking fault energy ($\gamma/Gb \sim 3.5\times10^{-3}$) and exhibits short range order after 100 hours of annealing at 600°C. Between 77K and 400K, conventional deformation tests lead to rather small activation volumes independent of strain, and a high "athermal stress" incompatible with the intersection mechanism (16).
- Cu-x at % Ge has a stacking fault energy of 35 and 15 mJ/m^2 for x = 2 and 6 respectively and shows between 77K and 400K a decrease in the flow stress with temperature, down to a high athermal plateau above 400K (17). Observations by high speed cinematography of slip band propagation with an optical microscope, allow a correct determination of the activation parameters to be made (18) which are connected with the by-passing of clusters of solute atoms of different strength with a high athermal stress (about 3/4 of the plateau stress).

Dynamic observations at 100K and 300K in Cu-Ge, exhibit homogeneous and heterogeneous glide for 2 at % and 6 at % Ge respectively (20), and heterogeneous glide in Ni-30 % Cr at 300K: large dislocation groups emitted by a single source, move on discrete slip planes, inducing a large local strain rate and a long range stress field, this process being responsible for most of the deformation (19). Individual dislocations are seldom observed to move, except in the vicinity of the active glide planes, because of the high stress field, which indicates strong anchoring forces. A careful inspection of the dislocations positions along an active slip plane shows that the dislocations pile up against each other thus concentrating the stress on hard points in the crystal and then separate after by-passing them. Obstacles to glide can therefore be inventoried and consist of: intersecting pile-ups, grain and twin boundaries, "invisible obstacles" probably bound to the oscillating internal stress field, or to fluctuations in alloy concentration. Since the long range stress field is weaker in thin foils (truncated dislocation segments), the latter explanation is more likely, although no direct evidence could be obtained yet. In addition, the gliding defects experience the intrinsic lattice resistance to glide, which maintains dislocations at unstable positions, when straining is interrupted. This resistance could possibly be higher on the lead dislocations of the groups, because of a glide softening process (21).

A quantitative estimation of the stress concentration at the pile-up head has been derived for the thin foil situation (20-21), the concentrated stress being as large as several times the flow stress.9 Critical stresses corresponding to the above obstacles can thus be measured.

4. Creep Under Irradiation of 316 Steel

The phenomenon of creep under irradiation has been considered by means of several models relying on few scattered experimental results. It has been simulated in the HVEM at 350°C and 450°C on an annealed steel (22).

Thermal creep occurs through heterogeneous glide as described above for face centered cubic alloys. Under irradiation conditions, first a low dose substructure is observed (below 0.35 dpa) consisting of small Frank loops (diameter smaller than 500 Å) parallel to the 4 {111} planes. Measurements of the loop densities in each of the 4 families indicate uneven loop populations: the higher the normal stress, the larger the density (care was taken to compare loop families submitted to equal surface forces). This confirms the hypothesis of Hesketh (23) and Brailsford and Bullough (24) in which the stress biases the nucleation of imperfect loops. For slightly higher doses unfaulting of the loop is observed, the perfect loops being identified as interstitial ones. This process is thermally activated and quite fast at these temperatures. Due to their small sizes, it was not possible to count the loop densities corresponding to the 6 possible Burgers vectors.

For higher doses, usually above 0.5 dpa, the irradiated area exhibits a "Corduroy contrast" similar to that reported by Kelly and co-workers (25) in neutron irradiated zirconium. This contrast is bound to crystal bands which are built up during irradiation (parallel to the basal plane in Zr, to a 111 plane in 316 steel) of alternating low and high loop densities. Since this anisotropy in loop condensation has a crystallographic origin in Zirconium (the 3 available Burgers vectors are in the basal plane), it is thought that in steel it results from the reaction of perfect loops, the Burgers vectors of which are distributed anisotropically, due to the influence of stress on the previous stages of the microstructural development. Therefore, in this latter stage, the climb of inhomogeneous loop population has to be taken into account to describe the creep rate.

Therefore, the successive microstructural stages could be clearly identified in such an experiment, for 316 steel between 350 and 450°C.

5. Conclusion

In both creep experiments, the dynamic properties of the microstructure do indicate which type of model should be developed to interprete the creep rate. In the deformation by glide of face centered cubic alloys, more quantitative results should be obtained and a correlation between the flow stress and the various measured obstacle strengths should be possible.

Acknowledgments

The author would like to thank D. Caillard, N. Clément, F. Monchoux and Prof. H. Neuhäuser for helpful comments on the manuscript.

References

(1) T. Imura, H.Saka, A.Nokara, N.Yukuwa and H.Kitagawa, in "Microscopie Electronique 1970" p.153 (P.Favard Ed.), SFME Paris (1970)
(2) H.G.F.Wilsdorf, A.S.T.M. Spec. Tech. Pub. 245, 43 (1958)
(3) T.Imura, in "Electron Microscopy and Structure of Materials" p.104 (G.Thomas Ed.) Univers. of California Press, Berkeley (1972)
(4) R.Sandström, B.Lehtinen and R.Lagneborg, Scand. J. Metal. 4, 17 (1975)
(5) H.Fujita, in "Electron Microscopy 1978" III, p.355 (J.M.Sturgess Ed.), Microscopical Soc. of Canada, Toronto (1978)
(6) E.P.Butler, Reports on Progress Physics 42, 833 (1979)
(7) L.P.Kubin and J.L.Martin, in "Strength of Metals and Alloys" 3, p.1639 (P.Haasen, V.Gerold and G.Kostorz Ed.) Pergamon N.Y. (1979)
(8) J.L. Martin and L.P.Kubin, Ultramicroscopy 3, 215 (1978)
(9) J.L. Martin and L.P.Kubin, Phys.Stat.Sol. a) 56, 487 (1979)
(10) M.M.Myshlyaev, in "Proceedings 4th I.C.S.M.A." 3 p.1037, E.N.S.M.I.M. Nancy (1976)
(11) D.Caillard, Doctorate Thesis, Toulouse 1980
(12) D.Caillard and J.L.Martin, in "Strength of Metals and Alloys", vol.2, p.1323 (P.Haasen, V.Gerold and G.Kostorz Ed.) Pergamon N.Y. (1980)
(13) S.F.Exell and D.H.Warrington, Phil.Mag. 26, 1121 (1972)
(14) M.M.Myshlyaev, D.Caillard and J.L.Martin, Scripta Met. 12, 157 (1978)
(15) J.Washburn, Appl. Phys. Letters 7, 183 (1965)
(16) N.Clément and P.Coulomb, Phil. Mag. 30, 683 (1974)
(17) H. Traub , H.Neuhäuser and Eh.Schwink, Acta Met. 25, 437 (1977)
(19) N.Clément, D.Caillard and J.L.Martin, in "Electron Microscopy 1978" vol.1, p.568, (J.M.Sturges Ed.) Microscop. Soc. of Canada, Ontario (1978)
(20) F.Monchoux and H.Neuhäuser, to be published
(21) N.Clément, F.Monchoux and J.L.Martin, in "Strength of Metals and Alloys" vol.2, p.1299 (P.Haasen, V.Gerold and G.Kostorz Ed.) Pergamon N.Y. (1979)
(22) D.Caillard, J.L.Martin and B.Jouffrey, Acta Met. 1980, under press
(23) R.V.Hesketh, Phil. Mag. 7, 1417 (1962)
(24) A.D. Brailsford and R.Bullough, Phil. Mag. 27, 49 (1973)
(25) P.M.Kelly, R.G.Blake and A.Jostons, J. Nucl. Mat. 59, 307 (1976)

STRAIN HARDENING AND DYNAMIC RECOVERY

H. Mecking*

Institut für Allgemeine Metallkunde und Metallphysik, RWTH Aachen, West Germany

Abstract

Work hardening and creep of pure metals are surveyed on the basis of the correlation between the main features of structure evolution and macroscopic behavior.

At low strains dislocation storage follows geometrical rules resulting in a stage of athermal hardening. In the subsequent stage the hardening coefficient becomes temperature dependent. This can be described consistently by the increasing influence of a thermally activated dynamic-recovery component which superimposes upon the athermal hardening component. In this regime dislocations arrange in a cellular pattern, i.e., cell walls at low temperatures and subboundaries at high temperatures.

As strain continues the development of the cellular structure appears to be disturbed by structural instabilities and inhomogeneities of deformation, such as dynamic recrystallization, micro-bands and shear bands. The influence of these processes on work hardening makes it necessary to distinguish four different stages of work hardening and has important consequences for the interpretation and evaluation of steady state flow.

Introduction

Due to the great variety of possible dislocation processes which occur during plastic flow, it will never be posisble on a purely theoretical basis to predict the evolution of the dislocation structure under different deformation conditions and to calculate the macroscopic properties of a material with sufficient accuracy. Rather, for a mechanistic understanding of mechanical properties, the common procedure is to define the phenomenological relationships experimentally, to correlate these with microscopic observations on the crystallography of slip and on the evolution of the dislocation structure and hence to deduce the physical nature of the controlling dislocation mechanisms.

Physical descriptions obtained in this way can be successfully checked by their ability to predict material behavior for different deformation conditions. For example calculations of creep curves from parameters measured in stress-strain tests are a useful tool to examine critically the competence of the physical concepts of work hardening. The accuracy to which stress-strain curves can be converted into creep curves and vice versa can also be considered as a quantitative measure of the degree of understanding of plasticity. Once the conversion rules for the results of these two simple test methods are known, material behavior under all other deformation modes can be described as long as they are unidirectional.

The intention of this paper is to discuss on the above basis our present knowledge about plastic behavior. The main elements of structure evolution will be correlated with macroscopic observations and an analysis will be made of the extent to which the rules deduced from this comparison can be used to convert the results of creep and stress-strain tests into each other.

It is well established (1) that several different stages of deformation can be distinguished as demonstrated by the stress-strain curves of <111>-oriented Ag single crystals tested at different temperatures (Fig. 1). At low strains the curves for all temperatures coincide indicating a range of athermal hardening (stage II). This will be discussed in detail in the next section. At medium strains above a certain stress level (τ_{III}) the curves become temperature dependent. The phenomenology of this range (stage III) appears to be rather complex.

*Now with Techn. Universität Hamburg-Harburg, West Germany.

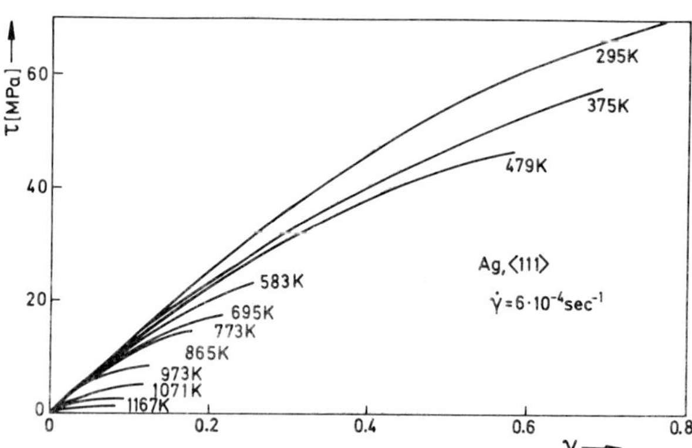

FIG. 1
Resolved shear stress versus shear strain for Ag single crystals oriented for polyslip. Wachholz (2).

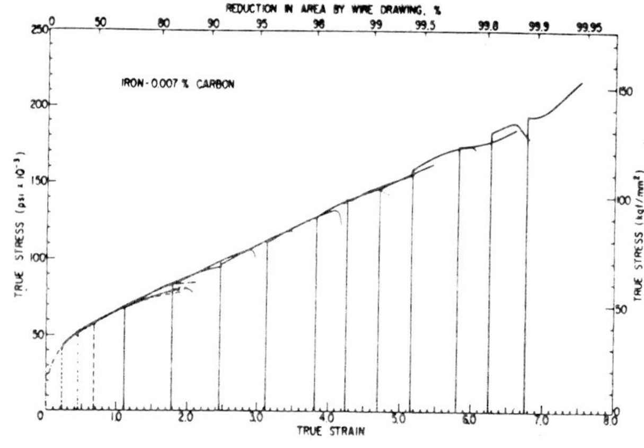

FIG. 2
Stress-strain curve of low carbon steel obtained from tensile tests after different amounts of strain by wire drawing. Langford and Cohen (5).

At large strains the curves flatten continuously towards a horizontal course which, however, is not reached in tension because, at low temperatures necking intervenes, and at high temperatures recrystallization occurs in most pure metals (3). Nevertheless, the position has often been taken that steady state flow is a typical behavior in the large-strain limit. In particular, the concept of steady state has extensively been applied in creep theories (4). There have been, however, many experimental observations of large-strain behavior, particularly after industrial forming operations, where a steady state condition has not been observed. Fig. 2 shows, as an example, the result of one of the early investigations (5) of this question, viz., the $\sigma_{0.2}$ proof-stress as a function of strain by wire drawing. The curve appears linear in the extremely large strain-range over a considerable strain interval (up to $\varepsilon = 8$) and exhibits no tendency for steady state flow. In this paper it will be discussed to what extent the concept of a steady state is a reasonable approach, what the important structural observations are and how this stage can be described phenomenologically.

Finally, some limitations have to be mentioned. This paper will deal with unidirectional deformation only, and only pure metals will be considered, although the general concepts can be applied directly, or after some modification, to alloys as well.

Dislocation Structure at Low Strains

The anisotropy of the dislocation mobility has a considerable influence on the dislocation pattern which is produced during deformation. For instance, the dislocation structure of b.c.c. metals deformed at low temperatures, where the edge dislocations have a much higher mobility than screw dislocations consists almost exclusively of screw dislocations (6). On the other hand in certain alloys which contain precipitates often only edge dislocations are observed (7) due to the higher mobility of the screw segments. These cases, however, can be considered as exceptional because in most cases the slip distances of edge and screw dislocations are of comparable magnitude (8). This is particularly true for pure metals with closed packed structures and for the b.c.c. metals deformed at high enough temperature (approximately above room temperature). In the following only these cases will be discussed.

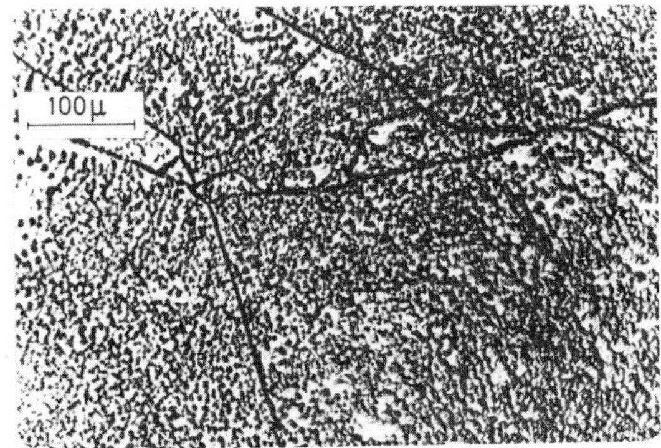

FIG. 3
Arrangement of forest dislocations in Cu single crystals after 4% shear in single slip. Mecking and Bulian (12).

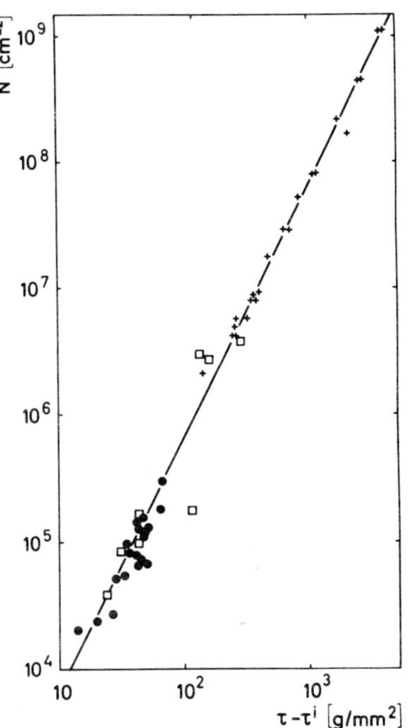

FIG. 4

FIG. 4
Etch pit density of forest dislocations as a function of flow stress for Cu single crystals deformed at room temperature. Replotted from data collected by Mecking and Bulian (12). From the flow stress a constant value of τ^i = 12 g/mm^2 is subtracted which is attributed to impurities.

Investigation of the dislocation structure at low strains have been extensively conducted by the etch-pit method (9-12). In Fig. 3 the arrangement of etch pits in a Cu single crystal oriented for single slip is shown for 4% shear at room temperature. The dislocations are arranged in a rather random pattern, and sometimes a slight tendency for cell formation is seen. The figure shows etch pits of secondary dislocations. Primary dislocations are often more regularly arranged along slip planes and tend to build polygonization walls which are, however, at least partly due to relaxation processes during unloading (12). In symmetrically oriented crystals, cells have been observed in the very early stages of deformation (13). However, since such investigations have always been conducted in the unloaded state it also remains questionable whether these cells are present during deformation.

Whatever the details of the dislocation arrangements may be, dislocation storage at low strains (and moderate temperatures) can be considered as a purely geometrical process (1). The dislocation density increases rapidly with strain by a shrinkage of the scale of the dislocation pattern but by keeping its geometry constant [principle of similitude (14)]. Therefore, depending on the orientation, the dislocations arrange in a more or less random pattern with a continuous decrease of the average dislocation spacing.

The flow stress in this deformation range as a function of the density of forest dislocations follows a simple square-root relationship while its correlation with the density of primary dislocations is rather peculiar (11, 12). Fig. 4 displays the forest dislocation density as a function of the flow stress demonstrating that the square root law is obeyed over 5 orders of magnitude of the dislocation density. In a more detailed discussion about the σ vs. $\sqrt{\rho}$ dependence in Ref. (17), where TEM data were also considered, it was concluded that the linear relationship between σ and $\sqrt{\rho}$ even holds for 7 orders of magnitude of the forest density.

Athermal Hardening

In the past many theoretical and experimental investigations of work hardening have been focussed on the low strain region (stages I and II). Disregarding the easy glide region (stage I) which is a characteristic of single slip, stage II behavior can be considered to be reasonably well understood, although, some controversies are still unsettled. It is not the aim of this contribution to revive the discussion, but rather to present the general physical background.

In agreement with the above discussed microscopical observations of the main geometrical features of structure evolution in this stage, hardening is athermal, i.e., the hardening coefficient Θ does not depend on temperature or strain rate. Originally the notation stage II hardening had been restricted to linear hardening (15), however, it has been suggested (1) that it should be extended to denote athermal hardening. This is because it has been shown that, when the stress strain curves are analyzed in their differentiated form, single crystals oriented for single or polyslip and polycrystals obey very similar relationships (1).

Furthermore, the Cottrell-Stokes law is obeyed in stage II of single and polycrystals of pure f.c.c. metals (16, 17). For example Fig. 5 shows the relative increase in stress due to strain rate changes plotted as a function of the flow stress for Ag single-crystals of different orientation and for polycrystals. The data follow closely a straight line through the origin at low stresses. The Cottrell-Stokes law is the macroscopic equivalent to the microscopic observation in Fig. 4 that the flow stress is controlled by the forest-dislocation density. Deviations from Cottrell-Stokes behavior which have been sometimes reported can be ascribed to contributions to the stress by impurities, lattice resistance, etc. (18).

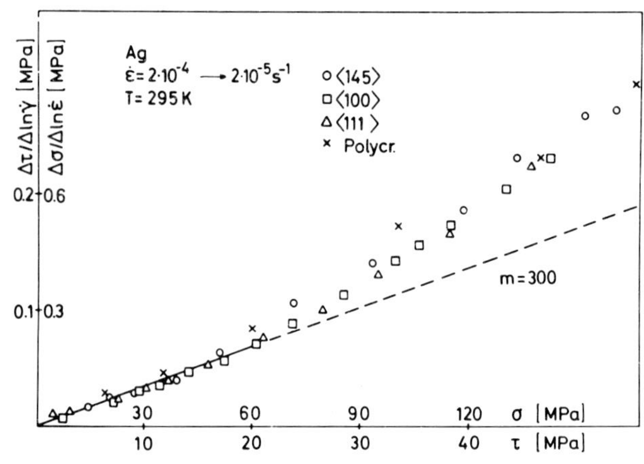

FIG. 5

The relative stress changes due to strain-rate changes as a function of the flow stress for Ag single crystals of different orientations and for polycrystals. Mecking and Kocks (17).

For an analytical description of plastic behavior often the differential form is preferred

$$d\tau = \left.\frac{\partial\tau}{\partial\gamma}\right|_{\dot\gamma} d\gamma + \left.\frac{\partial\tau}{\partial\ln\dot\gamma}\right|_{\gamma} d\ln\dot\gamma \ . \tag{1}$$

The stress τ necessary to enforce plastic flow changes because of two physical reasons, viz., due to changes of structure (work hardening) associated with an increase in strain γ and due to changes in strain rate $\dot\gamma$ at constant structure. (Resolved shear stresses and strains are always implied.)

Eq. (1) is generally assumed to be valid (19). To use it to describe plastic behavior it is, however, essential to know the two functions

$$\left.\frac{\partial\tau}{\partial\gamma}\right|_{\dot{\gamma}} \equiv \Theta \quad \text{and} \quad \left.\frac{\partial\tau}{\partial \ln\dot{\gamma}}\right|_{\gamma} \equiv \frac{\tau}{m} \tag{2}$$

According to the above considerations, Θ and m turn out to be very simple in stage II, viz., constants in a good approximation. Over the range in which stress strain curves appear to be straight it follows that creep curves should be also straight lines when plotted as $\ln\dot{\gamma}$ vs γ since for $d\tau = 0$ Eq. (1) yields

$$\left.\frac{\partial \ln\dot{\gamma}}{\partial\gamma}\right|_{\tau} = -\frac{m\Theta}{\tau} \tag{3}$$

For f.c.c. metals in the vicinity of room temperature the quantities m and Θ/τ range typically between

$$200 \leq m \leq 500 \quad \text{and} \quad 5 \leq \Theta/\tau \leq 100 \tag{4}$$

Here the range for Θ/τ varies with temperature and material due to the fact that $\Theta \simeq$ const. $= \mu/200$ (μ = shear modulus) but that the τ-range of stage II decreases with increasing temperature and stacking fault energy of the material (1). Hence, the range for the slope of the creep curves $(-m\,\Theta/\tau)$ in the athermal region lies between

$$5 \cdot 10^4 \geq -\partial\ln\dot{\gamma}/\partial\gamma \geq 10^3 \tag{5}$$

depending on the applied stress. The higher value in Eq. (5) corresponds to low stresses near the initial yield stress while the lower value corresponds to high stresses at the end of stage II, i.e., in the vicinity of τ_{III}. This estimate shows that creep in this range decelerates rapidly to rates below the observable range. For example, within 1% shear strain the creep rate falls by 200 orders of magnitude near the upper value of Eq. (5) and still by 4 orders of magnitude at the lower limit.

If Θ and m were exactly constant, integration of Eq. (3) would lead to a logarithmic creep law (20)

$$\dot{\gamma} = \dot{\gamma}_o/(1 + \dot{\gamma}_o \frac{\Theta m}{\tau} t) \quad \text{or} \quad \gamma = \frac{\tau}{m\Theta} \ln\,(1 + \dot{\gamma}_o \frac{\Theta m}{\tau} t) \tag{6}$$

where $\dot{\gamma}_o$ is the initial strain rate at zero time and the corresponding strain is assumed to be zero. Experimentally, in this equation it has been found that the exponent of time (t) is slightly higher than 1 and approximately 1.1 (21). These deviations from ideal logarithmic creep can be interpreted as a reflection of deviations from ideal stage II behavior of stress strain curves. It can easily be shown that an exponent slightly above 1 is predicted if Θ is not exactly constant but increases with strain or decreases with strain rate (1) (as has been observed to be typical for deformation in single slip) or if m increases with strain [as has been observed to be the standard deviation from Cottrell-Stokes behavior in metals containing solute impurities (18)] or if m increases with decreasing strain rate [which follows from theoretical considerations (22,23)]. Yield point effects which appear in connection with strain rate changes (19) should also lead to deviations from ideal behavior.

The final conclusions from this discussion is as follows: The dislocation processes underlying athermal hardening are very well explored. Since the dislocation pattern merely shrinks in its dimensions without changing its geometry the dislocation structure is sufficiently described by a single parameter such as the average dislocation-density. Equivalently, the mechanical state of the material is completely defined by a single macroscopic quantity such as the flow stress at a given strain rate and temperature. This together with the fact that the work hardening rate Θ and the strain-rate sensitivity m are rather insensitive to strain, allows an accurate conversion of stress strain data into creep data and vice versa.

Dislocation Structure at Large Strains

While the correlation of macroscopic behavior with microscopic observations has led to a consistent (though still contested) description of the athermal hardening regime, the situation is rather unsettled for larger strains where hardening becomes temperature dependent. In the following two sections a concept will be presented of what the important elements of structure evolution are and how they are interrelated with macroscopic behavior. The developed framework is based on data of pure copper and is not meant to be generally valid also for other metals, but may serve as a guideline for these as well.

It has been confirmed by many investigations that with continuing strain the dislocations tend to build up a cellular structure at every deformation temperature (25, 30). At high temperatures (around 0.7 T_M) well defined subgrains are formed (Fig. 6). Often the shape of the subgrains reflects slightly the deformation mode because they are respectively elongated or shortened in the tensile or compressive directions. They can, however, always be considered to be equiaxed if compared to the macroscopic shape changes imposed by the applied strain. This is a very important result since it shows that the subgrains do not represent a statically stable structure (3, 25) but that their actual size and shape is the result of a dynamical balance between two competing processes, namely subgrain formation and subgrain destruction. Therefore a steady state subgrain size will form (26) when both processes reach equilibrium. However, the equilibrium is not necessarily stable.

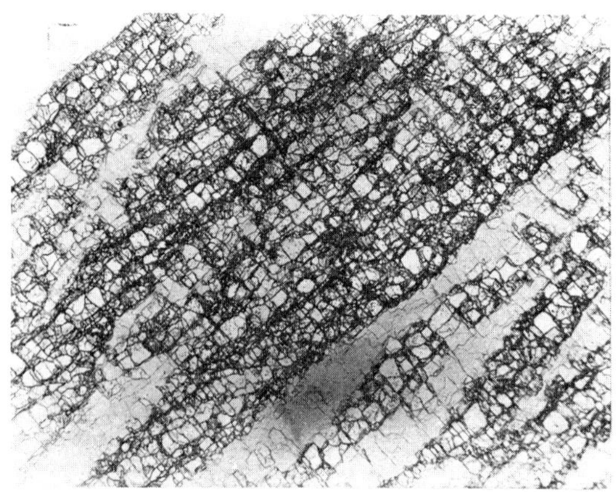

FIG. 6

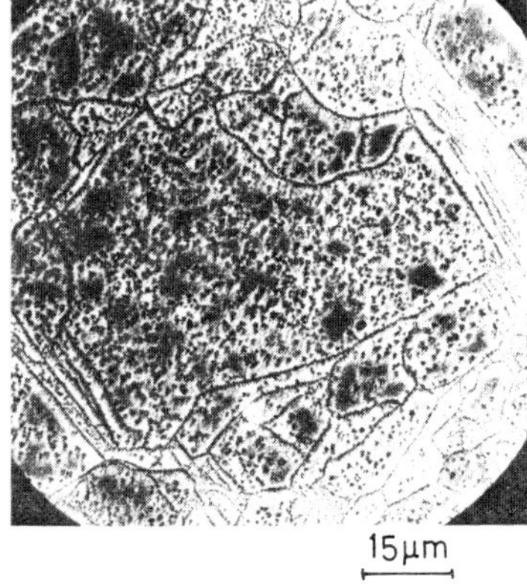

FIG. 7

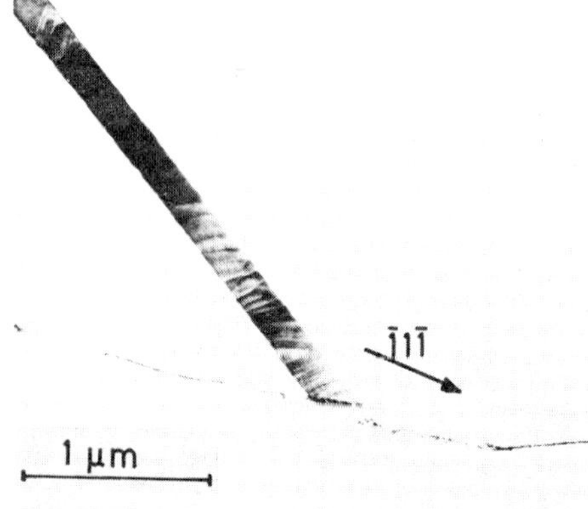

FIG. 8

FIG. 6. The subgrain structure of Cu single crystals deformed at 1180 K (Etch pit method). Hansen (24)

FIG. 7. Inhomogeneous subgrain growth in a Cu single crystal deformed to the beginning of dynamic recrystallization at T $\approx$ 1100 K. Karduck (29).

FIG. 8. Twin Lamella emerging from a sub-boundary in a Cu single crystal deformed to the beginning of dynamic recrystallization at moderate temperature. Karduck (29).

The Different Stages of Work Hardening

As already mentioned plots of the slope Θ versus the stress are better suited for a mechanistic interpretation of work hardening than the actual stress strain curves (Eq. (1)). Such plots correspond to presentations of creep curves in the form $d\ell n\dot{\gamma}/d\gamma$ versus strain rate $\dot{\gamma}$. Since the determination of these two quantities demands the first and second derivative of the creep curve (γ vs t) they are subject to large experimental scatter and hence are rarely used. Additionally, there are not many data on creep at large strains and low temperature. Therefore, the following discussion will be restricted to stress strain behavior.

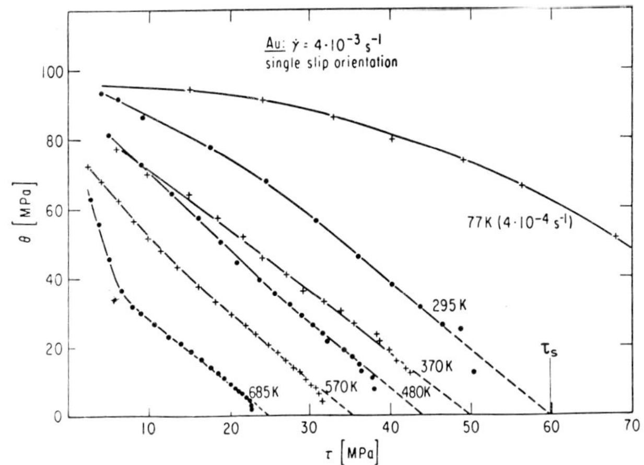

FIG. 12
Work hardening coefficient Θ versus the flow stress for Au single crystals. The dashed continuations represent the extrapolation to steady state. Kirch and Mecking (37,1).

Fig. 12 shows Θ versus τ as obtained for Au single crystals oriented for signle slip. All the curves start practically from the same Θ-level at low-stress values and decrease to low Θ-values with a slope that increases with temperature. This is the typical behavior in stage III where hardening becomes thermally activated. The curves do not exhibit any indication prohibiting an extrapolation beyond the necking point for the determination of a steady state stress τ_s at Θ = 0. They even develop rather straight branches at this stage and the extrapolation extends only over a relatively small Θ and τ range.

Regarding the correlation with structure evolution, the decreasing Θ-τ branches are in the range where the dislocations arrange in diffuse cell walls at low temperature and in subboundaries at high temperature. Consequently the τ_s values, as determined by the above extrapolation method, correspond to the stage where cell size (or subgrain size) would have come to a steady state. It is assumed to be reached when the processes of cell formation and cell destruction have come to an equilibrium.

Although the extrapolation beyond the necking point to Θ = 0 extends only over a small Θ - τ range the corresponding strain interval is infinitely large, since a linear approach to Θ = 0 in a Θ versus τ plot is identical with an exponential stress-strain relationship where a saturation stress is reached asymptotically at infinite strain (1). Therefore the large strain effects as discussed in the last section would become noticeable in this region. This explains the appearance of irregularities such as seen in Fig. 13 which displays Θ - τ data reported for the high-strain regime of a torsion test on Cu at room temperature. At Θ values considerably below the critical values for necking in tension, the curve changes abruptly from a steeply decreasing slope into a horizontal branch. This is clearly an indication of a change in the controlling mechanism (38).

A comparison with the microscopic observations suggests that the new stage with almost constant Θ is related to the appearance of microbands and the eventually decreasing part with shear bands.

It must be admitted that this is a somewhat speculative explanation since the microscopic observations were on rolled specimens while the curve in Fig. 13 was obtained in torsion. However, such behavior as shown in Fig. 13 seems to be typical for deformation of Cu at room temperature. Fig. 14 shows a stress strain curve for deformation by wire drawing (39) which quantitatively agrees well with the torsion data.

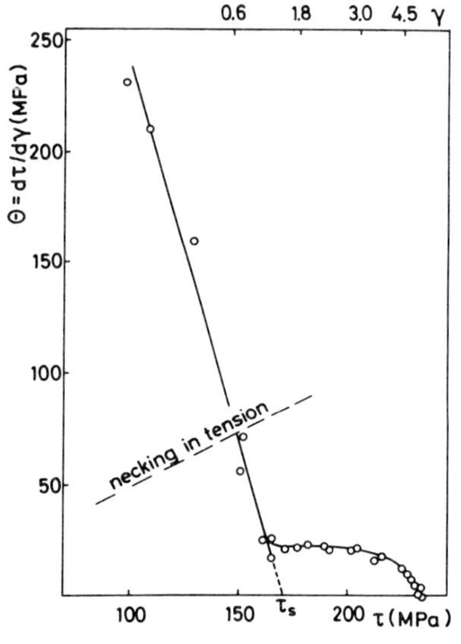

FIG. 13
Work hardening rate Θ versus the flow stress for very large shear strains in Cu deformed by torsion. Gil Sevillano, van Houtte and Aernoudt (38).

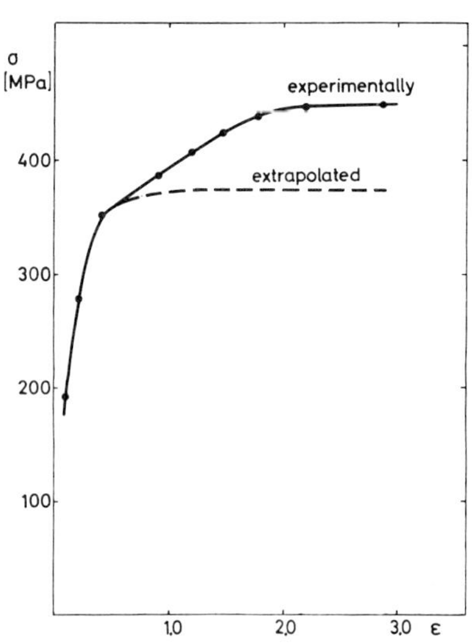

FIG. 14
$\sigma_{0.2}$ proof stress obtained from tensile tests after wire drawing. Lloyd and Kenny (39). For comparison with Fig. 13 note that $\sigma \simeq 2\tau$ and $\gamma \simeq 2\varepsilon$.

As a result of the comparison of macroscopic behavior with the microscopically determined features of structure evolution, it is concluded that four different stages of deformation should be distinguished as proposed in Ref. (38). Following common nomenclature these stages can be characterized by the following scheme displayed in Fig. 15.

Stage I is the easy glide region which only appears in deformation of single crystals oriented for single slip. It has not been considered in this paper. Stage II is the regime where hardening is athermal. Here the evolution of the structure follows geometrical rules. The dislocations are stored statistically in a random pattern.

Stage III is the subsequent temperature dependent part. It is usually explained by a continuous superposition of thermally activated softening processes (cross slip and climb) on the athermal hardening processes. Thereby the dislocation structure becomes cellular consisting of rather dislocation-free cells surrounded by diffuse cell walls. The dashed part in Fig. 15 represents the extrapolation of stage III behavior to large strains assuming that the underlying processes continue to operate. The asymptotic stress is the saturation stress σ_s corresponding to τ_s as defined in Fig. 12. Provided dynamic recrystallization does not occur, at high temperature possibly only stage III has to be considered, because the length of stage II practically becomes zero and the strains to generate a steady-state situation are small, less than a few percent. At low temperatures, however, where the corresponding strains are rather high new mechanisms become active.

Stage IV is here used as the collective term for these large-strain effects. It starts with an abrupt change of the characteristics of work hardening and smilutaneously new structural elements are observed microscopically. With the beginning of stage IV microbands become a dominant feature of the structure. They replace the diffuse cells of stage III. Intersecting microbands on different intersecting crystallographic planes ({111} planes) form elongated cells with sharp boundaries of high dislocation content. Possibly, the almost linear parts in the stress strain curves in Figs. 2, 13 and 14 are related to these processes. A characteristic of even larger strains is shear band formation which is possibly responsible for the flattening of the stress strain curves at the large-strain end. To what extent additional effects, such as discussed in the last chapter, influence stage IV behavior cannot as yet be deduced from published data. Also how this stage is affected by temperature and strain rate or by material parameters such as stacking fault energy and lattice structure is poorly known.

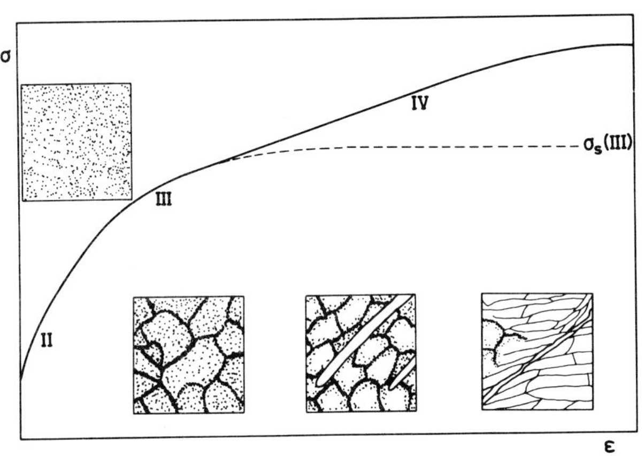

FIG. 15
Correlation of the four hardening stages proposed by Gil Sevillano, Van Houtte and Aernoudt (38) with structure evolution observed by Malin and Hatherly (35).

Phenomenology of Stage III

Physical concepts of the influence of temperature and strain rate on the work hardening and creep behavior at larger strains are almost exclusively obtained from data within the strain ranges typically reached in tensile tests. Therefore they refer to the stage III mechanisms but do not include the processes which have been denoted large-strain effects in this paper. It turns out that the phenomenology of stage III behavior is rather complex compared to the athermal region. Therefore only the basic physical features can be presented here; for further details the reader is referred to the original literature (1,17,40-42). For a proper description of stage III one needs to know how in Eq. (1) the work hardening coefficient $\Theta \equiv \partial\tau/\partial\gamma$, and the rate sensitivity of the stress $1/m \equiv \partial \ell n\tau/\partial \ell n\dot{\gamma}$, vary with deformation and also how they depend on temperature and strain rate.

Experimentally the dependences of Θ on $\dot{\gamma}$, T and τ have been very extensively analyzed and it is now well established that work hardening can be considered as being the result of two counteracting processes: strain hardening resulting from an approximately statistical storage of dislocations and dynamic recovery due to thermally activated rearrangements of dislocations. This situation can formally be described by composing the work hardening coefficient out of two terms

$$\Theta = \Theta_h - \Theta_r\ (\tau,\ \dot{\gamma},\ T) \tag{7}$$

Θ_h is the coefficient of athermal hardening of stage II which is a general constant in a first approximation. Θ_r denotes the softening effect of the recovery processes which increases (roughly linearly) with increasing flow stress; the increase is fastest at high temperatures and low strain rates (Fig. 12). Thus, experimentally the two contributions can be determined separately due to their very different dependence on temperature, strain rate and degree of deformation. To a first approximation Θ_h does not depend on T and $\dot{\gamma}$ nd only weakly on τ. From an analysis of data obtained in stress strain tests like those in Fig. 12, the empirical relation (1, 40-42) for the recovery term in Eq. (7)

$$\Theta_r = c\,\tau \left(\frac{\dot{\gamma}_o}{\dot{\gamma}}\right)^{1/n} \quad \text{with } n = n(T) \tag{8}$$

has been confirmed for a number of pure f.c.c. metals over a considerable temperature range.

The fundamental problem arising at this point is that the $\Theta_r - \tau$ relationship is not uniquely valid because it depends on the deformation path. This is immediately evident from the long transients in stress strain behavior appearing after changes in the strain rate (17,19). The path dependence means physically that Θ_r in Eq. (8) at a fixed $\dot{\gamma}$ and T is not defined sufficiently by the flow stress τ and, therefore, an additional parameter is needed.

In theoretical descriptions of stage III-hardening such an additional parameter is inherent in the balance between strain-hardening and dynamic recovery (41, 43) which by its very nature is path dependent. The path dependence of work hardening in stage III is in very good agreement with the observations of structure development. Stage III can be interpreted as a transient stage where the statistically arranged dislocation pattern of stage II is gradually converted into a cell structure. As such, the structure is bimodal consisting of cells interspersed with random tangles. Such a situation cannot be characterized by the specification of only one geometrical quantity such as cell size, but, in addition the density of dislocations in the tangles has to be specified (44). If both cells and tangles act as obstacles for dislocation glide (25, 45), the flow stress is not an unique parameter of the structure since the same flow stress can be achieved by manifold combinations of the two structural elements.

For a complete description of stage III-hardening not only the work hardening coefficient but also the other partial derivative in Eq. (1), the strain rate sensitivity m, has to be known. Like Θ, m also changes with deformation in stage III (Fig. 5) and becomes dependent on strain rate in addition to the temperature dependence which already exists in stage II (17, 46). This behavior has been explained by an interrelationship between m and Θ_r (17) of the form

$$\frac{1}{m_{III}} = \frac{1}{m_{II}} + F(T)\,\Theta_r \tag{9}$$

Here m_{III} is the strain rate sensitivity in stage III, m_{II} is the constant value in the Cottrell-Stokes regime (i.e., stage II), and F(T) is a semi-empirical function of temperature. This equation describes the increasing deviations from Cottrell-Stokes behavior with increasing deformation into stage III as shown in Fig. 5.

Summarizing it can be stated: (i) The phenomena which occur in stage III can be explained quantitatively by the concept of a superposition of strain hardening and dynamic recovery. (ii) The characteristics of the work hardening coefficient and the strain rate sensitivity can be related quantitatively to underlying physical processes. (iii) However, a generally applicable phenomenological description still has to be developed.

Regarding the possibility of the conversion of stress strain curves into creep curves is complex and at present, in the author's opinion, is an unsolved problem. However, the situation is very simple when a steady state exists. Then in both stress-strain and creep tests deformation proceeds under constant values of stress, temperature and strain rate and therefore if a sample is deformed up to this state, it cannot distinguish whether it is in a stress-strain or in a creep machine. It has been shown in this paper that the existence of a steady state is a realistic limiting assumption. It also has been confirmed experimentally that the relation between stress, strain-rate and temperature within the accuracy of the meausrements is identical for stress-strain and creep tests in this range.

Numerous publications have dealt with the question of the best analytical description of steady state flow, e.g., (4). The data analyses and their interpretation have been carried out in great detail in such investigations (48,49). However, the essential requirement for such sophisticated analyses is good and precisely defined data. Therefore it seems useful to draw attention to a few aspects following from the above presentation.

Firstly, theoretical and experimental investigations on steady state flow have almost exclusively dealt with the steady state limit of stage III as defined in this paper. For example, due to problems with necking in tension and barrelling in compression, the strain ranges considered in experiments are practically always below the range where the large strain mechanisms as discussed before can become effective. On the other hand, the precise steady state condition of stage III can only be determined by extrapolation. This has been considered in evaluating stress strain tests (1, 47, 50). For creep tests, however, it is common to take the range of minimum creep rate as steady state which corresponds to the UTS in stress strain tests.

Yet, the relative difference between σ_s, as determined by extrapolation, and σ_{uts} usually varies with deformation temperature. If, for example, the Θ - σ relationship is of the form (50)

$$\Theta = \Theta_o \left(1 - \frac{\sigma}{\sigma_s}\right) \tag{10}$$

then, by setting Θ_{UTS} according to the Considère criterion, one obtains

$$\frac{\sigma_s - \sigma_{UTS}}{\sigma_{UTS}} = \frac{\sigma_s}{\Theta_o} \; . \tag{11}$$

Since σ_s decreases strongly with increasing temperature while the variation with temperature of Θ_o is very weak (50), the relative difference between σ_s and σ_{UTS} increases continuously as the temperature decreases. (For example the values of σ_s/Θ_o for Cu are 0.1 at 473 K and 0.3 at 77 K (50).) Therefore also the parameters from a thermal-activation analysis will be affected systematically when the temperature dependence of the UTS (or minimum creep rate) is analyzed instead of the physically more relevant steady-state stress (or steady-state creep rate), to be determined by extrapolation. These differences must be taken into consideration in a detailed analysis.

Summary

The current understanding of the physical processes involved in work hardening and creep of pure metals is surveyed by attempting to correlate the observations of the evolution of the dislocation structure with macroscopic behavior. It is shown that four separate states of work hardening can be distinguished in each of which different processes are dominant.

Stage I, not treated in this paper, appears only in single crystals oriented for single slip and at very low strains.

In stage II work hardening is athermal because the storage of dislocations follows stat-statistical rules. Here the mechanical state is determined by a single parameter, namely the flow stress and, in the ideal case, both the work hardening coefficient and the relative strain rate sensitivity of the flow stress are independent of strain. Under these conditions creep is logarithmic.

In the subsequent stage III cell formation occurs. The dislocation structure is due to a balance between work hardening and dynamic recovery and therefore, the mechanical state is no longer determined anymore by a single parameter. The important macroscopic quantities, hardening coefficient and strain rate sensitivity, can be interpreted quantitatively by the concept of counteracting work hardening and recovery processes but a general phenomenological formalism is needed. Stage III has a steady state limit the phenomenology of which is well explored. Steady state is reached at moderate strains at high temperature provided dynamic recrystallization does not occur. At low temperature excessively large strains would be required.

Stage IV appears at low temperature before stage III has evolved to its equilibrium structure. The term stage IV includes several kinds of large-strain effects. The most important of these are instabilities of deformation which lead to microband and shear band formation at room temperature. The physical reasons for the appearance of the phenomena are poorly understood.

Acknowledgments

It is a pleasure to thank Dr. A. S. Malin for extensive, fruitful discussions, for many clarifying suggestions and for reading the manuscript. The continuous contribution of U.F. Kocks to the ideas developed in this paper is gratefully acknowledged.

References

1. H. Mecking, in "Work Hardening in Tension and Fatigue", A. W. Thompson, ed., p. 67, AIME, 1977.
2. F. Wachholz, Diploma Thesis, RWTH, Aachen (1968).
3. H. Mecking and G. Gottstein, "Recrystallization of Metallic Materials", F. Haessner, ed., p. 195, Dr. Riederer-Verlag, Stuttgart (1977).
4. W. D. Nix and B. Ilschner, Proc. ICSMA5, P. Haasen, V. Gerold, G. Kostorz, eds., p. 1503, Vol. 3, Pergamon Press (1979).
5. G. Langford and M. Cohen, Trans. ASM 62, 623 (1969).
6. B. Sestak, Proc. ICSMA5, P. Haasen, V. Gerold, G. Kostorz, eds., Vol. 3, p. 1461, Pergamon Press (1979).
7. H. Haberkorn, K. Hartmann and V. Gerold, Z. Metallkde 62, 200 (1971).
8. S. Mader, Z. Physik 149, 73 (1957).
9. J. D. Livingston, Acta Met. 10, 229 (1962).
10. F. W. Young, Jr., J. Appl. Phys. 33, 963 (1962).
11. Z. S. Basinski and S. J. Basinski, Phil. Mag. 9, 51 (1964). P. J. Jackson and Z. S. Basinski, Can. J. Phys. 45, 707 (1967).
12. H. Mecking and G. Bulian, Acta Met. 24, 249 (1976).
13. P. Ambrosi and Ch. Schwink, Proc. ICSMA5, P. Haasen, V. Gerold, G. Kostorz, eds., p. 29, Vol. 1, Pergamon Press (1979).
14. D. Kuhlmann-Wilsdorf, in "Work Hardening", J. P. Hirth and J. Weertman, eds., p. 97, Gordon and Breach (1968).
15. J. Diehl, Z. Metallkde. 47, 331 (1956).
16. H. Mecking, Habilitation Thesis, RWTH, Aachen (1973).
17. H. Mecking and U. F. Kocks, in preparation.
18. R. A. Mulford, Acta Met. 27, 1115 (1979).
19. U. F. Kocks and H. Mecking, this volume.
20. A. H. Cottrell, Dislocations and Plastic Flow in Crystals, p. 195, Oxford University Press, (1953).
21. M. Michelitsch, Z. Metallkde, 50, 548 (1959).
22. H. Kronmüller, "Moderne Probleme der Metallphysik," A. Seeger, ed., p. 126, Springer-Verlag (1965).
23. J. Friedel, Dislocations and Mechanical Properties of Crystals, Jon Wiley and Sons, New York (1957).
24. G. A. J. Hansen, Diploma Thesis, RWTH, Aachen (1971).
25. H. McQueen and J. J. Jonas, in "Treatise on Material Science", R. J. Arsenault, ed., p. 394, Academic Press, New York (1975).
26. H. Mecking and A. Grinberg, Proc. ICSMA5, P. Haasen, V. Gerold, G. Kostorz, eds., p. 289, Pergamon Press (1979).
27. G. Gottstein, D. Zabardjadi and H. Mecking, Metal Sci. 13, 223 (1979).
28. P. J. T. Stuitje and G. Gottstein, Z. Metallkde. 71, 279 (1980).
29. P. Karduck, Doctoral Thesis, RWTH Aachen (1980).
30. D. Embury, in "Strengthening Methods in Crystals", A. Kelly and R. B. Nicholson, eds., p. 331, Elsevier Publishing Company (1971).
31. D. H. Warrington, Proc. European Conf. on Electron Microscopy, Delft, Vol. 1, p. 354 (1960).
32. M. M. Myshlyaev and I. I. Khodos, Proc. ICSMA5, P. Haasen, V. Gerold, G. Kostorz, eds. p. 357, Vol. 1, Pergamon Press (1979).
33. H. M. Miekk-oja and V. K. Lindroos, in "Constitutive Equations" in Plasticity, A. S. Argon, ed., p. 327, MIT-Press (1975).

34. J. J. Jonas, Proc. ICSMA 1, Trans. Japan Inst. Met. Suppl. 9, 257 (1968).
35. A. S. Malin and M. Hatherly, Metal. Sci. 13, 463 (1979).
36. R. J. Asaro, Acta Met. 27, 445 (1979).
37. F. Kirch, Doctoral Thesis, RWTH Aachen (1970).
38. J. Gil Sevilano, P. van Houtte and E. Aernoudt, "Progress in Materials Science," B. Chalmers, J. W. Christian, T. B. Massalski, eds., Pergamon Press, in press.
39. D. J. Lloyd and D. Kenny, Scripta Met. 12, 903 (1978).
40. H. Mecking and K. Lücke, Acta Met. 17, 279 (1969).
41. K. Lücke and H. Mecking, in "The Inhomogeneity of Plastic Deformation", R. E. Reed-Hill, ed., p. 223 ASM (1973).
42. U. F. Kocks, J. Eng. Mat. Techn. 98, 76 (1976).
43. U. F. Kocks and H. Mecking, Proc. ICSMA 5, P. Haasen, V. Gerold, G. Kostorz, eds., p. 345, Pergamon Press (1979).
44. J. Hausselt and W. Blum, Acta Met. 24, 1027 (1976).
45. T. H. Alden, Met. Trans. 6A, 169 (1979).
46. E. W. Hart and H. D. Solomon, Acta Met. 21, 295 (1973).
47. B. Nicklas and H. Mecking, Proc. ICSMA 5, P. Haasen, V. Gerold, G. Kostorz, eds., p. 351, Vol. 1, Pergamon Press (1979).
48. O. D. Sherby, J. L. Lytton and J. E. Dorn, Acta Met. 5, 219 (1957).
49. H. Luthy, A. K. Miller and O. D. Sherby, Acta Met. 28, 169 (1980).
50. U. F. Kocks, J. Eng. Mat. Techn. (ASME H) 98, 76 (1976).

SOLUTION EFFECTS ON THE MECHANICAL BEHAVIOUR AND THE DISLOCATION MOBILITY IN SILICON CRYSTALS

K. Sumino, I. Yonenaga, H. Harada and M. Imai
The Research Institute for Iron, Steel and Other Metals
Tohoku University, Sendai 980, Japan

Abstract—Effects of interstitial oxygen atoms and substitutional phosphor atoms on the dislocation mobility and the mechanical behaviour in silicon crystals are investigated. Dissolved oxygen atoms do not affect the velocities of dislocations in the temperature range investigated, but develop atmospheres around slowly moving and static dislocations and lock them firmly. The locked dislocations lose the function of the multiplication centers and the mechanical behaviour of the crystals becomes similar to that of nearly dislocation-free crystals. Phosphor atoms increase the mobility of dislocations but have no appreciable locking effect. The behaviour of the effective stress, and, thus, the state of the collective motion of dislocations in the steady state of deformation both with and without the presence of the solute atoms are correctly described by the hypothesis of Sumino.

Introduction

In semiconductor crystals, the dislocation mobility is very sensitive to the temperature and rather insensitive to the stress. This feature in the dislocation mobility results in the mechanical behaviour of the crystals which is sensitive to both the temperature and the strain rate. The velocities of dislocations in this type of crystals are measured accurately as a function of the temperature and the stress by virtue of the viscous nature of the dislocation motion Thus, semiconductor crystals are suited for the investigation of a quantitative relationship between the dynamics of dislocations and the mechanical behaviour of crystals. From such a point of view, an extensive work has been developed on the dislocation-dynamical study on semiconductor crystals (1∿7). The present authors (6,7) have shown that experimental data on the mechanical behaviour of silicon crystals are described theoretically in a quite quantitative manner with the use of experimental data on the dislocation mobility and some reasonable assumptions for dislocation processes. Furthermore, it has been shown that the density and the velocity of dislocations participating to the deformation at a constant strain-rate take steady values in the deformation stage after the yielding. The characteristics in the collective motion of dislocations in such steady state of deformation have been successfully interpreted in terms of a hypothesis proposed by Sumino (5), i.e. the steady state of deformation is realized so that the component of the flow stress which is determined by the state of the collective motion of dislocations takes a minimum value to maintain the externally given strain-rate.

Meanwhile, it has been recognized in the electronic device technology that the mechanical strength of silicon crystals is fairly sensitive to the existence of interstitially dissolved oxygen atoms. The present work investigates the effects of dissolved oxygen atoms on the dynamical behaviour of dislocations and clarifies the mechanism of the solid solution hardening of silicon crystals due to oxygen atoms. It has also been known that the dislocation mobility in silicon crystals is increased by the existence of donor impurities. The present work confirms this effect and further investigates how such a change in the dislocation mobility is reflected in the mechanical behaviour and the collective motion of dislocations in the crystals.

The Mobility of Dislocations and Locking by Impurities

Velocities of isolated dislocations were measured by means of in-situ observations of the

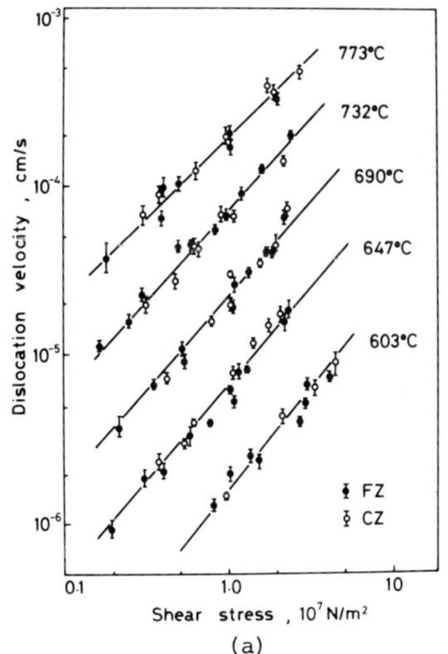

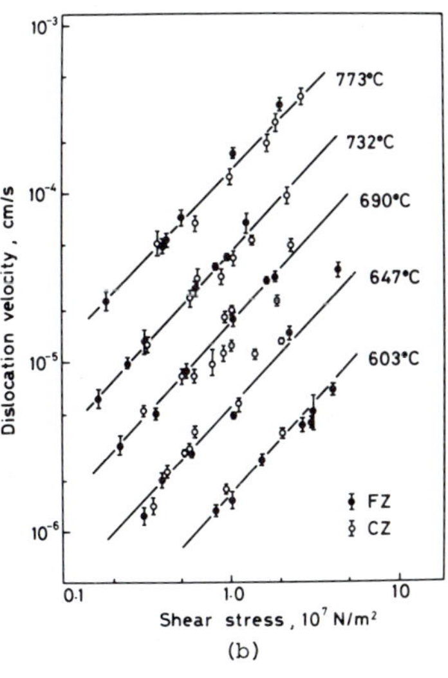

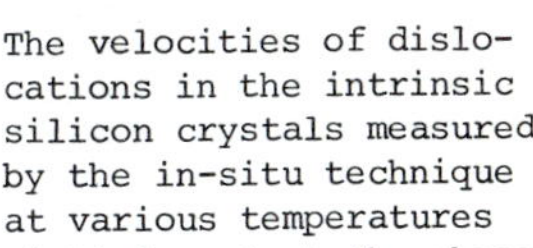
FIG. 1

The velocities of dislocations in the intrinsic silicon crystals measured by the in-situ technique at various temperatures plotted against the shear stress.
(a) 60° dislocations
(b) screw dislocations

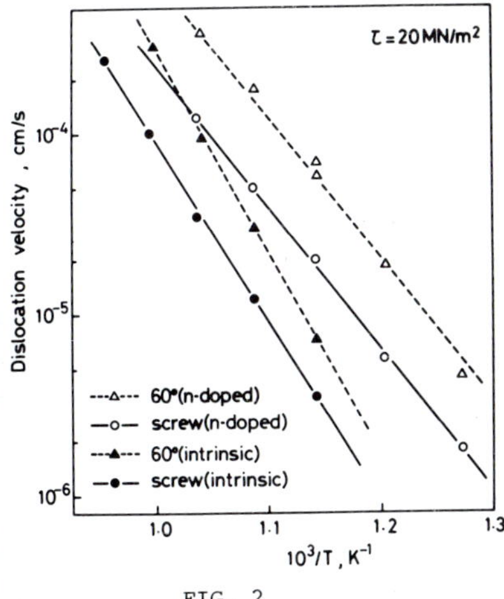

FIG. 2

The dislocation velocities for a shear stress 20 MN/m^2 plotted against the inverse temperature. Full marks are for the intrinsic crystals and open marks for the n-doped crystals.

dislocation motion with X-ray diffraction topography. A system consisting of a high power X-ray generator, a high temperature tensile stage and an X-ray T.V. system was employed in the measurements. Details of the system and of experimental procedure are described elsewhere (8). A big merit in the in-situ technique compared with the conventional one is to be able to see how slowly moving dislocations are immobilized by the development of impurity atmosphere.

Fig. 1(a) and (b) show velocities of 60° dislocations and screw dislocations, respectively, measured for intrinsic silicon crystals at various temperatures plotted against the shear stress. Open marks in the figures show data points for Czochralski crystals and full marks for float-zone crystals. The concentrations of interstitially dissolved oxygen atoms are 4×10^{17} cm^{-3} in the former and less than 10^{16} cm^{-3} in the latter. Oxygen atoms in the Czochralski crystals are supersaturated in the temperature range below 1200°C. In both types of crystals the logarithm of the velocity is linear to the logarithm of the stress at each temperature. In the float-zone crystals such a relation holds down to a small shear stress of order of 1.5 MN/m^2. The velocities in the Czochralski crystals are measured to be the same as those in the float-zone crystals in the stress range higher than about 3 MN/m^2. However, dislocations moving in such a stress range become immobile gradually when they are brought under stresses lower than 3 MN/m^2. Dislocations immobilized and kept for some period are not moved even if the applied stress is increased to an extremely high value. The above facts are true for both 60° and screw dislocations. It may be concluded that interstitially dispersed oxygen atoms do not interrupt the motion of dislocations under a high stress or at a high velocity in the temperature range concerned, but develop atmospheres around slowly moving dislocations and lock them firmly. It is interesting to note that the critical stress at which the immobilization takes place is insensitive to the temperature. Static dislocations in the Czochralski crystals are, of course, observed to be all locked.

Fig. 2 shows the dislocation velocities in intrinsic and highly doped n-type crystals under a shear stress of 20 MN/m^2 plotted against the inverse temperature. The doped crystals contain phosphor atoms of a concentration 1×10^{19} cm^{-3}. It is seen that the velocities in the n-doped crystals are much higher and the activation energies for the motion are lower compared with those in the intrinsic crystals in accord with the data so far published (9 ∿ 11).

In all the types of crystals, the mobilities of screw dislocations are observed to be lower than those of 60° dislocations. The velocity of screw dislocations is well expressed by an empirical equation ;

$$v = B(\tau/\tau_0)^m \exp(-E/kT) \qquad (1)$$

where τ is the shear stress, T the temperature, and $\tau_0 = 10$ MN/m^2. The magnitudes of B, m and E are given in TABLE I.

TABLE I

Magnitudes of B, m and E for screw dislocations

crystal	B (cm/s)	m	E (eV)
intrinsic	$(1.5 \pm 0.3)\times10^6$	1.1 ± 0.1	2.1 ± 0.1
n-doped	$(6.3 \pm 0.5)\times10^3$	1.2 ± 0.3	1.5 ± 0.1

Mechanical Behaviour and Collective Motion of Dislocations

The stress-strain behaviour of the crystals in [123] tensile deformation at high temperatures were investigated. The effective stress τ_{eff} was determined as a function of the strain, the strain-rate and the temperature by means of the strain-rate cycling tests with the use of Eq.1. Details of experimentals are described elsewhere (6).

Fig.3 shows stress-strain curves of a float-zone crystal and a Czochralski crystal which are of intrinsic type and are originally dislocation-free. The deformation takes place by the propagation of Lüders bands in the yield regions in both of the crystals. Though the flow stress in the Czochralski crystal is somewhat higher than the float-zone crystal throughout the whole deformation stage, the difference is rather small.

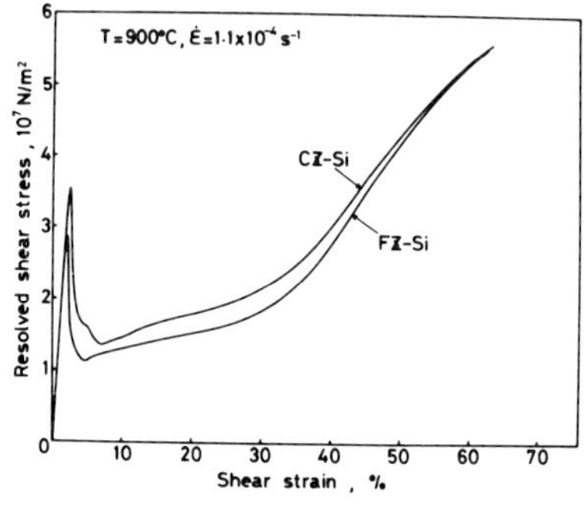

FIG. 3

Stress-strain curves for dislocation-free crystals of intrinsic type. FZ and CZ denote float-zone crystal and Czochralski crystal, respectively. [123] tensile deformation at 900°C. Shear-strain rate 1.1×10^{-4} s^{-1}.

Fig.4(a) shows stress-strain curves of float-zone crystals of intrinsic type having various values of the initial dislocation density N_0. All the crystals with $N_0 > 2\times10^4$ cm^{-2} undergo homogeneous deformation even during the yielding. The magnitude of the yield drop is strongly affected by the value of N_0. Such behaviour of the upper yield stress has been successfully described on the basis of a simple model of the dislocation multiplication (7). The high values of the lower yield stress in the crystals with low values of N_0 have been shown to be due to the activation of the secondary systems owing to the high flow stress around the upper yield point. The behaviour and the magnitude of τ_{eff} after the lower yield point are not affected by the magnitude of N_0 and the so-called steady state of deformation is realized after the middle of stage 0. Fig.4(b) shows stress-strain curves of solution-treated Czochralski crystals of intrinsic type having various values of N_0. All the crystals show marked yield drop phenomena and the deformation proceeds by the propagation of Lüders bands in the yield region for $N_0 < 10^5$ cm^{-2}. The comparison of the yield behaviour in Fig.4(b) with that in Fig.4(a) naturally leads to an idea that most dislocations initially contained in the Czochralski crystals are locked firmly by the oxygen atmospheres and only less than 1 % of them are able to act as multiplication centers in the deformation. The steady state value of τ_{eff} in the Czochralski crystals in Fig.4(b) is not affected by the magnitude of N_0 and coincides with that in the float-zone crystals in Fig.4(a). Also, the dependences of the steady state values of the effective stress, denoted as τ^*_{eff}, on the strain-rate $\dot{\varepsilon}$ and the temperature T approximately coincide between the float-zone crystals and the Czochralski crystals. These behaviours of τ^*_{eff} accord with those derived from the hypothesis proposed by Sumino : According to the theory, both the magnitude of τ^*_{eff} and its dependence on the deformation condition are determined by the expression for the velocity of individual dislocations, and, indeed, no difference in such expressions is found in the data in Fig.1 between the two types of crystals. It is noted that dislocations in the Czochralski crystals being deformed under a low strain-rate move under a stress somewhat lower than the critical stress for the immobilization mentioned in connection with Fig.1. During the deformation, oxygen atoms seem to be depleted inside slip bands to some extent due to the segregation to a high density of static dislocations accumulated, thus, the locking effect to moving dislocations becoming weaker. It is

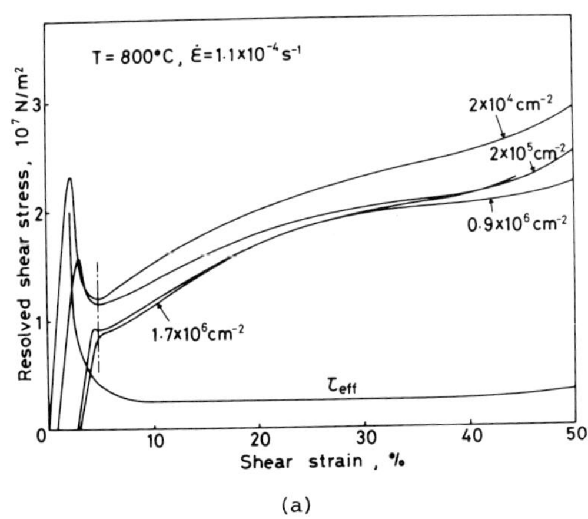

(a)

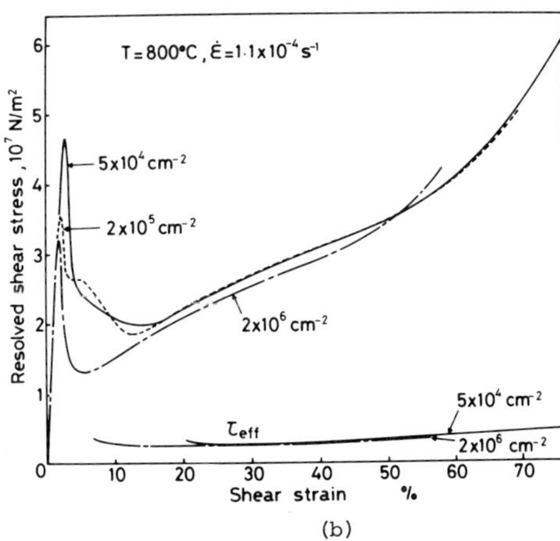

(b)

FIG. 4

Stress-strain curves for the intrinsic crystals containing dislocations. Numerals attached to the curves show the dislocation densities in the crystals before the deformation in unit of cm^{-2}. [123] tensile deformation at 800°C. Shear-strain rate $1.1 \times 10^{-4}\ s^{-1}$.
(a) Float-zone crystals, (b) Czochralski crystals

measured by the infra-red absorption method that the concentration of interstitially dissolved oxygen atoms in the Czochralski crystals decreases with the increase of the strain to a large extent during the deformation. Such a large reduction of dissolved oxygen atoms is not observed in a dislocation-free Czochralski crystal which is annealed at the temperature where the deformation experiments are conducted. These facts support the above interpretation.

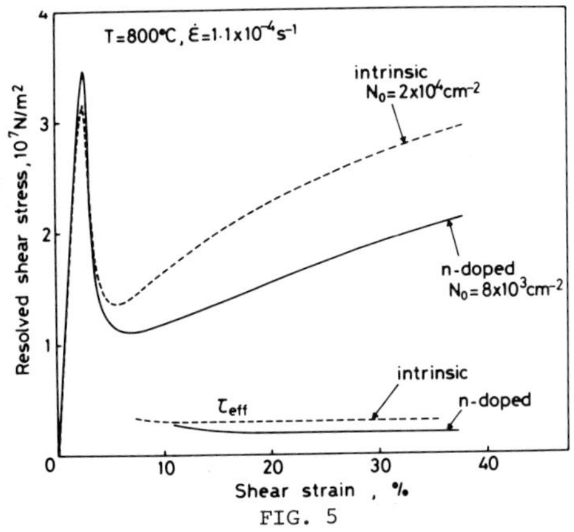

FIG. 5

Stress-strain curves of an intrinsic crystal and of an n-doped crystal both grown by float-zone technique. The initial dislocation densities are shown in the figure. [123] tensile deformation at 800°C. Shear-strain rate $1.1 \times 10^{-4}\ s^{-1}$.

Fig.5 shows stress-strain curves of an intrinsic crystal and an n-doped crystal, the latter containing phosphor atoms of a concentration $1 \times 10^{19}\ cm^{-3}$. Both crystals have been grown by float-zone technique. Though the upper yield stress of the n-doped crystal is somewhat higher than that of the intrinsic one owing to a lower value of N_o, the flow stress after the yield region is considerably lower in the former than the latter. τ^*_{eff} is lower in the n-doped crystal. The yield behaviour of the n-doped crystals is observed to be similar to that of the float-zone crystals of intrinsic type, showing that phosphor atoms have no appreciable locking effect on dislocations.

The dependences of τ^*_{eff} on the strain-rate and the temperature in both the intrinsic and the n-doped crystals are shown in Fig.6. In all the types of crystals, τ^*_{eff} is well expressed by a following equation ;

$$\tau^*_{eff} = A\ \dot{\varepsilon}^{1/n} \exp(U/kT) \qquad (2)$$

where A, n and U are constants. The experimental values of A, n and U are given in TABLE II. The hypothesis of Sumino leads to the parameters to be given by

$$A = (m^2 G^2 \tau_o^m b / 4\beta B)^{1/(m+2)}, \quad n = m+2, \quad U = E/(m+2), \qquad (3)$$

where G is the shear modulus and β a parameter characterizing the interaction force between moving dislocations of which magnitude is of order of $3 \sim 4$. Substitution of the magnitudes of B,

FIG. 6

The steady state values of the effective stress τ^*_{eff} as dependent on the deformation condition in the intrinsic and the n-doped crystals.
(a) strain-rate dependence
(b) temperature dependence

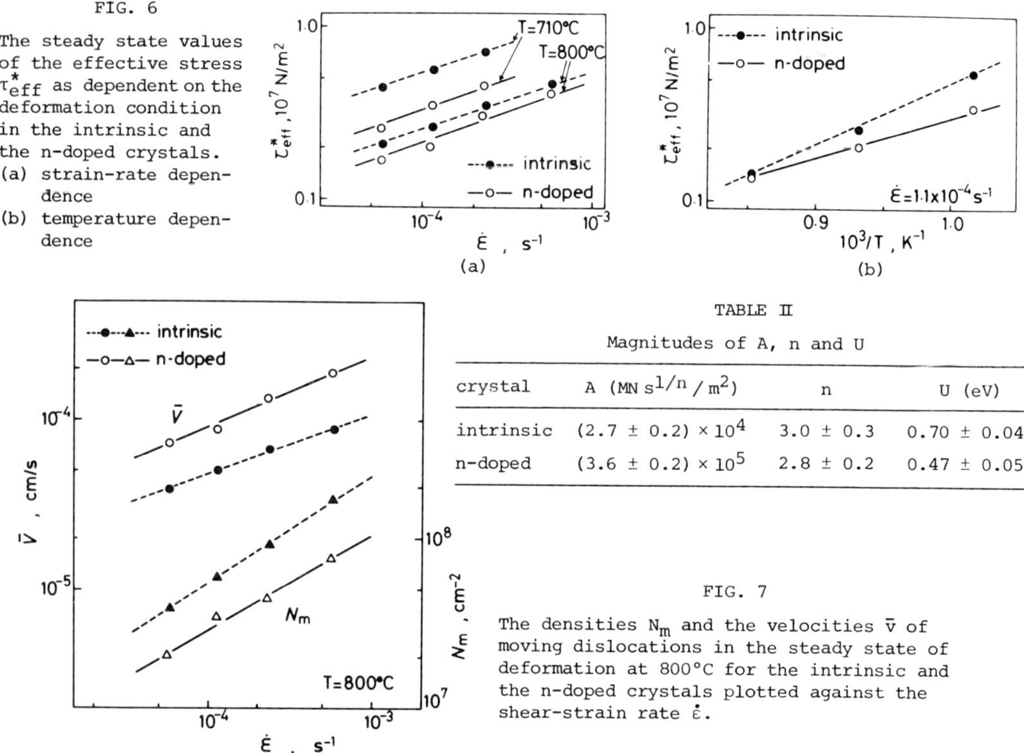

TABLE II

Magnitudes of A, n and U

crystal	A (MN $s^{1/n}$ / m^2)	n	U (eV)
intrinsic	(2.7 ± 0.2) × 10^4	3.0 ± 0.3	0.70 ± 0.04
n-doped	(3.6 ± 0.2) × 10^5	2.8 ± 0.2	0.47 ± 0.05

FIG. 7

The densities N_m and the velocities $\bar{v}$ of moving dislocations in the steady state of deformation at 800°C for the intrinsic and the n-doped crystals plotted against the shear-strain rate $\dot{\varepsilon}$.

m and E in TABLE I into Eqs.3 results in the magnitudes of A, n and U which are in good agreement with the experimental ones. Thus, the hypothesis of Sumino gives a good description for the collective motion of dislocations during deformation.

Fig.7 shows the densities N_m and the velocities $\bar{v}$ of moving dislocations for the steady state of deformation at 800°C in the intrinsic and the n-doped crystals as calculated by the substitution of the experimental values of τ^*_{eff} into Eq.1 plotted against the strain-rate. It is seen that N_m is lower in the n-doped crystals in which the dislocation mobility is higher. Since the density of static dislocations accumulated during the deformation is thought to be lower for lower N_m, this explains why the work-hardening is less remarkable in the n-doped crystals compared with the intrinsic crystals as seen in Fig.5.

References

1. H.Alexander and P.Haasen, Solid State Physics 22, 28 (1968)
2. K.Kojima and K.Sumino, Crystal Lattice Defects 2, 147 (1971)
3. K.Sumino and K.Kojima, Crystal Lattice Defects 2, 159 (1971)
4. K.Sumino, S.Kodaka and K.Kojima, Mater. Sci. Engng. 13, 263 (1974)
5. K.Sumino, Mater. Sci. Engng. 13, 269 (1974)
6. I.Yonenaga and K.Sumino, phys. stat. sol.(a) 50, 685 (1978)
7. M.Suezawa, K.Sumino and I.Yonenaga, phys. stat. sol.(a) 51, 217 (1979)
8. K.Sumino and H.Harada, Researches on High Temperature Materials, p.121, RIISOM (1980)
9. J.P.Patel and P.E.Freeland, Phys. Rev. Letters 18, 833 (1967)
10. V.N.Erofeev and V.I.Nikitenko, Sov. Phys.-Solid State 13, 116 (1971)
11. A.George and G.Champier, phys. stat. sol.(a) 53, 529 (1979)

THE NUCLEATION AND GROWTH OF DEFORMATION TWINS IN METALLIC CRYSTALS

S. Mahajan
Bell Laboratories
Murray Hill, New Jersey 07974

Abstract

The central elements of a generalized model for the formation of deformation twins in metallic crystals that does not require the existence of pole dislocations have been presented. Experimental evidence in support of this model has been developed by correlating the crystallographic features of twins with that of the emissary slip. Several unresolved problems pertaining to this model have also been delineated.

Introduction

Slip and mechanical twinning are the two principal modes by which metals and alloys can deform plastically. The propagation of slip from existing dislocations as well as the multiplication of dislocations during deformation have been studied extensively. Consequently, mechanistic details concerning these processes are relatively well understood. On the other hand, the corresponding situation regarding deformation twinning is rather ambiguous. Mechanistically, the twins can be envisaged as forming by the passage of twinning partials on the successive twin planes; the Burgers vector of these partials is always some fraction of the lattice vector. It is, however, not clear how these twinning partials evolve during deformation and what role slip may play in the twinning process.

The formation of a deformation twin can be visualized to occur in two stages: (i) nucleation of an embryo, and (ii) its subsequent growth into a macroscopically observable twin. Basically two different approaches have been adopted to rationalize the formation of deformation twins. Orowan (1) suggests that a twin embryo could nucleate homogeneously in a region of high stress concentration. On the other hand, Cottrell and Bilby (2), Thompson and Millard (3), Sleeswyk (4), Venables (5), Cohen and Weertman (6), Ogawa (7), Priestner and Leslie (8), Mahajan (9), Mahajan and Chin (10), Narita and Takamura (11), Mahajan (12), Sleeswyk (13), Fujita and Mori (14) and Vaidya and Mahajan (15) envisage that a twin could nucleate from a heterogeneity such as a particular dislocation arrangement.

The objective of this paper is twofold. Firstly, to present salient features of a generalized model for twin formation in B.C.C., F.C.C. and H.C.P. crystals. The central element of this approach is "Slip-Band Conversion" hypothesis (9). Secondly, to delineate problems which require a firmer theoretical base.

Generalized Model For Twin Formation, Supporting Experimental Evidence And Unresolved Problems

The basic elements of a generalized model for twin formation are: (i) micro-slip that occurs on the twinning plane and precedes twinning, and (ii) embryonic twins which result from either the dissociation of dislocations or the interaction of dislocations of different Burgers vectors constituting these micro-slip bands. The salient features of this approach can be best illustrated by taking a specific example. In the case of B.C.C. crystals, it is visualized that: (i) a twin grows from a region of planar slip that thickens with increasing plastic strain; (ii) positive and negative screw dislocations comprising the slipped region could, under stress, dissociate into either faults or three-layer twins as proposed by Sleeswyk (4); overlapping of the faults and twins occurs randomly at first and then gradually tends to approach the fault arrangement resembling that of a microscopic twin. Since the presence of stacking faults may produce long range electronic perturbations in the lattice (16), it is conceivable that faulted regions may tend to form in the vicinity of each other. This effect should enhance the probability of a particular slipped region, containing an interspersion of three-layer twins and faults, to evolve into a macroscopic twin.

Figure 1(a) or (b) schematically illustrates the dislocation arrangement envisaged at a non-coherent twin boundary of a three-layer twin. On the right side the three faults are bounded by twinning partials, while on the left in Fig. 1(a) they terminate on faults lying in the other two {112} planes which contain the twinning vector. Alternatively, in terms of the dislocation formalism, the left interface, Fig. 1(b), can be visualized as being composed of two $\frac{1}{6}\langle 1\bar{1}\bar{1}\rangle$ partials surrounding a $\frac{1}{3}\langle\bar{1}11\rangle$ complementary twinning dislocation; both these descriptions are equivalent. Under the influence of an applied stress, the twinning dislocations could glide away and thus extend the twinned region. Furthermore, an additional ramification of these descriptions is that the left interface would not exhibit contrast typical of that of dislocations for any operating reflection during the electron microscopy experiments.

Imagine a situation where the twinning partials gliding from the left encounter another three-layer twin lying in their glide path as shown in Fig. 1(c). When the gliding partials reach the right hand twin, dislocation annihilation as well as coalescence can occur, resulting in a four-layer twin as illustrated in Fig. 1(d). In addition, depending upon the relative vertical displacement between the two nuclei in Fig. 1(c), five or six-layer twins can result. Furthermore, a screw dislocation that has undergone multiple cross-slip, even over short distances, is capable of producing a large number of three-layer twins at different levels within a slip band. Consequently, the number of dislocations required to form a macroscopic twin will be substantially reduced. It is also conceivable that multiple cross-slip may occur as a result of the elastic interaction between screw dislocations gliding on neighboring but parallel planes. It is therefore envisaged that a stress pattern, conducive to cross slip, does not have to be generated by an externally applied stress.

The occurrence of slip on intersecting glide planes can also facilitate thickening of micro-scopic twins. Firstly, the passage of glide dislocations through twins whose Burgers vector is inclined to that the twinning partials produces steps in coherent twin interfaces as suggested by Sleeswyk and Verbraak (17). Under the influence of suitable applied stress, partials delineating these steps can glide and thus thicken the twin. Secondly, if twinning partials encounter a slip dislocation whose Burgers vector is not parallel to that of the partials, the movement of twinning dislocations through the slip dislocation would also result in steps (18). The movement of these steps would also lead the thickening. It has previously been hypothesized (4) that such an interaction would lead to spiralling of twinning partials as envisaged in the pole model (2). However, experimental evidence pertaining to the interaction of twinning partials with existing slip dislocations does not bear this out (19

Since micro-slip must precede twinning, an interesting question arises: what determines the transition from slip to slip accompanied by twinning. The situation regarding this question is not clear. Following Duesbery et al. (20), the transition is likely to depend upon how the dissociated core of the screw dislocation rearranges itself under an applied shear stress. If the constituting partials coalesce into a single-layer fault bounded by $\frac{1}{3}\langle 111\rangle$ and $\frac{1}{6}\langle 111\rangle$ dislocations, deformation will proceed exclusively by slip. On the other hand, the core may evolve into a multi-layer fault and then deformation will occur by slip and twinning because the edge dislocations will continue to move as total. With increasing strain additional length of pure screw dislocations will be generated. These may subsequently dissociate and thus add to the thickness of the existing twins. Furthermore, if the retarding influence of stacking faults resulting from the dissocation of screw dislocations is neglected, the partials should have higher mobility than that of the total dislocations because the former experience lower friction stress (21). Also, partial dislocations located on closely spaced parallel planes can by-pass each other at stresses lower than those required for bypassing of total dislocations. Thus the transition from micro-slip to micro-slip mixed with embryonic twins may be rationalized.

Figure 2 depicts a situation observed in B.C.C. Mo-35 at % Re alloy after tensile deformation at 77°K(9). This electron micrograph shows a well-developed faulted region F_1, a few tiny faults F_2, F_3, F_4 and F_5, and glide dislocations ahead of the main fault. Some of the small clear faults, i.e., F_3 and F_4, are still attached to a highly cusped slip dislocation.

It was ascertained by electron diffraction and trace analysis that F_1 is a thin twin having the (211) habit. It can also be deduced from Fig. 2 that the Burgers vectors of partials bounding F_1 and glide dislocations are $\pm \frac{1}{6}[\bar{1}11]$ and $\pm \frac{1}{2}[\bar{1}11]$, respectively. Comparing the projection of dislocations on the (111) plane with that of the $[\bar{1}11]$ vector, i.e., AA in Fig. 2, it is inferred that only those dislocations which are in screw orientation appear to dissociate into faults and twins. Furthermore, the contrast behavior of F_3 and F_4 implies that they are 3n-layer thick twins.

When the matrix is oriented along its $[1\bar{1}5]$ direction, i.e., as in Fig. 2(c), the twins with the $\pm \frac{1}{6}$ $[\bar{1}11]$ displacement vector should be oriented along their <111> directions. It it therefore suggested that the twins are strongly diffracting in Fig. 2(c) and the observed greyish contrast could result from this fact.

If the twins in Fig. 2 were to nucleate via the pole model of Cottrell and Bilby (2), the following dislocation reaction should govern their nucleation:

$$\frac{1}{2}[\bar{1}\bar{1}1] \longrightarrow \frac{1}{3}[\bar{1}\bar{2}1] + \frac{1}{6}[\bar{1}11] \;\ldots. \qquad (1)$$

The observed parallelism between the Burgers vectors of twinning partials and glide dislocations cannot be rationalized on the basis of this approach. However, the preceding observations are consistent with the occurrence of twin nucleation via the following reaction:

$$\frac{1}{2}[\bar{1}11]_{screws} \longrightarrow 3x\frac{1}{6}[\bar{1}11] \;\ldots. \qquad (2)$$

Since the traces of small twins with the foil surfaces are displaced from each other, these twins must be located at different levels within the slipped region. Again, this observation is consistent with one of the central elements of the generalized model for twin formation as discussed earlier. Furthermore, the more recent results on micro-twinning behavior of spinodally decomposed Fe-Cr-Co alloy has produced direct evidence for the agglomeration process evisaged above (22).

Conceptually, the formation of twins in F.C.C. crystals occurs in a manner analogous to that of B.C.C. crystals. In this case, three-layer twins may form by the interaction of two coplanar $\frac{1}{2}$<110> dislocations of different Burgers vectors. The reaction that governs the nucleation is the following (10):

$$\frac{1}{2}\langle\bar{1}10\rangle_{\{111\}} + \frac{1}{2}\langle\bar{1}01\rangle_{\{111\}} \longrightarrow 3x\frac{1}{6}\langle\bar{2}11\rangle_{\{111\}} \;\ldots. \qquad (3)$$

It has been suggested by several investigators (10, 23, 24) that, since this reaction is repulsive in nature, it must occur in the presence of stress concentration that may exist at the head of pile-ups. However, this reaction is observed to occur in the absence of pile-ups and very readily as well (23, 25). To rationalize these observations, a mechanism, schematically illustrated in Fig. 3, has been proposed (25). Using Thompson's notation (26), consider the interaction of dislocations with Burgers vectors $\vec{CA}$ and $\vec{CB}$ gliding on the plane (d). These dislocations could arise either from co-planar slip or from cross glide of one of the dislocations onto the glide plane of the other. Figure 3(a) shows a planar view of the situation. It can be shown that partials with Burgers vectors $\vec{C\delta}$ and $\vec{\delta B}$, resulting from the dissociation of $\vec{CB}$ dislocation, are repelled and attracted, respectively, by $\vec{\delta A}$ partial. Imagine a situation where a constricted portion of $\vec{CB}$ dislocation is contiguous to the dissociated dislocation $\vec{CA}$ as shown in Fig. 3(b). If $\vec{CB}$ dislocation dissociates in such a way that $\vec{\delta B}$ partial leads $\vec{C\delta}$ partial, the repulsive character of the interaction is changed into an attractive one. As suggested by Frank and Nicholas (27), $\vec{\delta B}$ and $\vec{C\delta}$ partials must, in turn, dissociate into $\vec{A\delta}+\vec{C\delta}$ and $\vec{\delta B}+\vec{\delta A}$ partials, respectively, to avoid anomalous faulting, Fig. 3(c). The resulting extrinsic fault will then blend in readily with the intrinsic fault to form a fault-pair as shown in Fig. 3(d). For the sake of clarity, the edge-on views of the various situations depicted in Figs. 3(a) → (d) are illustrated in Figs. 3(e) → h. It is emphasized that the fault-pair would likewise arise if $\vec{CB}$ and $\vec{CA}$ were gliding on adjacent planes instead of being co-planar. Furthermore, it is not clear how a fault-pair is transformed into a three-layer twin. It is likely that the conversion is induced by elastic interactions with other fault-pairs located within a deformed region.

Experimental evidence in support of the above nucleation hypothesis is presented in Figure 4(10). It shows a faulted region that terminates within a Co-9.5 wt %. Fe crystal (F.C.C.) deformed by bending. Slip is also observed and is not confined only to the region ahead of the fault. Furthermore, the traces generated by the habit plane of the fault and the glide planes of the slip dislocations are parallel, suggesting that the fault and the glide dislocations lie on parallel planes.

It was established by electron diffraction and contrast experiments that the faulted region is a twin having the (111) habit, and the Burgers vector of twinning partials bounding the non-coherent interface is $\pm\frac{1}{6}[\bar{2}11]$. Furthermore, it can be deduced from Fig. 4 that the Burgers vectors of dislocations P and Q are $\pm\frac{1}{2}[\bar{1}10]$ and $\pm\frac{1}{2}[\bar{1}01]$, respectively. In two instances dislocations P and Q interact to form dislocations R. Results of Fig. 4 are compatible with the assignation of a $\pm\frac{1}{2}[\bar{2}11]$ "effective" Burgers vector to dislocations R. These dislocations can evolve when the $\frac{1}{2}[\bar{1}01]$ and $\frac{1}{2}[\bar{1}10]$ dislocations interact according to reaction (3), and the "effective" Burgers vector is likely to be composed of three $\frac{1}{6}[\bar{2}11]$ shear vectors. There is also some indication of very short slip activity with $\pm\frac{1}{2}[01\bar{1}]$ vector (see dislocation S in Fig. 4(b)). The present example thus clearly demonstrates that the crystallography of the twin shown in Fig. 4 can be correlated with that of slip by reaction (3).

Recently, Vaidya and Mahajan (15) have interrelated the crystallography of $\{11\bar{2}1\}$ twinning with that of emissary slip in H.C.P. Co deformed by cavitation erosion. Figure 5 illustrates a typical situation where a twin T_1 has terminated within the crystal and generated a slip array ahead of it. Single surface trace analysis together with electron diffraction uniquely identify $(\bar{1}\bar{1}2\bar{1})$ as the habit plane of T_1. It is evident that the traces of T_1 and slip band emanating from the twin tip X are parallel. It has been ascertained by diffraction contrast experiments that the array consists of two sets of dislocations: (i) $\vec{C}+\vec{a}$ dislocations, identified by M in Fig. 5(a), have the $[\bar{2}113]$ Burgers vector and are visible in both Fig. 5(a) and (b) but disappear for $g=(\bar{2}11\bar{2})$; dislocations with Burgers vectors lying in the basal plane, i.e., $\vec{a}$ dislocations labelled N in Fig. 5(b), are out of contrast for $g=(0002)$ and visble for $g=(2\bar{1}\bar{1}0)$. Figure 5(c) is a dark-field image of the twin, and it delineates twin boundaries and emphasizes the jagged nature of the twin tip X. In addition, it shows evidence for deformation on $(01\bar{1}0)$ twin planes whose traces project parallel to LL′. A moire fringe pattern of non-uniform spacing and orientation is often visible within the twins as illustrated in Fig. 5(a).

The interrelationship between the twinning shear and emissary slip can be rationalized by hypothesizing that the dislocations delineating the non-coherent twin boundary of T_1 coalesce to generate matrix glide by the following reaction:

$$12 \times \frac{1}{12}\left\{\frac{1}{3}[\bar{1}\bar{1}26]\right\}_{(\bar{1}\bar{1}2\bar{1})} \longrightarrow 2 \times \left\{\frac{1}{3}[\bar{2}113]\right\}_{(\bar{1}\bar{1}2\bar{1})} + [1\bar{1}00]_{(\bar{1}\bar{1}2\bar{1})} \cdots \qquad (4)$$

It is implicit in this suggestion that the Burgers vector of the twinning dislocations has been approximated to $\frac{1}{36}\langle\bar{1}\bar{1}26\rangle$. The reaction is energetically unfavorable. It could, however, occur under the influence of the stress concentration existing at the head of a pile-up of twinning dislocations.

Experimental evidence (15) indicates that $\langle\bar{2}113\rangle$ and $\langle\bar{1}100\rangle$ glides occur on $\{11\bar{2}1\}$ planes prior to twinning. It is conceivable that these dislocations could, under an appropriate stress, interact to form a twelve-layer twin embryo according to a reaction which is reverse of equation (4). It is envisaged that the atomic displacements associated with the resultant $\frac{1}{3}\langle\bar{1}\bar{1}26\rangle$ zonal dislocation are spread over twelve twin planes, such that the Burgers vector of the zonal can be visualized as the sum of twelve individual $\frac{1}{36}\langle\bar{1}\bar{1}26\rangle$ twinning dislocations on adjacent $\{11\bar{2}1\}$ planes. Since the $\{11\bar{2}1\}$ interplanar spacing is only 1.197Å, the concept of a dislocation core extending over six twin planes on either side of the original location is not unreasonable. A microscopic twin could then evolve when such twin nuclei grow into each other under the influence of an applied stress.

Summary

A generalized model for the formation of twins in B.C.C., F.C.C., and H.C.P. crystals has been presented. The model involves the nucleation of embryonic twins at different levels within a micro-slip band. These twins nucleate either by the dissociation or by the interaction of dislocations of different Burgers vector which constitute the micro-slipped region. Under the influence of an applied stress, these embryonic twins could agglomerate to form a micro-twin. The passage of glide dislocations through existing dislocations could also facilitate thickening of embryonic twins.

References

1. E. Orowan, Dislocations in Metals, p. 116, AIME, New York (1954)
2. A. H. Cottrell, and B. A. Bilby, Phil. Mag., 42, 573 (1951)
3. N. Thompson, and D. J. Millard, Phil. Mag., 43, 422 (1952)
4. A. W. Sleeswyk, Phil. Mag., 8, 1467 (1963)
5. J. A. Venables, Phil. Mag., 6, 379 (1961)
6. J. B. Cohen, and J. Weertman, Acta Met., 11, 997 (1963)
7. K. Ogawa, Phil. Mag., 11, 217 (1965)
8. R. Priestner, and W. C. Leslie, Phil. Mag., 11, 895 (1965)
9. S. Mahajan, Phil. Mag., 26, 161 (1972)
10. S. Mahajan, and G. Y. Chin, Acta Met., 21, 1353 (1973)
11. N. Narita, and J. Takamura, Phil. Mag., 17, 1001 (1974)
12. S. Mahajan, Acta Met., 23, 671 (1975)
13. A. W. Sleeswyk, Phil. Mag., 29, 407 (1974)
14. H. Fujita, and T. Mori, Scripta Met., 9, 631 (1975)
15. S. Vaidya, and S. Mahajan, accepted for publication in Acta Met. (1980)
16. C. Nourtier, and G. Saada, Fundamental Aspects of Dislocation Theory, edited by J. A. Simmons. R. DeWit and R. Beellough, Nat. Bur. Stand. p. 1259. (1970)
17. A. W. Sleeswyk, and C. A. Verbraak, Acta Met., 9, 917 (1961)
18. G. B. Olson, and M. Cohen, Met. Trans. A, 7, 1897 (1976).
19. S. Mahajan, and G. Y. Chin, Acta Met., 22, 1113 (1974)
20. M. S. Duesbery, V. Vitek and D. W. Bowen, Proc. R. Soc. A332, 85 (1973)
21. J. P. Hirth and J. Lothe, Theory of Dislocations, p. 331, McGraw-Hill, New York, (1968)
22. S. Mahajan, S. Jin and D. Brasen, Acta Met., 28, 971 (1980)
23. P. C. J. Gallagher, Phys. Stat. Sol., 16, 95 (1966)
24. P. C. J. Gallagher and J. Washburn, Phil. Mag., 15, 969 (1967)
25. S. Mahajan, Met. Trans., 6A, 1877 (1975)
26. N. Thompson, Proc. Phys. Soc., B66, 481 (1953)
27. F. C. Frank and J. F. Nicholas, Phil. Mag., 44, 1213 (1953).

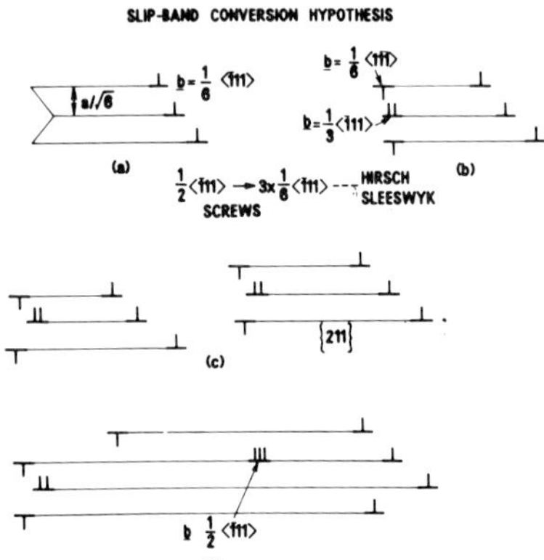

Fig. 1

(a, b) Schematic illustrations of the dislocation arrangements envisaged at a non-coherent twin boundary of a three-layer twin in B.C.C. crystals; (c) a possible configuration of two three-layer twins; (d) situation resulting from the coalescence of two three-layer twins shown in (c).

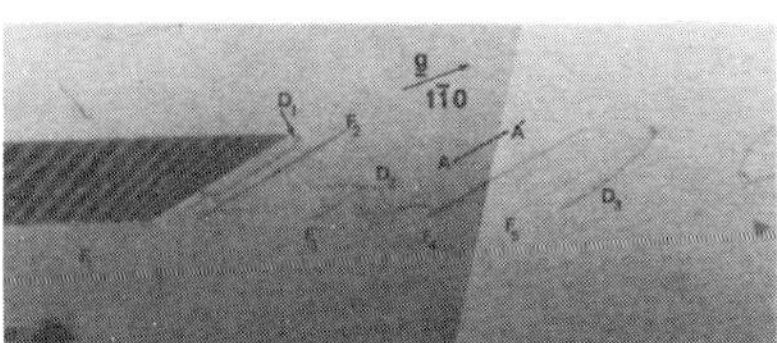

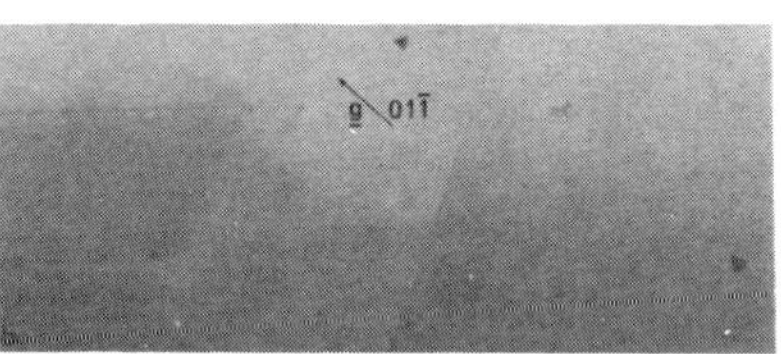

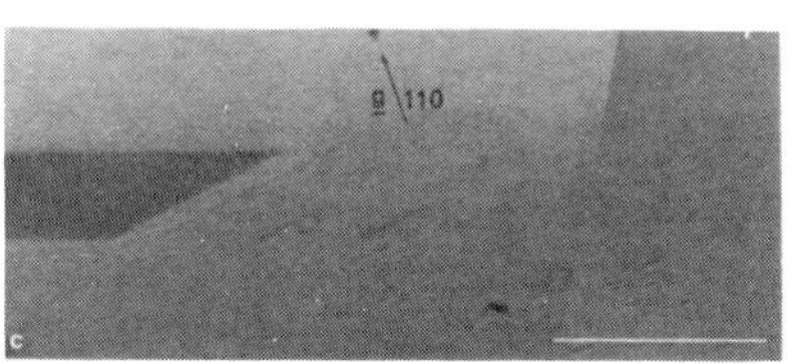

Fig. 2

Examination of the well-developed twin and emissary twins and slip dislocations observed in deformed Mo-35 at.% Re alloy (B.C.C.) The planes of the micrographs (a), (b) and (c), are ~ (111), ~ (111) and (11̄5), respectively. AA´ refers to the projection of the [1̄11] vector onto the (111) plane. The marker represents 1 μm.

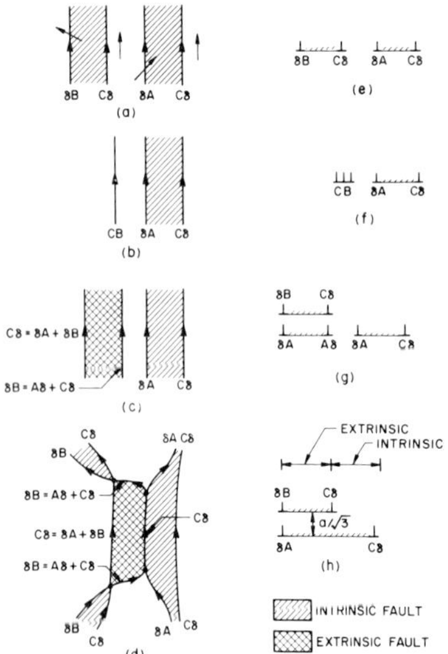

Fig. 3

Planar view of a situation that would result from the dissociation of $\vec{CA}$ and $\vec{CB}$ dislocations according to the Heidenreich-Shockley mode; these dislocations interact repulsively and are gliding on the plane (d); (b)-(d) Schematics of the fault-configurations which may evolve during the development of a fault-pair from the situation shown in (a); (e)-(h) Schematics of the corresponding edge-on views.

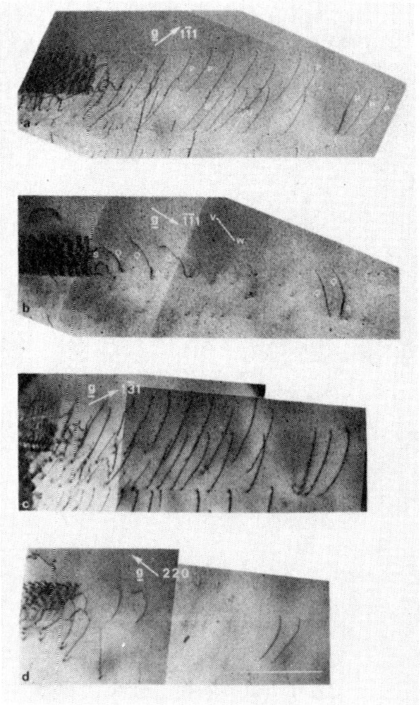

Fig. 4

Micrographs illustrating the contrast behavior of dislocations P, Q, R, and S observed ahead of a twin in deformed Co-9.5 wt% Fe alloy (F.C.C.). The planes of the micrographs are (a) ~ (011), (b) ~ (011), (c) ~ (112) and (d) ~ ($\bar{1}$14). TU and VW refer to the projections of the [1$\bar{1}$0] and [$\bar{1}$01] vectors onto the (011) plane. The marker represents 1 μm.

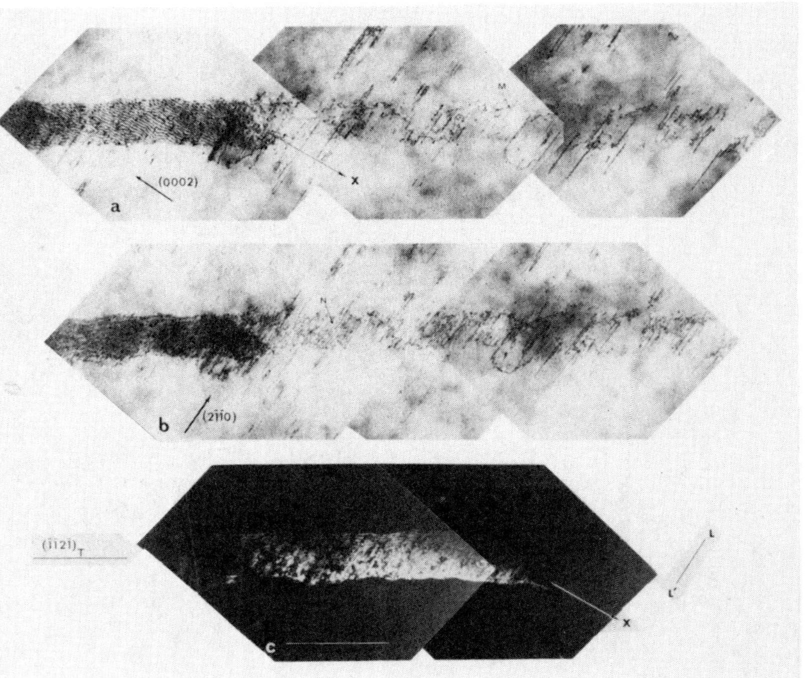

Fig. 5

Diffraction contrast behavior of dislocations emanating from the tip X of a ($\bar{1}\bar{1}2\bar{1}$)$_T$ twin T_1 observed in H.C.P. Co deformed by cavitation erosion. <$\bar{2}$113> and a dislocations are identified by M and N in (a) and (b), respectively. (c) Dark-field image of the twin tip. LL′ in (c) refers to the projection of (01$\bar{1}$0) planes within the twin onto the (0$\bar{1}$10) foil surface. The foil normal for these micrographs ~ [0$\bar{1}$10]. The marker represents 1 μm.

ADVANCES IN THE CONTINUUM ELASTIC THEORY OF DISLOCATIONS

David M. Barnett
Department of Materials Science and Engineering
Stanford University, Stanford, Ca. 94305

1. INTRODUCTION

Since more than eleven years has elapsed since the last major international conference devoted solely to dislocation theory and its applications, I am extremely flattered and pleased to have the opportunity to state what I believe to be the current state of continuum elastic dislocation theory. I shall try to point out the analytical and numerical machinery now available for performing calculations involving isolated dislocation lines or loops. The knowledgeable reader will probably be quick to note that the conceptual and analytical framework for these calculations was already available as early as 1967; I shall try, given current space limitations, to review this framework in a concise fashion and elucidate some of the clarifications and extensions which have since been made. For detailed derivations and formulae one would do well to refer to the original source articles cited. Two recent review articles [1, 2] may serve to furnish a more extensive bibliography and formulae collection than can be given here.

Much of what I shall describe as "advances" made since 1967 is colored by my own perception of topics which have struck me as being particularly interesting and in which I have been directly involved. I apologize for this rather parochial view, and I ask the reader's indulgence with it; it is simply easier for me to discuss topics with which I have some familiarity. I hope to be able to show convincingly that the theory in its present state allows us to study rather completely the elastic fields of isolated dislocations of arbitrary complex shapes in linear anisotropic solids of infinite extent; a truly remarkable feature of the theory is that the above-mentioned study can be successfully effected knowing only the elastic fields of infinite straight dislocations! When the theory is combined with expressions for Peach-Koehler forces and dislocation self-forces, equilibrium dislocation loop configurations may be computed using relatively simple numerical techniques.

I am keenly aware of the common criticism that dislocation theory has not lived up to its expectations, i.e., that the theory cannot be used to predict the macroscopic mechanical response of elastic solids containing large numbers of dislocations whose motion gives rise to plastic deformation. This criticism is valid and will, I suspect, remain valid. Of what use then is a sophisticated theory capable of dealing with only isolated (or maybe small numbers of) dislocations? Surprisingly, the dislocation theory I shall describe has proven to be indispensable for unravelling certain problems associated with the important, but seemingly unrelated, area of acoustic surface waves in solids [2, 3]. It takes little imagination to conjecture that dislocation theory may find great applicability in studies of microelectronic devices whose operating characteristics may be strongly affected by the presence of quite small numbers of crystalline imperfections.

Before entering the body of this presentation I should like to depart from tradition and acknowledge at the outset two gentlemen whose work has had a profound influence on dislocation theory in general and upon me, in particular. Professor J.D. Eshelby, FRS, needs no introduction to the audience of this conference. His contributions to the continuum theory of defects are legion, and I would simply like to thank him for making dislocation theory a respectable field in which to work and earn a living. I should also like to extend my deep gratitude to Professor Jens Lothe, who introduced me to many of the modern aspects of dislocation theory, kindly allowed me to collaborate with him on problems which he was certainly capable of solving alone, and who has remained a dear and valued friend. Of course, neither of these distinguished scientists bears any responsibility for errors in the text to follow.

2. STROH'S THEORY OF STRAIGHT DISLOCATIONS

As it turns out one of the cornerstones of modern dislocation theory is a knowledge of the elastic fields of an infinitely long and straight dislocation in an unbounded, homogeneous, anisotropic linear elastic solid. The late A.N. Stroh [4, 5] extended the previous work of Eshelby, Read, and Shockley [6] to provide a reasonably complete treatment of this class of problems. Most workers seem to be familiar with Stroh's 1958 paper; the 1962 article represents a more unified development, but it seems to have attracted much less attention, probably because of Stroh's untimely death following its publication. I would estimate that the ingredients for at least several Ph.D. dissertations are buried in the 1962 paper awaiting discovery by the curious. It is not my intention to review Stroh's theory in detail, but rather to highlight the results we shall need later. The notation to be used is that of Malén and Lothe [7] and Bacon, et al [1].

Consider an infinite straight dislocation in an anisotropic linear elastic solid of infinite extent. Let $\underset{\sim}{b}$ be the Burgers vector and $\underset{\sim}{t}$ be a unit vector tangent to the dislocation, and choose two unit vectors $\underset{\sim}{m}$ and $\underset{\sim}{n}$ normal to $\underset{\sim}{t}$ such that $(\underset{\sim}{m}, \underset{\sim}{n}, \underset{\sim}{t})$ form a right-hand rectangular Cartesian basis. If $\underset{\sim}{x}$ is the position vector measured from an origin on the dislocation line, Stroh showed that the displacement field of the dislocation is given by

$$u_i = \frac{1}{2\pi i} \sum_{\alpha=1}^{6} \pm A_{i\alpha} L_{s\alpha} b_s \, \ell n\{\underset{\sim}{m}\cdot\underset{\sim}{x} + p_\alpha \underset{\sim}{n}\cdot\underset{\sim}{x}\} \quad . \qquad (1)$$

$\underset{\sim}{A}$, $\underset{\sim}{L}$, and p are the eigenvectors and eigenvalues of the six-dimensional matrix eigenvalue problem represented by

$$[N] = p[\xi] \quad . \qquad (2)$$

The particular form of $[N]$ need not concern us here; the six eigenvalues p_α occur in complex conjugate pairs, and the $\pm$ sign in (1) is that of the imaginary part of p_α. Suffice it to say that, in general, one must solve eqn. (2) numerically, and that this is a simple task. It is easy to show that the dislocation stresses vary as the inverse normal distance from the dislocation line. In fact, if $\underset{\sim}{\tau}$ is any unit vector normal to $\underset{\sim}{t}$,

$$\sigma_{ij}(\pm\varepsilon\underset{\sim}{\tau}) = \pm\frac{\Sigma_{ij}(\underset{\sim}{b}, \underset{\sim}{t}; \underset{\sim}{\tau})}{\varepsilon} ; \qquad \varepsilon > 0 \tag{3}$$

and the energy per unit length of dislocation is merely

$$E = E(\underset{\sim}{t}) \ \ell n \frac{R}{r_0} \qquad (R, r_0 \text{ the usual cutoff-radii}) \tag{4}$$

where

$$E(\underset{\sim}{t}) = \frac{1}{2}\Sigma_{ij}\, b_i(\underset{\sim}{t} \times \underset{\sim}{\tau})_j \ . \tag{5}$$

$E(\underset{\sim}{t})$ is the pre-logarithmic dislocation energy factor and it depends upon $\underset{\sim}{t}$ but not on the choice of $\underset{\sim}{\tau}$. Σ_{ij} is linear in $\underset{\sim}{b}$ and is called the angular stress factor; Σ_{ij} is not independent of $\underset{\sim}{\tau}$. The importance of Stroh's analysis (at least for present purposes) is that Σ_{ij} may be easily computed. If we know Σ_{ij}, it turns out that we can perform analyses of curvilinear dislocation loops with ease.

Stroh also showed that the displacement field of a line of force coincident with the dislocation is

$$u_i = -\frac{1}{2\pi i}\sum_{\alpha=1}^{6} \pm A_{i\alpha} A_{s\alpha} f_s \ \ell n\{\underset{\sim}{m}\cdot\underset{\sim}{x} + p_\alpha \underset{\sim}{n}\cdot\underset{\sim}{x}\} \tag{6}$$

where $\underset{\sim}{f}$ is the line force strength (per unit length). Equation (6) is the fundamental ingredient for the two-dimensional Boundary Integral Method to be mentioned later.

3. THE INTEGRAL FORMALISM

Barnett and Lothe [8] have shown that it is possible to give alternative formulae for the elastic quantities associated with straight dislocations which circumvent the need for solving Stroh's eigenvalue problem. These alternate formulae involve three real matrices which may be computed by integration (usually these computations must be performed numerically). A detailed description and comparison of the Stroh and the "integral" formalisms and formulae is given in [1]. The choice of formalism to be used in straight dislocation calculations is of course left to the individual investigator. The integral formalism is useful when angular derivatives of dislocation quantities are required; it has proven to be indispensable for analyzing surface waves [2]. My experience has been that Stroh's method allows for a more rapid numerical determin tion of many of the straight dislocation field quantities. The integral formulae offer the advantage of never encountering degenerate cases; degenerate cases (repeated eigenvalues p_α) do occur in Stroh's method, but they can be dealt with rather easily by, say, "twiddling" the elastic constants a bit.

4. BROWN'S THEOREM AND THE INDENBOM-ORLOV FORMULA

As I see it, the basis for what I would call the modern theory of dislocations was provided in two papers published back-to-back in Philosophical Magazine (1967) by Lothe and Brown [9, 10].

I shall refer to the main result as Brown's theorem even though his paper is entitled "A Proof of Lothe's Theorem." (Perhaps the panel discussion at the end of this conference can help us decide exactly the number of and/or the order of names to be attached to this result.) At any rate, Brown's theorem is stated as follows.

Consider a plane curvilinear dislocation loop L (Burgers vector $\underset{\sim}{b}$) in an infinite anisotropic linear elastic solid. Let $\underset{\sim}{x}'$ be the position vector from some origin to an element of arc ds on the loop. Suppose we wish to evaluate the self-stress field of the loop at a point $\underset{\sim}{x}$ in the plane of the loop. Brown's theorem states that

$$\sigma_{ij}(\underset{\sim}{x}) = \frac{1}{2}\oint_L \frac{\Sigma_{ij}(\theta) + \Sigma''_{ij}(\theta)}{|\underset{\sim}{x}-\underset{\sim}{x}'|^2} \sin(\theta-\alpha)ds \ . \tag{7}$$

Figure 1 shows the relevant geometry for Brown's formula. θ is the angle measured counter-clockwise from a fixed arbitrary datum in the plane to the vector $\underset{\sim}{x}-\underset{\sim}{x}'$, and α is the angle measured from the datum to the local tangent to ds at $\underset{\sim}{x}'$. $\Sigma_{ij}(\theta)$ is the angular stress factor (in the loop plane) of an infinite straight dislocation along $\underset{\sim}{x}-\underset{\sim}{x}'$ (the direction θ) with the same Burgers vector as that of the loop L; $\Sigma''_{ij}(\theta)$ is the second derivative of $\Sigma_{ij}(\theta)$ with respect to θ. Elementary calculus can be used to show that

$$d\theta = \frac{\sin(\theta-\alpha)ds}{|\underset{\sim}{x}-\underset{\sim}{x}'|} \ , \tag{8}$$

so that another version of Brown's formula is the disconcertingly simple form

$$\sigma_{ij}(\underset{\sim}{x}) = \frac{1}{2}\oint_L \frac{\Sigma_{ij}(\theta) + \Sigma''_{ij}(\theta)}{|\underset{\sim}{x}-\underset{\sim}{x}'|} d\theta \ . \tag{9}$$

There are several striking features of Brown's formula. Obviously we need only know $\Sigma_{ij}(\theta)$ and its second angular derivative for all θ, $0 \leq \theta \leq 2\pi$, to construct, in principle, the in-plane stresses of a curvilinear loop (actually, since $\Sigma_{ij}(\theta) = \Sigma_{ij}(\theta+\pi)$, the straight dislocation data needed are for $0 \leq \theta \leq \pi$). This information is easily obtained using either the Stroh or the integral formalism, or both [1]. The really surprising feature of the formula is that the straight dislocation data used to build the loop solution is that for dislocations along $\underset{\sim}{x}-\underset{\sim}{x}'$, and not that for dislocations tangent to the loop, as one might have intuitively expected.

The next surprise arising from the theorem becomes apparent when one thinks of implementing the formula. The naive approach is to think of tabulating $\Sigma_{ij} + \Sigma''_{ij}$ and then numerically integrating (7) or (9) over the loop. The simplest numerical scheme would involve approximating the loop by polygonal segments. In fact, (7) or (9) can be integrated exactly for a finite line segment so that numerical integration can be entirely circumvented [1, 10]! I shall not present either the formula for the field of a segment or that for a polygonal loop; the importance of Brown's theorem for present purposes is that at this junction we have available some simple machinery for computing the in-plane stresses of planar dislocation loops.

I should digress briefly to mention that (7) is an alternate form of a line integral expression for dislocation loop stresses whose integrand involves derivatives of three-dimensional elastic Green's functions. The beauty of Brown's expression is that the integrands in (7) or (9) are directly related to two-dimensional dislocation solutions which are relatively simple quantities with which to work.

Indenbom and Orlov [11] provided the appropriate extension of Brown's theorem to non-planar loops (or to stresses out of the plane of a planar loop). Consider the non-planar curvilinear dislocation loop L (Burgers vector $\underset{\sim}{b}$) shown in Figure 2. We wish to compute the stress field of the loop at a point P in an infinite solid. We construct a cone with vertex at P by drawing straight lines from P to every point on L (these lines are the generators of the conical surface). Let ω be the angle measured from $\underset{\sim}{t}$, the unit tangent to L at $\underset{\sim}{x}'$, to the generator $\underset{\sim}{x}-\underset{\sim}{x}'$. Since $\underset{\sim}{x}-\underset{\sim}{x}'$ and $\underset{\sim}{t}$ lie in the tangent plane to the cone at $\underset{\sim}{x}'$, ω is an angle measured in the tangent plane. The Indenbom-Orlov formula states that

$$\sigma_{ij}(\underset{\sim}{x}) = \frac{1}{2} \oint_L \frac{\Sigma_{ij}(\omega) + \Sigma''_{ij}(\omega)}{|\underset{\sim}{x}-\underset{\sim}{x}'|^2} \sin\omega \, ds \, . \tag{10}$$

Here $\Sigma_{ij}(\omega)$ is the angular stress factor in the tangent plane for an infinite straight dislocation lying along the generator from $\underset{\sim}{x}'$ to P. $\Sigma''_{ij}(\omega)$ is the second derivative of $\Sigma_{ij}(\omega)$ _in the tangent plane_. It is clear that (10) reduces to (7) for planar loops with P in the loop plane. It is also clear that the stresses due to a loop of arbitrary shape can be built up from elementary infinite straight dislocation data.

The formulae of Brown and Indenbom-Orlov provide what is almost a purely geometrical representation of dislocation loop stress fields. Their simple appearance allows us to extract the singular behavior of these fields as the field point $\underset{\sim}{x}$ approaches a point on the loop, and this extraction is dealt with in the next section.

5. THE SINGULAR NATURE OF DISLOCATION LOOP STRESS FIELDS

Consider a plane loop of dislocation L, and a point P on this loop. The local unit tangent to the loop at P is $\underset{\sim}{t}$, and the unit principal normal to the loop at P is denoted by $\underset{\sim}{m}$ where

$$\kappa \underset{\sim}{m} = \frac{d\underset{\sim}{t}}{ds} \quad ; \qquad \kappa = \left|\frac{d\underset{\sim}{t}}{ds}\right| > 0 \tag{11}$$

κ ($\kappa > 0$) is the curvature of L at P, and ds is the element of arc at P (see Figure 3). We wish to determine the dominant singular part of the stress field of the loop at the point $P \pm \varepsilon\underset{\sim}{m}$ as $\varepsilon \to 0$. Gavazza and Barnett [12] and Barnett [13] have shown that as ε tends to zero ($\varepsilon > 0$)

$$\sigma_{ij}(P \pm \varepsilon\underset{\sim}{m}) \sim \pm \frac{\Sigma_{ij}(\underset{\sim}{t})}{\varepsilon} + \frac{\kappa}{2}\left(\Sigma_{ij}(\underset{\sim}{t}) + \Sigma''_{ij}(\underset{\sim}{t})\right) \ln \frac{8}{\kappa\varepsilon} \quad . \tag{12}$$

Thus, as we approach a point on the loop along the principal normal we encounter two types of singular behavior. The dominant singularity is an inverse distance singularity characteristic of an infinite straight dislocation tangent to the loop at P; this is to be expected, and, indeed, this term changes sign as we approach P from opposite sides of the loop. In addition, there is also a weaker curvature-dependent logarithmic singularity proportional to $\Sigma_{ij}(\underset{\sim}{t}) + \Sigma''_{ij}(\underset{\sim}{t})$, which is essentially a line tension term. To see this more clearly we may note that if we compute the "resolved stress" $\sigma_{ij} b_i n_j$, where $\underset{\sim}{n}$ is the normal to the plane of the loop, the coefficient of the logarithmic term in (12) is proportional to

$$\frac{1}{2} b_i n_j \{\Sigma_{ij}(\underset{\sim}{t}) + \Sigma''_{ij}(\underset{\sim}{t})\} = E(\underset{\sim}{t}) + E''(\underset{\sim}{t}) \tag{13}$$

which is the classical line tension factor for the "tangent dislocation". Furthermore, if we "average" the dislocation self-stress about P in the loop plane we find that near P

$$\frac{1}{2} \{\sigma_{ij}(P+\varepsilon\underset{\sim}{m}) + \sigma_{ij}(P-\varepsilon\underset{\sim}{m})\} \sim \frac{\kappa}{2} \{\Sigma_{ij}(\underset{\sim}{t}) + \Sigma''_{ij}(\underset{\sim}{t})\} \, \ell n \frac{8}{\kappa\varepsilon} , \tag{14}$$

since the coefficient of the curvature-dependent singularity does not change sign as we cross the loop.

Equation (12) was derived by Gavazza and Barnett [12] and Barnett [13] for circular and parabolic loops. I attempted in [13] to show that (12) is valid in general for smooth plane loops. My original proof lacks a bit of rigor, but I have constructed a better proof which demonstrates the general validity of (12).

At this point an obvious question to raise is "what is the singular nature of the self-stress of either a non-planar loop or of a planar loop approached normally along an arbitrary direction $\underset{\sim}{\tau}$?" One intuitively expects that we will still see the inverse distance singularity characteristic of the infinite straight dislocation tangent to the dislocation loop at the point in question. One also expects to see a curvature-dependent stress singularity; the problem is, "what curvature do we insert in equation (12)?", since in the more general case we are not approaching the loop along its principal normal (the direction of curvature, $\underset{\sim}{m}$). Most of us would guess that we should use the curvature resolved along the direction of approach to the loop. This is, in fact, correct, and I wish to give a heuristic argument leading to this result; the key to my argument is the Indenbom-Orlov formula.

Figure 4 depicts an arbitrary non-planar dislocation loop L as well as the Indenbom-Orlov cone to be used for evaluating the stresses at a point Q. We wish to determine the singular nature of the stresses at the point $Q = P \pm \varepsilon\underset{\sim}{\tau}$ as ε tends to zero, where $\underset{\sim}{\tau}$ is a unit normal to the loop at P, i.e., $\underset{\sim}{\tau}$ is a unit vector along $\vec{PQ}$ and thus $\underset{\sim}{\tau}$ lies in the tangent plane to the Indenbom-Orlov cone at P. The stress field at Q is given by equation (10), namely

$$\sigma_{ij}(Q) = \frac{1}{2} \oint_L \frac{\Sigma_{ij}(\omega) + \Sigma''_{ij}(\omega)}{|\underset{\sim}{x}-\underset{\sim}{x}'|^2} \sin\omega \, ds , \tag{15}$$

and we note that $\omega = \pi/2$ when $\underset{\sim}{x}'$ denotes the point P on L.

Now for a fixed Q, the vertex of the cone is fixed and we can regard the angle ω as being solely determined by position along the curve L, i.e., $\Sigma_{ij}(\omega) + \Sigma''_{ij}(\omega)$ as well as the other terms in the integrand in (15) are completely characterized or parmeterized by the arc length s. Hence we can view (15) as simply a mathematical formula and try to seek an expedient method for evaluating the integral.

Now a cone is a developable surface, and for the purpose of evaluating (15) we may imagine cutting the cone along any of its generators and developing the cone in the tangent plane to the cone at P; this situation is depicted in Figure 5. The process of development maps the non-planar real loop L forming the boundary of the base of the Indenbom-Orlov cone into the plane curve L^D in the tangent plane. Every point on the real curve L is mapped into a corresponding point on L^D. Furthermore, development of the cone preserves (point by point) arc length s, ω, and $|\underset{\sim}{x}-\underset{\sim}{x}'|$. Of course the developed curve L^D is not a physical dislocation; it is simply a mathematical construct. Since for fixed Q we regard $\Sigma_{ij} + \Sigma''_{ij}$ as a function of s in (15), the stress field of L^D at Q is simply

$$\sigma^D_{ij}(Q) = \frac{1}{2} \oint_{L^D} \frac{\Sigma_{ij}(s) + \Sigma''_{ij}(s)}{|\underset{\sim}{x}-\underset{\sim}{x}'|^2} \sin \omega \, ds \, . \tag{16}$$

Since the integrands in (15) and (16) are point-by-point identical, we conclude that the stresses at Q due to L are identical to those at Q produced by L^D, i.e., that

$$\sigma_{ij}(Q) \equiv \sigma^D_{ij}(Q) \quad . \tag{17}$$

As L^D is a planar curve, when $\varepsilon \to 0$ we can use (12) to deduce that

$$\sigma^D_{ij}(Q) \sim \pm \frac{\Sigma_{ij}(\underset{\sim}{t})}{\varepsilon} + \frac{\kappa_D}{2} \{\Sigma_{ij}(\underset{\sim}{t}) + \Sigma''_{ij}(\underset{\sim}{t})\} \, \ell n \frac{8}{\kappa_D \varepsilon} \tag{18}$$

where $\underset{\sim}{t}$ is the unit tangent to L^D (and also to L) at P, and κ_D is the curvature of L^D at P. There is a result from differential geometry [14] which states that the curvature κ_D of the developed curve L^D is merely the geodesic curvature of L in the tangent plane to the real Indenbom-Orlov cone at P, i.e., if κ and $\underset{\sim}{m}$ are the curvature and principal normal of L at P, respectively, then

$$\kappa_D \equiv \kappa |(\underset{\sim}{\tau} \cdot \underset{\sim}{m})| \quad . \tag{19}$$

Hence, the singular part of the stresses of a general loop (as we approach the loop normally along a direction $\underset{\sim}{\tau}$) is given by equation (12) with κ replaced by κ_D, i.e., the appropriate curvature is the component of curvature seen along the direction of approach. In addition, $\Sigma_{ij} + \Sigma''_{ij}$ must be evaluated in the tangent plane at P.

This result has one immediate interesting implication. If we approach the loop along a direction normal to the direction of curvature $\underset{\sim}{m}$, then κ_D vanishes and there is no apparent line-tension-like singularity. The formulae given by Kroupa [15] for a circular edge loop in an isotropic solid show this property when one carefully analyzes their limiting behavior. This leads me to believe that the somewhat heuristic proof outlined above is essentially correct, and I think this result has further implications about the concept of line tension, which I shall discuss in the following section.

6. THE SELF-FORCE AT A POINT ON A DISLOCATION LOOP

If we consider the generalized thermodynamic "force" at a point on a dislocation loop due to applied stresses, other sources of internal stress, and "image stresses", such forces always take the Peach-Koehler form

$$f_m = -\varepsilon_{jmn} b_i \sigma_{ij} t_n \tag{20}$$

where $\underset{\sim}{t}$ is the unit tangent to the dislocation at the point in question and σ_{ij} is the relevant stress field evaluated at this point. The self-force due to the dislocation self-stress cannot be obtained using (20) without some modification, since the self-stresses are singular at points on the loop. Brown [16] heuristically argued that (20) should apply for the self-force at a point P on a planar loop if, instead of the self-stress σ_{ij}, we use the "average" self-stress about P, i.e., if we use

$$\frac{1}{2}\{\sigma_{ij}(P+\varepsilon\underset{\sim}{m}) + \sigma_{ij}(P-\varepsilon\underset{\sim}{m})\} \quad .$$

We would typically choose ε to be on the order of $|\underset{\sim}{b}|$. Brown noted that there were some inconsistencies between this prescription and results obtainable from Kroupa's analysis [15]. Gavazza and Barnett [12] performed a careful analysis of the self-force by computing the energy variation during a small virtual in-plane change in the configuration of a planar loop. What they found was that Brown's prescription was essentially correct except for two missing terms which resolved the earlier noted inconsistencies; this is clearly a testimony to Brown's keen insight.

Apart from a contribution to the self-force from an integral over a tube of radius ε surrounding the dislocation (a discussion of which would require more space than I care to devote) the in-plane self-force at a point P on a planar loop whose local unit tangent, principal normal, and curvature are $\underset{\sim}{t}$, $\underset{\sim}{m}$, and κ, respectively, was shown in [12] to be

$$\underset{\sim}{f}\cdot\underset{\sim}{m} = -\kappa\, E(\underset{\sim}{t}) + \frac{1}{2}\, b_i\, n_j\{\sigma_{ij}(P+\varepsilon\underset{\sim}{m}) + \sigma_{ij}(P-\varepsilon\underset{\sim}{m})\} \tag{21}$$

where $\underset{\sim}{n}$ is the unit normal to the plane of the loop. $E(\underset{\sim}{t})$ is the pre-logarithmic energy factor for an infinite straight dislocation tangent to the loop at P. Clearly the first term on the right-side of (21) is a term missing from Brown's earlier prescription. Using (13) and (14), the dominant part of the in-plane self-force can be written as

$$\underset{\sim}{f}\cdot\underset{\sim}{m} \sim -\kappa\, E(\underset{\sim}{t}) + \kappa\{E(\underset{\sim}{t}) + E''(\underset{\sim}{t})\}\, \ell n\, \frac{8}{\kappa\varepsilon} \quad , \tag{22}$$

and one sees that the second term clearly displays the "line tension" contribution to the self-force. For an infinitely long straight dislocation, $\kappa = 0$ and the right-side of (21) vanishes, as it should. One can argue that (21) and (22) are unsatisfactory in the sense that the result depends on the cut-off parameter ε. This is the best one can do for "singular dislocations," and at least the derivation given in [12] is self-consistent.

If one attempts to redo the analysis of Gavazza and Barnett [12] in order to compute the self-force in an arbitrary direction $\underset{\sim}{\tau}$ normal to a dislocation loop element (the loop may be either planar or non-planar), one finds that many of the same arguments are valid for this more general case (we also need to use the results given in equations (17) - (19) of the last section). In particular, apart from a tube integral contribution

$$\underset{\sim}{f}\cdot\underset{\sim}{\tau} = -\kappa^* E(\underset{\sim}{t}) + \frac{1}{2} b_i n_j\{\sigma_{ij}(P+\varepsilon\underset{\sim}{\tau}) + \sigma_{ij}(P-\varepsilon\underset{\sim}{\tau})\} \tag{23}$$

where

$$\kappa^* = \kappa|(\underset{\sim}{\tau}\cdot\underset{\sim}{m})| \; ; \qquad \underset{\sim}{n} = \underset{\sim}{\tau}\times\underset{\sim}{t} \; . \tag{24}$$

The dominant part of the right side of (23) can be written as

$$\underset{\sim}{f}\cdot\underset{\sim}{\tau} \sim -\kappa^* E(\underset{\sim}{t}) + \kappa^*\{E(\underset{\sim}{t}) + E''(\underset{\sim}{t})\} \, \ell n \frac{8}{\kappa^*\varepsilon} \, , \tag{25}$$

and it is understood that $E''(\underset{\sim}{t})$ is the second angular derivative of $E(\underset{\sim}{t})$ in the plane whose normal is $\underset{\sim}{n}$. What we have essentially shown is that Gavazza and Barnett's formulae (21) and (22) carry over to the general case if we use the curvature component in the direction $\underset{\sim}{\tau}$ in which the self-force is to be computed.

At this point it seems appropriate to offer some comments on what I think is meant by the line tension of a dislocation. Many texts and papers consider the line tension to be given (roughly) by

$$\frac{\delta E}{\delta \ell} \, ,$$

i.e., the change in loop energy with a change in length. As Alan Head pointed out at the 1969 dislocation conference at Gaithersburg, such a definition of line tension is not satisfactory since one can easily conceive of different configurational changes involving the same $\delta\ell$, but different δE's. This simply states that it is unreasonable to expect that a dislocation loop with an infinite number of possible virtual changes in configuration can be totally characterized by one parameter, namely loop length, only. To me the more appropriate definition of local line tension is given by the second term on the right side of (25). This term is obtained by the imaginary process of "plucking" a dislocation element in a direction $\underset{\sim}{\tau}$ and computing the restoring force in that direction (if $E(\underset{\sim}{t}) + E''(\underset{\sim}{t}) < 0$ the force is not restoring, i.e., it aids rather than opposes the perturbation). If κ^* vanishes at the dislocation element, i.e., if we pluck the dislocation normal to its direction of curvature ($\underset{\sim}{m}$), there is no line-tension restoring force. Hence, a circular dislocation loop plucked normally to the loop plane experiences no line-tension restoring force according to the definition proposed here. I should point out that even though what I call line tension (the second term on the right side of (25)) vanishes when $\kappa^* = 0$, there are other terms in the self-force expression which do not vanish. These terms arise from the effects of the rest of the loop (these are the terms Gavazza and I denoted by $J[L]$ in [12]), but I do not see fit to classify these as local line tension effects.

7. SUMMARY AND CONCLUSIONS

I hope that I have been able to convey the feeling that at the present there is available a good deal of understanding about the stress fields of dislocations of complex shapes in linear anisotropic solids. It is, for example, possible to compute the equilibrium shapes of planar dislocation loops in crystals subjected to applied stresses in a straightforward way. Essentially one has to find a shape which balances the Peach-Koehler force due to the applied field and the dislocation self-force at each point on the loop. The simplest technique involves assuming a loop shape, computing the net force at discrete points on the loop (which may be approximated as a polygon to use the results of Brown's theorem), adjusting the loop to a relaxed configuration, recomputing the net force, and iterating until the the net force at all the points vanishes (to within some pre-assigned tolerance). The knowledge of the exact form of the self-stress singularity is an essential ingredient in performing this type of calculation. Gore [17] has performed such calculations using non-linear tension splines to facilitate accurate determination of the local loop curvature. He has included in his study loops in anisotropic materials for which $E(t) + E''(t)$ is negative over certain angular ranges (negative line tensions) so that the equilibrium loop shapes exhibit corners. Gore's relaxation scheme seems to work quite well; even ridiculous initial guesses for equilibrium loop shapes lead to convergent equilibrium loop shapes within about eight iterations. His results illustrate that the equilibrium loop shapes determined previously by Scattergood and Bacon [18] using a different self-force prescription are quite correct.

Even if very few investigators decide that it is worth their time to use the rigorous framework discussed here in actual calculations, the machinery I have presented may prove useful for deciding under what circumstances simpler approximations, such as the line tension approximation, may suffice. I hope the reader will be convinced that the framework is really quite simple. The extent to which this machinery is used will depend on the needs of the particular investigator. Numerous examples of the applications of the theory may be found in [1], since Bacon and Scattergood have been among the primary users of these methods, and the interested reader is referred to the relevant sections of that review.

I had initially planned to present some applications of a numerical method of stress analysis closely related to dislocation theory, but space limitations preclude this; I would like to at least mention the method in hopes of stimulating interest in its use. The method is known as the Boundary Integral Equation technique. It is simple to implement and can be used to study, among other topics, the elastic fields of arrays of inclusions and pressurized voids. G.W. Look and I [19] have used the method to predict the stability of void lattices in isotropic solids and cubic metals. According to Eshelby [20] the method dates back to Gebbia and Somigliana at the turn of this century. Altiero and Gavazza [21] have recently given a beautiful exposition of the technique and its relation to the theory of dislocations and misfitting inclusions.

In essence the method allows one to construct the solution to linear elastic problems (in two or three dimensions) using distributions of dislocations and/or point or line forces over the external and internal surfaces of <u>bounded</u> media; the dislocation and point/line force solutions needed are those for <u>infinite</u> solids, and these are easy to obtain. For any given problem one may choose a solution representation using either dislocations, localized forces, or both. The beauty of the method is that its implementation only requires a discretization of the boundary of the solid, whereas a method such as the finite element technique requires discretization of the interior of the solid as well. The method involves nothing more than solving integral equations by matrix algebra techniques.

It is my belief that the Boundary Integral Method provides a nice technique for coupling an exterior continuum to an interior region modelled as a discrete lattice. In my opinion such modelling could provide a better "flexible boundary" scheme for atomistic defect calculations than other schemes proposed previously. Perhaps I shall have the chance to raise this issue with the "lattice" talks following my presentation at this meeting. I might add that engineers are already coupling Boundary Integral and finite element methods in many problem areas, and I see no reason why the same pursuit cannot be developed with regards to atomistic-continuum coupling.

I should like to close by acknowledging my close association with Drs. R.J. Asaro, D.J. Bacon, S.D. Gavazza and R.O. Scattergood over the past ten years. Most of the extensions I have presented here grew out of our collaborative efforts, and I am sure that I have gleaned more knowledge from these colleagues than I have imparted to them in return. They know the theory at least as well as I purport to.

REFERENCES

1. D.J. Bacon, D.M. Barnett, and R.O. Scattergood, Prog. in Mat. Sci., v. 23, nos. 2-4, 51 (1978).
2. P. Chadwick and G.D. Smith, in Adv. in Appl. Mechanics, v. 17, 303, Acad. Press, N.Y. (1977).
3. D.M. Barnett and J. Lothe, J. Phys. F: Metal Physics, v. 4, 671 (1974).
4. A.N. Stroh, Phil. Mag., v. 3, 625 (1958).
5. A.N. Stroh, J. Math. and Phys., v. 41, 77 (1962).
6. J.D. Eshelby, W.T. Read, and W. Shockley, Acta Metall., v. 1, 251 (1953).
7. K. Malén and J. Lothe, Phys. Status Solidi, v. 39, 287 (1970).
8. D.M. Barnett and J. Lothe, Physica Norvegica, v. 7, 13 (1973).
9. J. Lothe, Phil. Mag., v. 15, 353 (1967).
10. L.M. Brown, Phil. Mag., v. 15, 363 (1967).
11. V.L. Indenbom and S.S. Orlov, Soviet Physics JETP, v. 6, 274 (1967).
12. S.D. Gavazza and D.M. Barnett, J. Mech. Phys. Solids, v. 24, 171 (1976).
13. D.M. Barnett, Phys. Status Solidi (a), v. 38, 637 (1976).
14. D. Laugwitz, Differential and Riemannian Geometry, 53, Acad. Press, N.Y. (1965).
15. F. Kroupa, Czech. J. Physics, v. 10B, 284 (1960).
16. L.M. Brown, Phil. Mag., v. 10, 441 (1964).
17. L.A. Gore, Ph.D. Thesis, Dept. of Mechanical Engineering, Stanford University (unpublished, 1980).
18. R.O. Scattergood and D.J. Bacon, Phys. Status Solidi (a), v. 25, 395 (1974).
19. G.W. Look and D.M. Barnett (to be published).
20. J.D. Eshelby, in Dislocations in Solids (ed., F.R.N. Nabarro), v. 1, ch. 3, North-Holland, Amsterdam (1979).
21. N.J. Altiero and S.D. Gavazza, Journal of Elasticity, v. 10, 1, (1980).

Addendum

Three comments have been kindly brought to my attention by the reviewers of this manuscript following the conference. Professor J.P. Hirth informed me that the newly revised edition of Theory of Dislocations by Hirth and Lothe will refer to the basic geometrical theorem as "Brown's Theorem"; this would seem to set an appropriate precedent. I should have mentioned that Willis [Phil. Mag. 21, 931 (1970)] and Mura and Mori [Phil. Mag. 33, 1021 (1976)] have given formulae for the elastic fields due to a dislocation segment; these formulae were derived without recourse to Brown's Theorem. Finally, Professor Hirth and Dr. Puls have noted that Dr. J. Sinclair has introduced a recent version of the flexible boundary scheme which bears a marked resemblance to the proposed coupling of a boundary integral technique with a lattice computation.

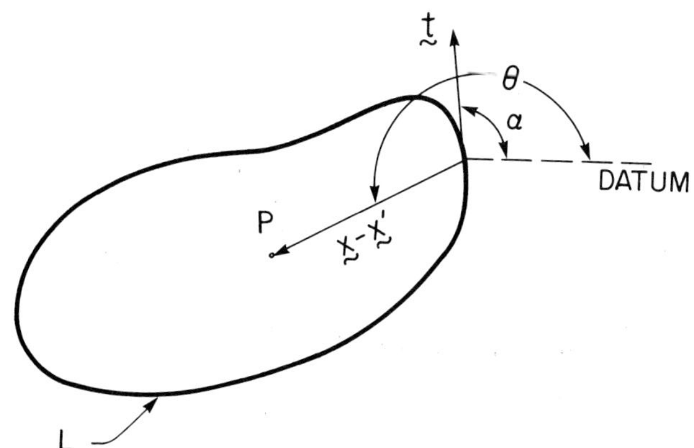

FIG. 1
The geometry appropriate for Brown's formula for a planar loop.

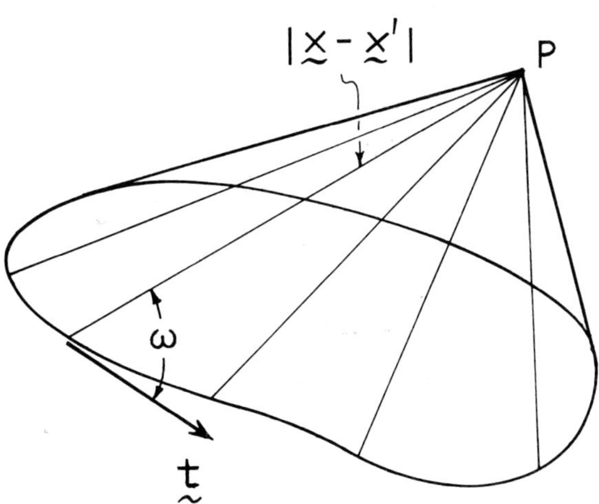

FIG. 2
The Indenbom-Orlov cone and associated geometry for a non-planar loop.

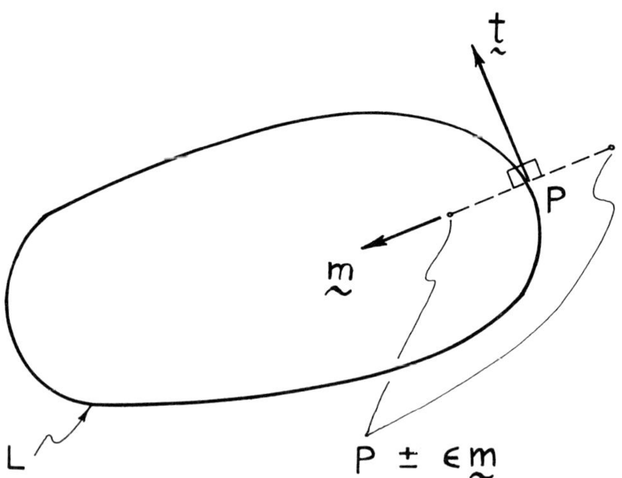

FIG. 3
The geometry associated with the asymptotic form of the stresses near a plane dislocation loop.

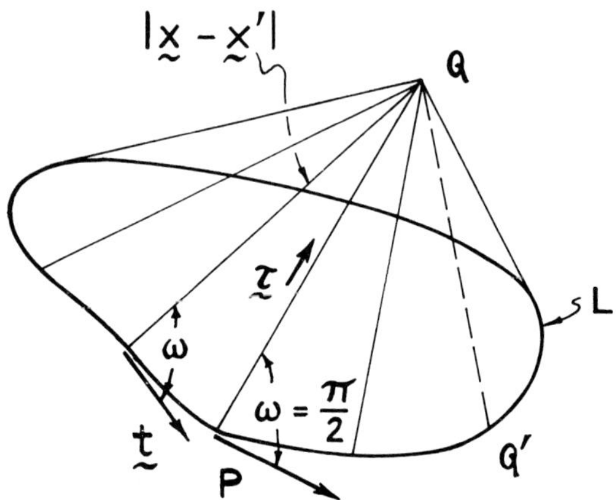

FIG. 4
The Indenbom-Orlov cone and associated geometry for deducing the asymptotic form of the stresses near a non-planar dislocation loop.

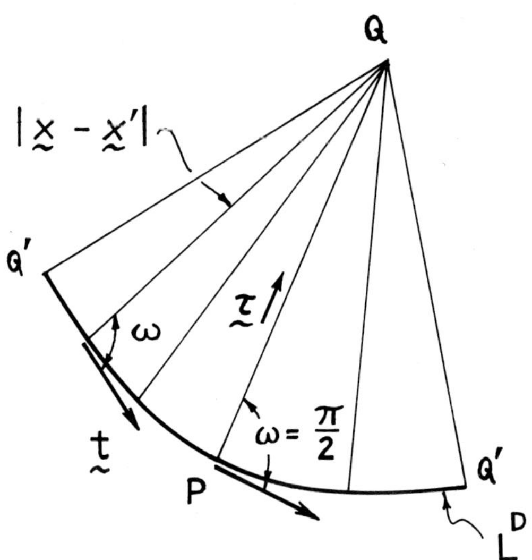

FIG. 5
The developed Indenbom-Orlov cone and the "developed" plane loop L^D in the tangent plane at P.

EDGE DISLOCATIONS AND DISCRETE ELASTIC THEORY

D. Evans and V. Celli
Department of Physics, University of Virginia

Introduction

An edge dislocation is usually described by continuum elastic theory, either isotropic or anisotropic. This theory is valid in regions well removed from the dislocation core, but serious difficulties arise if one tries to extend the continuum solution into the core region, because the strains go to infinity there. By excluding a small cylinder of material centered on the dislocation line, we can avoid having to deal with infinite stresses or strains. The strain energy of an elastic dislocation diverges as the size of the body containing the dislocation goes to infinity, or as the "excluded radius" goes to zero. The catastrophe as the excluded radius goes to zero results from unphysical assumptions in continuum theory, and is avoided in a discrete model. We show how to implement the discrete theory in a specific example; the difficulties are considerably greater than in the case of a screw dislocation.

We assume a two dimensional square lattice in which each atom interacts with its nearest neighbors and next-nearest neighbors by Hooke's law forces. Each atom is specified by a coordinate pair (m,n), where m and n are half-odd-integral. We introduce an edge dislocation dipole which extends along the y-axis from $n = 0.5 - N$ to $N - 0.5$. In the absence of a dislocation, the horizontal displacements u(m,n) and the vertical displacements v(m,n) satisfy:[1]

(1) $L_1 u(m,n) + L_2 v(m,n) = 0$, (2) $L_2 u(m,n) + L_3 v(m,n) = 0$, where for any function f(m,n)

$$(3)\quad L_1 f(m,n) = \alpha(f(m+1,n) + f(m-1,n) - 2f(m,n)) + \beta(f(m,n+1) + f(m,n-1) - 2f(m,n)) + \gamma(f(m+1,n+1) + f(m+1,n-1) + f(m-1,n-1) + f(m-1,n+1) - 4f(m,n))$$

$$(4)\quad L_2 f(m,n) = \gamma(f(m+1,n+1) - f(m+1,n-1) + f(m-1,n-1) - f(m-1,n+1))$$

$$(5)\quad L_3 f(m,n) = \alpha(f(m+1,n) + f(m-1,n) - 2f(m,n)) + \beta(f(m,n+1) + f(m,n-1) - 2f(m,n)) + \gamma(f(m+1,n+1) + f(m+1,n-1) + f(m-1,n-1) + f(m-1,n-1) - 4f(m,n))$$

According to isotropic continuum theory, a static edge dislocation satisfies the equations:

$$(6)\quad \partial^2 u/\partial x^2 + A\partial^2 u/\partial y^2 + (1-A)\partial^2 v/\partial x\partial y = 0$$

$$(7)\quad (1-A)\partial^2 u/\partial x\partial y + A\partial^2 v/\partial x^2 + \partial^2 v/\partial y^2 = 0,$$ where $A = \mu/(\lambda + 2\mu)$, with μ and λ being the Lame constants. If we want equations (1) and (2) to reduce to equations (6) and (7) in the continuum limit, we specify that $\alpha = (\lambda + 3\mu)/(2\lambda + 4\mu)$, that $\beta = (\mu - \lambda)/(2\lambda + 4\mu)$, and that $\gamma = (\lambda + \mu)/(4\lambda + 8\mu)$.

We now introduce the dislocation dipole, and consider its effect on the atoms in the column (0.5,n). If $|n| < N - 0.5$, then all bonds between the (0.5,n) atom and the atoms at (-0.5,n), (-0.5,n + 1), and (-0.5,n-1) are severed, and replaced by bonds with atoms at the newly introduced positions (0,n),(0,n-1), and (0,n + 1). We assume that, as m goes to zero from the right, u(m,n) goes to 0.5b, where b is the Burger s vector, and that, as m goes to zero from the left, u(m,n) goes to -0.5b. If $n = N - 0.5$, then the bonds between the (0.5,n) atom and the atoms at (-0.5,n) and (-0.5,n-1) are replaced by bonds with atoms at the (0,n) and (0,n-1) positions. The bond between the (0.5,n) atom and the (-0.5,n + 1) atom is severed, but not replaced. If n is $N + 0.5$, then we assume that the bonds between the (0.5,n) atom and the atoms at (-0.5,n) and (-0.5,n + 1) are left intact, while the bond between the (0.5,n) atom and the (-0.5,n-1) atom is severed but not replaced. The (0.5,N + 0.5) atom is closer to the (0,N-0.5) atom than it is to

[1] P. Dean, Proc. Camb. Phil Soc., 59, 383, (1963)

the (1.5,N-0.5) atom, so we must bond these two atoms. Unfortunately, the relative orientation of these two positions may not be that of a square lattice, so it is not immediately clear what strength to assign to this bond. We next employ a treatment similar to that described above for the (0.5,0.5-N) and (0.5,-0.5-N) positions, and for all the atoms in the (-0.5,n) column. This bond arrangement is illustrated in the following drawing.

FIG. 1

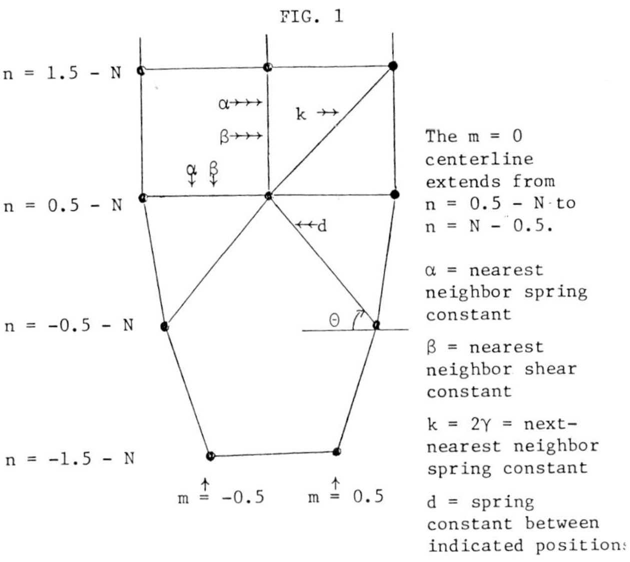

By using the procedure described above, we finally get the following pair of equations, analogous to (6) and (7):

$$(8)\quad L_1u(m,n) + L_2v(m,n) = (\delta(m,-0.5) - \delta(m,0.5))J(n)$$

$$(9)\quad L_2u(m,n) + L_3v(m,n) = (\delta(m,-0.5) + \delta(m,0.5))K(n)$$

From now on, we will let u(n) stand for u(0.5,n), v(n) stand for v(0.5,n), and w(n) stand for v(0,n). Also, u(-0.5,n) = -u(0.5,n), and v(-0.5,n) = v(0.5,n), allowing us to eliminate u(-0.5,n) and v(-0.5,n) from the expressions for J(n) and K(n). Let d be the strength of the spring connecting the (0.5,N+0.5) atom to the (0,N-0.5) atom, and let θ be the angle that this bond makes with the m-axis. We then define e to be $d\cos^2\theta$, e' to be equal to $d\cos\theta\sin\theta$, and e" to be $d\sin^2\theta$. Finally, we define h(n,N-0.5) to be one if $|n| < N - 0.5$, and zero otherwise, and we let $\delta(n,k)$ represent the standard Kronecker delta function. With these definitions, we can write the following expressions for J(n) and K(n):

$$(10)\quad J(n) = h(n,N-0.5)(0.5b + \alpha u(n) + \gamma u(n+1) + \gamma u(n-1) + \gamma v(n+1) - \gamma w(n+1) + \gamma w(n-1) - \gamma v(n-1)) + \delta(n,N-0.5)((\alpha + \gamma)(0.5b + u(n)) + \gamma(u(n+1) + u(n-1) + v(n+1) - v(n) + w(n-1) - v(n-1))) + \delta(n,N+0.5)(0.5eb + (\gamma - e)u(n) + \gamma u(n-1) + (\gamma - e')v(n) - \gamma v(n-1) + e'w(n-1)) + \delta(n,0.5-N)((\alpha + \gamma)(0.5b + u(n)) + \gamma(u(n+1) + u(n-1) + v(n) - v(n-1) + v(n+1) - w(n+1))) + \delta(n,-0.5-N)(0.5eb + (\gamma - e)u(n) + \gamma u(n+1) + (e' - \gamma)v(n) + \gamma v(n+1) - e'w(n+1))$$

$$(11)\quad K(n) = h(n,N-0.5)(\gamma(u(n+1) - u(n-1) + v(n+1) - w(n+1) + v(n-1) - w(n-1)) + \beta(v(n) - w(n))) + \delta(n,N-0.5)(-0.5\gamma b + \beta(v(n) - w(n)) + \gamma(u(n) + u(n+1) - u(n-1) + v(n+1) + v(n-1) - w(n-1) - v(n))) + \delta(n,N+0.5)(-0.5e'b + (e' - \gamma)u(n) - \gamma u(n-1) + (e'' - \gamma)v(n) + \gamma v(n-1) - e''w(n-1)) + \delta(n,0.5-N)(0.5\gamma b + \gamma(u(n+1) - u(n) - u(n-1) + v(n+1) - w(n+1) + v(n-1) - v(n)) + \beta(v(n) - w(n))) + \delta(n,-0.5-N)(0.5e'b + \gamma u(n+1) + (\gamma - e')u(n) + (e'' - \gamma)v(n) + \gamma v(n+1) - e''w(n+1))$$

Methods

We begin by solving for the Green functions associated with equations (8) and (9). There is a two-by-two Green function tensor whose elements satisfy the following equations:

(12) $L_1 GXX(m,n) + L_2 GYX(m,n) = \delta(m,0)\delta(n,0)$

(13) $L_2 GXX(m,n) + L_3 GYX(m,n) = 0$

(14) $L_1 GXY(m,n) + L_2 GYY(m,n) = 0$

(15) $L_2 GXY(m,n) + L_3 GYY(m,n) = \delta(m,0)\delta(n,0)$

We take the Fourier transform of both sides of equations (12) to (15), and we find that:

(16) $GXX(m,n) = (1/\pi^2) \int_0^{2\pi} \cos(mx)\cos(ny)\ G^{11}(x,y)\,dxdy$

(17) $GXY(m,n) = GYX(m,n) = (1/\pi^2) \int_0^{2\pi} \sin(mx)\sin(ny)\ G^{21}(x,y)\,dxdy$

(18) $GYY(m,n) = (1/\pi^2) \int_0^{2\pi} \cos(mx)\cos(ny)\ G^{22}(x,y)\,dxdy$

(19) $G^{11}(x,y) = N^{11}(x,y)/D(x,y)$; $G^{21}(x,y) = N^{21}(x,y)/D(x,y)$; $G^{22}(x,y) = N^{22}(x,y)/D(x,y)$

(20) $N^{11}(x,y) = \alpha(\cos(y) - 1) + \beta(\cos(x) - 1) + 2\gamma(\cos(x)\cos(y) - 1)$

(21) $N^{21}(x,y) = 2\gamma\sin(x)\sin(y)$

(22) $N^{22}(x,y) = \alpha(\cos(x) - 1) + \beta(\cos(y) - 1) + 2\gamma(\cos(x)\cos(y) - 1)$

(23) $D(x,y) = 2(\alpha(\cos(x) - 1) + \beta(\cos(y) - 1) + 2\gamma(\cos(x)\cos(y) - 1))(\alpha(\cos(y) - 1) + \beta(\cos(x) - 1) + 2\gamma(\cos(x)\cos(y) - 1)) - 8\gamma^2\sin^2(x)\sin^2(y)$

The integrals GXY(m,n) converge for all values of m and n, but the integrals GXX(m,n) and GYY(m,n) are unfortunately divergent. On the other hand, GXX(m',n') - GXX(m,n) and GYY(m',n') - GYY(m,n) are well behaved. The continuum equivalents of these Green functions are:

(24) $GXXC(x,y) = (1/4\pi)(((\lambda + 3\mu)/(2\mu))\ln(x^2 + y^2) + ((\mu + \lambda)/\mu)y^2/(x^2 + y^2))$

(25) $GXYC(x,y) = -(\mu + \lambda)xy/((4\pi\mu)(x^2 + y^2))$

(26) $GYYC(x,y) = (1/4\pi)(((\lambda + 3\mu)/(2\mu))\ln(x^2 + y^2) - ((\mu + \lambda)/\mu)y^2/(x^2 + y^2))$

Let S(m,n) stand for the right hand side of (8), and let T(m,n) stand for the right hand side of (9). From elementary Green's function theory, we know that:

(27) $u(m,n) = \sum_{j=-\infty}^{\infty}\sum_{k=-\infty}^{\infty}(S(j,k)GXX(m-j,n-k) + T(j,k)GXY(m-j,n-k))$

(28) $v(m,n) = \sum_{j=-\infty}^{\infty}\sum_{k=-\infty}^{\infty}(S(j,k)GXY(m-j,n-k) + T(j,k)GXY(m-j,n-k))$

Because of the $\delta(m,-0.5) \pm \delta(m,0.5)$ terms in S(m,n) and T(m,n), the infinite sum over j is entirely eliminated in (27) and (28), and the sum over k is reduced to a sum from -0.5 - N to N + 0.5. If we let m = 0.5 in (27) and (28), then the only necessary Green's functions are GXX(1,n) - GXX(0,n), GXY(1,n) ± GXY(0,n), and GYY(1,n) + GYY(0,n). Let GXX(n) be equal to GXX(1,n) - GXX(0,n), and let H(n) stand for GYY(1,n) + GYY(0,n). Clearly, GXX(n) is convergent, while H(n) is divergent, but H(n') - H(n) is finite. Now, GXY(0,n) is zero, so we can define GXY(n) by GXY(n) = GXY(1,n) + GXY(0,n) = GXY(1,n) - GXY(0,n). We know that u(-n) = u(n), v(-n) = -v(n), and w(-n) = -w(n), so we can write equations (27) and (28) in the form:

(29) $u(n) = \sum_{k=0.5}^{N+0.5}(J(k)(GXX(n-k) + GXX(n+k)) + K(k)(GXY(n-k) - GXY(n+k)))$

(30) $v(n) = \sum_{k=0.5}^{N+0.5}(J(k)(GXY(n-k) + GXY(n+k)) + K(k)(H(n-k) - H(n+k)))$

The continuum limit of S(m,n) is $-bh(y,L)d\delta(x)/dx$, where $\delta(x)$ is the Dirac delta function, and h(y,L) is equal to one if the absolute value of y is less than L (L = N times the lattice spacing) and zero otherwise. The continuum limit of T(m,n) is $wb(\delta(x)(\delta(y+L) - \delta(y-L)))$, where w involves various factors of e and γ. The continuum limits of equations (27) and (28) are

(31) $u(x,y) = b \int_{-L}^{L}(\partial GXXC(x,y-z)/\partial x)dz + wbGXYC(x,y+L) - wbGXYC(x,y-L)$

(32) $v(x,y) = b \int_{-L}^{L}(\partial GXYC(x,y-z)/\partial x)dz + wbGYYC(x,y+L) - wbGYYC(x,y-L)$

The continuum displacements u(x,y) and v(x,y) satisfy the equations (6) and (7), and the solutions to these two equations are known to be given by the following expressions, where a dipole configuration is assumed.[2]

(33) $u(x,y) = (b/2\pi)(\arctan((L-y)/x) + \arctan(L+y)/x) + ((\lambda + \mu)/(\lambda + 2\mu))(x(L-y)/(x^2+(L-y)^2) + x(L+y)/(x^2 + (L+y)^2)))$

(34) $v(x,y) = (b/4\pi)((\mu/(\lambda + 2\mu))\ln(x^2 + (L-y)^2)/(x^2 + (L+y)^2)) + ((\lambda + \mu)/(\lambda + 2\mu))(((L+y)^2 - x^2)/(x^2 + (L+y)^2) + (x^2 - (L-y)^2)/(x^2 + (L-y)^2)))$

If (31) and (32) are to lead to the continuum displacements, w must be equal to $\lambda/(\lambda + 2\mu)$. In the discrete case, this corresponds to e' equal in magnitude to β.

We now consider the equations obeyed by the atoms in the column m = 0. If the absolute value of n is less than N - 0.5, then the behavior of the atom at the position (0,n) is described by the equation:

(35) $\alpha(w(n+1) + w(n-1) - 2w(n)) - (4\gamma + 2\beta)w(n) + 2\gamma v(n+1) + 2\gamma v(n-1) + 2\gamma u(n+1) - 2\gamma u(n-1) + 2\beta v(n) = 0$

Due to the symmetries possessed by u(n),v(n), and w(n), there is no need to consider negative values of n in (35). If n = 0.5, equation (35) reduces to:

(36) $-2\gamma u(0.5) + 2\gamma u(1.5) - 2\gamma v(0.5) + 2\beta v(0.5) + 2\gamma v(1.5) - \alpha w(0.5) - (2\alpha + 4\gamma + 2\beta)w(0.5) + \alpha w(1.5) = 0$

If the position of the (0,N-0.5) atom is fixed arbitrarily, then equations (29),(30),(35), and (36) provide a complete and self-contained set of equations for the quantities u(n) and v(n) for n in the range 0.5 to N + 0.5, and also for w(n) for n between 0.5 and N - 0.5 inclusive. These equations can be solved by matrix inversion. Let us define a column vector Q(j) of rank 3N + 2 such that Q(j) = u(j-0.5) if j is between one and N + 1 , Q(j) = v(j-N-1.5) if j is between N + 2 and 2N + 2 inclusive, and Q(j) = w(j-2N-2.5) if j is greater than 2N + 2. Of course, Q(3N+2) corresponds to w(N-0.5), which has an arbitrarily assumed value. It is then a simple matter to write equations (29), (30), (35), and (36) in the form M(i,j)Q(j) = S(i), where M(i,j) is a 3N + 2 by 3N + 2 coefficient matrix, and S(i) is a column vector which does not contain any unknown terms. We take $M(3N + 2,j) = \delta(j,3N + 2)$ and S(3N + 2) = w(N-0.5). The matrix elements M(i,j) contain various combinations of the divergent Green functions H(n), but H(n) appears solely in expressions of the form H(n') - H(n), which are all finite. So, the matrix M(i,j) can be inverted, and all atomic positions determined, once the displacement of the (0,N-0.5) atom is assumed.

The atom at the (0,N-0.5) position must satisfy the following equation of equilibrium:

(37) $\alpha w(N-1.5) - (\alpha + 2\beta + 2\gamma + 2e'')w(N-0.5) + 2\gamma v(N-0.5) + 2\gamma v(N-1.5) + 2e''v(N+0.5) - 2\gamma u(N-1.5) + 2e'u(N+0.5) = 0$

By varying e,e', and e", one can relax the dislocation core into an equilibrium configuration which satisfies (37). Different values of the variable bond strengths correspond to different assumptions about the nonlinear interaction forces in the core.[3]

It can be seen from equations (29) and (30) that u(m,n) and v(m,n) for m > 0.5, or for $|n| > N + 0.5$ can be found from the already determined displacements u(n), v(n), and w(n) without any additional matrix inversion.

We now consider the strain energy. We start with a normal lattice, and cut all bonds crossing the line m = 0, and then calculate the work done in forcing the m = 0.5 and m = -0.5 columns to their final positions. At this point, the extra column at m = 0 is inserted, and the energies of the bonds connecting the atoms in the m = 0 column to the atoms in the m = 0.5 column and the m = -0.5 column must be added to the work previously done. Let $F_x(m,n)$ and $F_y(m,n)$ be the x and y components of the force acting on the atom at the position (m,n). The symmetries inherent in this dislocation configuration insure that $F_x(-m,n) = -F_x(m,n)$ and that $F_y(-m,n) = F_y(m,n)$, so the work done and the bond energies can be expressed in terms of u(0.5,n) and v(0.5,n), v(0,n), $F_x(0.5,n)$ and $F_y(0.5,n)$, with all other quantities eliminated. If $|m| > 0.5$, or $|n| > N + 0.5$, then all bonds are normal, and F = 0. We determine $F_x(0.5,n)$ and $F_y(0.5,n)$ by deleting all terms connecting the m = 0.5 column to the m = -0.5 column in (1) and

[2] J. Weertman and J. Weertman, Elementary Dislocation Theory, p. 36, MacMillan Co., N.Y. 1964

[3] D.M. Esterling, J. Appl. Physics, 47, 486 (1976)

(2). We take the m = 0.5 to be understood in all force expressions from now on. We find that, for $|n| < N - 0.5$:

$$(38)\quad F_x(n) = \alpha(u(1.5,n) - u(n)) + \beta(u(n+1) + u(n-1) - 2u(n)) + \gamma(u(1.5,n+1) + u(1.5,n-1) - 2u(n)) + \gamma(v(1.5,n+1) - v(1.5,n-1))$$

$$(39)\quad F_y(n) = \alpha(v(n+1) + v(n-1) - 2v(n)) + \beta(v(1.5,n) - v(n)) + \gamma(v(1.5,n+1) + v(1.5,n-1) - 2v(n)) + \gamma(u(1.5,n+1) - u(1.5,n-1))$$

By using equations (8) and (9), the two equations immediately above can be simplified considerably. We get, for $|n| < N - 0.5$:

$$(40)\quad F_x(n) = u(n) + \gamma w(n+1) - \gamma w(n-1) - 0.5b$$

$$(41)\quad F_y(n) = (\beta + 2\gamma)v(n) - \beta w(n) - \gamma w(n+1) - \gamma w(n-1)$$

By a similar procedure, we ascertain that:

$$(42)\quad F_x(N-0.5) = (\alpha + \gamma)(u(N-0.5) - 0.5b) + \gamma(v(N-0.5) - w(N-1.5))$$

$$(43)\quad F_y(N-0.5) = \gamma(-0.5b + u(N-0.5) - w(N-1.5)) + (\beta + \gamma)v(N-0.5) - \beta w(N-0.5)$$

$$(44)\quad F_x(N+0.5) = -0.5eb + eu(N+0.5) + e'v(N+0.5) - e'w(N-0.5)$$

$$(45)\quad F_y(N+0.5) = -0.5e'b + e'u(N+0.5) + e''v(N+0.5) - e''w(N-0.5)$$

Employing the results of equations (40) to (45), we obtain the following expression for the energy of the dislocation:

$$(46)\quad E = 2\sum_{k=0.5}^{N+0.5}(F_x(k)(u(k) + 0.5b) + F_y(k)v(k)) + 2\alpha\sum_{k=0.5}^{N-0.5}(u(k) - 0.5b)^2 + 8\gamma\sum_{k=0.5}^{N-1.5}(u(k) - 0.5b)^2 + 4\gamma(u(N-0.5) - 0.5b)^2 + 2d(u(N+0.5) - 0.5b)^2 + 2d(v(N+0.5) - w(N-0.5))^2 + 2\beta\sum_{k=0.5}^{N-0.5}(v(k) - w(k))^2 + 4\gamma\sum_{k=0.5}^{N-1.5}(v(k) - w(k+1))^2 + 4\gamma\sum_{k=0.5}^{N-0.5}(v(k) - w(k-1))^2$$

Maradudin derived analogous, although simpler, expressions for the displacements and energy of a screw dislocation.[4]

According to isotropic continuum elastic theory, the strain energy per unit length of an edge dislocation dipole is given by[5] (L = length of dipole, R = radius of cylinder in which the dislocation is assumed to reside, r = radius of exclusion at the core, ν = Poisson's ratio):

$$(47)\quad W = (\mu b^2/4\pi(1-\nu))\ln(L^2(R^2 - rL - L^2)/(r^2R^2)) \xrightarrow[R\to\infty]{} (\mu b^2/2\pi(1-\nu))\ln(L/r)$$

In the discrete model, it has been assumed that all bonds connected to atoms at positions (m,n) for $|m| > 0.5$ and/or $|n| > N + 0.5$ are undisturbed, thus eliminating all dependence on the size of the entire crystal. This corresponds to $R = \infty$ in the continuum limit. Note that E and W represent slightly different quantities: E is the total energy of the discrete dislocation, whereas W stands for the energy stored outside the excluded radius in the continuum case.

Results

We calculated the displacement components u(0.5,n), v(0.5,n), and v(0,n) for the case N = 32, b = 1, and $\lambda = -\mu/3$. These displacements are presented in the table at the top of the next page. The displacements are listed in order of increasing distance from the center of the dipole, with n = 0.5 being adjacent to the dipole center, and n = 32.5 being just past the end of the inserted center column. The appropriate continuum displacements are presented for purposes of comparison. In the continuum displacements u(x,y) and v(x,y), x is taken to be one, except for n = 32.5. The discrete displacements u(1.5,n) were also computed, and were found to converge to the continuum more rapidly than the displacements listed in the table.

The discrete strain energy E and the continuum strain energy W are found to be in good agreement if the excluded radius is taken to be 3.6b. However, W is the energy stored outside the excluded radius. The total continuum energy consists of the energy W stored outside a suitable excluded radius plus the (unknown) energy in the highly stressed material in the core. By contrast, in the discrete theory, there is no excluded radius, and there are no unknown energy values to be added.

4 A. A. Maradudin, J. Phys. Chem. Solids, 9, 1, (1958)

5 J. Hirth and J. Lothe, Theory of Dislocations, p. 138, McGraw-Hill Co., N.Y., 1968

TABLE ONE

n	Discrete u(n)	Continuum u(n)	Discrete v(n)	Continuum v(n)	Discrete w(n)	Continuum w(n)
32.5	-.11172	.14173	-.72014	-.34553		
31.5	.30120	.34775	-.31782	-.33483	.25000	-.46259
30.5	.38520	.43427	-.22524	-.31904	-.09242	-.35616
29.5	.43105	.45984	-.21547	-.29000	-.18244	-.30584
28.5	.45405	.47095	-.21236	-.26362	-.20404	-.27214
27.5	.46651	.47707	-.20664	-.24128	-.20495	-.24655
26.5	.47392	.48095	-.19871	-.22221	-.19890	-.22577
25.5	.47870	.48361	-.18966	-.20561	-.19038	-.20817
24.5	.48198	.48555	-.18019	-.19091	-.18101	-.19283
23.5	.48436	.48703	-.17068	-.17769	-.17147	-.17918
22.5	.48614	.48818	-.16131	-.16563	-.16203	-.16682
21.5	.48752	.48911	-.15218	-.15453	-.15282	-.15549
20.5	.48862	.48987	-.14331	-.14421	-.14387	-.14500
19.5	.48951	.49050	-.13470	-.13454	-.13520	-.13520
18.5	.49024	.49103	-.12636	-.12541	-.12680	-.12598
17.5	.49085	.49148	-.11827	-.11677	-.11866	-.11725
16.5	.49136	.49187	-.11041	-.10852	-.11075	-.10893
15.5	.49178	.49220	-.10276	-.10061	-.10306	-.10097
14.5	.49217	.49249	-.09531	-.09301	-.09557	-.09332
13.5	.49249	.49274	-.08803	-.08567	-.08826	-.08594
12.5	.49276	.49296	-.08090	-.07855	-.08111	-.07879
11.5	.49300	.49315	-.07393	-.07164	-.07411	-.07184
10.5	.49321	.49331	-.06709	-.06489	-.06725	-.06507
9.5	.49338	.49345	-.06036	-.05830	-.06050	-.05846
8.5	.49353	.49357	-.05373	-.05184	-.05385	-.05198
7.5	.49366	.49368	-.04718	-.04550	-.04729	-.04561
6.5	.49377	.49377	-.04072	-.03925	-.04081	-.03934
5.5	.49386	.49385	-.03432	-.03308	-.03440	-.03315
4.5	.49394	.49392	-.02798	-.02697	-.02804	-.02704
3.5	.49399	.49396	-.02168	-.02092	-.02174	-.02097
2.5	.49404	.49400	-.01543	-.01492	-.01548	-.01495
1.5	.49407	.49402	-.00924	-.00894	-.00928	-.00896
0.5	.49408	.49403	-.00311	-.00298	-.00309	-.00298

Conclusion

By assuming a discrete atomic lattice with nearest neighbor and next-nearest neighbor Hookean interactions, it is possible to solve for the displacements associated with an edge dislocation dipole, and these displacements are found to converge to the results of continuum isotropic elastic theory at points well removed from the dislocation core. The discrete model is an improvement over the continuum model in the core region because of the finite stresses and strains in the discrete model, but the assumption of Hookean forces near the core probably still leads to some overestimate of the strains and stresses.

By relaxing a larger number of interatomic forces in the core, it is possible to obtain a more realistic picture of the displacements. If we did not have to solve a self-consistent equation for the displacement field, the discrete elastic model would be highly advantageous, because only a few atomic bonds need to be relaxed and the matching of the core to the surrounding material could be achieved exactly as in the case of a screw dislocation.

ACCELERATING DISLOCATIONS: SOME RECENT RESULTS

Xanthippi Markenscoff
Department of Mechanical Engineering
Carnegie-Mellon University
Pittsburgh, PA 15213

I. Introduction

Accelerating, that is nonuniformly moving, dislocations have been first studied by Eshelby [1] and subsequently by several other investigators [2]-[4]. In a recent paper [5] the author analyzed in detail the stresses due to the general nonuniform motion of a screw dislocation starting from rest. By the same methodology the work has been extended to an edge dislocation starting from rest and moving thereafter nonuniformly. As a special case the stresses [5,8] and the radiated energy [6] from a dislocation starting from rest and moving with a constant velocity thereafter were obtained. With this treatment the solution to the stress field due to the general motion of a dislocation has been obtained in a usable form.

II. Solution to the General Transient Problem of a Nonuniformly Moving Dislocation

(a) Screw Dislocation

The problem is reduced to one in the half-space (Fig. 1) satisfying the differential equation

$$\frac{\partial^2 u_y}{\partial x^2} + \frac{\partial^2 u_y}{\partial z^2} = b^2 \frac{\partial^2 u_y}{\partial x^2}$$

and the boundary conditions at z=0:

$$u_y(x,0,t) = \frac{1}{2}\Delta u H(x) \qquad \text{for } t < 0$$

$$u_y(x,0,t) = \frac{1}{2}\Delta u H(x-\ell(t)) \qquad \text{for } t \geq 0$$

where H( ) denotes the step function, $x=\ell(t)$ the motion of the dislocation, and b the shear-wave slowness.

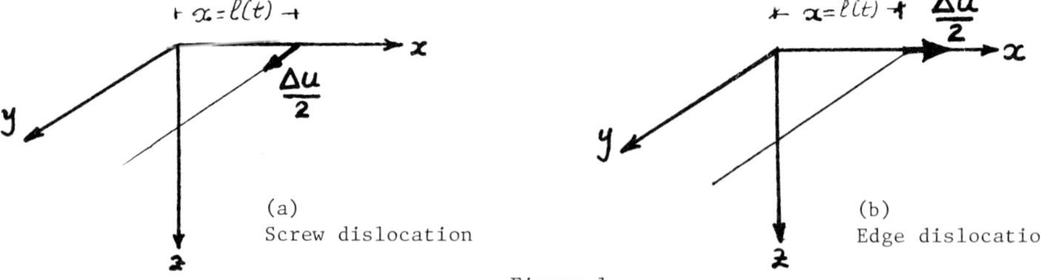

Figure 1

The problem is solved [5] by applying Laplace transforms in time and two-sided Laplace transforms in space. The inversion is carried out by Cagniard-de Hoop. The solution for $z \neq 0$ is obtained for the strain of interest $\frac{\partial u_y}{\partial z}$ in the form:

$$\frac{\partial u_y}{\partial z}(x,z,t) = \frac{\partial u}{\partial z}(x,z,t) = -\frac{\Delta u}{2\pi}\int_0^{\infty}\frac{(t-\eta(\xi))(x-\xi)^2 H(t-\eta(\xi)-rb)}{r^4[(t-\eta(\xi))^2-r^2b^2]^{1/2}}\,d\xi$$

$$+\frac{\Delta u}{2\pi}z^2\frac{\partial}{\partial t}\int_0^{\infty}\frac{(t-\eta(\xi))^2 H(t-\eta(\xi)-rb)}{r^4[(t-\eta(\xi))^2-r^2b^2]^{1/2}}\,d\xi - \frac{\Delta u}{2\pi}\frac{x}{x^2+z^2} \tag{1}$$

with $r^2 = (x-\xi)^2 + z^2$, and where $t = \eta(x)$ is the inverse function of $x = \ell(t)$. The function $\ell(t)$ can be any monotonically increasing function.

The argument of the step function $t - \eta(\xi) - rb = 0$ defines the limits of the integration. Naturally the solution is given in an integral form since the stress at any point will depend on the entire history of the motion.

The second term in (1) has to be brought to a form suitable for differentiating inside, since, as it is, carrying the differentiation inside the integral would yield a non-integrable singularity (see [5]). The solution is non-singular except, as expected, as one approaches the dislocation. For z=0 the solution has been derived independently and is found to be:

$$\frac{\partial u_y}{\partial z}(x,0,t) = \frac{\partial u}{\partial z}(x,0,t) = -\frac{\Delta u}{2\pi}\int_0^{\infty}\left\{\frac{(t-\eta(\xi))H(t-\eta(\xi)-|x-\xi|b)}{\sqrt{(t-\eta(\xi))^2-(x-\xi)^2b^2}\,(x-\xi)^2}\right.$$

$$\left.-\frac{(t-\eta(x)-\eta'(x)(\xi-x))H(t-\eta(x)-\eta'(x)(\xi-x)-|x-\xi|b)}{\sqrt{(t-\eta(x)-\eta'(x)(\xi-x))^2-(x-\xi)^2b^2}\,(x-\xi)^2}\right\}d\xi$$

$$+\frac{\Delta u}{2\pi}\frac{\operatorname{sgn} x}{t-\eta(x)}\sqrt{\left(\frac{t+\eta'(x)x-\eta(x)}{x}\right)^2-b^2}\;H(t-|x|b-\eta(x)+x\eta'(x)) - \frac{\Delta u}{2\pi}\frac{1}{x} \tag{2}$$

Here, because of the singular behavior of the integrand the singularity has been subtracted (and added) so that (2) is now integrable.

For the special case that the dislocation is starting from rest and moving with constant velocity, i.e. for $\eta(\xi) = \alpha\xi$ where α is the dislocation slowness, the solution is obtained in the form:

$$\frac{\partial u}{\partial z}(x,z,t) = \frac{\Delta u}{2\pi}\begin{cases}\dfrac{-x}{x^2+z^2} & \text{for } t < rb\\[2ex] \dfrac{-x}{x^2+z^2} + \dfrac{\left(b^2 - \dfrac{t^2x^2}{r_o^4} + \dfrac{z^2}{r_o^4}(t^2-r_o^2b^2)\right)\left(\alpha - \dfrac{tx}{r_o^2}\right) + \dfrac{2tx}{r_o^6}z^2(t^2-r_o^2b^2)}{\sqrt{t^2-r_o^2b^2}\left[\left(\alpha-\dfrac{tx}{r_o^2}\right)^2 + \dfrac{z^2(t^2-r_o^2b^2)}{r_o^4}\right]} & \text{for } t \geq rb\end{cases} \tag{3}$$

where $r_o^2 = x^2 + z^2$ and evaluating at z=0:

$$\frac{\partial u}{\partial z}(x,0,t) = \frac{\Delta u}{2\pi}\begin{cases}-\dfrac{1}{x} & \text{for } t < b|x|\\[2ex] -\dfrac{\sqrt{t^2/x^2 - b^2}}{\alpha - \dfrac{t}{x}}\dfrac{1}{|x|} - \dfrac{1}{x} & \text{for } t \geq b|x|\end{cases} \tag{4}$$

(b) Edge Dislocation

In the case of an edge dislocation the displacement is written in terms of the potentials ϕ and ψ, so that:

$$u_x = \frac{\partial\phi}{\partial x} - \frac{\partial\psi}{\partial z} \quad , \qquad u_z = \frac{\partial\phi}{\partial z} + \frac{\partial\psi}{\partial x}$$

and the problem reduces to the static one and superposition of the solution of

$$\frac{\partial^2\phi}{\partial x^2} + \frac{\partial^2\phi}{\partial z^2} = a^2\frac{\partial^2\phi}{\partial t^2}$$

$$\frac{\partial^2\psi}{\partial x^2} + \frac{\partial^2\psi}{\partial z^2} = b^2\frac{\partial^2\psi}{\partial t^2}$$

satisfying at z=0:

$$\left.\begin{aligned} u_x(x,0,t) &= \frac{1}{2}\Delta u\, H(\ell(t) - x) - \frac{1}{2}\Delta u\, H(x) \\ \sigma_{zz}(x,0,t) &= 0 \end{aligned}\right\} \text{ for all } t$$

when a and b are the dilatational and shear-wave slowness respectively. For the shear stress on the slip-plane, which is frequently of interest, it is obtained [8]:

$$\begin{aligned}\sigma_{xz} = \mu\Bigg[\frac{2\Delta u}{\pi b^2}\frac{\partial}{\partial t}\int_0^\infty & H(t-\eta(\xi)-ra)\Big(((t-\eta(\xi))^2z^2-((t-\eta(\xi))^2-a^2r^2)(x-\xi)^2) \\ & ((t-\eta(\xi))^2(x-\xi)^2-((t-\eta(\xi)^2-a^2r^2)z^2)-4(t-\eta(\xi))^2((t-\eta(\xi))^2-a^2r^2)(x-\xi)^2z^2\Big) \\ & \Big/\Big(r^8\sqrt{((t-\eta(\xi))^2-a^2r^2)}\Big)\,d\xi \\ - \frac{\Delta u}{2\pi b^2}\frac{\partial}{\partial t}\int_0^\infty & H(t-\eta(\xi)-rb)\Bigg(\Big(b^2 - \frac{2((t-\eta(\xi))^2(x-\xi)^2-(t-\eta(\xi))^2-b^2r^2)z^2)}{r^4}\Big)^2 \\ & - \frac{16(t-\eta(\xi))^2((t-\eta(\xi))^2-b^2r^2)(x-\xi)^2z^2}{r^8}\Bigg)\Big/\sqrt{((t-\eta(\xi))^2-b^2r^2)}\,d\xi\Bigg]\end{aligned}$$

In obtaining the above expression it was assumed that interchange of integration in λ and ξ was permissible, which happens to be the case for regions I and III (Figure 2). (See Markenscoff, 1980). Analysis similar to that for the screw dislocation must be performed for region II (Fig. 2). For region I (Fig. 2) the shear stress on the slip-plane z=0 is:

$$\begin{aligned}\sigma_{xz} = \mu\frac{2\Delta u}{\pi b^2}\frac{\partial}{\partial t}\int_0^\infty & \frac{-\sqrt{(t-\eta(\xi))^2-a^2(x-\xi)^2}}{(x-\xi)^4}(t-\eta(\xi))^2\,H(t-\eta(\xi)-(x-\xi)a)\,d\xi \\ - \mu\frac{\Delta u}{2\pi b^2}\frac{\partial}{\partial t}\int_0^\infty & \frac{(2(t-\eta(\xi))^2-(x-\xi)^2b^2)^2}{(x-\xi)^4\sqrt{(t-\eta(\xi))^2-(x-\xi)^2b^2}}\,H(t-\eta(\xi)-(x-\xi)b)\,d\xi\end{aligned}$$

In the first of the above integrals the differentiation in t can be carried inside, since the quantity $\sqrt{(t-\eta(\xi))^2-a^2(x-\xi)^2}$ appears in the numerator and there is no singularity involved. However, in the second integral there will be a term with $\sqrt{(t-\eta(\xi))^2-a^2(x-\xi)^2}$ in the denominator, for, if we write the numerator as

$$\left(2(t-\eta(\xi))^2-(x-\xi)^2b^2\right)^2 = \left[\left((t-\eta(\xi))^2-(x-\xi)^2b^2\right) + (t-\eta(\xi))^2\right]^2$$

and expand the square, then there will appear a term of the form:

$$\frac{\partial}{\partial t}\int_0^\infty \frac{(t-\eta(\xi))^4H(t-\eta(\xi)-(x-\xi)b)}{(x-\xi)^4\sqrt{(t-\eta(\xi))^2-(x-\xi)^2b^2}}\,d\xi$$

In order to bring this integral in a form suitable for differentiation inside, we write:

$$\frac{\partial}{\partial t}\int_0^\infty \frac{(t-\eta(\xi))^4\ H(t-\eta(\xi)-(x-\xi)b)}{(x-\eta)^4\ \sqrt{(t-\eta(\xi))^2-(x-\xi)^2b^2}}\ d\xi$$

$$= \frac{\partial}{\partial t}\int_0^\infty \frac{(t-\eta(\xi))^4-(x-\xi)^4b^4}{(x-\xi)^4\ \sqrt{(t-\eta(\xi))^2-(x-\xi)^2b^2}}\ H(t-\eta(\xi)-rb)\ d\xi + \frac{\partial}{\partial t}\int_0^\infty \frac{b^4H(t-\eta(\xi)-(x-\xi)b)d\xi}{\sqrt{(t-\eta(\xi))^2-(x-\xi)^2b^2}}$$

so that the first integral on the right-hand-side above is not singular any longer, and the second one may be integrated by parts as follows:

$$b^4\frac{\partial}{\partial t}\int_0^\infty \frac{H(t-\eta(\xi)-(x-\xi)b)d[(t-\eta(\xi))^2-r^2b^2]^{1/2}}{[-2(t-\eta(\xi))\eta'(\xi)-2(\xi-x)b^2]} = \frac{\partial}{\partial t}\left[\frac{b^4[t^2-r_o^2b^2]^{1/2}}{t\eta'(0)-xb^2}\right] H(t-r_ob)$$

$$+ b^4\frac{\partial}{\partial t}\int_0^\infty [(t-\eta(\xi))^2-r^2b^2]^{1/2}\ H(t-\eta(\xi)-rb)\ \frac{\partial}{\partial\xi}\frac{1}{\{(t-\eta(\xi))\eta'(\xi)+(\xi-x)b^2}\ d\xi$$

with $r_o^2 = x^2 + z^2$, where, now, the denominator does not vanish in the range of integration for subsonic motion and the differentiation in t can be carried inside the integral.

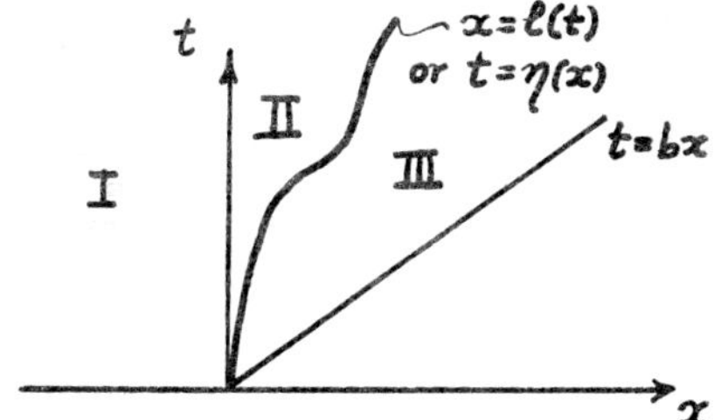

Figure 2

For the case of an edge dislocation starting from rest, and moving thereafter with constant velocity 1/d, for the stress σ_{xz} we have:

$$\sigma_{zx} = \mu(2\frac{\partial^2\phi}{\partial x\partial z} + \frac{\partial^2\psi}{\partial x^2} + \frac{\partial^2}{\partial z^2}) = 2\mu\frac{\Delta u}{\pi b^2}\cdot(\sqrt{(t^2-a^2r^2)}\ \ (r^6\ t\ x - d\ r^8)\ .\ ((t^4-a^2r^2t^2)z^4$$

$$+ (-6t^4 + 6a^2r^2t^2-a^4r^4)x^2z^2 + (t^4-a^2r^2t^2)x^4) + \sqrt{(t^2-a^2r^2)}\ \ (a^2r^8-r^6t^2)z((4t^3-2a^2r^2t)xz^3$$

$$+ (2a^2r^2t-4t^3)x^3z))/((a^2r^8-r^6t^2)^2\ z^2 + (t^2-a^2r^2)(r^6tx-dr^8)^2)$$

$$+ \frac{(\Delta u)\mu}{2b^2\ \pi}(((12t^5-24b^2r^2t^3 + 12b^4r^4t)x + 4dr^2t^4-8b^2dr^4t^2 + 4b^4dr^6)z^4 + ((8t^5-8b^2r^2t^3)x^3$$

$$+ (24b^2dr^4t^2-24dr^2t^4)x^2 + (4b^2r^4t^3-4b^4r^6t)x + 4b^2dr^6t^2-4b^4dr^8)z^2-4t^5x^5 + 4dr^2t^4x^4$$

$$+ 4b^2r^4t^3x^3-4b^2dr^6t^2x^2-b^4r^8tx + b^4dr^{10})/(\sqrt{(t^2-b^2r^2)}((r^6t^2-b^2r^8)z^2 + r^6t^2x^2-2dr^8tx + d^2r^{10}))$$

III. Radiation of Energy From a Nonuniformly Moving Dislocation

The radiation of energy out of the core of a screw dislocation being at rest and moving with a constant velocity, has been obtained by Clifton and Markenscoff [6] in a study of the effect of the energy radiation to the decay of the elastic precurser in plate-impact experiments.

If we consider an infinitesimal cylinder surrounding the dislocation and moving with it, the energy flux is given from [7], following the same notation,

$$\dot{W} = \lim_{S\to 0} \int [\sigma_{ij}\dot{u}_i n_j + U_n (\frac{1}{2} \sigma_{ij} u_{i,j} + \frac{1}{2} \rho u_i u_i)] dS$$

with

$$\sigma_{ij} = \sigma_{ij}^{applied} + \sigma_{ij}^{disl} ,$$

where σ_{ij}^{disl} is given by (3).

Performing the calculations we obtain

$$\dot{W} = \frac{-\mu(\Delta u)^2}{2\pi t} \left(\frac{1 - \sqrt{1 - U_d^2/C_2^2}}{\sqrt{1 - \frac{U_d^2}{C_2^2}}} \right) + \sigma^{appl} (\Delta u) U_d$$

where U_d denotes the velocity of the dislocation.

The left-hand side of this equation must be equated to the dissipated energy, while the first term in the right-hand side is the radiated energy and the second is the externally applied "driving force". It may be noticed that the same expression for $\dot{W}$ that is obtained for the circular contour is also obtained for a rectangular one. It was shown that $\dot{W}$ is independent of the shape of the contour as $S\to 0$.

For an edge dislocation the energy radiation as calculated both from a circular and a rectangular contour - (contour independence was also proven here) - is:

$$\dot{W} = \frac{-(\Delta u)^2 \mu}{2\pi t} [-2(1 - \frac{\alpha^2}{\beta^2}) + \frac{12 - 8\alpha^2}{\beta^2 (1-\alpha^2)^{1/2}} - \frac{12 - 20\beta^2 + 7\beta^4}{\beta^2 (1-\beta^2)^{3/2}}]$$

where $\alpha = \frac{a}{d}$, $\beta = \frac{b}{d}$ and the quantity inside the brackets was proven positive for all α and β for subsonic motion.

Conclusions

We presented here the solutions to the general transient motion of screw and edge dislocations and described the mathematical methodology used in the treatment of transient problems of moving singularities. Although here the motion was assumed subsonic, the motion of a supersonic dislocation was also obtained in [9]. An open problem that we hope to be able to solve in the future is the energy radiation for the general nonuniform motion of a dislocation.

Acknowledgments

Some of the algebraic computations were performed with the use of the MACSYMA program at M.I.T. This work was partially supported by NSF Grant CME-8007766 to Carnegie-Mellon University.

References

[1] J.D. Eshelby, Phys. Rev., 90, p248 (1953).

[2] J. Kiusalaas and T. Mura, in Recent Advances in Engineering Sciences, Vol I (ed. by A.C. Eringen) Gordon-Breach, NY, p543 (1964).

[3] T. Mura, in Dynamic Response of Structures (ed. by G. Herrmann and N. Perrone), Pergamon Press, p345 (1971).

[4] A. Roy, Int. J. Sol. Str., 14, p755 (1978).

[5] X. Markenscoff, J. Elasticity, Vol 10 (1980).

[6] R.J. Clifton and X. Markenscoff, (in preparation).

[7] L.B. Freund, J. Elasticity, 2, p341 (1972).

[8] X. Markenscoff and R. J. Clifton (in preparation)

[9] C. Callias and X. Markenscoff Quart. Appl. Mech., (in press).

ATOMIC MODELS OF SINGLE DISLOCATIONS

M.P. Puls
Materials Science Branch
Whiteshell Nuclear Research Establishment
Atomic Energy of Canada Limited
Pinawa, Manitoba Canada R0E 1L0

Abstract

The models and techniques used to simulate the atomic configuration in the core regions of single dislocations in metallic and ionic materials are reviewed. An assessment is made of the contribution that atomic dislocation models have made to our understanding of the interaction of dislocations with the lattice (dislocation splitting and the Peierls-Nabarro stress) and with point defects (solution hardening). These interactions are important elements in models of the mechanical strength of materials. It is seen that the techniques have evolved to the point where the principal limitation to the accuracy of the results resides in the reliability of the interatomic potentials. These are currently not very rigorously determined and reliance has mostly been placed on empirically based potentials derived from perfect lattice, macroscopic properties. This limitation is perhaps not so severe in simple ionic materials, where the bulk of the interaction is dominated by the Coulomb interaction. Further progress in this field must await the development of suitable potentials derived from rigorous quantum mechanical models.

Introduction

Accurate determinations of the core configurations of dislocations are important in quantitative evaluations of the interaction of dislocations with the lattice (dislocation splitting and the Peierls-Nabarro stress) and with point defects at close range (solution hardening). Elastic continuum models of dislocations are totally inadequate to give information about the above phenomena and recourse has to be taken to discrete or atomic models. Various models have been developed to take account of the atomic nature of the dislocation's core. These have recently been reviewed by Bullough and Tewary (1). In the following review, we will restrict our discussion to atomic dislocation models in which the core atoms are taken to interact via an explicit (non-linear) interatomic potential. In general, these types of models require the use of large, high-speed computers. Owing to the very explicit nature of the method of determining the core configuration, these models often give the appearance of greater realism compared to other, less direct approaches, such as the partially discrete or lattice statics methods. Although this view is to some extent justified, it is seen that some of the core properties are very sensitive to the assumed interatomic potentials. Thus caution has to be exercised in believing the quantitative conclusions derived from these atomic models.

In this paper we critically review the models and techniques used to simulate the atomic configuration in the core regions of single dislocations in metallic and ionic crystals. In view of the critical importance of the potential to the predictive capabilities of these models, this aspect will be given added attention. Surveying the applications, we will restrict ourselves mostly to work carried out from 1974 onwards. Much of the work up to that period has been discussed in various reviews by, for instance, i) Vitek (2), dealing exclusively with the bcc metals; ii) Perrin (3), emphasizing methods of calculation and applications to fcc metals; iii) Lidiard and Norgett (4), emphasizing models and methods in ionic crystals (but with applications only to point defects); and iv) Bullough and Tewary (1), emphasizing a broad spectrum of lattice models.

AECL 6929

The plan of the paper is as follows. We start with a section on methods of calculation which outlines the overall approach to the problem and assesses the methods developed to solve it. The second section deals with the important and vexing problem of choosing a suitable interatomic potential. The subsequent review on the applications is divided into a section on i) core structure analysis, ii) the Peierls stress, iii) calculations of the volume of expansion and iv) point-defect/dislocation interactions. Various methods of evaluating the latter three quantities are also critically reviewed in the appropriate sections. The paper concludes with an overall assessment of the limitations and successes of atomic dislocation models and the contributions that they have made to our understanding of the mechanical behaviour of materials.

Methods of Calculation

There is a standard approach to this problem. This involves dividing the crystal into two regions: the inner region (region 1) containing the defect where the atoms will be allowed to interact via a given non-linear potential and a boundary region (region 2) simulating the rest of the crystal and which should be sufficiently far from the defect center so that the atomic positions can be given by a harmonic or linear elastic approximation. From a computational efficiency point of view, it is desirable to make the size of region 1 as small as possible. On the other hand, the boundary region should not inhibit the relaxation of region 1. This can be accomplished in a variety of ways. The simplest approach is to make the size of region 1 very large, leaving region 2 fixed in its initial Volterra field configuration as region 1 is relaxed. This is called a rigid boundary method. It has the advantage of simplicity but does not fulfill the requirement of computational efficiency. To achieve the latter, the boundary atoms must be able to respond in some way to the new configuration achieved during the relaxation of region 1. This is done by means of so-called flexible boundary techniques, a variety of which have been developed over the last few years. These have been reviewed by Perrin (3) and Vitek (2). A method developed most recently is the Flex-II technique of Hoagland et al. (5). According to comparisons made by Puls and Woo (6) and by Sinclair et al. (7), this latter technique seems to be the best, with Sinclair's approach (called Flex-S by Sinclair et al. (7)) a close second.

In the Flex-II, as in the Flex-S method, region 2 is further subdivided into regions 2A and 2B. This is necessary because it is desired to monitor the forces on all boundary atoms which are within range of region 1 (region 2A). This requires the extension of region 2 to 2B to provide each atom in 2A with a full complement of neighbours. The Flex-II method uses an iterative scheme to adjust the positions of all the atoms (regions 1, 2A and 2B) by means of a Green's function after a rigid boundary relaxation of region 1 has occurred. The source forces of the anisotropic linear elastic Green's function are the forces on the atoms of region 2A. For a single dislocation, periodicity along the dislocation line makes it convenient to use a line force solution such as derived by Hirth and Lothe (8). For small source forces, the assumption of linear superposition is good and the resultant displacement of an atomic row due to a number of source forces is a linear combination of the displacements due to the individual source forces acting alone. Thus, if an exact Green's function can be found and the assumption of linear superposition is valid, a single application of this function should be able to remove the residual forces in region 2 without bringing in new forces in region 1. However, both of these conditions are not likely to be satisfied in an actual calculation, because an exact Green's function would be difficult to obtain, and furthermore, the residual forces in region 2 would not be small unless one used a very large region 1. Nevertheless, with a reasonable approximation to the Green's function and a region 1 that can comfortably contain the core, the residual forces in region 2 may generally be reduced without introducing large forces into region 1. When the new equilibrium positions of the region 1 atoms are computed again by minimization of the region-1 energy and forces with respect to the new boundary, the source forces imposed on region-2 atoms are generally smaller than those in the previous iteration. As the iteration goes on, the error introduced by the approximate Green's function and the assumption of linear superposition decreases steadily and in general, convergence is obtained after a few iterations when all forces in the entire crystal have been reduced below a specified limit. Further details of the method are given by Hoagland et al. (5), Sinclair et al. (7) and Puls and Woo (6).

A departure from the above linear elastic flexible boundary methods is the method developed by Petrasch and co-workers (9-11) based on a non-linear anisotropic elastic solution of the dislocation field (11). In common with the linear elastic solution, this field contains some constants not given by elasticity theory which must be obtained, iteratively, from the relaxed coordinates of region 1. The method used by them of obtaining these constants is essentially a non-linear variant of Flex-I, originally developed by Gehlen et al. (12) in the isotropic approximation, whereby the traction acting across a circular path lying within region 1 is calculated and elastic displacements are imposed on the outer region corresponding to these tractions. The Flex-I method has been found to be much less efficient than either Flex-S or Flex-II (7) and it thus seems likely that the same may apply to the non-linear technique. However, since no direct comparison has so far been carried out between these methods, it is not possible to state this with complete certainty.

There are two main methods used to relax the lattice in region 1. These are the static and the dynamic relaxation techniques. The static methods are basically gradient methods for minimizing a function of many variables such as the total energy of region 1. These have been reviewed by Perrin (3) and Lidiard and Norgett (4). For dislocations, the conjugate gradient methods turn out to be the most useful. A faster technique called the variable metric method is not really suited to dislocation core studies because of the considerable increase in storage requirements. In point-defect studies this can be reduced because of the greater symmetry around the defect. The dynamic relaxation technique uses the forces to solve the Newtonian equations of motion which are then periodically quenched out as the kinetic energy reaches some maximum value. The relaxation proceeds in an iterative fashion. The technique has the advantage that finite temperatures of the lattice can be simulated (by imposing a Maxwellian velocity distribution). It is also claimed to be less susceptible to finding metastable equilibrium states. However, there seems to be little concrete evidence to support this view. On the other hand, the dynamic method appears more cumbersome to use, on occasion requiring some fine tuning to maximize its efficiency (5).

Interatomic Potentials

Of crucial importance in the atomistic models is the interatomic potential. The problem of deriving suitable and reliable potentials has been a source of continuing difficulty in these calculations and, at the present time, represents their outstanding weakness. This weakness has been a problem throughout the history of these calculations and considerable effort has been devoted to the derivation of improved potentials. A number of reviews and discussions have appeared on this subject in recent years (1, 4, 13-17). The problem is somewhat different for the metals than for the ionics and, in the following, it will be convenient to discuss these separately. We will start with the metals. These represent the bulk of the materials studied in dislocation modelling, yet paradoxically, also present the greatest problem.

Metallic Crystals

The major difficulties are the lack of any direct experimental methods which can probe the full range of interactions needed in defect calculations. In addition, we have presently only a very limited ability to calculate interatomic potentials from first-principle quantum mechanical methods. The simplest way out of this dilemma involves, first of all, making two important simplifying assumptions: 1) atoms interact only by pair-wise forces and, 2) the range of the interactions can be limited to a few nearest neighbours. Based on these assumptions, one approach is to construct a pair potential having an assumed analytic form (which has some basis in atomic theory) and containing constants which are determined by relating the potential and its derivatives to various measured or calculated physical properties. Potentials of this type are the Morse, Lennard-Jones and Born-Mayer potentials or variants of these. They are often called semi-empirical potentials. Another approach is to allow the functional form of the potential to vary and to determine its actual form by fitting to various physical properties. This is a total empirical, curve-fitting method which has been popularized by Johnson (13-14). Potentials of this type are frequently called Johnson potentials and are most conveniently represented by composite polynominal expressions. Neither of these potentials considers the presence of the conduction electrons and their interaction explicitly. They are thus incapable of modelling the special features that distinguish metallic from covalent or ionic interactions.

One of the features of conduction electrons in metals, and one which makes a theoretical treatment so difficult, is that these electrons give rise to long-range many-body interactions and a volume-dependent potential energy. In the Johnson or semi-empirical potentials, the hope is that these electron gas effects have either been adequately factored into the curve or constant-fitting procedure, or properly separated out by including the free-electron contribution as a separate volume-dependent term. Various efforts have been made to include the effect of electrons in a more rigorous manner. The most useful of these is based on the pseudopotential theory of metals (18). The essence of this method is to replace the strong attractive potential experienced by the conduction electrons in the core region by a weak pseudopotential without changing the energies and the electronic distribution in the region outside the core. Because this bare-ion potential is much weaker than the original potential, it can be treated by perturbation theory. To second-order perturbation within the linear screening approximation for the electron-electron interaction, the total energy of the crystal can be written as a volume-dependent sum depending on the free electron energy and a second sum, also volume dependent (but to a lesser extent), which contains the changes in the electron energy due to the structure of the metal. This latter part can be represented as the sum of pair potentials. Thus the potential has a form similar to the above empirical and semi-empirical potentials and this result of second-order pseudopotential theory has been used to lend some theoretical validity to the empirical methods. Nonetheless, it sould be noted that the potential derived from pseudopotential theory has, in general, a slowly-decaying oscillatory tail (the "Friedel" oscillations) and can not really be represented by a short-ranged potential. In addition, a careful evaluation of the basic assumptions of the perturbative pseudopotential model, as applied to the vacancy formation energy, reveals some serious shortcomings of the method. These shortcomings and the reasons for them have been discussed by a number of workers recently (19-22). It appears that the method works fairly well for simple monovalent metals but becomes increasingly less reliable as the valency and the complexity of the electronic structure of the valence band increases. Prospects for improvement within the pseudopotential formalism for the transition metals (or even for a simple trivalent metal such as aluminum) applied to defects such as vacancies, dislocations, surfaces and cracks which produce large inhomogeneities in the electron distribution, appear not too promising (12-22). It must also be noted that the laudable efforts to place defect calculations on a firmer theoretical footing have resulted in potentials which have become exceedingly more complex (and long ranged!). Consequently they are ill-suited to defect calculations, requiring extensive and very non-symmetric rearrangements of atoms from perfect lattice positions such as are involved in most models of dislocation cores.

Ionic Crystals

Compared to the metals, the situation in ionic materials is much improved. In these materials, there are no free electrons and the major component of the total energy of the crystal comes from the Coulomb interaction of the atoms modelled as point ions. The problem is further simplified because the closed-shell nature of the electron distribution around each ion means that the bonding is fairly non-directional. Thus the short ranged ion core interactions can be given by simple central-force pair potentials such as the Born-Mayer and van der Waals potentials. This is one of the reasons that the first atomic model of a dislocation was applied to NaCl (23).

Although the functional form of the Coulomb potential is known accurately, its long range and the existence of charges on the ions can cause problems. Thus electrical polarization can be large and it is important that the potential reproduce the dielectric behaviour of the material accurately. Early models of ionic crystals were deficient in this description, but the introduction of shell models removed much of this difficulty (4). In the shell model, each ion is divided into a separate charged core and shell to describe ionic polarization. There is no Coulomb interaction between core and shell of the same ion, only an harmonic spring coupling. Because the interatomic forces act through the shells, the short range forces are modified by ionic polarization. This simple shell model is capable of describing both the elastic and dielectric properties of an ionic solid. Since it is a central-force, pair interaction model, its applicability is strictly limited to solids for which the Cauchy condition $C_{12}=C_{44}$ on the second-order elastic constants is obeyed. Deviation from this condition can be accounted for by the breathing shell model (24,25). The effective ion separation in this model is modified by the symmetric breathing deformation of the shells; the shell radius responds harmonically to the total effective overlap pressure of all surrounding ions. In this way, manybody forces may be simulated and the model need not satisfy the Cauchy condition.

Although attempts have been made to justify the shell models in terms of more rigorous quantum mechanical models (26), it should be kept in mind that the above are, like the metal potentials described earlier, essentially empirical models with parameters determined by fitting to various physical properties that can define these parameters mainly at close-to-perfect lattice spacing. Because the available physical data is always exceeded by the required number of parameters, a certain amount of arbitrariness creeps into the determination of the short range potentials. A frequent difficulty in many ionics is the attractive second neighbour anion-anion interaction, such as the F^--F^- interaction in the fluorides or the O^{2-}-O^{2-} interaction in the oxides, and discussions for improving the determination of these are given in a number of recent publications (27,28). These calculations are greatly aided by supplemental quantum mechanical calculations that determine the repulsive part of the second neighbour interaction. A further difficulty with an empirical potential model such as the shell model is that the limitations of the potential are generally difficult to pinpoint and quantify. As a result, it is difficult to estimate what influence these limitations will have on different defect properties. Nonetheless, the large body of recent work on point-defect properties (16,17,27,29,30) attests to the reliability of the shell model. This reflects the fact that the defect energies are dominated by the Coulomb and nearest-neighbour short range interactions and by lattice and electronic polarizations; these are all well described by the shell model.

The Core Structure

Body Centered Cubic Metals

By far the greatest volume of work on atomic models of dislocations has been carried out on bcc metals. This strong interest arose out of an effort to explain certain plastic deformation properties of bcc metals which differed considerably from those of the fcc metals; for instance, a rapid increase of flow stress with decreasing temperature and a complete breakdown of the Schmid law of critical resolved shear stress. The differences in the plastic properties of bcc as compared to fcc metals were conjectured to relate to differences in the core structure of dislocations in these metals. Observations suggested that screw dislocations were less mobile than edge dislocations. It was suggested that, at low temperatures, a Peierls stress mechanism was responsible for controlling the mobility of the dislocations. Hirsch (31) was the first to suggest that the core of the screw could be dissociated onto three intersecting {112} planes. As a result, these 1/2<111> dislocations were studied most extensively, although non-screw dislocations with the same Burgers vector were also investigated. In addition, because of its simple form and because it is thought to serve as a nucleus for a crack, calculations on the core structure of the sessile <100> edge dislocation were also carried out. The work done up to 1974 has been reviewed in detail by Vitek (2). Vitek concluded that the atomistic results indeed showed the unique role played by the 1/2<111> screw dislocation. This dislocation was seen to have a core, characteristic of the bcc structure, which extended into several non-parallel {110} and {112} planes of the <111> zone. As a result, the Peierls stress was always many times higher than that of the non-screw dislocations on {110} planes, which were consistently found to have very planar cores. All the Peierls stress calculations show that the screw dislocations are fairly sessile and must undergo large core changes before they can move. In addition, they possess an asymmetry connected with the twinning-antitwinning asymmetry of {112} planes. The above were observed to be features fairly independent of the potential used and thus could be construed as characteristic of the crystal structure. However, the exact details of the width and asymmetry of the core did depend on the details of the potential used, i.e., one might conclude that these are material-dependent properties. It is to be noted that the majority of the calculations reported were carried out with empirical, Johnson-type potentials, many of which were constructed to have a rather arbitrary form, i.e., no attempt was made to represent any particular type of metal. The sensible view was taken that no great significance would be attached to the absolute value of physical properties such as the Peierls stress derived by means of these potentials. Rather, an effort was made to distinguish those properties which were independent of the functional form of the potential used. In addition, many of the calculations on the 1/2<111> dislocations were carried out using a rigid boundary method. In the case of Peierls stress calculations, this method could lead to possible over-estimates of this quantity.

In the following, we will give a brief review of the work carried out since 1974. The major thrusts of this work have been: i) improvements in the potentials used, ii) improvements in the method of calculation, and iii) efforts at filling in some of the gaps left in the study of different dislocation types. There have also been a more detailed look at the orientation dependence of dislocation motion under a variety of applied stresses and a number of studies of point-defect/dislocation interactions. These latter topics will be treated in separate sections further on.

We start with the results for the 1/2<111> screw dislocation. It is worth noting that there are stable as well as unstable (metastable) locations for the center of the dislocation. The stable position is obtained when the sense of the screw dislocation and that of the immediately surrounding atoms is the same. The configuration (for the case when the ions are placed according to linear elasticity theory) is shown in Fig. 1 using the differential displacement method of Vitek et al. (32). In this method, relative displacements in the direction of the Burgers vector, b, of neighbouring atom rows are shown in appropriate projections of the crystal as arrows joining the corresponding rows. The arrow lengths are normalized to the range ±b/2 and scaled. This method is particularly useful in describing the core field of both screw and edge dislocations. As a result, most authors have tended to report their results in this form and our perceptions of their results, particularly in comparison between different models, will be based on this representation.

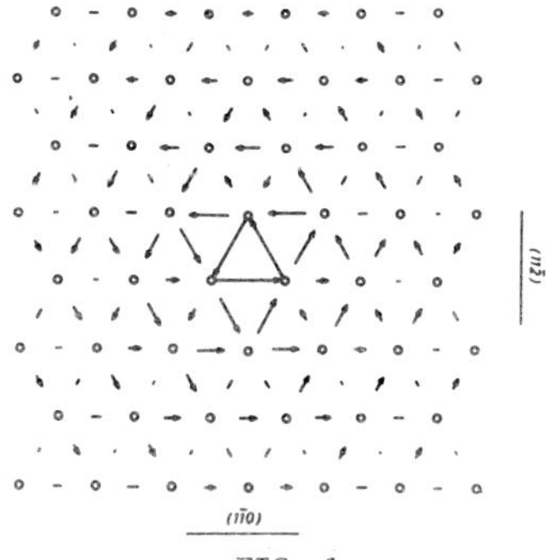

FIG. 1

Core structure of a 1/2<111> screw dislocation according to linear anisotropic elasticity theory; projection of {111} planes with atom rows shown as circles. Relative displacements, shown as arrows, represent actual displacements along <111>.

First of all, we report on studies by Hübner (33) and Masuda and Sato (34,35) which have attempted to use more rigorously based potentials. In the approach used by Hübner (33), the energy of the lattice is calculated within the second-order pseudopotential theory using a local pseudopotential representing various alkali metals. It is assumed that atom displacements perpendicular to the dislocation line can be neglected and that, moreover, the presence of the screw dislocation generates no dilatation of the lattice. This greatly simplified the calculation. The method of calculation was based on the rigid boundary, gradient relaxation techniques. As in calculations with empirical potentials (2), two energetically equivalent core configurations, shown in Fig. 2 were obtained. The core was very narrow but spread out over three {110} planes, similar to the configuration of Vitek et al. (32) obtained with the J_1-potential. The motivation of the work of Sato and Masuda (35) was to

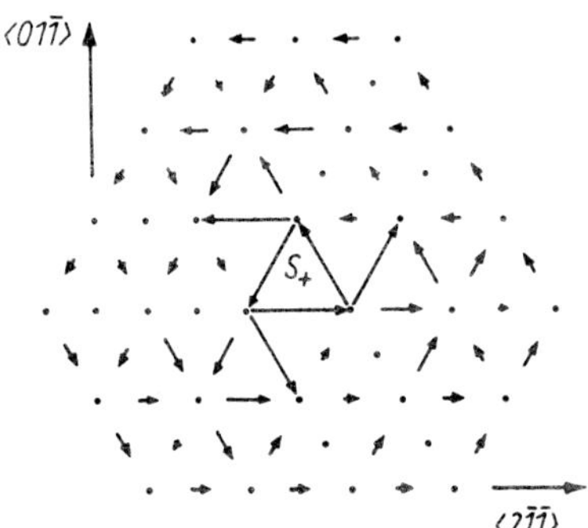

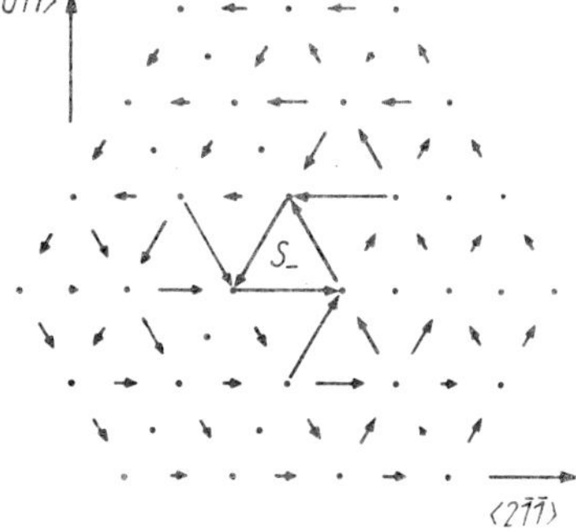

FIG. 2

Two energetically equivalent core configurations of a 1/2<111> screw dislocation calculated using an atomic model (33). The dots represent atom rows; same projection as in Fig. 1.

explain the difference between the orientation dependence of the Peierls stress for W, Mo and Ta compared to that for α-Fe. A possibility was that this difference could be attributed to the difference in the number of d-electrons between these metals (4.5 as compared to 7 for α-Fe). A (many-body) potential was developed which takes into account the density of states of the d-electrons. The repulsive core-core interaction energies were given by a Born-Mayer potential fitted to the lattice parameter and the elastic constants. As in Hübner's work (33), only displacements along the dislocation line were permitted. Moreover, the rigid boundary method made use of a fairly small region 1 (96 rows). The orientation differences calculated for the Peierls stress agree with experimental findings. However, the differences are small and probably less than the accuracy of the calculations. Agreement with experiment must therefore be considered as tentative. The only configuration given in the paper shows a core for α-Fe which is almost isotropically spread over a considerable distance as shown in Fig. 3. Only one stable configuration was found. This latter result is similar to that of Takeuchi (36) who used a series of empirical row potentials (no relaxation perpendicular to the dis-

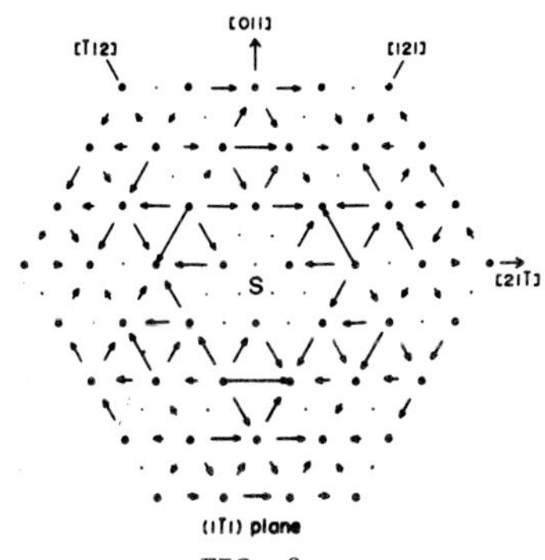

FIG. 3

Non-degenerate core configuration of a 1/2<111> screw dislocation calculated atomistically by Sato and Masuda (35) for α-Fe.

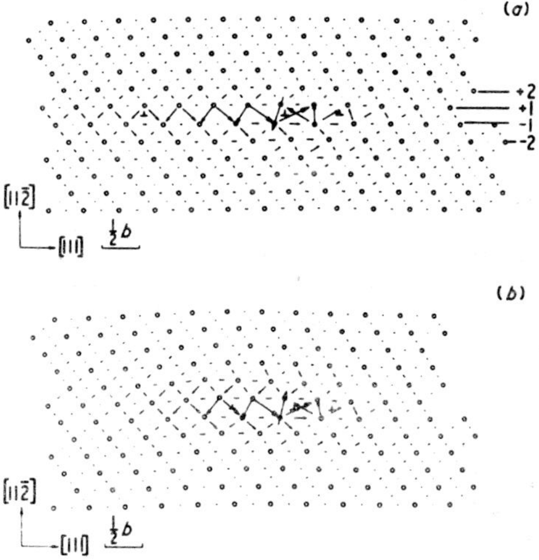

FIG. 4

Core configurations of the 1/2<111>{112} edge dislocation calculated atomistically by Yamaguchi and Vitek (40) using two different potentials [(a) and (b)]. Estimated positions of the fractional dislocations are indicated by ⊥.

location line) and a rigid boundary method with fairly large region-1 size. By varying the parameters in his potential he was able to obtain either doubly-degenerate (i.e., two energetically equivalent) cores, or a non-degenerate core. The non-degenerate core was always accompanied by the existence of a metastable trebly-degenerate split configuration. It is claimed that the deformation behaviour of α-Fe, in particular the asymmetry of slip on {112} planes, can be more easily explained by the presence of a non-degenerate core structure. An improvement in the method of calculation was implemented by Wüthrich (37) who used the Flex-S technique of Sinclair (38). As a result, he was able to estimate the volume expansion associated with this dislocation. This will be discussed further in a later section. The potential used was an empirical potential due to Chang and Graham (39). Again, a doubly-degenerate core split on three planes was obtained.

In addition to the 1/2<111> screw dislocation, various 1/2<111> non-screw dislocations have been investigated. Yamaguchi and Vitek (40) studied the core structures of non-screw dislocations on {112} planes using a series of empirical potentials and a rigid boundary method. It was found that non-screw dislocations on {112} planes are always planar and, like non-screw dislocations on {110} planes, appear to be dissociated into three fractional dislocations. The cores of the dislocations are spread over three parallel {112} planes and the twinning-antitwinning asymmetry for shear on {112} planes is reflected in the asymmetry of the core. This is shown in Fig. 4 for the edge dislocation. Using an empirical potential for α-Fe due to Johnson (41), Adams et al. (42) carried out a Flex-II calculation of the core of the 1/2<111>{110} edge dislocation. Comparison was made between the results of the Flex-II and the rigid boundary methods. The Flex-II method enabled a calculation of the volume expansion which is suppressed in the rigid boundary method. An interesting feature of this particular dislocation is the shearing of atomic rows along the dislocation line direction upon relaxation. These displacements are antisymmetric about the slip plane and have been observed in previous investigations based on the rigid boundary method (43-45). The Flex-II results show, however, that this field contains significant contributions from the region-2 relaxations. Aside from this result, however, differences between the core configurations based on the Flex-II and the rigid boundary method were small. Yamaguchi and Umakoshi (46-48) studied the <001> screw dislocation using various empirical potentials and the rigid boundary method. For the model CsCl-type ordered lattice, they reported a planar core, slightly dissociated into two fractional dislocations consisting of edge components. Slip occurs on {110} planes upon application of a shear stress.

Ionic Crystals

Compared to the metals (particularly the bcc metals), dislocation studies in ionic crystals have received far less attention. Despite this, new methods or advances in technique were frequently tested first on dislocations in ionic crystals, as can be seen by the following examples: i) the first atomic calculation of a dislocation core by Huntington et al. (23); ii) the first atomic calculation of point-defect/dislocation interaction energies by Bassani and Thomson (49); iii) the first application of lattice statics to dislocations by Maradudin (50); iv) the first attempt to use an atomic dislocation model to calculate the energy of kinks and jogs by Eisenblätter (51,52); v) the first application of a non-linear elastic representation of the boundary atoms in an atomic model by Petrasch and co-workers (9-11); vi) the first application of Flex-II by Hoagland et al. (5); and vii) the first calculation of the complete Peierls barrier for a straight dislocation under stress by Woo and Puls (53-55).

The main reason for this tendency to test new methods and models in these materials is that the potential interaction for ionic crystals is known with much greater certainty than for the metals (for a discussion see the section on potentials). A second, lesser reason is that the mechanical and physical properties of the simple ionic crystals used in the above studies have all been well characterized.

The most detailed and extensive work on dislocation modelling in ionic crystals in recent years has been carried out by Puls and co-workers (6,53-62). A prime objective of this work was the development of a model to calculate point-defect/dislocation interaction energies (59,60,62). However, in the course of the work leading up to this, a detailed comparison was made between the rigid boundary, Flex-S and Flex-II methods (6,58), various potential models were evaluated (56-58), and a new method was developed to calculate the Peierls stress (53-55). Similar work which emphasized comparison between boundary methods was carried out by Hoagland et al. (5) and Petrasch and co-workers (9,10).

All of the studies so far have concentrated on ionic crystals with the fcc structure and the majority have limited their efforts to the 1/2[110]{110} edge dislocation. Except for the original work of Huntington et al. (23), the work of Polisar et al. (63) and brief notes by Raupach (64) and Granzer et al. (9) there has been a dirth of studies on the 1/2[110] screw dislocation or other dislocation orientations. This reflects the greater difficulty of calculating Coulomb sums in cases where the dislocation is along crystal directions for which the ion rows are not alternately charged. We note that the 1/2[110]{110} edge dislocation consists of two extra half planes of {110} type and that the atom rows along the dislocation line direction are alternately charged. Because two planes are inserted to form the dislocation, there are two possible symmetric configurations of the dislocation core as shown in Figs. 5a and 5b. Unfortunately, there has been some confusion in the nomenclature of these symmetry types. Thus Huntington et al. refer to the configuration of Fig. 5a as type II and of Fig. 5b as type I, whereas in the papers by Granzer and co-workers and by Puls and co-workers, the opposite notation is used. The notation of Hoagland et al. (5) follows this latter trend, but he substitutes A for I and B for II. Whitworth (65), in his review article, uses the original designation of Huntington et al. (23). The designation shown in Figs. 5a and 5b seems to enjoy greater popularity and we will thus continue to use this in the following.

A fairly careful analysis was made of the core structure of the 1/2[110]{110} edge dislocation in KCl by Hoagland et al. (5) and in MgO by Woo and Puls (6,58). The results of Hoagland et al. are based on a simple point ion model (PIM) whereas those of Woo and Puls are obtained using the more elaborate breathing shell model (BSM). It is shown by Woo and Puls, however, that the core configurations obtained with the PIM and the BSM differ little as long as the potential is fitted to the shear moduli $1/2(C_{11}-C_{12})$ and C_{44} (in the case where the Cauchy condition is violated). In KCl, the Cauchy condition

FIG. 5

Ion arrangements for an, a) type I and, b) type II 1/2<110>{110} edge dislocation in an fcc ionic crystal. The small and large circles represent cations, and anions, respectively.

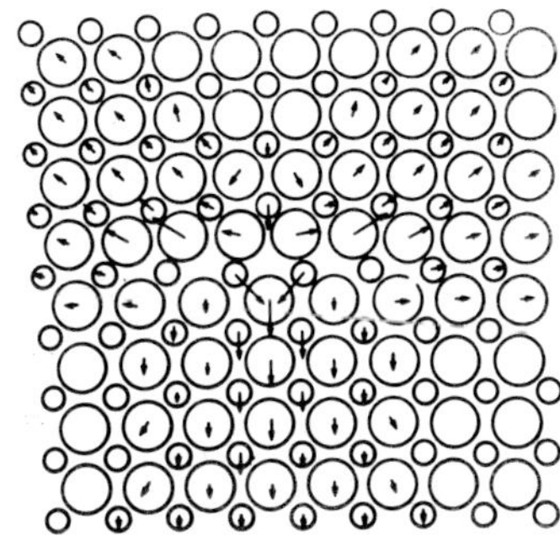

FIG. 6

Atomistically determined core configuration of a 1/2<110>{110} edge dislocation in MgO (type I symmetry) showing the atom row displacements (x5) due to relaxation from the initial anisotropic Volterra field (58).

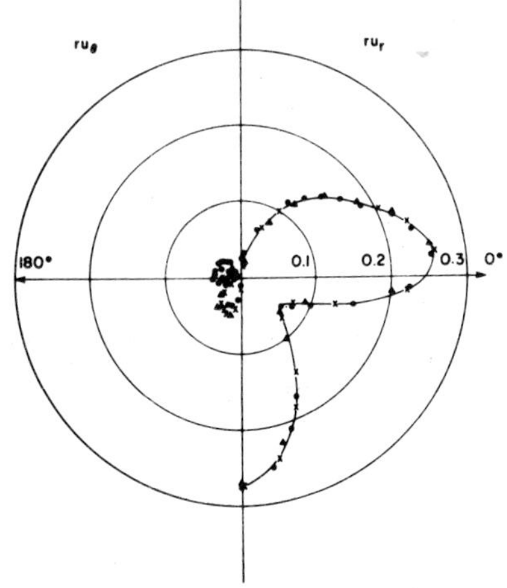

FIG. 7

Polar plots of the ru_r (right side) and ru_θ (left side) components of the core field (the difference between the atomistically relaxed and the initial linear elastic coordinates) for the same dislocation as shown in Fig. 6 (point ion model, Model 1 potential). The field was evaluated at different radii from the dislocation center: ▲, R=4b; X, R=5b; ●, R=6b. The axes are given in units of half the lattice parameter.

$C_{12}=C_{44}$ is fairly closely obeyed and consequently little error is anticipated in using the PIM. Both the KCl and the MgO calculations yield the result that type I symmetry is the stable configuration with type II situated on top of an energy barrier (the shape of this barrier is discussed in more detail in the section on Peierls stress). In representing the core configuration, changes in the structure from the Volterra configuration after the lattice has relaxed, as shown in Fig. 6, were emphasized. This was done because one of the objectives of the studies was a comparison between methods and potentials. A more mathematical representation of the core by means of ru_r and ru_θ plots is shown in Fig. 7 where u_r and u_θ are coordinate differences in the radial and circumferential directions, respectively, between the initial Volterra lattice and the final relaxed configurations; r is the distance from the dislocation's center. Such a plot is of some physical significance, as the ru_r component of the core field, integrated over $d\theta$, yields the area of expansion. An attempt was made to represent the nonlinear core expansion, consistent with the calculated volume expansion, analytically by means of a linear anisotropic elastic expansion field due to line dipoles. In the case of MgO, it was found that a rough representation of the core field could be obtained by two dipoles of unequal strength situated on the y-axis: one directed along the x-axis just slightly above the slip plane, the other directed along the y-axis and somewhat below the slip plane. Although differential displacement plots of the core configurations have not been given, all of the calculations appear to yield similar cores which are fairly narrow and planar. An indication of the planar nature of the core configuration can be seen in the plots of ionic polarization derived from the shell model calculations of Puls and Norgett (56) where the largest effect occurs for ions across the slip plane. With the exception of the work of Belzner and Granzer (66), all calculations also show no tendency for the dislocation to dissociate. This question was specifically addressed by Puls and So (61) for NaCl using a BSM with Flex-II. The calculations showed that a configuration of two 1/4<110> partials initially separated by either 4b or 2b would recombine into a single 1/2<110> edge dislocation.

Other Metals

There seem to be only two studies of dislocation cores carried out in recent years on other than bcc metals. One is on hcp metals, the other on the <100> screw in fcc Al (67-69). The objective of the study on hcp metals by Bacon and Martin (67) was to determine the reason for the difference between basal and prism slip by determining whether dissociation of $1/3[\bar{2}110]$ edge dislocations occurs for slip on these planes. A Lennard-Jones potential fitted to the average properties of various hcp metals with nearly ideal c/a ratios such as Mg, Zr, Ti, Be, etc. was used. It was found that the core on the prism plane was undissociated but fairly broad, whereas dissociation of the type $1/3[\bar{2}110] \rightarrow 1/3[\bar{1}010] + 1/3[\bar{1}100]$ occurred on the basal plane. A clear stacking fault could not be distinguished since the core widths of the partials were quite wide. The second study involved calculations of the <100> screw dislocation in Al by Esterling and Moriarty (69). This represents an interesting attempt to improve both on the potential and the method of treating the boundary region. Thus, the authors use a non-local first-principles pseudopotential to second order in the energy to determine the interactions in region 1. The rest of the crystal is positioned using the lattice statics approach. Unfortunately, a plot of the relaxed core configuration using the differential displacement method was not given, making it difficult to relate the result to other calculations.

The Peierls Stress

By itself, the core structure derived from atomic simulations of dislocations, although interesting, can only be used to provide qualitative information on how the dislocation can affect the mechanical properties of materials. As well, the structure is not presently accessible to direct experimental verification. A quantity strongly dependent on the core structure, however, which is to some degree accessible to experiment and which can be calculated using an atomic model, is the Peierls stress, i.e., the resistance of the lattice to dislocation motion under an applied shear stress. Most of the studies discussed in the previous section have consequently attempted to calculate this quantity.

There are basically two methods which have been used to accomplish this with an atomic model. In the first method, which we will call the linear interpolation method, the dislocation center is placed on various crystallographically non-equivalent positions and the energy of the core (region 1) is calculated as a function of position. This gives the Peierls energy barrier from which the Peierls stress can be derived by differentiation. In this method, relaxation of the lattice is only carried out at the minimum energy positions and intermediate configurations are obtained by assuming linear movement of all or some of the ions between the two minima.* The second method involves the application of an actual shear stress and monitoring the movement of the dislocation during relaxation. The stress is applied by means of a homogeneous shear strain derived from the stress according to linear anisotropic elasticity theory. The stresses are built up gradually until, at a critical stress which is the Peierls stress, the dislocation is observed to move. In the past, this method has mostly been used with a rigid boundary model. Obviously, region 1 must be chosen large enough so as not to inhibit the initial movement of the dislocation's center. Although this second method has the virtue of allowing non-linear structural redistributions of the core under stress, it does not yield the shape of the Peierls barrier. Recently a method was developed by Woo and Puls (53-55) which is capable of generating the full Peierls barrier with the dislocation under stress. The method takes advantage of the very stable iterative relaxation process provided by the combination of the Flex-II and the conjugate gradient relaxation techniques. Briefly, a stress is applied, slightly greater than the Peierls stress. During the rigid boundary relaxation, the dislocation center will move. It will experience two opposing forces, the lattice friction force and the rigid boundary force. The latter force will cause the dislocation to stop after the center has moved a certain distance. The next step is a relaxation of the rigid boundary force by the Green's function. This followed by another rigid boundary relaxation causing the center to move again. A plot is

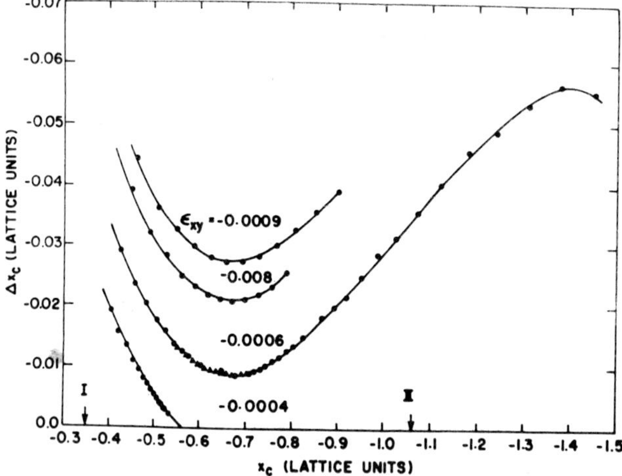

FIG. 8

Dislocation movement Δx_c versus dislocation position x_c for a 1/2<110> edge dislocation in MgO (55) for different applied shear strains ε_{xy}. The positions marked I and II indicate the points at which the dislocation assumes the symmetry configurations I and II, respectively.

* In some variants of this method only a few key core atoms plus the boundary atoms are linearly interpolated, the rest are subsequently relaxed to minimize the energy.

made of the extent of dislocation movement, Δx_c, between rigid boundary relaxations versus dislocation position, x_c, as shown in Fig. 8. For small movements of the dislocation's center with respect to the region's boundary, the rigid boundary force is independent of the dislocation's position, while the Peierls force varies according to its location in the lattice. Thus the variation of Δx_c versus x_c reflects the variation of the Peierls force. It is shown by Woo and Puls (55) that determination of two such curves (for two different values of applied strain) is sufficient to derive the complete Peierls stress barrier. The Peierls energy can then be determined from this by a simple integration. The method should work in a variety of materials, but to date has only been used on dislocations in ionic crystals (53-55,61). It is worth noting that an attempt by Hoagland et al. (5) to determine the Peierls stress using Flex-II with the dynamical relaxation technique was only moderately successful. In that particular case, the dynamical relaxation method either over- or under-responded to the perturbation imposed by the shear strain.

The results of the Peierls stress calculations have been summarized in Tables 1 and 2 for some bcc metals and fcc ionic crystals, respectively. Table 3 summarizes some of the experimental data. The detailed results for the bcc materials are very complex and only a broad summary of the numerical results has been given in Table 1. The results of Vitek and coworkers (2,70,71), employing a series of empirical potentials with varying strength of binding, are the most extensive. A wide selection of stresses, including uniaxial compressive and tensile stresses, were imposed on the dislocation and the structural modifications of the core monitored as the stress was increased. In the case of the 1/2<111> screw dislocation, complex structural changes were observed, an example of which is shown in Fig. 9, which evidently represent the changes required to transform an originally sessile core configuration into a glissile one. The observed changes depend on both the assumed interatomic forces and the type of external loading. Many of the results are qualitatively similar to those previously reviewed by Vitek (2) and will not be repeated here. A new result is the effect of uniaxial loading. Thus the geometry of dislocation motion is different for different loading even though the resolved shear stress is the same. In some cases only the Peierls stress varies.

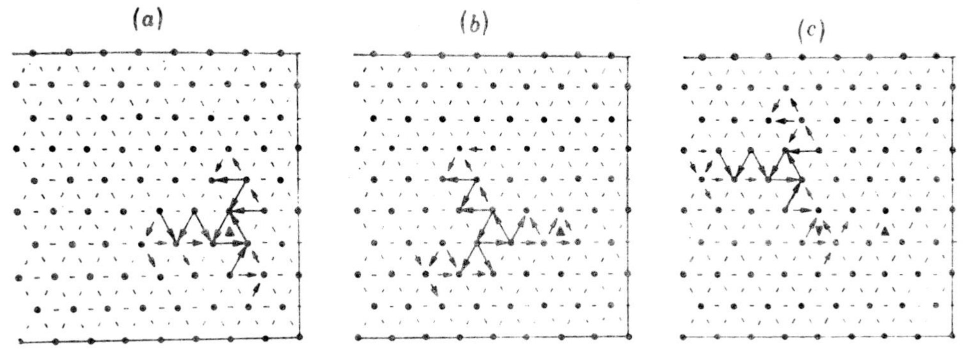

FIG. 9

Core structure determined atomistically (71) under [011] tension for increasing values of applied shear stress: a) 0.020μ; b) 0.029μ; c) 0.040μ where μ is the shear modulus along <111>.

For the same resolved shear stress, it usually falls lowest to highest depending on whether, respectively, tensile, shear or compressive stresses are applied. Vitek (71) also emphasized the important role played by atom-row relaxations lateral to the dislocation line; i.e., certain core structures may only be stable when no lateral relaxations are permitted. In the case of 1/2<111> non-screw dislocations on {112} planes (70), the core structure is strongly antisymmetric and this is also reflected in the asymmetry connected with the twinning, antitwinning sense of shear. Structural core changes after stress application are much more significant for non-edge dislocations on {112} as compared to {110} planes. Nonetheless, these changes are small compared to those of screw dislocations. Thus it is concluded that the observed asymmetries of low-temperature flow stress are governed primarily by the asymmetry of the core of the screw dislocation, with the non-screw dislocation playing only a secondary role.

TABLE 1

Results of Peierls Stress Calculations in bcc Metals for Various Dislocation Types

Material	Potential	Method	Peierls Stress* (10^{-3} μ)	Reference
		1/2<111> screw		
General bcc	empirical; J_2,J_1,J_3	Rigid boundary; shear on {110} and {112} planes	22-71	Vitek (71)
Na, K	pseudo-potential	same as above	1.74-6.15	Hübner (33)
General bcc	empirical; row-potential	rigid boundary; linear interpolation	Camel-humped Peierls energy barrier, maximum height: (2.5 - 10 x 10^{-3} μb^2)	Takeuchi (36)
α-Fe Nb W	tight-binding approximation; Born-Mayer for ion cores	rigid boundary; linear interpolation	24-32 71-92 18-26	Masuda and Sato (34)
α-Fe	same as above	rigid boundary; shear on {110} and {112} planes	46-69 20-34	Sato and Masuda (35)
		1/2<111>{112} non-screw		
General bcc	empirical; J_2,J_1 J_3	rigid boundary; linear interpolation	3.1-3.3 25	Yamaguchi and Vitek (70)
	J_2,J_1 J_3	rigid boundary; shear on {112} planes	1.6-2.9 10	
		<100> screw		
α-Fe CsCl model	empirical; J_0,J_3 empirical	rigid boundary; shear on (110)	18-20 7	Yamaguchi and Umakoshi (47, 48)
		1/2<111>{110} edge		
α-Fe	empirical; J_0	rigid boundary; shear on (110) plane	4	de Hosson (87)
α-Fe	empirical; J_0	rigid boundary; linear interpolation	6	de Hosson (83, 88)
General bcc	empirical; J_2,J_3 J_3	rigid boundary; shear on (110) plane	1.3-1.8 4.0	Yamaguchi and Vitek (70)

*μ is the shear modulus along <111>

TABLE 2

Results of Peierls Stress Calculations in Ionic Crystals for the 1/2<110>{110} Edge Dislocation

Material	Potential[†]	Method	Peierls Stress[#] (10^{-3} μ)	Reference
NaCl	various PIM	rigid boundary; linear interpolation	6.12-9.33*	Granzer, Wagner and Eisenblätter (89)
	PIM BSM	Flex-II, applied shear strain	3.47 0.751	Puls and So (61)
KCl	PIM	Flex-II, applied shear strain	2.04	Woo and Puls (53)[††]
	PIM		1.0-2.5	Hoagland et al. (5)
MgO	PIM and SM	rigid boundary; linear interpolation	1.61-3.81	Puls and Norgett (56)
	Model 1, SM and PIM	Flex-II, applied shear strain	0.92-0.94	Woo and Puls (54,55)[††]
	Model 2, SM		1.36	
	Model 1, BSM		1.50	
	Sangster's, BSM		1.37	

[†]PIM - point ion model; SM - shell model; BSM - breathing shell model

[††]The Peierls stresses calculated by Woo and Puls (53-55) are too low by a factor of two. This factor was erroneously omitted in relating shear stress to shear strain.

[#]The shear modulus, $\mu = 1/2(C_{11}-C_{12})$. Elastic constants correspond to the potential used in the calculations.

*These results are based on a Peierls barrier which has a periodicity of b/2 rather than b as in all the other results (b is the Burgers vector).

TABLE 3

Experimental Values of the Peierls Stress for bcc Metals and Ionic Crystals

Material	Peierls Stress (10^{-3} μ)	Method	Reference
K	2-4	Flow stress (4 K)	Hands and Rosenberg (90)
Mo	8	Flow stress (4 K)	Lawley and Gaigher (91)
Ta	7	Flow stress (4 K)	Mitchell and Spitzig (92)
α-Fe	7	Flow stress (4 K)	Altshuler and Christian (93)
Nb	6	Flow stress (4 K)	Kubin (94)
NaCl	0.5	Flow stress (4 K)	Tesh and Whitworth (95)
KCl	0.3-1.0	Bordoni Peak	Ikushima and Suzuki (96)
MgO	1.0[†]	Dislocation Velocity	Singh and Coble (97)
MgO	0.02-0.05	Bordoni Peak	Chang (98) and Ikushima and Suzuki (96)

[†]Based on the analysis of Woo and Puls (54)

It should be noted that the results of the atomic models are strictly applicable only near 0 K, whereas in reality, at higher temperatures, dislocation motion is thermally assisted, for instance, via double-kink formation. Consequently it is useful to obtain information on kink formation energies. Of course, the above results can provide qualitative guidelines on possible core transformations during double-kink formation. A more quantitative approach to determine kink formation energies using an atomic model was recently attempted by Wüthrich (72). Such a calculation, however, involves some serious computational difficulties. Moreover, the kink might actually be quite long, but to substantiate this would require a prohibitively large atomic model. An alternative approach might be to use the method of Woo and Puls (54) to determine the shape of the Peierls barrier atomistically and then to use this information to estimate the kink formation energy via the string model. This has not been done, so far, for bcc metals.

In ionic crystals, the most detailed and extensive work on Peierls barriers has been done by Puls and co-workers (53-55,61). The most reliable calculations, based on the method of Woo and Puls (53-55) show that the Peierls barrier has a sinusoidal shape and a periodicity of b (the Burgers vector). The maximum is situated on one of the two dislocation symmetry types. The calculations done so far indicate that type I is the minimum energy configuration for any potential which can be consistently used with the breathing shell model. Comparison with experimental results given in Table 3, shows agreement within a factor of two for NaCl and MgO. Agreement is also good for KCl, but the experimental results are based on Bordoni peak measurements and the interpretation of such experiments must be considered somewhat doubtful. For instance, for MgO Woo and Puls (54) show that the Bordoni peaks cannot be explained by a Peierls mechanism. Experimentally, it is observed that edge dislocations are more mobile than screw dislocations. This is similar to the situation in bcc metals. It would be interesting to determine the reason for this in ionic crystals by doing an atomistic calculation but the greater difficulty of handling the Coulomb sums has so far hindered development.

Overall, expressed in terms of the shear modulus μ, there is a remarkable similarity between the Peierls stresses of non-screw dislocations in both types of materials. Variations in the results due to the method of calculation and potential are generally (with a few exceptions) within a factor of 3. Thus the order of magnitude higher value for the Peierls stress of the 1/2<111> screw dislocation compared to the non-screw dislocations in the bcc metals is a clear-cut general result which can be attributed to the core structure of the screw dislocation. This result seems to be fairly robust to any uncertainties in potentials or methods. A persistent difficulty, however, is the order of magnitude disagreement between the calculated Peierls stress for the screw dislocation and the measured liquid helium temperature flow stresses given in Table 3.

Volume of Expansion

When a model of a dislocation obeys a non-linear force law, introduction of the dislocation into the perfect lattice generates an overall expansion. This is largely suppressed in rigid boundary models, since the boundary atoms act as a corset restricting the outward motion of atoms near the periphery of region 1. In flexible boundary models, however, this restriction is removed. As a consequence, models using flexible boundary techniques have been able to determine this dilatation and the results are summarized in Table 4. Unfortunately, most of the expansion volumes (given as volume/unit distance or area of expansion) listed in the table are incomplete, and thus too low, because, as recently pointed out by Sinclair et al., previous calculations neglected to include a member of subtle, but nevertheless significant, non-linear effects. Sinclair et al. (7) indicate that there are really three sources of dilatation:* i) due to the initial Volterra displacement field; ii) due to the boundary relaxation field; and iii) due to additional displacements that would be needed in the rest of the crystal to eliminate body forces and surface tractions originating from non-linear elastic displacements. Except for the results of Sinclair et al. (7), the values listed in Table 4 were determined only on the basis of the expansion which results from item (ii), the boundary relaxation field. There are two ways to determine this. One is from the Green's function source forces in region 2 when using Flex-II or alternately, when using Flex-S, from the coefficients of the higher order terms; the other is by comparing the difference in area of a circuit centered about the dislocation before and after relaxation. In either case, the result is relative to the initial expansion caused by the Volterra configuration which, due to non-linear geometric effects, will be different depending on how the dislocation is introduced into the

* There are actually four sources, but the fourth one can be eliminated if Eulerian rather than Lagrangian coordinates are used to position the atoms according to the Volterra field.

crystal. In some cases the dilatation due to this latter effect is as much as 50% of the relaxation terms, in others, it is practically negligible. On the other hand, the boundary non-linear dilatations (which, at present, can only be estimated in anisotropic elasticity) roughly double the expansion calculated due to the first two sources.

TABLE 4

Volume of Expansion

Material	Dislocation	Potential and Method	Volume change per unit length (b^2)	Reference
KCl	1/2<110>{110} edge	PIM, Flex-II	0.55	Hoagland et al. (5)
MgO	1/2<110>{110} edge	BSM, Flex-II	0.45	Woo and Puls (58)
MgO	1/2<110>{110} edge	PIM or SM, Flex-II	0.51	Woo and Puls (58)
NaCl	1/2<110>{110} edge	PIM, Non-linear Flex-I	0.50	Granzer et al. (9)
α-Fe	1/2<111> screw	empirical, Flex-S	0.42	Wüthrich (37)
α-Fe	1/2<111>{110} edge	J_o, Flex-II	0.62†	Adams et al. (42)
α-Fe	[100] edge	empirical; various flexible boundary methods	1.0#	Sinclair et al. (7)

† Volterra field contribution of 0.22 b^2 included.

\# Includes Volterra field and long-range non-linear contributions and surface relaxations (assuming effective crystal size of 10^4b).

Interactions of Point-Defects with Dislocations

Aside from the Peierls stress, another quantity which cannot be calculated with any confidence by a linear continuum model, and which to some degree is accessible to experimental verification, is the interaction energy between a point-defect and a dislocation at close point-defect/dislocation separation. To calculate this quantity accurately it is necessary to use an atomic model. Compared to atomic models of single dislocations or point-defects, the number of such calculations has, however, been few. This is probably because of their substantial cost. Adding a point-defect to an atomic dislocation model generally means that the lattice periodicity along the dislocation line can no longer be maintained (or the repeat distance must be substantially increased), changing an essentially two-dimensional problem to a three-dimensional one and, hence, considerably increasing the model size. Adding to the difficulty is that this three-dimensional problem is generally very unsymmetric. Partly for this reason, methods of flexing the boundary have not so far been developed for a combined point-defect/dislocation model. There is, however, also less reason for flexible boundaries in this case, since the strain field of the point-defect has a much shorter range than that of the dislocation. Thus, as far as the point-defect is concerned, rigid boundaries are unlikely to affect the relaxation in region 1 provided that this region is large enough for the purposes of generating an adequately relaxed dislocation core. Consequently, all the calculations done so far have used rigid boundary methods.

To determine the interaction energy, we need to evaluate the difference in the energy of formation of the point-defect in a perfect and a dislocated lattice. In metals, this is usually accomplished by a separate evaluation of the energy of, i) the point-defect, ii) the dislocation, iii) the perfect lattice, and iv) the point-defect/dislocation combination. In ionic crystals, a slightly different procedure is used (59). In these crystals, the long-range Coulomb interaction represents a substantial computational complication. The problem is made manageable by formulating it in terms of energy differences between the defected and the perfect crystal. The formalism was developed by Norgett (73) and incorporated into the computer program HADES. This is a very elegant and efficient program capable of evaluating point-defect formation energies in perfect ionic crystals. Using the same formalism, but a different method of calculating the electrostatic contribution to the energy (the Madelung sums), a program entitled PDINT, was developed (59) to calculate point-defect formation energies in either a dislocated or a perfect lattice, the choice depending on the chosen reference state. Thus, interaction energies evaluated with this program require only two energy determinations. Used with the perfect lattice as the reference state, PDINT can generate results identical to HADES, as was verified in one particular case (61). However, HADES, being dedicated to the study of point-defects only, is computationally about an order of magnitude more efficient. This greater efficiency is due to a more concise method of determining the Coulomb sums (the Ewald as opposed to the infinite ion string technique) and a faster minimization routine (variable metric versus conjugate gradient). As presently formulated, PDINT uses a rigid boundary method when calculating formation energies in the dislocated state. On the other hand, the reference dislocation core configuration is obtained from a Flex-II calculation. As well, in response to the charge of the defect in region 1, an additional harmonic displacement field based on the method of Mott-Littleton (74) is imposed on the boundary. In this method the ionic displacements and polarizations are determined by assuming that the crystal responds to the effective charge of the defect like a dielectric continuum.

For metals, recent atomistic studies of vacancy-dislocation interactions have been carried out by Schiffgens and Ashton (75) on 1/2<110> edge dislocations in copper and by Ingle and Crocker (76) on 1/2<111>{110} edge dislocations in α-Fe. Using empirical potentials the maximum interaction was always attractive and at the edge of the extra planes of atoms. The maximum binding energy varied with the potential used, ranging from 0.28 - 0.70 eV. The work on α-Fe covered numerous positions around the core and gave interactions which are attractive almost everywhere, with little angular variation. In terms of a simple elastic model, this was similar to the inhomogeneity interaction (77). Quantitatively, however, the agreement was not too good. Similar results were obtained some time ago by Bullough and Perrin (78) for the 1/2<100>{010} edge dislocation in α-Fe. These results contrast strongly with recent calculations on MgO by Puls and co-workers (59,60). In particular, Puls (60) found that the energy in a region approximately 2b by 4b surrounding the dislocation's center shown in Fig. 10 revealed a rather complex behaviour quite unlike that predicted by an isotropic linear elastic model. As with the metals, however, the maximum attractive interaction of $\sim$ 1.0 eV. occurred at the edge of the extra plane of atoms. This binding energy is in agreement with results inferred from internal friction experiments (79). Ignoring the angular dependence, the spatial variation of the binding energy has an approximately R^{-2} dependence (R is the defect-dislocation separation). This is in disagreement with an R^{-8} dependence inferred from internal friction studies (79). However, compared to the pinning energy, there is much greater uncertainty associated with this latter prediction. At the furthest point-defect/dislocation separation of 4b to 5b, comparison was also made by Puls (60) between values of the interaction energy derived from the atomic and the isotropic elastic models (the latter, based on the size effect and inhomogeneity interaction). It was found that, at this distance from the core, there is qualitative, but still not quantitative, agreement between the predictions of the two models.

Finally there have been a number of calculations by de Hosson and co-workers (80-85) on the interaction of carbon and helium with various dislocations in fcc and bcc metals. The calculations use mostly empirical or semi-empirical potentials and the rigid boundary method. In references (80) and (81) the diffusion drift paths of C to perfect <100> edge dislocations in Ni and γ-Fe were determined. It was found that inside about 5b from the dislocation's center, the drift paths differed considerably from the prediction based on isotropic elastic theory using only the size effect interaction. A surprising result on the interaction of interstitial He with a 1/2<111> screw dislocation in Mo determined by Heugten et al. (85,86) was the close agreement obtained between the atomic and elastic models within 1b of the dislocation's center. In fact, the agreement was at least as good as

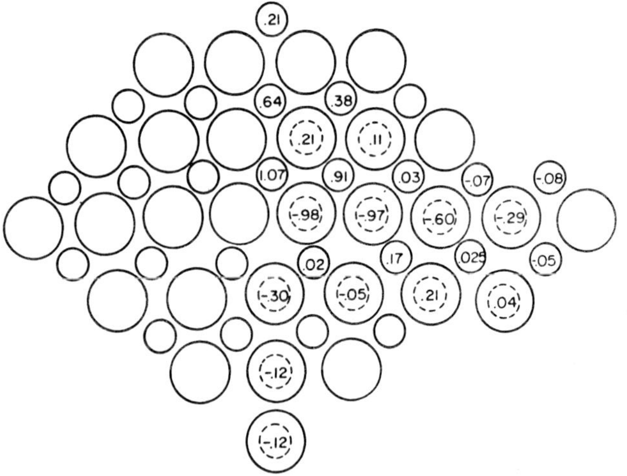

FIG. 10

Stable, type I core configuration of a 1/2<110> edge dislocation in MgO with values for the atomistically determined interaction energies of the cation vacancies superimposed (60).

the difference in binding energy of 1.3 - 1.6 eV. determined using two different sets of potentials.

This last point brings up an important general observation which concerns the overall accuracy of these atomistic calculations. In each of the cases (including ionic crystals) when calculations of binding energies were made with more than one potential, differences of up to 100% were found. This then represents the most optimistic limit for the accuracy of the atomistic calculations with the present set of potentials. It is, perhaps, surprising that such a low accuracy seems also to prevail for ionic crystals where we have claimed considerably greater confidence in the potentials, but evidently the interaction energies, representing less than 10% of the total formation energy of the defect, are more sensitive to the short-range interactions than one might initially suspect. Despite this limitation on the accuracy of the atomic models, they are nevertheless a considerable improvement over linear continuum or semi-discrete models which are hopelessly inadequate in the core region. This is particularly evident in the comparison done by Puls (60) on MgO.

Conclusions

The usefulness of atomic dislocation models is limited by the efficiency with which calculations based on these models can be carried out and by the accuracy of the chosen interatomic potential. To improve the efficiency without sacrificing accuracy, so-called flexible boundary methods have been developed. These are schemes in which the boundary atoms are positioned according to linear or non-linear elasticity theory in response to the atomistically determined core configuration. Of the various flexible boundary methods developed, comparisons (6,7) have shown that the Flex-II technique of Hoagland, Hirth and Gehlen (5) is the most efficient. Combined with the conjugate gradient method for relaxing the explicit atomic configuration of the core, there is now available an exceedingly efficient computational scheme for simulating an atomic dislocation model. The principal limitation of this method presently resides in our ability to provide for a realistic and computationally convenient interatomic potential. In the case of the metals, the current situation is that, except for monovalent simple metals, such a potential is unfortunately not available. The much touted pseudopotential method, which has been very successful in predicting various physical properties characteristic of the perfect lattice, has yielded potentials which have recently been shown to be inadequate for defects such as vacancies, dislocations, surfaces and cracks, which produce large inhomogeneities in the electron distribution. Consequently, it appears that we must continue to rely on empirical model potentials. It is important in this case to compare the results of atomic calculations by using a number of different potentials. Properties of the dislocation which vary significantly with the potential will obviously only be qualitatively useful.

In ionic crystals, the situation is, at first glance, much improved as the well-known electrostatic interaction between ions dominates the binding energy. The difficulties here are in finding an efficient method of summing the long range Coulomb interaction and in providing a model which describes the dielectric response of the material to the electric field produced by the defect. For dislocations, the former is accomplished via the infinite ion string method, whereas the latter is very adequately described by means of the shell model which, in its breathing modification, is capable of simulating many-body interactions. Attempts have been made to justify rigorously the shell models in terms of quantum mechanics. However, as used in defect calculations, they are essentially empirical models. The parameters of the models are determined from various measured physical properties and little physical significance should be attached to their actual values. Nevertheless, the models have been shown to be very successful in predicting numerous defect properties in simple ionic crystals. This suggests that, despite their somewhat empirical character, they are able to mimic the essential physics of the interaction very well.

Applications of atomic dislocation models in metals have concentrated mainly on crystals with the bcc structure. The objective has been to explain the plastic deformation properties which are characteristic of these metals, such as the rapid increase in flow stress at low temperature and the observed asymmetry in the slip geometry. It is believed that these features are controlled by the movement of the 1/2<111> screw dislocation. The modelling work was extensively reviewed in 1974 by Vitek (2). Work since then has concentrated on additional calculations using different or improved potentials, improved methods of calculations, other dislocation types and Peierls stress calculations with a greater variety of orientations of the maximum resolved shear stress. The overall assessment of these recent efforts is essentially the same as that made earlier by Vitek (2). That is, the atomic models point to the sessile and non-planar nature of the core of the screw dislocation as the dominant cause for the plastic deformation properties of the bcc metals. This result is characteristic of the of the bcc crystal. However, the exact details of the core configuration and the values of the Peierls stress vary with the potential used and would thus be expected to differ between different bcc metals. For the Peierls stress, the variation with potential is, however, surprisingly small (generally within a factor of 3). In agreement with experiment, the screw dislocation has the higher Peierls stress. On the other hand, low-temperature flow stress measurements give stresses which are generally an order of magnitude lower than those calculated for the screw dislocation. This difference continues to be a bothersome discrepancy between theory and experiment.

Atomic dislocation models in ionic crystals have concentrated exclusively on the fcc structure, emphasizing the 1/2[110]{110} edge. This emphasis reflects the greater difficulty in deriving efficient Coulomb summation methods for other dislocation types and orientations. Good agreement is found between calculated and measured values of the Peierls stress based on the most reliable methods and models. Except for one result, all the calculations yield a planar, non-dissociated core. This result casts some doubt on the interpretation of the pressure dependence of stage III flow stress in the alkali halides. This is interpreted in terms of a model providing for enhanced cross-slip of screw dislocations in stage III. Assuming that the dislocations are in a dissociated state, hydrostatic compression, by suppressing the dissociation, would favour this cross-slip mechanism. To clarify this point further, it would be useful to generate a corresponding atomic model of the 1/2<110> screw dislocation.

The effort placed on flexible boundary methods has led to an improved representation of the displacement field of the dislocation close to the core. Thus it is found that the edge dislocation can be represented as the sum of two linear fields: the Volterra field and a pair of orthogonal line-force dipole fields centred close to the dislocation line. Related to this, it is also possible (subject to the accuracy of the assumed potential) to give a fairly accurate estimate of the volume of expansion when a dislocation is formed in a non-linear medium. A recent careful analysis shows, however, that both of these results, based on an analysis of the relaxation field, depend, sensitively, on the initial conditions for the unrelaxed dislocation configuration.

Finally, there have been relatively few point-defect/dislocation interaction calculations based on atomic models. This is attributed to the greater cost and complexity of these calculations. The calculations that have been carried out show, as expected, that there are differences between the results based on an atomic as opposed to a linear elastic model. These differences are particularly marked in ionic crystals as indicated by a study of the interaction of a vacancy with the 1/2[110] edge dislocation in MgO. This work shows neither quantitative nor qualitative agreement with the elastic model. It is likely that ionic polarizations, induced by the local electric fields of the defects, contribute significantly to the total interaction energy. Such effects are, of course, completely absent from any model based solely on elastic interactions.

Acknowledgements

The author would like to thank C.H. Woo and C.B. So for useful discussions.

References

1. R. Bullough and V.K. Tewary, Dislocations in Solids, Vol. 2, F.R.N. Nabarro, ed., North-Holland, Amsterdam (1979), p.1.
2. V. Vitek, Crystal Lattice Defects 5, 1 (1974).
3. R.C. Perrin, J. Phys., Paris, Colloq. 35, C7-103 (1974).
4. A.B. Lidiard and M.J. Norgett, Computational Solid State Physics, F. Herman, N.W. Dalton and T.R. Koehler, eds., Plenum, New York (1972), p. 385.
5. R.G. Hoagland, J.P. Hirth and P.C. Gehlen, Phil. Mag. 34, 413 (1976).
6. M.P. Puls and C.H. Woo, Atomic Energy of Canada Ltd., Report AECL-5238 (1975).
7. J.E. Sinclair, P.C. Gehlen, R.G. Hoagland and J.P. Hirth, J. Appl. Phys. 49, 3890 (1978).
8. J.P. Hirth and J. Lothe, J. Appl. Phys. 44, 1029 (1973).
9. F. Granzer, V. Belzner, M. Bücher, P. Petrasch and C. Teodosiu, J. Phys., Paris, Colloq. 34 C9-359 (1973).
10. P. Petrasch and V. Belzner, J. Phys., Paris, Colloq. 31, C7-553 (1976).
11. A. Seeger, C. Teodosiu and P. Petrasch, Phys. Stat. Sol. (b) 67, 207 (1975).
12. P.C. Gehlen, J.P. Hirth, R.G. Hoagland and M.F. Kanninen, J. Appl. Phys. 43, 3921 (1972).
13. R.A. Johnson, J. Phys. F3, 295 (1973).

14. R.A. Johnson, Nucl. Met. 20, 1 (1976).
15. J. Grilhé and P. Beauchamp, J. Phys., Paris, Colloq. 35, C7-91 (1974).
16. C.R.A. Catlow, K.M. Diller and M.J. Norgett, J. Phys. C. 10, 1395 (1977).
17. J. Corish, B.M.C. Parker and P.W.M. Jacobs, Can. J. Chem. 54, 3839 (1976).
18. W.A. Harrison, Pseudopotentials in the Theory of Metals, W.A. Benjamin, Inc., New York, (1966).
19. R. Evans, Vacancies 76, R.E. Smallman and J.E. Harris, eds., The Metals Society, (1977), p.30.
20. R. Evans and M.W. Finnis, J. Phys. F6, 483 (1976).
21. C.B. So and C.H. Woo, unpublished, (1980).
22. D.M. Esterling and A. Swaroop, Phys. Stat. Sol. (b) 96, 401 (1979).
23. H.B. Huntington, J.E. Dickey and R. Thomson, Phys. Rev. 100, 1117 (1955).
24. U. Schröder, Solid St. Commun. 4, 347 (1966).
25. V. Nüsslein and U. Schröder, Phys. Stat. Sol. 21, 309 (1967).
26. H.P. Niedermann and M. Wagner, Phys. Stat. Sol. (b) 78, 615 (1976).
27. C.R.A. Catlow and A.B. Lidiard, Thermodynamics of Nuclear Materials 1974, Vol. II, Vienna; IAEA, (1975), p.27.
28. C.R.A. Catlow, I.D. Faux and M.J. Norgett, J. Phys. C. 9, 419 (1976).
29. C.R.A. Catlow, M.J. Norgett and T.A. Ross, J. Phys. C. 10, 1627 (1977).
30. W.C. Mackrodt and R.F. Stewart, J. Phys. C. 12, 431 (1979).
31. P.B. Hirsch, 5th Int. Conf. on Crystallography, Cambridge (oral communication) (1960), p.139.
32. V. Vitek, R.C. Perrin and D.K. Bowen, Phil Mag. 21, 1049 (1970).
33. U. Hübner, Phys. Stat. Sol. (b) 67, 113 (1975).
34. K. Masuda and A. Sato, Phil. Mag. B37, 531 (1978).
35. A. Sato and K. Masuda, Solid St. Commun. 29, 545 (1979).
36. S. Takeuchi, Phil. Mag. A39, 661 (1979).
37. C. Wüthrich, Phil. Mag. 35, 325 (1977).
38. J.E. Sinclair, J. Appl. Phys. 43, 5321 (1971).
39. R. Chang and L.F. Graham, Phys. Stat. Sol. 18, 99 (1966).
40. M. Yamaguchi and V. Vitek, J. Phys. F5, 1 (1975).
41. R.A. Johnson, Phys. Rev. A134, 1329 (1964).
42. B.L. Adams, J.P. Hirth, P.C. Gehlen and R.G. Hoagland, J. Phys. F7, 2021 (1977).
43. M. Yamaguchi and V. Vitek, J. Phys. F3, 523 (1973).
44. V. Vitek and M. Yamaguchi, J. Phys. F3, 537 (1973).
45. J. Th. M. de Hosson and A.W. Sleeswyk, Phys. Stat. Sol. (b) 71, 595 (1975).
46. M. Yamaguchi and Y. Umakoshi, Phys. Stat. Sol. (a) 31, 101 (1975).
47. M. Yamaguchi and Y. Umakoshi, Scripta Met. 9, 637 (1975).
48. M. Yamaguchi and Y. Umakoshi, Nucl. Met. 20, 763 (1976).
49. F. Bassani and R. Thomson, Phys. Rev. 102, 1264 (1956).
50. A.A. Maradudin, J. Phys. Chem. Solids 9, 1 (1959).
51. J. Eisenblätter, Phys. Stat. Sol. 31, 87 (1969).
52. J. Eisenblätter, Phys. Stat. Sol. 31, 71 (1969).
53. C.H. Woo and M.P. Puls, J. Phys. C. 9, 127 (1976).
54. C.H. Woo and M.P. Puls, Phil. Mag. 35, 1641 (1977).
55. C.H. Woo and M.P. Puls, Atomic Energy of Canada Ltd., Report AECL-5685, (1977).
56. M.P. Puls and M.J. Norgett, J. Appl. Phys. 47, 466 (1976).
57. M.P. Puls, Atomic Energy of Canada Ltd., Report AECL-5237, (1975).
58. C.H. Woo and M.P. Puls, Phil. Mag. 35, 727 (1977).
59a. C.H. Woo, M.P. Puls and M.J. Norgett, J. Phys., Paris, Colloq. 37, C7-557 (1976).
59b. M.P. Puls, C.H. Woo and M.J. Norgett, Phil. Mag. 36, 1457 (1977).
60. M.P. Puls, Phil. Mag. A41, 353 (1980).
61. M.P. Puls and C.B. So, Phys. Stat. Sol. (b) 98, 87 (1980).
62. M.P. Puls, J. Phys., Paris, Colloq., in press (1980).
63. L.M. Polisar, A.A. Predvoditelev and V.A. Skokov, Sov. Phys. Crystallogr. 20, 53 (1975).
64. W. Raupach, Phys. Stat. Sol. (a) 50, K81 (1978).
65. R.W. Whitworth, Adv. Phys. 24, 203 (1975).
66. V. Belzner and F. Granzer, Phys. Stat. Sol. (a) 39, 183 (1977).
67. J.W. Martin and D.J. Bacon, Nucl. Mat. 20, 298 (1976).
68. D.M. Esterling, J. Appl. Phys. 49, 3954 (1978).
69. D.M. Esterling and J.A. Moriarty, J. Appl. Phys. 49, 3960 (1973).
70. M. Yamaguchi and V. Vitek, J. Phys. F5, 11 (1975).
71. V. Vitek, Proc. Roy. Soc., London, A352, 109 (1976).

72. C. Wüthrich, Phil. Mag. 35, 337 (1977).
73. M.J. Norgett, AERE Harwell Report R7650, (1974).
74. N.F. Mott and M.J. Littleton, Trans. Faraday Soc. 34, 485 (1938).
75. J.O. Schiffgens and D.H. Ashton, J. Appl. Phys. 45, 1023 (1974).
76. K.W. Ingle and A.G. Crocker, Acta Met. 26, 1461 (1978).
77. R. Bullough and R.C. Newman, Rep. Prog. Phys. 33, 101 (1970).
78. R. Bullough and R.C. Perrin, Dislocation Dyanmics, A.R. Rosenfield, G.T. Hahn, A.L. Bement, Jr. and R.I. Jaffee, eds., McGraw-Hill, New York, (1968), p.175.
79. I.G. Ritchie and K.W. Sprungmann, Scripta Met. 7, 323 (1973).
80. J.Th.M. de Hosson and A.W. Sleeswyk, Phys. Stat. Sol. (b) 69, 417 (1975).
81. J.Th.M. de Hosson, Solid St. Commun. 16, 1231 (1975).
82. J.Th.M. de Hosson, Solid St. Commun. 17, 747 (1975).
83. J.Th.M. de Hosson, Nucl. Met. 20, 511 (1976).
84. F. van den Berg, W. van Heugten, L.M. Caspers and A. van Veen, Solid St. Commun. 27, 665 (1978).
85. W.F.W.M. van Heugten, F. van den Berg, L.M. Caspers, A. van Veen and J.Th. M. de Hosson, Phys. Stat. Sol. (b) 90, 643 (1978).
86. W.F.W.M. van Heugten, L.M. Caspers, J.Th.M. de Hosson, Phys. Stat. Sol. (b) 92, 199 (1979).
87. J.Th.M. de Hosson, Phys. Stat. Sol. (b) 87, 151 (1978).
88. J.Th.M. de Hosson, Scripta Met. 11, 571 (1977).
89. F. Granzer, G. Wagner and J. Eisenblätter, Phys. Stat. Sol. 30, 587 (1968).
90. B.A. Hands and H.M Rosenberg, Acta Met. 17, 455 (1969).
91. A. Lawley and H.L. Gaigher, Phil. Mag. 8, 1713 (1963).
92. T.E. Mitchell and W.A. Spitzig, Acta Met. 13, 1169 (1965).
93. T.L. Altshuler andJ.W. Christian, Phil. Trans. Roy. Soc. London A, 261, 253 (1967).
94. L.P. Kubin, Thesis, Université Paris-Sud, Centre d'Orsay, (1971).
95. J.R. Tesh and R.W. Whitworth, Phys. Stat. Sol. 39, 627 (1970).
96. A. Ikushima and T. Suzuki, J. Phys. Soc. Japan 18, Suppl. I, 163 (1963).
97. R.N. Singh and R.L. Coble, J. Appl. Phys. 45, 981 (1974).
98. R. Chang, J. Appl. Phys. 32, 1127 (1961).

COMPUTER SIMULATION OF DISLOCATION CORES IN HCP METALS

D.J. Bacon and J.W. Martin

Department of Metallurgy and Materials Science,
The University, P.O. Box 147,
Liverpool L69 3BX, England.

Abstract

The cores of the prism-edge, basal-edge and screw dislocations with $\underset{\sim}{b} = \frac{1}{3}\langle 11\bar{2}0\rangle$ have been simulated in two model hcp crystals with different basal stacking-fault energy γ. Three aspects are considered here. First, the prism edge is found not to dissociate, in contrast to several earlier predictions. Second, the basal edge and screw dissociate into Shockley partials of spacing close to that given by γ and linear elasticity, suggesting that the use of elasticity to estimate γ from observed partial spacing may not be as questionable as previous simulations had indicated. Finally, constriction and cross slip of the screw onto the prism plane under an applied shear stress have been simulated. They occur at the site of one of the partials only, and this can be explained by the asymmetry of Shockley partial dislocations.

1. Introduction

Numerous investigations by computer modelling of defect structures in metals have been reported in recent years, but little attention has been paid to the hexagonal close-packed (hcp) metals. This is somewhat surprising, for this class has considerable technological significance and the individual metals exhibit a wide range of properties. Of particular interest as far as dislocation behaviour is concerned are the crystallographic features of basal and prism slip, for information on these aspects is of importance for understanding the preference of some metals for basal slip and of others for prism slip.

Part of the reason for lack of study of these metals is the difficulty of obtaining potentials which accurately simulate the physical properties - such as c/a ratio, elastic constants, stacking fault energy (s.f.e.) - of individual metals, and, as far as is known, such potentials are not available at present. Thus, for defect modelling by computer simulation, it is necessary to use a potential which simply describes a stable hcp structure rather than a particular metal. This problem has been discussed at length (1), and the Lennard-Jones potential is quite suitable provided it is not truncated before the fifth-neighbour shell of atoms. In fact, the c/a ratio, elastic constants and phonon data of a Lennard-Jones crystal are similar to those of many real metals. By using truncations just beyond the fifth- and sixth-neighbour shells, which give the potentials designated ℓj5 and ℓj6 respectively in (1), it is possible to model crystals with high and low s.f.e. γ on the basal plane and almost identical elastic constants. The effect of changes in γ on dislocation processes can therefore be investigated.

These potentials have been used to simulate dislocations in unstressed and stressed crystals, and the results are presented in detail elsewhere (1,2). Here we discuss some of the features of interest for dislocation processes. The dislocations studied all have $\underset{\sim}{b} = \frac{1}{3}\langle 11\bar{2}0\rangle$ and are the 'prism edge' lying along [0001], the 'basal edge' along <1010> and the 'basal screw', which can slip on the basal and prism planes. The crystallites used contained up to approximately 9000 atoms, of which about 1000 were relaxable, and periodic boundary conditions were imposed along the dislocation-line direction. The block sizes were sufficiently large to prevent the boundaries affecting the core structures obtained.

2. Core of the Prism Edge

The structure of the prism edge dislocation is of importance for metals which slip predominantly on the prism plane, e.g. Ti, Zr and Hf. A number of authors have, in fact, speculated on the possibility of some form of dissociation occurring on this plane and enhancing the relative ease of prism slip. We have therefore examined the structure of the prism edge in crystals of the ℓj5 and ℓj6 potentials to determine the extent of dissociation. The computed atomic positions on two adjacent (0001) planes in a crystal with the ℓj6 potential are shown in fig.1(a).

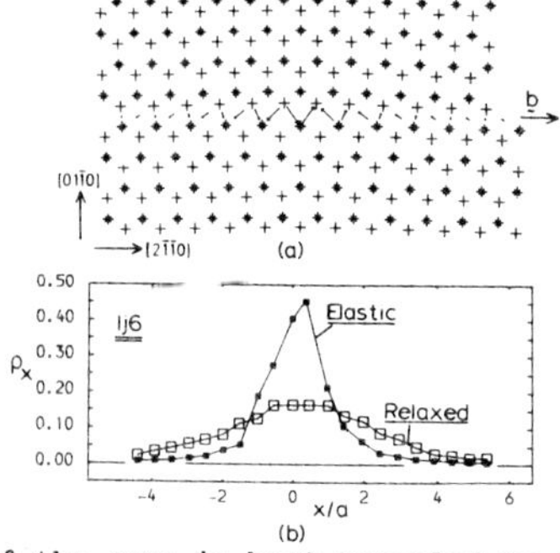

FIG. 1

(a) Relaxed atomic positions projected onto the basal plane for the prism edge dislocation(ℓj6 potential).

(b) Distribution of Burgers vector ρ_x along the slip plane.

The form of the core is best revealed by the plot of fig.1(b), which shows the distribution of Burgers vector as a function of position x along the $[2\bar{1}\bar{1}0]$ direction. (The distribution function ρ_x is defined to be $d(\Delta u_x)/dx$, where Δu_x is the difference in x displacement of atoms on opposite sides of the slip plane, as shown by the arrows in fig.1(a).) It is clear from fig.1(b) that, in comparison with the linear-elastic starting configuration, the atomic interactions allow the core to relax along the prism plane: a similar result holds for the ℓj5 crystal (1). There is no tendency for the dislocation to split into partials however, for this would result in a ρ_x curve having maxima corresponding to partial dislocations. Thus, the dissociation suggested by Regnier and Dupouy (3), namely

$$\frac{1}{3}\ \langle 11\bar{2}0\rangle \rightarrow \frac{1}{9}\ \langle 11\bar{2}0\rangle + \frac{2}{9}\ \langle 11\bar{2}0\rangle, \tag{1}$$

does not occur in Lennard-Jones crystals.

A second dissociation can be seen to be possible from the translation of adjacent $\{10\bar{1}0\}$ planes in a hard-sphere crystal (4,5): it gives the reaction

$$\frac{1}{3}\ \langle 11\bar{2}0\rangle \rightarrow \frac{1}{18}\ \langle 42\bar{6}3\rangle + \frac{1}{18}\ \langle 24\bar{6}\bar{3}\rangle. \tag{2}$$

Each partial has a screw component, which would require atomic displacements normal to the basal planes of fig.1(a). Upon giving small (<0.2a) c-axis displacements to atoms in the core region of our models, however, they always relaxed back to their parent basal plane. Reaction (2) is therefore not an appropriate description of the core, even with hard-core potentials such as ℓj5 and ℓj6.

Experimental evidence for prism-plane dissociation is at best indirect. Akhtar and Teghtsoonian (6,7) estimated values of prism-plane s.f.e. of 145 and 56 mJ/m^2 in Ti and Zr from measurement of the stress at the onset of stage III in single-crystal deformation. This method is open to considerable uncertainty (8), however, and could give a s.f.e. even when no true fault is possible, i.e. when the dislocation core is distributed smoothly rather than as two distinct partials. If dissociation does occur in real metals, our simulation experiments suggest it is unlikely to be reaction (2). Reaction (1) cannot be ruled out, but requires a crystal which is stable under the transformation of hcp prism planes to the stacking of {112} planes in bcc. The occurrence of this in computer modelling would therefore be a potential-dependent phenomenon, and further study will require development of suitable potentials. Weak-beam experiments in the t.e.m. on, say, Zr would be highly desirable.

3. Dissociation in the Basal Plane

Stable faults on the basal plane exist in crystals of both the ℓj5 and ℓj6 potentials (1), although the energies γ are somewhat lower than the experimental values quoted for the hcp metals, with the possible exception of Co. Perfect basal edge and screw dislocations were allowed to relax, and dissociation into two $\frac{1}{3}$ $\langle 10\bar{1}0\rangle$ Shockley partials occurred, as expected. It was found (1), however, that the partial spacing, defined by the peak-to-peak spacing on the ρ_x (edge component) and ρ_z (screw component) curves, was less than that calculated conventionally by using linear elasticity theory for dislocations with singular cores. On changing the starting configuration to two partial dislocations of spacing d, the unrelaxed and relaxed crystal energies decreased and then increased again as d was increased. The minimum energy occurred for partial spacings of approximately 3.5a and 10a for the basal screw in the ℓj5 and ℓj6 models, and 7.5a for the basal edge in the ℓj5 crystal. These values compare with those predicted by elasticity theory of 3a, 11a and 7a, respectively. Thus, the observed spacings are in good agreement with those given by elasticity for the same value of γ. This is despite the fact that the partial cores are considerably wider than their elastic counterparts, and the perfect fault width correspondingly smaller.

Two consequences arise from these observations. First, the partial spacing found in computer modelling may not correspond to the configuration of lowest energy if the intrinsic lattice resistance to partial motion is not negligible. Frequently in model crystals it is not (1), and this resistance can prevent the partials moving to the spacing of minimum total energy. The Shockley nature of the partials was clear in the present work and it was therefore possible to find the minimum-energy structure, but in many cases it may not be. Second, the good agreement observed between the partial spacings in the atomistic models and the estimates given by the continuum approximation is in contrast with the conclusion of Norgett et al. (9,10), who suggested that estimates of γ based on elasticity theory could be open to considerable error. They simulated dissociated edge and screw dislocations in a model copper crystal, and found from the partial spacings differences between the estimated and true γ of up to 50%. Their method for determining the equilibrium spacing was different from the one employed here, however, and further study of the relationship between partial spacing and γ is required in order for the most important method of determining this physical parameter to be assessed.

4. Cross Slip of the Screw Dislocation

Cross slip is an important process in the deformation of crystalline materials, particularly so in hcp metals where the number of slip systems is so limited. In the crystals of the present work, the screw dislocation in a stress-free crystal dissociates in the basal plane, as discussed above, and cross slip onto the first-order prism plane requires constriction. A similar mechanism must operate in the basal-slip metals such as Mg, Cd, Zn. It has been simulated here by applying an increasing shear stress on the prism plane (in the $\langle 11\bar{2}0\rangle$ direction) until constriction and cross slip occur.

Consider a right-handed screw dislocation to lie along the z axis with indices $[2\bar{1}\bar{1}0]$, and take y along $[0001]$ so that x points along $[0\bar{1}10]$. As the x axis is traversed through the centre of the core in the ℓj5 crystal, the relative displacement of pairs of atoms on the basal planes on either side of the slip plane (see fig.2(a)) varies as shown in fig.2(b); Δu_z is the difference in z (screw) displacement and Δu_x the difference in the x (edge) displacement. In the case

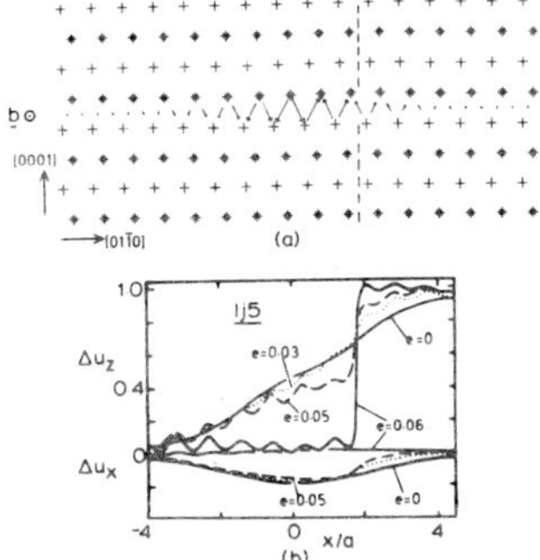

FIG. 2

(a) Atomic positions projected onto the $(2\bar{1}\bar{1}0)$ plane for the screw dislocation (ℓj5 potential). The dashed line denotes the cross-slip plane.

(b) Variation of relative displacement along the basal plane during the process of cross slip under applied strain $e \equiv e_{31}$.

of zero external strain ($e \equiv 0$), Δu_z increases from 0 to $a(\equiv b)$ with two maxima in gradient corresponding to the Shockley partials at $x/a \simeq -1.5$ and $+2$, and $|\Delta u_x|$ rises to a maximum in the middle of the fault. As the shear strain e acting on the prism planes is increased, the core is redistributed with the effect that the Burgers vector of the left-hand partial decreases and that of the right-hand partial increases, and the stacking fault tends to disappear. At a critical strain of 0.053, the constriction process is complete and the right-hand dislocation becomes perfect, and, as seen in the figure, at the higher strain of 0.06 prism slip has occurred leaving the original basal slip plane dislocation-free; the screw dislocation has slipped downwards in the -y direction. The cross-slip plane is denoted by a dashed line in fig.2(a), and the relative displacement of atoms on either side of this plane as a function of strain is shown in fig.3. It can be seen that the core is very narrow on the prism plane until just before constriction and cross slip occur. The process is similar in the ℓj6 case, except that γ is lower and the critical strain (0.092) therefore higher.

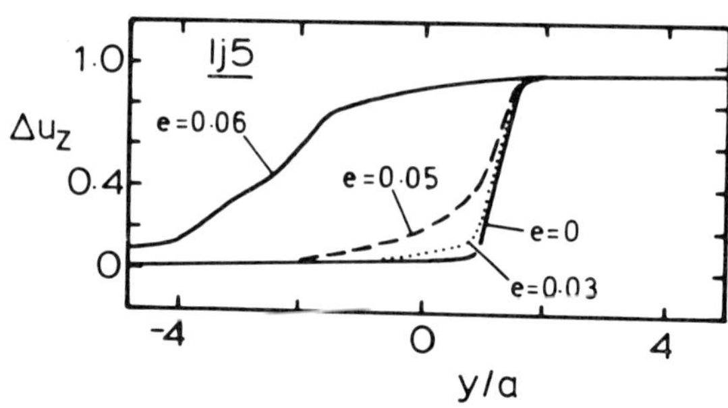

FIG. 3

Variation of relative displacement across the dashed prism plane of fig.2(a) during constriction and cross slip.

The inherent asymmetry of cross slip observed here has an explanation found in the structure of the dissociated dislocation (1). The two edge components are opposite in sign, that of the left-hand partial in fig.2 being negative and that of the right-hand being positive. The prism-plane spacing is therefore reduced below the partial on the left and above that on the right. This induces an asymmetry under shear on the prism plane in the sense that when the shear is such as to make the right-handed screw move down, it reduces the Burgers vector of the partial on the left and increases that of the partial on the right. The effect is reversed under reversal of the sense of shear. A similar process must occur in any structure in which the screw dislocation dissociates into partials with edge components.

This study therefore provides a clear picture of cross slip of a gliding screw dislocation. It does not take place by the partials moving together and constricting at the mid-point of the core, as commonly sketched, nor does constriction necessarily occur at the leading partial. It arises at the partial favoured by the edge-component asymmetry, and this may be the leading or trailing one depending on the sense of the shear stress acting on the cross-slip plane. This model strictly applies to an infinite, straight dislocation at 0 K, and therefore neglects the effects of double-kink nucleation and thermal activation. They must lower the critical strain for slip on any plane, but there seems to be no reason why the asymmetry should not persist.

Acknowledgement

This research was supported by a grant from the S.R.C.

References

1. D.J. Bacon and J.W. Martin, Phil. Mag. (in the press).
2. D.J. Bacon and J.W. Martin, ibid.
3. P.Regnier and J.M. Dupouy, Phys. Stat. Sol. 28, K55 (1968).
4. F.C. Frank and N. Thompson, Acta Met. 3, 30 (1955).
5. W.R. Tyson, Acta Met. 15, 574 (1967).
6. A. Akhtar and E. Teghtsoonian, Met. Trans. A 6, 2201 (1975).
7. A. Akhtar and E. Teghtsoonian, Acta Met. 19, 655 (1971).
8. T.E. Mitchell, Prog. Appl. Mater. Res. 6, 117 (1964).
9. M.J. Norgett, R.C. Perrin and E.J. Savino, J. Phys. F: Metal Phys. 2, L73 (1972).
10. R.C. Perrin, J. de Physique (Colloque C7) 35, C7-103 (1974).

THE LOW TEMPERATURE FLOW STRESS OF BODY-CENTRED CUBIC MATERIALS

Z.S. Basinski and M.S. Duesbery
Division of Physics, National Research Council of Canada
Ottawa, Canada, K1A 0R6

Abstract

A first-principles interatomic potential for potassium has been used to compute the Peierls stress for a rigid screw dislocation at OK as a detailed function of orientation for uniaxial tension and compression. The calculated relative orientation dependence is markedly different from that observed in body-centred cubic transition metals, but is in reasonable agreement with experiments performed on potassium. The calculated magnitude of the Peierls stress, however, is uniformly a factor of three too large. Some possible explanations for this discrepancy are discussed.

1. Introduction

The fundamental role played by the a/2 ⟨111⟩ screw dislocation in determining the low-temperature plastic properties of body-centred cubic metals is now quite well documented, both experimentally and theoretically; a number of excellent reviews exists (1-4) and the reader is referred to these for background. To summarize, the large, temperature dependent flow stress, its strong orientation dependence and the unusual slip geometry in bcc metals are all capable of explanation, in principle at least, by the hypothesis of a non-planar screw dislocation core, a concept deriving from an original suggestion by Hirsch (5). Quantitative agreement between theory and experiment has yet to be achieved and progress in this area has been one of the aims of the work reported in this manuscript.

Much of the theoretical work which has contributed to the present understanding has relied on computer modeling of the core of a rigid infinite screw dislocation at OK and its behaviour under stress (4). For the most part the various workers, using widely different interatomic central force models, are in broad agreement on the nature of the core structure (i.e., three-fold symmetric with total extension of 1-2b, where b is the magnitude of the total Burgers vector) and on the magnitude of the Peierls stress at OK in pure shear (reported values range from 0.02 μ to 0.04 μ, where μ is the ⟨111⟩ shear modulus). Unfortunately these figures are half to an order of magnitude larger than the typical experimental values of 0.002 μ in K (e.g. 6) and 0.007 μ in Fe (e.g. 7). This discrepancy is too large to be explained by neglect in the calculations of zero-point motion of the ions or of quantum-mechanical tunneling: possible quantum effects are barely observable in close packed metals (8,9), indicating that in bcc metals such processes will be significant only when the applied stress is very close to the Peierls stress. Similarly, spontaneous creation of kink pairs will be of importance only in this same high-stress regime (10).

Most of this computer modeling work utilizes pure shear strain to develop the constraint inducing dislocation motion, on the tacit assumption that the components of stress generated in addition to pure shear stress by anisotropic elastic couplings do not influence the core. The possibility that this might be responsible for the Peierls stress discrepancy was recognized some time ago and modeling calculations were performed, albeit still for the rigid OK case, for the physically realizable constraints of uniaxial stress by a few workers. Basinski, Duesbery and Taylor (11), using a first-principles interatomic potential appropriate for sodium, found Peierls stresses in tension for some orientations reduced by a factor of two from those calculated in pure shear strain for the same potential (12); due to a subsequently discovered program error (see section 2), these results are approximate only. Vitek (10) has also treated uniaxial stress, using a number of empirical potentials all with elastic constants appropriate for iron, showing a similar reduction by a factor of two in the Peierls stress for some orientations, but still with a lower limit three times larger than observed experimentally.

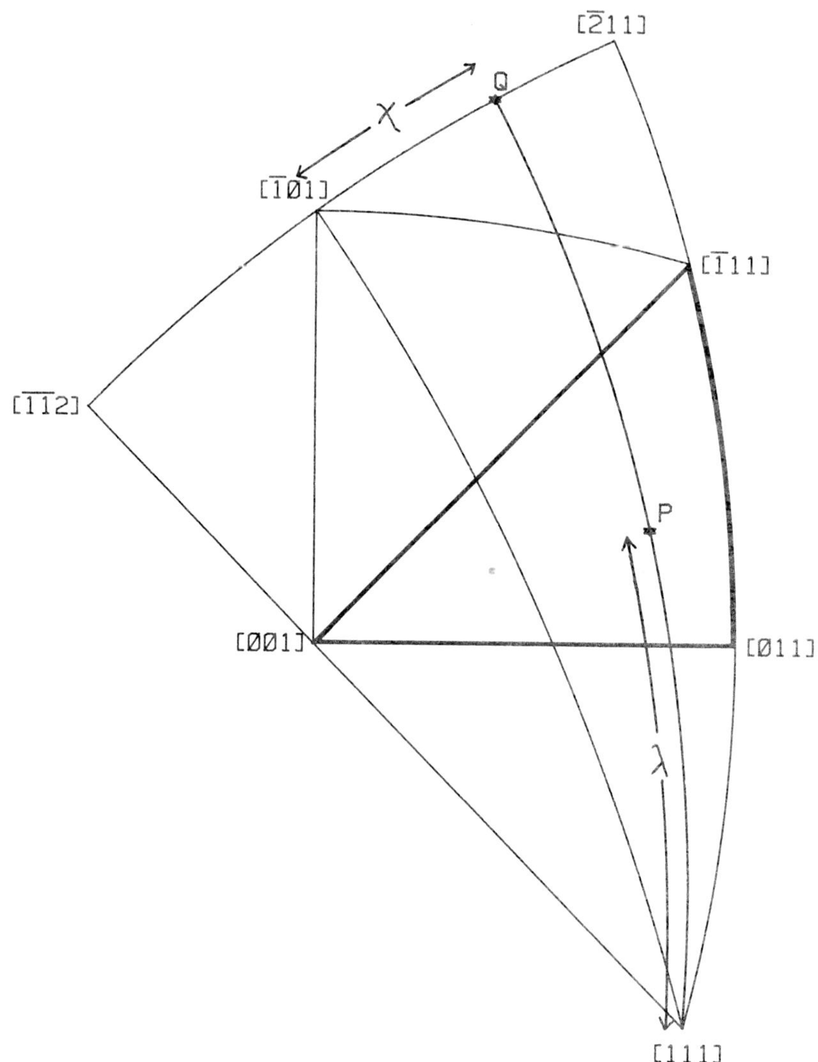

FIG. 1
Definition of orientation (P) via the pole (Q) of the plane bearing the maximum resolved shear stress.

Vitek's (13) approach was to treat a small number of orientations using a range of empirical interatomic potentials with carefully graded anelastic properties: this serves to give a broad overview of the likely effects to be encountered in uniaxial stressing, but does not give sufficient detail to span the entire accessible range of orientation. More extensive calculation would be desirable, for example, to determine whether the modeled core could fully account for the observed orientation dependence of the flow stress. A second aim of the work reported here has been to fill this gap by adopting an interatomic potential with a sound physical basis to compute the Peierls stress for a rigid screw dislocation at OK as a detailed function of uniaxial stress orientation.

2. Theoretical Background

In this section we discuss briefly the assumed interatomic potential, the numerical methods and procedures and the small amount of dislocation and elasticity theory necessary for an understanding of the later sections. Because of space limitations, extensive references are made to the literature.

2.1 The Interatomic Potential

Use has been made of a central (plus volume term) pair potential constructed by Dagens, Rasolt and Taylor (13). While the reader is referred to the original paper for details, it may be noted that the essential new physics involves folding into the central potential all two-body contributions from N-body terms to infinite order; in conventional pseudopotential theory these additional terms are ignored. The potential has been used successfully to account for a number of the physical properties of potassium, for example the phonon dispersion curves (13), thermodynamic properties (14,15), the liquid structure factor (16,17) point defect energies (18) and self-diffusion properties (19), electrical resistivity (20,21), thermal resistivity (22) and thermoelectric power (23). It is reasonable to state that within the limitations of current physics, there is little if anything that can be added to our knowledge of interionic forces in potassium. Thus if this potential cannot give a reasonable quantitative account of the Peierls stress and its orientation dependence, then serious doubts must arise about either the physics behind the potential or the validity of the screw dislocation core model.

2.2 Numerical methods and self-consistency

The numerical method has been described in earlier publications (11,12,24,25). For most calculations a fully relaxable block of dimensions ~ 20b x 20b normal to the dislocation line was used, embedded in a rigid elastic continuum, with periodic boundary conditions parallel to the dislocation line. This block size is substantially smaller than the 45b x 45b size used in the references cited above and results in an overestimate of the absolute Peierls stress by ~ 10%; use of the smaller size is considered justified because of the smaller error (~ 2%) in the relative Peierls stresses and because of the greatly enhanced throughput. The size dependence is discussed further in Section 4.

The uniaxial stress is applied by operating on the dislocated lattice block with the strains computed from the applied stress using the elastic compliances. A word of caution is appropriate here; in order to preserve the periodic boundary conditions, the [111] line direction must remain invariant, so that rotations are required to complement the strains. Earlier work (11) did not incorporate these rotations. A self-consistency check which rules out errors of this kind is to recompute the stresses directly from a strained perfect lattice (12) and compare these with the applied stress tensor.

2.3 Symmetry and Orientations.

The stable screw dislocation core in potassium is the threefold symmetric, twofold degenerate structure widely reported in the literature (4). To study fully the response of this dislocation to arbitrary uniaxial stress it is necessary to consider orientations spanning 1/12 of the stereographic projection and further to treat both tension and compression (26). Physically this can be understood as follows. Introducing a single dislocation into a perfect cubic lattice destroys the global fourfold ⟨100⟩ symmetry, so that the usual irreducible 1/48 of the stereographic projection (the "standard triangle") is replaced by an irreducible 1/12 corresponding to a point group of lower symmetry. The existence of the second core configuration doubles the new point group to give an irreducible 1/6 of the stereographic projection if only one sense of stress is considered, or 1/12 if both tension and compression are treated.

For reference we shall adopt the spherical triangle with vertices at [111] (also the Burgers vector, by convention) and at [$\bar{1}$01] and [$\bar{2}$11]. To define orientations the maximum resolved shear stress system will be used, referred to Cartesian axes $x_1 \equiv [\bar{1}2\bar{1}]$, $x_2 \equiv [\bar{1}01]$ and $x_3 \equiv [111]$. The reference triangle, axes and orientation definition are illustrated in

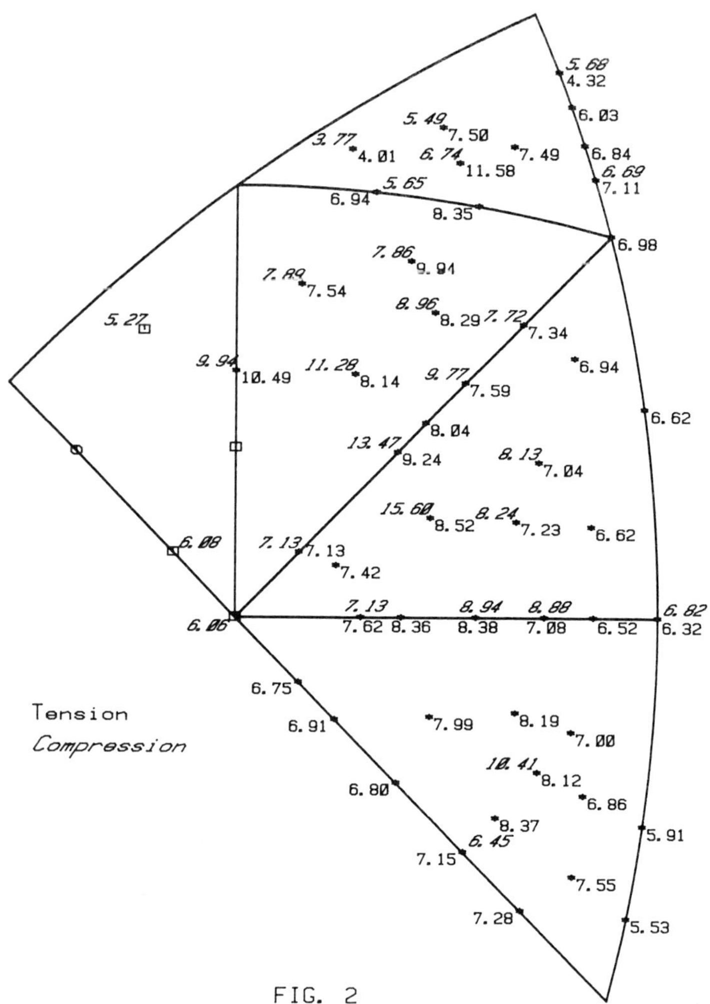

FIG. 2

Orientation dependence of the critical mrss in potassium calculated at 0K. Stresses in units of $10^{-3}\mu$.

fig. 1. The orientation P is defined completely by the orthogonal coordinates λ, the angle between the stress axis and the Burgers vector and χ, the angle between the x_2 axis, $[\bar{1}01]$, and the pole of the slip plane bearing the maximum resolved shear stress. The usual 1/48 of the projection, with vertices at [001], [011] and $[\bar{1}11]$ is indicated by heavy lines.

2.4 Elasticity

In a system of imposed stresses, the appropriate elastic constants are the compliances. Referred to the Cartesian axes defined in 2.3, the compliance tensor indicates anisotropic couplings of the glide strain to the applied stress via the components s_{15}, s_{25}, and s_{46} (where s_{ij} is the elastic compliance tensor in Voigt notation). For potassium and other materials with an anistropy factor $2s_{44} < s_{11} - s_{12}$, this means that a positive shear strain e_{13} is induced by a positive non-glide shear stress $(\sigma_{11} - \sigma_{22})$. It follows that a glide strain can be induced by non-glide stresses and that for orientations in which the signs of the glide stresses and of the shear stress $(\sigma_{11} - \sigma_{22})$ are opposite, there exists a regime for which the glide shear strain is zero or opposite in sign to the glide shear stress. These possibilities are discussed fully by Duesbery (26) and will be mentioned in a later section of this manuscript.

3. The Calculated Peierls Stresses

In Fig. 2 are shown the calculated Peierls stresses as a function of orientation. The numerical values represent the critical shear stresses for the first core instability, resolved in the plane of maximum resolved shear stress, in dimensionless units of 10^{-3} μ; erect font refers to uniaxial tension, while italicized values indicate results obtained in compression. In each case the initial core configuration has been chosen so that the calculated stress corresponds to the stress required for the identity translation rather than the equivalence translation (12) i.e. to the stress required for free propagation of the dislocation. The open square and circular symbols in Fig. 2 have a particular significance. For these orientations, in at least one sense of stress, the first instability of the core results not in dislocation movement but in the ejection of an incipient twinning partial dislocation, as observed for some empirical potentials in pure shear strain (25) and in uniaxial stress (10). The square symbols indicate those orientations for which this behaviour is observed in tension only, the circular symbols for both tension and compression. No numbers are given for these orientations since the partial dislocation ejection is an ill-defined process with no clear critical stress. It may be noted, however, that the mechanism operates at or below the stresses needed for normal glide. This behaviour is treated more fully elsewhere (26).

3.1 Orientations within the standard triangle

It is clear from Fig. 2 that the calculated Peierls stresses (resolved in the plane of maximum resolved shear stress) for potassium are lower than those reported in the literature for other materials or for empirical potentials, whether for applied shear strain or stress. Even within the standard triangle there is a substantial orientation dependence, with a maximum value in tension at $[\bar{1}25]$ some 30% higher than the minimum value at [011]; in compression the effect is much larger, the same $[\bar{1}25]$ orientation showing a critical stress more than twice that for [011]. There is a distinct trend for values in the centre of the triangle to be larger than at the boundaries. This is in contrast to experimental observations for transition bcc metals, in which the boundary orientations, in particular the triangle vertices, are harder than the central orientations (see (27,28) for example).

The calculated Peierls stresses in compression are uniformly larger than for the same orientation in tension and a small inverse asymmetry of slip is apparent. This latter observation agrees with calculations made for sodium in pure shear strain (12), but is contrary to all experimental evidence for transition metals, for which the Peierls stress for orientations having $\{112\}$ as the plane of maximum resolved shear stress is invariably smaller for the twinning direction than for the reverse.

The picture presented, in summary, gives promising new values for the Peierls stress, but disagrees with the established orientation dependence for transition metals. Because of this discrepancy, experiments were performed in this laboratory to measure the magnitude and orientation dependence of the Peierls stress in potassium single crystals deformed in tension at temperatures down to 1.5 K (29). The results show that there is an orientation dependence of the Peierls stress in quite close agreement with the calculated dependence of Fig. 2, such that the three corner orientations exhibit roughly equal Peierls stresses, while the central region of the triangle is about 30% harder. The experimental magnitudes are uniformly a factor of three less than the calculated values: this will be discussed in section 4.

3.2 Orientations outside the Standard Triangle

These orientations correspond on an experimental basis to secondary slip systems in that with an operational Schmid law there is always a crystallographically equivalent primary slip system (i.e. orientations within the standard triangle) for which the resolved shear stress is larger. Two striking features emerge from this part of the calculations. The first is the dependence of the Peierls stress on the λ-coordinate, which is at least as large as the χ-dependence. The importance of this dependence can be understood by noting that for a core model of splitting into pure screw partials, the Peierls stress should be independent of the λ-coordinate. This strong dependence is a clear indication that there is some non-screw character in the dislocation core.

The second feature of interest is the dramatic drop in the Peierls stress for λ approaching 90° and $\chi > 0$. Considering orientations lying in the zone from [111] to [$\bar{2}$11], for example, it can be seen that the Peierls stress increases smoothly to a maximum close to [$\bar{6}$55] and then drops sharply, tending asymptotically to zero at [$\bar{2}$11]. As closely as can be determined at present, the maximum value corresponds to an orientation for which the resolved glide shear strain due to the glide stress is balanced exactly by the (negative) shear strain induced by the anisotropic couplings to the stresses σ_{11} and σ_{22}, so that the net glide shear strain is zero. The regime of decreasing Peierls stress then has the unusual property that the glide shear strain is negative for positive glide shear stress.

The occurrence of vanishing Peierls stress for secondary slip systems brings to mind immediately the phenomenon of so-called anomalous slip in transition bcc metals (30,31), in which crystallographic slip on secondary slip systems bearing very low resolved shear stresses predominates for some orientations at low temperatures. This will be discussed more fully elsewhere (26).

4. Discussion

It is clear from the literature that the orientation dependence of the Peierls stress in bcc metals is complicated. The present work shows that for potassium at least, theory and experiment present a consistent picture in part. The indicated involvement of non-screw character in the dislocation core and of anisotropic elastic couplings are clear handles on the underlying mechanisms responsible for the orientation dependence; the tying in of anistropic elasticity is of particular interest with respect to the different orientation dependence observed in transition metals, since the elastic anisotropy is radically different in these metals. Overall, there is reason for optimism that a resolution of this problem is within our grasp.

4.1 The Magnitude of the Peierls Stress

The present results show that the calculated orientation dependence of the Peierls stress is in fair agreement with experiment. The calculated magnitudes, however, are uniformly a factor of three too large. As mentioned in section 2.1, the interatomic potential cannot be used as an excuse. To check the precise effects of size, a series of runs was done for the [011] orientation at steadily increasing block sizes up to 50b x 50b. The asymptotic value, saturating at a size of 40b x 40b, was $5.8 \times 10^{-3}\ \mu$, still a factor of 2.4 larger than the experimental Peierls stress. A further possible source of discrepancy lies in the method of experimental measurement. Due to the notoriously ill-defined yield process in bcc metals, it is customary to adopt a fixed plastic set as defining the initial flowstress - in (29), a figure of 0.1% was used; at this point the hardening is still largely elastic. If the stress is measured at an offset of 0.5%, the measured Peierls stress is increased only by ≈ 2%. Thus there is a real discrepancy between theory and experiment.

For reasons given earlier, it is unlikely that quantum effects or kink pair creation could have such a large effect. However, it has been demonstrated that non-glide stresses have a strong influence on the Peierls stress and can in extreme cases reduce it to zero. It is proposed, therefore, that apparent yield can take place below the nominal Peierls stress by activation at regions for which the Peierls stress is reduced by internal stress concentrations. Since the depression of the Peierls stress is nearly proportional to the Peierls stress (26), the gross features of the orientation dependence would be preserved. It would be expected, however, that experiment would show a subtantial scatter, as is indeed observed.

4.2 The Peach-Koehler Force

There is a distinct dichotomy associated with atomistic interpretations of dislocation behaviour, stemming from the Peach - koehler (32) notion of the thermodynamic force on a dislocation. In modeling mechanics, whether the applied stimulus is strain or stress, transmission of the constraint to the defect takes place by a coordinate transformation, i.e. a strain. In the absence of any other information, it would be natural to expect instability, i.e. slip to occur at a critical glide strain. The work of Peach & Koehler, however, shows that thermodynamics requires, instead, that the stress should be the driving tensor. The anisotropic couplings mentioned earlier afford an ideal opportunity to examine this. Glide strain can be applied directly, by a glide stress, or indirectly, in a positive or negative sense via the coupling constants to the orthogonal shear stresses (26). We have examined this by applying large glide strains via couplings to σ_{11} and σ_{22} only (no glide stress): slip does not occur until a glide stress is applied, regardless of the sense of the glide strain.

This confirmation that the atomistic models are consistent with thermodynamics is gratifying, but does not really resolve the issue. For example, at the orientation for which the glide shear strain is identically zero (see section 3.2), dislocation motion does occur at a sufficiently high stress. The physics is clear, but its application to the atomistic model is not; we do not yet have an adequate interpretation of modeling calculations.

5. Acknowledgments

The authors wish to acknowledge the indispensable contribution to this work of H.G. Champion for his tireless efforts in computer maintenance and hardware development.

References

1. J.W. Christian, 2nd Int. Conf. on Strength of Metals and Alloys, Asilomar, ASM, p. 31, (1970).
2. J.W. Christian, 3rd Int. Symposium "Reinstoffe in Wissenschaft und Technik", p. 263, Akademie Verlag, Berlin (1970).
3. B. Vestak, 3rd Int. Symposium "Reinstoffe in Wissenschaft und Technik", p. 221, Akademie Verlag, Berlin (1970).
4. V. Vitek, Crystal Lattice Defects, 5, 1 (1974)
5. P.B. Hirsch, 5th Int. Conf. on Crystallography, Cambridge (1960).
6. H.O.K. Kirchner, Acta Phys. Austriaca, 48, 111 (1978).
7. T.L. Altshuler and J.W. Christian, Phil. Trans. Roy. Soc. A261, 253 (1976).
8. Z.S. Basinski and M.S. Duesbery, unpublished work.
V.D. Natsik, V.P. Soldatov and V.I. Startsev, 5th Int. Conf. on Strength of Metals and Alloys, Aachen, Pergammon, Oxford (1979).
10. V. Vitek, Proc. Roy. Soc, Lond., A352, 109 (1976).
11. Z.S. Basinski, M.S. Duesbery and R. Taylor, Interatomics Potential and Simulation of Lattice Defects, Battelle Colloquium (eds. P.C. Gehlen, J.R. Beeler and R.I. Jaffee), P. 537, Plenum Press, New York.
12. Z.S. Basinski, M.S. Duesbery and R. Taylor, Can. J. Phys., 49, 2160 (1971).
13. L. Dagens, M. Rasolt and R. Taylor, Phys. Rev. B11, 2726 (1975).
14. S.S. Cohen and M.L. Klein, Phys. Rev. B12, 1984 (1975).
15. S.S. Cohen, M.L. Klein, M.S. Duesbery and R. Taylor, J. Phys. F, 6, 337; ib id, 6, L271 (1976).
16. G. Jacucci, M.L. Klein and R. Taylor, Solid State Comm., 24, 685 (1977).
17. G. Jacucci, I.R. McDonald and R. Taylor, J. Phys. F, 8, L121, (1978).
18. G. Jacucci and R. Taylor, J. Phys. F, 9, 1489 (1979).
19. A. DaFano and G. Jacucci, Phys. Rev. Lett., 39, 950 (1977).
20. R.C. Shukla and R. Taylor, J. Phys. F, 6, 531 (1976).
21. R. Taylor, C.R. Leavens and R.C. Shukla, Solid State Comm., 19, 809 (1976).
22. C.R. Leavens, J. Phys. F, 7, 163 (1977).
23. C.R. Leavens and R. Taylor, Thermoelectricity in Metallic Conductors (eds. F.J. Blatt and P.A. Schroeder), p. 131, Plenum Press (New York) (1978).
24. Z.S. Basinski, M.S. Duesbery and R. Taylor, Phil. Mag. 21, 1201 (1970).
25. M.S. Duesbery, V. Vitek and D.K. Bowen, Proc. Roy. Soc. Lond. A332, 85 (1973).
26. M.S. Duesbery, to be published.
27. D.K. Bowen, J.W. Christian and G. Taylor, Can. J. Phys., 45, 903 (1967).
28. A.S. Keh and Y. Nakada, Can. J. Phys. 45, 1101 (1967).
29. Z.S. Basinski, M.S. Duesbery and G.S. Murty, to be published.
30. M.S. Duesbery and R.A. Foxall, Phil. Mag., 20, 719 (1969).
31. A.J. Garratt-Reed and G. Taylor, Phil. Mag., 39, 597 (1979).
32. M.O. Peach and J.S. Koehler, Phys. Rev. 80 436 (1950).

ATOMISTIC STUDIES OF DISLOCATIONS AND STACKING FAULT-TYPE DEFECTS IN $L1_2$ ORDERED STRUCTURES

M. Yamaguchi*, V. Vitek and D. P. Pope

Department of Materials Science and Engineering
University of Pennsylvania
Philadelphia, PA 19104

ABSTRACT

The stability of possible planar faults on {111} and {100} planes is discussed using crystallographic symmetry considerations and the results of atomistic calculations of γ- surfaces. In general it is found that the APB on {100} planes and SISF on {111} planes are always stable while APB and CSF on {111} planes can be either stable or unstable. Results of atomistic calculations of the core structure of <110> screw dislocations dissociated on {111} planes are also presented. Different core configurations, some of them non-planar, are obtained, depending on the stability of APB and CSF. Possible implications of the different core structures for the plastic behavior of $L1_2$ alloys are discussed.

1. Introduction

Although the $L1_2$ structure is based on an f.c.c. lattice the plastic behavior of these alloys substantially differs from that of pure f.c.c. metals or disordered f.c.c. alloys. The most remarkable phenomenon, common to all $L1_2$ alloys, is the anomalous increase of the yield stress with increasing temperature observed at high temperatures (for a review see (1-3)). Furthermore, certain asymmetries of the yield stress have been predicted in this temperature region (4). Such plastic behavior is reminiscent of the low temperature plastic behavior of b.c.c. metals, which is governed by the cross-slip of screw dislocations the cores of which possess a special sessile configuration (5-7). It is, therefore, likely that the unusual features of the plastic behavior of $L1_2$ alloys are also related to the cross-slip of screw dislocations which is further controlled by their dissociation and/or core structure. It has, indeed, been proposed earlier that it is the cross-slip of screw dislocations from (111) to (100) planes and their low mobility on the latter planes which cause the anomalous increase of the yield stress at high temperatures (8).

In order to understand the cross-slip phenomenon we first need to know the possible modes of dislocation dissociation, core structure of the corresponding partials and/or the cores of total dislocations if dissociation into well defined partials does not take place. Although some guidelines for the trends of these properties may be obtained from a hard sphere model an unambiguous and complete investigation can only be performed by carrying out atomistic calculations. These studies can first reveal which metastable stacking fault-like defects, if any, may exist and thus indicate possible dislocation dissociations. This can be done by calculating the γ- surface** for corresponding crystallographic planes. Secondly, the dislocation core structure can be directly studied using these methods (6).

In the present paper we first summarize the results on the faults on (111) and (100) planes which have been obtained using a crystallographic analysis and by calculation of the γ-surfaces (10, 11). We further present results of the calculations of the core structure of superpartials of a $[\bar{1}01]$ screw dislocation dissociated on the (111) plane. These calculations have been made for several cases of varying stability of faults on the (111) plane which leads to significant

* On leave from Department of Materials Science and Engineering, Osaka University, Suita, Osaka, JAPAN

** The γ-surface is the energy of the stacking fault characterized by the displacement vector $\underset{\sim}{f}$ plotted as a function of f .

variations in the core structure. Possible implications of these different core structures for the glide and, in particular, the cross-slip of these dislocations are discussed.

2. Planar faults on {111} and {100} planes

Figure 1 shows the atomic arrangement on the (111) plane in an A_3B alloy with the $L1_2$ structure. Open and closed circles represent A and B atoms respectively. The layers labelled a, b and c constitute the repeat stacking along the [111] direction. Stacking faults can be formed by cutting the crystal, for example, between the a- and b- layers and displacing the upper half by a vector, $\underset{\sim}{f}$ with respect to the lower half. On the basis of a hard sphere model the anti-phase boundary (APB), complex stacking fault (CSF), and superlattice intrinsic stacking fault (SISF), characterized by vectors, $\underset{\sim}{b}_A$, $\underset{\sim}{b}_c$ and $\underset{\sim}{b}_s$, respectively, have been proposed to be the possible stable faults (1). However, since two different species of atoms are involved in this structure the application of the hard sphere model is not a priori justified and it is likely that similarly as in the case of b.c.c. lattices (6) predictions based on this model will be inadequate.

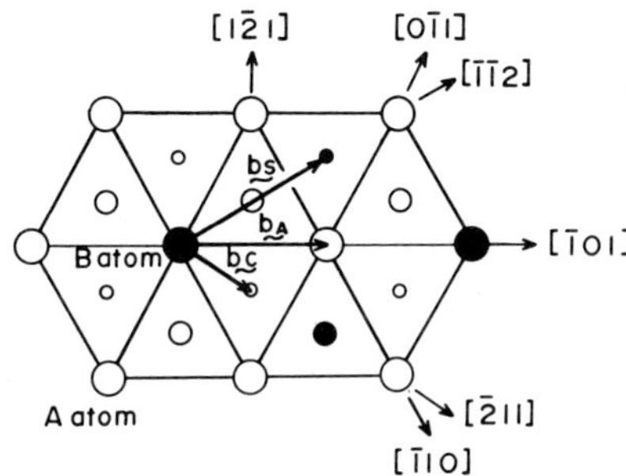

Fig. 1: Atomic arrangement on the (111) plane in an A_3B alloy with the $L1_2$ structure

Recently, we have carried out a systematic investigation of the stability of planar faults in ordered alloys using symmetry considerations and by calculating the γ-surfaces (10, 11). The symmetry arguments developed in the studies of grain boundaries in bicrystals (12) and our recent investigations (10, 11), lead to the following rules:

(i) Among various crystal symmetry elements, only mirror planes perpendicular to the fault plane affect the stability of planar faults.

(ii) If one mirror plane is preserved when displacing the two halves of the crystal by a vector $\underset{\sim}{f}$, the corresponding dependence of the fault energy on the displacement (γ surface) may, but need not, have an extremum for this value of $\underset{\sim}{f}$. In the latter case there is either no extremum in the vicinity of $\underset{\sim}{f}$ or a local extremum exists for the displacement vector, $\underset{\sim}{f} + \underset{\sim}{f}'$ where $\underset{\sim}{f}'$ is a vector parallel to the conserved mirror plane. A stable fault is found when the local extremum is a minimum. The case that $\underset{\sim}{f}' = 0$ is not, of course, excluded.

(iii) When displacement by $\underset{\sim}{f}$ conserves at least two intersecting mirror planes, the γ-surface must have a local extremum for $\underset{\sim}{f}$. If this extremum is a minimum, there is a stable fault for $\underset{\sim}{f}$ and its stability is guaranteed on symmetry grounds.

For the case of the (111) plane in the $L1_2$ structure, there are three mirror planes perpendicular to (111), namely ($\bar{1}$01), (0$\bar{1}$1) and ($\bar{1}$10), composed of A and B atoms. Obviously, the mirror plane is preserved by any displacement parallel to it and by those displacements perpendicular to it the magnitude of which is an integral multiple of the interplanar spacing between the corresponding mirror planes. Therefore, as seen from Fig. 1, the displacement by $\underset{\sim}{f} = \underset{\sim}{b}_s$ conserves all three mirror planes whilst displacements by $\underset{\sim}{f} = \underset{\sim}{b}_A$ and $\underset{\sim}{f} = \underset{\sim}{b}_c$ conserve only one of the mirror planes, ($\bar{1}$01) and (0$\bar{1}$1), respectively. It is clear from hard sphere considerations that if the γ-surface has extrema for $\underset{\sim}{f} = \underset{\sim}{b}_s$ and in the vicinity or at $\underset{\sim}{f} = \underset{\sim}{b}_A$ and $\underset{\sim}{f} = \underset{\sim}{b}_c$, they will be minima. Thus, the following conclusions regarding the stability of these three possible faults may be drawn:

(i) SISF is always stable. The γ-surface must have a local minimum for $\underset{\sim}{f} = \underset{\sim}{b}s$.

(ii) APB and CSF are either unstable or stable with $\underset{\sim}{f} = \underset{\sim}{b}_A + \frac{\chi_A}{6}[1\bar{2}1]$ and $\underset{\sim}{f} = \underset{\sim}{b}_c + \frac{\chi_c}{6}[2\bar{1}\bar{1}]$, respectively. The values of χ_A and χ_c depend on the interatomic interactions and cannot be determined on a crystallographic basis; in special cases they may be equal to zero.

On the {100} planes, there is only one possible fault, APB with $\underset{\sim}{f} = \underset{\sim}{b}_A$. It is always stable since the displacement by this vector conserves two orthogonal {100} minor planes.

γ-surfaces on (111) and (001) planes were calculated using several different series of central force potentials always representing a mechanically stable $L1_2$ lattice but leading to a large variation in the APB energy. They are described in detail in ref. (10). The results of calculations are in full agreement with the above predictions based on the symmetry arguments. For some potentials, stable APB and CSF were obtained with χ_A and χ_c varying from 0.1 to 0.2 In general the stability of APB and CSF decreases with an increase of their energies.

3. Configuration of <110> screw dislocations dissociated on {111} planes

Static relaxation method with fixed boundaries was used (for a review, see (6)). The crystallites simulated were rectangular parallelipipeds bounded by the planes, $(1\bar{2}1)$, (111) and $(\bar{1}01)$ containing about 1000 relaxed atoms. Dislocations lying parallel to the $[\bar{1}01]$ direction were introduced by originally displacing all the atoms in accordance with the corresponding anisotropic elastic displacements. Calculations were carried out for the following three cases:

(i) APB and CSF are stable, and either the four-fold dissociation (13) involving both these faults or the two-fold dissociation involving only APB are energetically favorable.

(ii) APB and CSF are stable, but either the six-fold dissociation (14) involving SISF together with APB and CSF or the two-fold dissociation involving only SISF are energetically favorable.

(iii) APB and CSF are unstable. In this case, only the two-fold dissociation with SISF is possible.

For case (i) the $[\bar{1}01]$ screw dislocation splits into well separated superpartials with the Burgers vectors $\frac{1}{2}[\bar{1}01] \pm \frac{\chi_A}{6}[\bar{1}2\bar{1}]$. The core structure of these dislocations can, in general, be interpreted as dissociated into partials bounding the CSF. When the four-fold dissociation is favored these partials are well separated while when the two-fold dissociation is favored they overlap. In both cases, however, purely planar splittings and/or cores occur.

In case (ii), the $[\bar{1}01]$ screw dislocation is split into well separated 1/3 $[\bar{2}11]$ and 1/3 $[\bar{1}\bar{1}2]$ superpartial dislocations. However, the core structure of these two dislocations is no longer planar.

Fig. 2 shows the differential displacement map (6) of the core of a 1/3 $[\bar{2}11]$ dislocation calculated using a potential which gives $\gamma_{APB} = 312\ mJ.m^{-2}$, $\gamma_{CSF} = 392\ mJ.m^{-2}$ and $\gamma_{SISF} = 15\ mJ.m^{-2}$. The arrows in the figure show the $[\bar{1}01]$, i.e. the screw components of the relative displacements of neiboring atoms. The edge components of displacements, parallel to $[1\bar{2}1]$, are confined to the (111) plane i.e. to the plane of the SISF; they are not shown in the figure. Therefore, the core structure of the 1/3 $[\bar{2}11]$ dislocation can be interpreted as dissociation into 1/2 $[\bar{1}01]$ screw and 1/6 $[\bar{1}2\bar{1}]$ edge dislocations. The core of the screw dislocation is non planar. Its center appears to be at the point marked S and it spreads principally onto the $(1\bar{1}1)$ plane which is inclined to the plane of the SISF. On the other hand the edge dislocation core is planar spread onto the plane of the SISF.

In case (iii), a $[\bar{1}01]$ screw dislocation is also split into well separated 1/3 $[\bar{2}11]$ and 1/3 $[\bar{1}\bar{1}2]$ superpartial dislocations whose cores are again dissociated into screw and edge components. In this scase, however, the core of the screw component is spread onto the (111) plan immediately above the SISF plane so that a stacking fault of extrinsic type is created in the core region. The core of the edge component is again spread onto the SISF plane. Fig. 3 shows an example of a core structure of this type. The most likely reason for this core structure is tha

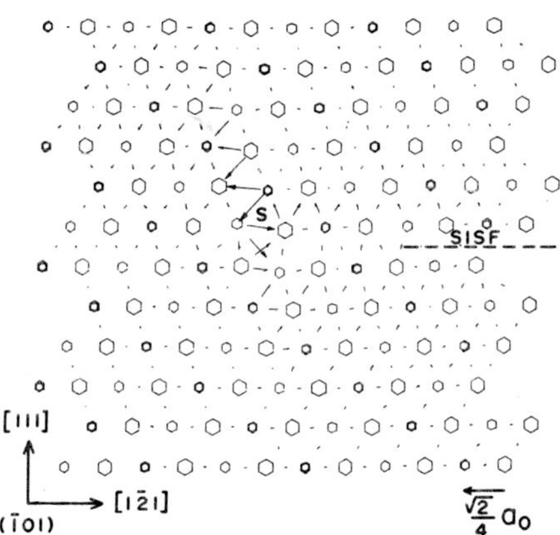

Fig. 2: Core structure of the $\frac{1}{3}$ [$\bar{2}$11] superpartial which bounds the SISF on the (111) plane. The hexagons of two different sizes represents atom in two adjacent ($\bar{1}$01) planes.

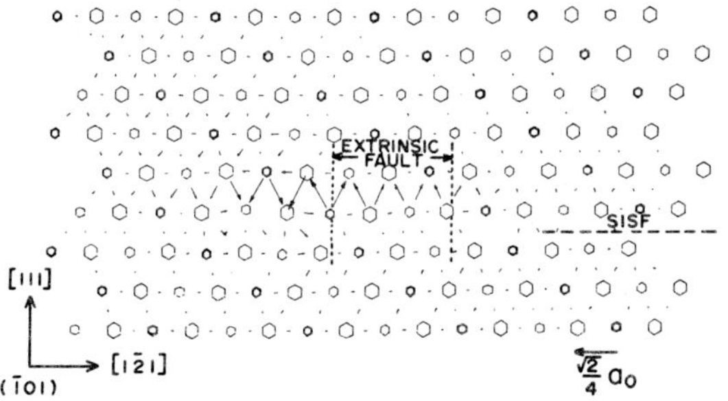

Fig. 3: Core structure of the $\frac{1}{3}$ [$\bar{2}$11] superpartial which bounds the SISF on the (111) plane. An extrinsic stacking fault is found inside the core.

the complex extrinsic stacking fault is locally stabilized although the APB and CSF are unstable. An analogous situation was found in the case of $\frac{1}{2}$ <111> screw dislocations in b.c.c. lattices where multilayer {112} faults are formed inside the cores (6).

4. Discussion

As predicted on the basis of the hard sphere model APB, CSF and SISF are all possible stable faults on {111} planes and APB can be stable on {100} planes. However, only the stability of SISF on {111} planes and APB on {100} planes is guaranteed on symmetry grounds and, therefore, these faults will be stable in every $L1_2$ alloy. On the other hand, the stability of APB and CSF on {111} planes decreases as their energy increases and thus, depending on the interatomic potentials used (i.e. in different alloys), they can be either stable or unstable. Naturally, the dissociation and core structure of <110> dislocations depend on the stability and energies of these faults.

When both the APB and CSF are stable and their energies are relatively low the <110> dislocation dissociates on a {111} plane into two superpartials the cores of which are spread onto the same plane. This configuration is well described by the four-fold splitting proposed in ref. (13). Movement of these dislocations will be confined to {111} planes and no significant Peierls (or lattice friction) stress can be expected.

When the energies of APB and CSF are high or the faults are unstable, the dissociation of the <110> dislocation on a {111} plane involves the SISF as originally suggested in ref. (14). However, the core of the corresponding superpartials is not in this case confined to the same {111} plane. When APB and CSF are stable the core is principally distributed in the {111} plane with the SISF and in another {111} plane inclined by 109.5° to the plane of the SISF. When both APB and CSF are unstable the core of the superpartials spreads into several {111} planes parallel with the plane of the SISF. In both cases a high Peierls stress can be expected. In particular in the former case the superpartials are sessile and their motion has to be associated with cross-slip. This is analogous to the case of screw dislocations in b.c.c. metals and, therefore, it is possible that this core configuration is responsible for the high low temperature strength in some $L1_2$ alloys.

The APB on {100} planes is always stable and its energy is generally low (10). Hence, on a {100} plane the <110> dislocation will always split into two well separated 1/2 <110> superpartials. Preliminary calculations of the core structure of these superpartials in screw orientation show that their cores are always spread on one of the two {111} planes intersecting in the direction of the dislocation line. Two energetically equivalent core configuraitons, therefore, exist. The slip on {100} planes, observed at high temperatures in some $L1_2$ alloys, may then be interpreted in terms of the continual cross-slip of 1/2 <110> screw dislocations on two intersecting {111} planes and the dislocation immobilization on {100} planes may be associated with this mechanism. A more detailed description of the core structure of dislocations dissociated on {100} planes as well as discussion of implications for the {100} slip and the anomalous increase of yield stress at high temperatures, will be presented in a future paper.

Acknowledgements

This work was supported by the National Science Foundation under Grant No. DMR79-05556.

References

1. M. J. Marcinkowski, Electron Microscopy and Strength of Crystals, Wiley, New York (1963), p. 333.
2. N. S. Stoloff and R. G. Davies, Prog. Mater. Sci., 13, (1966).
3. S. J. Liang and D. P. Pope, Acta Met. 25, 485 (1977).
4. C. Lall, S. Chin and D. P. Pope, Met. Trans. 10A, 1323 (1979).
5. J. W. Christian, Proc. 2nd Int. Conf. Strength of Metals and Alloys, ASM: Metals Park (1970), p. 31.
6. V. Vitek, Crystal Lattice Defects, 5, 1 (1974).
7. B. Sestak, Proc. ICSMA 5, Aachen, eds. P. Haasen et al., Pergamon Press: Oxford, p. 1461, (1979).
8. S. Takeuchi and E. Kuramoto, Acta Met., 21, 415 (1973).
9. V. Vitek, Phil. Mag. 18, 773 (1968).
10. M. Yamaguchi, V. Vitek and D. P. Pope, Phil. Mag. to be published (1980).
11. M. Yamaguchi, D. P. Pope, V. Vitek and Y. Umakoshi, to be published (1980).
12. R. C. Pond and W. Bollmann, Phil. Trans. Roy. Soc. Lond. 292, 449 (1979).
13. M. J. Marcinkowski, N. Brown and R. M. Fisher, Acta Met. 9, 129 (1961).
14. B. H. Kear, A. F. Giamei, J. M. Silcock and R. K. Ham, Scripta Met. 2, 287 (1968).

DESCRIPTION OF DISLOCATION DISTRIBUTIONS

E. Kröner

Institut für Theoretische und Angewandte Physik der Universität Stuttgart und Max-Planck-Institut für Metallforschung, Stuttgart, W-Germany

ABSTRACT

After a brief introduction into the general concept of defects in material structures this review concentrates on defects, mainly dislocations, in primitive solids, i.e. solids which consist of only one sort of structure-less particles. Many metals are an example hereof. Such materials are well described by the differential geometry of affinely connected spaces. Whereas Euclidean geometry is appropriate for compatibly distorted lattices, more general geometries are needed for the description of defect lattices. In particular, Cartan's torsion is the continuum version of the dislocation density and Riemann-Christoffel curvature is produced by foreign (=extra) matter on interstitial sites. Beside the strain space considered so far, a pertaining stress space is defined which is described by formally the same equations (of differential geometry) as the strain space.

The basic equations of the linearized theory are also given. They are particularly convenient for actual calculations. The paper concludes with a survey on solutions for the internal stress problem in terms of generalized Green functions where also a new solution of the incompatibility problem is presented.

1. INTRODUCTION

In this lecture we discuss distributions of defects, above all dislocations, in, mainly, crystalline structures. Defining defects implies generally that the state of matter which is declared as defect-free must be specified. Thus defect is not a thing in itself but related to some reference state. The defect-free state is also seen as an ordered state. In particular the undistorted perfect crystal is considered as the most ordered state of a material. But even the notion of order is relative. This is so because our intuition is formed by the Euclidean geomtery of that part of the universe into which we are born. It is certainly not against logic to imagine living beings in non-Euclidean spaces, for instance in so-called amorphous spaces. Such beings would have a completely different perception of what is order and disorder. These considerations show that it is not a priori meaningless to think of defects in structures which, with our Euclidean eyes, are specified as amorphous.

The foregoing considerations provide a strong motivation to look at defects within the mathematical framework of differential geometry because this branch of mathematics is perhaps more than others concerned with spaces. If, more accurately, one wishes to include the discrete, i.e. atomic or molecular, structure of matter, a difference geometry is the appropriate tool. However, since this geometry does not seem to be equally well developed as the differential geometry it is fortunate that for the numerical treatment of many important problems the discreteness as such of the matter is not of great relevance. Among such problems are found those of the interaction between defects over distances large compared with the elementary atomic distances. On the other hand, the form of the discreteness is indeed important for the development of static and dynamic continuum, or field, theories of crystal lattices and their defects.

Example: The defect and defect-free configurations of a primitive (Bravais) lattice, often realized in the case of metals, are completely specified by the positions of the atoms which for almost all purposes can be treated as mass-points. Lattices, like diamond, with two atoms in an elementary cell can, in some approximation, be handled like a primitive mass-point lattice, where the mass-points represent the center of gravity of the cells. However, proceeding in this way means to neglect all effects which have to do with distorsions of the cells themselves. For instance, these distortions are the causes of the optical branches in the vibration spectra of such materials. A more accurate field theory of such crystals has to take into account that the number of the macroscopic degrees of freedom of a lattice with n atoms in an elementary cell is n times that of the primitive lattice. An additional feature of covalent crystals which often possess a diamond structure is that here the atoms are not well described by the picture of structure-less mass-points because they have the well-known tetra-structure which gives rise to special effects like dangling bonds.

A useful division of defects in crystal lattices is that into topological and energetical defects. Also useful is the division into intrinsic (or eigen) and extrinsic defects. By topological defects we understand all intrinsic defects which can be defined by purely topological means, in particular by enclosure procedures, e.g. Burgers and Frank circuit. Most important examples are vacancies, eigen-interstitials, dislocations, stacking faults, twin and grain boundaries. Elementary excitations, which can include dangling bonds, leave the topology unchanged and are classified as energetical defects. Impurity atoms on interstitial and regular lattice sites are not included into the system but considered as external influences acting on the system. If at all the name defect is used, they belong to the extrinsic type. The foregoing distinction of defect types is of basic importance if one aims at an accurate theory of defects in crystals.

In this lecture the description of defects in material structures will be given both in the language of differential geometry which is indispensible for a deeper understanding of the concept of defect and in the language of small distortions which is very useful for the calculation of the stress and strain states which accompany the defects and give rise to an elastic energy which is an important or even decisive part of the free energy determining the behaviour of the material. Not described, however, will be the group theoretical approach to defects initiated by Rogula [1], by Toulouse and Kléman [2] and by Kléman, Michel and Toulouse [3]. This approach uses the homotopy group, and in particular, considers certain enclosure procedures by which defects can be defined for all sorts of structures. This subject, however, would deserve a lecture for itself and will be omitted here.

The differential geometry which applies to "primitive" materials, i.e. materials consisting of only one sort of structure-less particles, is that of the linearly or affinely connected spaces. It is typical for this geometry that it allows the particles the three degrees of freedom of translation. The most characteristic defect of this geometry is, in a sense, the dislocation. In other material structures, above all in the magnetic structures, the particles have the three degrees of freedom of rotation only. The differential geometry which describes such structures is that special case of Finssler's geometry in which the translational freedom has been removed. The most characteristic defect of this geometry is the disclination. The general Finssler geometry [4] allows the particles three degrees of freedom for both translation and rotation. The best known example of the pertaining material structures are the liquid crystals. Most characteristic for this geometry is the common appearance of dislocations and disclinations. More complex materials, for instance lattices with many atoms in an elementary cell can be described by the very general geometry of Kawaguchi [5] which, however, is so complicated that its usefulness has first to be shown.

In section 2 we consider the defect-free state of crystals, also called the vacuum state. The description of this state can be given with the help of Euclidean geometry. In section 3 dislocations are introduced. The comparison of the circuit of Burgers and Frank with that of Cartan leads to the identification of dislocation density as Cartan's torsion of the space. The needed geometry is now non-Euclidean, although the Riemann-Christoffel curvature vanishes. As shown in section 4, curvature is introduced if foreign (or extra) matter is dissolved on interstitial sites. In sections 5 and 6 it is explained very briefly that vacancies, self-interstitials, substitutional foreign atoms and amorphous primitive solids can be described by nonmetric affine differential geometry. Whether this is a useful approach has still to be shown. In section 7 we describe a far-reaching analogy between the basic equations of the strain space considered up to then and a suitably introduced stress state.

The linear version of the theory which is especially useful for calculations is presented in section 8. In section 9 we report on general solutions of the internal stress problem thereby also giving a new solution of the anisotropic incompatibility problem. Section 10 contains among others a brief remark explaining why the continuum theory of dislocations, though proven to be helpful in many problems, has, so far, not lead to a general macroscopic theory of plasticity.

2. CRYSTALS IN EUCLIDEAN STATES

In order to avoid the complications occuring in non-primitive lattices (which include also all alloys) we consider materials with only one sort of atoms which, at least for our present purpose, can be described as mass points. From the point of view of Euclidean observer the defect-free states are by definition the perfect crystal lattice states (perfect in contrast to defect). Among the perfect crystal lattice states there is one distinguished state called the rectilinear state. It represents what is commonly understood by the undisturbed crystal. The measurement of lengths in this crystal can be performed with the help of a rectilinear coordinate system (x^k) which has the coordinate lines along the crystallographic lattice lines.

The crystal which is compatibly distorted, for instance by external forces, is still defect-free by definition. Such distorted states are called Euclidean because we can use Euclidean geometry for their description. In fact, in the compatible case we can imagine that the original rectilinear coordinate system (x^k) is "dragged along" with the distortion of the crystal so that it becomes a distorted system $(x^{k'})$ with coordinate lines which coincide with the distorted atomic lattice lines. Apparently, x^k is numerically equal to $x^{k'}$ for $k = k'$. In applications to crystals the use of the rectilinear and the pertaining dragged-along coordinate system means that lenghts are measured by counting atomic steps in the three crystallographic directions. The prescription matches a fundamental feature of crystals. Coordinate systems of the just described sort will also be named crystallographic later-on.

We now give a more quantitative description of the situation of Euclidean lattices.[+] The coordinate transformation connecting the systems (x^k) and $(x^{k'})$ is written as

$$dx^k = A^k_{k'} dx^{k'}, \quad dx^{k'} = A^{k'}_k dx^k. \tag{1}$$

For our special choice of coordinates $A^{k'}_k = \delta^{k'}_k$ and $A^k_{k'} = \delta^k_{k'}$. However, the eqs. (1) remain meaningful also if other Euclidean coordinates are used in the rectilinear and (compatibly) distorted state. In both cases the Jacobian of the transformations (1) is characteristic of the distortion and has such a form that the eqs. (1) are integrable. Thus the coordinate systems considered are holonomic in the Euclidean space (the 3-dimensional physical space in which the crystal is embedded). In the rectilinear state lengths between two atoms are measured according to the formula

$$ds_o^2 = a_{kl} dx^k dx^l \tag{2}$$

where dx^k describes the relative position of the two points. The components a_{kl} of the Euclidean metric tensor are constant in the rectilinear system (x^k). In the compatibly distorted state lengths can be measured in the dragged-along system by

$$ds^2 = g_{k'l'} dx^{k'} dx^{l'}, \quad (g_{k'l'} = a_{kl} \text{ for } k' = k,\ l' = l) \tag{3}$$

(standpoint of an "inside observer") or by

$$ds^2 = g_{kl} dx^k dx^l, \quad g_{kl} = A^{k'}_k A^{l'}_l g_{k'l'} \tag{4}$$

(standpoint of an "outside observer"). The compatible strain in the distorted state is described by the difference $ds^2 - ds_o^2$ which can be expressed by means of the usual strain tensor

$$\varepsilon_{kl} \equiv (g_{kl} - a_{kl})/2. \tag{5}$$

[+] Here and later we use Ricci's symbolism of upper and lower indices in order to distinguish between co- and contravariant vectors. This, however, is done more for having the equations correct than for a general understanding which can be achieved also by ignoring this distinction.

As is well-known, the construction of the undisturbed lattice can be performed by repeated translations of a mass point in three crystallographic directions. The same result is achieved if, starting from a point, the three primitive lattice vectors are parallely displaced again and again. The continuum version of this procedure consists in the repeated parallel displacement of a vector, say v^l, along the vectors dx^m. The parallel displacement is performed with respect to an affine or linear connection $A_{ml}{}^k$ and defined by the law of parallel displacement

$$dv^h = -A_{ml}{}^h v^l dx^m, \quad A_{ml}{}^h \equiv a^{hk} A_{mlk} \,. \tag{6}$$

Here the affine connection has the special form of a Christoffel symbol

$$A_{mlk} = a_{mlk}, \quad a_{mlk} \equiv \frac{1}{2}(\partial_m a_{lk} + \partial_l a_{mk} - \partial_k a_{lm}). \tag{7}$$

For the compatibly distorted state the inside description is

$$dv^{h'} = -\Gamma_{m'l'}{}^{h'} v^{l'} dx^{m'}, \quad \Gamma_{m'l'}{}^{h'} = g^{h'k'} \Gamma_{m'l'k'}, \tag{8}$$

$$\Gamma_{m'l'k'} = g_{m'l'h'} \equiv \frac{1}{2}(\partial_{m'} g_{k'l'} + \partial_{l'} g_{m'k'} - \partial_{k'} a_{l'm'}) \tag{9}$$

and the outside description is

$$dv^h = -\Gamma_{ml}{}^h v^l dx^m, \quad \Gamma_{ml}{}^h = g^{hk} \Gamma_{mlk}, \tag{10}$$

$$\Gamma_{mlk} = g_{mlk} \equiv \frac{1}{2}(\partial_m g_{kl} + \partial_l g_{mk} - \partial_k g_{lm}). \tag{11}$$

$\Gamma_{m'l'h'}$ is the dragged-along connection belonging to A_{mlk}, thus

$$\Gamma_{m'l'k'} = A_{mlk} \quad \text{for} \quad k = k', \; l = l', \; m = m'. \tag{12}$$

Since in rectilinear coordinates the a_{lm} are constant, both A_{mlk} and $\Gamma_{m'l'k'}$ vanish in these coordinates. This is a manifestation of the fact that both the undistorted and the (compatibly) distorted lattice state are Euclidean.

The eqs. (10) and (11) represent a convenient description of the distorted lattice by the outside observer. Since $\partial_k g_{lm}/2$ is symmetric in l,m this part of the connection describes a change in length of the parallely displaced vector. The part $(\partial_l g_{mk} - \partial_k g_{lm})/2$, on the other hand, is antisymmetric in k,l and therefore describes a rotation of the vector. Both effects together represent the change of the lattice due to the considered distortion.

3. DISLOCATIONS AND CARTAN'S TORSION

It was the great discovery of Kondo [6] and of Bilby, Bullough and Smith [7] in 1952/54 that Cartan's torsion of a space is the continuum version of the notion of the crystal dislocation. This result was proved by a comparison of Cartan's circuit and Frank's Burgers circuit. Cartan's circuit implies that if an infinitesimal vector $\underline{u}(\underline{r})$ is parallely displaced along an infinitesimal vector $\underline{v}(\underline{r})$ and the vector $\underline{v}(\underline{r})$ is parallely displaced along the vector $\underline{u}(\underline{r})$ then the parallelogram formed by the undisplaced and displaced vectors does not close (figure 1) provided the affine connection $\Gamma_{ml}{}^{k}$ (or Γ_{mlk}) used for this displacement has a part antisymmetric in m,l. If we denote antisymmetrization by brackets [] around the concerned subscripts, then the condition for the existence of a closure failure reads $\Gamma_{[ml]}{}^{k} \neq 0$. The failure itself is a vector db^k and the result of Cartan's procedure is

$$db^k = \alpha_{ml}{}^{k} dS^{ml}, \quad \alpha_{ml}{}^{k} \equiv \Gamma_{[ml]}{}^{k}. \tag{13}$$

$\alpha_{ml}{}^{k}$ is called the tensor of dislocation density, here written as a (true) tensor of 3rd rank. db^k is the continuum version of the Burgers vector in crystals. When (13) is applied to crystals then, following Nye [8], db^k is to be interpreted as the resulting Burgers vector of all dislocations piercing through the area element $dS^{ml} \equiv e^{nml} dS_n$.

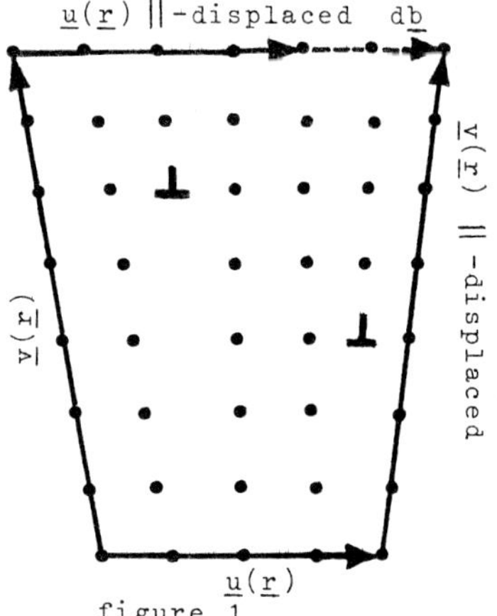

figure 1

It is a fundamental feature of the theory that the method of measuring distances by counting atomic steps is not destroyed by the introduction of dislocations. At first sight, this is surprising because the number of steps along two paths connecting the same distant atoms is not the same if the pertaining closed circuit (path 1 in forward direction, path 2 in backward direction) encloses one or more dislocations. As Kondo has remarked [9] this implies a principle of uncertainty of length measurements in a dislocated crystal, the uncertainty being of the size of one Burgers vector.

There exists an interesting analogy to the famous uncertainty in quantum mechanics. There the uncertainty refers to distances in the (momentum-position) phase space. This means that the uncertainty principle in quantum mechanics is in accord with a "crystalline" dislocated phase space with Burgers vector of the size of Planck's constant.

Returning to our problem of length measurements in a space with torsion we see that for $dx^m \rightarrow 0$ the uncertainty db^k itself vanishes like $dx^l dx^m$ whereas, on the other hand, it is the square of the distance which vanishes like $dx^l dx^m$. Thus db^k/ds is infinitesimal and can be omitted. In other words, the distance formula (4) can be used also in dislocated materials.

There is an interesting result referring to the crystallographic coordinate system which an inside observer would prefer to use. In a dislocated crystal a coordinate system can be connected with the rectilinear system by transformation formulae of type (1), but now the Jacobians have a form which makes the eqs. (1) non-integrable. Thus the crystallographic coordinate system of dislocated crystals is anholonomic: its coordinates $x^{k'}$ are defined only locally. This feature of dislocated crystals plays an important role in the works of Kondo [6], Marcinkowski [10] and Zorawski [11] (see also Schouten [12] for the more mathematical aspects). Here we shall not go into further details.

Since anholonomic coordinates are inconvenient to handle we shall prefer to use holonomic, namely Euclidean (or outside), coordinates. The disadvantage is, of course, that our coordinates x^k (which could be curvilinear but will now, for simplicity, be taken to be cartesian) are not connected with any non-Euclidean state of our solid.

It can be shown that the affine connection of the dislocated state has the form

$$\Gamma_{mlk} = g_{mlk} + h_{mlk} \tag{14}$$

where g_{mlk} is defined by (10). The tensor

$$h_{mlk} \equiv \Gamma_{[ml]k} + \Gamma_{[km]l} - \Gamma_{[lk]m} \tag{15}$$

is sometimes called the <u>contortion tensor</u>. It is easily verified that h_{mlk} is antisymmetric in l,k, thus describing a rotation during parallel-displacement of the vector v^l. This means that the lattice is curved by the presence of a dislocation density α_{mlk}. This curvature is also called Nye's curvature [8]. In his basic work on this subject Nye introduced two (axial) 2nd-rank tensors in order to describe dislocation density and dislocation curvature (contortion). These tensors are defined by

$$db^k = \alpha^{nk} dS_n, \quad d\omega^n = \kappa_m{}^n dx^m \tag{16}$$

where $d\omega^n$ is the infinitesimal rotation of the vector v^l parallely displaced along dx^m. The connection with our 3rd rank tensors is+

$$\alpha^n{}_k = e^{nml}\alpha_{mlk}, \quad \alpha_{mlk} = \frac{1}{2}e_{nml}\alpha^n{}_k \tag{17}$$

$$\kappa_m{}^n = -\frac{1}{2}e^{nlk}h_{mlk}, \quad h_{mlk} = -e_{nlk}\kappa_m{}^n. \tag{18}$$

Expressed by these tensors, Nye finds

$$\alpha^n{}_k = \kappa_k{}^n - \kappa_l{}^l\delta^n_k, \quad \kappa_k{}^n = \alpha^n{}_k - \frac{1}{2}\alpha^l{}_l\delta^n_k \tag{19}$$

+ Here and elsewhere the raising and lowering of indices is performed in the usual manner, namely by means of the metric tensor g_{lk}.

for the relation between the dislocation density and its curvature. (18) results from (15) by multiplication with $-e^{nlk}/2$ and subsequent use of (16).

Let now the vector v^l be parallely displaced (with respect to Γ_{mlk}) along the edge of an area element ΔS^{mn} which we consider infinitesimal but very large compared to the above used element dS^{lm}. The point is that, whereas Γ_{mlk} can be considered constant over dS^{lm} we now want to consider the derivatives $\partial_n \Gamma_{mlk}$ constant over ΔS^{mn}. The result of the parallel displacement is (see Schouten [12])

$$\Delta v^k = -\tfrac{1}{2}\Gamma_{nml}{}^k \, v^l \, \Delta S^{nm} \tag{20}$$

with

$$\Gamma_{nml}{}^k \equiv \partial_n \Gamma_{ml}{}^k - \partial_m \Gamma_{nl}{}^k + \Gamma_{np}{}^k \Gamma_{ml}{}^p - \Gamma_{mp}{}^k \Gamma_{nl}{}^p \tag{21}$$

as the Riemann-Christoffel curvature tensor formed with the affine connection Γ_{mlk}. It follows from our previous description that, when $\Gamma_{ml}{}^k$ represents a lattice, then a lattice vector remains a lattice vector under parallel displacement. If now the situation were such that $\Gamma_{nml}{}^k$ would not vanish than no lattice could be built up any more by the above described repeated parallel displacement of primitive lattice vectors. In fact, the result of the parallel displacement over distances of order Δx^m or larger would then be path-dependent. In other words, the continuum version of the notion of a crystal lattice is a space with distant parallelism as emphasized by Kondo [6] and by Bilby, Bullough and Smith [7]. In mathematical language we have

$$\Gamma_{nml}{}^k = 0 \quad \text{or} \quad \Gamma_{nmlk} = 0 \tag{22}$$

for crystals which may be dislocated or not.

Eqs. (22) means that the vector v^l when parallely displaced along the edge of ΔS^{mn}, returns unchanged in length and direction. Let us fix our attention to the rotation which takes place during the parallel displacement. As mentioned before, there are two contributions to the rotation, one coming from g_{mlk} and one from h_{mlk}. Accordingly we have

$$\Delta v^k = (\Delta v^k)_g + (\Delta v^k)_h \,. \tag{23}$$

Of course, the vanishing of Δv^k does not imply that the two contributions disappear separately. It has become popular to use the rotation failure $(\Delta v^k)_h$ for the definition of a line defect called <u>disclination</u> (see e.g. de Wit [13]) It is clear, however, that this defect line is determined by the presence of dislocations, for instance a grain boundary type of dislocation wall ending in the interior of the body. Therefore, such a disclination is <u>not</u> a genuine, independent new lattice effect. This remark applies also to the famous 60°-wedge disclination in hexagonal lattices where, if $(\Delta v^k)_h$ represents an angle of 60°, $(\Delta v^k)_g$ is -60°. There is, however, some ambiguity in this interpretation because in the case of 60° the dislocations are not seen anymore. This means that other interpretations are possible. The reader is referred to the work of Anthony [14]. Genuine, independent disclinations are bound to structures of particles with true rotational degrees of freedoms like spin systems or liquid crystals. I propose that this important fact is emphasized by calling disclinations in mass point lattices "improper", in contrast to "proper" disclinations in genuine rotational structures.

Every curvature tensor Γ_{nmlk} formed with an affine connection Γ_{mlk} is according to eq. (21) antisymmetric in the subscripts n,m. If the connection has the (still special) form (14,15) then it is also antisymmetric in l,k. In this case, the so-called Einstein tensor

$$\Gamma^{ij} \equiv \frac{1}{4} e^{inm} e^{jlk} \Gamma_{nmlk} \tag{24}$$

contains the same information as the curvature tensor. For many purposes, the Einstein tensor is more convenient to handle than the Riemann-Christoffel tensor. In particular, certain identities on the curvature tensor can be written as identities for the Einstein tensor, namely

$$\nabla_i^* \Gamma^{ij} = 0 \qquad \text{(Bianchi identity)} \tag{25}$$

$$\nabla_i^* \alpha^i{}_k = -e_{mlk} \Gamma^{ml} \qquad \text{(torsion identity)} . \tag{26}$$

Here the symbol ∇_i^* is defined by

$$\nabla_i^* \equiv (\mathrm{div}_\Gamma)_i = \nabla_i + 2\,\Gamma_{[ij]}{}^j \tag{27}$$

where ∇_i is the usual symbol of covariant differentiation. Apparently, ∇_i^* denotes the divergence operation in a geometry with torsion. Since in our dislocated crystal $\Gamma^{ij} = 0$ we see that dislocations cannot end in its interior as was discussed in particular by Bilby, Bullough and Smith [7].

4. FOREIGN ATOMS ON INTERSTITIAL SITES (EXTRA MATTER)

Consider a defect-free crystal in its rectilinear state and cut this crystal into small pieces which are called volume elements δV. Desolve foreign atoms on the interstitial sites of the δV's such that the concentration is homogeneous in δV but varies slowly from one δV to its neighbour thereby changing correspondingly the atomic distances of the original lattice. If lengths within each δV are still measured by atomic steps along the original atoms then we expect that the metric tensor is constant in each δV but varies slowly from one δV to its neighbour. It is then sensible to comprise many δV's in one larger element dV over which Γ_{mlk} can be considered constant and many dV's in one even larger element ΔV over which Γ_{nmlk} is constant.

We now consider the successive limiting processes $\delta V \to 0$, $dV \to 0$, $\Delta V \to 0$. Thereby we distinguish two cases. In the first case the volume elements dV fit together so that a compact body can be formed by them without applying any constraint. In this case the foreign atoms act like an external force density and the resulting lattice can be considered compatibly distorted from the rectilinear state so that it is still Euclidean. Then our former results apply accordingly. A new situation, however, arises when the elements dV do not fit in the (mentioned) limit. The resulting distribution of the original atoms does no longer represent a global crystal lattice but rather an assembly of different lattices each belonging to one element dV.

From the considerations on the Euclidean states of crystal lattices it follows that all configurations of the original atoms which differ by a compatible distortion have the same metric tensor whose components can be formed in a set of coordinate systems which are all compatibly connected. If, however, the mentioned disconnected assembly of lattices is fitted together by applying the necessary constraint, then the metric is not only altered by the originally distributed extra matter in the disconnected lattices dV through the change of atomic distances but also by the fitting procedure which corresponds to an incompatible distortion of the volume elements. For both reasons the metric tensor now becomes a function of position which is similar to the case when matter is introduced into the empty universe (vacuum).

The present situation is clearly Riemannian so that the Riemann and Einstein tensors do not vanish. Instead of eqs. (22) we then have the Einstein equations

$$\Gamma^{ij} = T^{ij} \tag{28}$$

and T^{ij} is a tensor which somehow characterizes the introduced extra matter. This tensor will therefore be called "tensor of extra matter" or, simply, "matter tensor".

Assume we know how to calculate T^{ij} from the information about the extra matter. We have then to solve eq. (28) in order to find the metric tensor. Like in general relativity theory the metrics are determined by the Einstein equations only up to a compatible strain. Unlike in relativity, however, here exists an additional condition, namely that there are no external forces (if we are interested in such a situation).

If the eqs. (28) are introduced into (25) and (26) we obtain

$$\overset{*}{\nabla}_i T^{ij} = 0, \quad \overset{*}{\nabla}_i \alpha^{i}_{\cdot k} = -e_{mlk} T^{ml} \tag{29}$$

where the latter equations express the fact that dislocations can end inside the body provided the matter tensor has an antisymmetric part. Such a situation is conceivable if the extra matter consists, for instance, of diatomic molecules on interstitial sites which exert a torque on their environment. Since this is certainly not a normal situation we shall not consider it further but rather assume that the matter tensor is symmetric.

5. VACANCIES AND SELF-INTERSTITIALS

At sufficiently high temperatures and also under other conditions, crystal lattices can contain an appreciable number of vacancies and self-interstitials. Since the now considered configurations* of the same particles which also form the defect-free lattice we consider the introduction of vacancies and self-interstitials as a distortion mode. If we maintain our fundamental concept that length measurements in (global or local) lattices are performed by counting steps always from one atom to its neighbour then we are in trouble at all places where the counting path touches a vacancy or an interstitial atom. The effect, of course, increases with the concentration of these defects and is not infinitesimal for finite concentrations. In other words, vacancies and self-interstitials destroy the intrinsic metrics of the lattice and therefore, have to be assigned to the non-metric part of a general connection. However, since this concept has not been worked out in any detail, we shall not pursue it further.

*consist

6. COMBINATION OF DEFECTS, AMORPHOUS STATE

Sofar we have treated separately the various defects such as dislocation, foreign atom on intersitital sites, vacancy and self-interstitial. Foreign atoms on substitutional sites not only represent extra matter but, contrary to those on interstitial sites, also destroy the metrics. Therefore we expect that they correpsond to a non-vanishing curvature tensor which is symmetric in the subscripts l,k. Other defects like stacking faults, twin boundaries and grain boundaries can probably be represented as conglomeratcs and superpositions of the previously discussed elementary defects. So it seems that all known topological defects in primitive mass-point lattices have their place within the differential geometry with the general affine connection. The same seems to be true for amorphous bodies. These certainly correspond to a non-metric situation. In this case the affine connection will be a function of position which is more or less random on a small scale.

7. THE STRESS STATE OF A CRYSTALLINE SOLID WITH DEFECTS

We now restrict ourselves to metric spaces which are described by the Einstein equations (28) and eqs. (29). This means that we consider only the (intrinsic) defect dislocation and the (extrinsic) defect foreign atom on interstitial site. Of these two sorts of defects only the former represents a distortion mode. Any such mode leads to a specific response of the body which in our case has the dimension of a couple stress. To understand this consider the special case of a homogeneous dislocation density which in the continuum limit is not accompanied by elastic strains, but is equivalent to the discussed Nye curvature. Clearly such a state which is realized for instance in plastically bent, annealed materials, leads to an increase of energy, therefore to a response of the body. The problem is also related to the energy of small angle grain boundaries which are formed by dislocation walls.

If we accept that, in general, the response of a dislocated body consists of ordinary (or force) stresses σ^{ij} and of couple stresses $\tau^{i}{}_{k}$ then the equilibrium equations in the volume of the body have the form

$$\overset{*}{\nabla}_i \sigma^{ij} = 0, \quad \overset{*}{\nabla}_i \tau^{i}{}_{k} = -e_{mlk}\sigma^{ml} \qquad (30)$$

Here we have used the divergence symbol $\overset{*}{\nabla}_i$ introduced in section 3 in order to emphasize that the equilibrium conditions would apply not only to a medium in an Euclidean space but even to a medium in a general metric space, i.e. in a space with torsion and curvature. This can be concluded from the facts that (i) these equations are of a local nature and (ii) each space with a metric affine connection has locally a Euclidean tangent space [12]. Now the eqs. (30) have exactly the same structure as the eqs. (29). It is therefore tempting to consider σ^{ij} and $\tau_{mlk} \equiv e_{nml}\tau^{n}{}_{k}/2$ respectively as the matter tensor and the torsion tensor of a metric space which we call the <u>stress space</u> in order to distinguish it from the sofar considered spaces which now will also be named <u>strain spaces</u>. For the stress space we write Einstein equations

$$\chi^{ij} = \sigma^{ij} \qquad (31)$$

where the Einstein tensor X^{ij} is formed with an affine connection

$$X_{mlk} = \frac{1}{2}(\partial_m X_{lk} + \partial_k X_{ml} - \partial_l X_{km}) + \tau_{mlk} + \tau_{kml} - \tau_{lkm}. \quad (32)$$

It is interesting to note that χ_{lk} is identical with the well-known 2nd-order stress-function tensor which is used to satisfy the equilibrium conditions for the force stresses in the absence of external forces. Airy's stress function is just one component of this tensor. The eqs. (30)-(32) are formally identical with the eqs. (29), (28), (14) and (15).

The unexpected symmetry between stress and strain space, discovered in 1964 [15] is not yet fully understood. At first sight, there seems to be an imcompleteness of this symmetry insofar as the stress space metric tensor χ_{lk} is not observable but rather calls for a gauge condition, similar to the vector potential of electrodynamics to which it has a far-reaching analogy. On the other hand, the strain space metric tensor g_{lk} is usually considered an observable quantity. As we have seen, however, a compatible strain is not observable by an inside observer. Therefore, he, too, can propose a gauge condition. Of course, it is only consistent not to consider compatible distortions in the strain space if we asuume the validity of the Bianchi identities (30_1) as the force equilibrium equations in the stress space. This means that the above mentioned apparent incompleteness of the symmetry is removed.

8. LINEARIZED EQUATIONS

A fundamental problem in the theory of defects consists in the calculation of the internal stress and strain fields produced by defect distributions of which the single defect is a special case. The most important tool for this purpose is the linear theory of elasticity. Our interest is now directed towards such a description of defects which allows us a convenient application of the tools of linear elasticity theory. Of course, this implies that we need a linearized version of the sofar discussed nonlinear equations.

We are not interested in stress and strain due to external forces but rather in internal, i.e. self- (or eigen-) stress and strain. It is merely a technical problem to include the boundary conditions which can always be treated by standard methods. Therefore, we can restrict ourselves to bodies without boundaries, i.e. to infinite bodies.

Most problems of the indicated kind have been treated by means of isotropic elasticity theory. Nevertheless, a trend towards calculations which take into account anisotropic elasticity can be observed. Such calculations are indispensible in many cases if one wants to have quantitatively correct results. For this reason we shall now give the relevant equations in an invariant form so that they apply to the anisotropic theory as well.

Point defects. These are conveniently described by forces $f^{(\nu)}$ which they exert on the neighbouring atoms here labelled by (ν). These forces, sometimes called Kanzaki forces [16], were first introduced by the lecturer [17] who in the spirit of fundamental work of Eshelby [18] defined the (symmetric) elastic dipole tensor

$$P_{ij} = \sum_{\nu} x_i^{(\nu)} f_j^{(\nu)} \quad (33)$$

for the quantitative description of point defects. Here we are interested in a dipole density $p_{ij} = dP_{ij}/dV$ which produces a stress-free (or extra) strain q_{kl} so that $p_{ij} = C_{ijkl}q_{kl}$ where C_{ijkl} is the 4th rank tensor of elastic moduli which has the symmetry properties

$$C_{ijkl} = C_{jikl} = C_{ijlk} = C_{klij}. \tag{34}$$

In this and in the next section we shall consider q_{kl} as the sources of the internal stress and strain in the case of point defects. We shall see that an antisymmetric part of q_{kl} does not contribute to the stresses in the present theory which takes into account only a symmetric force stress tensor.

Dislocations. For their description we need an equation connecting the tensor $\underline{\alpha}$ of dislocation density defined in (16_1) with the quantities of elasticity theory. To find this equation we define the elastic distortion tensor β_{kl} by means of the elastic relative displacement du_l of two points which are separated by dx_k:

$$du_l = \beta_{kl}dx_k. \tag{35}$$

The symmetric part of β_{kl} is the strain tensor ε_{kl} and the antisymmetric part the rotation tensor ω_{kl} of the linear theory of elasticity. Since the distortion is incompatible in the presence of dislocations we know that the vector

$$\underline{b} = \int_C d\underline{r} \cdot \underline{\beta} = \int_S (d\vec{S} \times \nabla) \cdot \underline{\beta} = \int_S d\vec{S} \cdot \mathrm{curl}\underline{\beta} \tag{36}$$

is not zero for all closed circuits C. In (36) we have used Stokes' formula, therefore S is any surface with edge line C. Clearly, $\underline{b}$ is the resulting Burgers vector of all dislocations piercing through the surface S. Since (36) applies to infinitesimal circuits as well we have with (16_1)

$$e_{ijk}\partial_j\beta_{kl} = \alpha_{il} \quad \text{or} \quad \mathrm{curl}\underline{\beta} = \underline{\alpha} \tag{37}$$

as the fundamental relation between elastic distortion and dislocation density. A better understanding of this equation is achieved if we go back to differential geometry and (for simplicity in linear approximation and cartesian coordinates) redefine the affine connection by the equation

$$d\beta_{lk} = d\varepsilon_{lk} + d\omega_{lh} \equiv -\Gamma_{mlk}dx^m \tag{38}$$

which is consistent with the fact that a parallely displaced vector v^l undergoes a strain and a rotation. The vanishing of the curvature tensor is the integrability condition for eq. (38) which, when integrable, can be transformed into

$$\Gamma_{mlk} = -\partial_m\beta_{lk}. \tag{39}$$

The existence of β_{lk} guaranted by (39) implies distant parallelism, i.e. the existence of a lattice. In the linearized theory this was assumed a priori by the definition of β_{kl}. The part antisymmetric in m,l of eq. (39) is equivalent to (37). Thus we see that in order to retain a lattice inspite of the introduction of dislocations, elastic distortions must satisfy (37).

Incompatibilities. According to a theorem of Reissner [19] internal stresses in infinitely extended linearly elastic bodies can develop only in the presence of an incompatibility tensor η_{ij} so that de St. Venant's compatibility equations have to be extended to the form

$$-e_{ikm}e_{jln}\partial_k\partial_l\varepsilon_{mn} = \eta_{ij} \quad \text{or} \quad \text{inc}\underline{\varepsilon} = \underline{\eta}. \tag{40}$$

Since the stresses produced by defects are internal it must be possible to connect η_{ij} with the above introduced densities q_{kl} and α_{kl}. The corresponding formulae are well-known and read

$$\gamma_{ij} = e_{ikm}e_{jln}\partial_k\partial_l q_{mn} \quad \text{or} \quad \underline{\eta} = -\text{inc}\underline{q} \tag{41}$$

for a symmetric q_{mn} and

$$\eta_{ij} = \frac{1}{2}(e_{jkl}\partial_l\alpha_{ik} + e_{ikl}\partial_l\alpha_{jk}) \text{ or } \underline{\eta} = (\underline{\alpha}\times\nabla)_{sym} \tag{42}$$

where the symbol "sym" denotes symmetrization.

9. FUNDAMENTAL SOLUTIONS OF THE INTERNAL STRESS PROBLEM

Let there be given internal stress sources in the form q_{rs}, α_{js} or η_{ab}. Both q_{rs} and α_{js} may have antisymmetric parts. The antisymmetric part of q_{rs}, however, does not contribute to the stresses as noted before. The fundamental solutions referring to an infinite medium can be found by means of certain kernels in the form

$$\sigma_{kl}(\underline{r}) = \int L_{klrs}(\underline{r}-\underline{r}')q_{rs}(\underline{r}')dV' \tag{43}$$

$$\sigma_{kl}(\underline{r}) = \int M_{kljs}(\underline{r}-\underline{r}')\alpha_{js}(\underline{r}')dV' \tag{44}$$

$$\sigma_{kl}(\underline{r}) = \int N_{klab}(\underline{r}-\underline{r}')\eta_{ab}(\underline{r}')dV' \tag{45}$$

The problem has been treated extensively by Simmons and Bullough [20] at a previous conference on Fundamental Aspects of Dislocation Theory (Washingtron 1969). Here we only give their results and a new (simpler) expression for the kernel $\underline{N}$. Simmons and Bullough find, using comma notation for differentiation,+

$$L_{klrs} = -C_{klrn}\delta_{ns}\delta(\underline{r}-\underline{r}') - C_{klmn}G_{nq,mp}(\underline{r}-\underline{r}')C_{pqrs}. \tag{46}$$

Here $G_{nq}(\underline{r}-\underline{r}')$ is the Green tensor of anisotropic elasticity theory, defined by

$$-C_{pqms}G_{nq,pm} = \delta_{ns}\delta(\underline{r}-\underline{r}') \tag{47}$$

+ Instead of extra strain (or -distortion) q_{rs} which describes the point defects, Simmons and Bullough use the corresponding elastic distortion β_{rs} in eq. (43). This implies a change of the sign.

This substituted in (46) gives the alternative form

$$L_{klrs} = -2[C_{klmn}C_{pqrs}]_{[mr]} G_{nq,mp}(\underline{r}-\underline{r}') \tag{48}$$

where [mr], as usual, denotes antisymmetrization. If (48) is substituted into (43) one finds after a partial integration with respect to ∂_m and subsequent comparison with (37) and (44) the fundamental result of Mura [21] and Willis [22], namely

$$M_{kljs} = [C_{klmn}C_{pqrs}]e_{mrj}G_{nq,p}. \tag{49}$$

Except for a convention (Simmons and Bullough use the transposed tensors instead of our tensors β_{kl} and α_{kl} and have a different sign convention for the dislocation density) their "Mura-Willis-equation" (4.1) agrees with our eq. (44) if (49) is inserted into it.

In order to find the kernel $\underline{N}$ in eq. (45) we introduce, following Fredholm [23], the differential operators D_{ik}, D^*_{km} and f, where $D_{ik} = C_{ijkl}\partial_j\partial_l$, D^*_{km} is the tensor of 2x2 subdeterminants of D_{ik}, thus a 4th order differential operator and f is the 3x3 determinant of D_{ij} so that $f\delta_{im} = D_{ik}D^*_{km}$. If the Green function $U(\underline{r}-\underline{r}')$ belonging to the 6th order differential operator f is introduced by

$$f(\nabla)U(\underline{r}-\underline{r}') = \delta(r-r') \tag{50}$$

then the Green function defined by (47) is

$$G_{nq}(\underline{r}-\underline{r}') = D^*_{nq}(\nabla)U(\underline{r}-\underline{r}'), \tag{51}$$

a known result. We now introduce an auxiliary tensor field Ψ_{ab} by [24]

$$\Psi_{ab}(\underline{r}) = \int U(\underline{r}-\underline{r}')\eta_{ab}(r')dV' . \tag{52}$$

The Ψ_{ab} are called <u>4th order stress functions</u> because, as we shall see, the stresses are derived from them by a 4-fold differentiation. We write

$$\sigma_{kl}(\underline{r}) = Y_{klab}\Psi_{ab} = \int Y_{klab}U(\underline{r}-\underline{r}')\Psi_{ab}(\underline{r}')dV' \tag{53}$$

and show below how $\underline{Y}$ can be found. Setting

$$\eta_{ab}(\underline{r}') = e_{apr}e_{bms}\partial_{p'}\partial_{m'}q_{rs}(\underline{r}') \tag{54}$$

we have

$$\sigma_{kl}(\underline{r}) = \int Y_{klab}e_{apr}e_{bms}U(\underline{r}-\underline{r}')\partial_{p'}\partial_{m'}q_{rs}(r')dV'. \tag{55}$$

On the other hand, eq. (43) with (48), (51) and the partial integrations with respect to $\partial_{m'}$ and $\partial_{p'}$ gives

$$\sigma_{kl}(\underline{r}) = - \int 2[C_{klmn}C_{pqrs}]_{[mr]} D^*_{nq}(\nabla)U(\underline{r}-\underline{r}')\partial_{p'}\partial_{m'}q_{rs}(r')dV' \tag{56}$$

The comparison of the last two equations yields

$$Y_{klab}e_{apr}e_{bms}\Big|_{(pm)(rs)} = -2\{[C_{klmn}C_{pqrs}]_{[mr]}\}_{(pm)(rs)}D^{*}_{nq}(\nabla) \qquad (57)$$

where the parentheses {} indicate that the antisymmetrization [mr] comes before the symmetrizations (pm) and (rs). Eq. (57) can be solved for $\underline{Y}$ by multiplication with $e_{cpr}e_{dms}$ because $(e_{apr}e_{bms})_{(pm)(rs)}e_{cpr}e_{bms} = 6(\delta_{ac}\delta_{bd} + \delta_{ad}\delta_{bc})$. If at the end of this calculation c,d is replaced again by a,b we obtain the result

$$Y_{klab} = -\frac{1}{6}e_{apr}e_{bms}\{[C_{klmn}C_{pqrs}]_{[mr]}\}_{(pm)(rs)}D^{*}_{nq}(\nabla). \qquad (58)$$

Inserting (58) into (53) and comparing with (45) leads to

$$N_{klab} = -\frac{1}{6}e_{apr}e_{bms}\{[C_{klmn}C_{pqrs}]_{[mr]}\}_{(pm)(rs)}G_{nq}(\underline{r}-\underline{r}'). \qquad (59)$$

It is interesting to note that the kernels appearing in eqs. (43)-(45) contain in a sequence $G_{nq,mp}$, $G_{nq,p}$ and G_{nq}. In this respect, the integral (45) is simpler than the other two. It is known that D^{*}_{nq} contains the elastic moduli only in terms of 2x2 subdeterminants of the Voigt matrix (C_{ik}). It seems, but still has to be proved that the same is true for the factor of D^{*}_{nq} in (58). If this is so then it is very likely that the elastic moduli in $\underline{Y}$ occur as 4x4 subdeterminants of (C_{ik}) which, according to a well-known law of the theory of determinants can be replaced by the 2x2 subdeterminants of the inverse of Voigt's matrix thus leading to a considerable simplification. The belief in this likeliness stems from the fact that the 2-dimensional anisotropic stress problem has successfully been formulated in terms of stress functions by Voigt [25]. The method was later used by Seeger and Schöck [26] for the solution of certain dislocation problems. The point is that the anisotropic elastic constants enter Voigt's formulation only in the form of 2x2 subdeterminants of the compliance tensor.

It should be mentioned that the integrals (43)-(45) and the stress function method simplify greatly when elastic isotropy is assumed. This, however, has already become material of many monographs such as Friedel's [27], Lardner's [28] and Nabarro's [29] so that it need not be treated here.

10. CONCLUSION

In this lecture we have discussed the concept of a crystal as a material space which can be considered as a continuum if one is interested in phenomena, such as defect interactions and their consequences, which occur on a scale sufficiently large compared to the atomic scale. We explained that defect-free states require a description by Euclidean geometry whereas Cartan's torsion is needed to handle dislocations. Foreign matter on interstitial sites produces Riemann's curvature. There are other defects that with increasing concentration destroy more and more the intrinsic metricity of the crystal so that the possibility of measuring length by counting atomic steps breaks down at sufficiently high concentration. The amorphous solid is an extreme case hereof.

We have argued that the affine differential geometry applied here is really good only for solids consisting of structureless mass-points. If the particles of the body have rotational degrees of freedom rather than translational ones then the geometry of Finsler is best suited. A promising attempt in this direction has been undertaken by Amari [30] who treated spin structures in magnetic materials. Again another geometry is needed for the dealing with line structures as polymers and flux line lattices in superconductors. This problem has been treated extensively by Anthony [31].

If now the differential geometry is so fundamental to the phenomenon "defect" then the question arises why has it been so unpopular among the researchers concerned with the influence of defects on material properties? The main reason is <u>not</u> that differential geometry is considered unjustifiedly an especially difficult mathematical discipline. It is rather the fact that no good proposals have been made how to solve practical problems with dislocations by the help of differential geometry. The point is that the actual dislocation distributions in real materials almost never correspond to a more or less continuous torsion of the space, i.e. dislocations almost never form smooth line densities, as it would be needed to apply differential geometry successfully. What one observes are rather irregular networks. In principle, such a network can be described by a dislocation density tensor $\alpha^{nk}(\underline{r})$ if one assumes that α^{nk} degenerates to a delta distribution

$$\alpha^{nk}(\underline{r}) = t^{n}(\underline{r})b^{k}(\underline{r}), \quad \int_{c} t^{n}(\underline{r})dc = 1 \tag{60}$$

where the tangential vector $t^{n}(\underline{r})$ vanishes outside the dislocation core and is very large in the core so that its integral over the core cross section gives unity.

Due to the irregularity of the network a statistical description seems to be adequate. The important quantities for such a description are the ensemble average dislocation density $\langle\alpha^{nk}(\underline{r})\rangle$, which now should be identified with that of Nye [8], and, even more important, the correlation functions

$$\langle\alpha^{nk}(\underline{r}_1)\alpha^{pl}(\underline{r}_2)\rangle, \quad \langle\alpha^{nk}(\underline{r}_1)\alpha^{pl}(\underline{r}_2)\alpha^{qm}(\underline{r}_3)\rangle \ldots \tag{61}$$

of which there exists an infinite number. Only the inclusion of all these functions does, principally, permit a sharp prediction of the expectation values for experiments with dislocations such as plasticity experiments. In practical calculations we can only afford to consider a small number of such "internal variables". This is the reason why, in my opinion, there is no hope to develop a general theory of inelastic behaviour on the basis of dislocation or defect theory. The term "general theory" is here used in the sense that all inelastic phenomena are described and predicted with a reasonable accuracy. I am afraid that this remark also applies to the remarkable work of Augusti, Baldacci and Capurro [32] who develop a "microrelativistic dislocation theory" in which inside and outside observers are introduced similar to ours. It seems that at least for some time, we have to live with models of a rather limited range of validity. The success of such models will largely depend on the choice of a small number of internal variables. In this connection it is worthwhile to note that the invariant part

$<\alpha^n{}_k(\underline{r})\alpha_n{}^k(\underline{r})>$ of the 2-point correlation function represents the total length of all dislocations within a unit volume [33]. This internal variable has been used with great success in many models.

Whereas the differential geometry of dislocations and other defects has certainly increased our insight in the very nature of the defects, it was the linearized theory which was most successful in handling the important problems of internal stresses produced by defects. Here we have, in particular, presented the stress function method for calculating the stresses from given distributions of defects. The reason was not only that they are comparatively little known and, therefore, deserve some popularization, but also that they play the role of the metric tensor of the stress space, thus being involved in a peculiar symmetry between stress and strain space. Finally, recent work of Saada and coworkers [34,35] shows once more that, at least for certain distributions of dislocations, the stress functions are useful tools.

We have seen that internal stresses can always be related to the incompatibility tensor as a source tensor for these stresses. The formalism resulting from this fact has a great value even if internal stresses are produced not by defects but by other causes such as a varying field of temperature or magnetostriction. Formally, these stress fields can also be thought to be produced by suitably distributed dislocations so that the just mentioned stress sources are sometimes called quasi-dislocations.

In this report I have tried to sketch both the developments which are important for the understanding of the basic concepts and those which provide useful tools for the solution of practical problems. I hope that this was done in a sufficiently objective manner. The limited space did not permit to mention all the work which is important for our subject. The authors of such work are requested to accept my apology.

REFERENCES

1. D. Rogula in Trends in Applications of Pure Mathematics to Mechanics, Conf. Univ. Lecce, ed. by G. Fichera, p. 311, Pitman Publ., London 1976
2. G. Toulouse and M. Kléman, J. de Phys. Lettres 37, L-149 (1976)
3. M. Kléman, L. Michel and G. Toulouse, J. de Phys. Lettres 38, L-195 (1977)
4. H. Rund, The Differential Geometry of Finsler spaces, Springer-Verlag, Heidelberg 1959
5. A. Kawaguchi, Proc. Imp. Acad., Tokyo 7, 211 (1931), 8, 340 (1932), 9, 347 (1933) and M. Kawaguchi, RAAG Memoirs of the Unifying Study of the Basic Problems in Engineering Sciences by Means of Geometry, Vol. III, p. 717, Tokyo, Gakujutsu Bunken Fukyu-Kai 1962
6. K. Kondo ibid. Vol. I (1955), pp. 453-562
7. B.A. Bilby, R. Bullough and E. Smith, Proc. Roy. Soc. London A 231, 263 (1955)
8. J.F. Nye, Acta Met. 1, 153 (1953)
9. K. Kondo, RAAG Memoirs (sec ref. 5) Vol. II (1958), p. 227
10. M.J. Marcinkowski, Unified Theory of the Mechanical Behaviour of Matter, J. Wiley and Sons, New York 1979

11. M. Zorawski, Theorie Mathématique des Dislocations, Dunod, Paris 1967
12. J.A. Schouten, Ricci Calculus, Springer-Verlag, Heidelberg 1954
13. R. deWit, J. Research Nat. Bur. Stand. 77A, 49, 359, 607 (1973)
14. K.-H. Anthony, Arch.Rat.Mech.Anal. 37, 161 (1970), 39, 43 (1970) and 40, 50 (1971)
15. E. Kröner, Ann. der Physik VII 11, 13 (1963)
16. H. Kanzaki, J. Phys. Chem. Solids 2, 24 (1957)
17. E. Kröner, Z. Naturforschung 11a, 969 (1956)
18. J.D. Eshelby, Phil. Trans. Roy. Soc. London A 244, 87 (1951)
19. H. Reissner, Z. Ang. Math. Mech. 11, 1 (1931)
20. J.A. Simmons and R. Bullough in Fundamental Aspects of Dislocation Theory, Nat. Bur. Stand. (U.S.) Spec. Publ. 317, Vol I.,p. 89 (1970)
21. T. Mura, Phil.Mag. 8, 843 (1963)
22. J. Willis, Int. J. Eng. Sc. 5, 171 (1967)
23. I. Fredholm, Acta Math. 23, 1 (1900)
24. E. Kröner, Z. Phys. 141, 386 (1955)
25. W. Voigt, Lehrbuch der Kristallphysik, Teubner, Berlin 1928
26. A. Seeger and G. Schöck, Acta Met. 1, 519 (1953)
27. J. Friedel, Dislocations, Pergamon Student Editions, London 1967.
28. R.W. Lardner, Mathematical Theory of Dislocations and Fracture, Univ. Toronto Press 1974
29. F.R.N. Nabarro, Theory of Crystal Dislocations, Univ. Press Oxford 1967
30. S. Amari, RAAG Memoirs (see ref. 5) Vol. III, p. 193
31. K.-H. Anthony, Feldtheorie physikalischer Linienstrukturen, Habilitationsschrift Univ. Stuttgart, 1974
32. V. Augusti, R. Baldacci and M. Capurro, Atti Acc. Naz. Linc., Cl. Sci. fis. mat. nat. VIII, 15, 22 (1979)
33. E. Kröner in Inelastic Behaviour of Solids, ed. by M.F. Kanninen, W.F. Adler, A.R. Rosenfield and R.I. Jaffee,McGraw Hill, New York 1970
34. G. Saada, Acta met. 27, 921 (1979)
35. C. Rey-Villain, Contribution a l'Etude de l'Influence des Interfaces sur la Plasticité a Froid des Métaux Purs, Thèse Doctorat d'Etat, Univ. Paris-Nord-Villetaneuse 1980

THE DISPLACEMENT FIELD OF A DISLOCATION DISTRIBUTION

Roland de Wit
Fracture and Deformation Division
National Bureau of Standards
Washington, DC 20234

ABSTRACT

Burgers' formula gives the displacement field due to a discrete dislocation line of arbitrary shape in an isotropic, linearly-elastic, infinitely extended, homogeneous body. This paper derives the analogous expression for a continuous distribution of dislocations in an anisotropic body. Special cases give the displacement due to a discrete dislocation line in anisotropic elasticity and due to a continuous dislocation distribution in isotropic elasticity. These expressions all represent generalizations of Burgers' original formula.

1. Introduction

This paper develops general expressions for the displacement field of a continuous distribution of dislocations in a linearly-elastic, infinitely extended, homogeneous body.

Various expressions exist for the displacement of a discrete dislocation line in a continuous medium, the best-known being Burgers' formula (1), which was quite influential in the development of dislocation theory. These expressions can be quite useful in applications to cases where the displacement field is needed for further calculations, e.g. to interpret the diffraction effects from a crystal containing a dislocation.

However, no formulas seem to be available for the case where the dislocations are continuously distributed. In applications to deformed crystals a dislocation tangle could be represented by such a continuous distribution. It is the purpose of this paper to develop such formulas, i.e. to generalize Burgers' formula to a dislocation distribution.

The general approach in this paper will be to derive the desired formulas from results that already exist in the literature.

In section 2 we review the result that was obtained by Burgers for an isotropic solid. We write it in convenient tensor notation. Burgers' formula contains essentially two terms: one is a line integral along the dislocation line and the other is a term proportional to the solid angle subtended by the dislocation line.

Section 3 addresses the question of writing the solid angle as a line integral. Though it is defined as a surface integral, the important fact for this paper is that it can also be written as a line integral. Hence, Burgers' formula is essentially a line integral along the dislocation line. This makes a generalization to a continuous distribution conceptually easy.

In section 4 we discuss the generalization of Burgers' formula to a continuous dislocation distribution in anisotropic elasticity. We start from a well-known volume integral expression for the displacement interms of Green's tensor function and the plastic distortion of a deformed body.

We introduce a useful auxiliary function to help simplify the derivation. We then derive the general expression for the displacement of a dislocation distribution, i.e. the generalization of Burgers' formula.

We close the section by specializing the result to a discrete dislocation line, thus also generalizing Burgers' formula for a discrete dislocation line to anisotropic elasticity.

Section 5 specializes the results derived so far to isotropic elasticity. A general formula for the displacement of a dislocation distribution is found. If this result is specialized to a discrete dislocation line, then Burgers' original formula is recovered.

In this paper we shall freely refer to the results of two previous publications (2,3), which will be denoted as II and III for convenience. These references may also be consulted for more information on the notation in this paper.

2. Burgers' Formula

Burgers derived his formula for a single closed dislocation loop of arbitrary shape in an infinite, elastically isotropic body. It gives the displacement, u_n, at a point, r, due to the discrete dislocation along the curve L, as follows (1,4)

$$u_n(\mathbf{r}) = -b_n\Omega/4\pi\ (8\pi)^{-1}\oint_L [\epsilon_{pnl}R_{,mm} - \epsilon_{pml}R_{,nm}/(1-\nu)]\, b_l dL'_p \tag{2.1}$$

Here b_n is the Burgers vector of the dislocation line, Ω is the solid angle subtended at **r** by the closed curve L, ϵ_{ijk} is the permutation symbol, and ν is Poisson's ratio. The relative radius vector, **R**, is defined by

$$\mathbf{R} = \mathbf{r} - \mathbf{r}' \tag{2.2a}$$

or, in component notation

$$X_n = x_n - x'_n \tag{2.2b}$$

where **r** or x_n is the field point and $\mathbf{r}'$ or x'_n the source point, i.e. the integration variable on the dislocation line. The Einstein summation convention is implied for repeated indices, and a subscripted comma followed by an index indicates partial differentiation, e.g.

$$R_{,n} = X_n/R \tag{2.3}$$

$$R_{,nm} = \delta_{nm}/R - X_nX_m/R^3$$

where δ_{nm} is the Kronecker delta. If the partial differentiation is with respect to x'_n, a prime is attached to the index.

The integral in (2.1) is taken along the dislocation line, L. It has been recognized intuitively for a long time by workers in dislocation theory that such an integral can easily be generalized to a continuous distribution by replacing $b_l dL'_p$ by $\alpha_{pl}(\mathbf{r}')dV'$ and changing the line integral to a volume integral. This amounts to approximating the dislocation density, α_{pl}, by a collection of discrete dislocation lines with Burgers vector b_l and line direction x_p. Hence, to generalize Burgers' formula, (2.1), to a continuous distribution it is helpful to find a line integral expression for the solid angle, Ω.

3. The Solid Angle

The solid angle, Ω, at a point, **r**, subtended by the surface, S, is defined by (4)

$$\Omega(\mathbf{r}) = \int_S X_m R^{-3} dS'_m \tag{3.1}$$

Burgers pointed out that this surface integral could be replaced by a line integral but did not express it as such. Peach and Koehler (5) stated that the problem of replacing the surface integral by a line integral does not have a unique solution. They chose a symmetric solution which enabled them to write out the components of the displacement as line integrals in full and rather cumbersome detail. Eshelby (6) and Mossakowska (7) pointed out that a much more compact expression for the line integral can be obtained. To see how this is done recall that there exists a general relation between surface and line integrals, namely, Stokes' theorem

$$\int_S \epsilon_{ijk}T_{,j}\, dS_i = \oint_L T dL_k \tag{3.2}$$

Hence, to write the solid angle as a line integral it is only necessary to find a solution to the equation

$$\epsilon_{pmk} A_{k,p}(\mathbf{r}) = x_m/r^3 \tag{3.3}$$

that is, to find the vector potential, A_k, of the field $\mathbf{r}/r^3$. It is straightforward though tedious to check that a solution is given by

$$A_k(\mathbf{r}) = \frac{\epsilon_{ijk}x_i e_j}{r(r+\mathbf{r}\cdot\mathbf{e})} \tag{3.4}$$

where e_j is an arbitrary but constant unit vector, which embodies the lack of uniqueness mentioned by Peach and Koehler.

The solid angle (3.1) now becomes

$$\Omega(\mathbf{r}) = \int_S \epsilon_{pmk} A_{k,p}(\mathbf{R})\, dS'_m$$
$$= \oint_L A_k(\mathbf{R})\, dL'_k \tag{3.5}$$

by (3.3) and (3.2). Hence, it is now simple to generalize Burgers' formula (2.1) to a continuous distribution of dislocations, as mentioned in the previous section. The result, given by (5.10) below, is however derived from the more general formula.

4. Anisotropic Elasticity

4.1 *The Displacement due to a Plastic Distortion*

We start the derivation from the general relation for the displacement in an anisotropic body that has experienced a plastic deformation. The result has been derived before (2) and is given by (II3.15)[1]

[1]The symbol (II3.15) will be used to refer to Eq (3.15) in reference (2).

$$u_n(\mathbf{r}) = -\int C_{ijkl} G_{jn,i}(\mathbf{R})\, \beta^P_{kl}(\mathbf{r}')\, dV' \tag{4.1}$$

Here the C_{ijkl} are the anisotropic elastic constants, G_{jn} is Green's tensor function, and β^P_{kl} is the plastic distortion. The integration is taken over all space, i.e. the whole body.

Green's tensor function, $G_{jn}(\mathbf{r})$, represents the displacement in the x_j-direction at the field point, **r**, arising from a point force in the x_n-direction at the origin. For an infinitely extended body it is defined by (II2.7)

$$C_{ijkl} G_{jn,ik}(\mathbf{r}) + \delta_{ln}\delta(\mathbf{r}) = 0 \tag{4.2}$$

where $\delta(\mathbf{r})$ is the three-dimensional Dirac delta function.

4.2 *A Useful Auxiliary Function*

To proceed with the derivation it was found useful to introduce the auxiliary function, J_{klmn}, defined by

$$J_{klmn}(\mathbf{r}) \equiv \int C_{ijkl} G_{jn,i'm'}(\mathbf{r}')\, (4\pi R)^{-1}\, dV' \tag{4.3}$$

This function bears some resemblance to the incompatibility source tensor (II4.25), introduced by Simmons and Bullough (8) to solve the incompatibility problem. It is also a type of potential function. It's usefulness stems mainly from the fact that it satisfies two relationships that facilitate the derivation.

The first relationship is found by taking the divergence with respect to the third index

$$J_{klmn,m}(\mathbf{r}) = \int C_{ijkl} G_{jn,i'm'}(\mathbf{r}')\, (4\pi R)^{-1}_{,m}\, dV'$$
$$= \int C_{ijkl} G_{jn,i'}(\mathbf{r}')\, (4\pi R)^{-1}_{,mm}\, dV'$$
$$= -\int C_{ijkl} G_{jn,i'}(\mathbf{r}')\, \delta(\mathbf{R})\, dV'$$
$$= -C_{ijkl} G_{jn,i}(\mathbf{r}) \tag{4.4}$$

In the above relations the first equality follows from carrying out the differentiation under the integral (4.3), where R is the only function that depends on r, the second equality follows by a partial integration, where the surface integrals vanish at infinity, the third equality from the relationship

$$(4\pi r)^{-1}_{,mm} + \delta(\mathbf{r}) = 0 \tag{4.5}$$

and the fourth equality by the basic property of the delta function (IIB3).

The second relationship for the auxiliary function is found by taking the divergence with respect to the first index

$$J_{klmn,k}(\mathbf{r}) = \int C_{ijkl} G_{jn,i'm'}(\mathbf{r}')\, (4\pi R)^{-1}_{,k}\, dV'$$
$$= \int C_{ijkl} G_{jn,i'k'}(\mathbf{r}')\, (4\pi R)^{-1}_{,m}\, dV'$$

$$= - \int \delta_{ln}\delta(\mathbf{r}')(4\pi R)^{-1}_{,m}\, dV'$$
$$= - \delta_{ln}(4\pi r)^{-1}_{,m}$$
$$= \delta_{ln}x_m/4\pi r^3$$
$$= \delta_{ln}\epsilon_{pmk}A_{k,p}(\mathbf{r})/4\pi \tag{4.6}$$

Here the second equality follows by a partial integration, the third from (4.2), and the sixth from (3.3).

4.3 *Continuous Distribution of Dislocations*

For a continuous distribution the dislocation density, α_{pl}, is defined in terms of the plastic distortion, β^P_{kl}, by (II3.4)

$$\alpha_{pl} \equiv - \epsilon_{pmk}\beta^P_{kl,m} \tag{4.7}$$

or, equivalently (II3.5)

$$\epsilon_{pmk}\alpha_{pl} \equiv \beta^P_{ml,k} - \beta^P_{kl,m} \tag{4.8}$$

The following derivation then shows how the displacement (4.1) can now be written in terms of the dislocation density

$$u_n(\mathbf{r}) = \int J_{klmn,m}(\mathbf{R})\,\beta^P_{kl}(\mathbf{r}')\, dV'$$
$$= \int J_{klmn}(\mathbf{R})\,\beta^P_{kl,m'}(\mathbf{r}')\, dV'$$
$$= \int J_{klmn}(\mathbf{R})\,[\beta^P_{ml,k'}(\mathbf{r}') - \epsilon_{pmk}\alpha_{pl}(\mathbf{r}')]\, dV'$$
$$= \int J_{klmn,k}(\mathbf{R})\,\beta^P_{ml}(\mathbf{r}')\, dV' - \int \epsilon_{pmk}J_{klmn}(\mathbf{R})\,\alpha_{pl}(\mathbf{r}')\, dV'$$
$$= (4\pi)^{-1} \int \delta_{ln}\epsilon_{pmk}A_{k,p}(\mathbf{R})\,\beta^P_{ml}(\mathbf{r}')\, dV' - \int \epsilon_{pmk}J_{klmn}(\mathbf{R})\,\alpha_{pl}(\mathbf{r}')\, dV'$$
$$= (4\pi)^{-1} \int A_k(\mathbf{R})\,\epsilon_{pmk}\beta^P_{mn,p'}(\mathbf{r}')\, dV' - \int \epsilon_{pmk}J_{klmn}(\mathbf{R})\,\alpha_{pl}(\mathbf{r}')\, dV'$$
$$= - (4\pi)^{-1} \int A_k(\mathbf{R})\,\alpha_{kn}(\mathbf{r}')\, dV' - \int \epsilon_{pmk}J_{klmn}(\mathbf{R})\,\alpha_{pl}(\mathbf{r}')\, dV' \tag{4.9}$$

Here the first equality follows from (4.4), the second by a partial integration, the third from (4.8), the fourth from a partial integration again, the fifth from (4.6), the sixth by a partial integration once more, and the seventh by (4.7).

Equation (4.9) is the basic result of this paper. It gives the displacement field, u_n, of a continuous dislocation distribution, α_{pl}, in a linearly elastic, infinitely extended, homogeneous body, in terms of the potential functions A_k and J_{klmn} given by (3.4) and (4.3). It represents the complete generalization of Burgers' formula (2.1) to a continuous dislocation distribution in anisotropic elasticity.

4.4 *The Discrete Dislocation Line*

We now wish to show the result that the previous relation (4.9) reduces to for a discrete dislocation line along the curve L. For this case the dislocation density is given by (II3.23)

$$\alpha_{pl}(\mathbf{r}) = \delta_p(L)b_l \tag{4.10}$$

where (IIB4)

$$\delta_p(L) = \oint_L \delta(\mathbf{R})\, dL'_p \tag{4.11}$$

is the Dirac delta function for a curve. If these two relations are substituted in (4.9) and the integration over dV' carried out (IIB11), the result is

$$u_n(\mathbf{r}) = - (4\pi)^{-1} \oint_L A_k(\mathbf{R})\, b_n dL'_k - \oint_L \epsilon_{pmk}J_{klmn}(\mathbf{R})\, b_l dL'_p \tag{4.12}$$

This relation represents the generalization of Burgers' formula (2.1) for a discrete dislocation line L to anisotropic elasticity.

We now investigate the consistency of (4.12) with previously published results. From Stokes' theorem (3.2)

$$u_n(\mathbf{r}) = - (4\pi)^{-1} \int_S \epsilon_{pmk}A_{k,p}(\mathbf{R})\, b_n dS'_m$$

$$- \int_S [J_{klmn,m}(\mathbf{R})\, b_l dS'_k\, J_{klmn,k}(\mathbf{R})\, b_l dS'_m]$$
$$= \int_S C_{ijkl} G_{jn,i}(\mathbf{R})\, b_l dS'_k \tag{4.13}$$

by (4.4) and (4.6). This is the well-known relation (II3.30) for the displacement of a discrete dislocation by an integral over the surface S that spans the line L. The conventional way of deriving this relation is to substitute $\beta^P_{kl}(\mathbf{r}) = -\delta_k(S) b_l$ into (4.1).

5. Isotropic Elasticity

5.1 *Evaluation of the Auxiliary Function*

To specialize the results of section 5 to the case of isotropic elasticity we first find the isotropic form of the auxiliary function. For an isotropic material the elastic constants are given by (III2.1)

$$C_{ijkl} = 2G\,[\delta_{ik}\delta_{jl} + \delta_{ij}\delta_{kl}\,\nu/(1-2\nu)] \tag{5.1}$$

where G is the modulus of rigidity and ν is Poisson's ratio. Furthermore Green's tensor function is given by (III2.2)

$$G_{jn}(\mathbf{r}) = (8\pi G)^{-1}\,[\delta_{jn} r_{,qq} - r_{,jn}/2(1-\nu)] \tag{5.2}$$

Hence we find from (4.3) that the auxiliary function reduces to

$$J_{klmn}(\mathbf{r}) = \int [\delta_{ln} r'_{,k'q'q'm'} + \delta_{kn} r'_{,l'q'q'm'}$$
$$-r'_{,k'l'm'n'}/(1-\nu) + \delta_{kl} r'_{,n'q'q'm'}\nu/(1-\nu)]\,(32\pi^2 R)^{-1} dV' \tag{5.3}$$

From the integral (III2.6)

$$\int r'_{,q'q'}\, R^{-1}\, dV' = -4\pi r \tag{5.4}$$

it then follows by partial integration that

$$\epsilon_{pmk} J_{klmn}(\mathbf{r}) = -(8\pi)^{-1}\,\epsilon_{pmk}\,[\delta_{kn} r_{,lm} + \delta_{kl} r_{,nm}\nu/(1-\nu)] \tag{5.5}$$

where the terms in (5.3) that are symmetric in (km) vanish because of the antisymmetry of ϵ_{pmk}. This is the desired result.

5.2 *Continuous Distribution of Dislocations*

We now derive several versions of the displacement field for a dislocation distribution in isotropic elasticity. If we substitute (5.5) into (4.9) we obtain

$$u_n(\mathbf{r}) = -(4\pi)^{-1} \int A_k(\mathbf{R})\,\alpha_{kn}(\mathbf{r}')\,dV'$$
$$+ (8\pi)^{-1} \int [\epsilon_{pmn} R_{,lm} + \epsilon_{pml} R_{,nm}\nu/(1-\nu)]\,\alpha_{pl}(\mathbf{r}')\,dV' \tag{5.6}$$

This equation represents one generalization of Burgers' formula (2.1) to a continuous dislocation distribution in isotropic elasticity.

The relation (5.6) can be written in a form that more closely resembles the original Burgers' formula by using the identity

$$\epsilon_{pmn} R_{,l} \equiv \epsilon_{mnl} R_{,p} - \epsilon_{pnl} R_{,m} + \epsilon_{pml} R_{,n} \tag{5.7}$$

Then we find for (5.6)

$$u_n(\mathbf{r}) = -(4\pi)^{-1} \int A_k(\mathbf{R})\,\alpha_{kn}(\mathbf{r}')\,dV'$$
$$+ (8\pi)^{-1} \int [\epsilon_{mnl} R_{,pm} - \epsilon_{pnl} R_{,mm} + \epsilon_{pml} R_{,nm}/(1-\nu)]\,\alpha_{pl}(\mathbf{r}')\,dV' \tag{5.8}$$

This relation can be simplified. A consequence of the definition for the dislocation density, (4.7), is that α_{pl} satisfies the continuity condition (II3.6)

$$\alpha_{pl,p} = 0$$

(5.9)

Hence, by a partial integration the first term in the second line of (5.8) vanishes. We therefore have

$$u_n(\mathbf{r}) = -(4\pi)^{-1} \int A_k(\mathbf{R})\, \alpha_{kn}(\mathbf{r}')\, dV' \\ -(8\pi)^{-1} \int [\epsilon_{pnl} R_{,mm} - \epsilon_{pml} R_{,nm}/(1-\nu)]\, \alpha_{pl}(\mathbf{r}')\, dV' \tag{5.10}$$

This equation represents another generalization of Burgers' formula (2.1) to a continuous distribution of dislocations in isotropic elasticity.

5.3 *The Discrete Dislocation Line*

We now come full circle and show that the last result is consistent with Burgers' original formula (2.1). This is easily seen if we substitute (4.10–11) into (5.10), carry out the integration over dV', and substitute for the solid angle from (3.5).

6. Summary

We have generalized Burgers' formula for the displacement of a single discrete dislocation line in isotropic elasticity, eq(2.1), to a displacement formula for a continuous distribution of dislocations in anisotropic elasticity, eq(4.9). The result is a volume integral over the dislocation density in terms of two useful functions, a vector potential (3.4) and an auxiliary function (4.3). The former is not unique and the latter is expressed as a volume integral over the elastic constants and Green's tensor function.

We have specialized the above result to a discrete dislocation line in anisotropic elasticity, eq(4.12), and to a continuous dislocation distribution in isotropic elasticity, eq(5.10). These are also generalizations of Burgers' formula. Finally we have shown that the results reduce to Burgers' original formula as a special case.

As a corollary we have also reviewed the compact line integral expression for the solid angle subtended by a closed curve, eq(3.5). It is given in terms of the non-unique vector potential (3.4).

What is the physical significance of the non-unique displacement? We believe that it represents the fact that different displacement fields can be associated with the same dislocation distribution. In other words one could visualize a given dislocation distribution as being created by many different possible deformation histories, giving different displacement fields for the same dislocation distribution. It is natural for this uncertainty in the displacement to be manifested through the solid angle term, because this term embodies the essential property of a dislocation line, namely, that a Burgers circuit around the dislocation produces the Burgers vector. In fact, the original expression for the solid angle, (3.1), is not unique, because the surface S is arbitrary. It can be chosen anywhere, as long as it is bounded by the curve L. On passing through the surface S the solid angle $\Omega(\mathbf{r})$ changes discontinuously by an amount 4π.

The results have been derived for a linearly-elastic, infinitely extended, homogeneous body. Nevertheless it is hoped that some of these results may be useful in application to dislocation distributions in real crystals. A dislocation tangle could be modeled by a continuous dislocation distribution and the resulting displacement field could be used in expressions for the diffraction from such a crystal.

References

1. J. M. Burgers, Proc. Koninkl. Ned. Akad. Wetenschap. **42**, 293 (1939).

2. R. deWit, J. Res. Nat. Bur. Stand. (U.S.) **77A**, 49 (1973). In this paper referred to as II.

3. R. deWit, J. Res. Nat. Bur. Stand. (U.S.) **77A**, 359 (1973). In this paper referred to as III.

4. R. deWit, Sol. State Phys. **10**, 269 (1960).

5. M. O. Peach and J. S. Koehler, Phys. Rev. **80**, 436 (1950).

6. J. D. Eshelby, Private Communication (1972).

7. Z. Mossakowska, Private Communication (1972).

8. J. A. Simmons and R. Bullough, in *Fundamental Aspects of Dislocation Theory*, ed. by J. A. Simmons, R. deWit, and R. Bullough, Nat. Bur. Stand. (U.S.) Spec. Pub. 317, Vol. I (1970), p. 89.

THE STRESS FIELD OF DISLOCATION ARRAYS

G. SAADA

Laboratoire P.M.T.M., C.S.P. Université Paris Nord, Avenue J.B. Clément
93430 Villetaneuse, France

Abstract

Many physical situations occur in which periodic planar dislocation arrays play a key role : low angle grain boundaries, plastic deformation of crystals, epitaxial growth. A general method based upon Kröner's theory of internal stresses is shown to give simple mathematical formulae allowing a clear physical discussion of the relevant parameters.

The general features of the method are first given as well as the necessary formulae. Surface effects are explicitly discussed. Specific applications are made to low angle grain boundaries, kink bands, bicrystal plastic deformation, epitaxial growth. More sophisticated problems such as the influence of elastic anisotropy or inhomogeneity are briefly discussed.

THE STRESS FIELD OF DISLOCATION ARRAYS

G. SAADA

Laboratoire P.M.T.M., C.S.P. Université Paris Nord, Avenue J.B. Clément
93430 Villetaneuse, France

1. Introduction

The description of the stress field of dislocation arrays is a necessary step toward the understanding of many physical phenomena : strain hardening, epitaxial growth, stability of interfaces. In principle the description of the elastic field of dislocation arrays can be obtained by summing the contribution of individual dislocations [1,2,3]. The method may be painful however. When the array is made of dislocation segments (honeycomb network), the resulting formulae are quite involved [4,5,6,7], the relevant physical quantities are obtained through tedious numerical calculations. In any case a good deal of the physical meaning of the results is darkened by the mathematical complexity of the formulae.

As pointed out by Mura [8], the elastic field of periodic planar dislocation networks is easily derived from the knowledge of the dislocation density tensor $\underline{\alpha}$ through the approach developped by Kröner [9,10], Nye [11], Bilby, Bullough and Smith [12]. Recently Rey and the author [13,14,15,16,17] have developped the method and applied it to various physical situations : low angle grain boundary analysis, description of complex boundaries.

In this paper we shall give a detailed account of the method and describe some applications, restricting ourselves to results which are rigorous and obtained through simple calculations. There is strictly speaking neither new concepts nor new theory in this paper. It is rather a way of describing simply situations of physical interest and this paper could have been written 30 years ago. It is based mostly on Kröner's approach. Since it is now quite popular we shall only quote the few usual concepts and equations relevant to our purpose (§ 2). General formulae for dislocation arrays in an infinite medium are given in § 3 and surface effects discussed in § 4. Examples of applications are reviewed in § 5. More sophisticated problems arising from elastic anisotropy, elastic inhomogeneity, non linear elasticity are shortly discussed in § 6.

2. Self stress theory [9,10,11,12,18]

2.1. Geometry

In the case of compatible elasticity the following quantities are defined : the displacement $\underline{u}$, the distorsion $\underline{\beta}$, the strain $\underline{e}$, rotation $\underline{\omega}$ and the bend twist $\underline{K}$:

$$\underline{u} = \overrightarrow{P_I P_F} \qquad 1a$$

$$du_j = \beta_{ij} dx_i \qquad 1b$$

$$ds_{II}^2 - ds_I^2 = 2\, e_{ij} dx_i dx_j \qquad 1c$$

$$e_{ij} = \frac{1}{2} (\beta_{ij} + \beta_{ji}) \qquad 1d$$

$$\omega_{ij} = \frac{1}{2} (\beta_{ij} - \beta_{ji}) \qquad 1e$$

Figures 1

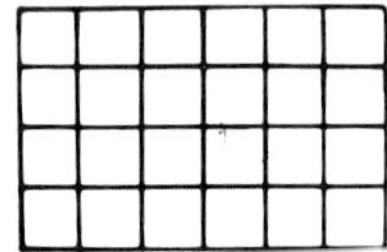

1a

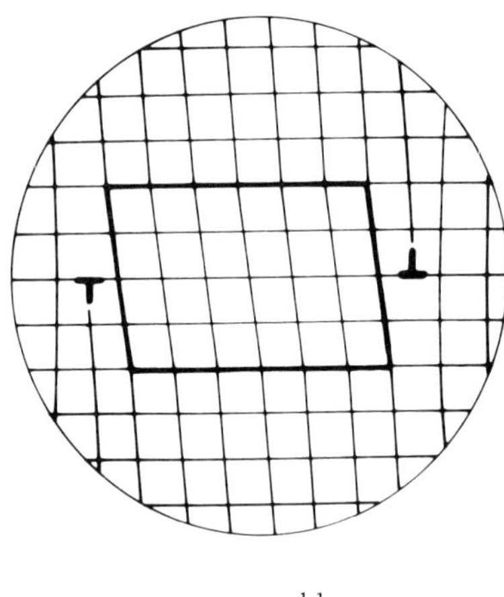

1d

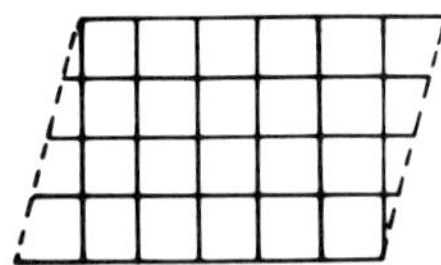

1b

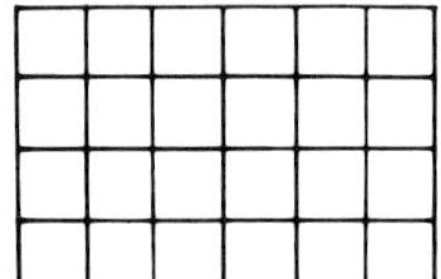

1c

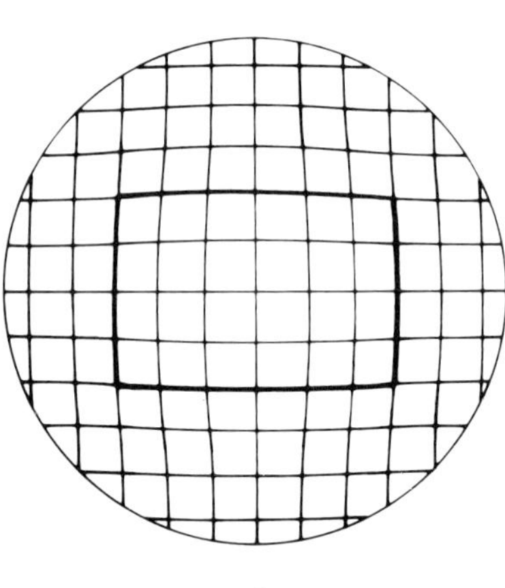

1e

1a initial state

1b homogeneous stress free plastic shear

1c homogeneous stress free dilatation

1d heterogeneous stress free plastic shear elastically accomodated

1e heterogeneous stress free dilation elastically accomodated.

$$\omega_k = \frac{1}{2} \varepsilon_{ijk}\omega_{ij} \qquad \text{1f}$$

$$d\omega_l = K_{lm}dx_m \qquad \text{1g}$$

In these formulae the symbols I and F refer to the initial and final state respectively, ds is the infinitesimal distance between two corresponding points in either state, ε_{ijk} is the permutation tensor (= 0 if two indices are equal and to ± 1 following the parity of the permutation). $\underline{\omega}$ is the rotation vector associated to the antisymmetrical rotation tensor $\underline{\omega}$. The Einstein convention is used in this paper. The distortions are assumed to be small.

Whenever equations 1b and 1g can be integrated the solid is said to be in a compatible state. Then $\underline{\beta}$ and $\underline{K}$ are the gradients of the displacement and rotation vector. The condition for compatibility is that the so called incompatibility tensor $\underline{\eta}$ vanishes :

$$\eta_{ij} = - \varepsilon_{kli}\, \varepsilon_{mnj} \frac{\partial^2 e_{nl}}{\partial x_m \partial x_k} \qquad \text{1h}$$

Then K_{lm} is given by :

$$K_{lm} = \varepsilon_{knl} \frac{\partial e_{nm}}{\partial x_k} \qquad \text{1i}$$

When $\underline{\eta}$ differs from zero the solid is said to be in an incompatible state. Formally the existence of non zero incompatibility implies a state of self stress : the stress field vanishes on the outside surface of the medium but differs from zero inside the medium. Equivalently cut an elastic medium into small pieces : if no piece deforms <u>elastically</u> after cutting, there are no self stresses and the medium is in a compatible state. If some pieces deform elastically, there are self stresses and the medium was, before cutting in an incompatible state. Try then to weld the pieces, cavities and overlap will occur which are a measure of incompatibility.

Elastic incompatibility is often a consequence of inhomogeneous stress free transformation β^P : fig. 1a,b,c pictures stress free homogeneous plastic shear or stress free dilatation. If the transformation is not homogeneous (Fig. 1d,1e) the solid has to accommodate elastically the deformation to avoid cavities or matter overlapping. The total deformation $\underline{\beta}^T$ is then compatible, but the stress free transformation $\underline{\beta}^P$ and the elastic transformation $\underline{\beta}$ are not compatible. The body is then self stressed [10,18].

To describe the state of the body Kröner [9,10] introduces the dislocation density tensor $\underline{\alpha}$:

$$\alpha_{ij} = \varepsilon_{ikl} \frac{\partial}{\partial x_k} \beta_{lj} = - \varepsilon_{ikl} \frac{\partial \beta^P_{lj}}{\partial x_k} \qquad \text{2a}$$

$$\frac{\partial \alpha_{ij}}{\partial x_i} = 0 \qquad \text{2b}$$

Consider a surface S bounded by a curve C. From 2a and 2b it is easily shown that :

$$\int_S \alpha_{ij} n_i dS = + b_j \qquad \text{2c}$$

$\underline{b}$ is the total Burgers vector of the dislocations threading the surface. Equation 2b simply express the fact that dislocations cannot end inside the crystal.

From the knowledge of $\underline{\alpha}$ two important tensors, the incompatibility tensor $\underline{\eta}$ and the contortion $\underline{\kappa}$ are derived :

$$\eta_{ij} = \frac{1}{2}\left[\varepsilon_{jmn}\frac{\partial\alpha_{im}}{\partial x_n} + \varepsilon_{imn}\frac{\partial\alpha_{jm}}{\partial x_n}\right] \qquad 2d$$

$$\kappa_{ij} = \frac{1}{2}\alpha_{mm}\,\delta_{ij} - \alpha_{ij} \qquad 2e$$

since $\underline{\omega}$ is defined everywhere there exist a tensor $\underline{K}'$ such that :

$$\frac{\partial\omega_1}{\partial x_m} = K'_{1m} \qquad 2f$$

On the other hand, since $\underline{\eta} \neq 0$, 1g cannot be integrated if $\underline{K}$ is defined by equation 1i. Kröner [9] has shown that :

$$\underline{K}' = \underline{K} + \underline{\kappa} \qquad 2g$$

2.2. Statics

In the absence of body forces the stress field obeys the well known equation :

$$\frac{\partial\sigma_{ij}}{\partial x_j} = 0 \qquad 3a$$

In isotropic elasticity to which we specialize from now on :

$$\sigma_{ij} = 2\mu(e_{ij} + \frac{\nu}{1-2\nu}e_{mm}) \qquad 3b$$

where μ is the shear modulus and ν the Poisson ratio.

From equations 2f, 2g, 1i, 3b, 3a, it is easily shown that :

$$\vec{\nabla}^2\omega_1 = \frac{\partial\kappa_{1m}}{\partial x_m} \qquad 3c$$

Which allows the determination of the rotation field when $\underline{\alpha}$ is known. To determine the stress field Kröner [9] defines a stress function $\underline{\chi}$ such that :

$$\sigma_{nk} = -\varepsilon_{pnm}\,\varepsilon_{ijk}\frac{\partial^2\chi_{mj}}{\partial x_p\,\partial x_i} \qquad 3d$$

$$\frac{\partial\chi'_{ij}}{\partial x_j} = 0 \qquad 3e$$

$$\chi_{ij} = 2\mu(\chi'_{ij} + \frac{\nu}{1-\nu}\chi'_{kk}\delta_{ij}) \qquad 3f$$

And shows that :

$$\nabla^4\underline{\chi}' = \underline{\eta} \qquad 3g$$

which allows the determination of stress together with suitable boundary conditions. In our case the boundary condition is :

$$\sigma_{ij}n_j = 0 \text{ on the external surface} \qquad 3h$$

The interaction energy of two dislocation distributions can be calculated either by standard formulae or by Kröner's formula :

$$E_{12} = \int \eta_{ij}^{(1)} \chi_{ij}^{(2)} dx_1 dx_2 dx_3 \qquad 3i$$

which is valid for an infinite body.

2.3. Fictitious and actual dislocations

As we have seen in § 2.1. formula 2a, an incompatible transformation $\underline{\beta}^P$ results in the existence of a dislocation density tensor $\underline{\alpha}$. On the other hand the Burgers vectors of dislocations must obey strict crystallographic conditions. This is always the case when $\underline{\beta}^P$ occurs by true plastic strain. However if one considers the case of inhomogeneous heating or the case of coherent matching of different crystals on an interface, the self stress can still be described by the above equations but the dislocations do not exist physically. We call them fictitious dislocations and represent the corresponding dislocation density tensor as $\underline{\alpha}'$.

In many case the stress arising from $\underline{\alpha}'$ is so large as to induce plastic deformation which results in creating actual dislocations, the corresponding dislocation density tensor being $\underline{\alpha}$. Then the source of stress is the sum of $\underline{\alpha}' + \underline{\alpha}$. Low angle grain boundaries and misfit dislocations have been described as actual dislocations relaxing the stress of fictitious dislocations or allowing good matching along an interface.

3. Periodic planar dislocation arrays

3.1. General equations

Periodic planar dislocation arrays occur in various physical situations : low and large angle grain boundary, misfit dislocations plastic deformation of bicrystals, kink bands.

The purpose of this section is to outline the common features of all these arrays without insisting on their specific aspects. We therefore describe such an array in plane $x_1 O x_2$ by expressing the dislocation density tensor $\underline{\alpha}$ as :

$$\underline{\alpha}(\underline{r}) = \underline{a}(\underline{\rho})\ \delta(x_3) \qquad 4a$$

$\underline{a}$ is the surface dislocation density tensor, $\underline{r}$ stands for a point (x_1, x_2, x_3) in the 3 dimensional space ; similarly $\underline{\rho}$ stands for a point (x_1, x_2) of the plane of the array.

To take full advantage of the periodicity of the dislocation array, $\underline{a}$ must be expressed as a Fourier sum. Then all previous equations are integrated in a very simple way. We therefore write [13] :

$$\underline{a}(\underline{\rho}) = \sum_{\vec{g}} a^{\underline{g}}(\underline{\rho}) \exp 2\Pi i\ \underline{g}\ \underline{\rho} \qquad 4b$$

Where $\underline{g}$ is the set of the reciprocal vectors of the periodic dislocation array, $a^{\underline{g}}$ is the corresponding Fourier component. General formulations of $\underline{a}^{\underline{g}}(\underline{\rho})$ have been calculated [13,16,17] and are summarized below :

I. One set of parallel equidistant infinite dislocations (Fig. 2a)

Let h be the distance between dislocations, $\underline{t}$ the unit vector characterizing each dislocations, $\underline{h}$ be the vector $h\ (-t_2, t_1, 0)$, $\underline{g}$ be the vector $\underline{h}/h^2$, then the set of the reciprocal vector is $n\underline{g}$ (n being a whole number) and :

$$a_{ij}^{(n)} = g\ t_i\ b_j \qquad 4c$$

$$a_{ij} = \sum_{n=-\infty}^{n=+\infty} g t_i b_j \exp 2\Pi i\ n\ \underline{g}\ \underline{\rho} \qquad 4d$$

Figures 2

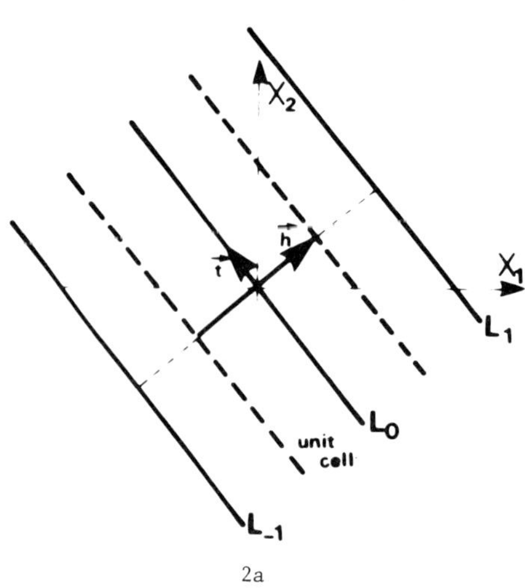

2a

Figure 3

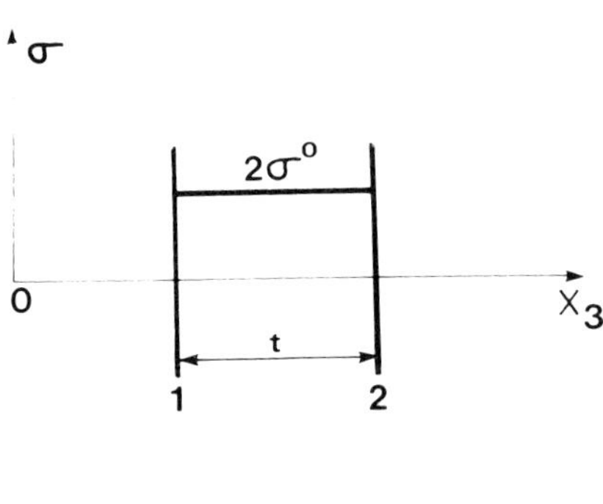

3a

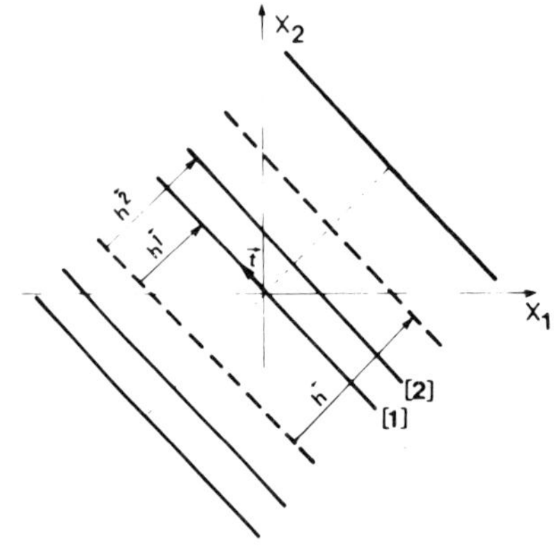

2b

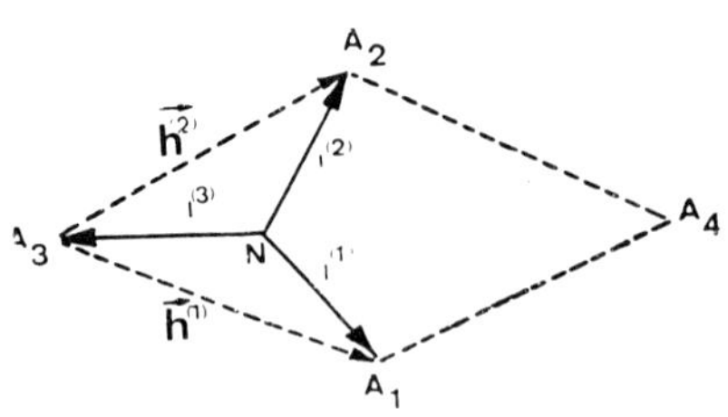

2c

Figures 2 : Cell structure of periodic planar dislocation arrays

2a One set of parallel infinite dislocations. The spacing of dislocations is h, the common Burgers vector of the dislocations is $\underline{b}$, the unit vector along dislocations lines is $\underline{t}$.

2b Same case when there are more than one dislocation per unit cell.

2c Honeycomb network.

Figure 3

3a Parallel planar arrays (1 and 2) in an infinite medium. σ, σ° sketch the behaviour of the stress components.

Figures 3

Figures 4, 5, 6

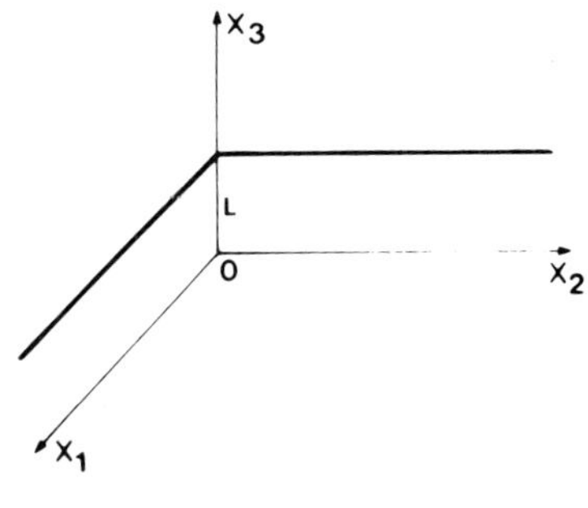

3b

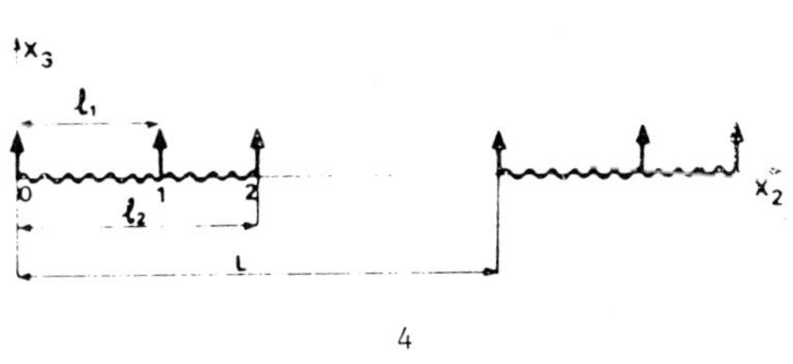

4

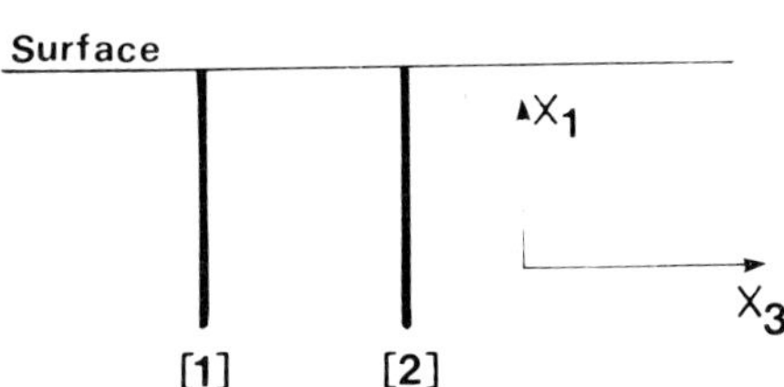

3c

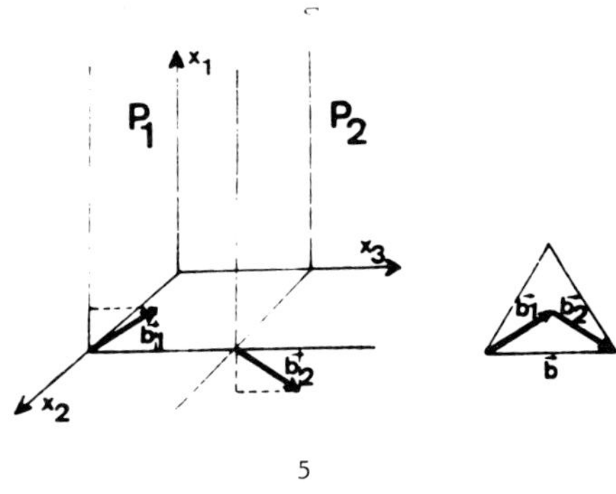

5

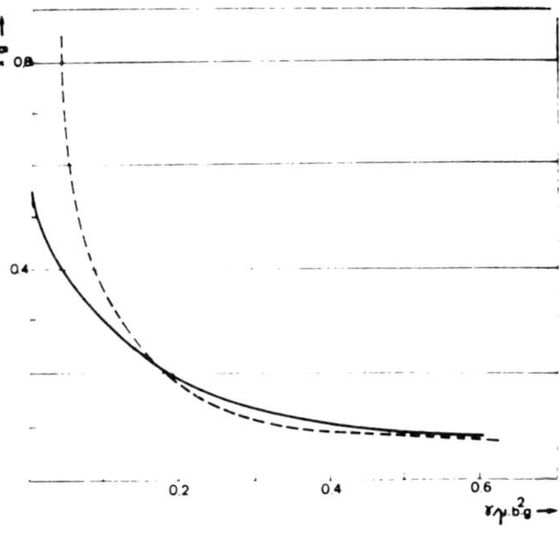

6

Figures 3

3b Planar array parallel to a planar surface The array is in the plane $x_3=0$, the surface is the plane $x_3=1$.

3c Boundary perpendicular to a surface. Case of a slab. $\sigma_{11}^{(o)}=0$ outside the slab and differs from zero inside the slab. The surface relaxation induces a stress $e_{11} \sim -\sigma_{11}^{o}/2\mu$.

Figure 4

Sketch of a tilt boundary whose unit cell is made of three Frank partial dislocations.

Figure 5

Splitting of dislocations outside from the boundary.

Figure 6

Variation of the splitting width of dislocations with respect to the stacking fault energy.

In that case it is useful to introduce $\underline{\lambda}$:

$$\underline{\lambda} = g\underline{t} \qquad \text{4e}$$

II. Case where the unit cell contains more than one dislocation (Fig. 2b)

In that case :

$$a_{ij}^{(n)} = g\, t_i \sum_{p=1}^{p=m} b_j^{(p)} \exp - i\, n\, \delta_p \qquad \text{4f}$$

$$\delta_p = 2\Pi\, \underline{g}\, \underline{h}_p \qquad \text{4g}$$

In formulae 4f, 4g, m refers to the number of dislocations in the unit cell, p labels the pth dislocation and $\underline{h}_p$ characterizes the position in the unit cell.

III. Sets of intersecting dislocations

The components of $\underline{a}^{(n)}$ are obtained by summing the components for each set.

The term $\underline{a}°$ equal to the average < a > of $\underline{a}$ in the plane $x_1\, Ox_2$ is of particular interest [17] :

$$< a_{ij}^{(p)} > = \sum_{p=1}^{p=m} \lambda_i^{(p)}\, b_j^{(p)} \qquad \text{4h}$$

4. Honeycomb network (Fig. 2c)

The unit cell is made of three dislocation segments $\underline{1}^{(p)}$ (p = 1,2,3) with Burgers vectors $\underline{b}^{(p)}$ satisfying the node condition :

$$\underline{b}^{(1)} + \underline{b}^{(2)} + \underline{b}^{(3)} = 0 \qquad \text{4i}$$

The basis is defined by the position of the node N characterized by the vector $\underline{1}$ joining the origin of the unit cell to N.

The Bravais lattice is defined by the vectors $\underline{h}^{(1)}$, $\underline{h}^{(2)}$ and the reciprocal lattice by the vectors $g^{(1)}$, $\underline{g}^{(2)}$:

$$\underline{g}^{(i)}\, \underline{h}^{(j)} = \delta_{ij} \qquad \text{4j}$$

Then :

$$\underline{a}(\underline{\rho}) = \sum_{\underline{g}} a^{(\underline{g})} \exp 2\Pi i\, \underline{g}\, \underline{\rho} \qquad \text{4k}$$

And :

$$a_{ij}^{(g)} = \frac{\exp -2\,\Pi i\, \underline{g}\, \underline{1}}{2\, i\Pi\, |h^{(1)} \times h^{(2)}|} \sum_{p=1}^{p=3} \frac{1_i^{(p)}\, b_j^{(p)}}{\underline{g}(\underline{h}^{(p)} - \underline{1})} \qquad \text{4l}$$

With :

$$\underline{1}^{(p)} = - \underline{1} + \underline{h}^{(p)} \qquad \text{4m}$$

$$\underline{h}^{(3)} = 0 \qquad \text{4n}$$

Note that :

$$a_{ij}^{(o)} = \sum_{p=1}^{p=3} \frac{l_i^{(p)} b_j^{(p)}}{|\underline{h}^{(1)} \times \underline{h}^{(2)}|} \qquad 4o$$

Another useful expression for $a_{ij}^{(o)}$ is [17] :

$$a_{ij}^{(o)} \doteq \sum_{p} \lambda_i^{(p)} b_j^{(p)} \qquad p = 1,2 \qquad 4p$$

$$\underline{\lambda}^{(p)} = \underline{h}^{(p)} / |\underline{h}^{(1)} \times \underline{h}^{(2)}| \qquad 4q$$

General methods for describing more complicated networks are given in [13,17].

3.2. Contortion tensor and rotation field

From 2e, 4a, 4b :

$$\kappa_{ij}(\underline{r}) = k_{ij}(\underline{\rho})\ \delta(x_3) \qquad 5a$$

$$k_{ij} = \frac{1}{2} a_{mm} \delta_{ij} - a_{ij} \qquad 5b$$

Whence :

$$\frac{\partial \kappa_{ij}}{\partial x_j} = \left(\frac{\partial k_{i1}}{\partial x_1} + \frac{\partial k_{i2}}{\partial x_2} \right) \delta(x_3) + k_{i3}\ \delta'(x_3) \qquad 5c$$

Equation 3c is then integrated without trouble by developing in Fourier sum and taking the Fourier transform :

$$\omega_j = \frac{1}{2} \sum_{\underline{g}}{}' \left\{ \left[k_{j3}^{\underline{g}} \text{ sgn } x_3 - i(g_1\, k_{j1}^{\underline{g}} + g_2\, k_{j2}^{\underline{g}})/g \right] \exp 2\,\Pi i\ \underline{g}\ \underline{\rho} \exp - 2\Pi g|x_3| \right.$$

$$+ \omega_{\bar{j}} + k_{j3}^{(o)}\ Y(x_3) \qquad 5d$$

In formula 5d the symbol Σ' means as usual that $\underline{g} = 0$ is excluded from the sum ; sgn x_3 is the generalized function $|x_3|/x_3$.

The sum is easily calculated in all cases of interest. It decreases exponentially with the distance $|x_3|$ from the plane of the array.

$\omega_{\bar{j}}$ is a constant which cannot be determined from the above equations ; it defines in fact the reference lattice and can be taken equal to zero in most cases.

$Y(x_3)$ is the Heaviside step function (equal to zero for $x_3 < 0$ and to 1 for $x_3 > 0$).

3.3. Incompatibility tensor, stress potential, stress field

From formulae 2d and 4a the incompatibility tensor is :

$$\underline{\eta}(r) = \underline{A}(\underline{\rho})\ \delta(x_3) + \underline{B}(\underline{\rho})\ \delta'(x_3) \qquad 6a$$

$$\underline{A} = \left\| \begin{matrix} -\dfrac{\partial a_{13}}{\partial x_2} & \dfrac{1}{2}\left[\dfrac{\partial a_{13}}{\partial x_1} - \dfrac{\partial a_{23}}{\partial x_2}\right] & \dfrac{1}{2}\left[\dfrac{\partial a_{11}}{\partial x_2} - \dfrac{\partial a_{12}}{\partial x_1}\right] \\ \dfrac{1}{2}\left[\dfrac{\partial a_{13}}{\partial x_1} - \dfrac{\partial a_{23}}{\partial x_2}\right] & \dfrac{\partial a_{23}}{\partial x_1} & \dfrac{1}{2}\left[\dfrac{\partial a_{21}}{\partial x_2} - \dfrac{\partial a_{22}}{\partial x_1}\right] \\ \dfrac{1}{2}\left[\dfrac{\partial a_{11}}{\partial x_2} - \dfrac{\partial a_{12}}{\partial x_1}\right] & \dfrac{1}{2}\left[\dfrac{\partial a_{21}}{\partial x_2} - \dfrac{\partial a_{22}}{\partial x_1}\right] & 0 \end{matrix} \right\| \qquad 6b$$

$$\underline{B} = \left\| \begin{matrix} a_{12} & \dfrac{1}{2}[a_{22} - a_{11}] & 0 \\ \dfrac{1}{2}[a_{22} - a_{11}] & -a_{21} & 0 \\ 0 & 0 & 0 \end{matrix} \right\| \qquad 6c$$

$\underline{A}$ et $\underline{B}$ are then developped as Fourier sums with the help of 4b and by taking the Fourier transform of equation 3g one obtains easily [13].

$$\underline{\chi}' = \sum_{\underline{g}}{}' \underline{\chi}'^{(g)}(x_3) \exp 2\,\Pi i\ \underline{g}\ \underline{\rho} + \frac{1}{4}\underline{B}^{(o)}\ x_3^2 \operatorname{sgn} x_3 \qquad 6d$$

$$\chi'^{(\underline{g})}(x_3) = \left[\left(\frac{1+2\Pi g|x_3|}{32\Pi^3 g^3}\right) \underline{A}^{\underline{g}} - \frac{1}{8\Pi} x_3 \frac{\underline{B}^{\underline{g}}}{g} \right] \exp - (2\Pi g|x_3|) \qquad 6e$$

When all dislocations are parallel to the same direction $\sum_{\underline{g}}'$ is replaced by $\sum_{n=-\infty}^{n=+\infty}{}'$ and $\underline{\chi}'^{\underline{g}}$ by $\underline{\chi}'^{(n)}$:

$$\underline{\chi}'^{(n)} = \left[\left(\frac{2\Pi g|nx_3|+1}{32\Pi^3 g^3 |n|^3}\right) A^{(n)} - \frac{x_3}{8\Pi g|n|} B^{(n)} \right] \exp - 2\Pi g|nx_3| \qquad 6f$$

From formulae 6d, 6e, 6f, the stress field is easily obtained with the help of 3d and 3f :

$$\underline{\sigma} = \sum_{\underline{g}}{}' \ \underline{\sigma}^{\underline{g}}(x_3) \exp 2\Pi i\ \underline{g}\ \underline{\rho} + \underline{\sigma}^{(-)} + \underline{\sigma}^{(o)}\ Y(x_3) \qquad 6g$$

In 6g the term under Σ' comes from the corresponding term under Σ' in 6d. It is easily calculated in all cases of interest and decreases exponentially with the distance $|x_3|$ from the boundary. $\underline{\sigma}^-$ is a constant which cannot be determined only by the above equations ; $\underline{\sigma}^{(o)}$ is defined as :

$$\underline{\sigma}^{(o)} = 2\mu \left\| \begin{matrix} \dfrac{a_{21}^{(o)} - \nu a_{12}^{o}}{1-\nu} & \dfrac{1}{2}(a_{22}^{(o)} - a_{11}^{(o)}) & 0 \\ \dfrac{1}{2}(a_{22}^{(o)} - a_{11}^{(o)}) & \dfrac{\nu a_{21}^{(o)} - a_{12}^{(o)}}{1-\nu} & 0 \\ 0 & 0 & 0 \end{matrix} \right\| \qquad 6h$$

Let us discuss formulae (6g, 6h) in the following cases :

3.3.1. The boundary is isolated in an infinite medium

Then $\underline{\sigma}$ can be taken equal to zero and $\underline{\sigma}^{\circ}$ has to be zero in order to obtain a sensible solution. This implies that :

$$a_{21}^{(o)} = 0 \qquad 6i$$

$$a_{12}^{(o)} = 0 \qquad 6j$$

$$a_{11}^{(o)} - a_{22}^{(o)} = 0 \qquad 6k$$

which are well known conditions for the stability of a planar array.

3.3.2. Two parallel boundaries in an infinite medium (Fig. 3a)

In that case a sensible solution exists provided the long range stress fields of both arrays cancel : i.e. the components $a_{21}^{(o)}$, $a_{12}^{(o)}$, $a_{11}^{(o)}-a_{22}^{o}$, are of opposite sign in both arrays. Let then $x_3 = 0$ and $x_3 = 1$ be the position of the arrays. One has :

$$\underline{\sigma} = \begin{cases} 2\,\underline{\sigma}o & 0 < x_3 < 1 \\ 0 & \text{everywhere else} \end{cases} \qquad 6l$$

where $\underline{\sigma}^{o}$ is calculated for the network located at $x_3 = 0$.

The strain energy due to this term is equal to 4wl with :

$$w = \mu\left[\frac{1}{2}\,(a_{22}^{(o)} - a_{11}^{(o)})^2 + \frac{a_{21}^{(o)^2} + a_{12}^{(o)^2}}{1-\nu} - \frac{2}{1-\nu}\,a_{12}^{(o)}\,a_{21}^{(o)}\right] \qquad 6m$$

3.3.3. Boundary parallel to a planar surface (Fig. 3b)

Let the boundary be parallel to the surfaces of a slab at distance H and l from the two surfaces. Assume that the boundary is made of homogeneously distributed dislocations. Then only the terms $\underline{\sigma}^{-}$ and $\underline{\sigma}^{o}\,Y(x_3)$ have to be considered in 6g. Since σ_{3j}^{o} is zero one may choose the stress field as :

$$\underline{\sigma} = \begin{cases} \phi\underline{\sigma}^{o} & 0 < x_3 < 1 \\ (1-\phi)\underline{\sigma}^{o} & -H < x_3 < 0 \end{cases} \qquad 6n$$

ϕ has to be choosen in order to minimize the total elastic energy. With the help of 6 m :

$$\phi = \frac{1}{1+(l/H)} \qquad 6o$$

4. Surface effects

The elastic field of dislocations arrays has been studied by various methods. Mura [20,21] has given general formulae, the use of which however is not very convenient. In this paper we shall restrict ourselves to the case of dislocations arrays parallel to the free surface. This case is relevant to the experimental situation corresponding to epitaxial growth. A few remarks on the case of arrays not parallel to the surface are given at the end of this section. Let us consider a periodic planar dislocation network in the plane $x_1\,Ox_2$, the free surface being in the plane $x_3 = 1$.

From equations 3d, 6d, 6e, 6g, $\underline{\sigma}^g(x_3)$ can be expressed as :

$$\underline{\sigma}^g(x_3) = \left[\underline{F}^g(\underline{a}^g)x_3 + \underline{E}^g(\underline{a}^g)\right] \exp - 2\Pi g|x_3| \qquad 7a$$

$\underline{F}^g$ and $\underline{E}^g$ are second rank tensors depending linearly on a^g. Then :

$$\underline{\sigma}^g(0) = \underline{E}^g(a^g) \qquad 7b$$

Let us then consider an image distribution [18,19] $\underline{I}$:

$$\underline{I} - \underline{y}(x_1, x_2)\ \delta(x_3-1) \qquad 7c$$

$$\underline{y} = \sum_g{}' \underline{y}^g \exp 2\Pi i\ \underline{g}\ \underline{\rho} \qquad 7d$$

Here the image distribution is a surface distribution at $x_3 = 1 + \delta 1$ where $\delta 1$ tends towards 0. $\underline{y}^g$ is determined by writing that $\underline{\sigma}^g$ is zero on the surface :

$$\left[F^g_{3j}(\underline{a}^g)\ 1 + E^g_{3j}(a^g)\right] \exp - 2\Pi g 1 + E^g_{3j}(\underline{y}^g) = 0 \qquad 7e$$

From which y^g is easily determined in the general case.

The general formulae would be tedious to write and one has rather to calculate them in cases of interest. For example let all the dislocations of the $\underline{a}$ distribution be parallel to ox_1. After simple calculations the above method gives :

$$y^{(n)}_{11} = - a^{(n)}_{11} \exp - 2\Pi g|n|1 \qquad 7f$$

$$y^{(n)}_{12} = \left[2\Pi i\ ngl\ a^{(n)}_{13} - (1-2\Pi|n|gl)\ a^{(n)}_{12}\right] \exp - 2\Pi g|n|1 \qquad 7g$$

$$y^{(n)}_{13} = \left[-\ (1+2\Pi|n|gl)\ a^{(n)}_{13} + 2\Pi i\ ngl\ a^{(n)}_{12}\right] \exp - 2\Pi g|n|1 \qquad 7h$$

In these formulae notice that the terms $- a^{(n)}_{1j} \exp - 2\Pi g|n|1$ correspond in fact to the mirror distribution. Since $\underline{a}^o$ gives no contribution to the stress on the free surface, $\underline{y}^o$ may take any value, we take it equal to zero.

When the network intersects the surface, more elaborate calculations [20,21] are necessary and the results are quite involved. However in the case of a dislocation network in a slab of thickness t two remarks can be added :

. As is well known [22] the stress field of the dislocations is restricted to a cylinder of radius t surrounding the dislocations, therefore the above method does not apply if the average distance between dislocations is larger than t, i.e. if $gt \lesssim 1$.

. If however $gt \gtrsim 1$ and 6i,j,k are not fulfilled, the stress at the free surface (as calculated by equations 6) will exhibit a jump at the intersection of the free surface and the plane of the array. Relaxation of this stress to fulfill the boundary conditions will induce additional strain. An example is sketched on fig. 3.

5. Applications

5.1. Low angle grain boundary

A low angle grain boundary can be described as a dislocation array whose elastic field reduces to a given rotation field at infinity, i.e. such that its stress field at infinity be zero. From equations 5c,d and 6i,j,k, $\underline{a}^o$ or equivalently $\underline{a}$ is completely determined. This is in fact the basis of Frank's analysis of low angle grain boundaries. It applies to low angle grain boundaries made of lattice dislocations as well as to arrays of so called secondary dislocations.

When analyzing such arrays observed experimentally one generally knows the dislocation directions and spacings i.e. the $\underline{\lambda}$ vectors defined by equations 4e,q. Writing equations 5c,d and 6i,j,k in terms of the $\underline{\lambda}$ vectors is simple and has been detailed in [17]. It allows necessary conditions to which the Burgers vectors must satisfy, in some cases it allows a complete determination of the latter.

In the case of a low angle boundary, the Burgers vectors satisfy crystallographic conditions : either they are lattice vectors or well known partial dislocations. In most case they can be completely known by contrast analysis. One has to check however that conditions 5c,d and 6i,j,k are satisfied. If they are not, the observed array having long range stress field is not stable : either it is stabilized by image force, or it is stabilized by an other array which has escaped the observation. Examples of such situations are quite frequent [17].

The situation is more involved in the case of secondary dislocation networks. Then diffraction contrast analysis does not give clear cut informations in all cases of interest and the Burgers vectors need no longer be crystallographic. The O lattice approach [25] is generally used and gives indeed good hints though the following remarks must be made :

- The O lattice approach does not provide in all cases the condition that the stress field be zero at infinity.
- It is also a geometrical theory. It cannot go beyond predicting $< \underline{a} >$.
  Although the Burgers vectors predicted by this approach are the most probable other Burgers vectors may be stable in grain boundaries.

5.2. Stability of dislocation arrays. Elastic theory

5.2.1. General remarks

As pointed out in the previous section condition 6i,j,k are not enough to resolve the cell structure of dislocation networks. To go beyond that one has to make effective energy calculations. The method developped by Rey and the author [13] allows for simple calculations which are summarized below.

5.2.2. Twist boundary in (III) plane of FCC crystals. Fig. 2c

As is well known two simple configuration have been observed for such boundaries : intersecting infinite dislocations or honeycomb network made of screw dislocations of equal length. Application of the method gives for the energy per unit surface for the honeycomb network [16] :

$$E = \frac{9\mu b^2}{4\Pi^3 l} f(x) \qquad \text{8a}$$

$$x = r_o/l \qquad \text{8b}$$

$$f(x) = \sum_{\substack{n_1 \neq 0 \\ n_2 \neq 0}} \frac{N^{3/2} \sin^2[\Pi(n_1+n_2)/3] \exp[-(4/3)\Pi N^{1/2} x]}{(2n_1-n_2)^2 (n_1-n_2)^2 (n_1+n_2)^2} \qquad \text{8c}$$

$$N = |n_1^2 + n_2^2 - n_1 n_2| \qquad \text{8d}$$

r_o is the core size and n_1, n_2 are integers.

For less symmetrical situations one would have less symmetrical expressions for f(x). The latter is easily shown to behave logarithmically when x tends toward 0. One finds finally :

$$E = - 0.14 \mu b \omega \operatorname{Log} (4.5 \frac{r_o}{b} \omega) \qquad \text{8e}$$

ω is the twist angle.

The method is more general and more precise than those already published [26,27].

For the case of intersecting dislocations one finds the well known formula of Read and Shockley [28].

The honeycomb network is found to have a smaller energy than a network made or parallel dislocations when ω is smaller than 5.10^{-2} radians.

5.2.3. Tilt boundaries

The case of tilt boundaries is by far more simple since in most cases all dislocations are parallel to the same direction which we take to be ox_1. The boundary is in the plane $x_1\ Ox_2$. For the simple tilt boundary one finds for the energy the classical formula [29,13] :

$$E = - \frac{\mu\ b\ \omega}{4\Pi(1-\nu)} \operatorname{Log} \frac{2\Pi r_o}{eb}\ \omega \qquad 9a$$

The case of complex planar tilt boundaries with undissociated dislocations can be calculated [17] as :

$$E = \frac{\mu K \omega}{4\Pi \bar{b}_3} \left[- \operatorname{Log} \omega + \frac{\overline{b_2^2} + \overline{b_3^2}}{(1-\nu)K} + \operatorname{Log} \frac{m\ \bar{b}_3}{2\Pi r_o} + \varepsilon \right] \qquad 9b$$

$$\varepsilon = - \frac{1}{2mK} \sum_{p \neq q} K_{pq} \operatorname{Log} 2\left[1-\cos(\delta_p - \delta_q)\right] \qquad 9c$$

$$K_{pq} = b_1^{(p)}\ b_1^{(q)} + (b_2^{(p)}\ b_2^{(q)} + b_3^{(p)}\ b_3^{(q)})/(1-\nu) \qquad 9d$$

$$K = \frac{1}{m} \operatorname{Tr} \underline{K} \qquad 9e$$

m is the number of dislocations in the unit cell, δ_p has been defined in 4g. Formulae 9 apply to boundaries having no long rang stresses i.e. satisfying equations 6i,j,k. The bar symbols indicate averages on the dislocations of the unit cell. Comparison of 9b with 9a confirms that the energy of the boundary is lowered by splitting dislocations and emphasizes the importance of a good knowledge of the core size and core energy. An application of formula 9b to the observations make by Bourret [29,30] has been worked out explicity [17]. The geometry of the boundary is shown fig. 4 : it is a tilt boundary with < $0\bar{1}1$ > tilt axis in a (III) plane of Ge. The unit cell is made of 3 Frank partials and the energy is easily calculated by using formula 9b and adding the stacking fault energy [17]. An interesting result is that half of the boundary surface is covered by stacking fault when :

$$\omega = 6(1-\nu) \frac{(\gamma_1 \gamma_2)^{1/2}}{\mu b} \qquad 9f$$

γ_1 and γ_2 are the intrinsic and extrinsic stacking fault energies.

ω varies between 5.10^{-3} and 5.10^{-2} radians following the value of the stacking fault energy, i.e. one may observe widely extended dislocations in high stacking fault energy crystals.

Splitting of dislocations perpendicular to the boundary (Fig. 5) can also be studied. Generally each dislocation wall made of partial dislocations is such that its long range stress field $\underline{\sigma}_o$ is not zero. However the total long range stress field is zero as it should be. A specific case has been worked out for a tilt boundary in a prismatic plane in an hexagonal structure [13], see fig. 5.

Let x be the width of splitting. Then :

$$\gamma = \frac{\mu b^2 g}{24} \left[- 1 - \frac{2}{\exp 2\Pi gx - 1} + \frac{6\Pi gx}{(1-\nu)\{ch(2\Pi gx)-1\}} \right] \qquad 9g$$

Which for :

$$2\Pi gx << 1$$

gives the well known formula |31| :

$$\gamma \sim \frac{\mu b^2}{24\Pi x} \frac{2+\nu}{1-\nu} \qquad 9h$$

Fig. 6 represents the results of formulae 9g,h. The complete formula 9g gives results differing from those obtained by 9h when :

$$\omega > 5 \frac{\gamma}{\mu b} \qquad 9i$$

which compares with 9f.

More important, due to the term $\underline{\sigma}^o$ the width of splitting remains finite when the stacking fault tends to zero. This allows the boundary to remain stable when there is a phase transformation.

Notive that splitting of dislocations perpendiculary from the boundary needs more energy than splitting in the plane of the boundary.

5.2.4. Zig zag boundaries

Given a tilt boundary whose rotation axis is parallel to ox_1 equations 6i,j,k write :

$$\sum_{p=1}^{p=m} b_1^{(p)} = 0 \qquad 10a$$

$$\sum_{p=1}^{p=m} b_2^{(p)} = 0 \qquad 10b$$

$$g \sum_{p=1}^{p=m} b_3^{(p)} = - \omega \qquad 10c$$

Since the Burgers vectors must be crystallographic there are values of ω for which m has to be quite large. In most cases [17] the boundary is not stable, due to short range stresses. A zig zag boundary may then form. If it is made of Nye walls the misorientation of each wall must be the same [17] which implies restrictions so drastic that in most cases zig zag tilt walls are not stable. If the Nye walls have not the same misorientation their junction can be described as quasidisclinations giving rise to long range stresses. The latter can be relaxed by building a third tilt boundary. The misorientation of the latter is quite small however and it may escape the observation.

5.2.5. Simple tilt boundary parallel to a free surface

For a tilt boundary parallel to a free surface $a_{13}^{(n)}$ is equal to gb. Then formulae 7f,g,h give the image distribution :

$$y_{12}^{(n)} = 2\Pi i n g^2 1b \exp - 2\Pi g|n|1 \qquad 11a$$

$$y_{13}^{(n)} = - (1+2\Pi|n|g1) \; gb \exp - 2\Pi g|n|1 \qquad 11b$$

Closed expression of $\underline{y}$ are easily obtained :

$$y_{12} = 2gb\ L(\frac{dC}{dX_2})_{X_3=L} \qquad 11c$$

$$y_{13} = -\ 2gb\ (C - L\ \frac{dC}{dX_3})_{X_3=L} \qquad 11d$$

$$C = \frac{\cos X_2 - \exp - |X_3|}{2(\mathrm{ch}|X_3| - \cos X_2)} \qquad 11e$$

$$X_i = 2\Pi g\ x_i \qquad 11f$$

$$L = 2\Pi g l \qquad 11g$$

From the above equations the stress field can easily be calculated. We restrict our interest to the calculation of the elastic energy of the network by evaluating the work done on the cut surfave from $x_3 = -\infty$ to $x_3 = 0$. The relevant component of the stress field is σ_{23} which is the sum of two contributions :

$$\sigma_{23}(\underline{a}) = \frac{\mu g b}{1-\nu} \sum_{n=1}^{n=\infty} 2\Pi n g\ x_3 \cos 2\Pi n g\ x_2 \exp 2\Pi n g\ x_3 \qquad 11h$$

$$\sigma_{23}(\underline{y}) = \frac{\mu g b}{1-\nu} \sum_{n=1}^{n=\infty} -\ 2\Pi n g \left[x_3 + 4\Pi n g l(x_3-l)\right] \exp 2\Pi n g(x_3-2l) \qquad 11i$$

Let then S(x) be defined as :

$$S = \sum_{n=1}^{n=\infty} \frac{1}{n} \exp - 2\Pi g n x \qquad x > 0 \qquad 11j$$

The energy per unit surface is then :

$$E = \frac{\mu g b^2}{4\Pi(1-\nu)} \left\{ (S - r_o \frac{dS}{dx})_{x=r_o} - \left[S - (r_o+2l) \frac{dS}{dx}\right]_{x=r_o+2l} - 2l(r_o+l)(\frac{d^2S}{dx^2})_{x=r_o+2l} \right\} \qquad 11k$$

The first term is the self energy of the boundary in an infinite medium, the second term gives the contribution of the mirror distribution.

Together with :

$$S = -\ \mathrm{Log}(1 - \exp - 2\Pi g x) \qquad 11l$$

$$\frac{dS}{dx} = -\ 2\Pi g(\exp 2\Pi g x - 1)^{-1} \qquad 11m$$

$$\frac{d^2S}{dx^2} = (2\Pi g)^2 .\ \exp(2\Pi g x)\ .\ (\exp 2\Pi g x - 1)^2 \qquad 11n$$

since the surface effects are more important when l is smaller than the spacing h of the dislocations, we restrict to that case and obtain :

$$E = \frac{\mu g b^2}{4\Pi(1-\nu)} \left[\mathrm{Log}\ (1 + \frac{2l}{r_o}) - 2l\ \frac{(r_o+l)}{(r_o+2l)^2} \right] \qquad 11o$$

If besides $l >> r_o$:

$$E \sim \frac{\mu g b^2}{4\Pi(1-\nu)} \text{Log} \frac{2l}{e^{1/2} r_o} \qquad 11p$$

5.3. Misfit dislocations and epitaxy

Let us consider two crystals of different lattice parameters the substrate S and the film F matching coherently along the plane $x_1 Ox_2$; F is of thickness l, le substrate is in the negative half space. The stress free transformation writes :

$$\underline{\beta}^P = \underline{E}\, Y(x_3) \qquad 12a$$

where $\underline{E}$ is the uniform misfit strain.

The elastic field can be described by an homogeneous distribution $\underline{a}'$ of fictitious surface dislocations. From equation 2a the non zero components of $\underline{a}'$ are :

$$a'_{1j} = + E_{2j} \qquad 12b$$

$$a'_{2j} = - E_{1j} \qquad 12c$$

Notice that $\underline{E}$ is a symmetrical tensor and E_{3k} is zero.

Such a distribution satisfies the boundary conditions provided the plane $x_1 Ox_2$ can be considered as infinite. This definition of misfit agrees with that given by Matthews [32].

It leads to a strain energy per unit surface W_s which is easily evaluated through equations 6m,n,o :

$$W_s = \frac{\mu H l}{(H+l)} \left[\frac{1}{2} (a'_{22} - a'_{11})^2 + \frac{a'^2_{21} + a'^2_{12}}{1-\nu} - \frac{2\nu}{1-\nu} a'_{12}\, a'_{21} \right] \qquad 12d$$

where H is the thickness of the substrate.

The strain energy can be relaxed by adding a periodic network of lattice dislocations on the plane $x_1 Ox_2$ such that :

$$< \underline{a} > + \underline{a}' = 0 \qquad 12e$$

A general treatment including surface effects can then be given using the results of section 2, 3, 4. For sake of simplicity let us restrict first to the classical case where $\underline{a}$ is made of a square grid of equally spaced edge dislocations to see how the complete treatment compares with previous ones. Let then :

$$E_{11} = E_{22} = - f \qquad 12f$$

be the non zero components of $\underline{E}$ and :

$$a^{(o)}_{12} = - a^{(o)}_{21} = + f \qquad 12g$$

Whence :

$$gb = |f| \qquad 12h$$

b being the common length of the Burgers vectors.

Figures 7

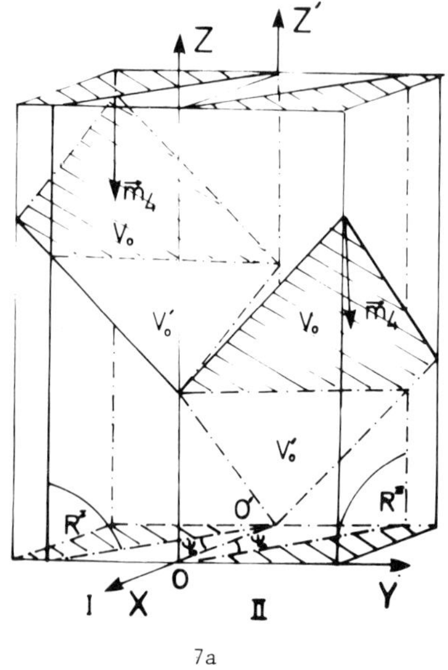

7a

Figure 8

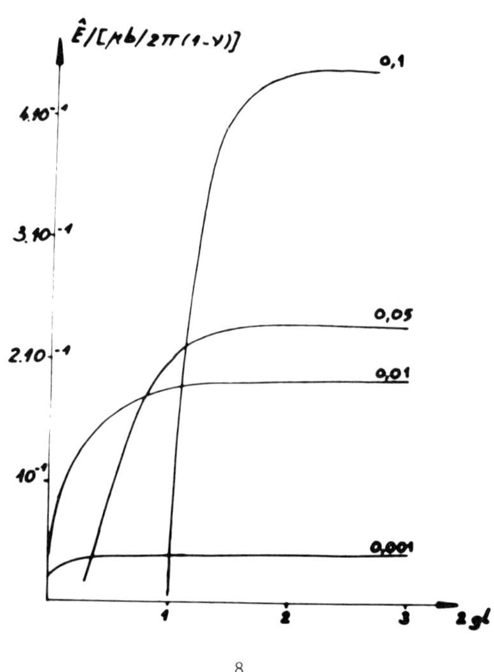

8

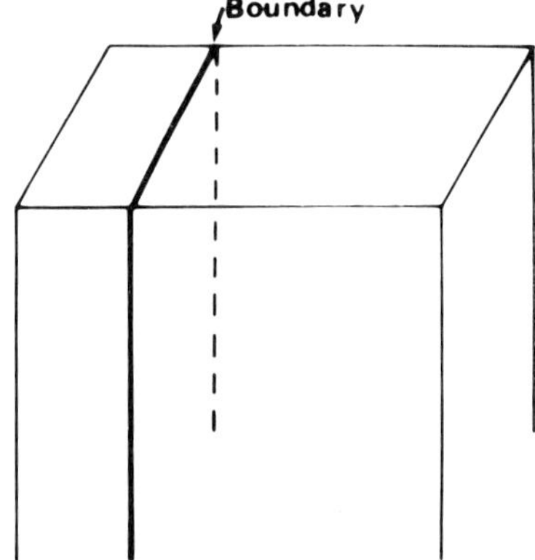

7b

Figures 7

7a Schematic illustration of the propagation and blocking of glide fronts during plastic deformation.

7b Bicrystal with grain boundary close to the surface.

Figure 8

Plot of the elastic energy per unit surface of a square grid of dislocations. Vertical Axis $E/[\mu b/2\Pi(1-\nu)]$. Horizontal axis 2gbl. Curves correspond to f = 0.1 ; 0.05 ; 0.01 ; 0.001 .

Let us consider first the set of dislocations parallel to ox_1 of Burgers vector (0, b, 0). The corresponding surface dislocation density tensor is from formula 4d :

$$a_{12} = gb \sum_{n=-}^{n=+} \exp 2\Pi ing\, x_2 \qquad 12i$$

The Fourier components of the image surface dislocation density tensor are (from formulae 7f,g,h) :

$$y_{12}^{(n)} = -(1- 2\Pi|n|gl)\, gb \exp - 2\Pi g|n|l \qquad 12k$$

$$y_{13}^{(n)} = (2\Pi ingl)\, gb \exp - 2\Pi g|n|l \qquad 12l$$

Notice that $\underline{a}$ is on the plane $x_3 = 0$ and $\underline{y}$ on the plane $x_3 = 1$. Similar formulae are obtained for sets of dislocations parallel to ox_2.

Since there is no elastic interaction between both sets of dislocations the strain energy per unit surface may be calculated as twice the strain energy of one set of dislocation by using the same method as in section 5.2.5. One obtains :

$$E = \frac{\mu|f|b}{2\Pi(1-\nu)}\left\{(S+r_o \frac{dS}{dx})_{x=r_o} - \left[S+(2l+r_o)\frac{dS}{dx}\right]_{x=2l+r_o} - \left[2l(l+r_o)\frac{d^2S}{dx^2}\right]_{x=2l+r_o}\right\} \qquad 12m$$

Which for $r_o << l << h$ reduces to :

$$E \quad \frac{\mu|f|b}{2\Pi(1-\nu)} \, \text{Log} \, \frac{2l}{e^{1/2}b} \qquad 12n$$

These equations are somewhat different from those published elsewhere and deserve a full discussion which is still under progress. Fig. 8 only gives the plot of the elastic energy per unit surface in the case where all the misfit has been relaxed. Obviously when the misfit is too large (f ~ 0.1) the results have no physical meaning when l is to small and core problems become more important.

There is no difficulty in studying either more complicated networks or layered networks.

5.4. Plastic deformation

Plastic deformation does not create infinite periodic planar dislocation networks. However consider a planar network of size L and dislocations whose average spacing is $\underline{h}$. If L >> h one may consider 3 regions in the crystal following the distance from the network [17] :

- . D ≤ h. The network may be considered as infinite and the above equations apply.
- . h ≤ D ≤ L. The above equations still apply and indicate that the stress is of the order of μb/h and does not vary with distance.
- . D >> L. The network behaves as if it were a single dislocation. These considerations apply of course if the network is more or less regular and not in the case for example of pile ups. Having these remarks in mind a few situations can be reviewed rapidly.

. Plastic deformation of bicrystals [33,34]

During the plastic deformation of a bicrystal the grain boundary can be considered to be planar and infinite indeed. Intersection of the glide systems with the boundary provide a distribution of dislocations which may be averaged as a surface dislocation density tensor $\underline{a}$. Necessary conditions for the existence of no long range stress field $\underline{\sigma}^\circ$ defined by equation 6h helps explaining the observed glide structure. Other phenomena, for example the asymmetry of the glide system on both faces of the bycristal can be explained by taking into account the fact that a glide front (see fig. 7a) can be blocked by the grain boundary, making a secondary boundary. Increasing the plastic strain makes the glide front to rotate and generate long range stress field which has to be taken into account to explain the appearance of new glide systems.

In some cases the glide front may have steps. Experiments on bicrystals in which the boundary is closer of one surface (Fig. 7b) instead of being at the center of the bicrystal would be worthwile.

. Kink bands [35]

Kink bands are a nice example of the relaxation of the long range stress field $\sigma°$ on a boundary by the building of a parallel boundary compensating $\underline{\sigma}°$. As pointed out at the end of section 4 when the kink band is perpendicular to the surface of observation one should observe a strain parallel to the normal of the surface as a consequence of surface relaxation. This has been observed in sodium chloride [36].

. Dislocation cells

Secondary creep or stage III of plastic deformation induces the building of more or less defined dislocation cells. It would be interesting to check to which extent the edges of dislocation walls satisfy compatibility conditions as far as rotations are concerned. Another remark is worth mentionning : the rotation field of such walls may be quite large, e.g. larger than 10^{-2} radians, on the other hand the long range stress $\underline{\sigma}°$, if any, should be smaller than $10^{-3}\mu$. Since both quantities are proportional to the dislocation density it is obvious that those dislocations which produce the driving force for the transformation of the substructure are a small proportion of the total number of dislocations. This is probably why they generally escape the observation [17].

. Secondary glide

A building glide line can be considered as a set of roughly equidistant parallel dislocations. Then $\underline{\sigma}°$ is not zero and is comparable with the external applied stress. This is probably responsible for secondary glide occuring during stage I of plastic deformation.

6. Miscellaneous remarks and conclusion

The above results are easily generalized to the case of anisotropic crystals provided one may neglect the variation of the elastic constants with orientation. One must then apply the Mura Willis formula for the stress [37] and the results have the same structure as those developped above [14]. Note however that formula 3c is no longer valid. One has to add $\varepsilon_{knl} \frac{\partial e_{nm}}{\partial x_k}$ to the right member of the equation. In the case of two media of different elastic moduli a detailed calculation shows that again the structure of the solution (i.e. homogeneous long range stress field in each half space and exponentially decreasing field) is not changed [38]. The formulae however are quite involved and in some cases, for example for the case of epitaxial growth replacing the shear modulus by $2\mu_1\mu_2/(\mu_1 + \mu_2)$ is probably a good approximation [32]. More difficult is the situation arising from too narrowly spaced dislocations. One must then distinguish between local and true Burgers vector [12], besides problems of non linear elasticity occur and they mix with core problem. However the general solution given above is correct at distances larger than the spacing of dislocations. Direct observations by high resolution electron microscopy should allow the determination of both the local and the true Burgers vector.

The authors is glad to thank Dr Faivre, Rey and Zaoui for helpful discussion and comments.

REFERENCES

1 BURGERS J.M., Proc Kon Ned Akad Wenschap. 42 293 1939.
2 COTTRELL A.H., Dislocations and plastic flow in crystals, Oxford University Press Fair Lawn N.J. (1953) p. 94.
3 LI J.C.M., Electron Microscopy and Strength of Crystals (ed. G. Thomas and J. Washburn) Interscience, New York (1963) p. 713.
4 HOKANSON J.L. and WINCHELL P.G., J. Appl Phy 39 (1968) 3311.
5 YOFFE E.H., Phil Mag 5 (1960) 161.
6 BASSIM M.N. and KUHLMANN-WILSDORF D., Phys Stat Sol a 15 (1973) 725
7 BASSIM M.N. and KUHLMANN-WILSDORF D., Phys Stat Sol a 16 (1973) 241
8 MURA T., Proc Roy Soc A 280 (1964) 528.
9 KRONER E., Kontinuums theorie der Verstetzungen und Eigenspannungen. Springer Verlag, Berlin 1958.

10 KRONER E., Theory of crystal defects Academia, Prague 1966.
11 NYE J.F., Acta Met 1 (1953) 53.
12 BILBY B.A., BULLOUGH R. and SMITH E., Proc Roy Soc A 231 263 (1955).
13 REY C. and SAADA G., Phil Mag 33 5 (1976) 825.
14 SAADA G., Phil Mag 34 4 (1976) 639.
15 SAADA G., Proc ICSMA 4, Conf Nancy (1976) 368.
16 REY C. and SAADA G., J. Phys 38 (1977) 721.
17 SAADA G., Acta Met 27 (1978) 921.
18 ESHELBY J., Solid State Physics, Vol 3, Acad Press New York (1956) p. 79.
19 ESHELBY J., Phil Trans R. Soc. A 244 (1951) 87.
20 MURA T., J. Appl Phys 38 5 (1967) 1999.
21 MURA T., J. Appl Phys 43 1 (1972) 182.
22 ESHELBY J.D. and STROH A.N., Phil Mag 7 42 (1951) 335.
23 WINTNER E., KARNTHALER H.P., HAZZLEDINE P.M., Proc ICSMA 4, Conf. Nancy (1976) 927.
24 FRANK F.C., Report on the conference of defects in crystalline solids, London Physical Society (1954).
25 BOLLMANN W., Crystal defects and crystalline interfaces, Berlin, Heidelberg, New York, Springer (1970).
26 HOKANSON J.L. and WINCHELL P.G., J. Appl Phys 39 (1968) 331.
27 YOFFE E.H., Phil Mag 5 (1960) 161.
28 READ W.T. and SHOCKLEY W., Phys Rev 78 (1950) 275.
29 BOURRET A. and DESSEAUX J., Phil Mag 39 4 405 (1979)
30 BOURRET A. and DESSEAUX J., Phil Mag 39 4 419 (1979)
31 READ W.T., Jr Dislocations in crystals (1953) New York, Mac Graw Hill.
32 MATTHEWS J.W., Dislocations in Solids, vol. 2, North Holland, Amsterdam, New York, Oxford (1979) p. 461.
33 REY C. Thèse, Villetaneuse 1980.
34 REY C., ZAOUI A., Acta Met 6 28 687 (1980).
35 NABARRO FRN Theory of crystal dislocations. Clarendon press, Oxford (1967) 100.
36 STOWELL M.J., Phil Mag 7 (1962) 677.
37 MURA T. (1971) J. Appl Mech 38 865.
38 HIRTH J.P., BARNETT D.M. and LOTHE J., Phil Mag 40 1 39 (1979).

THE DYNAMICS OF THE GROUP OF DISLOCATIONS AND THE INSTABILITY OF PLASTIC FLOW

A. Korbel and A. Pawelek
Institute for Metal Working and Physical Metallurgy
Academy of Mining and Metallurgy
Krakow, Poland

Abstract

The pile-up model was used to explain the mechanism of slip band propagation and such features of non-uniform deformation as the strain rate in the band, the width of the band and the velocity of the band propagation. On the basis of the Rosenfield and Kanninen model the dynamical properties of a group of dislocations emitted by a source were numerically found and the conditions under which load and flow instabilities may occur were specified.

The model predicts that the width of the slip band and the velocity of band propagation are functions of the strain and decrease as the strain increases. The model seems to be especially adequate for a non-uniform deformation of the Portevin-LeChatelier type.

Introduction

The mechanism believed to be responsible for the non-uniform deformation of metals is the dislocation - solute atom interaction. The origin of this viewpoint is the fact that non-uniform deformation of Lüders or Portevin-LeChatelier type occurs in alloys in which solute atoms strongly strengthen the solid solution[1].

It is known, however, that Lüders deformation can occur even in very pure metals. Basinski and Jackson [2,3] have shown that unstable flow occurs in very pure copper single crystals if predeformed on slip systems secondary with respect to that which is being tested. The authors found that this phenomenon results from an "alien" dislocation distribution produced by prior deformation. Similar effects of prior deformation on subsequent tensile characteristics have been reported by other workers [4,5,6]. This implies that strain localization may occur in a repeated manner(Portevin-LeChatelier deformation)provided that the strain can itself reproduce the "alien dislocation distribution".

Another very important feature of Lüders deformation in single crystals is the formation of coarse slip lines. From structural observations [7,8] it is known that they consist of a few hundred coplanar dislocations emitted by a source which somehow activates the neighbor sources on adjacent slip planes, so that the slip band is formed. This process has the dynamical features of the pile-up formation and the rate of growth of the slip band is determined by the mutual interaction of the dislocations [8,9]. All these attributes of the phenomenon are even more distinct in alloys, especially those with a low stacking fault energy [7,10,11]. The collective behaviour of dislocations in such an alloy was recently observed by Clement et al. in HVEM [12].

It is seen, therefore, that the flow instability is closely connected with the dynamical properties of groups of dislocations. This is why we have performed calculations of the dynamics of a pile-up formation in the hope of explaining some features of non-uniform deformation. Particular attention has been paid to those features which ensure instability of flow and which have not yet been explained by any other model e.g. the width of the slip band and its dependence upon the strain.

Procedure of Calculation

The model used was that of Rosenfield and Kanninen [13]. The most important assumptions were that: 1) the emission of the leading dislocation takes place when the applied stress becomes equal to the yield stress level of the source, which is determined by a source's dimension $R_o = \frac{A}{\sigma_a}$. R_o is simply the distance of the leading dislocation from the source when it is ready to emit the second dislocation. 2) The emission of subsequent dislocations occurs every time the effective stress on the source is reduced to zero, or

$$\sigma_a - \sum_{i=1}^{N} \frac{A}{x_i} = 0 \tag{1}$$

where σ_a is the applied stress (equal to the yield stress of the source), A is the interaction constant, N is the number of dislocations emitted and x_i is the distance of the i-th dislocation from the source.

The velocity of i-th dislocation is assumed to follow the relationship

$$V_i = B\sigma^m_{eff,i} \quad , \quad X_i \leq 0 < L \tag{2}$$

where the effective stress on i-th dislocation is given by

$$\sigma_{eff,i} = \sigma_a \left[1 + \frac{A}{\sigma_a} \sum_{\substack{j=0 \\ j \neq i}} \frac{1}{X_i - X_j} \right] \tag{3}$$

where B and m are material constants, and L is the pile-up length. It is also assumed that the effect of thermal activation of dislocation motion is taken care of by the form of the velocity - stress relationship. This seems to be justified for cooperative movement of dislocations, which is only little influenced by thermal fluctuations. The calculated values are: 1) time dependent velocities of dislocations in the group during emission and pile-up formation, and 2) the time dependent stress concentration at a strong obstacle which permanently stops the leading dislocation. This is given by

$$\frac{\sigma_c}{\sigma_a} = 1 + \frac{A}{\sigma_a} \sum_{i=1}^{N} \frac{1}{L - X_i} \tag{4}$$

The calculations were performed using the numerical Runge-Kutta method [23] for the normalized time parameter, $\theta = \frac{B\sigma_a^m}{L}$, so that the solutions do not depend upon the particular σ_a, B, and L values [14].

Results

In Fig. 1a and b are shown the relative velocities of particular dislocations in a group ($\frac{v_k}{v_a}$, where v_a is the velocity of a dislocation under the applied stress only) during the emission and pile-up formation, for different values of strain rate sensitivity parameter m.

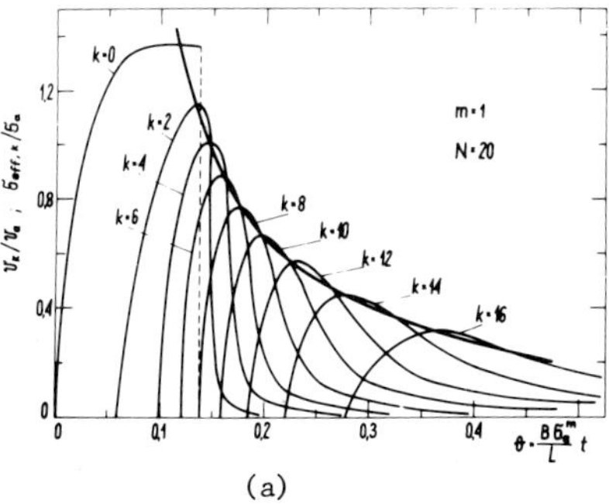

(a)

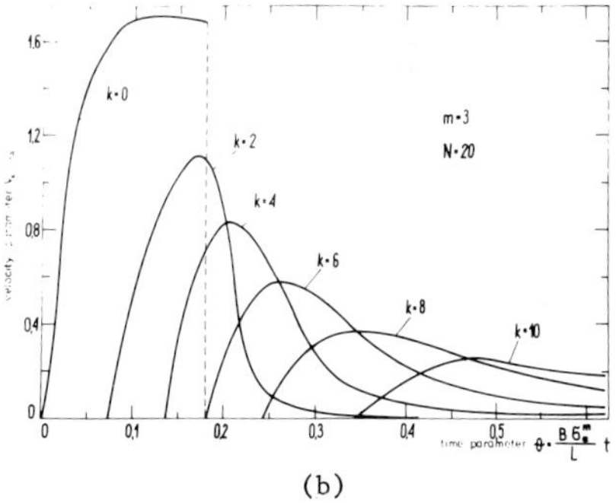

(b)

Fig. 1. The relative velocities of the dislocations during the parametric time θ of the pile-up formation, for (a) m = 1 and (b) m = 3.

The strain rate in the slip plane is proportional to the sum of the velocities of all dislocations in the group. The profiles of the strain rate change during the formation of the pile-up for different total numbers of dislocations emitted and different values of m are shown in Fig. 2 a and b respectively.

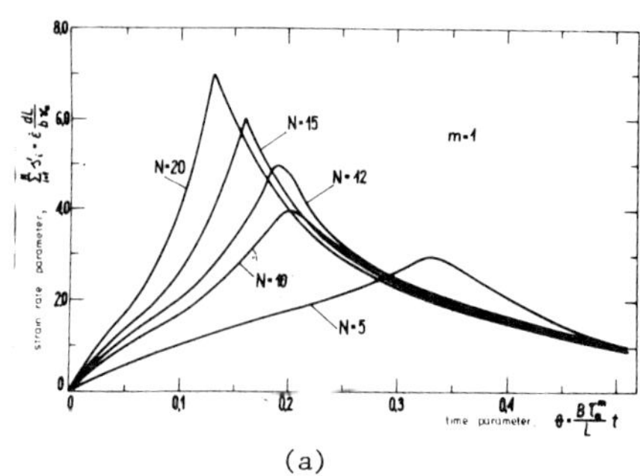

(a)

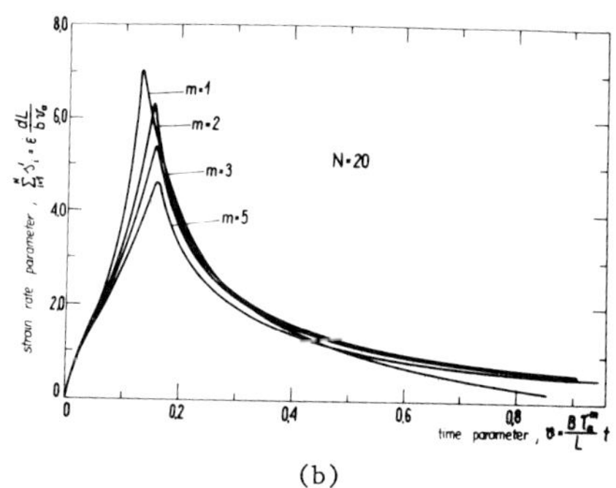

(b)

Fig. 2. The profiles of the strain rate in the slip plane during the pile-up formation for different numbers of dislocations in the pile-up (a) and different m values (b).

It is seen that there is an acceleration of the strain rate in the early stages of emission. The effect depends upon the m and B values as well as on the yield stress of the dislocation source (σ_a) and the pile-up length L (distribution of strong obstacles) and is the biggest for the parametric time of the order of 0.1-0.2.

The acceleration of the strain rate is a necessary condition for the load and flow instability during the deformation [15,16,17,18]; however, there remains the question of whether this local and momentary strain rate acceleration can lead to macroscopic flow and load instability of Lüders or Portevin-LeChatelier types.

The problem is closely connected with the slip band growth during deformation. The propagation of the band throughout the specimen suggests that there are stress concentrations at the front of the band which help to keep the band moving at a smaller stress than at nucleation. The pile-up is possibly the best known example of an internal stress concentration in metals.

The magnitude of the stress at an obstacle located on the front of the band is the function of the time of pile-up formation. Fig. 3 shows the increase of the stress concentration σ_c/σ_a vs. the parametric time θ for different numbers of dislocations emitted. Each asymptote denotes an extreme value of the stress concentration for a given number of dis-

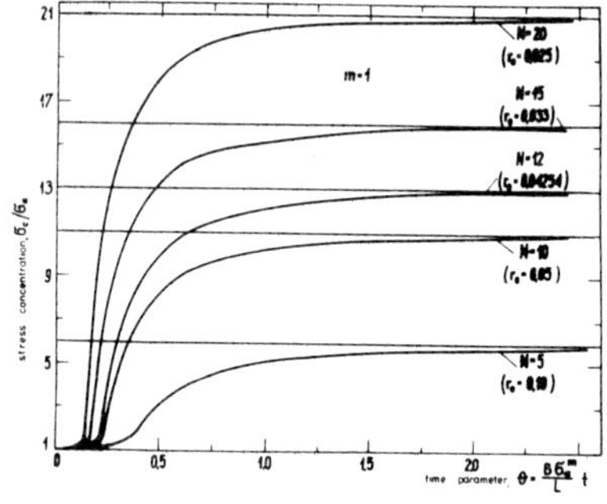

Fig. 3. The stress concentration at an obstacle vs. the parametric time θ for different numbers of dislocations in the pile-up.

locations in the static pile-up. These values exceed the applied stress by many times. On the other hand, inspection of the stress-strain curve in the vicinity of the instability point (Fig. 4) suggests that the stress at which the band propagates is only a few percent smaller than at nucleation. The difference reflects the strength of the stress concentrations at the front of the band which is sufficient to activate slip on the neighbour planes. If the time necessary to produce this concentration is a little shorter than that at the maximum strain rate in the slip plane (parametric time $\theta \approx 0.1$ Fig. 2) then the local instantaneous strain rate will be maintained during the slip band formation. This can lead to the load instability if the strain rate in the slip plane exceeds the strain rate imposed by the tensile system [18].

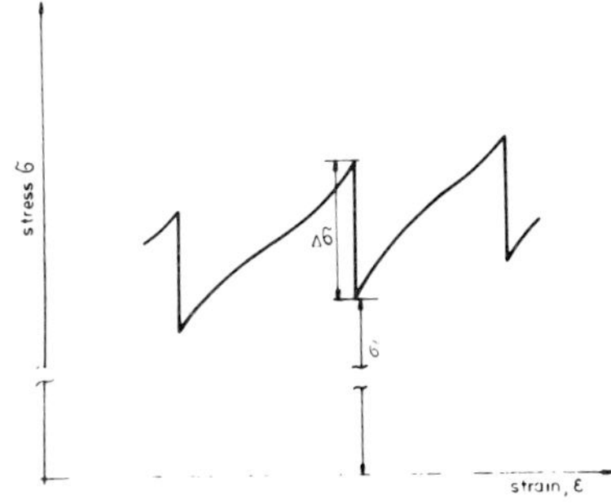

Fig. 4. A portion of the stress-strain curve recorded for Cu-Zn alloy under P-L deformation conditions. The difference between the stress at which the slip band nucleates and that at which it then propagates is shown.

Discussing the stress concentration at the obstacle, it seems to be reasonable to assume that the extreme possible concentration ($N\sigma_a$) is never reached and that probably it exceeds only slightly the critical one at which slip propagates, unless the pile-up is still not completed yet. This is because of the pairing of the opposite dislocations from adjacent slip planes and the stress relaxation due to activation of secondary dislocation sources along the head of the pile-up. This last process was postulated by Hirsch and Mitchell [19] and has been experimentally confirmed in the low stacking fault energy Cu-Al alloy [7]. Moreover, what seems to be especially important, the process produces an alien dislocation distribution with respect to the system which carries the deformation, so that the structure behind the slip band is unstable for the following deformation. Therefore the instability of the load may occur in repeated manner, like the P-L deformation.

There are also other consequences of this model. Because the time to get the critical stress concentration at the front of the band, t_c, is much shorter than the time of the pile-up completion t_s, the deformation is continued at the back of the front and the band has a width which is proportional to the pile-up length multiplied by the ratio t_s/t_c, or

$$d_B = L\,\frac{\theta_s}{\theta_c} \tag{5}$$

Since L decreases with deformation, while θ_s/θ_c is assumed to be constant, the width of the band should diminish for subsequent bands. This result agrees with the observations [11,20].

Similarly, taking

$$V_B = \frac{L}{t_c} \tag{6}$$

where

$$t_c = \frac{\theta_c L}{B\cdot\sigma_a^m}$$

it may be found that the average velocity of the band propogation V_B is

$$V_B = \frac{L^2}{\theta_c B\cdot\sigma_a^m} \tag{7}$$

and for constant θ_c it decreases with deformation, which again is supported by experiments [11,20].

Finally it is necessary to emphasize that all the features of the group of dislocations strongly depend upon the stability of the planar arrangement of the dislocation and any process which destabilizes the group will favour uniform deformation. This explains why non-uniform deformation readily occurs in low stacking fault energy alloys or in those where the pile-up formation is very fast, so that there is no chance of distroying the group of moving dislocations by cross slip or climb processes [21,22].

References

1. B. J. Brindley, P. J. Worthington, Metall. Rev., 145, 101, (1970).
2. Z. S. Basinski, P. J. Jackson, Phys. Stat. Sol., 9, 805, (1965).
3. Z. S. Basinski, P. J. Jackson, Phys. Stat. Sol., 10, 45, (1965).
4. J. U. Sharp. M. J. Makin, Can. Jour. Physics, 45, 519, (1961).
5. J. Washburn, G. Murty, Can. Jour. Physics, 45, 523, (1967).
6. A. Korbel, M. Szczerba, to be published.
7. J. W. Mitchel, J. C. Chevrier, B. J. Hockey, J. P. Monaghan, Jr., Can. Journ. Physics, 45, 453, (1967).
8. H. Neuhäuser, R. Rodloff, Acta Met., 22, 375, (1974).
9. M. J. Makin, J. U. Sharp, Phys. Stat. Sol., 9, 109, (1965).
10. B. J. Brindley, D. J. H. Corderoy, R. W. K. Honeycombe, Acta Met., 10, 1043, (1962).
11. A. Korbel, J. Zasadzinski, Z. Sieklucka, Acta Met., 24, 919, (1976).
12. N. Clement, D. Caillard, J. L. Martin, Proceeding of the Ninth Intern. Congress on Electron Microscopy, Toronto, Canada, Vol. 1., 1978.
13. A. R. Rosenfield, M. F. Kanninen, Phil. Mag., 22, 143, (1970).
14. A. Pawelek, A. Korbel, Metallurgy and Foundary Practice, Vol. 5, No. 4, 507, (1979).
15. E. W. Mort., Acta Met., 15, 351, (1967).
16. J. J. Jonas, R. A. Holt, C. E. Coleman, Acta Met., 24, 911, (1976).
17. V. F. Kocks, J. J. Jonas, H. Mecking, Acta Met., 27, 419, (1979).
18. P. Penning, Acta Met., 20, 1169, (1972).
19. P. B. Hirsch, T. E. Mitchel, Can. Journ. Physics, 45, 663, (1967).
20. A. Wijler, A. Van den Benkel, Scripta Met., 4, 705, (1970).
21. A. Korbel, H. Dybiec, A. Pawelek, Materials of Polish-French Colloqium on Strain Localization and Recrystallization, Krakow, September 1979, will be published in Archiwum Hutnictwa.
22. A. Korbel, H. Hybiec, submitted for publication in Acta Met.
23. A. Ralston, A first Course in numerical analysis, McGraw-Hill, New York, 1968.

DISLOCATION STRUCTURES PRODUCED BY CROSS-SLIP

by

P.J. Jackson, O.L. de Lange and P.D.K. Nathanson*
Physics Department, University of Natal, P.O. Box 375,
Pietermaritzburg 3200, Republic of South Africa

ABSTRACT

The cross-slip of single dislocations has often been discussed. However, plastic deformation is usually inhomogeneous, and it is probable that dislocations frequently move in groups rather than glide singly. We argue that this may affect the process of cross-slip. We consider the successive cross-slip of a procession of dislocations and describe a mechanism which produces dislocation structures like those observed in crystals which work-harden in three stages. Evidence for the operation of this mechanism is found in neutron irradiated copper crystals, where similar structures are clearly linked to cross-slip (1,2). We propose a re-interpretation of the role of cross-slip in three-stage hardening; namely, that cross-slip occurs throughout deformation, that it induces secondary slip, and that it is responsible for the growth of structures known as 'carpets' (3) or 'versetzungsschichten' (4). We attribute Stage III to the onset of primary dislocation multiplication by cross-slip, rather than to the onset of cross-slip itself.

1. MODELS OF CROSS-SLIP

Consider a procession of dislocations of mixed character which cross-slips over an obstacle. In general this is a complex process and the outcome depends on a variety of factors such as the screw-edge character of the procession (5), the possibility of primary dislocation multiplication on neighbouring planes (6), and whether the dislocations cross-slip over the obstacle singly or in groups (see below). For example, in crystals containing a dispersion of hard particles, arrays of loops are formed by cross-slip (7) at a particle. One loop (or dipole) per dislocation in the procession is formed if the procession is edge-like, two, (one on either side of the particle) if it is screw like. In a detailed analysis, Humphreys and Hirsch (8) consider dislocations that cross-slip singly over the particle, presumably because the spacing of dislocations in the procession is greater than the particle dimensions. The particle acts as a template for cross-slip. After cross-slip, the loops or dipoles which form repel each other and collect in arrays where their spacing ≈ the loop height. One does not expect secondary slip to be associated with the stresses of these arrays (9) and, in fact, no secondary slip is observed.

More complex structures may result when several dislocations move simultaneously over an obstacle. This is likely whenever at least one dimension of the obstacle is large compared to the dislocation spacing in the procession. We describe here a model of the cross-slip of a group of dislocations. This model, which is based on an analysis (10) of the structures formed by slip in neutron irradiated copper crystals (1,2) has features which appear to be pertinent to work-hardened crystals (see discussion).

Figure 1 shows the encounter of a procession of mixed character with a planar obstacle (0) which is elongated perpendicular to $\underset{\sim}{b}$. The encounter is shown at a stage when several dislocations (group A) have cross-slipped twice and their 'upper' segments are gliding over the obstacle. The first dislocation to arrive cross-slips at the obstacle; successive dislocations cross-slip at points such as S, which are a distance d_s from previously formed jogs. The region close to the line of an array of jogs such as JJ is a favourable site for cross-slip; here the stress in the cross-slip plane is large (10). The length of the jogs increases with distance from the obstacle and the upper segments glide on different primary planes (see Fig. 2), if successive dislocations cross-slip over previously formed jogs. We have assumed for simplicity that segment interchanges, which are common in dispersion hardened crystals (7), do not occur here. If cross-slip is accompanied by interchanges the details of the process are more complicated, but the essential features (1 - 5, listed below) remain.

The structure acts as a template for cross-slip and the two sets of jogs (JJ and J'J') lie close to a common cross-slip trace. The set JJ is trapped at the obstacle: the set J'J' can glide away with the procession. When the procession is close to screw orientation the jogs J'J' may also become trapped, if the segments Q cross the line of the jogs.

A key feature of this process is that a region of overlap develops between the upper segments and the segments EE on the primary plane and spreads out from the jogs JJ (cf. the stippled region, Fig. 1). In this region the structure approximates two overlapping sheets of edge-like dislocations of opposite sign, in which the average spacing is $\bar{d}$. The regions between such sheets are particularly favourable sites for the production of secondary slip. When the length of overlap along $\underset{\sim}{b}$ is greater than the separation of the sheets, h, the internal stress between the sheets is predominantly a tensile stress in the primary glide direction (9). This stress, which is proportional to $(\bar{d})^{-1}$, acts only on secondary systems and can cause slip if $\bar{d} \lesssim h$ (9).

* Present address: Metallurgy Dept., Atomic Energy Board, Private Bag X256, Pretoria 0001.

If secondary slip does occur, edge segments can become trapped by interactions with secondary dislocations and a pair of tangled dislocation layers forms: primary dislocation multiplication by double cross-slip is inhibited.

Dislocations in the advancing procession (such as the group B) encounter an obstacle which now includes this pair of tangled layers. The original obstacle has become a structure of dislocation layers growing by a process we call multiple cross-slip; i.e., the cross-slip of groups of dislocations accompanied by secondary slip, in the manner described above. Secondary slip plays the crucial role of stabilising the structure. As the process continues, the layers grow parallel to $\underset{\sim}{b}$ and become more widely separated because successive groups of dislocations must cross-slip over the regions of secondary slip (Fig. 2). The growing structure, which continues to act as a template for cross-slip, is bounded by an array of jogs approximately parallel to a cross-slip trace.

Growth of the structure perpendicular to $\underset{\sim}{b}$ will depend on the screw-edge character of the procession. If the procession is close to edge orientation, cross-slip at T can terminate growth as the dislocations wrap around the obstacle and screw segments form near T. Therefore, structures produced by the multiple cross-slip of an edge procession are unlikely to grow perpendicular to $\underset{\sim}{b}$. If the procession is close to screw orientation the structure may develop as a ribbon along the primary edge direction, as secondary slip spreads from the jogs JJ (cf. the stippled region, Fig. 1).

Growth parallel to $\underset{\sim}{b}$ can initiate additional processes: cross-slip may occur at several points in front of the structure if screw segments accumulate here. The upper segments formed by such separate cross-slip events will in general lie in different primary planes. Jogs may form where these upper segments meet, as they sweep over the structure. If these jogs provide the nucleus of a new obstacle above the previously formed sheets, a new structure, also consisting of tangled layers inclined to the primary plane, may grow around this obstacle. Such features have been observed in neutron irradiated copper (1,2). We conclude that the initial obstacle serves only to initiate the process - once multiple cross-slip of a screw-like procession has started, complex structures may develop.

We note the following.

1. A basic feature of structures formed by multiple cross-slip is sheets of edge-like dislocations of opposite sign which have been converted by secondary slip into *tangled layers of dislocations*. The layers *form in pairs* and may be either *parallel to or inclined to the primary glide plane* (Fig. 2).
2. The long-range stresses of the accumulating dislocations become screened when secondary slip traps the upper segments. Between the sheets the internal stress on the primary system is small (9).
3. In f.c.c. crystals the most highly stressed systems between edge sheets are AVI and CI (9). These equally stressed systems can operate to form Lomer-Cottrell locks in the sheets, which may cause the growing structure to develop along a front inclined at 60° to $\underset{\sim}{b}$.* *Lomer-Cottrell dislocations should be common in the layers.* If AVI and CI operate equally, the lattice between the layers will develop a *tilt about an axis parallel to the primary edge direction*: the edge sheets are converted into a pair of asymmetric tilt boundaries of opposite sign.
4. If the jogs J'J' (Fig. 1) are not trapped, the upper segments P continue to glide to the right on a set of closely spaced planes. These segments, moving in echelon, can glide polygonise to form a *short wall perpendicular to the primary plane.*
5. The general appearance of slip traces formed when such a group escapes through a crystal surface is that of *cross-slip traces linking the ends of primary slip lines.* A stair-case like set of traces results if the procession which cross-slips has come from a region such as P (Fig. 1). The corresponding pair of layers are parallel and inclined to the primary plane (Fig. 2b).
6. Multipoles and dipoles may form by this mechanism when the internal stresses are insufficient to cause secondary slip (e.g., if $d/h \leq 1$, (9)). Segment interchange (7) is likely in this case. *Multipoles and dipoles* formed in this way *should terminate in a common cross-slip trace.* Alternatively, the segments which have cross-slipped may bypass the segments on the primary plane and act as sources of primary dislocations (6).

Evidence for the operation of the mechanism we have described is provided by observations of secondary slip in copper crystals strained after neutron irradiation (1,2). In these crystals 'islands' of forest dislocations are observed on the primary slip plane. The islands are bounded by cross-slip traces. Viewed in section, they are seen to consist of inclined layers of dislocations, as in the model. Particularly convincing evidence for the operation of the model is provided by the presence of regions swept clear of radiation damage just where these are

* This feature is found in neutron irradiated crystals (1,2).

expected (cf. C, Fig.2 with C, Fig. 5a (2)). Sequences of islands are also observed. These can develop by a variety of outflanking processes which affect termination of the structures (1,10). The sequences are elongated in a direction which is believed to be the direction of motion of the slip front behind which they form, rather like the stepped trails sometimes left behind screw dislocations (6).

2. DISCUSSION

A survey of the extensive literature shows that the items listed above in italics feature prominently in crystals which display three-stage work hardening behaviour (for reviews see 4,11,12,13). In particular, the growth of dislocation layers nearly parallel to the main glide plane ('carpets' or 'versetzungsschichten') is considered to be an important aspect of work-hardening (13). The growth of these layers is generally attributed to encounters between groups of dislocations in the two-way dislocation traffic that accompanies deformation. It has been suggested that layers grow when dislocations are trapped by the stress-field of a pile-up on a neighbouring plane (12), and that Lomer-Cottrell networks are produced when secondary slip spreads out by double cross-slip between groups of primary dislocations of opposite sign (14). No detailed mechanism for juxtaposing such groups has been proposed, nor has the formation of dislocation walls been associated with the growth of layers. Wilkens (15) has made the interesting suggestion that "systematic interactions of an autocatalytic, rather than of an accidental nature" are responsible for layer formation. He also remarks that "some special kind of secondary slip ---- plays the crucial role", but attributes this slip to the stress field of a pile-up. We note that the mechanism of multiple cross-slip, described here, is of an autocatalytic nature and that it involves secondary slip of the kind which is observed. We suggest that dislocation 'carpets' and walls are formed by this mechanism during three-stage hardening[†], probably by the cross-slip of *screw-like* processions, since it is these processions that can develop into extensive structures, like those observed in work-hardened crystals. We propose the following re-interpretation of the role of cross-slip in the three stages of hardening.

Stage I. At low applied stresses cross-slip occurs at obstacles such as small jogs produced by screw annihilation. The pairs of edge arrays which result are too closely spaced for secondary sources to operate between them (9). In a material with a low friction stress the arrays dissociate into individual dipoles which are swept through the crystal by dislocation traffic to collect in bundles (17) and the only indication of their origin is termination in a common cross-slip trace. In an alloy the arrays may be stabilised by a high friction stress (18), producing multipoles which terminate along a cross-slip trace (see Fig. 10 of (19)). The edge sheets formed by multiple cross-slip at larger obstacles such as dipole bundles (e.g., pure copper), or closely spaced multipoles (e.g., copper-based alloys), are sufficiently widely separated for secondary slip to occur between them, and inclined layers begin to form when such obstacles have developed. In alloys this stage may be delayed until the crystal is filled with multipoles (20), because these are sessile and cannot be swept together to form bundles; Stage I is therefore extended in alloys. The transition to Stage II takes place when a network of inclined layers fills the crystal.

Stage II. Throughout Stage II the formation of new structures by this mechanism continues, thereby reducing the scale of the distribution. In addition, dislocations continue to accumulate at existing inclined layers (21) and the dislocation density in the layers increases. During Stage II multiple cross-slip does not produce sources of dislocations on new glide planes, because the edge segments of dislocations which cross-slip are trapped by secondary slip. We note that in Stage II secondary dislocations multiply profusely without leaving visible slip traces. While multiple cross-slip is more likely to leave visible slip traces (such traces are observed in Stage I, even at 4.2K (21)), it may also produce traces on a scale which is too fine to be detected.

We remark that a mechanism such as multiple cross-slip which traps dislocations in stress-cancelling arrays, also allows a source to operate repeatedly before any back stress is sufficient to stop it, and this is consistent with the observation that slip lines are formed in small strain intervals (see, for example, (12)).

To summarise, we propose that multiple cross-slip is responsible for the basic cell structure which characterises the dislocation distribution from the end of Stage I.

Stage III. The model suggests a criterion for the onset of Stage III; namely, that this occurs when the multiple cross-slip of screw-like processions ceases to inhibit primary dislocation multiplication by double cross-slip. Cross-slip can then assume its traditional role of transferring glide to new planes and producing slip bands. We suggest that this transition

† We suppose that the cross-slip of our mechanism commences at low applied stresses: cross-slip has in fact been observed in copper at an applied stress of 10^{-5} μ (16).

occurs because secondary slip becomes difficult between newly formed sheets, and new layers cease to form (22). A quantitative treatment of a dynamic situation such as that shown in Fig. 1 is hardly feasible and we offer the following arguments for the purpose of discussion.

Slip on the cross-slip plane occurs predominantly in response to the internal stress, so that the proximity of cross-slip to the outermost jog (cf. the distance* d_s, Fig. 1) is probably determined by the stresses which the structure produces at S. We therefore suppose that d_s is relatively insensitive to the applied stress σ_a; specifically that it varies less rapidly than σ_a^{-1}. We also suppose that following the cross-slip of a screw-like procession, the distance d_s sets the scale for the average spacing, $\bar{d}$, which develops as overlapping segments form (see Fig. 1). An estimate of the stress which multiplies secondary dislocations between these segments is (9) $\sigma_i \approx \mu b/\bar{d}$. The average flow stress on secondary systems, σ_s, is approximately proportional to σ_a. Therefore, the above remarks suggest that the ratio σ_s/σ_i increases with strain and eventually exceeds unity, at which point secondary slip between the overlapping segments becomes difficult. We identify this transition with the onset of Stage III.†

3. CONCLUSION

We conclude that an examination of the role played by the cross-slip of groups of dislocations during deformation may provide a simple basis for understanding the development of the gross dislocation structures observed in crystals which work-harden in three stages. It is encouraging to note that structures of the type produced by multiple cross-slip are observed in deformed b.c.c. crystals only when screw dislocations are as mobile as edges (13).

REFERENCES

1. P.D.K. Nathanson, Ph.D. Thesis, University of Natal (1979).
2. P.D.K. Nathanson, P.J. Jackson and D.R. Spalding, *Acta Metall.*, In Press (1980).
3. J.W. Steeds, *Proc. Roy. Soc.* A 292, 343 (1966).
4. B. Šesták and A. Seeger, *Z. Metallk.* 69, 195 (1978); ibid. 355; ibid. 425.
5. P.B. Hirsch, *J. Inst. Metals* 86, 7 (1957).
6. W.G. Johnston and J.J. Gilman, *J. appl. Phys.* 31, 632 (1960).
7. F.J. Humphreys and J.W. Martin, *Phil. Mag.* 16, 927 (1967).
8. F.J. Humphreys and P.B. Hirsch, *Proc. Roy. Soc.* A 318, 73 (1970).
9. O.L. de Lange, P.J. Jackson and P.D.K. Nathanson, *Acta Metall.*, In Press (1980).
10. C.J. Young, O.L. de Lange and P.J. Jackson, to be published.
11. F.R.N. Nabarro, Z.S. Basinski and D.B. Holt, *Adv. in Phys.* 13, 193 (1964).
12. P.B. Hirsch, *The Physics of Metals*, Vol. 2 (Edited by P.B. Hirsch) Cambridge University Press, p. 189 (1975).
13. B. Šesták and A. Seeger, *Phys. Stat. Sol.* 43, 433 (1971).
14. H. Strunk, *Mat. Sci. Engng.* 26, 231 (1976). See also references in this article.
15. M. Wilkens, *Can. J. Phys.* 45, 567 (1967).
16. F. Minari, B. Pichaud and L. Capella, *Phil. Mag.* 31, 193 (1974); ibid. A41, 281 (1980).
17. J.V. Sharp and M.J. Makin, *Phil. Mag.* 10, 1011 (1964).
18. P.M. Hazzledine, *J. de Physique* 27, C3-210 (1966).
19. C.S. Pande and P.M. Hazzledine, *Phil. Mag.* 24, 1039 (1971).
20. J.W. Mitchell, J.C. Chevrier, B.J. Hockey and J.P. Monaghan, Jr., *Can. J. Phys.* 45, 453 (1967).
21. Z.S. Basinski and S.J. Basinski, *Phil. Mag.* 9, 51 (1964).
22. D. Kuhlmann-Wilsdorf, in *Work Hardening* (Edited by J.P. Hirth and J. Weertman) Gordon and Breach, New York, p. 97 (1968).
23. R. Berner, *Z. Naturforsch.* 15a, 689 (1960).

* This distance, which also depends on the temperature, stacking fault energy and strain-rate, should decrease as cross-slip becomes more difficult.

† This criterion provides a rough estimate of $\bar{d}$ at the end of Stage II: if $\sigma_i \approx \sigma_s$ when $\sigma_s \approx \sigma_{III}$, then $\sigma_{III} \approx \mu b/\bar{d}$. For copper at room temperature $\sigma_{III}/\mu = 6 \times 10^{-4}$ (23), which means $\bar{d}$ is of order 0.4 μm in arrays which form at the end of Stage II.

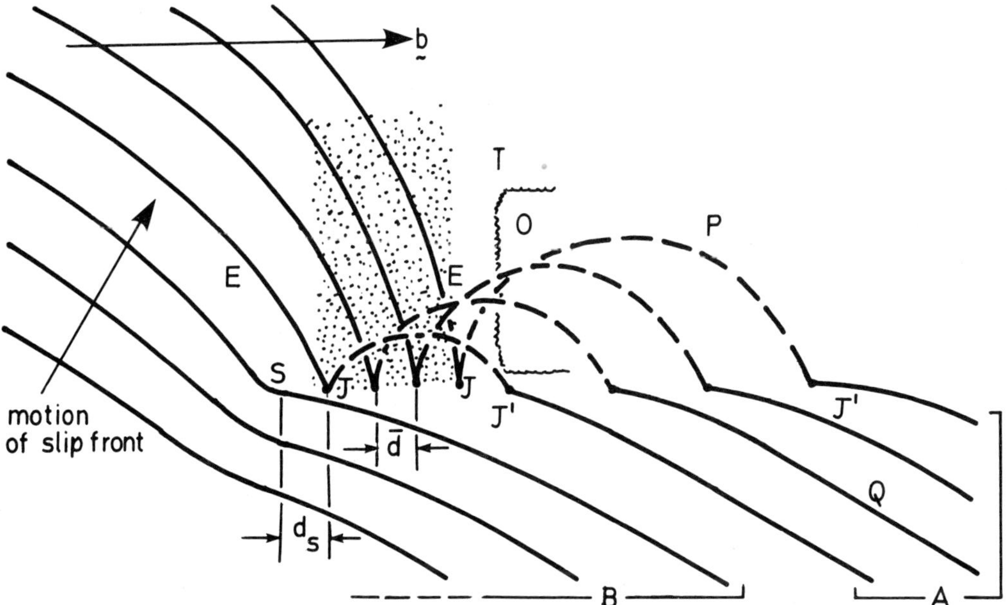

Figure 1. A procession of dislocations of mixed character on a single glide plane which encounters an obstacle O. The dashed lines are segments which have cross-slipped over the obstacle (or previously formed jogs) and are gliding on successively higher 'upper' planes (U, Fig. 2). The screw segment S is about to cross-slip at a distance d_s from the growing structure. The jogs JJ and J'J' lie in the cross-slip plane.

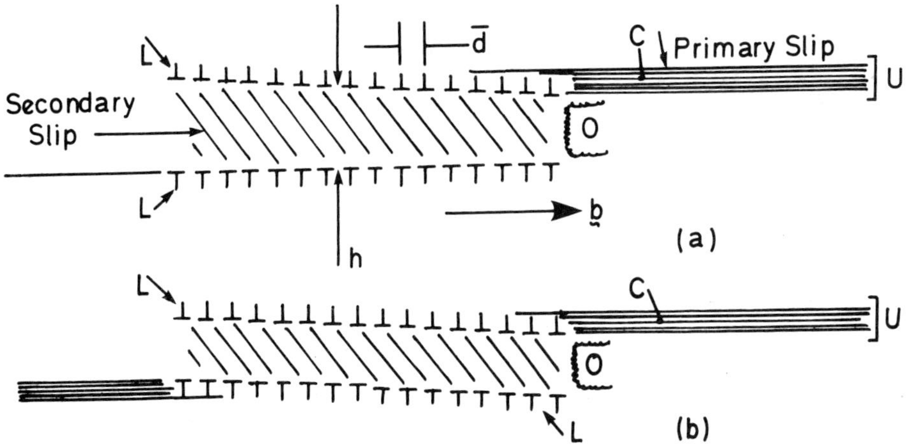

Figure 2. (a) Section through a structure which has formed by multiple cross-slip. The original obstacle is O. The section is normal to the glide plane and contains b. The structure consists of overlapping layers of dislocations, L.

(b) As Fig. 2a, except that the procession which has cross-slipped was gliding on a set of closely spaced glide planes.

THE DISLOCATED STATE

M.J. Marcinkowski
Department of Mechanical Engineering,
and Engineering Materials Group
University of Maryland
College Park, Maryland 20742

Introduction

The first general theory of dislocations was developed by Volterra (1) in 1907. He referred to them as "distorsioni" which was later changed to dislocation by Love (2). Although Volterra developed the concept of a dislocation solely in terms of an elastic entity, Orowan (3), Polyani (4) and Taylor (5) almost simultaneously in the early thirties made the concept of a dislocation nearly synonymous with crystal plasticity. This in turn fixed in the minds of nearly everyone the notion of a dislocation which possessed a quantized Burgers vector whose magnitude was more or less equal to that of the interatomic spacing within a crystal. This limited view of a dislocation thus kept it from being utilized to its fullest potential, as will be shown shortly.

Another development which tended to obscure the real meaning of dislocations was the introduction of the image dislocation by Head (6) who borrowed this idea from the field of electrostatics. In fact, much of the present theoretical background associated with dislocations is based upon the notion of a point force analogous to a point charge and the resulting Green's function representation of the displacement arising from the point force. This approach is in general very complex and the mathematics are such as to obscure the physical aspects of the problem, which in many respects is the most important aim of any successful theory.

In 1979, a unified theory of the mechanical behavior of matter was introduced by the present author (7) in which it was shown that any distortion, whether occasioned by an interal or an external stress, and whether in a solid or in a liquid, or within the universe itself for that matter, could be described in terms of some suitable distribution of surface dislocations. In principle, it thus now seems possible to unify and quantify that vast area of science which deals with the mechanical behavior of matter in much the same way as was done for the disciplines of statics, dynamics and classical thermodynamics, to name only a few. In what follows, this theory will be developed in detail and applied to a whole host of important mechancial phenomena.

Surface Dislocations

Let us consider an edge type crystal lattice dislocation within the center of a body such as shown in Fig. 1a. In order to make the surfaces of this body stress free, a continuous distribution of surface dislocations, each with infinitesimal Burgers vector, must be placed on the surfaces of the body. One such surface dislocation is shown in enlarged form in the inset toward the right in Fig. 1a. The surface dislocation is seen to act as the transition zone which connects the stress-free surface, as characterized by vertical and horizontal segments, to the elastically strained interior of the crystal. These surface dislocations were first postulated by Marcinkowski and Das (8) in 1974 in connection with surface cracks. In order to form a proper Burgers circuit (7,9), it is necessary that a path be taken about the entire outer surface of the body in the manner shown in Fig. 1a. The closure failures associated with the crystal lattice dislocation and surface dislocations are shown dotted. It is necessary that the Burgers vector of the lattice dislocation b^{ℓ} and the sum of all the Burgers vectors associated with the surface dislocations Σb^{s} obey the following powerful conservation law:

$$b^{\ell} + \Sigma b_i^s = 0 \tag{1}$$

where as usual (10)

$$b_i = \oint \frac{\partial u_i}{\partial x_j} dx_j \tag{2}$$

with a similar expression for b_i^s, where u_i is the displacement. In view of equation (2) it is apparent that equation (1) also states that the sum of all the displacements associated with a stress-free surface should add up to zero. It should be pointed out here that the surface dislocation defined above is different than that given by Bilby (11), wherein he defined such a defect as a surface with an invariant Burgers vector associated with it. In the present case, the surface contains dislocation lines of arbitrary dislocation density so that the Burgers vector at any one point is function of position.

In the construction of Fig. 1a, the extra half plane comprising the lattice dislocation is shown squeezed to a point. This constitutes an extremely high strain energy which may be relieved by having an asymmetric crack form just below the extra half plane such as shown in Fig. 1b. Equilibrium is attained when the increase in surface energy associated with the crack is just balanced by the decrease in strain energy. In order to insure stress-free boundary conditions on the surface of the crack, it is necessary to distribute surface dislocations upon it, one of which is shown in enlarged form in the inset to the right of Fig. 1b. It is interesting to note that only the core of the lattice dislocation is altered by this modification, as can be seen by the identical Burgers circuits of Figs. 1a and 1b.

If the asymmetric crack in Fig. 1b is allowed to move to the bottom surface of the body, say by allowing the surface energy to become sufficiently small, the configuration shown in Fig. 1c obtains. This may be referred to as a perfectly torn configuration (7) in which all of the elastic strains are removed. In terms of dislocations, we may describe this state as one in which the surface dislocations in Fig. 1a have annihilated with the crystal lattice dislocation, as indicated by the dislocation pair of equal but opposite sign in Fig. 1c. Similarly, the dislocation shown in Fig. 1a could have moved to the rightmost surface by glide to annihilate with the surface dislocations so as to generate the step shown in Fig. 1d. The configurations shown in Figs. 1c and 1d may be referred to as states of pure plastic distortion (7) since they have no elastic distortion associated with them and give rise to a permanent shape change in the outer surface (12). By the same token, the configurations shown in Figs. 1a and 1b may be designated as partially plastic since the lattice and surface dislocations have not yet combined to form a surface step.

In order to demonstrate the method whereby the distribution of surface dislocations may be determined, let us consider an infinite half space such as shown in Fig. 2a where a screw type lattice dislocation is located at a distance d from the free surface. In order to satisfy the stress-free boundary conditions, surface dislocations, which obey the conservation law of equation (1), are distributed on the vertical free surface as indicated. Let us now assume the surface dislocations to be continuously distributed or smeared out such that $b^{\ell}f(y)dy$ corresponds to the total amount of Burgers vector lying between y and y + dy. The stress at some point y due to all the surface dislocations, except those in the immediate neighborhood of y, is given by the Cauchy principal value integral (13,14)

$$P \int_{-\infty}^{+\infty} \frac{\mu b^{\ell}}{2\pi} \frac{f(y')dy'}{y-y'} \tag{3}$$

For the stress to vanish on the free surface, we require that the stress from the surface dislocations, as given by equation (3), plus that due to the lattice dislocation vanish, i.e.

$$\frac{\mu b^{\ell}}{2\pi} P \int_{-\infty}^{+\infty} \frac{f(y')dy'}{y-y'} - \frac{\mu b^{\ell}}{2\pi} \frac{y}{(d^2+y^2)} = 0 \tag{4}$$

which is a singular integral equation whose solution can be found to yield (14)

$$f(y) = \frac{1}{\pi} \frac{d}{(d^2+y^2)} \tag{5}$$

When substituted back into the stress field equations, it is found that the stress field due to the surface dislocation array together with that of the lattice dislocation array vanish in the half space for $x > 0$ in Fig. 2a, i.e. the vacuum, but becomes finite within the body, i.e. $x < 0$. This can be seen more clearly in Fig. 3 which shows the equi-stress contours (14) corresponding to the dislocation configuration shown in Fig. 2a.

The continuous dislocation distribution discussed in the previous paragraph allowed an exact solution to be obtained for the singular integral equation given by equation (4). However, for finite bodies, such as shown in Figs. 1a and 1b, or else irregularly shaped surfaces, such solutions are extremely difficult, if not impossible to obtain. It is then more convenient to replace the continuous surface dislocation array by a discrete array and write down the total energy of the system as follows:

$$E_T = \sum_{i=1}^{N} \frac{\mu b_i^2}{4\pi} \ln\left(\frac{4R}{b_i}\right) + \sum_{i=1}^{N} \sum_{\substack{j=1 \\ i \neq j}}^{N} \frac{\mu b_i b_j}{2\pi} \ln\left(\frac{R}{y_i - y_j}\right) + \frac{\mu (b^\ell)^2}{4\pi} \ln\left(\frac{4R}{b^\ell}\right) + \sum_{i=1}^{N} \frac{\mu b_i b^\ell}{2\pi} \ln\left(\frac{R}{[d^2 + y_i^2]^{\frac{1}{2}}}\right) \tag{6}$$

where the first and second terms are the self and interaction energies respectively associated with the surface dislocations, while the third and forth terms are the self energy of the lattice dislocation and interaction energy between it and the surface dislocations respectively. The quantity R is a constant which has dimensions on the order of the size of the crystal. In addition, in accordance with equation (1), we must have

$$b^\ell + Nb_i = 0 \tag{7}$$

The total energy E_T given by equation (6) can now be minimized with respect to the positions of the surface dislocations using very powerful numerical techniques (15). Mathematically, this means that equation (6) becomes

$$F_i = \frac{\Delta E_T}{\Delta y_i} = \sum_{\substack{j=1 \\ i \neq j}}^{N} \frac{\mu b_i b_j}{2\pi} \frac{1}{y_i - y_j} - \frac{\mu b_i b^\ell}{2\pi} \frac{y_i}{(d^2 + y_i^2)} = 0 \tag{8}$$

where F_i is the force on the dislocation at y_i due to the stress fields of all the other dislocations. Since force and stress are related by the following equation (16):

$$dF_k = \varepsilon_{ijk} d\ell_i b_\ell \sigma_{j\ell} \tag{9}$$

where $d\ell_i$ is the dislocation line length, it is clear that dividing through equation (8) by b_i gives the discrete counterpart of equation (4). Thus, the continuous dislocation distribution method involving the solution of singular integral equations is in principle equivalent to the discrete dislocation method in which energy minimization techniques are employed. As we shall see later, the latter technique is much more powerful.

In the case of an edge type lattice dislocation located near a free surface, the surface dislocation array is somewhat more complicated, as can be seen from Fig. 2b. In addition to a surface dislocation array, which satisfies equation (1) and which we shall term the primary array, it is also necessary to add a secondary array of dislocations whose Burgers vectors are perpendicular to that of the lattice dislocation (8,14). These secondary surface dislocations are seen to occur in pairs of opposite sign so that the law of conservation of Burgers vectors is satisfied. The secondary array accounts for only about ten percent of the energy of the system and is thus of secondary importance and is related to the warping of the slip plane by the lattice dislocation (7). It should also be pointed out here that the surface dislocation model employed here is a physically as well as mathematically perfect concept as compared to the image dislocation model which requires additional stress functions of some unknown physical origin to fix up what the image stresses alone could not accomplish.

Application of Surface Dislocations

a) Cracks

Of all the areas dealing with the mechanical behavior of matter, none is in such a need of a sound theoretical basis as that of fracture. Because of its economic importance, a great deal of pressure exists to obtain empirical results, but because of the lack of theory, these results have very little connection to any simple physical concepts. The situation is further complicated by those from within the disciplines of continuum mechanics who construct huge mathematical edifices with almost zero physical insight into the problem. It will now be shown that the problem of fracture can in principle be solved by the methods of surface dislocations.

Consider first the tensile or mode I crack such as shown in Fig. 4a. Such a crack can obviously be viewed in terms of an array of surface dislocations which are associated with the ledge structure on both faces of the crack. Others (17,18) have recognized that a crack could be viewed in terms of dislocations, i.e. the so-called crack dislocations, but this idea was fuzzy since the concept of surface dislocations was not then known. An array of lattice dislocations corresponding to the tensile crack is shown in Fig. 4a', but it is physically distinct from the crack in Fig. 4a. The crack takes its elliptical shape from the repulsive nature of the surface dislocations on opposite faces of the crack. For completeness (7), Figs. 4b and 4c show a shear, or mode II crack and a mode III crack respectively. Their corresponding lattice dislocation configurations are shown in Figs. 4b' and 4c' respectively. It is apparent that the mode III crack is comprised of screw type surface dislocations.

We can, without any loss of generality, concern ourselves with the tensile crack. The one shown in Fig. 4a is a special case of an elastic crack since, upon removal of the load, it will vanish, i.e. all of the surface dislocations will annihilate with one another. A much more general type of tensile crack is the plastic crack shown in Fig. 5a. In this case, as the crack propagates, some of the crack dislocations leave the crack at its tip and become lattice dislocations, which in turn generates a plastic zone comprised of dislocations represented by solid symbols (19,20). For simplicity, the surface dislocations comprising the crack are represented in terms of crack dislocations, as indicated by the dotted symbols. Cottrell, Bilby and Swinden (21) were the first to treat a plastic crack in terms of dislocations, but it was a very special and oversimplified model without general applicability as compared to the configuration shown in Fig. 5a. Similar to the discrete surface dislocation methods used to treat the behavior of a screw type dislocation near a free surface, the total energy associated with the dislocations of a plastic crack of given radius R can be written as (19,20)

$$E_T^D = E_S^c + E_S^\ell + E_I^c + E_I^\ell + E_I^{c\ell} + E_f - E_\sigma^c - E_\sigma^\ell \tag{10}$$

where E_S^c and E_S^ℓ are the self energies of the crack and lattice dislocations respectively, E_I^c and E_I^ℓ the interaction energies between the crack and the lattice dislocations respectively, $E_I^{c\ell}$ the interaction energy between the crack and lattice dislocations, E_f the frictional energy expended in moving the lattice dislocations, while E_σ^c and E_σ^ℓ are the energies of the applied stress in moving the crack and lattice dislocations respectively. The positions of the crack and lattice dislocations can be obtained by minimizing E_T^D given by equation (10) with respect to these positions (19,20). The total energy then of a crack of given length R is then

$$E_T = E_T^D + E_\gamma \tag{11}$$

where E_γ is the surface energy of the crack which can be written as $R\gamma$ where γ is the surface energy per unit area. An important physical significance can be obtained from equation (11) by noting that dE_T^D/dR is the elastic energy released per unit length growth of the crack. When this quantity is greater than $dE_\gamma/dR = 2\gamma$ then crack growth can occur. This is termed the crack extension force G (22) or the J (23) integral which we may write as

$$G \equiv J = -dE_T^D/dR$$

It is apparent that the Griffith condition corresponds to

$$dE_T/dR = 0 \qquad \text{or} \qquad dE_T^D/dR = -dE_\gamma/dR = -2\gamma$$

When E_T given by equation (11) is plotted against R for a fixed stress σ, a maximum is obtained as shown schematically by the dotted curve in Fig. 6. Again, the maximum is simply related to the Griffith condition and is the energy which must be supplied to form a crack of critical size in a perfect crystal. In general, such energies are not available, either through thermal activation or by other means. It is therefore necessary to assume the presence of preexisting voids within the crystal which form unavoidably during processing. Consider such a preexisting void of size corresponding to A in Fig. 6. Upon application of a load σ_1, the energy drops precipitously owing to the formation of crack and lattice dislocations. Further crack growth at this stress level results in an increase in energy and thus cannot occur. If the stress is further increased to σ_2, further energy decrease occurs and in addition the crack can grow until it reaches the energy minimum, at which point it is stabilized. A further increase in load to σ_3 causes the crack to propagate catastrophically. When a larger preexisting void size is considered, such as that indicated by point B in Fig. 6, application of a load σ_1 causes the crack to propagate indefinitely with a decrease in energy. An increase in the lattice dislocation friction stress also alters the curves shown in Fig. 6 in a manner which makes crack propagation more favorable. It is with energy curves of the type shown in Fig. 6 that the complete range of fracture behavior, extending from fully brittle to fully ductile, can be completely understood (24,25).

When the plastic crack shown in Fig. 5a is unloaded the configuration shown in Fig. 5b obtains. In the first place the crack dislocations are seen to annihilate with one another. However, the crack cannot disappear since the lattice dislocations which comprise the plastic zone are prevented from moving back into the crack by the frictional stresses which act on them. In order to satisfy the stress-free boundary conditions on the surface of the unloaded crack due to the presence of the lattice dislocations, it is necessary to place surface dislocations on the surfaces of the crack. Since these are of opposite sign to the crack dislocations they have been termed anti-crack dislocations (26). With exact models of the loaded and unloaded cracks, such as illustrated in Fig. 5, it is possible to formulate a very accurate model of the fatigue process in terms of energy curves similar to those shown in Fig. 6 (27).

Perhaps the severest criticism that can be leveled at the classical continuum mechanics approach to the solution of crack problems is that it is completely unable to treat irreversible crack behavior. Since only the purely elastic crack behaves reversibly, all other crack behavior is path dependent. In the present discrete dislocation approach to the study of crack behavior, where energy methods are employed, no such problem exists, since it is always possible to calculate the irreversible contributions to the energy, i.e. those due to lattice friction (28,29). For example, if we consider the total frictional energy associated with a single lattice dislocation within the plastic zone up to some given crack size and designate it by E_f^R, then the new accumulated frictional energy when the dislocation moves a distance $(x-x^R)$ from its previous reference point is given by

$$E_f = |\tau_f b(x-x^R)| + E_f^R \tag{12}$$

where the absolute value symbol insures that the frictional energy is always positive. The present unified theory of the mechanical behavior of matter can thus be readily extended to formulate a unified theory of fracture (25).

b) Grain Boundaries

One of the most intractable problems in the area of materials science to date has been that concerning the structure of a grain boundary of arbitrary misorientation angle. We will now show how that problem can be solved using the surface dislocation method. Consider first the grain boundary in Fig. 7a of misorientation angle 36.9°. There are no elastic distortions associated with this boundary, and in the context of Fig. 1d, we may refer to such a boundary as being in the fully torn or uncoalesced state. We may also say that the dislocations at the grain boundary are fully compensated by surface dislocations to generate the ledges. Most theories of grain boundaries of arbitrarily large misfit are based upon such rigid models (30). If the boundary shown in Fig. 7b is allowed to coalesce, beginning at the contact points, so as to reduce its surface energy, the configuration in Fig. 7b obtains. Here the coalescence is shown to be complete, which is equivalent to saying that the surface dislocations (not shown) have uniformly distributed themselves along the grain boundary and are in fact responsible for the bending of the vertical faces so that coalescence can occur. On the other hand, the grain boundary dislocations are discretely distributed as shown in the figure. Figs. 7a and 7b obviously represent the extreme cases of high surface energy and high elastic strain energy respectively. In real materials a compromise is reached as shown by the partially coalesced configuration of Fig. 7c (31). The coalescence has caused a portion of the ledge to become uncompensated, as is indicated by the material in between the dashed line and grain boundary free surface.

In particular, if the strength of the ledge in Fig. 7a is b, while that in Fig. 7c is b', the uncompensated dislocation at the grain boundary has a strength (b-b') and would be reflected in the difference in the closure failures in this figure (32). Just as the dislocation configuration associated with Fig. 1a could lower its energy by distributing itself over a free surface in the form of surface dislocations, the uncompensated dislocations of Fig. 7c, which are in fact surface dislocations, could redistribute themselves over the grain boundary free surfaces as shown by the surface steps in Fig. 7d. In reality, a compromise is struck between the configurations shown in Figs. 7c and 7d, and the surface dislocations of strength (b-b') distribute themselves over all the grain boundary free surfaces (32). Thus, the surface dislocation model portrays the grain boundary as a type of periodic array of asymmetric cracks. It is important to emphasize here that in order for the Burgers circuit in Figs. 7c and 7d to be complete, they must be taken about the hole surfaces. For the purposes of clarity this has not been done in Fig. 7d.

A common or coincidence site lattice of edge length a_{oc} can be constructed for any of the grain boundary configurations shown in Fig. 7, as illustrated in Fig. 8 (31). This coincidence site lattice can in turn be subdivided into a smaller sublattice of unit cell length a_{ocs} as shown by the dotted lines in Fig. 8. For the purposes of clarity, only the vertical faces of the sublattice are shown and represent the closest interplanar spacing of atoms.

The 36.9° grain boundary of Fig. 7a can be converted to one of misorientation angle θ = 43.6° by the suitable introduction of lattice dislocations into the former (31) as illustrated in Fig. 9a. Two distinct types of ledge may now be identified, one that makes contact and the other which does not. Upon coalescence, the contacting ledges will become more severely distorted than the non-contacting ones. The grain boundary of Fig. 9a may also be depicted in terms of the coincidence site sublattice of the 36.9° boundary from which it was derived as illustrated in Fig. 9b. In terms of this picture, the grain boundary is comprised of pairs of dislocations of strength a_{ocs} measured in terms of the old coincidence site lattice sublattice spaced at intervals of a_{oc} in terms of the new coincidence site lattice. It is apparent, from comparison of Figs. 9a and 9b, that the coincidence site model of the grain boundary is not as complete as the lattice dislocation model since it does not deal fully with the ledge structure which is ultimately related to the lattice dislocations (31). It thus appears that when one observes the fine dislocation structure within grain boundaries, he is in fact observing the effects of distortions at the point of contact associated with the coincidence site lattice. Since the surface dislocation model portrays the grain boundary as an array of asymmetric elastic cracks, its energy may be written in a form given by equation (11).

c) Externally Stressed Bodies

Consider next the case of a hole subjected to simple tension such as shown in Fig. 10a. The surfaces of the hole are seen to be covered with surface dislocations (7,8), and it is clear by comparison of this figure with that of Fig. 4a that the crack is simply the limiting case of a hole as the thickness of the hole approaches zero. If, on the other hand, the thickness of the hole approaches infinity, then what remains is a body under uniform tension such as shown in Fig. 10b. Both vertical faces of the body must now be covered with a uniform array of surface dislocations in order that the stress-free boundary conditions be satisfied (7,33). This concept can be extended to show that any finite body subjected to an arbitrary state of external stress can be described in terms of some suitable distribution of surface dislocations (34). From what has been shown in the previous sections, an even stronger concept may be formulated, namely that any state of distortion, whether external or internal, can be represented in terms of some suitable and unique distribution of surface dislocations. It should be pointed out, however, that the components of stress lying within the free surface need not vanish. This in turn causes the stress field within the vacuum to not always vanish (8,34). However, these regions never contribute to the energy of the system since no dislocations are ever associated with them and so they may be neglected. Such a case occurs for the hole in Fig. 10a, within which the stress does not vanish.

Effects of Surface Tension

When the surface dislocation distribution of Fig. 2b was determined earlier using energy methods, no consideration was given to the surface energy associated with forming the ledges. When this energy contribution is included, the effect is to reduce the number of surface dislocations so that the conservation law given by equation (1) no longer holds (35). It is clear that for a surface energy that is sufficiently high, all ledge formation will be suppressed, or equivalently, all surface dislocations will vanish and the surface will possess non-vanishing surface tractions. This concept is expected to have important consequences with respect to catalysis.

In the previous paragraph we have considered the effects of surface tension in response to an internal stress. Should not surface tension effects be present even in an unstressed body? The answer is yes and the results are shown in Fig. 11a where surface energy effects lead to a decrease in surface area, which in turn gives rise to an array of surface dislocations (36). The density of surface dislocations in Fig. 11a is determined by a balance between the surface energy decrease and the strain energy increase occasioned by these dislocations. These same concepts can be readily extended to a liquid, as shown in Fig. 11b. However, in the case of a liquid, the surface can further reduce its surface energy by altering its shape to that of a sphere. An earlier attempt was made to relate surface tension to dislocations (37), but not in terms of the unified surface dislocation approach employed here.

Not only does the removal of the surface dislocations associated with Fig. 1a give rise to surface tension, but it makes the finite body become identical to that which would be present if it were imbedded in an infinite medium. This means that surfaces in Fig. 1a interact with its counterparts in the imbedded medium. The concept of surface tension can thus now be made equivalent to a stress generated by the surface. In the case where two surfaces are close together such as at the crack tip in Fig. 4a, or within the grain boundary of Fig. 7, their mutual interaction will be such as to either reduce or eliminate the number of surface dislocations in these regions.

Interphase Boundaries

Let us consider a coherent two-phase interface in which the lattice constant of each phase is identical. This insures that no strains are associated with such an interface. Suppose now that a screw dislocation is located at a distance -d from the interface and that the lattice constants μ_1 and μ_2 of each phase are different. The configuration shown in Fig. 12c then obtains in which the interface is seen to consist of two surface dislocation arrays, one associated with each phase. The interface could also be visualized as being formed by considering two infinite bodies, one with surface dislocation distribution $f_1(y)$ as shown in Fig. 12a, and the other with surface dislocation distribution $f_2(y)$, as shown in Fig. 12b. In order to be able to cut along the x = 0 plane and join the two half-spaces together to generate Fig. 12c, it is first necessary that the stress σ_{xz} and the displacement μ_z across the interface be continuous (38). The first condition is satisfied by writing

$$\frac{\mu_1 b^\ell}{2\pi} \, P \int_{-\infty}^{+\infty} \frac{f_1(y')dy'}{y-y'} - \frac{\mu_1 b^\ell}{2\pi} \frac{y}{(d^2+y^2)} = \frac{\mu_2 b^\ell}{2\pi} \, P \int_{-\infty}^{+\infty} \frac{f_2(y')dy'}{y-y'} \tag{13a}$$

whereas the second is satisfied by writing

$$b^\ell f_1(y) + b^\ell f_2(y) = \frac{b^\ell}{\pi} \frac{d}{(d^2+y^2)} \tag{13b}$$

which can be solved to yield

$$f_1(y) = \frac{1}{\pi}\left(\frac{\mu_2-\mu_1}{\mu_2+\mu_1}\right)\frac{d}{(d^2+y^2)} \tag{14a}$$

and

$$f_2(y) = \frac{2}{\pi}\left(\frac{\mu_1}{\mu_2+\mu_1}\right)\frac{d}{(d^2+y^2)} \tag{14b}$$

When $\mu_2 = 0$, equations (13a) and (14a) reduce to those results associated with a dislocation near a free surface, i.e. equations (4) and (5) respectively, thus illustrating the generality and self consistency of the present surface dislocation formulation. The above analysis could also have been carried out using the image dislocation approach (39); however, the same limitations are encountered here as were met with in connection with free surfaces. Another point of interest is that upon integration of equation (13b), we obtain

$$b_1 + b_2 = b^{\ell} \tag{15}$$

where

$$b_1 = \left(\frac{\mu_2-\mu_1}{\mu_2+\mu_1}\right) b^{\ell} \tag{16a}$$

and

$$b_2 = \left(\frac{2\mu_1}{\mu_2+\mu_1}\right) b^{\ell} \tag{16b}$$

where b_1 and b_2 represent the total Burgers vector of the interface dislocations in phases 1 and 2 respectively. It is apparent that unlike the case for a free surface embodied in equation (1), the conservation law for a lattice dislocation and its corresponding interface dislocations at the two-phase interface does not hold. A similar result was also encountered when the free surface was allowed to possess surface tension. Thus, the stresses on such a free surface and those on the boundary of a coherent two-phase interface are in principle the same.

Another common type of interface is that which forms when two crystals of differing lattice constant are forced into a common coincidence such as occurs along the dotted line in Fig. 13a. The crystal to the left of the dotted line is $\frac{5}{4}$ larger than that towards the right of this line. It was first recognized by the author that such an interface could be represented in terms of dislocations (40), but they were termed "virtual" dislocations since no evidence of their presence, i.e. extra half planes, could be detected. However, it was subsequently found that if a common or coincidence lattice were constructed for the two crystals, the presence of the interface dislocations were immediately revealed, as can be seen in Fig. 13b (41, 42). Still more insight into the interphase boundary of Fig. 13b can be obtained by considering the special case where $\mu_1 = \mu_2$. Under these conditions, equations (15) and (16) yield $b_1 = 0$ and $b_2 = b^{\ell}$. This means that the interface in Fig. 13b can be described by considering the edge dislocation counterpart of the interface shown in Fig. 12c, where the lattice dislocation is allowed to move to the boundary and dissociate into a uniform array of interface dislocations. In the case where the elastic constants of both phases are identical, each phase will contain equal numbers of interface dislocations. On the other hand, when $\mu_2 = \infty$, equations (15) and (16) yield $b_2 = 0$ and $b_1 = b^{\ell}$, which is equivalent to placing all of the interface dislocations in the leftmost phase of Fig. 13b. Another important point to be made is that when the surface dislocations are included in the construction of a two-phase interface, as shown by the steps in Fig. 13b, the conservation of Burgers vectors is seen to be obeyed, i.e. (43,44)

$$b^{\ell} + \Sigma b_i^S + \Sigma b_i^I = 0 \tag{17}$$

where Σb_i^I is merely the sum of all the interface dislocations. This equation is simply a more generalized version of equation (1). Equation (17) may also be applied separately to each phase comprising the interphase boundary. When the interface dislocations move to the surface of the crystal they combine with the surface dislocations to generate ledges such as illustrated in Fig. 13c in much the same way as was the case in Figs. 1d and 7a.

When an originally coherent interface comprised of two phases of the same lattice constant but differing elastic constants is subjected to an externally applied stress in the manner depicted in Fig. 14, the resultant distortions can also be described in terms of a suitable surface dislocation array (43,44). In this case $\mu_2 > \mu_1$ so that the leftmost phase is stretched further. The surface dislocation arrays in the leftmost and rightmost faces of Fig. 14 are similar to those shown in Fig. 10b. At the interface, however, the dislocations consist of an array of dipoles of infinitesimally small separation so that they generate no stress field. However, upon separation of the two phases without relaxation, the dipoles become uncoupled and thus generate stress fields. Again, the surface and interface arrays shown in Fig. 14 obey a conservation law of the type given by equation (17) whether the phases are considered separately or together. The surface dislocation concept can also be extended to the case of a liquid droplet in contact with a solid as shown in Fig. 15 (36). This allows the contact angle between the liquid and solid to be readily determined.

For the purposes of clarity and in order to introduce the basic principles involved, all of the distortions considered earlier have been kept relatively simple. It follows that more generalized distortions involving endless combinations of these basic distortions can also be formulated in terms of the discrete surface dislocation approach and solved numerically by minimizing the total energy of the system. In this manner, all of the relevant problems dealing with the mechanical behavior of matter can be solved with a degree of accuracy, physical insight and simplicity hitherto unknown.

Summary and Conclusions

It has been shown that any state of stress, arising from either an internal or an external source can be described in terms of some suitable distribution of surface dislocations. When formulated in terms of discrete dislocations, the total energy of the system can be readily minimized using computer methods so as to obtain the surface dislocation distributions. This approach in turn has allowed such formerly intractable problems as plastic crack behavior, the structure of grain boundaries and two-phase interfaces in both liquids and solids to be solved with a degree of physical insight which until now has not been possible.

The basis of the surface dislocation analysis is extremely simple as can be seen by considering an infinite body A such as shown in Fig. 16. If within this infinite body we consider a finite body B within which exists an internal source of stress I, then the surface tractions on the surface S of the finite body can be made to vanish by applying forces F_i over this surface which are equal in magnitude but opposite in sign to those initially present. The strain energy released by this process is simply

$$E_e = \tfrac{1}{2}\int_S u_i dF_i \tag{18}$$

where u_i is the displacement on the surface accompanying the stress relaxation there. Since $dF_i = \sigma_{ij} ds_j$, equation (18) can be written as (45)

$$E_e = \tfrac{1}{2}\int_S u_i \sigma_{ij} ds_j = \tfrac{1}{2}\int_V \sigma_{ij} u_{i,j} dv = \tfrac{1}{2}\int_V \sigma_{ij} e_{ij} dv \tag{19}$$

where the surface integral has been converted into a volume integral by means of Stokes' theorem. Now it is quite clear from what we have done earlier that the surface tractions are reduced to zero by an array of surface dislocations. Furthermore, if a Burgers circuit is made about an infinitesimal element of the surface, such as shown at point P in Fig. 16, then it is clear from a relation of the type given by equation (2) that $b_i = u_i$, so that equation (19) can be written as

$$E_e = \tfrac{1}{2}\int_S b_i \sigma_{ij} ds_j \tag{20}$$

This relation is simply another form of equation (6) which was used to determine the surface dislocation distribution (19).

Acknowledgment

The present research effort was supported by the U.S. Department of Energy under Contract No. EY-76-S-05-3935.

References

1. V. Volterra, Ann. Éc. norm. (Sér. 3), Paris, 24, 401 (1907).
2. A.E.H. Love, A Treatise on the Mathematical Theory of Elasticity, Fourth Edition, Dover Publications, New York (1927).
3. E. Orowan, Z. Physik, 89, 634 (1934).
4. M. Polanyi, Z. Physik, 89, 660 (1934).
5. G.I. Taylor, Proc. Roy. Soc., A145, 369 (1934).
6. A.K. Head, Phil. Mag., 44, 92 (1953).
7. M.J. Marcinkowski, Unified Theory of the Mechanical Behavior of Matter, a Wiley-Interscience Publication, John Wiley and Sons, New York (1979).
8. M.J. Marcinkowski and E.S.P. Das, Int. J. Fract., 10, 181 (1974).
9. M.J. Marcinkowski, Acta Mech., 33, 207 (1979).
10. J.P. Hirth and J. Lothe, Theory of Dislocations, McGraw-Hill Book Company, New York (1968).

11. B.A. Bilby, Report of the Conference on Defects in Crystalline Solids, published by the Physical Society, London, p. 124 (1954).
12. M.J. Marcinkowski, Acta Mech., 34, (1979).
13. R.W. Lardner, Mathematical Theory of Dislocations and Fracture, University of Toronto Press, Toronto (1974).
14. K. Jagannadham and M. J. Marcinkowski, Phys. Stat. Sol., a50, 293 (1978).
15. M.J. Marcinkowski, Advances in Materials Research, Vol. 5, edited by H. Herman, John Wiley & Sons, Inc., New York (1971).
16. E. Kröner, Kontinuumstheorie der Versetzungen und Eigenspannungen, Springer-Verlag, Berlin (1958).
17. B.A. Bilby and J.D. Eshelby, Fracture, Vol. 1, edited by H. Liebowitz, p. 99, Academic Press, New York (1968).
18. J. Friedel, Les Dislocations, p. 214, Gauthier-Villars, Paris (1956).
19. M.J. Marcinkowski and E.S.P. Das, Phys. Stat. Sol., a8, 249 (1971).
20. K. Sadananda, K. Jagannadham and M.J. Marcinkowski, Phys. Stat. Sol., a44, 633 (1977).
21. B.A. Bilby, A.H. Cottrell and K.H. Swinden, Proc. Roy. Soc., A272, 304 (1963).
22. G.R. Irwin, Fracturing of Metals, p. 147, ASM, Cleveland (1948).
23. J.R. Rice, Fracture, Vol. 2, p. 191, Academic Press, New York (1968).
24. K. Jagannadham and M.J. Marcinkowski, Mater. Sci. Eng., submitted for possible publication.
25. K. Jagannadham and M.J. Marcinkowski, Unified Theory of Fracture, Academic Press, New York, submitted for possible publication.
26. K. Jagannadham and M.J. Marcinkowski, Int. J. Fracture, in publication.
27. K. Jagannadham and M.J. Marcinkowski, Mat. Sci. Eng., in publication.
28. K. Sadananda and M.J. Marcinkowski, J. Appl. Phys., 43, 2609 (1972).
29. K. Sadananda, Computer Simulation for Materials Applications, Nuclear Metallurgy, Vol. 20, p. 252 (1976).
30. W. Bollmann, Crystal Defects and Crystalline Interfaces, Springer-Verlag, Berlin (1970).
31. M.J. Marcinkowski and K. Jagannadham, Phys. Stat. Sol., a50, 601 (1978).
32. K. Jagannadham and M.J. Marcinkowski, J. Mat. Sci., in publication.
33. M.J. Marcinkowski, Phys. Stat. Sol., b93, 703 (1979).
34. K. Jagannadham and M.J. Marcinkowski, Mater. Sci. Eng., 38, 259 (1979).
35. K. Jagannadham and M.J. Marcinkowski, J. Mat. Sci, 14, 1052 (1979).
36. K. Jagannadham and M.J. Marcinkowski, J. Mat. Sci. 14, 1717 (1979).
37. C. Herring, Physics of Powder Metallurgy, edited by W. Kingston, p. 143, McGraw-Hill, New York (1951).
38. K. Jagannadham and M.J. Marcinkowski, J. Mat. Sci., in publication.
39. J. Weertman and J.R. Weertman, Elementary Dislocation Theory, The MacMillan Company, New York (1964).
40. M.J. Marcinkowski, Fundamental Aspects of Dislocation Theory, p. 531, NBS Special Publication No. 317, Vol. 1, edited by J.A. Simmons, R. deWit and R. Bullough (1970).
41. M.J. Marcinkowski, K. Sadananda and W.H. Cullen, Jr., Acta Cryst., A31, 292 (1975).
42. M.J. Marcinkowski, J. Mat. Sci., 14, 205 (1979).
43. M.J. Marcinkowski and K. Jagannadham, J. Mater. Sci. Eng., 41, 75 (1979).
44. K. Jagannadham and M.J. Marcinkowski, J. Mater. Sci. Eng., 41, 81 (1979).
45. R. deWit, Solid State Physics, Vol. 1 edited by F. Seitz and D. Turnbull, p. 249, published by Academic Press, New York (1960).

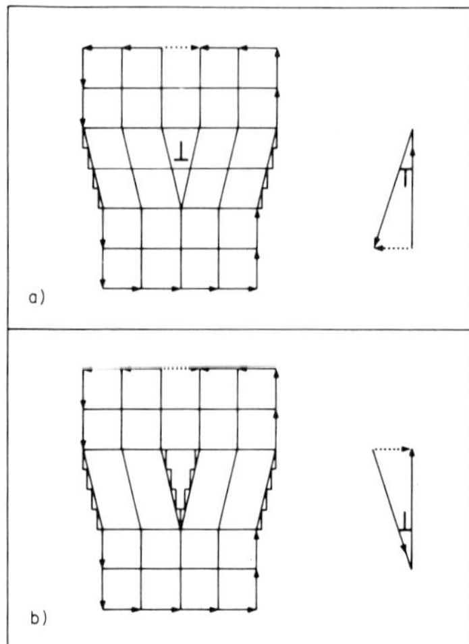

FIG. 1

a) Edge dislocation in a finite crystal with surface dislocations present. b) same as in a) but after relaxation of dislocation core by formation of an asymmetric crack. Insets in a) and b) correspond to magnified views of surface dislocations on outer and internal surface of a) and b) respectively.

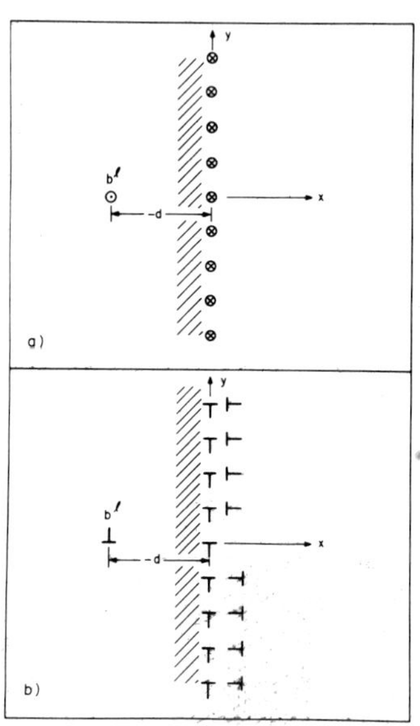

FIG. 2

Surface dislocation arrays associated with a a) screw b) edge type lattice dislocation situated near a free surface.

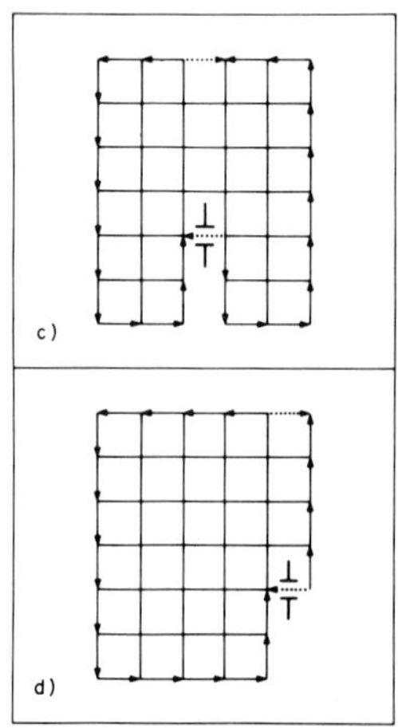

FIG. 1

(Continued) c) Fully torn configuration of the dislocated state shown in a). d) Configuration obtained after dislocation in a) has moved to surface by glide.

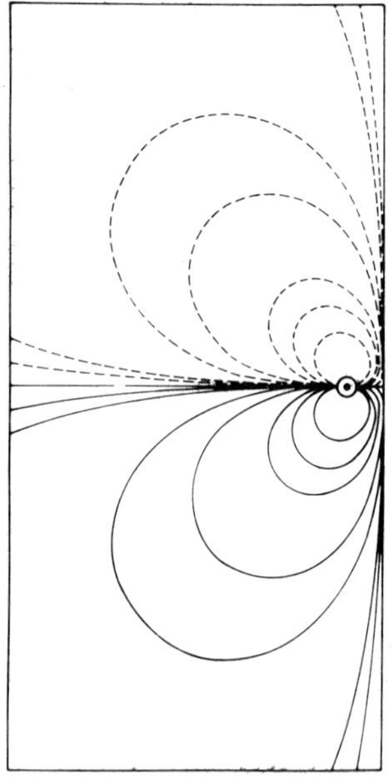

FIG. 3

Equi-stress contours corresponding to the dislocation configuration shown in Fig. 2a.

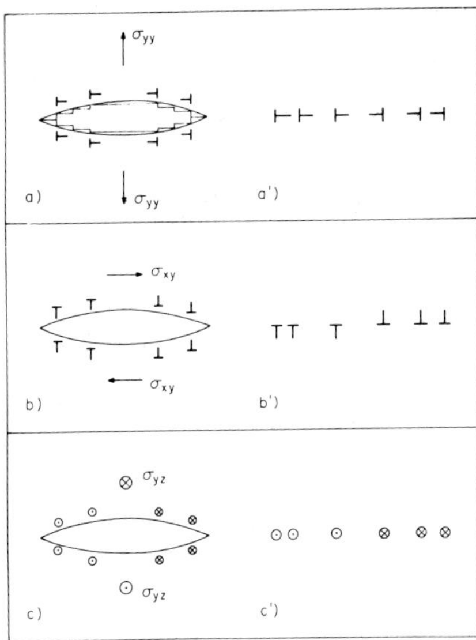

FIG. 4

Cracks of Mode a) I b) II and c) III and corresponding dislocation arrays a'), b'), c').

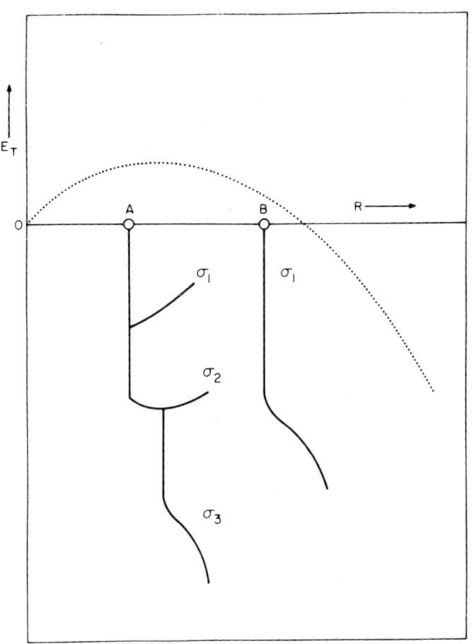

FIG. 6

Effect of preexisting crack size on plastic crack growth.

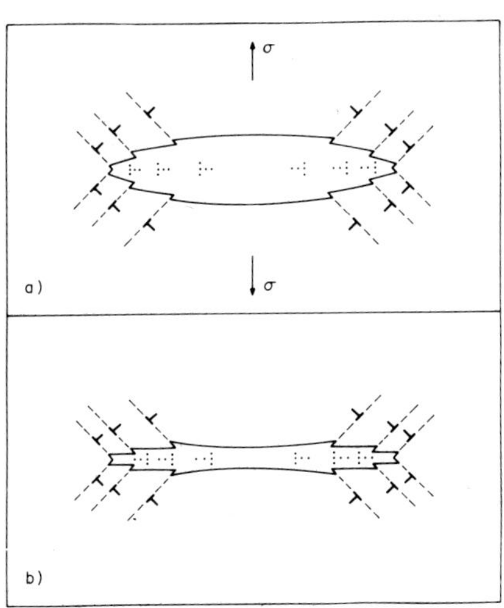

FIG. 5

a) Loaded b) Unloaded plastic tensile crack.

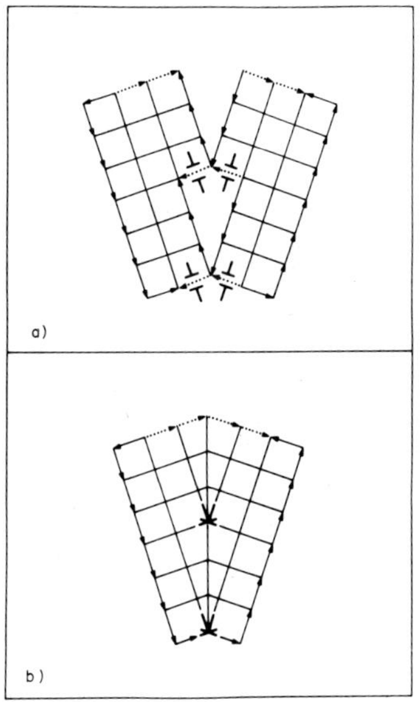

FIG. 7

Completely a) Torn or uncoalesced b) Coalesced grain boundary of misorientation angle 36.9°.

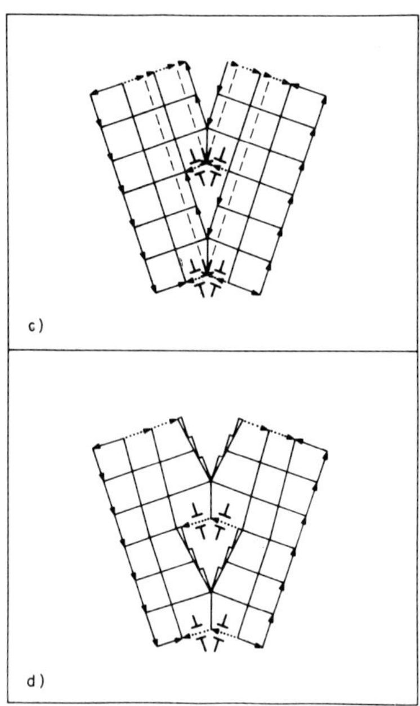

FIG. 7

(Continued) Partially coalesced grain boundary of misorientation angle 36.9° with two different surface dislocation configurations.

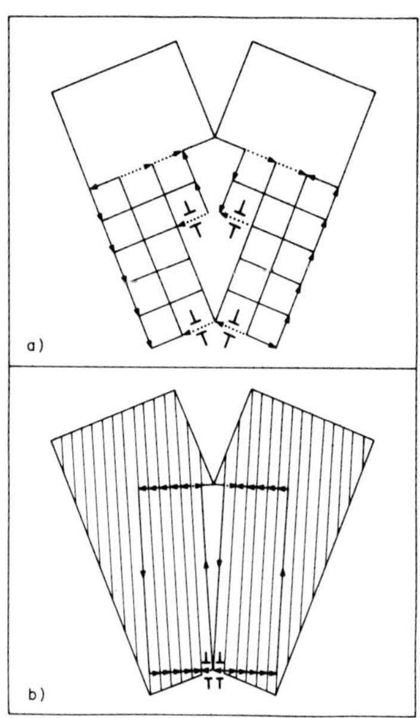

FIG. 9

Creation of a tilt boundary of misorientation angle $\theta = 43.6°$ from the 36.9° boundary of Fig. 7a illustrated in terms of the a) crystal lattice b) coincidence site sublattice.

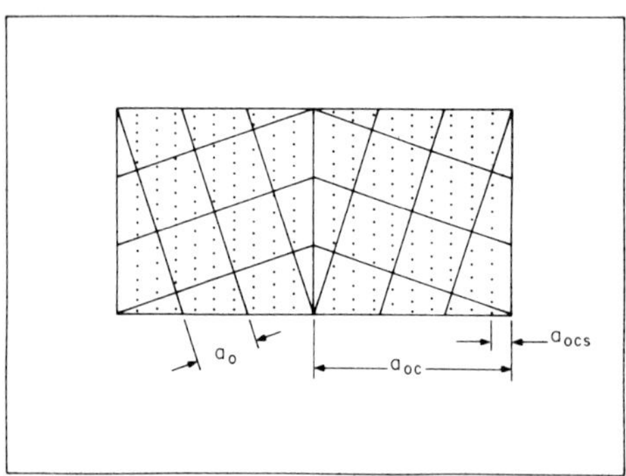

FIG. 8

Coincidence site lattice associated with the grain boundary in Fig. 7.

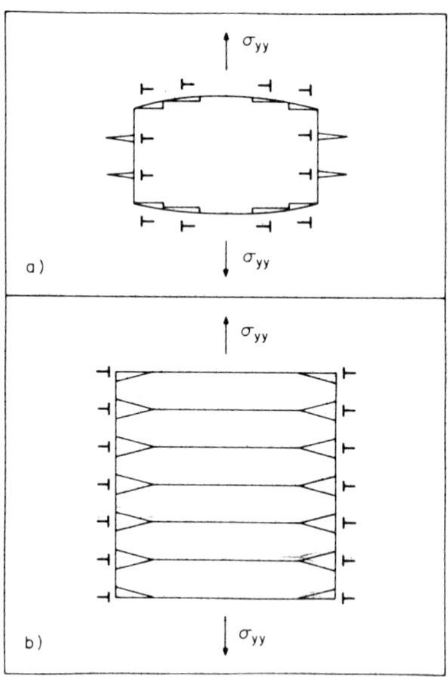

FIG. 10

a) Hole b) Finite body under simple tension.

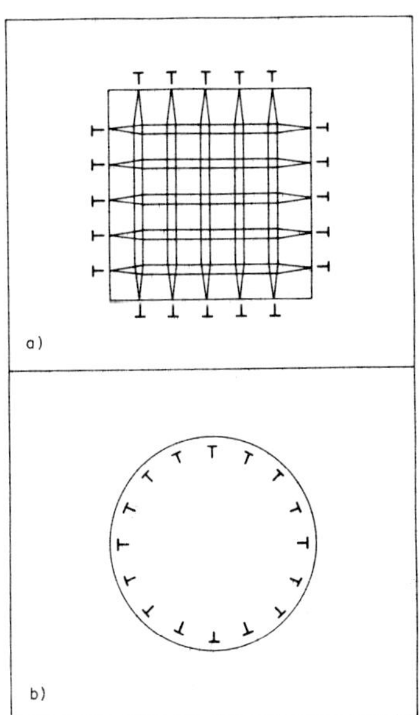

FIG. 11

Creation of surface dislocations as a result of surface tension in a a) solid b) liquid.

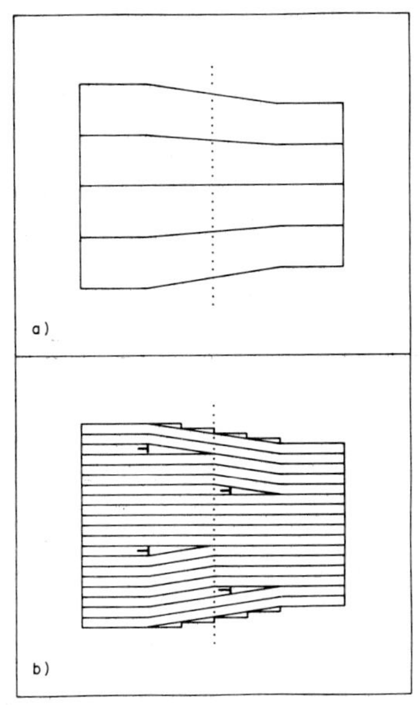

FIG. 13

Coherent interphase boundary drawn in terms of a) the original crystal lattice b) a common or coincidence lattice.

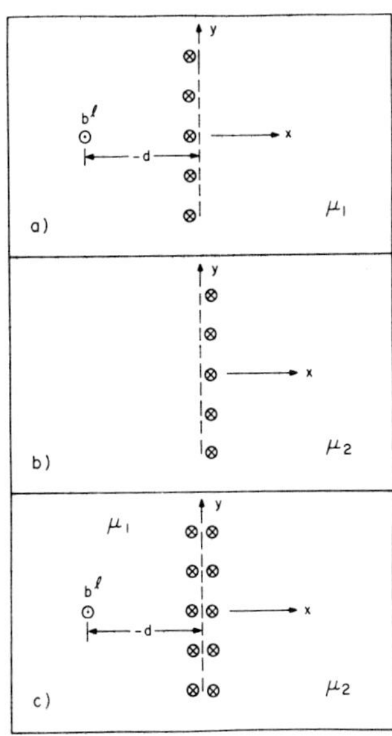

FIG. 12

Surface dislocation arrays associated with a dislocation near a two-phase interface.

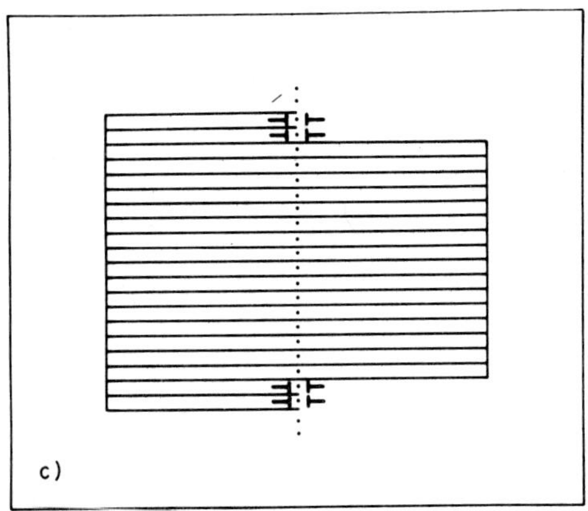

FIG. 13

(Continued) c) Completely torn or incoherent counterpart of b).

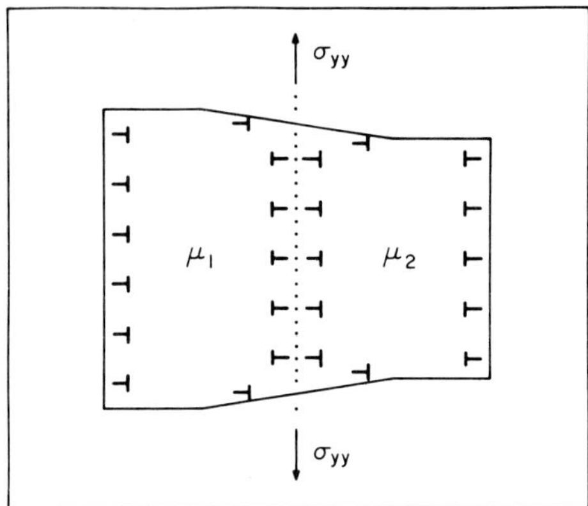

FIG. 14

Surface dislocation configuration associated with a two-phase body subjected to an externally applied stress.

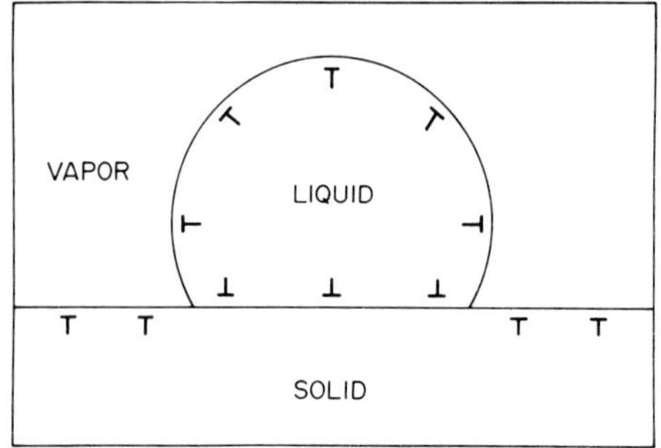

FIG. 15

Interface between solid, liquid and vapor described in terms of surface dislocations.

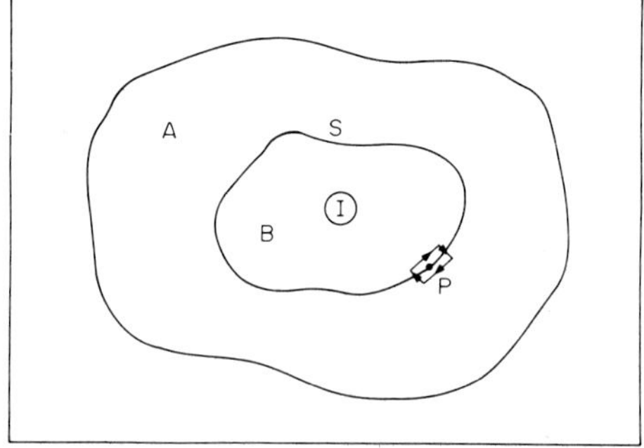

FIG. 16

Strain energy release in finite body B containing an internal stress I due to introduction of surface dislocations on surface S of finite body.

DISLOCATIONS AND INELASTICITY OF SOLIDS*

T. Mura

Dept. of Civil Engineering and Materials
Research Center, Northwestern University
Evanston, Illinois 60201

Continuum Plasticity and Dislocations

The stress field in continuum plasticity is the sum of the applied elastic stress and the dislocation stress. It will be shown here that the dislocation density tensor α_{hi} can be expressed in terms of the stress components if the impotent dislocations are excluded and the material is infinitely extended.

As shown by Mura (1), the stress field due to α_{hi} in an infinite body is

$$\sigma_{ij}(\underset{\sim}{x}) = C_{ijk\ell} \int_{-\infty}^{\infty} \varepsilon_{\ell nh} C_{pqmn} G_{kp,q}(\underset{\sim}{x} - \underset{\sim}{x}')\alpha_{hm}(\underset{\sim}{x}')d\underset{\sim}{x}' \tag{1}$$

where $C_{ijk\ell}$ are the elastic moduli, $\varepsilon_{\ell nh}$ is the permutation tensor, and G_{kp} is Green's function. According to Kröner (2), α_{hi} is expressed by

$$\begin{aligned} \alpha_{hi} &= -\varepsilon_{h\ell j}\beta^{p}_{ji,\ell} \\ &= \varepsilon_{h\ell j}\beta_{ji,\ell} \\ &= \varepsilon_{h\ell j}e_{ji,\ell} + \varepsilon_{h\ell j}\omega_{ji,\ell} \end{aligned} \tag{2}$$

where

$$\begin{aligned} u_{i,j} &= \beta_{ji} + \beta^{p}_{ji} \\ \beta_{ji} &= e_{ji} + \omega_{ji}. \end{aligned} \tag{3}$$

In the above equations e_{ij} is the elastic strain, $\omega_{ij} = -\omega_{ji}$ the rotational tensor, β^{p}_{ji} the plastic distortion, β_{ji} the elastic distortion and u_i is the displacement. It can be shown that the second term in (2) is impotent. When $\varepsilon_{hab}\omega_{bm,a}$ is substituted into α_{hm} in (1), we have

$$\begin{aligned} \sigma_{ij}(\underset{\sim}{x}) &= C_{ijk\ell} \int C_{pqmn} G_{kp,q}(\underset{\sim}{x} - \underset{\sim}{x}')(\omega_{nm,\ell}(\underset{\sim}{x}') - \omega_{\ell m,n}(\underset{\sim}{x}'))d\underset{\sim}{x}' \\ &= -C_{ijk\ell} \int C_{pqmn} G_{kp,qn}(\underset{\sim}{x} - \underset{\sim}{x}')\omega_{\ell m}(\underset{\sim}{x}')d\underset{\sim}{x}' \\ &= C_{ijk\ell} \int \delta(\underset{\sim}{x} - \underset{\sim}{x}')\omega_{\ell k}(\underset{\sim}{x}')d\underset{\sim}{x}' = 0, \end{aligned} \tag{4}$$

where $\delta(\underset{\sim}{x} - \underset{\sim}{x}')$ is the Dirac delta function, $C_{ijk\ell}\omega_{\ell k} = 0$ and $C_{pqmn}G_{kp,qn}(\underset{\sim}{x} - \underset{\sim}{x}') = -\delta_{mk}\,\delta(\underset{\sim}{x} - \underset{\sim}{x}')$. Therefore, the potent part of α_{hi} can be written as

$$\alpha_{hi} = \varepsilon_{h\ell j}e_{ji,\ell}. \tag{5}$$

---

*This work was supported by the U.S. Army Research Office under Grant No. DAA 29-77-G-0042 and partly by the National Science Foundation to the Materials Research Center of Northwestern University under Grant No. 7601057.

Since $e_{ji} = e_{ij} = C^{-1}_{ijmn}\sigma_{mn}$, (5) becomes

$$\alpha_{hi} = C^{-1}_{ijmn}\,\varepsilon_{h\ell j}\,\sigma_{mn,\ell}. \tag{6}$$

The above result indicates that the potent dislocation distributions can be obtained if the stress distributions are known in an infinite body.

As an example, Vilmann and Mura (3) obtained a distribution of dislocations at a crack tip as shown in Fig. 1. The dislocation density tensor is obtained from (6) by use of the stress field in the slip line theory. The dislocations are distributed only in the fan-shaped plastic

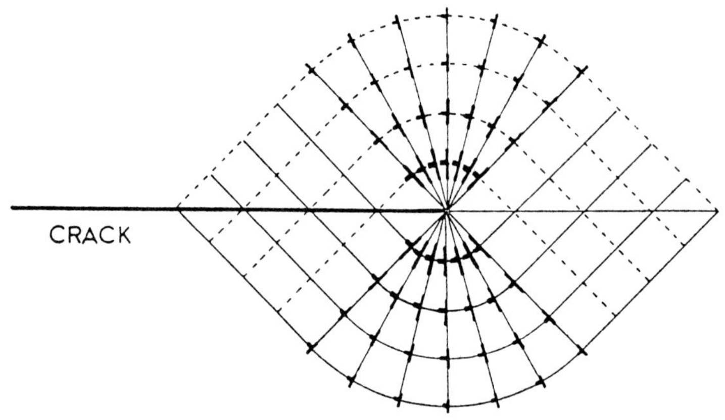

FIG. 1

Dislocation distribution at the crack tip.

domain and the Burger vector is radial. The density is

$$\alpha_{3r} = (1-\nu)2k/\mu r, \tag{7}$$

where ν is Poisson's ratio, k the shearing yield stress, μ the shear modulus, and r is the distance from the crack tip. Mura and Vilmann (4) considered its application to a propagating fatigue crack and obtained the dislocation distribution as shown in Fig. 2. Since the cyclic loading (tension-zero-tension) is applied in the fatigue, the dislocation distribution after an unloading changes the sign. Thus, the sum of the dislocation distributions of loading and unloading constitutes a dipole distribution of dislocations. The strength of the dipoles increases by the repeating loading. After some number of cyclic loadings, the crack tip advances by δa. The domain of the dipole distribution increases with the growth of the crack.

Dislocation Model for Fatigue Crack Initiation

The initiation of fatigue cracks is one of the most important stages in the fatigue fracture processes of metals. Recently Tanaka and Mura (5) have proposed a dislocation dipole model (two adjacent layers of dislocation pile-ups) to explain the ratcheting phenomena of cyclic loading, extrusions and intrusions of a smooth specimen. This model can also yield the Coffin-Manson type law for the crack initiation and the Petch type equation for the grain size dependency of fatigue strength.

Figure 3 shows the model of dislocation pile-ups. Dislocations are piled up on layer I by the loading and on layer II by the unloading. The signs of the dislocations are opposite in the two layers. We will calculate the plastic strain, number of dislocations, associated energy and the stress field caused by the dislocations. Under the first loading τ_1 greater than the frictional stress k, the dislocation distribution with density $D_1(x)$ produced on layer I is obtained from the equilibrium condition:

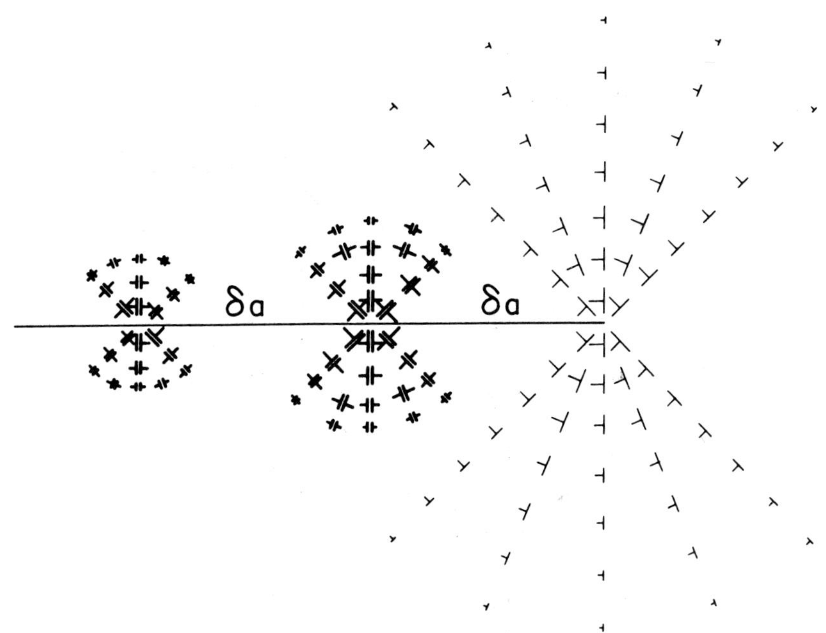

FIG. 2

Dislocation dipole distribution of a propagating fatigue crack.

$$\tau_1^D + \tau_1 - k = 0, \tag{8}$$

where τ_1^D is the dislocation stress (the back stress) given by

$$\tau_1^D = A \int_{-a}^{a} \frac{D_1(\xi)d\xi}{x - \xi} . \tag{9}$$

The domain of the dislocation distribution is $-a < x < a$, and $A = \mu b/2\pi(1-\nu)$. b is the Burgers vector. The plane $x = 0$ is assumed to be a free surface. The unbounded solution at $x = \pm a$ is obtained as

$$D_1(x) = (\tau_1 - k)x/\pi A(a^2 - x^2)^{\frac{1}{2}}. \tag{10}$$

The total number of dislocations existing between $x = 0$ and a is

$$N_1 = \int_0^a D_1(x)dx = (\tau_1 - k)a/\pi A \tag{11}$$

and the average strain caused by these dislocations is

$$\gamma_1 = \int_{-a}^{a} bD_1(x)\, x\, dx = (\tau_1 - k)ba^2/2A. \tag{12}$$

The stored energy of dislocations is

$$U_1 = -\frac{1}{2} \int_{-a}^{a} \tau_1^D\, bD_1(x)\, x\, dx = \gamma_1(\tau_1 - k)/2. \tag{13}$$

In the above expressions, a is supposed to be a grain size.

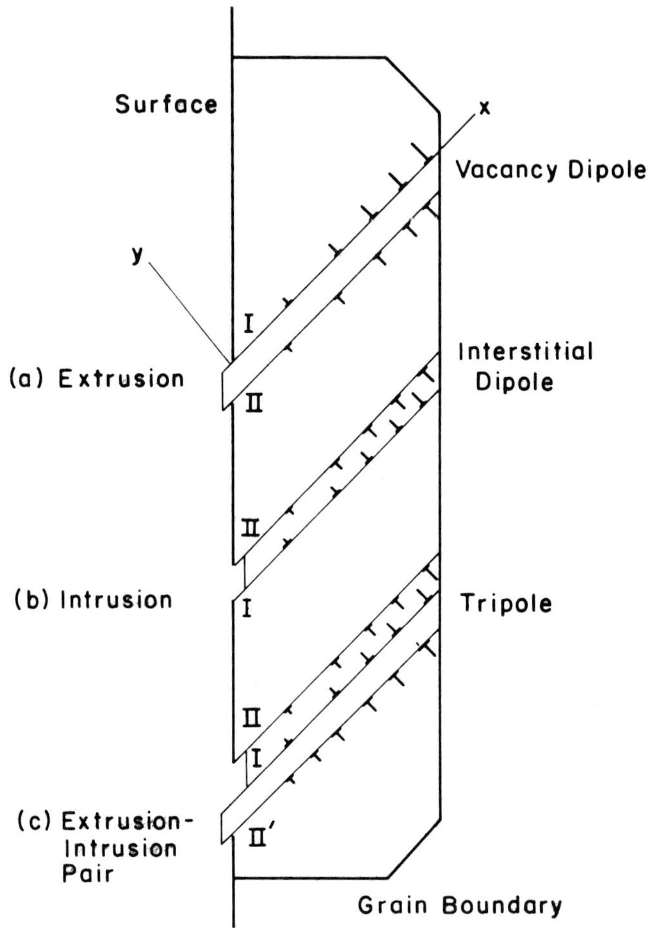

FIG. 3

Dislocation distribution in a most favorably oriented grain and the formation of extrusion and intrusion.

When the applied stress is reversed from τ_1 to τ_2, the dislocations with opposite signs are piled up on layer II. The distance between layers I and II is assumed to be very small compared with the pile-up length. It is also assumed that the dislocations in layer I do not move by the unloading. This assumption of the irreversibility of dislocation motion can be justified in some extent by the experiment of Keith and Gilman (6). By denoting the density of dislocations on layer II by $D_2(x)$ and the back stress due to $D_2(x)$ by τ_2^D, the equilibrium condition on layer II is expressed as

$$\tau_2^D + \tau_1^D + \tau_2 + k = 0. \tag{14}$$

τ_1^D on layer II is assumed to be the same as on layer I. The dislocation density on layer II, its total number of dislocations and the plastic strain increment due to the new dislocations on layer II are obtained as

$$D_2(x) = -(\Delta\tau - 2k)x/\pi A(a^2 - x^2)^{\frac{1}{2}}$$
$$N_2 = -(\Delta\tau - 2k)a/\pi A \qquad (15)$$
$$\gamma_2 = -(\Delta\tau - 2k)ba^2/2A$$

where $\Delta\tau = \tau_1 - \tau_2$ is the applied stress amplitude. The stored energy of dislocations $D_2(x)$ becomes

$$U_2 = -\gamma_2(\Delta\tau - 2k)/2. \qquad (16)$$

The pile-up of opposite dislocations on layer II causes a positive back stress on layer I. This back stress enhances the pile-up of dislocations on layer I during the next loading. Similarly, the dislocations on layer I enhance the pile-up of dislocations on layer II during the subsequent unloading. This is the ratcheting mechanism. After repeated loading and unloading, the extrusion or intrusion is created on the surface of the specimen as shown in Fig. 3. The corresponding quantities D_k, N_k, γ_k, and U_k at the k-th step of the loading and unloading processes are obtained in a similar manner. U_k is found to be a constant equal to U_2 for any k except k = 1.

We assume that when the stored strain energy due to the dislocations accumulated after n cycles of loading becomes a critical value $4aW_s$ which may be a surface energy, the layers of pile-up dislocations are transformed into a free surface (crack). The life of the crack initiation n_c is now defined by

$$2n_cU_2 = 4aW_s. \qquad (17)$$

The last equation (17) leads to

$$n_c(\Delta\gamma)^2 = 2bW_sa^3/A \qquad (18)$$

where $\Delta\gamma$ is the plastic strain amplitude given by $|\gamma_2|$. This result is similar to the Coffin-Manson law: $n_c(\Delta\gamma)^2$ = constant. Equation (18) can also be rewritten as

$$\Delta\tau = 2k + (8AW_s/bn_c)^{\frac{1}{2}} a^{-\frac{1}{2}} \qquad (19)$$

which is of the Petch type equation for constant initial life (n_c = constant and a is assumed to be a grain size). The Petch type relation has been reported for the long-life fatigue of several metals [Taira et al. (7)].

Relaxation, Creep and Internal Friction

Let us consider an elastic body which contains harder particles than the matrix. Under an applied stress σ^0_{31}, the interface between the particle and the matrix will be deformed into the dotted curve as shown in Fig. 4 if the particle had no resistance for the deformation. The resistance for this deformation, however, will be decreased if diffusion takes place and mass in region I is transformed to region V in Fig. 4. The misfit between the particle and the matrix would disappear and the internal stress and the associated energy will decrease by this diffusion.

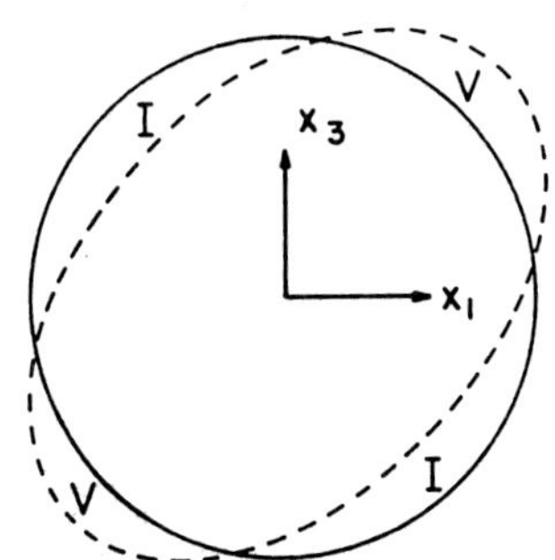

FIG. 4

When an inhomogeneous particle is harder than a matrix, the interface boundary of the matrix (dotted line) and the particle (solid line) deforms as shown here by σ^0_{31} if no constraint exists from the surroundings.

Similarly, the diffusional relaxation takes place when the matrix is subjected to plastic deformation and the inclusion is not. The misfit between the particle and the matrix can be represented by the Orowan loops surrounding the particle. The relaxation can be expressed by the climb of the Orowan loops as shown in Fig. 5. If atoms flow from the left hand side of an

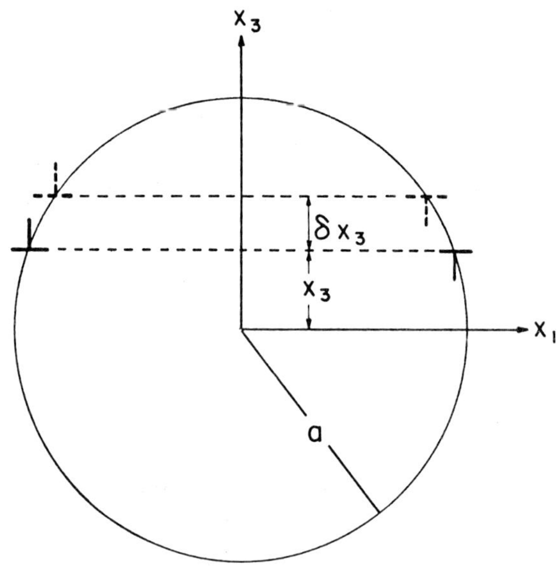

FIG. 5

The climb process of the Orowan loop.

Orowan loop to the right hand side through its own pipe, the Orowan loop with the Burgers vector $\underset{\sim}{b}$ climbs from x_3 to $x_3 + \delta x_3$. The climb rate calculated by Mori and Tokushige (8) is

$$V = \frac{dx_3}{dt} = Ax_3(a^2 - x^2)^{-2} \tag{20}$$

with

$$A = \alpha\mu\Omega h^2 D_p/kT \tag{21}$$

where α is a constant of the order of unity, μ the shear modulus, Ω the atomic volume, h^2 the cross section of the dislocation pipe, D_p the pipe diffusion constant along the dislocation, and k is the Boltzmann constant.

Annealing

Suppose that the matrix deforms with a strain rate of $\dot{\varepsilon}^p_{31} = \dot{\gamma}_p/2$. If slip occurs uniformly the Orowan loop density distribution N increases by $\dot{\gamma}_p N_p/b$ per unit time, N_p being the number of the inclusions per unit volume. At the same time, the loop density decreases with a rate of $-\partial(NV)/\partial x_3$. Thus, we have the equation for the dislocation climb as

$$\frac{\partial N}{\partial t} = \dot{\gamma}_p N_p/b - \frac{\partial(NV)}{\partial x_3} . \tag{22}$$

When $\dot{\gamma}_p = 0$, we have

$$\frac{\partial N}{\partial t} = - \frac{\partial(NV)}{\partial x_3} \tag{23}$$

which corresponds to the annealing case. It can be shown that the solution of (23) with the initial condition $N = N_0$ at $t = 0$ is obtained by Mori and Mura (9) as

$$N(x_3,t) = N_0 \frac{z(a^2 - x_3^2)^2}{x_3(a^2 - z^2)^2} \tag{24}$$

with

$$At = a^4 \log(x_3/z) - (x_3^2 - z^2)(4a^2 - x_3^2 - z^2)/4. \tag{25}$$

The back stress decrease (flow stress decrease) of a work hardened alloy by annealing can be calculated by the last result (24).

According to Mori and Tanaka (10), the average back stress in the matrix is

$$<\sigma_{31}>_M = -f <\sigma_{31}>_I \tag{26}$$

where f is the volume fraction of particles and $<\sigma_{31}>_I$ is the average value of σ_{31} inside one inclusion when it exists alone in the infinite medium. Annealing time t and annealing temperature T determine N in (24), where D_p is assumed to be $b^2\nu \exp(-Q/kT)$ with $\nu = 10^{12}/s$. σ_{31} is the eigenstress caused by the eigenstrain $\varepsilon^p_{31} = -Nb/2N_p$ inside the inclusion. In the calculation, (24) is approximated by a polynomial, and the formula of Asaro and Barnett (11) is used. The theoretical prediction of the back stress after the annealing agrees reasonably with experiments.

Stationary creep

If the Orowan loop density distribution is in balance between loss due to climb and gain due to the accumulation by plastic deformation, one would expect stationary creep. The Orowan loop density distribution corresponding to a stationary creep rate $\dot{\gamma}_p$ is obtained from

$$\frac{\partial(NV)}{\partial x_3} = \dot{\gamma}_p N_p/b \tag{27}$$

since $\partial N/\partial t = 0$ in (22). Since V is given by (20), the solution of (27) is obtained as

$$N = \dot{\gamma}_p N_p (a^2 - x_3^2)^2/Ab. \tag{28}$$

The eigenstress σ_{31} is calculated from the eigenstrain given by

$$\varepsilon^p_{31} = -Nb/2N_p = -\dot{\gamma}_p(a^2 - x_3^2)^2/2A. \tag{29}$$

The back stress $<\sigma_{31}>_M$ can be evaluated from (26) after calculating $<\sigma_{31}>_I$. The applied stress needed to maintain the stationary creep is given by

$$\sigma^A_{31} = \sigma^D_{31} - <\sigma_{31}>_M \tag{30}$$

where σ^D_{31} is the flow stress if no particles exist in the material. Equation (30) was first derived by Mori and Osawa (12). Evaluation of (30) shows that the stationary creep rate is found to depend linearly on the applied stress and is also proportional to the inverse fourth power of the particle size. Mori and Osawa also showed that their theory agrees with the experiments using Al-Si alloy.

Internal friction

As mentioned in the beginning of this section, even if an applied stress is small enough to not produce any plastic flow, the material could be subjected to an inelastic deformation when the surface diffusion takes place at the particle interfaces. Thus, an internal friction is caused in the material under an oscillating applied stress field. We will review the work of Mori, Okabe and Mura (13).

When the material is subjected to an elastic applied stress $\sigma^0_{31} = \sigma^0$ at infinity, the stress disturbance caused by a spherical particle with shear modulus μ^* can be simulated by the eigenstress of the inclusion with eigenstrain ε^*_{31} given by

$$\varepsilon^*_{31} = (\mu - \mu^*)\sigma^0/2\mu g \tag{31}$$

with

$$g = 2(\mu^* - \mu)S_{3131} + \mu \tag{32}$$

where the shear modulus and Poisson ratio of the matrix are μ and ν, S_{3131} is the Eshelby tensor

$S_{3131} = (4 - 5\nu)/15(1 - \nu)$. The effective shear modulus (unrelaxed modulus) of the composite containing particles with volume fraction f can be evaluated as

$$\bar{\mu} = \mu\{1 - f(\mu - \mu^*)/g\}. \quad (33)$$

If an additional eigenstrain $\Delta\bar{\varepsilon}^p_{31}$ given by

$$\Delta\bar{\varepsilon}^p_{31} = \sigma^0/2\mu(1 - 2S_{3131}) \quad (34)$$

takes place in the particles, the stress inside the particles becomes zero and the Gibbs free energy of the system can be minimized. Then, the effective shear modulus (relaxed modulus) of the composite becomes

$$\bar{\mu}_\infty = \mu\{1 - f/(1 - 2S_{3131})\}. \quad (35)$$

The inelastic strain (34) can be achieved by the surface diffusion on the particles if a sufficient time is allowed. For a short period of time δt, the inelastic strain is achieved only by $\delta\Delta\varepsilon^p_{31}$. The number of the transported atoms to produce $\delta\Delta\varepsilon^p_{31}$ is calculated as

$$\delta N = 2V_0\delta(\Delta\varepsilon^p_{31})/\pi\Omega \quad (36)$$

where V_0 is the volume of a particle and Ω is the atomic volume. The corresponding change in the Gibbs free energy is calculated as

$$\delta\Delta W = 4\mu\mu^*(1 - 2S_{3131})V_0(\Delta\varepsilon^p_{31} - \Delta\bar{\varepsilon}^p_{31})\delta(\Delta\varepsilon^p_{31})/g. \quad (37)$$

The drift velocity of an atom, v, is found by the Einstein equation

$$v = -(D/kT)F \quad (38)$$

with

$$F = -(\delta\Delta W/\delta N)/(\pi a/2) \quad (39)$$

where D is the diffusion constant. Equation (38) yields

$$v = \frac{D}{kT}\,\frac{4\mu\mu^*(1 - 2S_{3131})\Omega}{ag}\,(\Delta\bar{\varepsilon}^p_{31} - \Delta\varepsilon^p_{31}). \quad (40)$$

During the time interval of δt, δN can also be written as

$$\delta N = (\pi^2 a/2)h\,v\,\delta t/\Omega \quad (41)$$

where h is the thickness of the interface layer. Equations (36) and (41) yield

$$\frac{d(\Delta\varepsilon^p_{31})}{dt} = (\Delta\bar{\varepsilon}^p_{31} - \Delta\varepsilon^p_{31})/\tau \quad (42)$$

with

$$\tau = \frac{kT}{Dh}\,\frac{4\{2(\mu^* - \mu)S_{3131} + \mu\}a^3}{3\mu\mu^*(1 - 2S_{3131})\pi^2\Omega}. \quad (43)$$

The last equation (42) indicates that the particle bearing specimen behaves like a standard linear solid. The relaxation strength $\Delta\mu$ is given by

$$\Delta\mu = (\bar{\mu} - \bar{\mu}_\infty)/\bar{\mu} = f\mu^*/(1 - 2S_{3131})g. \quad (44)$$

Okabe, Mori and Mura (14) examined their theory by measuring the internal friction of Al-Si alloys by the use of an inverted torsion apparatus. Figure 6 shows the logarithmic decrement and the square of the angular frequency against temperature. Single crystal specimens were quenched from 823K after 24 h solution annealed and aged at 573K ~ 603K for various times. Si particles, thus formed, had shapes approximately spherical. Since a single crystal of Al does not show the internal friction peak in the experimental condition similar to the present study [Kê, (15)], it is concluded that the peak such as observed in Fig. 6 is caused by the presence of Si particles. This conclusion is further substantiated by knowing that the peak height is

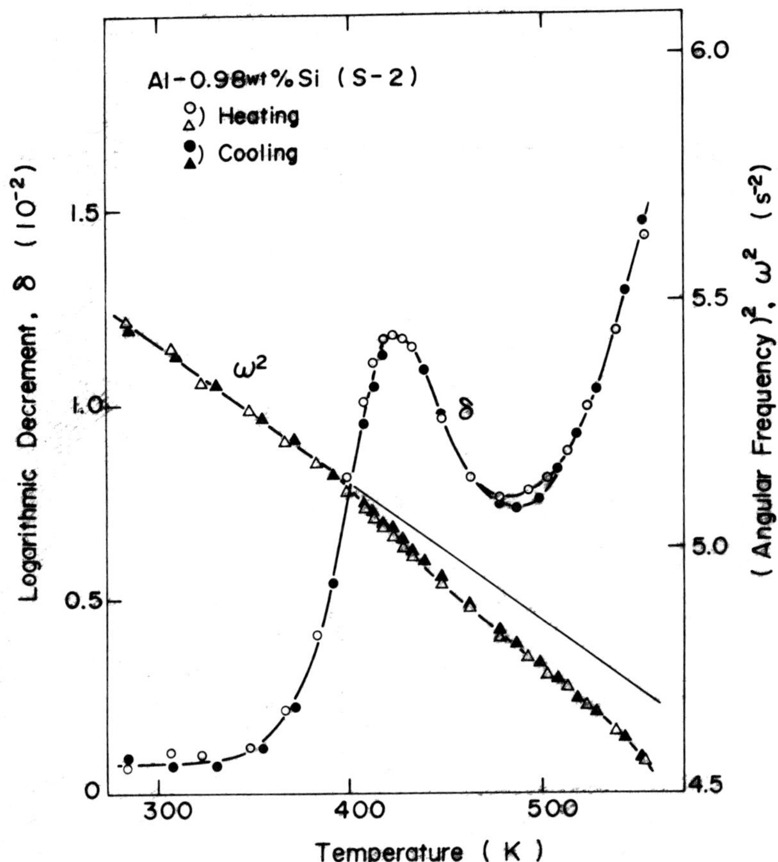

FIG. 6

Logarithmic decrement and square of frequency vs temperature.

proportional to the amount of the Si particles.

Since the elastic constant of Al linearly depends on the temperature in the present experimental condition [Kê, (15)], the linearly extrapolated line from the low temperature region of the square of the frequency is taken to represent the unrelaxed modulus μ. The completely relaxed modulus $\bar{\mu}_\infty$ is taken from the experimental point at the temperature at which the internal friction peak merges into the background and the relaxation is considered to be completed.

Denoting T_p and ω_p as the temperature and the frequency at the peak of internal friction, we have the relation

$$\omega_p \tau = 1 \tag{45}$$

where τ is the relaxation time defined by (43). Assuming that

$$D = D_0 \exp(-Q/kT_p), \tag{46}$$

where Q is the activation energy, equation (45) yields

$$\omega_p T_p \exp(Q/kT_p) = Aa^{-n} \tag{47}$$

where n is 3 and A is a constant. The above relation is well demonstrated by Fig. 7. The particle size dependency n is certainly close to 3.

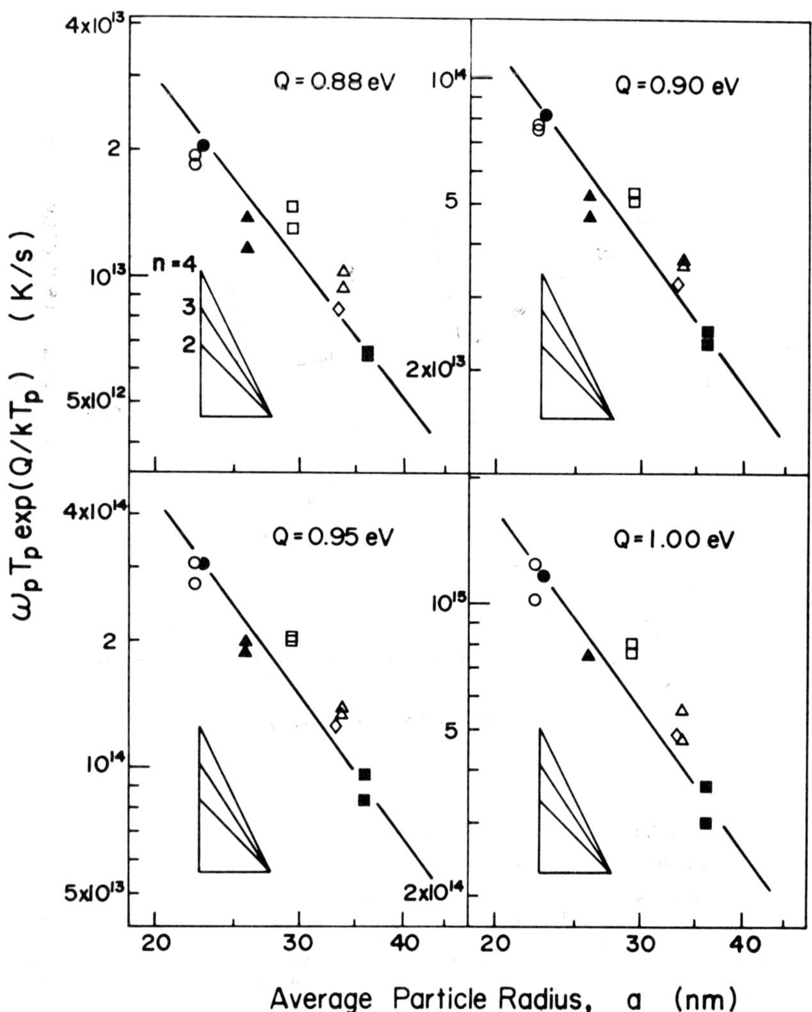

FIG. 7

Particle size dependence of peak temperature.

References

1. T. Mura, Phil. Mag. 8, 843 (1963).
2. E. Kröner, Annal. Phys. 11, 13 (1963).
3. C. Vilmann and T. Mura, J. Appl. Mech. 46, 817 (1979).
4. T. Mura and C. Vilmann, to be published (1980).
5. K. Tanaka and T. Mura, to be published (1980).
6. R. E. Keith and J. J. Gilman, ASTM Special Tech. Publ., No. 237, 3 (1959).
7. S. Taira, K. Tanaka and M. Hoshina, ASTM Special Tech. Publ., No. 675, 135 (1979).
8. T. Mori and H. Tokushige, Act. Metall., 25, 635 (1977).
9. T. Mori and T. Mura, Act. Metall., 26, 1199 (1978).
10. T. Mori and K. Tanaka, Act. Metall., 21, 571 (1973).
11. R. J. Asaro and D. M. Barnett, J. Mech. Phys. Solids, 23, 77 (1975).
12. T. Mori and T. Osawa, Phil. Mag., 40, 445 (1979).
13. T. Mori, M. Okabe and T. Mura, Act. Metall., 28, 319 (1980).
14. M. Okabe, T. Mori and T. Mura, to be published (1980).
15. T. S. Kê, Phys. Rev. 71, 535 (1947).

INFLUENCE OF TEXTURE AND SLIP MODE ON CREEP CONSTITUTIVE BEHAVIOR OF HEXAGONAL METALS

Brent L. Adams* and John P. Hirth**

*Hanford Engineering Development Laboratory, Richland, Washington
**The Ohio State University, Columbus, Ohio

Abstract

Steady-state creep behavior in hexagonal metals is modeled using a uniform strain-rate upper-bound estimate for power-law behavior. Single crystal creep is assumed to occur on prismatic, basal, and pyramidal slip systems in proportion to the resolved shear stress raised to a power. Polycrystalline behavior is deduced from the range of crystallite behavior by averaging using the crystallite orientation distribution function (CODF).

Two textures of engineering importance are modeled: a radial sheet rolling texture and a bimodal tubing texture. Predicted multiaxial creep behavior is mapped and compared with the isotropic idealization. Power-law behavior for stress exponents, n, in the range $1 \leq n \leq 6$ are compared with the Taylor-Bishop-Hill bound for rigid/perfectly plastic material ($n=\infty$). Results for the bimodal tube texture are compared with creep data obtained at 673°K for highly cold-worked Zircalo tubing.

Substantial creep anisotropy is predicted in non-isotropic polycrystals when one slip mode is preferred over the others. Comparisons of the model with 673°K cold-worked Zircaloy tube data suggest a high reliance on basal slip to maintain microcompatibility in low $\hat{c}/\hat{a}$ hexagonal metals. This requirement is observed to relax in recrystallized Zircaloy tubing of nominally the same crystallographic texture, suggesting hardening of prismatic slip due to prior cold-working.

Introduction

The observation of anisotropic creep in hexagonal metals under multiaxial loading suggests a strong link between inelastic behavior and crystallographic texture. Hexagonal structural forms frequently exhibit strong preferred orientations arising due to crystalline rotations during thermomechanical processing. Texture impacts creep through an interaction of the stress-state with allowable crystalline deformation modes. In this paper, theoretical predictions of steady-state creep are presented based on constitutive equations which embody slip on prism, basal, and pyramidal systems in the hexagonal crystallite. Single crystal behavior is averaged over the range of orientations exhibited in the polycrystal using the crystallite orientation distribution function (CODF). Theoretical predictions are compared, in the case of the tubing texture, with data obtained under biaxial loading conditions for Zircaloy 4.

Crystal Slip Creep Theory

Generally, hexagonal metals exhibit slip on prism and basal planes at lower shear stresses than those required for slip (or twinning) on pyramidal planes (1). Since the combined operation of prism and basal slip modes supplies only four linearly independent strains, whereas the von Mises criterion requires a minimum of five for an arbitrary incompressible strain, pyramidal activity has a controlling effect on general inelastic deformation. In hexagonal crystals any strain in the $\hat{c}$ direction must be accommodated by either pyramidal slip or twinning. This mode can supply as many as five additional linearly independent slip systems.

Pyramidal deformation is particularly important in low $\hat{c}/\hat{a}$ lattices, where prism slip is dominant and basal slip is difficult . . . basal slip not commonly being observed except at higher temperatures (>700-800K in Zr) and in regions of stress concentration in the crystallite (2,3). In such materials, pyramidal twinning normally occurs at lower temperatures, and pyramidal slip shows increased activity with increased temperature (4,5). In this paper, deformation is assumed to occur exclusively by slip on $\{1\bar{1}00\}\langle 11\bar{2}0\rangle$ prism, $\{0001\}\langle 11\bar{2}0\rangle$ basal, and $\{11\bar{2}2\}\langle\bar{1}\bar{1}23\rangle$ pyramida systems.

The fundamental relation linking slip system geometry with single crystal slip strain is μ_{ij}

where for the kth slip system

$$\mu_{ij}^{(k)} = \frac{1}{2}\left(b_i^{(k)}n_j^{(k)} + b_j^{(k)}n_i^{(k)}\right) \quad [1]$$

In equation [1] $\vec{b}$ and $\vec{n}$ represent unit vectors in the Burger's and slip plane normal directions respectively. The crystal stress, σ_{ij}, is connected to the resolved shear stress on the kth slip system, $\tau^{(k)}$, by the slip tensor:

$$\tau^{(k)} = \sigma_{ij}\mu_{ij}^{(k)}. \quad [2]$$

Similarly, if $\dot{\gamma}^{(k)}$ is used to represent the shear-rate of the kth slip system, the crystal strain rate, $\dot{\varepsilon}_{ij}$, is obtained by summing over the active slip-systems using:

$$\dot{\varepsilon}_{ij} = \sum_k \mu_{ij}^{(k)}\dot{\gamma}^{(k)} \quad [3]$$

Typically, materials exhibit a power-law creep dependence on stress over a wide range of stresses for temperatures typically greater than one-third the melting point (6). Following the work of Hutchinson (7,8) on isotropic polycrystals, we assume that the shear rate, $\dot{\gamma}$, is linked to the resolved shear stress, τ, through a power-law relation.

$$\dot{\gamma}^{(k)} = A\left|\tau^{(k)}/\tau_o^{(k)}\right|^{n-1}\tau^{(k)}/\tau_o^{(k)}. \quad [4]$$

Here, A is the reference creep-rate, also containing the primary temperature dependence, and τ_o is the reference shear-stress prescribing the ease of slip at any moment during the strain history. The stress exponent, n, is determined from polycrystalline behavior and is assumed to apply independently to each active slip mode.

At a constant temperature during steady-state creep it is reasonable to assume that τ_o is constant for each slip mode since strain-hardening and recovery processes are balanced. The validity of this assumption over a range of variable strain-rates must be verified by experiment.

A convenient upper-bound will be applied in the present analysis. If a uniform strain-rate is applied to the specimen boundary, we apply a uniform strain-rate upper-bound (USRUB) analysis to power-law creep through equation [3]. The strain-rate perceived at the boundary is extended uniformly throughout the specimen. Strain-rate compatibility is automatically preserved with this restriction, but crystal stresses are allowed to be discontinuous from one crystallite to another. The solution is achieved by solving for the stress in each crystallite, and then averaging over the range of preferred orientations. If $w(\theta,\psi,\phi)$ is the CODF describing the probability of a crystallite occurring at an orientation θ,ψ,ϕ (where θ,ψ,ϕ are the Euler angles) with respect to the specimen axes, then the average polycrystalline stress is $\{\sigma_{ij}\}$, where

$$\{\sigma_{ij}\} = \iiint \sigma_{ij}\cdot w(\theta,\psi,\phi)\cdot d\theta d\psi d\phi. \quad [5]$$

Obviously, σ_{ij} must be referred to specimen coordinates in equation [5].

The rigid/perfectly plastic analysis of Taylor (9), Bishop and Hill (10) provides a helpful comparison with the USRUB analysis. The method consists of minimizing the internal work $\Sigma\tau_i\gamma_i$ or maximizing the external work $\Sigma\sigma_{ij}\varepsilon_{ij}$ subject to the requirement that slip operates only for resolved shear stresses greater than or equal to an assumed critical resolved shear stress. A comparison with the USRUB analysis is possible by connecting ratios of the reference shear-stress to ratios of critical resolved shear-stress in the rigid plastic analysis. Hutchinson has shown (7) that the USRUB analysis converges to the rigid plastic prediction in the limit $n=\infty$.

Description of Texture

The CODF, $w(\theta,\psi,\phi)$, is usually determined from a set of direct pole-figure data, as a series of generalized spherical harmonic functions using the method of Roe (11) and Bunge (12). We denote θ and ψ as the polar and azimuthal angles, respectively, referring the $\hat{c}$ direction of the hexagonal crystal to principal specimen working directions, and ϕ as the rotational angle representing crystal rotations about the $\hat{c}$ pole. The CODF is expressed as

$$w(\theta,\psi,\phi) = \sum_{\ell=0}^{\infty}\sum_{m=-\ell}^{\ell}\sum_{n=-\ell}^{\ell} W_{\ell mn}Z_{\ell mn}(\theta)e^{-im\psi}e^{-in\phi}; \quad [6]$$

where $W_{\ell mn}$ are series coefficients computed from a set of pole-figure data. $Z_{\ell mn}(\theta)$ denotes the set of Augmented Jacobi Polynomials. Equation [6] is applied as a truncated series since coefficients $W_{\ell mn}$ for values of $\ell \gtrsim 18$ are obscured by the accuracy of the x-ray data. Taking advantage of hexagonal symmetry elements and the orthogonal biaxial symmetry observed in pole-figures, all coefficients $W_{\ell mn}$ for $\ell < 18$ are obtained using only three pole figures (13).

Figures 1 and 2 show basal, prism and pyramidal pole-figures for the transverse bimodal tube texture, and a predominantly radial sheet texture. They were obtained from composite specimens prepared by the method of Elias and Heckler (14), with a Siemens texture goniometer, using the Schultz reflection method (15). CuKα radiation was used for both materials. In the case of the tubing texture, tubing segments were flattened by bending (zero thickness reduction) to facilitate specimen lamination. This process was determined to have a negligible effect on the texture. Texture coefficients $W_{\ell mn}$ for the two textures have been reported elsewhere (16).

Results and Discussion

The USRUB analysis was applied to the tube and sheet textures, shown in Figures 1 and 2, over a range of strain-rate ratios $-1 \leq \dot{\varepsilon}_\theta / \dot{\varepsilon}_r \leq 1$; $\dot{\varepsilon}_\theta$ is the hoop strain-rate in the tubing and the transverse strain-rate in the sheet; $\dot{\varepsilon}_r$ is the tubing or sheet thickness strain-rate. The tubing axial strain-rate, $\dot{\varepsilon}_z$, (coincident with the rolling direction in the sheet) is connected to the hoop (transverse) and thickness strains through the incompressibility condition ($\dot{\varepsilon}_z + \dot{\varepsilon}_\theta + \dot{\varepsilon}_r = 0$). A wide range of reference shear stress ratios were investigated for stress exponents $1 \leq n \leq 6$. For comparison, the rigid/perfectly plastic ($n = \infty$) analysis was also performed, using linear programming techniques, by equating critical resolved shear-stress and reference shear stress ratios. In this case, absolute strains are imposed, in a strict sense, rather than strain-ratios.

The results are shown in Figures 3-6 for several cases of interest. Stress ratios $\{\sigma_\theta\}/\{\sigma_z\}$ were determined by texture averaging over 486 discrete texture orientations using equation [5]. The isotropic curve, corresponding to a uniform distribution of crystallite orientations, remains invariant regardless of the choice of reference shear stresses. Notably, the influence of the stress exponent, n, was found to be second-order compared to the effect of texture. Over the range $1 \leq n \leq 6$ only slight differences were observed in the predicted curves. While measurable differences are shown (Figures 3 and 4) between the USRUB (n=6) and rigid plastic ($n = \infty$) predictions, the general curvature trends are preserved between the two models. The notation $\tau_B = \tau_C > 10\tau_A$ suggests that the reference shear-stress for slip on prism systems (A) was selected to be at least a factor of ten times smaller than those for basal (B) and pyramidal (C) slip.

Figures 3 and 4 highlight the impact of low reference shear stresses (critical resolved shear stresses) in the prism and basal slip modes. In all cases where pyramidal slip is restricted (large τ_C) substantial deviations from isotropy were evident. Isotropic behavior was observed in both analyses when reference (critical resolved) shear stresses were equal among the slip modes ($\tau_A = \tau_B = \tau_C$). Figures 5 and 6 compare the impact of slip mode reference stress on creep anisotropy for easy slip on prism, basal and pyramidal planes for the USRUB analysis only. Significant differences in creep anisotropy due to varied slip mode activity are predicted.

Figure 6 shows steady state creep data (17) obtained at 673°K on highly cold-worked (CW-SR) Zircaloy 4 tubing. (Texture shown in Figure 1.) Variable biaxial stress states were achieved by combinations of uniaxial tension and internal pressurization. Strains were monitored continuousl using hoop and axial mechanical extensometry. Notably, for this data set the creep anisotropy is best predicted by the condition of relaxed basal slip ($\tau_A = \tau_C > 10\tau_B$). This result suggests that, based on USRUB predictions, basal slip is more important than had previously been expected. In this material, at temperatures below 800°K, prism slip is observed to be dominant (2,3). Since very high levels of cold-work (~80% final reduction) were present in CW-SR tubing, we suggest that hardening of the prism slip mode, due to cold-work, is strongly impacting the creep constitutive behavior at these temperatures.

Figure 7 shows further experimental data obtained by Stehle, et al. (18), on Zircaloy tubing at varying levels of cold-work. Tube species 1,2 are highly cold-worked and exhibit textures similar to CW-SR tubing. Tube species 5, by contrast, received a partial recrystallization anneal after the final reduction pass, although the basal pole texture remains substantially unaltered. Consistent with the prism mode strain hardening hypothesis, tube species 5 shows anisotropic creep behavior nearly constistent with USRUB predictions for relaxed prism slip while tube species 1,2 show behavior similar to the CW-SR tubing. Evidently, the strain hardening constraints present in the cold-worked material have been at least partially removed by the recrystallization anneal.

Since the Taylor-Bishop-Hill analysis allows explicit determination of shear magnitudes on each slip system, it is instructive to compare the average shear magnitude in each slip mode as a function of critical resolved shear-stress ratio and imposed strain ratio. Denoting average shear strains in prism, basal, and pyramidal modes as $\bar{\gamma}_A$, $\bar{\gamma}_B$, and $\bar{\gamma}_C$ respectively, where, for example

$$\bar{\gamma}_A = \iiint \gamma_A \cdot w(\theta,\psi,\phi) \cdot d\theta d\psi d\phi. \qquad [7]$$

Figure 8 shows the ratio of prism to basal shear activity, $\bar{\gamma}_A/\bar{\gamma}_B$, over a range of strain-ratios for the CW-SR tube texture. Although basal slip is relaxed ($\tau_A=\tau_C>10\tau_B$), the rigid plastic analysis predicts that prism slip still plays an important role over the entire range of strain ratios. The activity of basal slip exceeds that for prism slip only for $\epsilon_\theta/\epsilon_r \lesssim -0.15$. This importance of prism slip, even under the constraint of a large critical resolved shear stress τ_A, can be attributed to a geometric dependence on prism slip to maintain compatibility in the tubing texture. The ratios $\bar{\gamma}_A/\bar{\gamma}_C$ and $\bar{\gamma}_B/\bar{\gamma}_C$ were relatively insensitive over the range investigated. This is consistent with the observation (8) that pyramidal slip (or twinning) is required for an arbitrary strain in hexagonal crytallites.

Acknowledgements

The authors would like to thank J. W. Hutchinson, J. S. Kallend, and K. L. Murty for valuable discussions. This work was sponsored by the Exploratory Research Division of Babcock and Wilcox Company.

References

1. G. W. Groves and A. Kelley, Phil. Mag. 8, 877 (1963).
2. R. E. Reed-Hill, "Deformation Twinning," p. 295, Gordon and Breach, New York (1964).
3. J. L. Martin and R. E. Reed-Hill, Trans. AIME 230, 780 (1964).
4. E. Tenckhoff, Z. Metallke 63, 192 (1972).
5. J. A. Jensen and W. A. Backofen, Can. Met. Quarterly II, 39 (1972).
6. M. F. Ashby and H. J. Frost, "Constitutive Equations in Plasticity," p. 114, MIT Press, Cambridge, MS (1975).
7. J. W. Hutchinson, Proc. R. Soc. Lond. 348A, 101 (1976).
8. J. W. Hutchinson, Met. Trans. 8A, 1465 (1977).
9. G. I. Taylor, J. Inst. Metals, 62, 307 (1938).
10. J. F. W. Bishop and R. Hill, Phil. Mag., 42, 414 and 1298 (1951).
11. R. J. Roe, J. Appl. Phys., 35, 2024 (1965).
12. H. J. Bunge, Z. Metallke, 56, 872 (1965).
13. P. R. Morris and A. J. Heckler, Trans. AIME, 245, 1877 (1969).
14. J. A. Elias and A. J. Heckler, Trans. AIME, 239, 1237 (1967).
15. L. G. Schultz, J. Appl. Phys., 20, 1030 (1949).
16. B. L. Adams, PhD Thesis, The Ohio State University (1979).
17. K. L. Murty, unpublished data.
18. H. Stehle, E. Steinberg and E. Tenckhoff, "Zirconium in the Nuclear Industry," ASTM STP 633, p. 486, 1977.

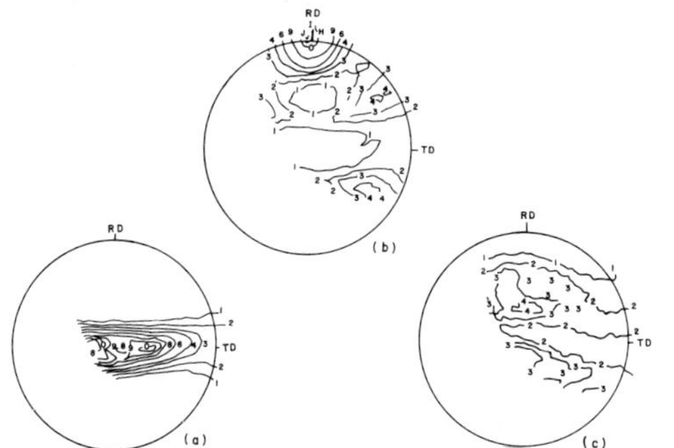

FIGURE 1: Pole-figures for transverse bimodal tube texture. (a) - (0002) basal; (b) - (10$\bar{1}$0) prism; (c) - (10$\bar{1}$2) pyramidal. (RD and TD represent longitudinal and hoop directions in the tubing.)

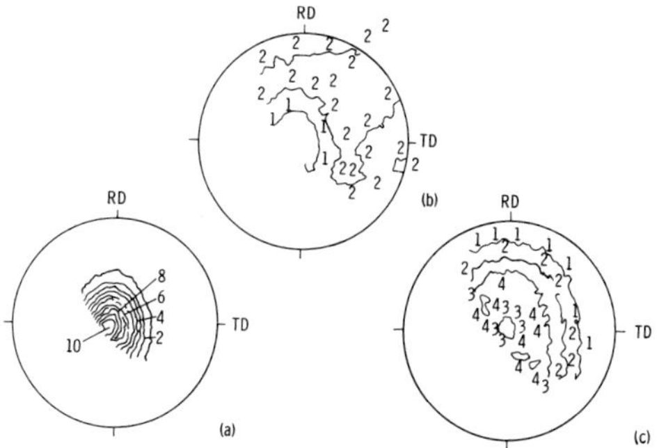

FIGURE 2: Pole-figures for radial sheet texture. (a) - (0002) basal; (b) - (10$\bar{1}$0) prism; (c) - (10$\bar{1}$2) pyramidal. (RD and TD represent rolling and transverse directions in the sheet.)

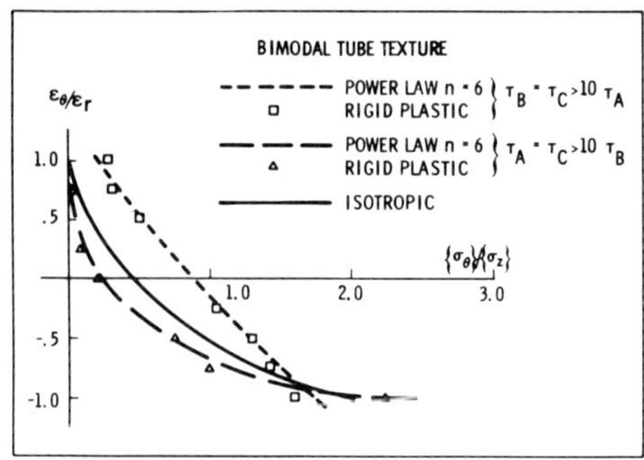

FIGURE 3: Power-law (n=6) and rigid plastic predictions for bimodal tube texture. $\tau_B=\tau_C>10\tau_A$ indicates ease of prism slip. $\tau_A=\tau_C>10\tau_B$ indicates ease of basal slip

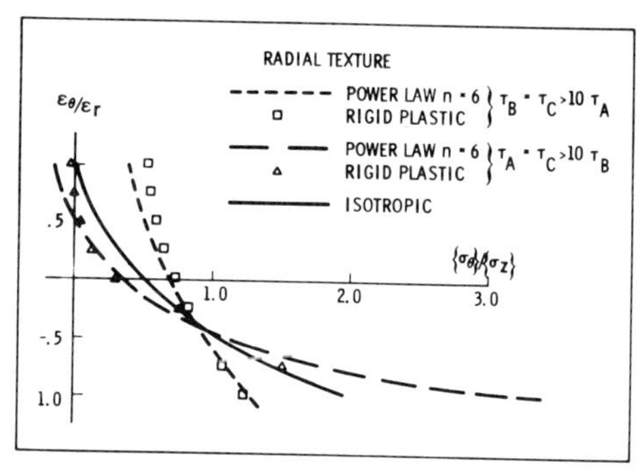

FIGURE 4: Power-law (n=6) and rigid plastic predictions for the radial sheet texture. $\tau_B=\tau_C>10\tau_A$ indicates ease of prism slip. $\tau_A=\tau_C>10\tau_B$ indicates ease of basal slip.

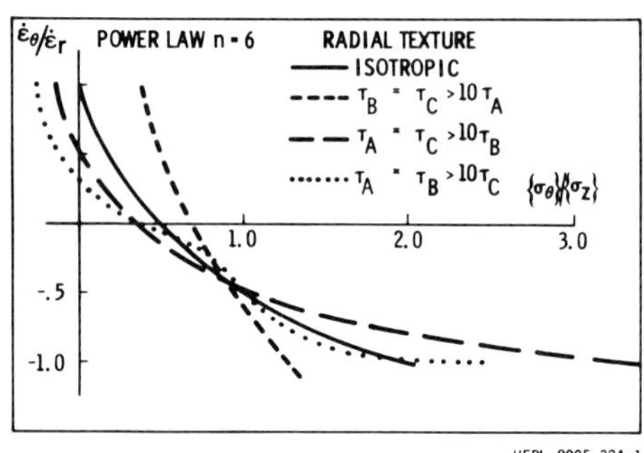

FIGURE 5: Power-law (n=6) predictions for the radial sheet texture. ($\tau_B=\tau_C>10\tau_A$), ($\tau_A=\tau_C>10\tau_B$), and ($\tau_A=\tau_B>10\tau_C$) indicate easy slip on prism, basal and pyramidal slip systems respectively.

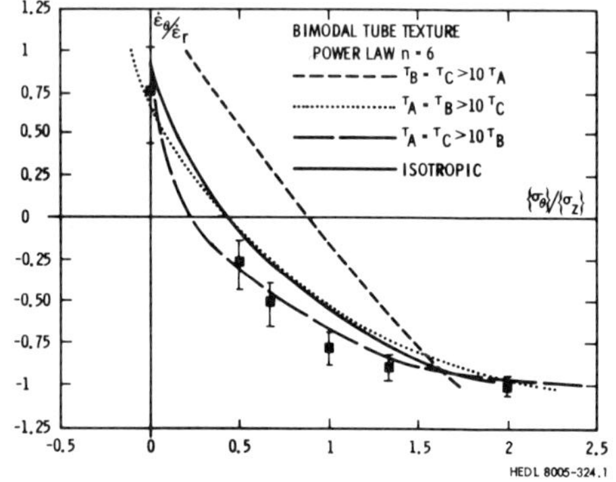

FIGURE 6: Power-law (n=6) predictions for the bimodal tube texture. ($\tau_B=\tau_C>10\tau_A$), ($\tau_A=\tau_C>10\tau_B$), and ($\tau_A=\tau_B>10\tau_C$) indicate easy slip on prism, basal and pyramidal slip systems respectively. Steady-state creep data are show for CW-SR Zircaloy 4 tubing (texture shown in Figure 1) at 673°K.

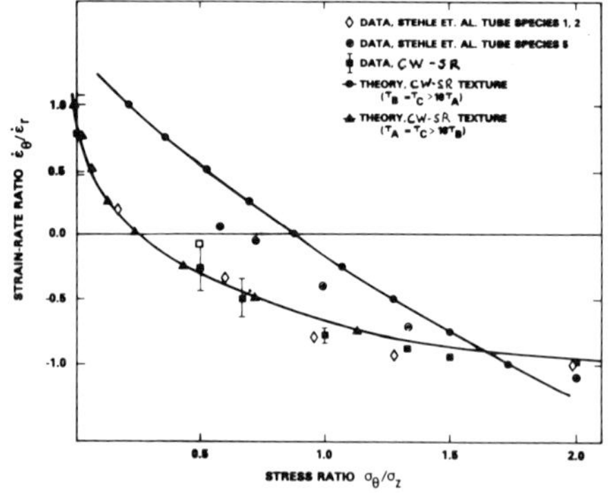

FIGURE 7: Comparison of CW-SR and Stehle data (17) for steady-state creep at 673°K, with power-law (n=6) predictions for easy prism slip ($\tau_B=\tau_C>10\tau_A$) and easy basal slip ($\tau_A=\tau_C>10\tau_B$).

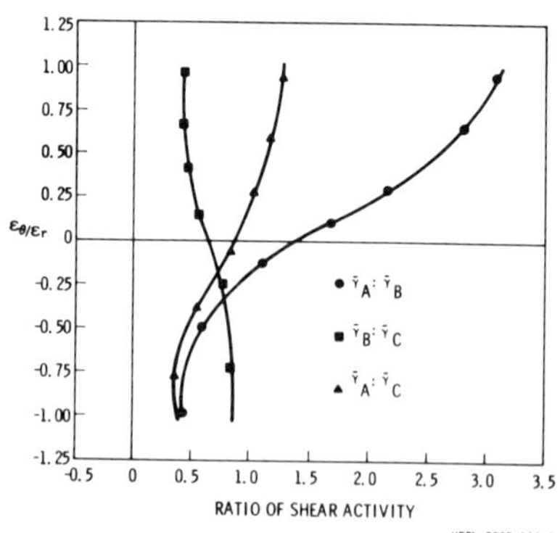

FIGURE 8: Functional dependence of shear acti- vity, calculated using Taylor-Bishop-Hill rigi plastic model, for prism, $\bar{\gamma}_A$, basal, $\bar{\gamma}_B$, and pyramidal, $\bar{\gamma}_C$, shears in bimodal tube texture.

PHENOMENOLOGICAL PLASTICITY AND DISLOCATION MODELS OF DEFORMATION

C. S. Hartley
Department of Materials Science & Engineering
State University of New York at Stony Brook
Stony Brook, New York 11794

and

M. A. Eisenberg
Department of Engineering Sciences
University of Florida
Gainesville, Florida 32611

Abstract

Two particular decompositions of the total distortion rate into elastic and inelastic components provide the bases for development of parallel theories of plasticity utilizing the thermodynamic properties of a continuum on one hand and properties of a continuum containing dislocations on the other. The equivalence of the two viewpoints manifests itself through the introduction of a term coupling the elastic distortion rate and the plastic distortion rate. This term arises naturally from the introduction into the Helmholz function of coupling coefficients related to the internal stress field of the current dislocation distribution.

Introduction

Although dislocations form the physical basis for macroscopically observable plastic deformation of crystalline materials, mathematical theories of continuum plasticity rarely include explicit reference to dislocation mechanics. This circumstance does not diminish the utility of continuum theories as tools for the solution of a wide class of problems which consider complex thermomechanical deformation behavior in engineering structural applications. However, it does obscure the underlying correspondences which must exist between dislocation behavior and phenomenological material parameters appearing in continuum theories. Advances in both the understanding of inelastic behavior of materials and the application of dislocation theoretic concepts to the solution of engineering problems require the development of a common point of view for the two approaches.

Research on various aspects of this problem has been conducted by scientists in a variety of disciplines for over twenty years. Eisenberg's (1) use of continuum dislocation concepts (2-5) to construct a continuum theory of plasticity that parallels Green and Naghdi's (6) rate-independent theory illustrates an effort to develop a connection between continuum plasticity and dislocation mechanics. Recently Eisenberg, Lee and Phillips (7) explored the restrictions inherent to the assumption of insensitivity of the elastic relations to inelastic deformation in the context of a theory of rate-dependent thermoplasticity. These authors showed that there exist alternate decompositions of the distortion tensor into elastic and inelastic components, the ambiguity arising from the structure of the Helmholz free energy. This paper interprets the decompositions in terms of a plastic component of distortion and an elastic component containing contributions from the stress fields arising from internal and external sources.

Decomposition of Total Distortion Rate

The total distortion rate, $\dot{\beta}^T_{ij}(\bar{r})$, denotes the gradient of the continuous velocity field, $\dot{u}^T_i(\bar{r})$, at a field point designated by the position vector, $\bar{r}$*. When $\dot{\beta}^T_{ij}(\bar{r})$ consists of elastic and inelastic components there exists a convenient decomposition which emphasizes the contribution of dislocation motion to the deformation,

*A superscript dot refers to differentiation with respect to time.

$$\dot{\beta}^T_{ij}(\bar{r}) = \dot{\beta}^L_{ij}(\bar{r}) + \dot{\beta}^D_{ij}(\bar{r}) \tag{1}$$

where the superscripts L and D refer to lattice and dislocation, respectively. A generalized form of Hooke's law relates the symmetric part of the lattice distortion rate, the lattice strain rate, $\dot{\varepsilon}^L_{ij}(\bar{r})$, to the local stress rate, $\dot{\sigma}^L_{k\ell}(\bar{r})$, at the field point. The local stress field consists of components due to externally applied loads, $\sigma^E_{ij}(\bar{r})$, and to internal sources of stress, $\sigma^I_{ij}(\bar{r})$, so that

$$\sigma^L_{ij}(\bar{r}) = \sigma^E_{ij}(\bar{r}) + \sigma^I_{ij}(\bar{r}) \tag{2}$$

relates the components*.

The rate of work done by the local stresses, per unit volume, during deformation**,

$$\sigma^L_{ij}\dot{\varepsilon}^T_{ij} = \sigma^L_{ij}\dot{\varepsilon}^L_{ij} + \sigma^L_{ij}\dot{\varepsilon}^D_{ij}, \tag{3}$$

contains a non-dissipative component, $\sigma^L_{ij}\dot{\varepsilon}^L_{ij}$ and a dissipative component, $\sigma^L_{ij}\dot{\varepsilon}^D_{ij}$. Subsequent developments will reveal how the dissipative term depends explicitly on the motion of dislocations participating in the deformation process.

Since applications of classical plasticity theory generally take the form of problems for which loads and displacements on external surfaces specify the boundary conditions, stress components used to develop constitutive theories arise only from external sources. Material parameters appearing in constitutive relations subsume the stress fields arising from internal sources. The decomposition of total distortion rate appropriate to this point of view takes the form

$$\dot{\beta}^T_{ij}(\bar{r}) = \dot{\beta}^E_{ij}(\bar{r}) + \dot{\beta}^P_{ij}(\bar{r}). \tag{4}$$

In this case the generalized form of Hooke's law connects the symmetric part of $\dot{\beta}^E_{ij}(\bar{r})$ to $\dot{\sigma}^E_{ij}(\bar{r})$.

Eisenberg, Lee and Phillips (7) developed a plasticity theory based on the existence of the functions

$$\Psi = \Psi(\varepsilon^E_{ij}, \varepsilon^P_{ij}, k^{(t)}, \theta) \tag{5a}$$

$$s = s(\varepsilon^E_{ij}, \varepsilon^P_{ij}, k^{(t)}, \theta) \tag{5b}$$

$$q_k = q_k(\varepsilon^E_{ij}, \varepsilon^P_{ij}, k^{(t)}, \theta, \theta_{,k}) \tag{5c}$$

$$F = F(\sigma^L_{ij}, \theta, \varepsilon^P_{ij}, k^{(t)}) \tag{5d}$$

$$\kappa = \kappa(\theta, \varepsilon^P_{ij}, k^{(t)}) \tag{5e}$$

relating the specific Helmholz free energy, Ψ, the specific entropy, s, and the outward heat flux vector, q_k, to the thermodynamic temperature, θ, the spatial gradient of temperature, $\theta_{,k}$, the stress tensor, σ^L_{ij} and components ε^E_{ij} and ε^P_{ij} of the total strain tensor, ε^T_{ij}. The parameters $k^{(t)}$, t = 1,2 . . . N, represent a finite set of measures of the effect of prior deformation history on the material response which obey evolution-type constitutive relations of the form

$$\dot{k}^{(t)} = g^{(t)}(\sigma^L_{ij}, \varepsilon^P_{ij}, k^{(t)}, \theta, \dot{\varepsilon}^P_{ij}), \tag{6}$$

homogeneous of degree one in $\dot{\varepsilon}^P_{ij}$. Under these assumptions F and κ remain invariant under all purely elastic deformation processes. The functions F and κ define the elastic domain, $F \leq \kappa$, and the current boundary of the elastic domain, $F = \kappa$, called the flow surface. A purely elastic process corresponds to a state of stress that satisfies the inequality or

*In a deformed material which does not exhibit yielding during the removal of external loads, stress fields due to internal sources produce no resultant forces or moments on the surface of the body.

**In the subsequent discussion all components of the field quantities and other quantities defined in terms of them pertain to the general field point.

$$F = \kappa; \left(\frac{\partial f}{\partial \sigma^L_{k\ell}}\right) \dot{\sigma}^L_{k\ell} + \left(\frac{\partial f}{\partial \theta}\right) \dot{\theta} \leq 0 \tag{7}$$

Changes in the state of stress that cause the loading function, F, to remain on the boundary or in the interior of the current flow surface correspond to neutral loading or unloading, respectively, and result in purely elastic processes. Components of the total strain rate tensor obey the conditions that $\dot{\varepsilon}^P_{ij} = 0$ and $\dot{\varepsilon}^E_{ij}$ is a linear function of $\dot{\sigma}_{ij}$ and $\dot{\theta}$ for purely elastic processes.

The definitions given in equation 5 and the Clausius-Duhem inequality lead to the relations

$$s = -\frac{\partial \Psi}{\partial \theta}, \tag{8}$$

and

$$\sigma^L_{k\ell} = \rho \frac{\partial \Psi}{\partial \varepsilon^E_{k\ell}} \tag{9}$$

with ρ as the mass density of the material. Requiring that the elastic incremental stress-strain-temperature relation be insensitive to the history of inelastic deformation restricts the Helmholz function to the form (8)

$$\Psi = \Psi'(\varepsilon^E_{mn}, \theta) + \Psi''(\varepsilon^P_{mn}, k^{(t)}) + \varepsilon^E_{k\ell}\phi_{k\ell}(\varepsilon^P_{mn}, k^{(t)}) - \left[\theta - \theta_a(\varepsilon^P_{mn}, k^{(t)})\right] s''(\varepsilon^P_{mn}, k^{(t)}), \tag{10}$$

whence follow the incremental relations

$$\dot{s} = -\left(\frac{\partial^2 \Psi'}{\partial \theta \partial \varepsilon^E_{mn}}\right)\dot{\varepsilon}^E_{mn} - \left(\frac{\partial^2 \Psi}{\partial \theta^2}\right)\dot{\theta} + \left(\frac{\delta s''}{\delta \varepsilon^P_{mn}}\right)\dot{\varepsilon}^P_{mn}, \tag{11}$$

and

$$\dot{\sigma}^L_{k\ell} = \rho \left[\left(\frac{\partial^2 \Psi'}{\partial \varepsilon^E_{k\ell} \partial \varepsilon^E_{mn}}\right)\dot{\varepsilon}^E_{mn} + \left(\frac{\partial^2 \Psi'}{\partial \varepsilon^E_{k\ell} \partial \theta}\right)\dot{\theta} + \left(\frac{\delta \phi_{k\ell}}{\delta \varepsilon^P_{mn}}\right)\dot{\varepsilon}^P_{mn}\right], \tag{12}$$

employing the operator

$$\frac{\delta}{\delta \varepsilon^P_{mn}} = \frac{\partial}{\partial \varepsilon^P_{mn}} + \sum_{t=1}^{N} \frac{\partial}{\partial k^{(t)}} \left(\frac{\partial g^{(t)}}{\partial \dot{\varepsilon}^P_{mn}}\right). \tag{12}$$

Assuming that Hooke's Law with temperature-dependent elastic moduli, $C_{k\ell mn}(\theta)$, governs the elastic response leads to the expressions

$$\rho \frac{\partial^2 \Psi}{\partial \varepsilon^E_{k\ell} \partial \varepsilon^E_{mn}} = C_{k\ell mn}(\theta) \tag{14}$$

and

$$\rho \frac{\partial^2 \Psi}{\partial \varepsilon^E_{k\ell} \partial \theta} = \eta_{k\ell}(\theta) \tag{15}$$

for the elastic moduli and the thermoelastic coupling coefficients, $\eta_{k\ell}(\theta)$. Defining the compliance tensor, $S_{k\ell mn}$

$$S_{k\ell mn} C_{k\ell pq} = 1/2 \left(\delta_{mp}\delta_{nq} + \delta_{mq}\delta_{np}\right) \tag{16}$$

permits the inversion of equation (12) to obtain

$$\dot{\varepsilon}^{E}_{pq} = S_{pqk\ell}\dot{\sigma}^{L}_{k\ell} - S_{pqk\ell}\eta_{k\ell}\dot{\theta} - \rho S_{pqk\ell}\left(\frac{\delta\phi_{k\ell}}{\delta\varepsilon^{P}_{mn}}\right)\dot{\varepsilon}^{P}_{mn}, \tag{17}$$

where the use of rates emphasizes the incremental nature of the constitutive relation. Inspection of equation 17 reveals that the strain rate $\dot{\varepsilon}^{E}_{mn}$ and the stress rate $\dot{\sigma}^{L}_{k\ell}$ obey Hooke's Law during unloading or neutral loading, both corresponding to $\dot{\varepsilon}^{P}_{mn} = 0$, but not during loading.

The decomposition of distortion rates and stress components given in equations 1, 2 and 4 lead to the relations

$$\dot{\varepsilon}^{L}_{pq} = S_{pqk\ell}\,\dot{\sigma}^{L}_{k\ell} \tag{18}$$

and

$$\dot{\varepsilon}^{D}_{pq} = \left[\delta_{pm}\delta_{qn} - \rho S_{pqk\ell}\left(\frac{\delta\phi_{k\ell}}{\delta\varepsilon^{P}_{mn}}\right)\right]\dot{\varepsilon}^{P}_{mn} \tag{19}$$

so that

$$\dot{\varepsilon}^{E}_{mn} = \dot{\varepsilon}^{L}_{mn} - \rho S_{mnk\ell}\left(\frac{\delta\phi_{k\ell}}{\delta\varepsilon^{P}_{pq}}\right)\dot{\varepsilon}^{P}_{pq} \tag{20}$$

and

$$\dot{\varepsilon}^{P}_{mn} = \dot{\varepsilon}^{D}_{mn} + \rho S_{mnk\ell}\left(\frac{\delta\phi_{k\ell}}{\delta\varepsilon^{P}_{pq}}\right)\dot{\varepsilon}^{P}_{pq}. \tag{21}$$

Equation 20 suggests the decomposition

$$\dot{\varepsilon}^{E}_{mn} = \dot{\varepsilon}^{L}_{mn} - \dot{\varepsilon}^{I}_{mn} \tag{22}$$

where

$$\dot{\varepsilon}^{I}_{mn} = \rho S_{mnk\ell}\left(\frac{\delta\phi_{k\ell}}{\delta\varepsilon^{P}_{pq}}\right)\dot{\varepsilon}^{P}_{pq} \tag{23}$$

represents the component of the lattice strain rate due to internal sources. Then by equation 16

$$\dot{\sigma}^{I}_{ij} = \rho\left(\frac{\delta\phi_{ij}}{\delta\varepsilon^{P}_{pq}}\right)\dot{\varepsilon}^{P}_{pq} = C_{ijmn}\dot{\varepsilon}^{I}_{mn} \tag{24}$$

represents the component of the local stress rate due to internal sources. Using equations 2, 23 and 24, equation 17 yields

$$\dot{\varepsilon}^{E}_{pq} = S_{pqk\ell}(\dot{\sigma}^{E}_{k\ell} - \eta_{k\ell}\dot{\theta}). \tag{25}$$

showing that Hooke's Law connects $\dot{\varepsilon}^{E}_{pq}$ and the stress rate due to external sources.

An elastic material possesses a homogeneous, stress-free natural state for which there exists a finite neighborhood characterized by the existence of a one-to-one correspondence between the Eulerian stress tensor and the Almansi strain tensor. A Green-elastic or hyperelastic material possesses a scalar specific strain energy function, which depends on the strain tensor formed with respect to the natural state, such that the rate of change of the specific strain energy function equals the rate of work done by the stresses. Hypoelastic materials possess a homogenous linear relationship between the components of the stress rates and the distortion rates (9).

Equation 17 illustrates that hyperelastic behavior characterizes the relation between $\dot{\varepsilon}^E_{mn}$ and $\sigma^L_{k\ell}$, while equation 25 demonstrates that hypoelastic behavior describes the connection between $\dot{\varepsilon}^E_{mn}$ and $\dot{\sigma}^E_{k\ell}$. However the nature of experimental measurements of stress on bulk material renders this distinction purely formal. Since the resultant forces and moments due to $\sigma^L_{k\ell}$ vanish on the external surfaces, no distinction is possible between $\sigma^L_{k\ell}$ and $\sigma^E_{k\ell}$ on the basis of measurements of forces and moments on the external surfaces; both fields possess the same resultants.

Calculation of the coupling coefficients, $\phi_{k\ell}$, follows from the identification of $\dot{\varepsilon}^D_{ij}$ with the symmetric part of the distortion rate due to dislocation motion,

$$\dot{\beta}^D_{ij} = \sum_n b^{(n)}_i \dot{A}^{(n)}_j , \tag{26}$$

where $\dot{A}^{(n)}_j$ represents the rate at which dislocations having Burgers vector, $b^{(n)}_i$, sweep out area by motion on the nth slip system, and the sum extends over all slip systems active at the point under consideration.* The component of the lattice strain rate at $\bar{r}$ due to a flux of sources distributed throughout the body follows from

$$\dot{\varepsilon}^I_{ij}(\bar{r}) = \int_R C_{mnk\ell}\, G_{k(i,j)\ell}(\bar{r}-\bar{r}')\,\dot{\varepsilon}^D_{mn}(\bar{r}'), \tag{27}$$

with $G_{k(i,j)\ell}(\bar{r}-\bar{r}')$ identified as the components, symmetric with respect to i and j, of the second derivatives of the Green's function appropriate to the material and boundary conditions.

Equating the expressions for $\dot{\varepsilon}^I_{mn}$ given in equations 23 and 27 gives

$$\rho S_{mnk\ell}\left(\frac{\delta\phi_{k\ell}(\bar{r})}{\delta\varepsilon^P_{pq}(\bar{r})}\right)\left(\dot{\varepsilon}^D_{pq}(\bar{r}) + \dot{\varepsilon}^I_{pq}(\bar{r})\right) = C_{ijk\ell}\int_R G_{k(m,n)\ell}(\bar{r}-\bar{r}')\,\dot{\varepsilon}^D_{ij}(\bar{r}')\,dv' \tag{28}$$

where the decomposition

$$\dot{\varepsilon}^P_{pq}(\bar{r}) = \dot{\varepsilon}^D_{pq}(\bar{r}) + \dot{\varepsilon}^I_{pq}(\bar{r}) \tag{29}$$

appears on the left hand side. Contracting both sides of equation 28 with C_{rsmn} and using equation 27 leads to

$$\int_R \left[\delta_{ip}\delta_{jq}K_{rspq}(\bar{r}')\,\delta(\bar{r}-\bar{r}') + K_{rspq}(\bar{r})\,C_{ijk\ell}G_{kp,q\ell}(\bar{r}-\bar{r}')\right]\dot{\varepsilon}^D_{ij}(\bar{r}')\,dv' =$$

$$\int_R C_{rspq}C_{ijk\ell}\,G_{kp,q\ell}(r-r')\,\dot{\varepsilon}^D_{ij}(\bar{r}')\,dv' \tag{30}$$

where

$$K_{rspq}(\bar{r}) = \rho\left[\frac{\delta\phi_{rs}(\bar{r})}{\delta\varepsilon^P_{pq}(\bar{r})}\right]^S \tag{31}$$

with S designating the components symmetric with respect to the indices r and s. Since equation 30 must hold for arbitrary $\dot{\varepsilon}^D_{ij}$,

$$K_{rspq}(\bar{r}')\,\delta_{ip}\delta_{jq}\,\delta(\bar{r}-\bar{r}') + \left[K_{rspq}(\bar{r}) - C_{rspq}\right]F_{pqij}(\bar{r}-\bar{r}') = 0. \tag{32}$$

where

$$F_{pqij}(\bar{r}-\bar{r}') = C_{ijk\ell}G_{kp,q\ell}(\bar{r}-\bar{r}') \tag{33}$$

---

*The discussion applies only to conservative motion of dislocations.

represents one of the projection operators defined by Simmons and Bullough (10). Integrating equation 32 with respect to the source point over the entire region yields

$$K_{rspq}(\bar{r}) \left[\delta_{ip}\delta_{jq} + F_{pqij}(\bar{r})\right] = C_{rspq}F_{pqij}(\bar{r}) \tag{34}$$

with

$$F_{pqij}(\bar{r}) = \int_R F_{pqij}(\bar{r}-\bar{r}')\, dv'. \tag{35}$$

Defining $M_{uvij}(\bar{r})$ by

$$M_{uvij}(\bar{r}) \left[\delta_{ip}\delta_{jq} + F_{pqij}(\bar{r})\right] = 1/2(\delta_{up}\delta_{vq} + \delta_{uq}\delta_{vp}) \tag{36}$$

permits an explicit solution for $K_{rspq}(\bar{r})$

$$K_{rsuv}(\bar{r}) = C_{rspq}F_{pqij}(\bar{r})M_{uvij}(\bar{r}). \tag{37}$$

A suitable integration of equation 31 gives

$$\rho\phi_{rs} = K_{rspq}\varepsilon^{P}_{pq} \tag{38}$$

whence

$$\varepsilon^{E}_{rs}\phi_{rs} = \frac{1}{\rho} K_{rspq}\varepsilon^{E}_{rs}\varepsilon^{P}_{pq} \tag{39}$$

represents the contribution to the Helmholz function due to the coupling of elastic and inelastic strain components.

Thermodynamic Considerations

The preceding interpretation of the strain rate components shows that the decomposition of the work done by external forces, equation 3, represents a statement of the first law of thermodynamics for deformation of a body containing moving dislocations. The rate of increase in internal energy due to a change in lattice strain arises from the term $\sigma^{L}_{ij}\,\dot{\varepsilon}^{L}_{ij} = \dot{U}$. Since dislocation motion is a purely dissipative process, the term $\sigma^{L}_{ij}\,\dot{\varepsilon}^{D}_{ij} = -Q$ represents the heat output of the system.* The rate at which the stresses do work on the system represents the power input, $P = \sigma^{L}_{ij}\,\dot{\varepsilon}^{T}_{ij}$. Conservation of energy requires that (11)

$$\dot{U} = P + Q, \tag{40}$$

neglecting changes in the total energy due to changes in kinetic energy.

Identification of the inelastic distortion rate with the dislocation distortion rate permits the representation of the dissipative term in two equivalent forms, one of which emphasizes the role of dislocation motion. By equation 26

$$\dot{\beta}^{D}_{ij} = \sum_n b^{(n)}_i \dot{A}^{(n)}_j = \sum b^{(n)}_i\, \varepsilon_{jk\ell}\rho^{(n)}_k v^{(n)}_\ell \tag{41}$$

where $\rho^{(n)}_k$ represents the projection on the x_k axis of the total length of dislocations per unit volume moving in the nth slip system and $v^{(n)}_\ell$ represents the component in the x_ℓ direction of a weighted average velocity of dislocations moving on the nth slip system. The dissipation term becomes

$$-Q = \sigma^{L}_{ij}\,\dot{\beta}^{D}_{ij} = \sum_n \varepsilon_{jk\ell}\sigma^{L}_{ij}b^{(n)}_i\rho^{(n)}_k v^{(n)}_\ell = \sum_n f^{(n)}_\ell v^{(n)}_\ell \tag{42}$$

with

$$f^{(n)}_\ell = \varepsilon_{jk\ell}\,\sigma^{L}_{ij}b^{(n)}_i v^{(n)}_\ell \tag{43}$$

*The minus sign arises from the convention that heat input is positive.

defined as a thermodynamic body force density due to the Peach-Koehler force on dislocations moving on the $n^{\underline{th}}$ slip system. The duality of the force-flux pairs $(\sigma^L_{ij}, \dot{\beta}^D_{ij})$ and $(f^{(n)}_\ell, v^{(n)}_\ell)$ permits the formulation of constitutive relations in either formalism. Notice that no such duality exists between σ^E_{ij} and $\dot{\beta}^P_{ij}$.

The approach adopted for the preceding development employs an incremental, field theoretic formalism. Solution techniques for problems involving real structures must specify boundary conditions appropriate to the situation analyzed. These can affect significantly the values of the $K_{ijk\ell}$ as these arise from an integration of the second derivatives of the Green's function over the volume of the deforming region. Distortions of a global body, corresponding to measurements of shape changes made on the external surfaces, represent volume averages of the field in the interior. In particular the plastic distortion obtained by measuring the change in external shape of a body includes a compatible component due to dislocations which have passed out of the material. This component of the dislocation distortion does not affect the current internal state of a deforming material, which depends only on the instantaneous dislocation content, and does not appear explicitly in the preceding development.

Conclusions

The connection between elastic constitutive relations relating material behavior to external loads and local material behavior to the local state of stress arises from the component of the local stress field due to internal sources. The dependence of elastic constitutive relations on loading conditions illustrates the differences which can arise as a result of the two levels of description. Interatomic force laws normally used in the description of crystal behavior dictate a linear relation between the local incremental lattice strain and the local state of stress due to internal and external sources. In contrast constitutive equations relating the local stress field to those components of the local strain field due to external forces depend on loading conditions as shown by equation 17.

Definition of a convenient natural state from which to measure material response to stresses due to internal and external sources complicates the comparison of these viewpoints. On the microscopic scale only a material containing no internal sources of stress can exist in the natural state. In the presence of internal stresses, linear response of the material to external loads can occur for loading states lying within or on the current flow surface, however the material does not return to a stress-free natural state upon removal of external loads. Furthermore, if loading occurs, followed by unloading, the state to which the material returns upon the removal of external loads does not generally correspond to that in which it existed before loading because of the possibility of changing the internal stress distribution as a result of loading. Consequently constitutive laws relating material response to external loads are not elastic in the same sense as that which characterizes local elastic behavior at points in the interior unless no internal stress field exists. Lack of a unique natural state from which to measure the response to external loads of material containing internal sources of stress renders the use of the term "elastic" non-rigorous in this context.

Interpretation of inelastic distortion rates in terms of dislocation motion permits the construction of parallel constitutive theories utilizing microscopic or macroscopic descriptions of inelastic bheavior. This circumstance provides a framework for the use of dislocation models to interpret phenomenological parameters associated with formulations based on continuum plasticity. Successful application of the incremental field-theoretic formalism to the solution of problems involving finite bodies relies on the careful formulation of appropriate boundary conditions.

Acknowledgements

The authors wish to acknowledge helpful discussions with R. Bullough and L. E. Malvern.

References

1. M. A. Eisenberg, Int. J. Eng. Sci., 8, 261 (1970).
2. T. Mura, Int. J. Eng. Sci. 1, 371 (1963).
3. E. Kröner, "Kontinuumstheorie der Versetzungen und Eigenspannungen", Springer (1958).
4. E. Kroner and G. Rieder, Z. Phys. 145, 424 (1956).
5. F. R. N. Nabarro, "Theory of Crystal Dislocations", Oxford University Press (1967).
6. A. E. Green and P. M. Naghdi, Arch. Ration, Mech. Analysis, 18, 251 (1965).

7. M. A. Eisenberg, C. W. Lee and A. Phillips, Int. J. Solids Structures, 13, pp. 1239-1255 (1977).
8. W. E. Dalby, Phil Trans. Royal Soc. London, A22, 117 (1921).
9. Y. C. Fung, "Foundations of Solid Mechanics", Prentice-Hall, Inc. Englewood Cliffs, N.Y. 445 (1965).
10. J. A. Simmons and R. Bullough, "Fundamental Aspects of Dislocation Theory", NBS Special Publication 317, Volume I, National Bureau of Standards, Washington, D.C., 89, (1970)
11. L. E. Malvern, "Introduction to the Mechanics of a Continuous Medium", Prentice-Hall, Inc. Englewood Cliffs, N.Y., 338-340, (1969).

A THEORY ON THE SHAPE OF SATURATED HYSTERESIS LOOPS*

D. J. Quesnel and J. C. Tsou

Materials Science Program,
Department of Mechanical and Aerospace Sciences,
University of Rochester,
Rochester, New York 14627 USA

Abstract

A model is proposed for the distribution of mesh lengths available for reversed deformation based on the structural parameter R, the statistical fraction of dislocation-obstacle interactions that allows reverse motion of the dislocation on stress reduction. This model is applied to the case of cyclically stable fatigue where well defined cell structures are present. It is shown that the average spherical cell size determined from quantitative electron metallography should be directly proportional to the critical mesh length and hence, inversely proportional to the saturation flow stress at constant cell wall structure. It is also shown that at a given critical mesh length and hence flow stress there is a unique stabilizing cell size which will allow continued cyclically stable deformation. Experimental measurements of the average cell size as a function of critical mesh length taken from the literature are used to estimate the value of the structural parameter R during cyclically stable fatigue as R=0.052. Normalized reversed stress-strain curves are calculated for a range of R and comparisons are made with experimentally obtained results. It is concluded that when the role of thermal activation and internal stresses are included in the present model, the calculated reverse stress-strain behavior will closely approximate the experimental results.

Introduction

The inverse linear relation between the flow stress and the largest dislocation mesh length, or critical mesh length as it is often called, is well documented. This relationship has been used to measure the local stress after deformation (1). An inverse linear relation has also been observed between the saturation flow stress obtained during moderate to large amplitude fully reversed plastic strain controlled fatigue deformation and the average dislocation cell diameter obtained by electron metallographic techniques (2-5). This relationship has been used to estimate the local stress amplitude near the tips of cracks (6). Clearly, a parametric linear relation exists between the critical mesh length and the average cell size. To extend this statement to say that a parametric relationship exists between the average mesh length and the average cell size requires an examination of the distribution of mesh lengths and the distribution of cell sizes. Figure 11 of reference 1 shows two of the few measured distributions of dislocation mesh lengths present in the literature for an FCC metal after fully reversed fatigue deformation (1,7). Examination of these distributions show that any relation between the mean (linearly averaged) mesh length and the critical mesh length will depend on the functional form of the distribution of mesh lengths if the details of the distribution

* This work supported by the National Science Foundation under Grant No. DMR-7922028.

depend on the critical mesh length. In other words, the proportionality constant between the average and critical mesh lengths depends on the critical mesh length. This is more clearly illustrated in Fig. 8b of reference 5, which shows the average cell size as a function of the average mesh length for cyclically saturated cell structures in copper (5). Figure 8a from reference 5 shows that the distribution of dislocation cell sizes also depends on the flow stress and hence critical mesh length, perhaps with the same functional form and critical mesh length dependence as the mesh length distribution. The nonlinearity of the average cell size plotted against the average mesh length as illustrated in Fig. 8b of Ref. 5, which is apparent despite the similarity in functional form of the mesh length and cell size distributions, appears to preclude this possibility. Statistically larger sample sizes are needed to reflect the distributions of the parent populations with sufficient detail to explain this effect. This brief review serves to illustrate that knowledge of the detailed functional form of the mesh length distribution plays a key role in understanding the interrelationships between deformation parameters at both the microscopic and macroscopic scale.

In the present work we shall investigate the implications of a particular functional form for the dislocation mesh length distribution. Since we are concerned with fully reversed fatigue deformation, the mesh length distribution which is most useful is the distribution of mesh lengths available for reverse deformation which arise from a single cell shuttling (8) event. We propose

$$F(\ell,\ell_c,R) = \sum_{n=1}^{\infty} (1-R)R^{n-1}\left[\frac{\ell^{n-1}}{\ell_c^n(n-1)!}\sum_{j=0}^{n}\frac{n!}{j!(n-j)!}(-1)^j H(\ell-j\ell_c)\right] \quad (1)$$

where: $F(\ell,\ell_c,R)$ is the fraction of meshes with length between ℓ and $\ell+d\ell$ available for reversed deformation excluding meshes already present in the cell wall.

H is the Heavyside function,

ℓ is the mesh length,

ℓ_c is the critical mesh length,

R is the repulsive fraction (see below), and,

n and j are summation indices.

In brief, its development is based on the realization that all dislocation interactions with the cell wall do not anchor the dislocation to the cell wall. Statistically, a certain fraction, R, of the interactions serves only to stop the dislocation from moving into the cell wall, yet allows the dislocation to, at least locally, reverse its path when the stress is reduced. The simplest example of such an interaction is the repulsion between dislocations of the same Burgers vector on the same slip plane. In that sense, we call R the repulsive fraction of those dislocation interactions with the cell wall which stopped the dislocation during the previous half cycle. Consequently, R is a microstructural parameter characterizing the interaction of the cell wall with incoming dislocations. The functional form for the distribution of mesh lengths available for reverse deformation represented by equation 1 follows immediately from this line of thought. A graphical representation of the distribution function is shown for R=0.5 in Fig. 1. The functional shape has been verified with statistical calculations based on random interactions of dislocations with cell walls for a range of repulsive fractions from 0.02 to 0.97. The detailed development of equation 1 is lengthy and is published elsewhere (9).

Numerical Considerations

Numerical evaluation of equation 1, which we shall subsequently refer to as F, requires that the summation over n be truncated at finite n. Physically, the largest mesh length cannot be greater than the cell diameter, d, so that if cells are present the summation should be truncated at ℓ=d. This implies that F should also be a function of d but in a very complicated way since d is itself distributed in a manner which is mathematically unknown at present. Both of these truncations will reduce the upper limit for the integration of F so that the integrated fraction of meshes will be less than unity. Figure 2 presents the $\int_0^\infty F d\ell$ where the summation over n in the numerical evaluation of F has been carried out up to and including n=10 as a function of R. All numerical calculations involving F reported in this work are based on n=10 as the upper limit of the summation. From Fig. 2 we conclude that the truncated terms do not play a significant role in determining the distribution for R≤0.5. As we shall see, physically meaningful values of R for structures containing well defined dislocation cells are typically less than this value.

Fig. 3 presents the mean mesh length as a function of R. Notice that the ordinate is in units of ℓ_c. Consequently at fixed R the average mesh length will be directly proportional to ℓ_c. The proportionality constant between $\bar{\ell}$ and ℓ_c is unity at R=0.5 and decreases smoothly to 0.5 at R=0. It can be easily shown that the function

$$\bar{\ell} = \frac{\ell_c}{2(1-R)} \tag{2}$$

represents the result if all terms were retained in the evaluation of F. The smooth line represents this function while the result based on our computer evaluation of F retaining 10 terms is indicated by the open circles. The error is 0.5% for R=0.5 and decreases with decreasing R.

It can also be shown from equation 1 that the fraction of meshes, f, with mesh lengths greater than ℓ_c is given by

$$f = \left(1-\frac{1}{R}\right) e^R + \frac{1}{R} = \int_{\ell_c}^{\infty} F d\ell. \tag{3}$$

f is shown as a function of R by the solid line in Fig. 4. The results obtained by integration of our computer evaluation of F, as a function of R, are shown as open circles in Fig. 4. The error caused by not retaining the higher order terms in F is less than 0.5% at R=0.5 and again decreases with decreasing R.

These numerical considerations allow us to state that our computer evaluations of F retaining the first 10 terms adequately represent the behavior of F for R≤0.5. We shall now explore the implications of the proposed dislocation mesh length distribution function with respect to the inter-relationships between deformation parameters as a function of the structural parameter R.

Phenomenological Considerations

The structure parameter, R, in principle applies to any obstacle array where repulsive dislocation interactions are possible including dislocation cell structures, persistent slip band (PSB) structures, precipitation hardened structures, dispersion hardened structures, etc. Consequently, equations 1-3 should also apply to these structures. To calculate the stress strain behavior during reversed deformation up to the stress applied in the forward direction from F we need to assign a mean free path to the dislocation meshes as they become critical with increasing reversed stress. The incremental flow stress is taken from its inverse dependence on the current critical mesh length (7) as longer meshes are exhausted. The strain is taken as the sum of the strain per dislocation bow out event times the number of bow out events that occur during a given stress increment. The strain per bow out event is a structural parameter, S_b, representing the expected value of the mean free path (area) of a bowing segment coupled with its strain generating efficiency. While in general S_b will change during deformation, at least conceptually the reverse strain as a function of decreasing instantaneous critical mesh length, ℓ_c, will be given by

$$\varepsilon_i(\ell_{ic}) = \int_{\infty}^{\ell_{ic}} FS_b d\ell \tag{4}$$

where: i is an increment index,
ℓ_{ic} is the instantaneous critical mesh length,
ε_i is the strain after the ith increment,

and

$$\sigma_i = \frac{\alpha Gb}{\ell_{ic}} \tag{5}$$

represents the parametric dependence of the reversed stress on the instantaneous critical mesh length. α is a constant which contains the orientation factor relating σ and τ and the dislocation character factor (7) for the $\tau = Gb/\ell_c$ relation. The expectation value for α is about 4 for statistically mixed character dislocation segments bowing out in a non-textured polycrystal and about 3(2.2) for primary edge(screw) character dislocations bowing out in single crystals oriented for single slip. As we shall see, the exact value of α will not affect the shape of normalized stress strain curves computed parametrically from equations 4 and 5.

Cyclic Stability Considerations

For ease in calculating S_b, we shall restrict the remainder of this paper to the case of materials which deform to produce well defined equiaxed cellular dislocation structures such as pure Al and Cu single and polycrystals which are deforming under strong multiple slip conditions. In such a structure, the expectation value of the mean free path (area) for cell shuttling is related to the average cell size in a manner which depends on the distribution of cell sizes and shapes. Considering only spherical cell shapes, this implies that there is a particular cell size whose average strain per cell shuttling event is identical to the expectation value of the mean free path(area) for the entire cell size distribution including the effects of off diameter shuttling. We call this the effective cell size, d_e. Under the restriction that the cell size distribution is symmetric about its mean, d_e is the cell size obtained from linear intercept analysis, ie. 1.5 times the mean intercept distance (10). Using the 3.06 Taylor factor (11), we can then estimate S_b for cell structures in random polycrystals as

$$S_b = \frac{3.06\sqrt{3}}{\sqrt{2}} \frac{b}{d_e} \tag{6}$$

Since S_b depends only on the effective cell size which is statistically invariant during cyclically stable deformation we now proceed to examine the dependence of cell size on the structural parameter R.

When a single mesh bows out, it is eventually stopped by the cell wall after traversing the cell interior. This dislocation has a total length which is statistically $\pi/4$ times the cell perimeter at its diametral plane. Consequently, the total length of dislocation produced by N bow out events is just $N\pi^2 d_e/4$ less the mesh length which bowed out. Assuming the bowing mesh length is small in comparison to the cell perimeter, $N\pi^2 d_e/4$ represents the total mesh length available for reverse deformation. The total number of meshes produced is then just $N\pi^2 d_e/4\bar{\ell}$. If the material is to deform in a cyclically stable manner with respect to stress and cell size at constant plastic strain amplitude, then the number of meshes produced which can bow out under a reversed stress equal in magnitude to the forward stress must be identically equal to the number of meshes which bowed out under the forward stress. Meshes already in the cell wall would have bowed out on the previous cycle. Mathematically, this is stated as

$$\frac{N\pi^2 d_e}{4\bar{\ell}} f = N \tag{7}$$

or using the exact expressions for $\bar{\ell}$ and f in terms of ℓ_c and R derived earlier

$$d_e = \frac{2\ell_c}{\pi^2} \frac{R}{(1-R)-(1-R)^2 e^R} \tag{8}$$

This is the cell size which will allow cyclically stable behavior. It is important to note that it is independent of N, the number of bow out events per average cell, and hence is not a direct function of the strain amplitude. The stabilizing cell size is directly proportional to the critical mesh length showing that the cell size is inversely related to the saturated flow stress at constant R. In other words, if the cell walls have the same character of interaction with the incoming dislocations, R, then the cell size for cyclically stable deformation is inversely proportional to the flow stress since it is directly proportional to the critical mesh length. The dependence of the stabilizing cell size on the structural parameter R is shown in Fig. 5. The solid line represents equation 8 while the open circles represent equation 7 when f and $\bar{\ell}$ are taken from the numerical calculation of F. As we have seen, the error is ignorably small for $R \leq 0.5$ and decreases with decreasing R.

Recent experimental determinations of $\bar{d}$ from diameter measurements (5) as a function of saturated flow stress during cyclically stable deformation of copper suggest that $\bar{d}/\ell_c$ is 5.44. Since this technique underestimates (10) the spherical cell size by 2/3, we estimate d_e/ℓ_c from their measurements as 8.14. Earlier measurements of d_e have been compiled after correction for the quantitative metallographic technique employed in the original works by Staker and Holt (4). Employing Mughrabi's (7) definition of $\ell_{FR}=\ell_c$ for mixed character bow out events, d_e/ℓ_c from this compilation is 8.07. Consequently the experimental value for d_e/ℓ_c is approximately 8. Under the present model then we find the corresponding value of R, the structure parameter representing dislocation cell wall interaction, to be approximately R=0.052. This suggests that in a cell wall environment an incoming dislocation has a one in twenty chance of finding a repulsive interaction. Considering that if all Burgers vectors on all slip systems are equally represented in the cell wall, the chance of a direct repulsive interaction is one in twenty-four, this value of R is not unreasonable. A complete analysis of the physical significance of R is, however, beyond the scope of this paper.

Normalized Stress Strain Curves

Fig. 6 presents the normalized stress strain curves for reversed deformation calculated by employing equations 4 and 5 with the numerical evaluation of F for several values of R. The curve with R=0.05 has been emphasized since this corresponds most closely to the physically significant value of R obtained above. The curves have been normalized so that the constant α and the value of S_b do not affect their shapes. Since these curves assume the cell size is constant during each curve, they should be compared with the shape of a "leg" of a normalized hysteresis loop. Such a comparison, however, shows that the work hardening rate of the calculated stress strain behavior is too large in comparison with normalized experimental results. Fig. 7 presents such a comparison with data from polycrystalline aluminum (99.998% Al, 175 micron grain size, tested in ultra-low dew point nitrogen) after plastic strain controlled low cycle fatigue into the saturated stage at $\Delta\varepsilon_p/2=0.005$ (12). Since both F and S_b are implicitly dependent on the distribution of internal stresses and deformation temperature, the results of such a comparison are not surprising. While variations in S_b will have not effect on the curves in normalized form, variations in F will have a significant effect. It is easy to visualize that the effect of including the distribution of internal stresses and the probability of thermal activation of sub-critical meshes will serve to smear average F over a range of mesh lengths. This will, in general, decrease the work hardening rate at larger stresses making the model more realistic. Such an analysis is in progress (13). For the present paper, the stress-strain results are those that would be obtained at 0°K neglecting the effects of the distribution of internal stresses. The results regarding the relationship between d_e and ℓ_c and its R dependence are not subject to these restrictions.

Acknowledgment

This work is supported by the National Science Foundation under Grant No. DMR-7922028. The helpful hint from Prof. A. Simon with respect to LaPlace transforms is gratefully acknowledged. The patient encouragement shown by our spouses is most appreciated.

References

1. H. Mughrabi, J. Microsc. Spectrosc. Electron, vol. 1, 571, (1976).
2. J. E. Pratt, J. of Materials, vol. 1, No. 1, 77, (1966).
3. J. E. Pratt, Acta Met., vol. 15, 319, (1967).
4. M. R. Staker and D. L. Holt, Acta Met., vol. 20, 569, (1972).
5. F. Prinz and A. S. Argon, Phys. Stat. Sol. (a), vol. 57, 741, (1980).
6. D. L. Davidson, J. Lankford, T. Yokobori, and K. Sato, Int. J. of Fract., vol. 12, No. 4, 579, (1976).
7. H. Mughrabi, in Constituative Equations in Plasticity, ed. by A. S. Argon, MIT Press, Cambridge, Mass. and London, England, 199, (1975).
8. C. E. Feltner and C. Laird, Acta Met., vol. 15, 1633, (1967).
9. D. J. Quesnel and J. C. Tsou, to be published.
10. Metals Handbook, 8th ed., vol. 8, ASM, Metals Park, Ohio, 47, (1973).
11. W. F. Hostford, Jr. and W. A. Backofen, in Fundamentals of Deformation Processing, Syracuse Univ. Press, 259, (1964).
12. J. T. Tsou and D. J. Quesnel, unpublished data.
13. D. J. Quesnel and J. C. Tsou, to be published.

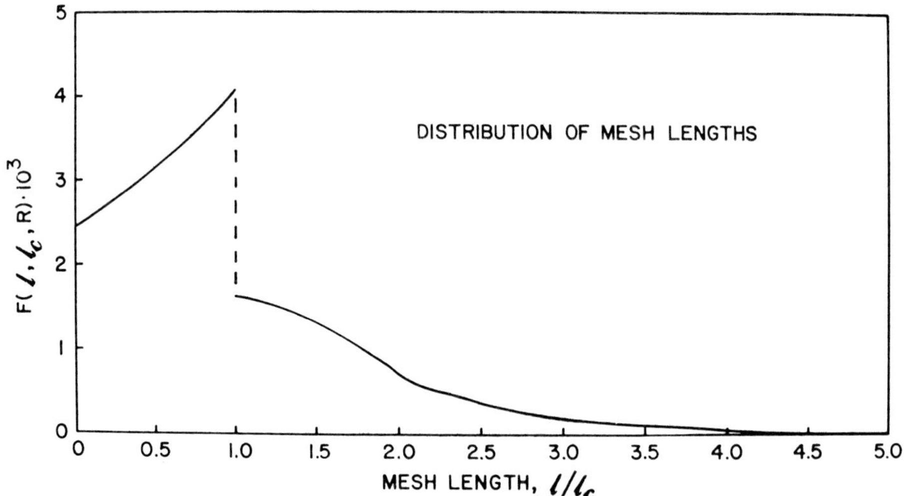

Fig. 1: Numerical Evaluation of equation 1, the distribution of dislocation mesh lengths available for reversed deformation, evaluated for the structural parameter R=0.5, retaining the first 10 terms in the series. Units correspond to the probability of finding a mesh in an interval between ℓ/ℓ_c and $\ell/\ell_c + 0.05$.

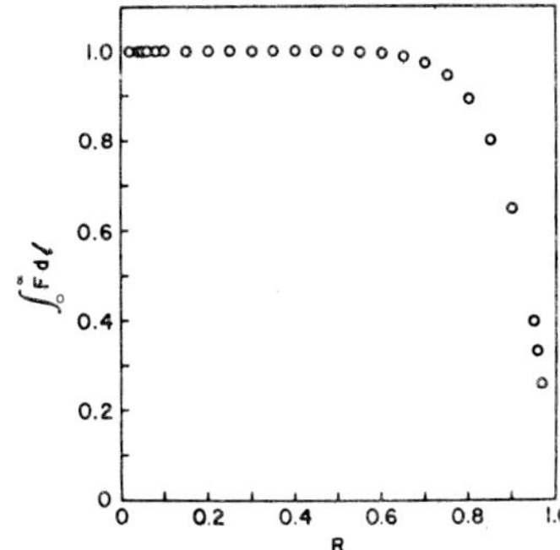

Fig. 2: Numerical integration of F as a function of R retaining the first ten terms of the infinite series. More terms must be retained for R greater than 0.6 since the integral should be identically unity.

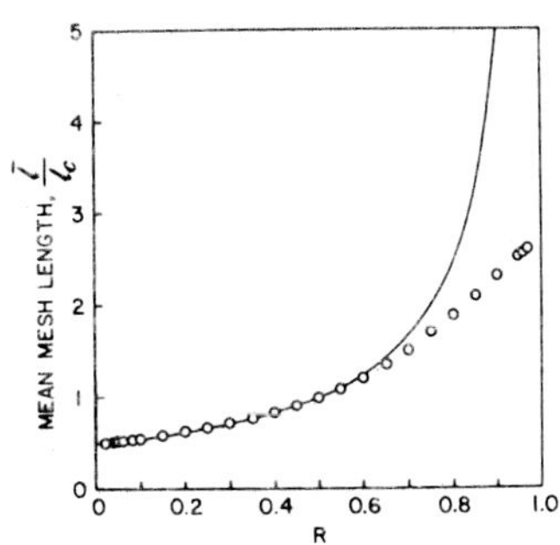

Fig. 3: Mean mesh length determined from the proposed distribution as a function of R. Open circles represent numerical calculations while the solid line represents equation 2.

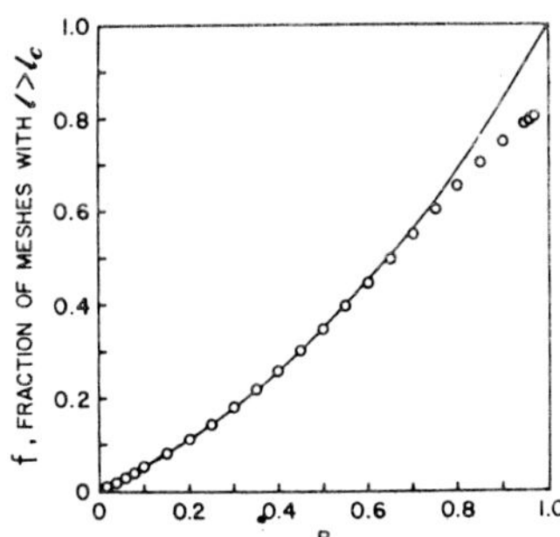

Fig. 4: Fraction of meshes with ℓ greater than ℓ_c as a function of R.

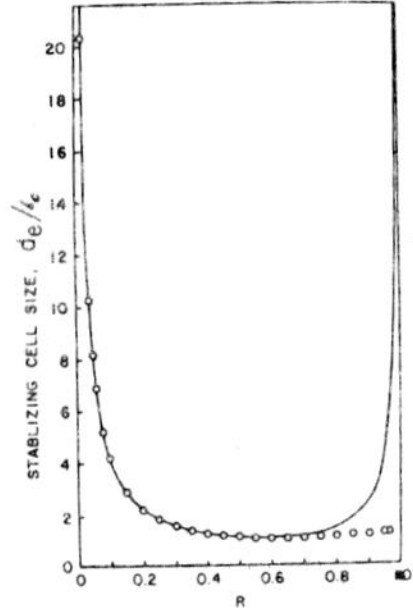

Fig. 5: Stabilizing cell size vs the structure parameter, R. d_e is proportional to ℓ_c .

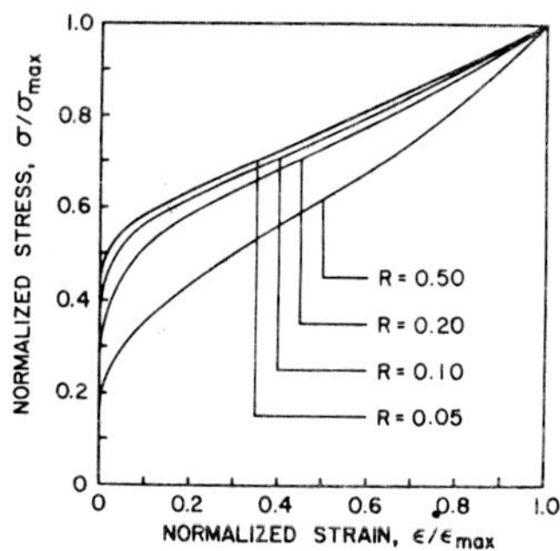

Fig. 6: Normalized stress strain curves obtained by numerical integration of the numerical evaluation of equation 1.

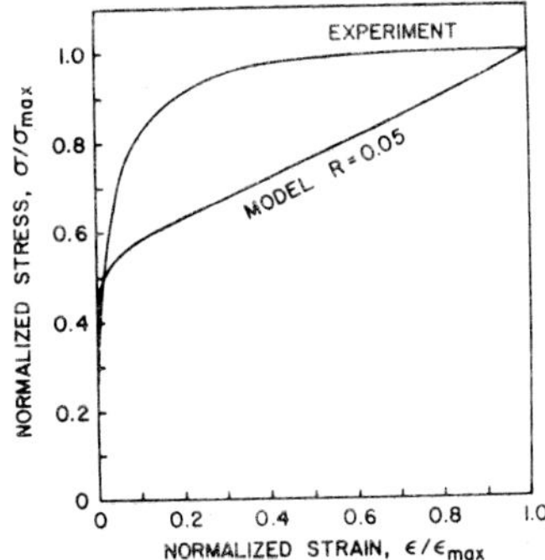

Fig. 7: Comparison of cyclically saturated experimental data for 99.998% pure Aluminum.

VOID STRENGTHENING[†]

R. O. Scattergood* and D. J. Bacon**

*Materials Science Division, Argonne National Laboratory, Argonne, IL 60439
**Department of Metallurgy and Materials Science, University of Liverpool, Liverpool, UK

Abstract: The strengthening effect due to voids is calculated using anisotropic elasticity theory.

Introduction

In recent years, the advent of high-speed digital computers combined with certain improvements in the continuum linear-elastic dislocation theory has made it possible to model complex problems in which dislocations play a governing role. In particular, dislocation strengthening mechanisms can be treated such that elastic anisotropy and dislocation self interaction can be taken into account. The success in dealing with these problems lies, first, in the application of the geometric-theorem approach for curvilinear dislocation stress fields in infinite homogeneous linear-elastic anisotropic media. This approach requires basic input data which need be calculated only for infinite straight dislocations. Second, the self interaction, due to the action of the dislocation's own stress field upon itself, can be included by using a proper definition of the dislocation self force. This has recently been obtained from rigorous energy minimization arguments. The line tension approximation precludes a proper treatment of self interaction for the sake of mathematical simplification. The only arbitrary parameter in the calculations then becomes the dislocation core cutoff r_0 required for linear-elastic strain energy evaluation. The continuum theory background and a treatment of self force have been summarized in the review paper by Barnett (1), and in more detail elsewhere(2).

In the results reported here, a representative example of strengthening mechanism calculations will be given. The critical stress necessary for a dislocation to cut through a row of equi-spaced voids was determined using anisotropic elasticity and the complete dislocation self force. The void-dislocation boundary conditions were treated by an approximate method. Based on the results, a simple correlation describing the critical stress was developed, and elastic anisotropy was included by means of effective values for the shear modulus and Poisson ratio. Dislocation self interaction is manifest as a critical line-tension cusp angle and a logarithmic cutoff parameter for the line energy. Voids are strong obstacles to dislocation motion, with critical stress values about 70% or more of the Orowan stress.

Outline of the Calculations

The computational scheme is similar to the one developed earlier by Scattergood and Bacon (3) for their analysis of the Orowan mechanism. An infinite equi-spaced row of circular voids of diameter D and spacing L shown in Fig. 1a was used, and the Burgers vector b is oriented by the angle β shown. The glide self force acting at a point P on the planar cutting dislocation was determined using anisotropic elasticity by summing contributions from the circular arc segment through P and the two adjacent points, and from straight segments joining all remaining points in a finite-segment representation of the line shape. Appropriate finite-segment self force formulae were given earlier (3), and the underlying rationale for the self force method has recently been discussed (4). The geometric-theorem approach permits a ready evaluation of the necessary dislocation stresses, with basic input data being the Fourier coefficients in Fourier expansions of the orientation factors and their derivatives in the line energy for infinite straight dislocations. These were obtained from separate calculations reported elsewhere (5). Elastic anisotropy is handled as an extremely simple feature of the calculations through the Fourier coefficients, with isotropy available for simple checks. The dislocation self interaction is properly taken into account, and, for example, the dipole-like attraction between the two dislocation branches incident on a given void will be manifest in the critical

†Work supported by U. S. Department of Energy.

stress values. For the actual calculations, it was found that two dislocation loops on either side of the one shown in Fig. 1a were sufficient to account for all interactions on the center loop.

The image-stress interaction between the void and dislocation is difficult to treat on an exact basis. As an approximation, we have used Lothe's (6) solution for a straight dislocation terminating at the free surface of an elastically anisotropic half-space. Let $E = e \ln(R/r_o)$ denote the energy per unit length of an infinite straight dislocation in an infinite homogeneous linear-elastic anisotropic body where e is the orientation factor and R and r_o are, respectively, the outer and core cutoff parameters. Taking e and its derivative e' with respect to Burgers vector orientation as vectors along and normal to the line direction at its termination on the free surface, respectively, the Lothe solution requires that the in-surface vector components balance. For the case at hand, this means that the dislocation incident on the void surface will be made to have a line direction such that the components of e and e' resolved in the tangent plane at the void surface will balance. As shown in Fig. 1b, the direction of e for the $\beta = 90°$ case will lie on the forward side of the tangent-plane normal n while e for $\beta = 0°$ (dashed) lies on the opposite side. This results because e' acts in effect as a torque tending to rotate the line direction into the lower energy screw configuration, so that it coincides more closely with the direction of b. The deviations from the tangent-plane normal can be as high as 30° for large screw-edge energy differences; in other words, large values of the Poisson ratio ν for isotropic elasticity.

Equilibrium is obtained when the applied force σb balances the self force at all points P along the dislocation line. σ is the applied shear stress in the glide plane resolved along the direction of b. In the calculations, σ is increased incrementally until no further stable equilibrium shapes can be found, and the largest equilibrium σ is the critical stress. Equilibrium shapes were found using a modified version of the finite-segment scaling procedure and curvature relaxation algorithm developed earlier (3). In order to assess the effect of the numerical representation and boundary conditions, systematic changes were made which included varying the number of finite segments on a bowing loop, and the length of the segment incident on the void surface required to lie in the orientation given by the Lothe solution. The critical stress decreased linearly as the incident segment length decreased, and with at least 30 segments on a bowing loop and an incident segment length less than D/10, the addition of more segments and/or extrapolation to zero incident segment length produced less than a 2% change in results. These conditions were maintained for all calculations, and an overall accuracy of 2% was maintained in the force balance and incremental approach to the critical stress.

Results

For each set of conditions, critical stress calculations were made over the range $10 \le D/b \le 1000$ and $100 \le L/b \le 3200$ such that D/L varied systematically from 0.03 to 0.56 for 16 different D, L pairs. A range of Poisson ratio values $0 \le \nu \le 0.4$ was used for the isotropic cases, while the anisotropic cases included Cr, Nb, Ni and Cu. The slip system for Cr and Nb was $(11\bar{2})$ 1/2[111] and for Ni and Cu it was $(1\bar{1}1)$ 1/2[110].

Fig. 2 shows typical equilibrium shapes for Cu, Ni, Nb and Cr (top-to-bottom) with $\beta = 0°$, 45° and 90° (left-to-right) for D/b = 180 and L/b = 1000. The increasing bowouts for each sequence correspond to increasing σ up to the last shape displayed, which is the critical one. The influence of elastic anisotropy and the orientation β are evident in the figures. The shapes for $\beta = 0°$ and 90° are centrosymmetric because of the slip system symmetry. Each sequence of shapes shown required about 15 seconds computing time on an IBM model 195 computer.

Fig. 3 shows typical isotropic critical stress values in units of Gb/L, where G is the shear modulus, plotted against the logarithm of the harmonic mean cutoff distance $\bar{D}/r_o = (D^{-1} + L^{-1})^{-1}/r_o$ for $\nu = 0$ with $\beta = 0°$ or 90° (Fig. 3c), $\nu = 0.4$ with $\beta = 90°$ (Fig. 3b) and $\nu = 0.4$ with $\beta = 0°$ (Fig. 3a). For each case, the four solid lines through the computed points correspond to $D/L \le 0.1$, 0.18, 0.32 and 0.56 (top-to-bottom). The heavy dashed lines are the Orowan stress correlation obtained earlier (3), and all Orowan stress values for the conditions shown would lie on these lines. As D/L increases, the void critical stresses decrease relative to the Orowan stress. For given D/L, as ν increases the differences become smaller for $\beta = 90°$ and larger for $\beta = 0°$. Results for the anisotropic calculations show similar behavior, and may be plotted in an analogous fashion using effective values of G and ν given by

$$Gb^2/4\pi = e_s \quad ; \qquad 1 - \nu = e_s/e_e , \tag{1}$$

where e_s and e_e are the orientation factors for infinite straight screw and edge dislocation energies. Effective values of ν for Cr, Nb, Ni and Cu are, respectively, 0.103, 0.278, 0.362 and 0.431. For D/L less than about 0.2, the void critical stress values are at least 70% of the Orowan stress values for similar conditions, over the range of parameters used here.

Discussion

The primary intent of the analysis of the results will be to develop a simple correlation to describe the calculated values of the critical stress. For simplicity, we restrict the discussion to the symmetric cases $\beta = 0°$ and $90°$. Fig. 3 shows that the Orowan stress correlation (dashed) is unsuitable for voids because of the dependence on D/L and ν. This is not surprising since the cutting dislocation moves around the void periphery, and the effective void size and spacing must differ from D and L for larger D/L (Fig. 1a). Likewise, deviations are expected for increasing ν because the incident line direction increasingly diverges from the tangent plane normal (Fig. 1b).

Fig. 4 shows line tension construction that will be used to develop and rationalize the correlation. E and E' represent the line energy and its derivative with respect to Burgers vector orientation for a dislocation cutting a void. These may be viewed as a tension E and torque E' force acting along and normal to, respectively, the incident line direction at the void surface. The Lothe-solution boundary condition requires a balance of E and E' forces in the tangent plane defined by θ; alternatively, the same requirement follows in a line tension model by assuming a discontinuous change in E such that $E \equiv 0$ within the void. For a centro-symmetric bowing loop, the total applied force in the forward direction $\sigma b\ell$ must balance the components of E and E' resolved in this direction, where ℓ is the spacing between terminations on the voids, as indicated. With ϕ denoting the dislocation cusp angle subtended at the void and the subscripts 0 and 90 the values $\beta = 0°$ and $90°$, the isotropic elastic line energy for the incident line direction is

$$\begin{aligned} E_0 &= e_0 \ln(R/r_o) = (Gb^2(1 - \nu\sin^2\phi)/4\pi(1 - \nu)) \ln(R/r_o) \\ E_{90} &= e_{90} \ln(R/r_o) = (Gb^2(1 - \nu\cos^2\phi)/4\pi(1 - \nu)) \ln(R/r_o) \; . \end{aligned} \tag{2}$$

The void-dislocation boundary conditions give the relations

$$\begin{aligned} \cos\theta &= (e_0\sin\phi + e_0'\cos\phi)/(e_0^2 + e_0'^2)^{1/2} \quad \text{for } \beta = 0, \quad \text{and} \\ \cos\theta &= (e_{90}\sin\phi + e_{90}'\cos\phi)/(e_{90}^2 + e_{90}'^2)^{1/2} \quad \text{for } \beta = 90° \end{aligned} \tag{3}$$

where the primes imply differentiation with respect to ϕ. Finally, the forward force balance gives

$$\begin{aligned} \sigma_0 b\ell &= 2(e_o - e_0'\tan\phi) \cos\phi \ln(R/r_o) \\ \sigma_{90} b\ell &= 2(e_{90} - e_{90}'\tan\phi) \cos\phi \ln(R/r_o) \end{aligned} \tag{4}$$

where the prelogarithmic factors may be written as $2\Gamma \cos\phi$ with Γ being an effective line tension orientation factor. Note that e_0' is a negative quantity in these expressions.

The calculated critical stress values were fitted by least-squares to equations (2) - (4), using the ν and G values defined in equation (1), with the aim of obtaining a correlation in terms of a critical cusp angle ϕ_c and an effective value of R. Among the various fits tried, the best overall correlation emerged as the one shown in Fig. 5. The critical cusp angle can be expressed as a function only of ν for a given β, as is shown for all anisotropic cases (open symbols) and isotropic cases (filled symbols). For each ϕ_c, the comcomitant value of θ_c is obtained from equation (3). The value of the cutoff distance R was determined as a function of D and L, and $R/r_o = 4.7\,\bar{d}/r_o = 4.7(d^{-1} + \ell^{-1})^{-1}/r_o$ where $d = D\cos\theta$ and $\ell = L(1 + D(1 - \cos\theta)/L)$ are effective void sizes and spacings measured between terminations on the void surfaces (Fig. 4). Using the values of ϕ_c given in Fig. 5, the critical stresses obtained from equations (1) - (4) deviate from the calculated critical stresses by at most 5%, with the deviations generally being much less than this. The isotropic results showed somewhat less deviation than the anisotropic results. Equations (1) - (4), when used in conjunction with the ϕ_c values in Fig. 5 and the effective value of R, give a concise and accurate correlation for the void critical stress values. The only arbitrary parameter in this correlation is the core cutoff r_o, which, presumably, is the order of b.

A few remarks are in order with regard to the void strengthening mechanism. The computer calculations indicated that the onset of instability with increasing σ results from the pinching-together of the two dislocation branches incident on a given void. This occurs when the dipole-like attraction between them becomes strong enough to complete the cut around the void periphery. It is not possible to disclose this complex effect in a line tension approximation, and only a proper treatment of the dislocation self interaction will account for the true behavior. The line tension construction shown in Fig. 4 can, however, be used aposteriori to help identify some important features. When a critical value of the cusp angle ϕ is reached, the dislocation configuration becomes unstable because of the dipole attraction. As ν increases with $\beta = 0°$, the void-dislocation boundary condition requires that the incident line direction lie further behind the tangent-plane normal n (Fig. 1b), and larger values of ϕ_c, hence lower critical stress, suffice to move the dislocation to an unstable position on the void surface. For $\beta = 90°$, the line direction lies on the forward side of n, and a decrease in ϕ_c and increase in stress should result for increasing ν. Fig. 3 confirms this behavior, however, it is somewhat less pronounced for $\beta = 90°$; this is reasonable because ϕ_c is already small and with further decrease dipole extension can contribute to instability, as in the Orowan mechanism (3). It appears, therefore, that the image-stress interaction between the void and dislocation, approximated here by the Lothe-solution boundary condition, is a significant factor in determining the effect of ν on the critical stress values.

The effective R already mentioned is also a manifestation of the self interaction, and represents the appropriate cutoff distance in the line energy terms contributing to an effective force on the void. For smaller D/L, R is proportional to the distance between the dislocation branches incident on the void. This again seems reasonable because this appears to be a natural screening distance for the dominant dislocation stresses.

The elastic anisotropy has been incorporated by means of effective ν and G values, which reduce to the familiar ones for isotropy. Although all possible anisotropies have not been examined in this investigation, the cases used were selected to provide a wide range of ν values, and we expect the correlation to hold in general. Values of ν and G for various crystals and slip systems can be obtained from tabulated data (5).

In summary, the work reported here has demonstrated the extent of modeling capabilities that are available within the framework of the continuum linear-elastic dislocation theory. Moreover, we have determined that voids are strong obstacles to dislocation motion and, for D/L ratios of about 0.2 or less (void volume fractions less than about 5%), their critical stress values are at least 70% of the Orowan stress values for similar conditions. Further details and more extensive void calculations will be reported elsewhere (7).

Acknowledgements

Numerous discussions with Dr. U. F. Kocks, Materials Science Division, Argonne National Laboratory, are gratefully acknowledged.

References

1. D. M. Barnett, see contribution to this conference.
2. D. J. Bacon, D. M. Barnett and R. O. Scattergood, Prog. in Mat. Sci., edited by B. Chalmers, J. W. Christain and T. B. Massalski, Pergamon, Oxford 23, 51 (1978).
3. R. O. Scattergood and D. J. Bacon, Phil. Mag. 31, 179 (1975).
4. R. O. Scattergood, submitted for publication.
5. D. J. Bacon and R. O. Scattergood, J. Phys. F: Metal Phys. 4, 2126 (1974).
6. J. Lothe in "Fundamental Aspects of Dislocation Theory", edited by J. A. Simmons, R. deWit and R. Bullough, p. 11, NBS Spec. Pub. 317, 1 (1970).
7. R. O. Scattergood and D. J. Bacon, to be published.

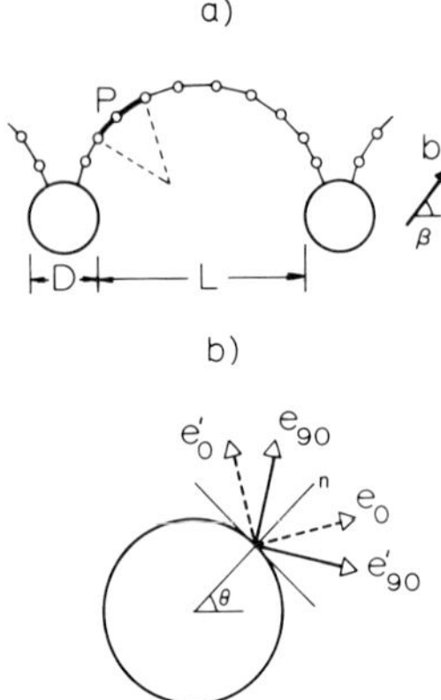

Fig. 1. (a) Single bowing loop in the infinite row of loops. (b) Lothe-solution boundary conditions. e is along the incident line direction and the components of e and e' balance in the tangent plane at θ. Subscripts denote β values.

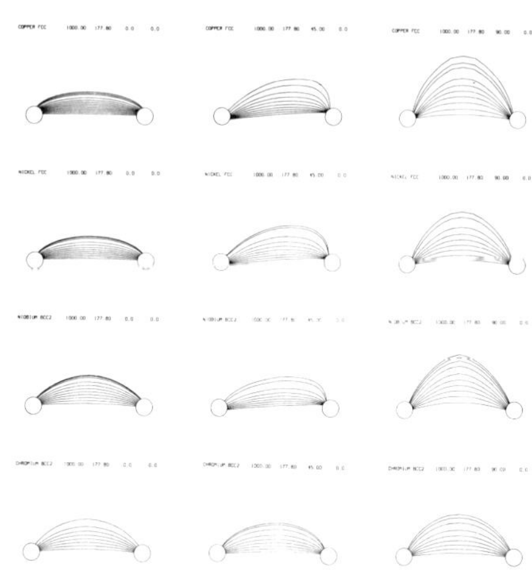

Fig. 2. The equilibrium shapes shown are for single void pairs from the infinite row.

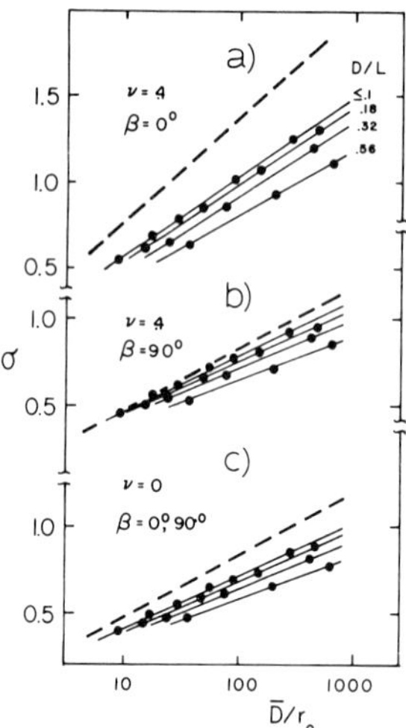

Fig. 3. Critical stress σ in units of Gb/L vs. $\ln(\bar{D}/r_o)$.

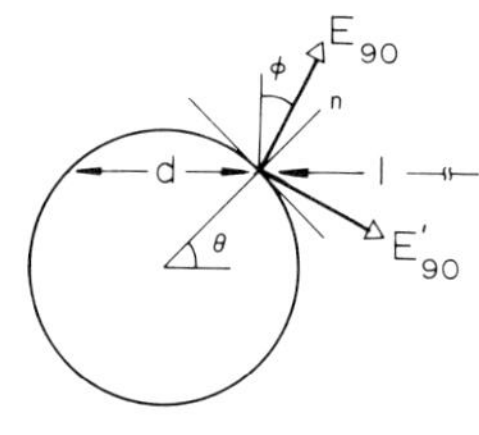

Fig. 4. Line tension construction for $\beta = 90°$. The diagram for $\beta = 0°$ follows from Fig. 1b with E replacing e.

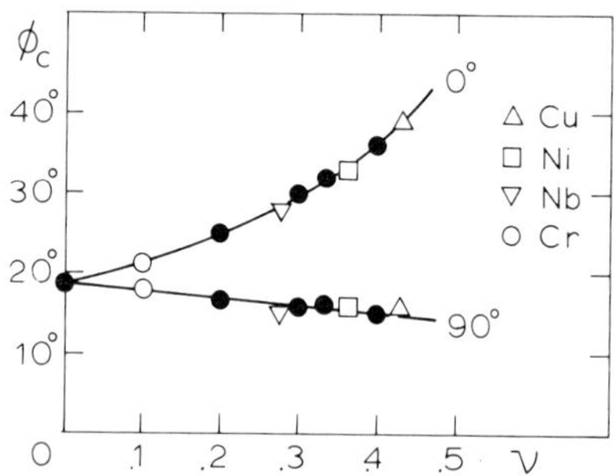

Fig. 5. Fitted values of the critical cusp angle ϕ vs effective ν for the values of β shown. Isotropic results are filled circles while anisotropic results are as shown in the figure.

DISLOCATIONS IN NON-CRYSTALLINE MEDIA?

A. S. Argon
Massachusetts Institute of Technology, Cambridge, Mass. 02139

ABSTRACT

Dislocations are the basic ingredient of crystal plasticity. The corresponding question for non-crystalline matter is to inquire about the mechanism of plasticity in such materials. As in the case of crystal plasticity Bragg's soap bubble model, appropriately generalized has been an invaluable analog medium for the study of plasticity in glassy media. Quantitative dynamic analysis of sheared glassy soap bubble rafts has established that the mechanism of inelastic deformation in this analog medium (and by extension in three-dimensional glassy media) is by the nucleation of a broad distribution of shear transformations occurring in small clusters of bubbles (or atoms), usually centered around sites of large free volume. It is assumed that the distributed nature of the process seen in the disordered bubble raft corresponds to the spectral distributions of transformation free energies that have been measured experimentally in metallic glasses.

On the basis of this understanding it is possible to answer questions on the stability of line dislocations in glassy media as a special case of the dissipation of long range residual stresses. Thus, if a group of dislocations were artificially introduced into a glassy medium it would be stable for long periods of time at very low temperatures but would readily dissipate above 0.5 T_g by a combination of shear transformations and free volume redistribution. Since this dissipation is a rate phenomenon itself it is possible to foresee the role of mobile groups of continuously distributed dislocations in shear fronts in low temperature, high strain rate deformation, where the presence of long lived atomic dilatancy will cause shear localization. The analysis of the kinetics of the deformation process, whether in homogeneous or heterogeneous form, however, is best understood by concentration of attention to the repeated nucleation of shear transformations which is the basic mechanism of deformation.

I. INTRODUCTION

The subject of dislocations in glassy media, whether they can exist or not, and whether the concept is even useful, has been a source of controversy both in its own right and in connection with the mechanism of plastic flow in such materials. The problem is related directly to the difficulty of describing the atomic structure of glasses - even the simplest kinds that can be idealized by a collection of hard spheres. It becomes even more complex in glasses with strong directional bonds such as oxide glasses with space networks and polymers with chain molecules. It is for this reason that lattice defects and their well understood role in diffusion and plastic flow have proved to be so attractive to many investigators to furnish quick explanations to the corresponding phenomena in glassy media. Such attempts, however, have generated as many new questions as the questions that were to be answered. They have, nevertheless, been of great value not only in clarifying where the important differences lie in the structure-property relationships between crystalline and glassy media, but also in the better understanding of the basic properties of vacancies and dislocations as lattice defects. Since vacancies and dislocations find their <u>raison d'être</u> in the phenomena of diffusion and plastic flow of crystalline matter, we find it useful to discuss first the likely mechanism of plastic flow in simple glasses and delay a specific discussion of the question of whether or not line dislocations can exist in glassy media. Although the mechanism of diffusion in simple glasses will also enter our discussion as it is directly tied to the shear localization process, the emphasis will be on the mechanism of shear flow and any statements on diffusion must be taken as not definitive. Finally, we will concentrate our attention to idealized hard sphere glasses that should be good representations of metallic glasses, and avoid additional complexities of the oxide glasses and polymers.

II. PLASTIC FLOW IN HARD SPHERE GLASSES

2.1 The Disordered Soap Bubble Raft as a Quantitative Analog.

Much of the recent stimulus for the modeling of the structure of simple glasses has come from the dense, randomly packed, hard sphere model of liquids investigated by Bernal (1) and the many equivalent models of others constructed as computer exercises of the spatial packing of touching hard spheres. Such models have been invaluable in the characterization of atom positions in simple glasses to explain their radial distribution functions, and to visualize the free volume distribution in them. Since there are no attractive forces between touching hard spheres and since the repulsive forces rise almost discontinuously rapidly, hard sphere models are not suitable to study inelastic deformation mechanisms in glassy media. Computer models (2), based on the use of pseudo-potentials between points, can overcome some of these difficulties but suffer from different artifacts. Unless a mixture of different force laws is used between points or other artificial constraints are introduced, under simulated thermal motion such models are likely to undergo crystallization. In addition, under an applied shear stress these models can only give the stable portions of the cluster relaxation paths while a more complete understanding and modeling of the plastic flow process also requires dealing with the unstable portions of the relaxation paths. The Bragg soap bubble model used very successfully by Bragg and his co-workers to study the motion of dislocations and the structure of low angle tilt boundaries in close packed crystals (3,4), and by some other investigators to study the structure and sliding of large angle boundaries (5,6), has far fewer shortcomings. As was shown by Lomer (7), small soap bubbles of equal size floating on the surface of water interact with each other through capillary attractive forces and through re - pulsive forces governed by their internal pressure. For certain sizes of bubbles (c.a. 1.2 mm diameter), these forces give rise to an inter-bubble potential that resembles closely the shape of interatomic potentials between close packed metal atoms. This, and the fact that all bubble motions are critically damped by viscous forces resulting from their partial immersion into water, make the Bragg bubble model a very useful quantitative analog in which plastic flow can be studied cinematographically. Argon and Kuo (9) have discussed how the use of the soap bubble raft can be generalized to successfully model the structure and properties of glassy media by mechanically mixing two somewhat different sizes of soap bubbles, with an overall average diameter of c.a. 1.2 mm, to preserve the resemblence of the inter-bubble potential to that of close packed metals . Radial distribution functions of bubble centers determined from such two-dimensional bubble rafts resemble very closely the radial distribution functions determined from three-dimensional metallic glasses by X-ray diffraction techniques - provided certain constraints imposed by the two-dimensional nature of the bubble raft are recognized. An immediate benefit of the glassy soap bubble raft is the visualization and measurement of the distribution of free volume in a glass. Defining elements of free volume as the excess interstitial hole size between bubbles over and above the interstitial hole size in a close packed reference bubble raft, Argon and Kuo have obtained the distribution shown in Fig. 1, where ΔA is the excess interstitial area, and a the radius of an average size soap bubble. This distribution which is not directly measurable in a real metallic glass bears a close resemblence to the interstitial hole size distributions obtainable from hard sphere models (10). We take it to be an accurate model of the free volume distribution in an idealized hard sphere glass.

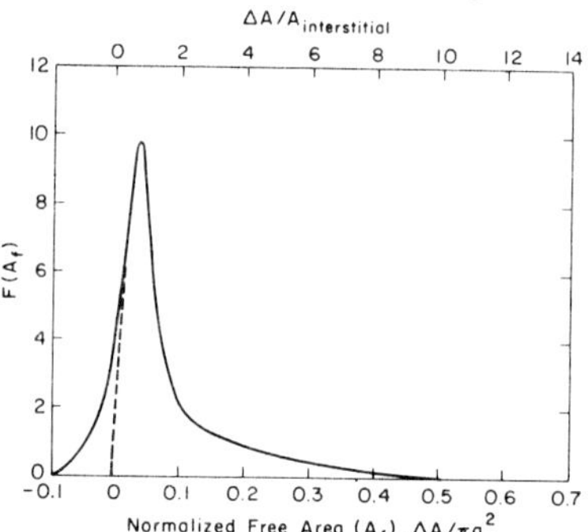

FIG. 1. Distribution of free volume obtained from analysis of a disordered bubble raft with 1.2 mm average bubble diameter. The portion on the left of the dotted line is an artifact of sampling error and should be ignored. (From Argon and Kuo, (9), courtesy of Elsevier, S.A.)

The most useful application of the generalized glassy bubble raft, however, has been in clarifying the nature of inelastic deformation in glassy media. When the relative bubble displacements were analyzed systematically from the bubble rearrangements recorded on motion picture film it became clear that inelastic shear is accomplished by means of shear transformations in small clusters of bubbles, typically five bubble diameters across. These transformations were of a distributed nature and ranged from flat configurations consisting of two nearly close packed short rows of bubbles translating past each other giving rise to a large transformation shear strain in a small volume element to more equiaxed configurations involving more complex internal bubble rearrangements leading to relatively small net shear strains over these larger volume elements. No significant alterations could be found in the bubble rafts after considerable shear, beyond a slight increase in the coordination number of bubbles from 5.346 to 5.385. This indicates that, as expected, a random and disordered structure will transform into another and equivalent random and disordered structure upon shear. Two important aspects of inelastic deformation in simple glasses are also readily determinable from the sheared bubble rafts. First, examinations of the prior state of the bubble raft in the regions that undergo shear transformations have disclosed that those shear transformations that result from the more complex rearrangements of bubbles in the relatively equiaxed regions are initiated from portions of the raft with large free volume, while the more intense transformations that occur from the relative translation of two short close packed rows of bubbles do not require much free volume. Second, a high incidence was found for all shear transformations to be accompanied by significant amounts of readily measurable dilatation.

The shear transformations observed in the disordered bubble raft reported above have also been found in two dimensional computer models of a glass by Maeda and Takeuchi (2) and indicate that even though such computer models might be criticized for tendencies to crystallization, they reflect properly the modes of deformation of the structure.

Thus, if these observations made on the sheared bubble rafts are taken seriously and are extrapolated to real three-dimensional metallic glasses, it must be concluded that inelastic deformation in the latter also is a result of local shear transformations occurring in small atom clusters preferentially around regions of large free volume. Furthermore, the largest dimension of these clusters is roughly five atom diameters across, but both the shapes of the transforming regions as well as their transformation shear strains vary considerably from one element of volume to another. Finally, the shear transformations are usually accompanied by a certain amount of dilatation that is a direct result of the hard sphere nature of the atomic interactions.

2.2 Kinetics of Shear Transformations in Metallic Glasses

The distributed nature of shear transformations in metallic glasses has been measured directly by Argon and Kuo (11) by means of a series of anelastic creep and step recovery experiments at low stress levels where the behavior of a glass is linearly viscoelastic, lending itself readily to operational analysis techniques. A typical result for $Pd_{80}Si_{20}$ is shown in Fig. 2 and indicates that the free energy barriers ΔF to shear transformations in atom clusters of volume Ω_f are distributed over a range of energy values where the maximum energy is more than a factor of two larger than the smallest. It is interesting to note that the shape of this free energy spectrum is a mirror image of the free volume spectrum measured in the generalized Bragg bubble model, and shown in Fig. 1. Based on the analysis of the plastic shear process in the bubble model of a glass that has been outlined in the preceding section, the kinetics of the plastic shear strain rate in a metallic glass can be described by utilizing the well developed technique of transition state theory of thermally activated processes (12). Detailed analysis of the

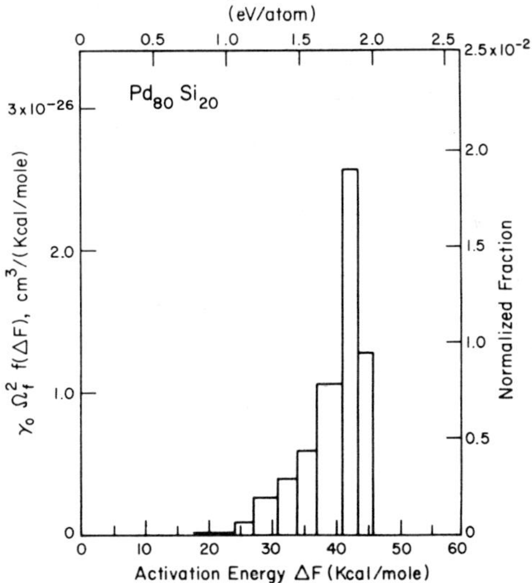

FIG. 2. Histrogram of free energies for inelastic shear transformations in a $Pd_{80}Si_{20}$ glass obtained from the analysis of recovery creep curves (from Argon and Kuo (11), courtesey of North Holland Publishing Co.).

activation configurations of atom clusters under an applied shear stress σ gives a plastic shear strain rate $\dot{\gamma}$ that is (13)

$$\dot{\gamma} = \alpha\gamma_o\nu_G \exp\left(-\frac{\Delta G^*(\sigma,\hat{\tau})}{kT}\right) \quad (1)$$

with the activation free enthalpy ΔG^* being usually of the form (13),

$$\Delta G^* = \hat{\tau}\Omega_f\beta\left(1 - \frac{\sigma}{\hat{\tau}}\right)^2 \quad (2)$$

where α is a constant of order unity, representing the volume fraction of potentially activable units, γ_o is the transformation shear strain in the atom cluster of size Ω_f , ν_G is the normal mode frequency of shear oscillation of the atom cluster in the activation path, $\hat{\tau}$ is the athermal threshold strength in shear of the atom cluster, and β is a numerical constant dependent upon the shape of the transforming region and the magnitude of the transformation shear strain. At low stress levels, where a significant reverse flux of configurations exists against the applied stress, a simplified analysis of the deformation process gives an expression of (13)

$$\dot{\gamma} = \alpha\gamma_o\nu_G \exp\left(-\frac{\Delta F}{kT}\right)\sinh\frac{\sigma\gamma_o\Omega_f}{kT} \quad . \quad (3)$$

where ΔF is a short hand description of a function of $\hat{\tau}$, Ω_f and β ; and represents the effective activation barrier to the shear transformation in the low stress limit where Eqn. (3) leads to linear behavior. As already discussed above, the structural variations in a glass result in a relatively wide distribution of values of barrier heights - with the results for $Pd_{80}Si_{20}$ glass given in Fig. 2 being typical. How the distributed nature of the shear transformations naturally gives rise to viscoelastic behavior at low stress levels over short periods of observation, and how it results in non-recoverable inelastic strains over longer periods of time or under stress levels near the mechanical threshold $\hat{\tau}$ has been discussed in detail elsewhere (13) and can not be repeated here. Let it suffice to state that this representation of the plastic shear process has successfully described the temperature and strain rate dependence of the flow stress in a typical metallic glass such as $Pd_{80}Si_{20}$.

The observation of a dilatation accompanying the shear transformations in the glassy bubble raft has indicated the existence of dilatancy at the atomic level that has been proposed by some investigators in the past as being the cause of shear localization (13-15). It can be expected that the deformation induced excess free volume can decay diffusively at elevated temperatures near T_g but is stable at low temperature. If it is assumed that the result of such retained excess free volume is a reduction in either the free energy barrier ΔF or the threshold plastic resistance $\hat{\tau}$ for further shear transformations in the area already sheared, it is clear that localization of shear deformation will result in those regions in which shear transformations are initiated. How rapidly such shear localization occurs at low temperatures can be calculated by elementary considerations of shear bifurcation based on reasonable assumptions (13). Experimental measurements of Neuhäuser (16) by high speed cinematography are in agreement with these theoretical estimates. It has been this intense shear localization in metallic glasses that has stimulated the discussion of the possible presence of dislocations in glasses.

In summary, we re-state the basic features of the mechanism of plastic deformation in metallic glasses as being a result of local shear transformations in small atom clusters with a relatively wide distribution of free energies of activation that is directly related to the distribution of free volume in the glass. Furthermore, we expect that shear transformations are accompanied by dilatations that can be retained as excess free volume at low temperatures, catalyzing further shear into rapid localization, while at higher temperatures near T_g , where the excess free volume can diffuse away as rapidly as it is produced, no shear localization should occur.

III. DISLOCATIONS IN GLASSY MEDIA

Dislocations have been used often in the past to explain the deformation processes in glassy media. Of these attempts, the more noteworthy ones were those of Gilman (17), Li (18), and Bowden and Raha (19), who have found the familiar concept of a dislocation and its mobility in a crystal useful to explain aspects of the rate dependence of plastic flow, and possible shear localization phenomena in glasses of all types. In most of these discussions, the possible existence and mobility of dislocations was based on analogy of processes in glasses to the more familiar and explained phenomena in crystalline substances. More recently, the possible existence of vacancies and dislocations in glassy media has been considered from the point of view of their stability in a disordered medium. Thus, Spaepen (20) and Bennett et al (21) have shown that vacancies in both glassy, hard sphere models and Lennard-Jones glasses fractionate in the presence of thermal motion and become dispersed as free volume, giving considerable insight into the possible mechanism of diffusion in glasses by thermally activated free volume redistribution. In a similar but more elaborate exercise, Chaudhari et al (22) have considered the stability of both edge and screw dislocations in glassy cylindrical solids. They have stated that after the introduction of such dislocations into the cylinders by the classical cutting and welding steps, relaxation of the glassy solid to re-equilibrate the atoms results in the dispersal of the edge dislocation but not the screw dislocation. They have attributed this unsymmetrical result to the ready dispersal of dilated material by a re-distribution of free volume in the case of the edge dislocation.

We will consider the stability of dislocations in glasses and the possible role of internal stress fields of the type that dislocations represent in the light of the mechanism of plastic flow that we presented in Section II above. In this context, the question can be stated much more broadly and is: "can self-equilibrated, long range internal residual stresses be dissipated by inelastic deformation in glasses in very small volume elements?" Clearly, if small volume elements mean small atom clusters in the immediate surroundings of regions of free volume, then the answer is a qualified "yes". Thus, based on our understanding of the mechanism of plastic deformation as a set of shear transformations having a wide distribution of free energy barriers, we expect that at very low temperatures any dislocation line with a Burgers vector equal to the mean atom spacing in a glassy medium should be stable for very long periods of time. If the athermal shear resistance of the hardest component of the glass is taken as $\hat{\tau}$, we expect that the only effect of relaxation at low temperatures around the dislocation will be the spreading out of the core over a distance $r_o = \mu b/2\pi\hat{\tau}$ as is well known from the classical analysis of Peierls and Nabarro. At higher temperatures, but still below T_g, however, over long periods of time, all elastic shear strains around the dislocation can be relaxed by a multitude of shear transformations occurring in the strain field of the dislocation. This becomes possible because the elastic strain field of a line dislocation is of much longer range than the size of the volume element that can admit an independent shear transformation. The experiments of Argon and Kuo (11) have shown that the distributed nature of processes makes it possible to have significant amounts of inelastic shear relaxation at temperatures above $T_g/2$ at stresses $\sigma/\mu = O(10^{-3})$ over periods of time of the order of 10-100 hours. This indicates that over such time periods, the shear stresses around a dislocation can be relaxed effectively to distances of the order of 100 atom dimensions from the core of a pre-existing dislocation line, whether that be of edge or screw nature. Complete relaxation of stresses to larger distances requires longer periods or higher temperatures. This is independent of the free volume redistribution that can effectively relieve dilational strain fields around edge dislocations at a rate dependent on the kinetics of such redistribution, which is still not clear at this writing. The early phases of this type of dissipation of the strain field around a line dislocation have been directly observed by Cherns, et al (23) in crystalline quartz under electron irradiation in a transmission electron microscope, where the radiation damage due to the high energy electrons occurs preferentially around the highly strained material around the center of the dislocation, producing vitrification, and subsequent dissipation of the strain field. Hence, we conclude that since plastic flow in simple glasses occurs independently over small atom clusters typically having dimensions of only five atom diameters, long range internal stress fields that line dislocations or groups of line dislocations typify can be effectively relaxed at temperatures above $T_g/2$, but that such stress fields are stable over very long periods of time at low temperatures.

The above discussion, however, is not complete in answering the possible role of entities resembling dislocations, and the mobility of shear fronts during high rates of deformation at low temperatures. In fact, both the mechanism of plastic flow by shear transformations with associated dilatancy effects presented above and the stability arguments just discussed lead us to conclude that at low temperatures, where shear localization occurs by dilatancy, deformation in the softened zone is not likely to be by rigid translation of one layer over another, but by repeated initiation of shear patches. At any one stage of the deformation process a shear front, formally describable as a continuous distribution of dislocations, will exist, and can either propagate in a self-similar way or can dissipate by fragmentation of the front. At low temperatures, where the stresses associated with the front will be long lived, the shear stress concentration due to the superposition of several concurrent transformations in the softened region will propagate under stress by ready nucleation of additional shear transformations preferentially in the highly stressed zone in the same plane, ahead of the front—in a way envisioned in the theory of Becker and Orowan (24). Under these conditions at very low temperatures the accentuated stresses of the shear fronts, initiated from surface flaws or other macro inhomogeneities, will overwhelmingly concentrate the nucleation of shear transformations to the periphery of the shear front, to the near complete exclusion of other portions of the material. At higher temperature, however, where excess free volume decays rapidly, there will be no important dilatancy associated with the shear transformations, and shear localization should not occur. Under these conditions, if a group of dislocations were suddenly introduced into the structure under stress, it too would favor shear transformations around the dislocation group, but over a much wider fan of angles in which the shear stresses are high enough to initiate some of the broad range of transformations that now become possible at these temperatures. This should level down the shear stress concentration around the front of the dislocation group much as in the plastic blunting of a shear crack under stress. Thus, we expect that shear fronts will not propagate in a concentrated form at elevated temperatures.

It is important to note, in this connection, a very fundamental distinction between crystal plasticity by dislocation motion and the plasticity of glassy media. In the former, at stress levels under which dislocations become mobile, they rarely ever nucleate as isolated loops in the perfect crystal. A new dislocation line is generated by convoluting a portion of an existing line at source configurations, which are normally not widely encountered. Thus, propagation or motion of dislocation lines over long distances in the perfect crystal always produces a heterogeneous and "non-local" form of plasticity. In glassy media, a complementary form of plasticity exists. The disorder of the structure inhibits easy mobility of a dislocation line—if one were to exist. On the other hand, as the structure can support far larger stresses, a very large fraction of the volume is fertile for initiating local shear transformations independent from each other. Hence, there is no shortage of transformation sources. The elementary shear transformations, however, do not propagate outward—being apparently hindered by the structural disorder, so that large strain production requires the repeated nucleation of new shear transformations throughout the stressed volume. Barring dilatancy effects at low temperatures, this leads to a very homogeneous and "local" form of plasticity.

ACKNOWLEDGEMENT:

The research of the author in plasticity of glassy media is supported by the National Science Foundation (MRL) through the Center for Materials Science and Engineering at M.I.T. under Grant No. DMR-78-24185.

REFERENCES

1. J. D. Bernal, Proc. Roy. Soc.(London), A280, p. 299. (1964).
2. K. Maeda and S. Takeuchi, Phys. Stat. Sol., 49, p. 685. (1978).
3. L. Bragg, and J. F. Nye, Proc. Roy. Soc. (London), A190, p. 474. (1947).
4. L. Bragg, and W. M. Lomer, Proc. Roy. Soc. (London), A196, p. 171. (1949).
5. M. F. Ashby, Surf. Sci., 31, p. 498. (1972).
6. Y. Ishida, in Grain Boundary Structure and Properties, edited by G. A. Chadwick, and D. A. Smith, Academic Press: New York, p. 93. (1976).
7. W. M. Lomer, Proc. Camb. Philos. Soc., 45, p. 660. (1949)
8. A. W. Simpson, and P.H. Hodkinson, Nature, 237, p. 320. (1972).
9. A. S. Argon, and H. Y. Kuo, Materials Science and Engineering, 39, p. 101. (1979).
10. H. Frost, "Hole Statistics in Dense Random Packing," Tech. Rept. No. 6, Div. Appl. Sciences, Harvard University: Cambridge, Mass. (1978).
11. A. S. Argon, and H. Y. Kuo, "J. Non-Cryst. Solids, 37, p. 241. (1980).
12. U. F. Kocks, A. S. Argon, and M. F. Ashby, Thermodynamics and Kinetics of Slip, Progress in Materials Science, edited by B. Chalmers et al, Pergamon Press: Oxford, vol. 19, (1975).
13. A. S. Argon, Acta Met., 27, p. 47. (1979).
14. J. J. Gilman, J. Appl. Phys., 46, p. 1625. (1975).
15. F. Spaepen, Acta Met., 25, p. 407. (1977).
16. H. Neuhäuser, Scripta Met., 12, p. 471. (1978).
17. J. J. Gilman, in Dislocation Dynamics, edited by A. R. Rosenfield et al, McGraw-Hill: New York, p. 3. (1968).
18. J.C.M. Li, in Frontiers in Materials Science, edited by L. E. Murr and C. Stein, Marcel Dekker: New York, p. 527. (1976).

20. F. Spaepen, J. Non-Cryst. Solids, 31, p. 207. (1979).
21. C. H. Bennett, P. Chaudhari, and V. Moruzzi, Phil. Mag., 40, p. 485. (1979).
22. P. Chaudhari, A. Levi, and P. Steinhardt, Phys. Rev. Letters, 43, p. 1517. (1979).
23. D. Cherns, M. L. Jenkins, and S. White, in Electron Microscopy and Analysis 1979, Inst. Physics: London, p. 121, (1980).
24. R. Becker, and E. Orowan, Zeitschrift f. Physik, 79, p. 566. (1932).

DISLOCATION MODEL OF SOLIDIFICATION AND THE FORMATION OF AMORPHOUS SOLIDS

Hideji Suzuki
Department of Physics, Faculty of Science
University of Tokyo, Bunkyo-ku, Tokyo 113, Japan

The solidification and the formation of amorphous solids are discussed on the basis of a dislocation model of liquids, which was recently developed by the present author. We suppose that the solidification takes place when dislocation networks in a liquid are swept out of it. Then the rate of crystal growth and its anisotropy are calculated in terms of the mobility of dislocation networks. An amorphous solid is obtained when a liquid is cooled to a temperature, at which dislocation networks cannot move, before the dislocation networks are swept out of it. The Peierls-Nabarro stress and the interaction with solute atoms resist the motion of dislocations, and conduce the formation of amorphous solids.

Introduction

The atomic process of solidification has been discussed already by many authors. Jackson (1,2) discussed in detail the interface structure specifying each atom in the interface as a solid-like atom or a melt-like atom. Temkin (3) extended Jackson's theory considering an interface of many atomic layers thickness. Cahn (4,5) examined the theory of crystal growth for a diffuse solid-liquid interface. A different diffuse interface model was discussed by Fletcher (6). Ookawa and Kuroda (7,8) have discussed the crystal growth on the basis of a dislocation model of liquids in order to explain the origin of dislocations in crystals grown from melt. These theories, however, could not predict the preferred orientation of dendrites, nor the condition of the formation of amorphous solids.

The present author (9,10) has developed a dislocation theory of liquids in order to discuss various important problems related to the liquid phase, which had not been discussed successfully. In this paper we will discuss the process of crystallization from liquids, the relation between the rate of crystal growth and the supercooling, the orientation dependence of the growth rate, and also the condition of the formation of amorphous solids on the basis of the dislocation theory of liquids.

Dislocation model of solidification

According to the dislocation theory of liquids (10) dislocations in a liquid form networks with a mesh size of 3 to 4 atomic distances together with randomly distributed wedge disclinations with a density of the order of one per cent of that of dislocations. Dislocation networks in normal liquids repeat local unknitting and knitting to change their configuration, namely, the dissociation of a dislocation segment into two and the union of two dislocation segments into one by the thermal motion. The geometrical properties of dislocations are preserved in normal liquids, except for the annihilation and the creation of dislocations at disclinations.

A crystal grows when dislocations in the liquid are swept out of it. The deriving force to move dislocations is the gradient of the chemical potential of a dislocation. The crystal growth takes place when the temperature at the solid-liquid interface becomes lower than the melting point. In the steady growth, the released latent heat of fusion must be removed by a heat process. This process plays an important role in the rate of crystal growth (11). We will, however, not discuss the problem caused by the released latent heat of fusion in the course of solidification.

The free energy of a supercooled liquid may be shown by the curve in Fig. 1 as a function of the dislocation density. It has a minimum for high density of dislocations. The minimum free energy is that of the supercooled liquid, and should be higher by ΔF than that of the solid which corresponds to the zero of the dislocation density. We approximate the free energy per mol in the vicinity of the minimum by

$$F = \Delta F + \alpha(n - n_o)^2 \qquad (1)$$

referred to that of the solid, where α is a constant independent of the dislocation density and the temperature, and of the order of $3b^4R$ (10). Here b is the magnitude of the Burgers vector, and R the gas constant. If the total number of dislocations is conserved, the free energy in the vicinity of the interface becomes the minimum by increasing the dislocation density in the liquid adjacent the interface to n_1, so that the region of solid increases. The dislocation density, n_1, is given by

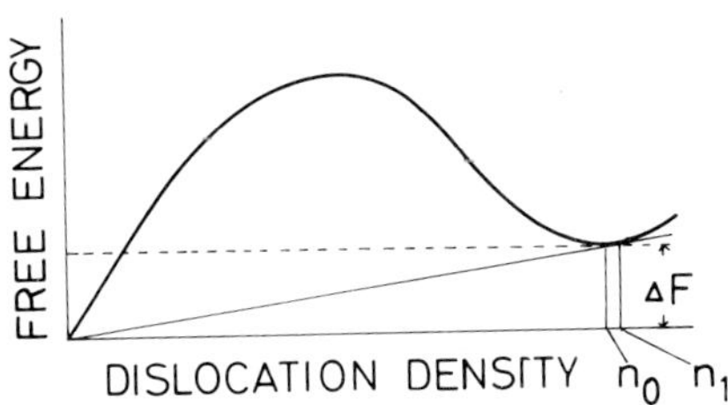

FIG. 1. Free energy plotted against the dislocation density.

$$(\partial F/\partial n)_{n_1} = F(n_1)/n_1. \qquad (2)$$

Dislocations tend to annihilate when the density exceeds the metastable equilibrium value n_o. The major part of dislocations annihilate at dislocations.

The dislocations in liquids form networks, and the motion of a dislocation network requires climb usually. If the unknitting of the network takes place, however, the whole of dislocations in the network can move only by glide. An unknitted part of a dislocation network is free to glide in the surface containing itself and its Burgers vector. A hexagonal dislocation network, therefore, can glide in any direction in the plane containing three Burgers vectors of dislocations meeting at each node. Dislocation networks in a liquid repeat the unknitting and the knitting to change the configurations of networks. The drift velocity of dislocation networks only by the glide motion is considerably higher than that by the climb motion. The rate of crystal growth, therefore, depends on the velocity of dislocation networks by the glide motion with the aid of the unknitting.

Growth rate of a crystal from liquid

Let us suppose an interface normal to the x-axis moving to the direction, x, at a constant velocity, v. The change in the dislocation density, which is defined as the total length of dislocations, per mol is then given by

$$\frac{\partial n}{\partial t} = v\frac{\partial n}{\partial x} - \beta(n-n_o) - \frac{\partial J}{\partial x}, \qquad (3)$$

in the moving coordinate fixed to the interface. Here β is the coefficient denoting the annihilation rate of dislocations, and J is the flow of dislocations and given by

$$J = \ell\frac{Df_\nu}{kT}n = -\ell^2\frac{D}{kT}\frac{\partial n}{\partial x}\frac{\partial F}{\partial n}, \qquad (4)$$

where ℓ is the cosine of the angle between the normal to the interface and the plane of the Burgers vectors, D is the diffusion constant of dislocation segments and is related to the viscosity of liquid, η, by

$$D = \frac{5kT}{\eta b}, \qquad (5)$$

and f_ν is the component of driving force in the direction of glide motion of the dislocation network. For the steady motion of the interface, we have

$$\frac{\partial}{\partial x}[\ell^2\frac{D}{kT}\frac{\partial n}{\partial x}\frac{\partial F}{\partial n} + vn] = \beta(n-n_o). \qquad (6)$$

The growth rate of the crystal is given by

$$v = -\frac{\ell^2}{n_1}\frac{D}{kT}\left(\frac{\partial n}{\partial x}\right)_{x=o}\left(\frac{\partial F}{\partial n}\right)_{n=n_1}. \qquad (7)$$

Then we need only the solution of eq.(3) which is correct in the limit of x = 0. The dislocation density is n_1 for x = 0, and decreases with increasing x, so that we assume

$$(n-n_o) = (n_1-n_o)e^{-\lambda x}, \qquad (8)$$

in the vicinity of x = 0. When we substitute eq.(8) into eq.(6), we obtain

$$4\alpha\ell^2 D(n_1-n_o)\lambda^2 - vkT\lambda - \beta kT = 0, \qquad (9)$$

in the limit of x = 0. From eqs.(7), (8) and (9) we obtain

$$v = [\{2\alpha\beta\ell^2 D(n_1-n_o)^3\}/\{kTn_1(n_1+n_o)\}]^{1/2}. \qquad (10)$$

From the consideration of annihilation of dislocations at disclinations, we have (10)

$$\beta = 5kT/27\eta b. \qquad (11)$$

We may use an approximate relation

$$\Delta F = \alpha(n_1-n_o)(n_1+n_o) = S_m\Delta T \simeq R\Delta T, \qquad (12)$$

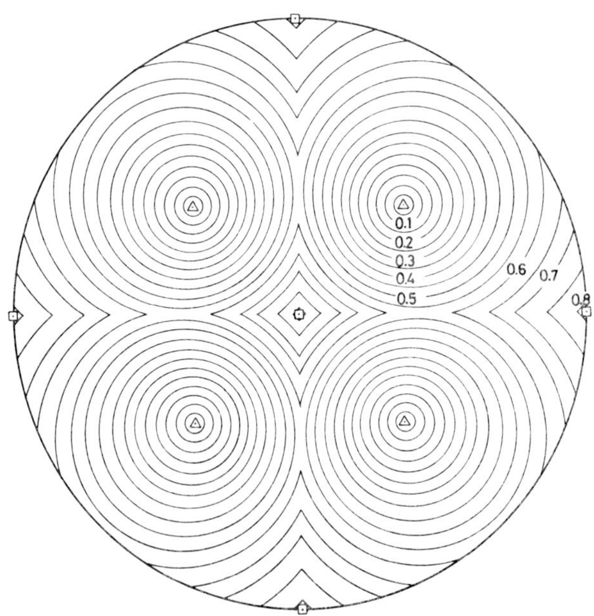

Fig. 2. Orientation dependence of the growth rate of the fcc crystal. Curves drawn in the standard projection indicate the velocity of crystal growth in an arbitrary unit.

where S_m is the entropy of fusion, ΔT the supercooling. Since $n_1 - n_o \ll n_o$, we approximate $n_1 + n_o \simeq 2n_o$. Combining eqs.(5), (10), (11) and (12)

$$v = \ell\eta^{-1}V^{-2.5}T^{0.5}(\Delta T)^{1.5}\times 5\times 10^{-14}m^{8.5}mol^{-2.5}\ Pa\ K^{-2}, \qquad (13)$$

where V is the molar volume of the liquid. Numerical values given by eq.(13) are of the order of the observed values of the growth rates of dendrites. However, it must be noted that the observed values are not given for the true supercooling at the solid-liquid interface so that we must estimate the true supercooling, solving the heat transport equation.

Equation (13) shows that the velocity is proportional to ℓ. Dislocations around a disclination tend to form networks with the same plane of the Burgers vectors, so that the plane of the Burgers vector is the same over a distance between neighbouring disclinations. The velocity of the interface is determined by the slowest part of dislocation networks. The plane of the Burgers vectors in a fcc crystal is {111}, so that the orientation dependence of the growth rate is readily obtained as shown in Fig. 2. The growth rate in the <111> directions is zero, because we neglected the velocity of the climb motion of dislocation networks. The growth rate in the <100> directions is the maximum as well known by observations (12). In a bcc crystal hexagonal networks are formed by $(a/2)[1\bar{1}\bar{1}]$, $(a/2)[111]$, $a[\bar{1}00]$ dislocations. The minimum growth rate of a bcc crystal is in the <110> directions, and the growth rate is the maximum again in the <100> directions as known by experiments (12).

Formation of amorphous solids from liquid

If the cooling rate of a liquid is higher than the rate to sweep the dislocations out of the liquid, we can cool it to a temperature at which the unknitting and the knitting of networks are frozen and dislocations cannot disappear. Some glide motions of dislocation segments in networks may take place to shorten their lengths, because the line tention of a dislocation increases with decreasing temperature. We suppose that amorphous solids are the state of liquid in which the unknitting and the knitting of dislocation networks are frozen. The rate to sweep the dislocations out of the liquid is slow when the Peierls-Nabarro stress is high or the interaction between a solute atom and a dislocation is very strong.

A typical example of high Peierls-Nabarro stress is SiO_2, which is in amorphous state even if it is cooled slowly. It is well known that the viscosity of fused silica is very high at the melting point. Equation (13) shows that the growth rate of a crystal is inversely proportional to the viscosity. The activation energy of the viscosity of fused silica is about 6 eV (13), which is about a few tens times that of liquid metals as normalized by the melting point. This size of the activation enthalpy is reasonably understood if we consider the extra energy required to escape a dislocation segment from the Peierls potential in a quartz crystal, which gives the Peierls stress of the order of 10^{10} Nm^{-2}.

Meanwhile, we cannot obtain amorphous Si and Ge by rapid cooling from the liquid phase, though these materials reveal high Peierls stresses in solids. The melting and related properties of Si and Ge are reasonably interpreted under the assumption that liquid Si and Ge are in metallic state with the simple body-centred tetragonal structure packed with dislocations (14). The Peierls-Nabarro stress of metals in this structure is supposed to be negligible.

It is easily seen that the basic structure in liquid phase should be found in the solid phase under normal pressure or under high pressures, because the strain energy of dislocations is small in a close packed crystal with a small elastic constant. High pressure measurements have not shown that Se and Te change their structure into another which is suitable for the basic structure in the liquid phase. We may suppose therefore that the basic crystal structure in liquid Se and Te are the same with that in the solid phase. It is well known that we can obtain amorphous Se and Te by rapid cooling from the liquid phase and also the Peierls-Nabarro stress in solid Te seems to be high (15).

Amorphous metals are obtained by adding large amounts of solute atoms, which would interact with dislocations by the size interaction or some chemical interactions. The diffusion of solute atoms in a liquid alloy usually takes place by the thermal motion of dislocations. The diffusion by the vacancy mechanism is usually slower than that caused by dislocation motions. The formation of solute atmosphere is, therefore, impossible at high temperatures. But at low temperatures the thermal motion of dislocations may be disturbed by the fluctuation of solute atoms and then atmospheres are formed more perfectly with decreasing temperatures. Actually, the viscosity of the liquid alloy increases very rapidly at temperatures lower than a temperature. In order to obtain amorphous metals we need, therefore, to add solute atoms which interact with dislocations strongly.

Amorphous solids may be ductile, if dislocations can move rather easily. In SiO_2, however, the force required to move a dislocation against the Peierls-Nabarro stress at room temperature is estimated to be about 5×10^9 Nm^{-2}, and we cannot expect the ductility in amorphous SiO_2 or silica glass. Meanwhile, in a metallic glass the locking force due to the solute atmosphere may be smaller than 5×10^8 Nm^{-2}, so that the resistance to the motion of dislocations in mainly caused by the pinning at the nodes of dislocations and also by the elastic interactions with other dislocations. Taking into account the decrease in the line tention of dislocations with high density we can obtain the yield stress of metallic glass of the order of observed values.

References

1. K.A. Jackson, Liquid Metals and Solidification, p.174, ASM, Cleveland (1958).
2. K.A. Jackson, Crystal Growth, ed. H.S. Peiser, p.17, Pergamon, Oxford (1967).
3. D.E. Temkin, Crystallization Process, p.15, Consultant Bureau, New York (1966).
4. J.W. Cahn, Acta Met., 8, 554 (1960).
5. J.W. Cahn, W.B. Hillig and G.W. Sears, Acta Met., 12, 1421 (1964).
6. N.H. Fletcher, J. Cryst. Growth, 28, 375 (1975).
7. A. Ookawa and T. Kuroda, Solid State Physics, 4, 185 (1969).
8. T. Kuroda and A. Ookawa, J. Cryst. Growth, 24/25, 403 (1974).
9. H. Suzuki, Bull. Japan Inst. Metals, 13, 773 (1974)
10. H. Suzuki, to be published.
11. M.E. Glicksman, R.J. Schaefer and J.E. Ayers, Met. Trans., 7A, 1747 (1976).
12. J.W. Christian, The Theory of Transformations in Metals and Alloys, p.566, Pergamon Press, Oxford (1963).
13. J.D. Mackenzie, Chem. Rev., 56, 455 (1956).
14. T. Matsuo and H. Suzuki, to be published.
15. L.C. Lovell, J.H. Wernick and K.E. Benson, Acta Met., 6, 716 (1958).

SOME RECENT RESULTS OF THE THEORY OF DISCLINATIONS AND DISPIRATIONS

Tsu-Wei Chou
Mechanical and Aerospacc Engineering Department
University of Delaware
Newark, Delaware 19711

Abstract

The complete theory of line imperfections in crystals should include dislocations, disclinations and dispirations. These are defects due to discontinuity in translational, rotational and screw displacements. This paper first reviews the developments in the theory of disclinations and dispirations. The concept of disclinations recently has been applied to treat the problem of rotational fault movements in geology. A model for fault surfaces involving both disclinations and dislocations has been established and it is proposed that this model can assist in defining the nature of fault planes. The concept of dispirations can be applied to study the deformation and contraction of some biological systems assuming the form of cylindrical crystals. The passage of a dispiration along a helical plane in a cylindrical crystal can induce one of two types of shape changes. The magnitude of these contractions as well as the local elastic deformation due to dispiration dipoles have been examined. Future topics of research are also discussed.

Introduction

The development of the continuum theory of line imperfections can be traced back to the turn of this century. It was in 1901 when Weingarten (1) first postulated the theorem which forms the basis for the study of line defects in solids. Volterra (2) later discussed Weingarten type imperfections in detail for the case where the elastic body is a circular cylinder. The Weingarten-Volterra type of line imperfections were first given the term dislocations by Love (3), who defined six orders of dislocations due to discontinuities in translational and rotational displacements.

Translational dislocations are traditionally known simply as dislocations while the term disclinations has been accepted by researchers for rotational dislocations. A third type of line imperfection is known as dispirations. From the continuum viewpoint a dispiration can be regarded as the combination of its dislocation and disclination components. However, when the crystallographic nature is taken into consideration it becomes evident that dispirations form a new and distinct category of line imperfections.

This paper surveys the recent advancement of the theory of disclinations and dispirations in physics, geology and biology. Future topics of research are also discussed.

Disclinations

Rotational dislocations were first given the name disclinations by Nabarro (4). Nabarro classified disclinations as screw and edge types where the axes of rotation are parallel and perpendicular to the imperfection line, respectively. However, researchers generally prefer to adopt the more descriptive terms of "twist" and "wedge" in place of edge and screw (5,6).

Rotational dislocations were traditionally neglected by metallurgists because these defects were considered to be high energy configurations. However, in recent years disclinations have been observed in liquid crystals (7), flux-line lattices (8), block wall lattices (9), pseudo-pentagonal twins (10), surface crystals (11), and carbonaceous mesophase and graphite (12). Bubble raft crystals containing disclinations also have been demonstrated (13).

Considerable efforts have been made to study the elastic field of disclinations in isotropic (14,15) and hexagonal close packed (16-18) media. Quantitative evaluations of the interactions of a disclination with dislocations (19,20), point defects (21), non-homogeneities (22, 23) and other disclinations (24) are also available. The concept of disclinations has been adopted by researchers to model the behavior of a wide range of materials including metals (25), polymers (26) and biological materials (27,28).

The concept of disclinations may have provided a new dimension to the understanding of the residual deformation fields due to faulting in geology (29). Rotational fault movements are defined as blocks rotating relative to each other about an axis perpendicular to the fault. Lahee (30) reports that rotational faults are also referred to as hinge, pivotal or scissors. Several examples of rotational faults have been identified in geological literature. Fig. 1a shows an example of rotational fault movement, a pivotal fault in Osage County, Oklahoma (31). Detailed structural mapping of the Edison spodumene mine in the Black Hills, South Dakota has revealed that the east and central faults appear to be rotational, with the central fault being further classified as a hinge fault (32). Galligan (33) presented a picture of an anticline which has been formed through a folding operation. Very large rotational displacement exists in such a deformation. Other examples of rotational faults exist (34), but the faults are the result of at least two episodes of fault block movement.

Many scientists have employed the basic concept of dislocation theory in studying specific earthquakes (29). Although much work has been done on the application of statical elasticity theory of dislocations to faulting, none has been directed at disclinations. Previous analyses only permitted strike or dip slip movement. A unified approach recently has been developed for the elastic field solutions of twist disclination and glide dislocation loops. The method is based upon the Green's functions (35-37) in elasticity and is valid for loops of arbitrary shapes and orientations beneath a free surface and permits any rake of net slip, as defined by Dennison (38). Fig. 1b depicts the calculated vertical displacement field in the vicinity of a twist disclination loop intersecting a free surface. The resemblance in the nature of the deformation fields is evident. Fig. 2 presents the displacements measured in the direction paralel to the fault plane in the Tango earthquake in Japan. Also shown are the theoretical predictions based upon rectangular (39) and circular dislocation loop models. The difference in predictions is due only to the difference in the assumed loop sizes.

Although no rotational faulting has been identified in monitored, earthquake-prone areas, the introducing of the concept exposes the researcher to the possibilities of such fault mechanics. The theoretical model can assist in characterizing both translational and rotational faults in geological structures by inspection of the displacement contours.

Dispirations

When the Burger's vector $\bar{b}$ of a dislocation and the rotation vector $\bar{\omega}$ of a disclination are combined it is possible for a singular line to result with no surface defects attached, even when neither $\bar{b}$ nor $\bar{\omega}$ is, by itself, a symmetry operation. This occurs when the lattice has screw symmetry and $\bar{b}$ and $\bar{\omega}$ are components of a screw symmetry operation. The resulting singular line is termed a dispiration (40). Dispirations are classified as twist and wedge types when both $\bar{b}$ and $\bar{\omega}$ are normal and parallel to the line directions, respectively.

The concept of dispirations is valid, provided the combination of the dislocation and disclination components of a dispiration is energetically favorable. It has been found (41) that the decomposition of twist and wedge dispiration lines into component partial dislocations and disclinations is opposed by the stacking fault energy. Mixed dispiration lines and dispiration loops with oblique screw axes are unstable configurations and the separation distance between the dislocation and disclination components is determined by the balance between interaction energy and stacking fault energy.

Currently, a major application of the concept of dispirations has been in the modelling of the contractile mechanisms in some biological systems (41,42). The understanding of the shape changes and movements of biological systems such as viruses, bacteria and striated muscles is of great interest to researchers. It is believed that the deformation of these biological systems is the result of movement of defects in the packing of the structural units. The nature of the defects in biological systems can be easily understood by using the concept of crystal defects. The most important defects that may appear in biological structures are in fact dispirations.

Consider, for instance, the dispiration structures in the tail sheath of T-even bacteriophages and tobacco mosaic viruses. These systems assume the shape of hollow rods with molecular units packed in stacked disks, and simple or double helix. Continuous structural rearrangement results in the neighborhood of a moving dispiration. The glide line of the defect is usually a long-pitched helix. The magnitude of the translational and rotational displacements in a cylindrical crystal can be related to the lattice structural parameters.

Dispiration movements are responsible for the contraction of cylindrical crystals. There exist two types of deformation due to dispirations in cylindrical crystals. In the first type, only the disclination component of the dispiration contributes to contraction, whereas in the second type, both the dislocation and disclination components contribute to contraction. Fig. 3 demonstrates these two types of deformations where R is the mean radius of a circular cylindrical crystal. Estimates of the magnitude of contraction have been made in terms of the dimensional and structural parameters of the cylindrical crystal. The dispiration movements in both types of contractions are conservative. There is no net change in mass of the cylindrical crystal as a result of the movements. Contractions of cylindrical crystals discussed by Harris & Scriven (41) correspond to the second type.

Fig. 4 illustrates a cylindrical crystal with a dispiration dipole situated on a helical slip plane A'ADD'. The screw axis of the dispirations is colinear with the cylinder axis. It can be shown that for the first type of deformation discussed above, the elastic field of the dispiration dipole associated with the helical slip is equivalent to those of the dipoles associated with the circumferential cut (AEE'A') and axial cut (EDD'E'). Each of these two dispiration dipoles can in turn be decomposed into its dislocation and disclination components. These components are illustrated in Figure 5 for the dipole EDD'E' where the free body diagram is drawn by taking a horizontal cut halfway through the segment CB. The complete elastic field of the dispiration dipole situated on a helical plane has been found in closed form by an application of the shallow shell theory (41-43).

The deformation field of the second type of contraction mentioned above as well as the transition between the first and second types of contraction deserves further examination. The biological factors governing these changes should also be studied.

Finally, it should be pointed out that problems involving large rotation and translation due to the defects, the treatment of the deformation field based upon linear elasticity breaks down. Analysis of the non-linear aspect of deformation should then be pursued. The topological aspect of disclinations and dispirations may be another fruitful area of future research and the knowledge gained in its study could benefit, in particular, the field of micro-biomechanics.

References

1. G. Weingarten, Atti Accad, naz. Lincei R., 5, 57 (1901).
2. V. Volterra, Annls. scient. Ec. norm. sup., Paris, 24, 401 (1907).
3. A.E.H. Love, A Treatise on the Mathematical Theory of Elasticity, 3rd edition, Cambridge University Press, London (1920).
4. F.R.N. Nabarro, Theory of Crystal Dislocations, Oxford University Press, London (1967).
5. K. Anthony, J. Essmann, A. Seeger and H. Traüble, Mechanics of Generalized Continua (Edited by E. Kröner), Springer-Verlag, New York (1967).
6. J.A. Simmons, R. deWit and R. Bullough, eds., Fundamental Aspect of Dislocation Theory, National Bureau of Standards, Washington, D.C., Special Publication No. 137 (1970).
7. G.H. Brown, Am. Scient., 60, 64 (1971).
8. A. Seeger, Metall. Trans., 1, 2987 (1970).
9. J.P. Hirth and R.G. Wells, J. Appl. Phys., 41, 5250 (1970).
10. R.L. Schwoebel, J. Appl. Phys. 37, 2515 (1966).
11. W.F. Harris, Defects in Surface Crystals, Ph.D. thesis, University of Minnesota, Minneapolis (1970).
12. J.E. Zimmer and J.L. White, ACS Symposium Series, No. 21, Petroleum Derived Carbons, American Chemical Society (1975).
13. Y. Ishida and S. Iyama, Act. Met., 24, 417 (1976).
14. W. Huang and T. Mura, J. Appl. Phys., 41, 5175 (1970).
15. G.C.T. Liu and J.C.M. Li, J. Appl. Phys., 42, 3313 (1971).
16. T.W. Chou and Y.C. Pan, J. Appl. Phys., 44, 63 (1973).
17. T.W. Chou and Y.C. Pan, J. Appl. Phys., 44, 3889 (1973).
18. T.W. Chou, Scripta Met., 7, 151 (1973).
19. R.J. DeWit, J. Appl. Phys., 42, 3304 (1971).
20. T.W. Chou, J. Appl. Phys., 46, 523 (1975).
21. T.W. Chou and Y.C. Pan, J. Appl. Phys., 46, 528 (1975).
22. T.W. Chou, J. Appl. Phys. 44, 3889 (1973).
23. T.W. Chou and T.L. Lu, Mat. Sci. Engr. 12, 163 (1973).
24. T.W. Chou and Y.C. Pan, J. Appl. Phys., 46, 4641 (1975).
25. J.C.M. Li, Surface Science, 31, 12 (1972).

26. J.C.M. Li and J.J. Gilman, ASM Materials Science Seminar "Relationships between Structure and Mechanical Behavior of Polymeric Solids", American Society for Metals (1973).
27. F.R.N. Nabarro, Physics of Strength and Plasticity, edited by A.S. Argon, MIT Press, Cambridge, Mass. (1969).
28. W.F. Harris, Phil. Mag. 32, 37 (1975).
29. T.W. Chou and H.J. Malasky, J. Geophys. Res., 84, 6083 (1979).
30. F.H. Lahee, Field Geology, P. 215, McGraw-Hill, New York (1941).
31. C.M. Nevin, Principles of Structural Geology, 3rd ed., p. 84, John Wiley, New York (1942).
32. L.R. Page, et al., U.S. Geol. Surv. Prof. Pap., 247, 104 (1953).
33. J.M. Galligan, Nature, 240, 144 (1972).
34. F.A. Donath, Geol. Soc. Amer. Bull. 73, 1 (1962).
35. Y.C. Pan and T.W. Chou, J. Appl. Mech., 43, 608 (1976).
36. Y.C. Pan and T.W. Chou, Int. J. Enging. Sci., 17, 545 (1979).
37. Y.C. Pan and T.W. Chou, J. Appl. Mech., 46, 551 (1979).
38. J.M. Dennison, Analysis of Geologic Structures, p. 151 and 180, W.W. Norton, New York (1968).
39. M.A. Chinnery, Bull. Seismol. Soc. Amer., 51, 355 (1961).
40. W.F. Harris, Phil. Mag., 22, 949 (1970).
41. T.W. Chou and M. Ichikawa, J. Biomechanics, 12, 321 (1979).
42. M. Ichikawa and T.W. Chou, J. Theor. Biol., 78, 129 (1979).
43. M. Ichikawa and T.W. Chou, J. Appl. Phys., 49, 4737 (1978).

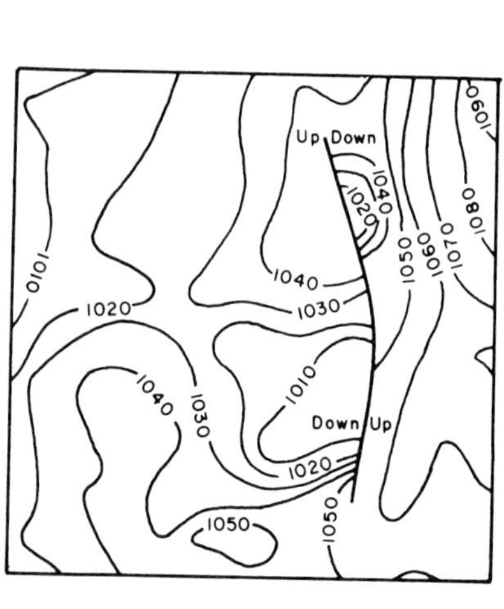

FIG. 1a

Contour map of a rotational fault in Osage County, Oklahoma [after Ref. 31].

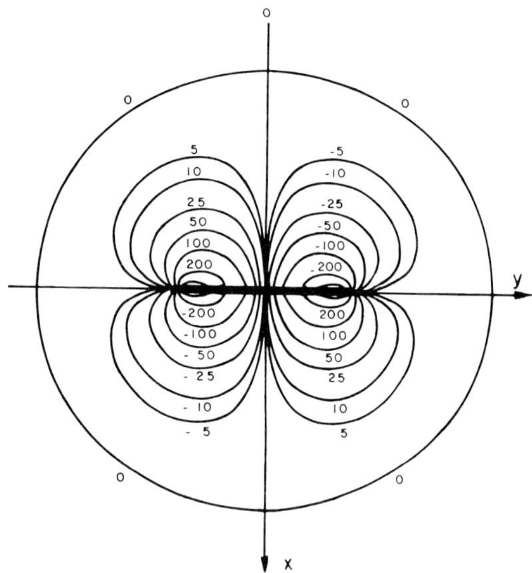

FIG. 1b

Vertical displacement of a half circular twist disclination loop centered at the free surface. Contour values are in the units of $10^{-3}\omega a$ where a is the loop radius.

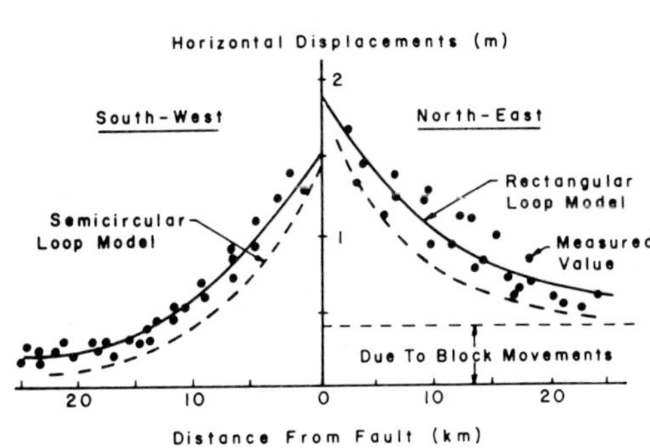

FIG. 2

Falloff of the horizontal displacements with distance in the Tango earthquake [after Ref. 29].

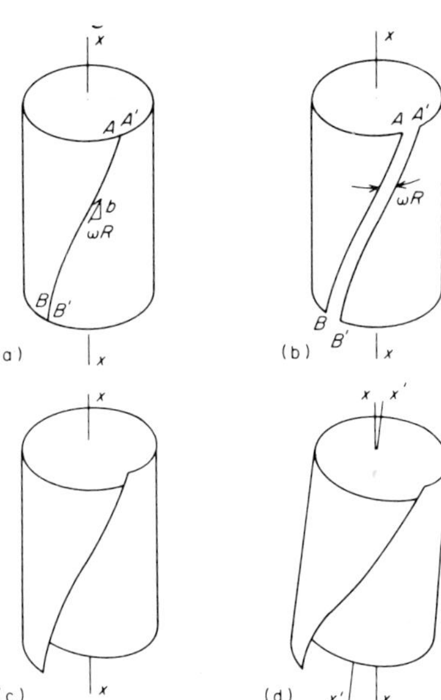

FIG. 3

(a) The application of shear displacement along a helical plane, (b) the deformation due to the rotational component ω, (c) the first type of deformation due to the helical slip, (d) the second type of deformation due to the helical slip [after Ref. 42].

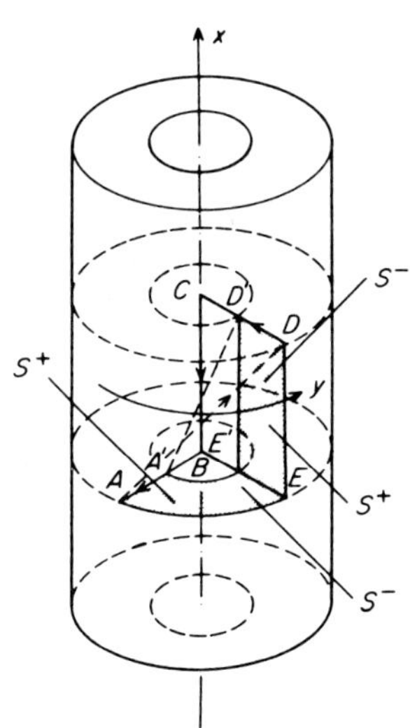

FIG. 4

The dispiration dipole DCBA produced by screw displacement on the vertical (DD'E'E) and horizontal (AA'E'E) cuts [after Ref. 42].

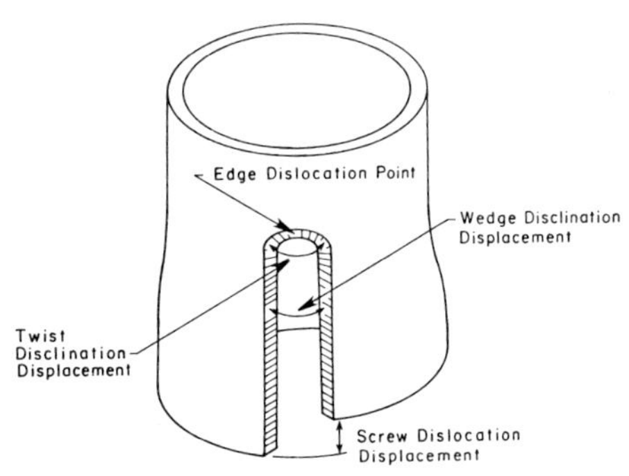

FIG. 5

Thc three-dimensional view of a dispiration dipole [after Ref. 41].

APPLICATION OF INTERNAL FRICTION MEASUREMENTS TO THE STUDY OF DISLOCATIONS AND PLASTIC FLOW*

Ricardo B. Schwarz

Materials Science Division, Argonne National Laboratory, Argonne, IL 60439

Abstract

Most of the available information on solution strengthening in alloys has been deduced from macroscopic flow-stress measurements. Amplitude-dependent internal-friction losses also depend on the dislocation-solute interaction, and this technique should, in principle, convey the same information. The interpretation of these measurements requires, however, a suitable model. The previous internal friction models are briefly reviewed. The recent flow-stress and internal-friction calculations based on the Labusch-Schwarz computer modeling method are discussed in more detail. In these, the dislocation-obstacle interaction has a smooth profile of finite range, the obstacles are distributed at random on the dislocation glide plane, and the dislocation motion is also opposed by viscous forces. The quantitative correspondence found between the flow-stress and the amplitude-dependent internal friction calculations is discussed. Experiments in which the internal-friction method has been used to investigate solution strengthening are discussed. These include: (a) the measurement of the force-distance law for the dislocation-solute interaction, (b) the study of the decrease in strength of superconducting alloys upon going from the normal to the superconducting state, and (c) the study of dynamic strain-aging in dilute Cu-Al alloys. The experimental results are used to evaluate the proposed models.

1. Introduction

Most of the available information on solution strengthening in alloys has been deduced from macroscopic flow-stress measurements. Amplitude-dependent internal-friction losses also depend on the dislocation-solute interaction, and this technique should, in principle, convey the same information. These measurements have the advantage that, during the tests, the dislocations need to overcome only the weakest obstacles (solutes). This is in contrast with the macroscopic flow-stress experiments, where the dislocations are forced to overcome stronger obstacles as well (forest dislocations, for example), which are often out of experimental control. An added advantage is that the internal-friction measurements do not change the "state" of the system and hence repeated experiments can be conducted with the same sample in order to examine the effects of annealing and thermal treatments.

In internal-friction measurements, a sample is subjected to a periodic applied stress of amplitude σ_0, and the energy ΔW dissipated internally during one cycle is measured as a function of the amplitude of the applied stress, its frequency, temperature, time, etc. It is convenient to measure ΔW in units of twice the elastic energy W_0 stored in the sample when $\sigma(t) = \sigma_0$. This defines the decrement

$$\delta = \Delta W/(2W_0) \quad (1)$$

Although not necessary, the sample is usually driven into one of the mechanical resonant modes compatible with its shape. An equivalent definition of δ then follows from the logarithmic rate of decay of the oscillations with the driving force removed.

Contributions to δ can arise from a variety of structural and lattice defects in the solid (1). Of concern here are only the contributions due to the cyclic motion of the dislocations. We further require that these glide over distances large compared to the intersolute spacing in the alloy, for only then will the dislocations be forced to overcome a solute environment similar to that encountered by dislocations in macroscopic flow-stress experiments.

*Work supported by the U. S. Department of Energy.

Since the underlying aim is to achieve a better understanding of the mechanisms involved, rather than phenomenology, we follow the usual approach of interpreting experimental results in terms of models. In Sections 2 and 3 we summarize the models, and the calculations based on these, that have been proposed to interpret internal friction measurements in alloys. Certainly this will not be a detailed review on the subject since excellent ones have been written (1,2). The recent numerical calculations based on computer-modeling (3-5), which are not included in these reviews, will be described in some greater detail. In section 4 we present experiments in which the internal-friction method, as outlined in Section 2.2.2, has been used to investigate alloy properties. These results also allow for an evaluation of the proposed models.

2. Internal Friction Calculations for Fixed Solutes

2.1 Decrement for a random distribution of obstacles on a line.

Granato and Lücke (6) first interpreted internal friction losses in terms of a hysteretic-type motion of dislocations. Their model portrays the dislocations in terms of a string analogy introduced by Koehler (7), which has been used since. The description of the alloy involves the following simplifying assumptions: (a) The weak obstacles (solutes) are distributed at random along straight line segments. (b) Strong (impenetrable) obstacles are located at both ends of each segment, and (c) The range of the interaction force between the dislocations and the weak obstacles is assumed equal to zero (point-like obstacles).

The large number of experimental investigations which have tried to exploit the potentialities of the internal friction method have confirmed the general picture drawn from the Granato-Lücke (GL) model: (a) for small values of the stress amplitude σ_0 the dislocations bow out between the obstacles giving an amplitude-independent decrement contribution δ_i. (b) For larger values of σ_0 the dislocations overcome some of the obstacles conducing to an amplitude-dependent contribution δ_h. Thus, the theory allows an estimate to be obtained of the stress required to initiate the transition from very restricted dislocation motion to dislocation displacements restricted only by dislocation network nodes or other strong anchoring points. However, the simplifying assumptions concerning the obstacle distribution should make this theory most applicable to situations in which solutes have been able to diffuse to stationary dislocations. We will return to this point in Section 4.1. The several models later devised to generalize the GL model (8-11) were mostly based on this model and have resulted in only some minor modifications to the GL model.

2.2 Decrement for a random distribution of obstacles on a plane.

2.2.1 Analytic methods

The major difficulty in calculating the internal losses for a dislocation interacting with a random planar distribution of obstacles is that of obtaining the required statistical averaging. Friedel (12) used a steady-state model in which the dislocations are either at rest or moving at a steady-state thermally activated velocity. More recently, Schlipf (13) used a similar model based on the statistical flow-stress calculations of Labusch and coworkers(14-15). While, as is indeed found (Ref. 5; see also Section 4.3), steady-state models yield reasonably accurate results for stress amplitudes far exceeding the macroscopic flow stress, they cannot predict the value of the decrement at lower stress amplitudes, where the amount of plastic deformation in each half-cycle is strongly dependent on the previous loading under a shear stress of the opposite sign. It is precisely this stress regime -- slightly below the macroscopic flow stress -- that we want to address.

2.2.2 Numerical methods

Within the last years, the advent of fast digital computers has opened the possibility of developing models that give a better description of the real situation but are far too complicated to be treated by analytical methods alone. Labusch and Schwarz (LS), (3,4) developed a computer code that solves the trajectory of a dislocation moving through a random distribution of weak obstacles on a plane. The dislocation-obstacle interaction has a smooth profile of finite range, and the dislocation motion is also opposed by viscous forces. These added features of the model require the specification of a large number of parameters. However, for weak obstacles (substitutional solutes in fcc alloys), LS have shown that by a proper variable normalization, this number can be drastically reduced. This method has been applied to the calculation of the critical stress necessary for the long-range motion of a dislocation (3,4) and recently, to the calculation of the amplitude-dependent internal-friction losses under a periodic applied stress (5). The use of the same alloy model in both calculations allows a quantitative comparison to be made between the flow stress and the internal friction calculations. Some results are summarized in Figs. 1-3. The normalized stress S measures the physical stress σ in units of the Friedel stress $2\Gamma f_o^{3/2}/(b\ell_s)$, while the drag constant γ measures the usual drag constant B in units of the quantity $B^* = (4\Gamma m f_o)^{1/2}/\ell_s$. Here Γ is the dislocation line tension, f_o is the strength of the dislocation-obstacle interaction in units of 2Γ, b is the magnitude of the Burgers vector, $\ell_s \approx b/\sqrt{c}$ is the average spacing between the solutes on the glide plane (c is the solute molar fraction), and m is the dislocation mass density.

Figure 1 shows the critical values of the applied stress for the long-range motion of the dislocation. The S_c and S_L curves correspond to different initial conditions for the dislocation motion. The S_c curve, also labeled "macroyield stress", gives the critical value of a slowly increasing applied stress at which the dislocation is found to move freely through the whole obstacle array. The S_L curve gives the critical value of a slowly decreasing applied stress at which an already moving dislocation stops. For $\gamma > 2$, the critical stresses are independent of the initial conditions for the dislocation motion and $S_c = S_L$. The physical reason behind this is that in the presence of a sufficiently high viscous drag, each breakaway event is approached quasistatically. The dislocation then percolates through the obstacles, and, independently of the initial conditions, it always finds the strongest configuration of obstacles in the random obstacle array. The regime $\gamma < 2$ is characterized by a decrease in S_c and S_L, with $S_L < S_c$. Here the viscous drag is sufficiently low for the dislocation to develop high velocities and influence the overcoming of some of the obstacles on account of its stored kinetic energy. These have been called "inertial effects". We will return to this subject in Section 4.3. For the internal friction calculations it is also important to know the stress at which a dislocation, placed at an arbitrary position on the glide plane, begins to surmount some of the obstacles, without being yet able to glide macroscopically. This "microyield stress" is also shown in Fig. 1, as a segmented line (*). Since the dislocation breaks away from the first obstacle from essentially a rest position, the "microyield stress" is independent of γ.

Figures 2 and 3 are the corresponding low-frequency internal-friction calculations (5). Figure 2 shows the amplitude-dependent decrement δ (**) as a function of the normalized stress amplitude S_o for a planar distribution of weak obstacles and three values of the normalized viscous drag. As may have been expected, for all three values of γ the decrement begins to increase when S_o becomes larger than the "microyield stress" and asymptotically approaches a vertical line as S_o approaches the "macroyield stress", S_c. The stress-amplitude/viscous-drag regime where the decrement is amplitude-dependent corresponds therefore to the shaded area in Fig. 1.

Figure 3 shows $\delta(S_o)$ for two of the previous values of γ but for a dislocation interacting with a superposition of two kinds of obstacles: the same planar distribution of weak obstacles that give rise to the curves in Fig. 2, plus a much smaller density of strong (impenetrable) obstacles. As in the previous case, δ begins to increase as S_o exceeds the "microyield stress" but, since the strong obstacles limit the dislocation displacements, δ remains finite for $S_o > S_c$. The decrement reaches a broad peak, the location and height of which can be predicted analytically: assuming that for $S_o >> S_c$ the effects of the weak obstacles can be averaged into a frictional stress of value $(S_c + S_L)/2$, it is obtained (5) that the maximum in δ occurs at $S_o = S_c + S_L$ and the height of the peak is $\delta_{max} = \Lambda L^2/12$. Here Λ is the total dislocation density and L is the physical separation between the strong obstacles on the dislocation. These results are within 15% from those deduced from the computer simulations. We further notice that while the location of the peak depends on the properties of the weak obstacles alone, the height of the peak reflects only those of the strong obstacles.

As a result of the LS normalization (3,4), the parameters y_o (range of the interaction force between the dislocation and a weak obstacle), ℓ_s, and f_o are not independent but combine to define $\eta_o = y_o/(\ell_s\sqrt{f_o})$. All the results shown in Figs. 1-3 are for $\eta_o = 0.5$, which, with the reasonable values $y_o = 2b$ and $f_o = 0.02$, can be seen to correspond to a rather dilute alloy. For higher η_o values (corresponding, for example, to more concentrated alloys) these curves retain their basic shape but are shifted in parallel fashion to higher S values.

For the case that the glide plane contains only identical obstacles and the dislocation is overdamped (see Fig. 2, curve for $\gamma = 2$) it was empirically found (5) that for different η_o

---

*For a given obstacle array the macroyield stress has a well defined value. On the other hand, the microyield stress has a large variance, depending on the initial dislocation position.

**From here on we omit the subindex h and, unless otherwise noted, δ is meant to be the amplitude-dependent component of the total decrement.

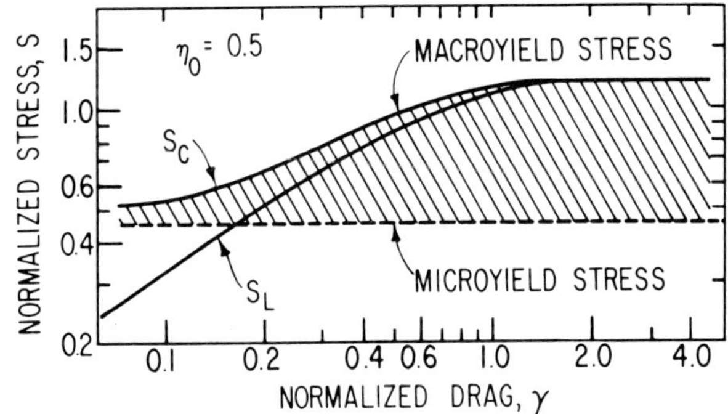

FIG. 1

Critical stresses S_C and S_L as a function of the normalized drag, γ, for a normalized interaction range η_o = 0.5 (from Ref. 4).

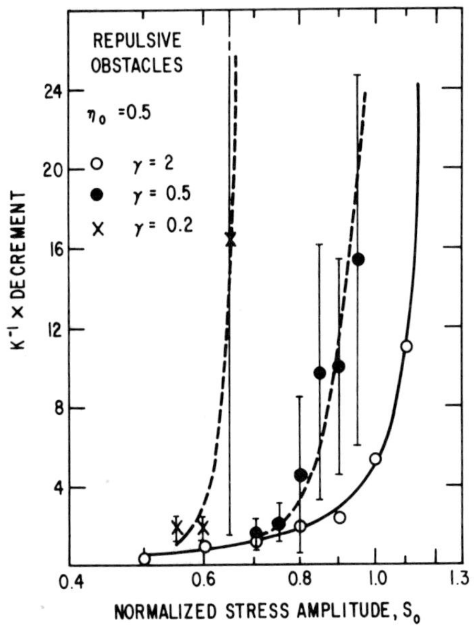

FIG. 2

Stress-amplitude dependence of the decrement for identical obstacles of normalized interaction range η_o = 0.5 and three values of the normalized viscous drag γ. Each point is an average over five independent computer simulations. The bars denote $\pm$ one standard deviation. $K = \Lambda \ell_s^2/f_o$ (see the text).

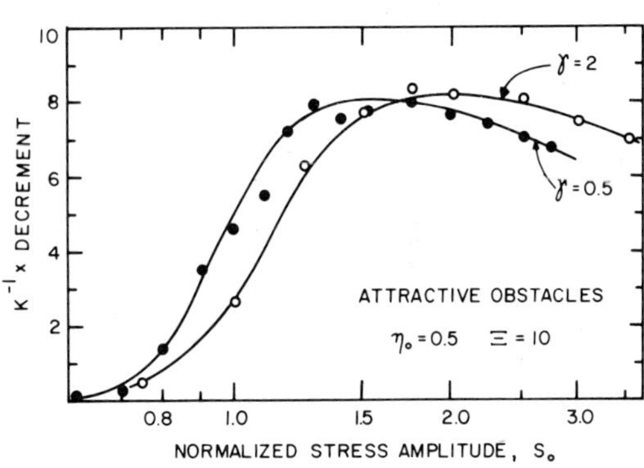

FIG. 3

Stress-amplitude dependence of the decrement for a superposition of weak and strong obstacles and two values of the normalized viscous drag γ. The weak obstacles have a normalized interaction range η_o = 0.5. $\Xi = L\sqrt{f_o}/\ell_s$ and $K = \Lambda \ell_s^2/f_o$ (see the text).

values the decrement has the simple analytic form

$$\delta_{\gamma > 2} = \left(\frac{\Lambda \ell_s^2}{f_o}\right) \delta' \left[\frac{\sigma_o}{\sigma_c(n_o)}\right] . \tag{2}$$

A direct implication of this equation is that the stress amplitude $\sigma_o(\Delta)$ necessary to give a constant value Δ to the product $\delta f_o/(\Lambda \ell_s^2)$ is proportional to the critical stress σ_c. Therefore, changes in σ_c can be conveniently studied by measuring $\sigma_o(\Delta)$. A similar relationship was previously deduced by GL (2) for their linear obstacle-distribution model. In this case a proportionality was found between $\sigma_o(\delta)$, the stress amplitude necessary to maintain a constant value δ of the amplitude-dependent component of the decrement, and σ_B, the statistically averaged breakaway stress of a dislocation from a random linear row of point obstacles. In the present case, where the solutes are randomly distributed on a plane, an equivalence between σ_c and σ_o for constant δ, instead of constant $\Delta = \delta f_o/(\Lambda \ell_s^2)$, is an approximation that seems justified only if a large value of δ is chosen, since then $\sigma_o d\delta/d\sigma_o >> 1$ (Fig. 2) and $\sigma_o(\delta)$ is rather insensitive to the value of δ. Furthermore, since for $\gamma < 2$ (underdamped regime) the decrement is an even stronger function of σ_c (Fig. 2), measurement of $\sigma_o(\delta)$, as described above, are a good measure of σ_c for all values of γ. This method has the advantage that measurements of $\sigma_o(\delta)$ can be obtained using a very simple experimental technique (16).

Temperature can affect the dislocation motion mainly in two ways:(*) (1) by an increase in the viscous forces, as a result of the increase in the phonon density, and (2) by a reduction in the stress necessary to surmount the obstacles due to thermal fluctuations. The results of Figs. (1-3) include the presence of viscous forces, but preclude the effect of thermal fluctuations. The effects of thermal fluctuations are presently being included in the simulations(17). These effects have been estimated (4) for dilute alloys and low temperatures, based on the assumption that, in this regime, the effect of increasing the temperature is equivalent to that of lowering the strength of all the obstacles in the array by an amount such as to maintain, on the average, the same waiting time for thermal activation at each equilibrium configuration. This indicates that in the presence of thermal fluctuations, Eq. (2) retains its form, where f_o and σ_c are now functions of temperature and the frequency ω of the applied periodic stress. Therefore, the implication of Eq. (2) as to the equivalence between σ_c and $\sigma_o(\Delta)$ remains unchanged (5).

3. Internal Friction Calculations for Mobile Obstacles

There is ample experimental evidence that at temperatures too low for bulk diffusion to occur, solutes are already mobile in the distorted crystal region adjoint to the dislocation core (2). Even though there is no lattice diffusion, the concentration of solutes that are effective pinners to a stationary dislocation can largely increase. Two mechanisms can be envisaged: (a) If the dislocation is initially curved or in a zig-zag configuration, it will tend to drag the solutes it is interacting with until becoming straight. In the process, the dislocation sweeps small regions of the glide plane, collecting extra solutes. (b) Since the strength of the dislocation-solute interaction is expected to decrease rapidly with the increase in the distance between the solutes and the glide plane, for most calculations it is sufficient to only consider the solutes lying on the two atomic planes adjacent to the glide plane (18). The rearrangement of the solutes near the core region will transform the weak obstacles into strong ones, increasing the number of effective obstacles.

At higher temperatures, a stationary dislocation also attracts solutes from distant regions in the lattice. The solute concentration near the dislocation increases until reaching a saturation value

$$c' = c \cdot \exp(W/kT) \tag{3}$$

where c is the solute concentration in the perfect lattice and W is the dislocation-solute interaction energy at the point where c' is evaluated. Equation (3) is valid, provided $c' << 1$.

A periodic applied stress of low amplitude can change this solute configuration. In a simplified picture the solutes can either move with the dislocation (in a path perpendicular to the dislocation line) or along the dislocation (pipe diffusion). Calculations of the decrement have been performed for these two extreme cases (see Review articles, Refs. (1), (2)) under the assumption that the total number of solutes interacting with the dislocation remains constant

*The temperature dependence of the elastic constants is here neglected.

(no breakaway events nor encounters with new obstacles). These calculations apply therefore to low stress amplitudes which is not the case of primary concern here. This topic will be discussed by Brailsford at this conference.

No internal-friction calculations have been made for higher stress amplitudes, taking into account both breakaway and diffusion. The available measurements show that for $T \lesssim 0.5\ T_m$ the decrement versus stress-amplitude curves retain the basic shape found at low temperatures, where the solutes are known to be fixed in the lattice. The observed rapid increase in δ for large stress amplitudes must thus be associated with the breakaway of the dislocations from the solute environment, whatever it may be. Measurements of $\sigma_o(\delta = \text{constant})$[see Eq. (2) and following discussion] therefore constitute a good gauge of the breakaway stress. The results are, however, meaningful provided the solute environment can be characterized. This is not only history-dependent, but may be affected by the measurements as well. Measurements using a reliable characterization are present in Section 4.4.

4. Experimental Observations

In this section we discuss experiments in which amplitude-dependent internal friction measurements have been applied to the study of solution strengthening in alloys. This discussion will also allow us to compare and evaluate the models that were presented in Section 2.

4.1 The force-distance law for the dislocation-solute interaction

The strengthening that the solutes can provide at low temperatures depends in principle on the interaction force F(s) between a dislocation and a single solute placed a distance s apart. The exact form of F(s) is not known. However, F(s) is expected to have the general features of the expression derived by Cottrell (19) using continuum linear elasticity. At very low temperatures, thermal fluctuations are only able to activate the breakaway of dislocations stopped near the top of F(s). It then follows from very general physical arguments (20) that, for a constant probability of overcoming the F(s) barrier, the applied stress and the temperature must be related by

$$\sigma = \sigma_m(1 - AT^{2/3}), \tag{4}$$

where σ_m is the stress necessary for the mechanical breakaway at T = 0 and A is proportional, among others, to the curvature of F(s) at its maximum. Measurements of the temperature dependence of the flow stress in constant-elongation-rate tests have often been fitted by theoretical curves based on different assumed force distance laws for the dislocation-obstacle interaction, but the measurements have not been sufficiently precise to distinguish between the theories or to determine the temperature law (21). The $T^{2/3}$ dependence of Eq. (4) has been clearly observed in amplitude-dependent internal friction measurements. Figure 4 shows measurements of σ_o (δ = constant) as a function of temperature (see Section 2.2.2) for crystals of Al, Cu and NaCl with a low density of residual impurities (22). The measurements show that the interaction profile increases in sharpness in the given order.

For lower stresses (higher temperatures), the dislocation does not anymore feel only the round top of the barrier, and the $\sigma(T)$ relationship deviates from that of Eq. (4). (See measurements for NaCl in Fig. 4.) The extent to which the method may be applied to map F(s) is determined by the transition from thermally activated events involving one obstacle to events involving several obstacles, in which case $\sigma(T)$ reflects statistical properties of the obstacle distribution rather than the features of a single obstacle (23). Estimates for this transition have been obtained for the case of a colinear obstacle distribution (24). For a planar distribution, likely to be encountered at low temperatures (see Section 4.2), the transition temperature may be obtained through dynamic computer simulations (including thermal activation) using a known profile F(s). The potentiality of the method has not been fully exploited.

4.2 Distribution of Solutes Interacting with the Dislocation

The numerical internal-friction calculations (5) described in Section 2.2.2 consider the dislocations interacting with a random distribution of obstacles on a plane that is likely to reflect that encountered by fresh dislocations at low temperatures. On the other hand, the GL model assumes a linear distribution of obstacles, which is more likely to represent that of aged dislocations, to which solutes have been able to segregate. A simple way to compare these calculations with experiments is to make use of the so-called GL plots of log $(\sigma_o\delta)$ versus

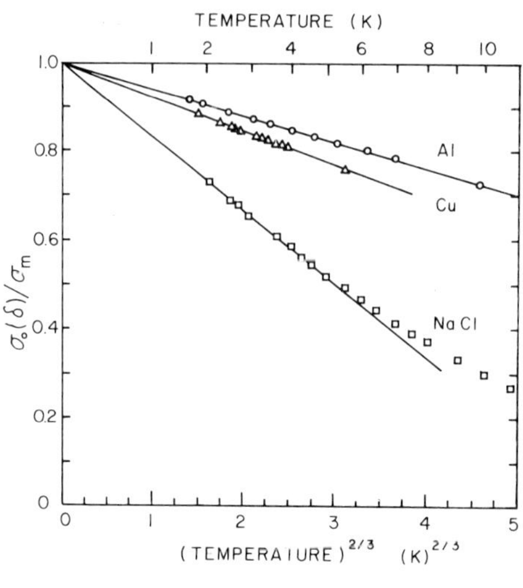

FIG. 4

Stress-amplitude as a function of (temperature)$^{2/3}$ required to produce a constant value δ of the amplitude-dependent decrement. The stress-amplitude measurements are normalized by the stress σ_m required for the breakaways at T = 0.

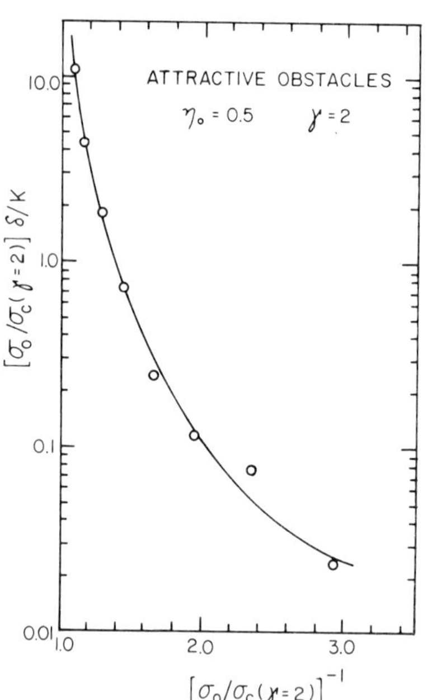

FIG. 5

GL-type plot of the computer-modeling internal-friction calculations (5) for dislocation motion through a random planar distribution of identical obstacles (normalized interaction range η_o = 0.5) and high viscous forces (γ = 2). $\sigma_c(\gamma = 2)$ follows from the flow stress calculations of Fig. 1. $K = \Lambda \ell_s^2/f_o$ (see the text).

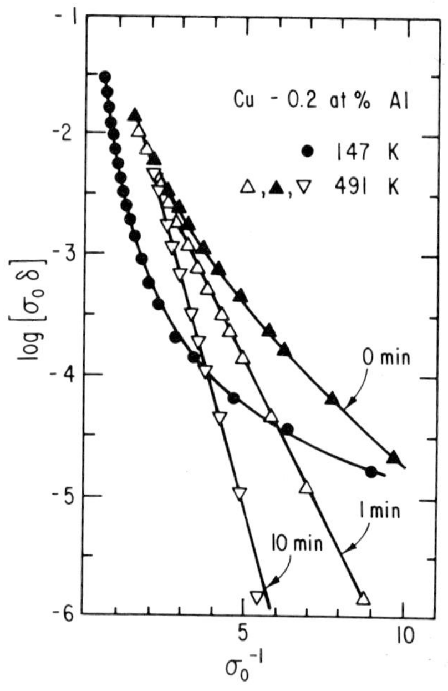

FIG. 6

GL-type plot of internal-friction measurements in a Cu-0.2 at.% Al single crystal at two temperatures, following thermal annealing periods of 0, 1 and 10 min. The stress-amplitude measurements are in arbitrary units.

σ_o^{-1}.(*) In such a plot the GL theory predicts the measurements to give a straight line. Figure 5 shows the calculation results for $\gamma = 2$ (also shown in Fig. 2, open symbols) plotted in the GL fashion. For comparison, Fig. 6 shows measurements (25) in a Cu-0.2 at.% Al single crystal made at two temperatures and for different annealing conditions, also plotted in the GL fashion. The measurements at 147 K were very reproducible and stable, showing no time-dependent effects. They present a concave curvature of the same nature as that shown by the calculations of Fig. 5. A similar concave curvature was observed at 491 K, provided the measurements were taken immediately following the straining of the sample to a stress amplitude higher than that used for the subsequent measurements (see also Section 4.4). When an annealing period under zero applied stress of one minute was interposed between the high-amplitude straining and the taking of the measurements, the data could be well fitted by an almost straight line. For an annealing period of ten minutes, the fitting curve is perfectly straight. This change in behavior reflects changes in the solute environment the dislocation interacts with: the curvature in the data for T = 147 K and T = 491 K (immediately following the high initial straining) is indicative of a planar solute distribution; the shift with annealing time of the curves for 491 K towards higher stresses, together with a straightening of these curves, is indicative of solute segregation to the stationary dislocations. This mechanism also provides an explanation for the difference in behavior observed between other alloy systems (Ref. 1, p. 368).

4.3 Flow stress at very low temperatures -- Dislocation dynamic effects

Within the last decade it has been realized that a superconducting alloy kept at a constant temperature is mechanically weaker when in the Superconducting (S) state than in the Normal (N) state. The many studies made since the initial reports of Kojima and Suzuki (26) and Pustovalov et al. (27) have been reviewed by Kostorz (28).

At present there is a rather general consensus that, as first suggested by Granato (29), Suenaga and Galligan (30), and Kamada and Yoshizawa (31), this decrease in strength reflects changes in the dynamic behavior of the dislocations which occurs with the lowering of the viscous drag constant B below a critical value. In superconductors this always happens when the material is taken from the N to the S state, because of the drastic reduction in the electronic contribution to B (32). The viscous drag changes as well with temperature (phonon contribution), and the anomalous T dependence of the flow stress found at very low temperatures in many dilute alloys (31, 33-35) has also been shown to correspond to dynamic dislocation effects (36).

Most of the theories and calculations that have addressed this problem within the framework of dynamic dislocation effects (29-31, 37, 38) have as a common feature that the decrease in alloy strength is attributed to the mechanical overshoot by a moving dislocation of the obstacles ahead on its trajectory. Except for the calculations by Kocks et al. (37), the above theories deal with point obstacles (interaction force of zero range).

It was recently shown by Schwarz (39) that the only quantity that decides whether a dislocation dynamically overcomes a point obstacle is its local velocity v before impact. If $v > v_s f_o$, where v_s is the sound velocity in the solid, the moving dislocation continues its motion unperturbed, as if the point obstacle had not been there. Thus, the flow stress for a dislocation in motion at $v > v_s f_o$ (for substitutional alloys, $f_o \approx 0.02$) is purely viscous, independent of the concentration of point obstacles and of the distribution in which the obstacles are located on the glide plane.

The computer-modeling calculations of Labusch and Schwarz (4) show that "overshoot" effects can explain only one aspect of the problem: As shown in Fig. 1, the decrease in B below the critical value decreases not only the minimum stress S_L that maintains a dislocation in motion, but also the stress S_c necessary to start a dislocation moving from rest. While the decrease in S_L is caused by dynamic "overshoot" events, along the lines first discussed in (29-31), the decrease in S_c is produced by dynamic "unzipping" events in which a dislocation initially at rest breaks away catastrophically from all the obstacles along its length. Because this requires a trigger event (the first breakaway), S_c is expected to be sensitive to thermal activation. (See Ref. 4, Eq. (32).) This is contrary to the case of S_L.

*The fact that the GL theory is for point-like obstacles, while the numerical calculations (5) are for extended obstacles, should play no role as long as we are dealing with very dilute alloys (Friedel limit).

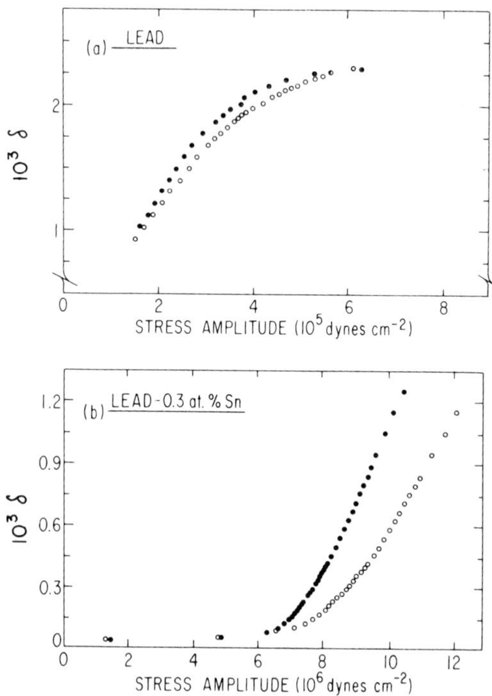

FIG. 7

Stress dependence of the decrement for (a) pure lead and (b) Pb-0.3 at.% Sn single crystals, in the normal (open circles) and superconducting (closed circles) states at 4.2 K (from Ref. 40).

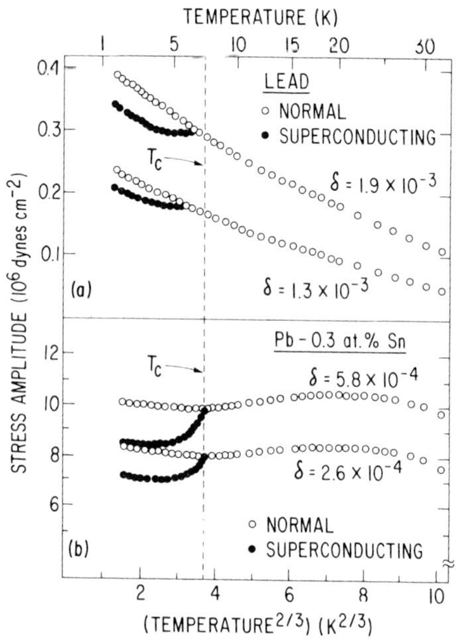

FIG. 8

Temperature dependence of the stress amplitude required to maintain constant decrement levels in (a) lead and (b) Pb-0.3 at.% Sn.

No theory has been advanced to predict the value of the flow stress measured in a constant-strain-rate experiment in terms of S_c and S_L. From the computer simulations (4) it is apparent that the measured flow stress should correspond primarily to S_c, with S_L affecting the flow stress only through the rate sensitivity of S_c. (The results of Fig. 1 do not include thermal activation events and thus S_c is rate-independent.)

Dislocation dynamic effects have also been studied through amplitude-dependent internal friction experiments. Figures 7 and 8 show internal-friction measurements by Isaac, Schwarz, and Granato (40) on Pb and Pb-0.3 at.% Sn alloy single crystals. These measurements were used to evaluate the several theories that have been proposed to explain flow-stress changes in sueprconducting materials. It was concluded that while "overshoot" effects could justify the general features of the measurements, a model based on dynamic "overshoot" events alone could not explain the fact that, as shown in Fig. 8, at the lowest temperatures $\sigma_0(\delta = \text{constant})$ decreases with the 2/3 power of T in both the N and S states and for both the pure and alloyed crystal. This is a characteristic of thermally activated rate processes (see Section 4.1).

The decrement versus stress-amplitude measurements of Figs. 7 and 8 can be directly compared to the results of the computer-modeling internal-friction calculations (5) shown in Figs. 2 and 3, which intrinsically include "overshoot" as well as "unzipping" events. Figure 7(a) shows the measurements in Pb (trace impurities, $c < 20$ ppm). In both the S and the N state $\delta(\sigma_0)$ presents a convex curvature reaching a peak (*). This behavior, which appears to be typical of the $\delta(\sigma_0)$ curves for very pure crystals (41), is due to the presence of strong obstacles (most probably forest dislocations) which limit the dislocation displacements to a low number of intersolute spacings. These measurements compare very favorably with the curves of Fig. 3 obtained with a superposition of a random planar distribution of weak obstacles plus fewer strong (impenetrable) ones. In agreement with these, the peak in the S state (lower viscosity) is shifted toward lower σ_0 values with respect to the peak in the N state (higher viscosity), while the height of the peak remains essentially unchanged.

In the Pb-0.3 at.% Sn alloy crystal the dislocations encounter mostly the solutes. The $\delta(\sigma_0)$ measurements of Fig. 7(b) should be compared with the curves of Fig. 2, calculated with a planar distribution of identical (weak) obstacles. As can be deduced from Fig. 8(b), the dislocations in the Pb-Sn alloy at 4.2 K in the N state are partially underdamped (i.e., $\gamma < 2$). [This conclusion is explained in great detail in Ref. (36).] Upon taking the sample to the S state, there is a further reduction on γ. The measurements of Fig. 7(b) should therefore be compared with the two segmented curves of Fig. 2, calculated with $\gamma \approx 0.5$ and 0.2, respectively.

The good qualitative agreement found between the internal friction calculations and the measurements, in particular that the decrease in the viscous drag affects the high-stress regions of the $\delta(\sigma_0)$ curves while, in both the Pb and the Pb-Sn alloy, the stress amplitude that denotes the onset of breakaway events remains largely unaffected by this change (see "microyield stress" in Fig. 1), supports the previous claim (Section 2.2.2) that $\sigma_0(\delta = \text{constant})$ measurements with large δ values are proportional to σ_c. Therefore, the observed $T^{2/3}$ dependence of the $\sigma_0(\delta)$ measurements in the S state of both Pb and Pb-Sn can be easily explained. A complementary explanation for this temperature dependence has been recently advanced by Isaac and Granato in terms of a rate theory which incorporates inertial-overshoot effects. Aspects of this work will be presented later in this conference.

4.4 Flow stress at high temperatures -- Diffusing solutes

The most recent interpretations of dynamic-strain-aging in alloys (42) assume that the dislocation kinetics in the presence of diffusing solutes is similar to that for fixed solutes: the dislocations spend times t_w waiting for thermally activated breakaway events (collecting extra solutes in the process) and times t_r, where $t_r \ll t_w$, gliding between these configurations. Therefore, a desirable quantity for dynamic-strain-aging studies is the stress necessary to remove a dislocation from the solute environment that develops after a waiting time t_w. This quantity can be measured by internal-friction techniques (Section 2.2.2), provided the solute environment can be initialized into a state resembling that encountered by glisile dislocations in macroscopic flow-stress tests. An obvious problem is that in the internal-friction tests the dislocations move back and forth over the same glide-plane area.

---

*The measurements in Ref. 40 were taken using the method described in Ref. 16 in which the oscillations are stable only if $d\delta/d\sigma_0 > 0$. Measurements beyond the peak could not be taken but the presence of the peak was experimentally established by the fact that the system became unstable, with the stress amplitude rapidly increasing by more than an order of magnitude.

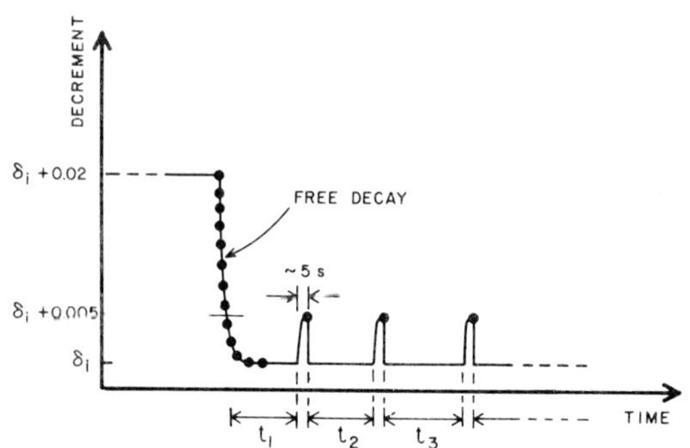

FIG. 9

Histogram of the internal friction measurements taken at a constant temperature on a Cu-0.2 at.% Al single crystals subjected to flexural vibrations. Each dot represents a measurement. During the free-decay of the oscillations, the stress amplitude is measured once every cycle (resonant frequency $\approx$ 1.3 KHz).

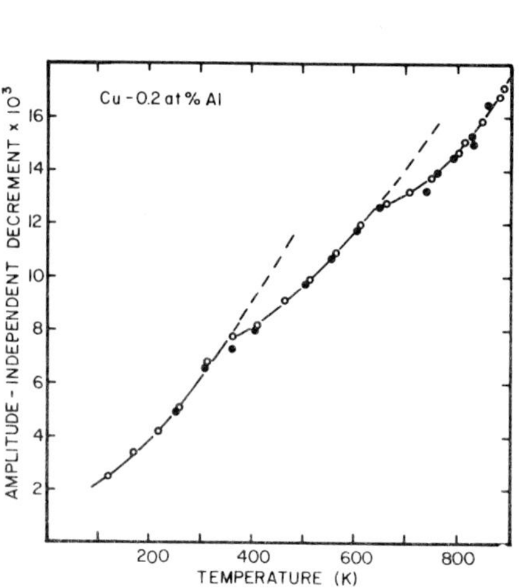

FIG. 10

Temperature dependence of the amplitude-independent decrement in a Cu-0.2 at.% Al single crystal measured from the free-decay of the oscillations starting from a large initial value. Open/solid circles correspond to increasing/decreasing temperature. (see text and Fig. 9)

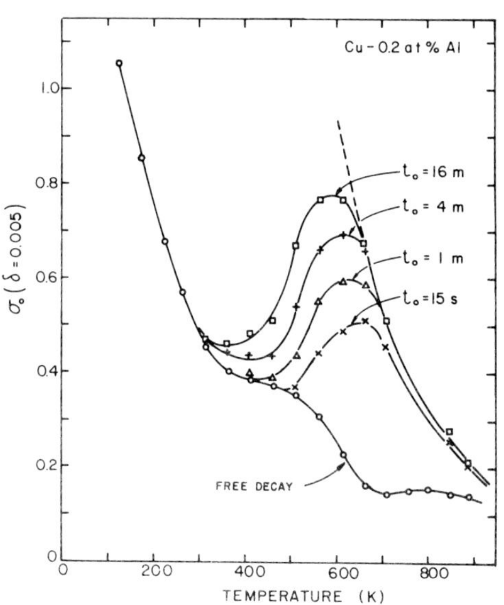

FIG. 11

Stress amplitude (in arbitrary units) required to maintain a value δ = 0.005 of the amplitude-dependent decrement in a Cu-0.2 at.% Al single crystal as a function of temperature and annealing times t_o (see the text).

The following set of internal-friction measurements, taken in a Cu-0.2 at.% Al single crystal between 100 and 900 K, are part of an ongoing study of alloy hardening in the presence of diffusing solutes. In these measurements we have chosen to initialize the samples by the application -- for a sufficiently long time -- of a stress amplitude such as to produce an amplitude-dependent decrement $\delta = 0.02$ (Fig. 9). This requires a stress-amplitude much larger than that of the subsequent measurements. At the end of this high-stress period the driving force is removed and the decrement versus stress-amplitude curve is measured during the free decay of the oscillations. At the lowest stress-amplitudes, the $\delta(\sigma_o)$ curve gives the amplitude-independent decrement, δ_i. Interpolating this "decay" curve with the value $(\delta_i + 0.005)$ produces the result $\sigma_o(\delta = 0.005$, decay). The free decay is followed by annealing periods of duration t_i during which the amplitude σ_o of the applied stress is extremely low and the dislocations are, for all practical purposes, stationary. After cumulative annealing times t_o, readings of $\sigma_o(\delta = 0.005, t_o)$ are obtained. At the end of this sequence the sample is again driven into large amplitude oscillations (compatible with $\delta = 0.02$) and the temperature is changed. Upon achieving a new stable temperature, the measurement sequence is repeated. Further details will be published separately (25).

Figure 10 shows the temperature dependence of the amplitude independent decrement δ_i, deduced from the free decay of the oscillations from the high initial amplitude. According to the GL theory (6), $\delta_i \propto B\ell^4$, where ℓ is the average length of the dislocation segments defined by the solutes (*). The average slope of the $\delta_i(T)$ curve reflects the increase in B with increase in temperature (larger phonon-drag component). In addition, $\delta_i(T)$ presents two sharp changes in slope at approximately 350 and 650 K, respectively. These are associated with changes in ℓ. On decreasing T, the measurements followed the same path, except near the first break in slope ($\sim$350 K), where the δ_i values sometimes jumped upward onto the extrapolation of the measurements below the first break.

The $\sigma_o(\delta, t_o)$ measurements in Fig. 11 show three distinct regimes: (a) For $T < T_1 = 300$ K, $\sigma_o(\delta, t_o)$ is independent of t_o indicating that the solutes are immobile everywhere. (b) For $T_1 < T < T_2(t_o)$, $\sigma_o(\delta, t_o)$ increases with increasing t_o reflecting the segregation of solutes to the stationary dislocations. (c) For $T > T_2(t_o)$, $\sigma_o(\delta, t_o)$ reaches, after time t_o, a curve that is again independent of t_o, indicating now that the solute concentration near the dislocations has achieved thermodynamic equilibrium with the lattice. The decrease in $\sigma_o(\delta, t_o)$ for $T > T_2$ must be attributed to a combination of thermal activation and the "evaporation" of the solute cloud according to Eq. (3).

Of great importance is the measure of the flow stress for a moving dislocation, which we consider equivalent to $\sigma_o(\delta, t_o \to 0)$. Our closest approach to this quantity are the $\sigma_o(\delta$, decay) measurements. The temperature dependence of $\sigma_o(\delta$, decay) in Fig. 11 shows two broad peaks (centered around 500 and 800 K, respectively) superimposed onto what would have otherwise been a smooth monotonic decay. These measurements were taken at an average rate of increase in temperature $\langle dT/dt \rangle \gtrsim 1$ K/m. (It takes $\sim$1 hour to reach a constant temperature and take a series of measurements with different t_o.) In order to further investigate the nature of this curve, $\sigma_o(\delta$, decay) measurements were also taken with $\langle dT/dt \rangle \gtrsim 3$ K/m (Fig. 12). At the faster rate, the first peak is seen to decrease in amplitude and to shift to higher T values. It seems safe to speculate that an even higher value of <dT/dt> would make this peak disappear, as indicated by the segmented curve on Fig. 12. On the other hand, the second peak remained unchanged, both for increasing and decreasing temperature changes. On decreasing T below 600 K, the $\sigma_o(\delta$, decay) curve took a different course (independent of dT/dt), joining the low-temperature $\sigma_o(\delta$, decay) curve at a stress amplitude $\sigma_o \gtrsim 0.6$. Abrupt drops in $\sigma_o(\delta$, decay) were sometimes observed near this σ_o value. These measurements, together with the $\delta_i(T)$ results of Fig. 10, suggest that the anomalies in $\sigma_o(\delta$, decay) for $300 < T < 600$ K are a consequence of the accumulation of solutes on the dislocation glide planes. This requires some further discussion. Let us consider first the measurements under increasing T. For $T < 300$ K, the dislocations encounter a planar distribution of solutes characteristic of the random alloy. For $T \gtrsim 300$ K, solute trapping begins to occur, most probably while the dislocations are at rest (annealing periods t_o), and involves only solutes very near the glide planes, diffusing through the distorted lattice that surrounds the dislocation. During the high-amplitude oscillations ($\delta = 0.02$) that precede the measurements at the next higher temperature, the dislocations are moving back and forth, but the excess solutes on the glide plane cannot diffuse

*This result was deduced for a linear obstacle distribution. However, the analytic dependence of δ_i should be preserved for other distributions. See also Brailsford (this conference).

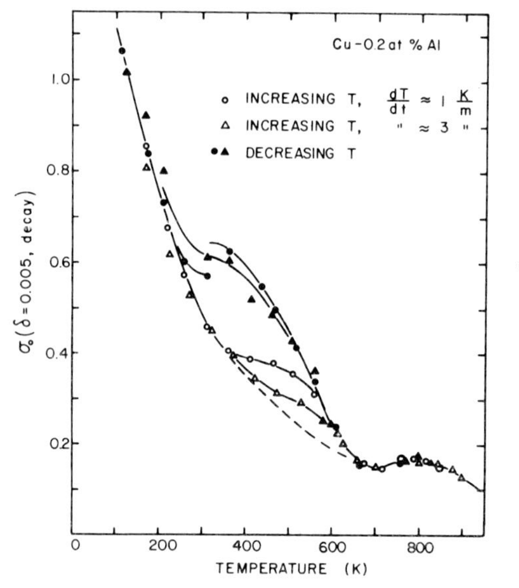

FIG. 12

Temperature dependence of the stress-amplitude (in arbitrary units) required to maintain a value $\delta = 0.005$ of the amplitude-dependent decrement in a Cu-0.2 at.% Al single crystal, obtained from the free-decay of the oscillations starting from a decrement value $\delta = 0.02$ (see Fig. 9 and the text).

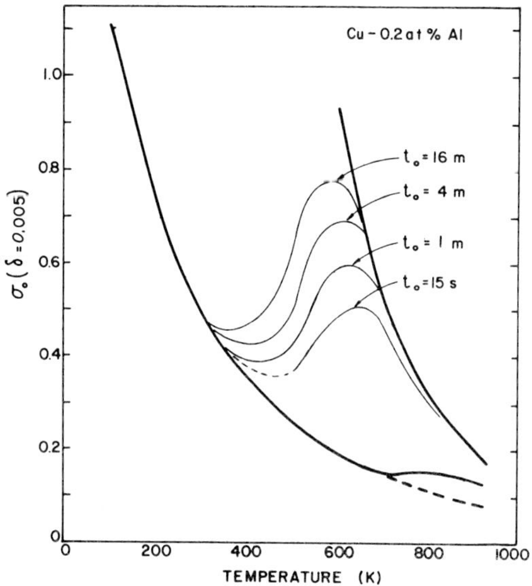

FIG. 13

Expected temperature dependence of the solution-hardening component of the flow stress for dislocations in Cu-0.2 at.% Al alloys. Lower heavy-trace curve: flow stress for a continuusly moving dislocation. Upper heavy-trace curve: stress required to remove a dislocation from a solute cloud in thermodynamic equilibrium with the lattice.

back into the lattice. They begin to leave the glide planes when $T \sim 500$ K, with the advent of lattice diffusion. For the measurements taken under decreasing T, the effects of the solute trapping on the glide planes are more accentuated. (Fig. 12) Since the solutes cannot escape once $T \lesssim 500$ K (this value is rather speculative), the $\sigma_o(\delta$, decay) curve should continue along a curve parallel to that for increasing temperature but shifted to higher stress, corresponding to a higher solute density. However, as T is further decreased, the stress amplitude required for $\delta = 0.02$ becomes so high that the dislocations cross-glide or new ones are generated somewhere else. This also explains the random and abrupt jumps observed in the otherwise very smooth $\sigma_o(\delta$, decay) and $\delta_i(T)$ curves. After this happens, the $\sigma_o(\delta$, decay) curve joins the low-temperature curve characteristic of fresh dislocations on the random alloy.

We conclude therefore that the anomalies in the T dependent of $\sigma_o(\delta$, decay) for $250 < T < 600$ K are characteristic of the internal friction method but are not representative of the flow stress for a dislocation in a *continuous* unidirectional motion. However, a similar glide-plane solute-trapping may occur in this temperature regime for arrays of dislocations moving consecutively on the same glide plane.

Based on our previous discussion, Fig. 13 shows qualitatively the flow stress for a dislocation in a Cu-0.2 at.% Al single crystal, as a function of T and t_o. The lower heavy-trace curve gives the flow stress for a dislocation in continuous motion. We cannot presently ascertain if the peak at 800 K is an intrinsic effect or another manifestation of the periodic nature of the glide. The upper heavy-trace curve gives a measure of the stress required to free a dislocation surrounded by a solute cloud in thermodynamic equilibrium with the rest of the lattice. The light-trace curves for different t_o, give the transition between these two extremes. These curves should help attain a better understanding of both static and dynamic strain aging effects in alloys.

The present internal-friction measurements show that in dilute CuAl alloys, localized diffusion starts at a temperature $T_1 \gtrsim 300$ K. Therefore, experiments that purport to evaluate the different solution-strengthening theories derived for fixed solutes (18) should be ideally performed at temperatures below T_1. This has not been always the case (18). The present results indicate that in the temperature regime $300 < T \lesssim 500$ K, stationary dislocations capture solutes from a prismatic region parallel to the core, the cross section of which increases rapidly with increasing temperature. Therefore, in this T regime, the c dependence of the flow-stress in constant elongation-rate tests should be approximately linear, rather than $c^{\frac{1}{2}}$, as expected on theoretical grounds for dilute alloys with fixed solutes. A linear concentration dependence in the flow stress of dilute CuAl alloys at 473 K has been measured by Basinski, Foxal, and Pascual (43).

5. Conclusions

Under certain conditions, internal friction measurements can be advantageously used to study solution strengthening in alloys. The interpretation of these measurements has been greatly improved by the recent calculations based on computer-modeling, which have established a quantitative correspondence between the amplitude-dependent internal-friction losses and the solution-strengthening component of the flow stress. This equivalence means that, provided the solutes are randomly distributed on the glide planes, the information gained by the internal-friction method from the motion of the dislocations through the relatively weaker regions of the glide planes (regions with a lower obstacle density) is sufficient to characterize the strongest regions as well.

The proportionality between $\sigma_o(\Delta)$ and σ_c that follows from Eq. (2) is based on computer-modeling calculations that do not include thermal fluctuations. It has been claimed (Section 2.2.2) that this relation should also hold for dilute alloys (Friedel limit) at low temperatures. For higher alloy concentrations (Mott limit) and higher temperatures, this proportionality may change, especially at low stress amplitudes, the reason for it being that in the weaker glide-plane regions the dislocations may be able to overcome the solutes one at a time, while glide through the denser regions may require thermal-activation events over groups of obstacles. Only if the dislocations are forced to glide macroscopically would they "feel" these more athermal effective obstacles.

References

1. R. DeBatist, in Internal Friction of Structural Defects in Crystalline Solids, Vol. V. of Defects in Crystalline Solids, edited by S. Amelinckx, R. Gevers and J. Nihoul (North Holland, New York, 1972).
2. R. W. Balluffi and A. V. Granato in "Dislocations in Solids", edited by F. R. N. Nabarro (North Holland, Amsterdam, 1980), Vol. 4, chap. 13.
3. R. Labusch and R. B. Schwarz, in Computer Simulation for Materials Applications, Proc. of International Conference held at NBS, Gaithersburg, MD, edited by R. J. Arsenault, J. R. Beeler, Jr., and J. A. Simmons, Nuclear Metallurgy 20, 650 (1976).
4. R. B. Schwarz and R. Labusch, J. Appl. Phys. 49, 5174 (1978).
5. R. B. Schwarz, submitted to Acta Metall.
6. A. V. Granato and K. Lücke, J. Appl. Phys. 27, 583 (1956).
7. J. S. Koehler, in Imperfections in Nearly Perfect Crystals, edited by W. Schockley, J. H. Hollomon, R. Maurer, and F. Seitz (John Wiley & Sons, Inc., New York,1952), Ch. 7, p. 197.
8. J. C. Swartz and J. Weertman, J. Appl. Phys. 32, 1860 (1961).
9. P. Peguin and H. K. Birnbaum, J. Appl. Phys. 39, 4428 (1968).
10. D. G. Blair, T. S. Hutchinson and D. H. Rogers, J. Appl. Phys. 40, 97 (1969); Can. J. Phys. 48, 2943,2955 (1970); Can. J. Phys. 49, 633 (1971).
11. D. G. Blair, J. Appl. Phys. 43, 37 (1972).
12. J. Friedel, in The Relation Between the Structure and Mechanical Properties of Metals, Proceedings of Conference held at the National Physical Laboratory, Teddington, England (Her Majesty's Stationery Office, London, 1963), p. 410.
13. J. Schlipf, in Internal Friction and Ultrasonic Attenuation in Solids, Proceedings of the Sixth International Conference held in Tokyo, edited by R. R. Hasiguti and N. Mikoshiba (University of Tokyo Press, 1977), p. 91.

14. R. Labusch, Acta Met. 20, 917 (1971).
15. R. Labusch, G. Grange, J. Ahearn and P. Haasen, in Rate Processes in Plastic Deformation of Materials, Proceddings of the John E. Dorn Symposium, edited by J. C. M. Li and A. K. Mukherjee (Amer. Soc. for Metals, Metals Park, Ohio, 1975), p. 26.
16. R. B. Schwarz, Rev. Sci. Inst. 48, 111 (1977).
17. R. Labusch and R. B. Schwarz, to be published.
18. P. Haasen, in "Dislocations in Solids", edited by F. R. N. Nabarro (North Holland, Amsterdam, 1980), Vol. 4. ch. 15.
19. A. H. Cottrell, in "Report of a Conference on Strength of Solids", H. H. Wills Physical Laboratory, University of Bristol (The Physical Society of London, England, 1948), p. 30.
20. N. F. Mott, Phil. Mag. 1, 568 (1956).
21. K. Ono, J. Appl. Phys. 39, 1803 (1968).
22. R. B. Schwarz and A. V. Granato, Phys. Rev. Letters 34, 1174 (1975). The data for Cu and NaCl is unpublished work by the same authors.
23. U. F. Kocks, A. S. Argon and M. F. Ashby, in "Thermodynamics and Kinetics of Slip", edited by B. Chalmers, J. W. Christian and T. B. Massalski (Pergamon Press, New York, 1975), p. 141.
24. L. J. Teutonico, K. Lücke, F. W. Heuser, and A. V. Granato, J. Acoust. Soc. Amer. 45, 1402 (1969).
25. R. B. Schwarz and L. Funk, to be published.
26. H. Kojima and T. Suzuki, Phys. Rev. Lett. 21, 869 (1968).
27. V. V. Pustovalov, V. I. Startsev, D. A. Didenko and V. S. Fomenko, Fiz. Metal. i Metalloved. 23, 312 (1967).
28. G. Kostorz, Phys. Stat. Sol. (b) 58, 9 (1973).
29. A. V. Granato, Phys. Rev. Lett. 27, 660 (1971); Phys. Rev. B4, 2196 (1971).
30. M. Suenaga and J. M. Galligan, Scr. Met. 5, 829 (1971).
31. K. Kamada and I. Yoshizawa, J. Phys. Soc. Japan 31, 1056 (1971).
32. A. V. Granato, in "Internal Friction and Ultrasonic Attenuation in Solids", Proc. of 6th International Conference held in Tokyo, edited by R. R. Hasiguti and N. Mikoshiba (University of Tokyo Press, 1977), p. 81.
33. R. E. Jamison and F. A. Sherill, Acta Metall. 4, 197 (1956).
34. T. Suzuki and T. Ishii, in "Physics of Strength and Plasticity", edited by A. S. Argon (The MIT Press, Cambridge, 1969), p. 159.
35. Z. S. Basinski, R. A. Foxall, and R. Pascual, Scripta Met. 6, 807 (1972).
36. R. B. Schwarz, R. D. Isaac and A. V. Granato, Phys. Rev. Lett. 38, 554 (1977).
37. See Ref. 23, p. 79.
38. V. L. Indenbom and Yu Z. Estrin, J. Low Temp. Phys. 19, 83 (1975).
39. R. B. Schwarz, Phys. Rev. B (in press; ∿June 1980).
40. R. D. Isaac, R. B. Schwarz and A. V. Granato, Phys. Rev. B 18, 4143 (1978).
41. See, for example, measurements by S. Takahashi, J. Phys. Soc. Japan 11, 1253 (1956).
42. A. van den Beukel, Acta Metal. (in press). See also references therein.
43. Z. S. Basinski, R. A. Foxal and R. Pascual, unpublished data shown in F. R. N. Nabarro, Z. S. Basinski and R. Pascual, Scripta Met. 12, 931 (1978).

A RATE THEORY OF DISLOCATION MOTION INCORPORATING INERTIAL OVERSHOOT EFFECTS*

R. D. Isaac† and A. V. Granato
Materials Research Laboratory and Department of Physics
University of Illinois at Urbana-Champaign
Urbana, Illinois 61801

Abstract

A rate theory of dislocation motion which incorporates inertial overshoot effects is developed by extending the results for Brownian motion of a particle in a force field. The theory provides a probability distribution function $\rho(x,v,t)$ that the system is at a position x with velocity v at time t. The distribution is centered about the purely mechanical solution. This provides a satisfactory theoretical basis for the observed mechanical-thermal dichotomy in experimental plastic flow and internal friction results. When the damping constant is changed suddenly the changes are mechanical-like. For changes with temperature, the distribution effects dominate, and the behavior is thermal or rate-like. For small damping coefficients B, the advance distance per thermally activated event is strongly dependent on B. For moderate B, the usual rate theory is obtained. For large B, the frequency factor ν is reduced. Thus a phase diagram separating different regimes and behavior is obtained.

Theories of deformation and internal friction of dilute alloys have developed independently to a large degree, each emphasizing different aspects of dislocation behavior. For example, the role of viscosity, as measured by the drag constant B in the familiar relation

$$Bv = b\sigma \ , \tag{1}$$

plays a dominant role in internal friction theories, but does not appear in traditional theories of plastic flow. In Eq. 1, v is the dislocation velocity, b the Burgers vector and σ is the applied stress. But in both fields, the basic interaction is that of dislocations with point defects and phonons, and there must exist close relations between the effects.

The first hint that viscosity plays an important role in deformation came with the discovery by Kojima and Suzuki (1) and Pustovalov, et al. (2) of a drop in flow stress of lead entering the superconductivity state. The only thing that changes during the transition is the dislocation viscosity provided by the scattering of normal electrons.

The basic notions used in describing plastic flow can be illustrated with Fig. 1. The plastic strain rate $\dot{\varepsilon}$ is given (3) by

$$\dot{\varepsilon} = \Lambda b v \qquad (2)$$

where Λ is the mobile dislocation density and v is the average velocity. The motion is impeded by obstacles, such as point defects, which can be represented either as positive or negative (repulsive or attractive) potential wells. The effect of an external stress is represented by a tilting of the line between obstacles by an amount proportional to the applied shear stress σ.

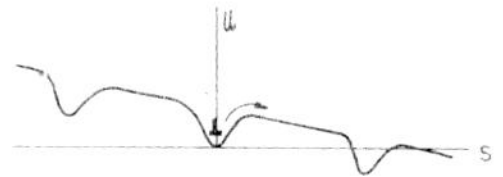

FIG. 1

Schematic representation of dislocation motion under stress through a crystal containing obstacles.

In the standard theories of plasticity (4), it is supposed that the rate-limiting step is provided by the overcoming of the obstacles by thermal fluctuations. There are two implicit assumptions made: (1) that the process is a quasistatic one in which the dislocations do not overshoot the obstacles by reason of their inertia, and (2) that the viscous drag provided by the medium is not so large that the time required to travel between the obstacles is greater than that to overcome them. Overshoot should occur if the viscous drag provided by phonons, electrons, and other point defects is very small, and viscoelastic behavior should occur if it is very large.

A classification of plasticity effects according to the magnitude of the viscosity, with examples and discussion of each type, has been given by Granato (5).

An example of the "large drag" (viscoelastic) case is given by measurements of deformation of ice by Joncich, Holder and Granato (6). Here one has for the total strain $\varepsilon = \varepsilon_{e\ell} + \varepsilon_d$, so that

$$\dot{\varepsilon} = \frac{\dot{\sigma}}{G} + \frac{\Lambda b^2 \sigma}{B} \qquad (3)$$

where $\varepsilon_{e\ell} = \sigma/G$, G is the elastic shear modulus, and Eq. (1) has been used in Eq. (2).

For the "medium drag" (quasistatic) case, the dislocation velocity is given by rate theory as

$$v = d\nu \exp[U(\sigma)/kT] \qquad (4)$$

where d is the distance between obstacles, ν is the effective attack frequency, $U(\sigma)$ is the activation energy required to overcome the obstacle and kT has its usual meaning. For low temperature and relatively pure materials, it was shown by Granato and Lücke (7) that the stress-temperature law for constant strain-rate deformation should have the form

$$\sigma = \sigma m - AT^{2/3} \qquad (5)$$

This has been observed by Schwarz and Granato (8). The significance of this relation is that its observation is evidence that deformation is initiated by thermal breakaway of a dislocation from a single pinning point.

For the "small drag" (inertial) case, evidence is obtained from measurements of the stress change in superconductors entering the superconducting state and from the anomalous maximum in the temperature dependence of the flow stress at low temperatures of dilute alloys.

Two types of theories have been offered to explain the superconducting effect. One by Natsik (9), modified by Suzuki (10), and by Estrin (11), supposes that the frequency factor ν in the rate theory expression in Eq. (4) decreases in the normal state. In a second theory (the mechanical inertial model)(12-15) it is supposed that the drag constant B at low temperatures is small enough so that dislocations can mechanically overshoot the obstacles inertially. The mechanical inertial model is analogous to a loaded spring in a viscous medium falling under a force corresponding to the applied stress on a dislocation. For large viscosity the dislocations approach their static equilibrium positions against obstacles exponentially, while for viscosities less than a critical value, the dislocations may overshoot this static equilibrium value and cut through the obstacles at a stress lower than that required for high viscosity. This simple mechanical model accounts for many of the observed changes in mechanical properties of superconductors entering the superconducting state. In particular, it predicts that the stress change at constant strain rate will go to zero below the superconducting transition temperature for sufficiently pure materials. In addition, it predicts that an anomalous maximum in the flow stress of dilute alloys will occur at temperatures where the phonon density is large enough to overdamp the dislocations. This behavior was subsequently observed by Schwarz, Isaac, and Granato (16) in lead and copper. The Natsik model cannot account for these facts.

However, this mechanical model cannot explain the fact that the flow stress is temperature dependent with a characteristic $T^{2/3}$ rate theory temperature dependence at low temperatures in both the normal and superconducting states. This temperature dependence, illustrated in a figure in the paper by Schwarz (17) at this conference, suggested to Schwarz, Isaac and Granato (16) that in the presence of inertial effects, dislocation motion is initiated by a thermally activated process and continues by the overcoming of additional obstacles inertially. We thus have a dilemma in which changes of viscosity at constant temperature appear to be mechanical-like, while changes with temperature at constant viscosity appear to be thermal like.

We have recently developed a generalized rate theory in which the effects of viscosity are taken into account explicitly and which resolves the above mechanical-thermal dichotomy. It is found that the frequency factor ν in Eq. (4) is not dependent on the damping constant B for the range of B expected, but that the advance distance d per thermally activated event in Eq. (4) is strongly dependent on B. The theory is an extension to the many degrees of freedom corresponding to dislocation motion of some earlier results found for Brownian motion of a particle in a force field by Kramers (18), Chandrasekhar (19), and Wang and Uhlenbeck (20).

A dislocation moves in a multidimensional potential. When a thermally activated transition is made from one stable position to another, the reaction path goes over the lowest energy saddle point dividing the stable positions. This reaction path can be considered to be an independent coordinate corresponding to a particular normal mode of the system. The remaining normal modes influence the activation rate only through an entropy factor. The problem is thus reduced to consideration of a system moving in a one-dimensional potential together with a determination of the normal modes of the system.

The potential including the effects of the applied external stress, the obstacles, and the dislocation line tension is then as indicated in Fig. 2 for repulsive obstacles. The reaction coordinate may be thought of as the average dislocation displacement. This is in contrast to the displacement at the obstacle position, which has often been used in other considerations. In this representation the curvature at the bottom of the well is given by the lowest normal mode frequency ω_A, and at the top of the barrier is called ω_C. With this potential one can determine a probability distribution function $\rho(x,v,t)$ that the system is at a position x with velocity v at time t. There are no thermal equilibrium distributions for this potential. With this distribution one can calculate the rate ν at which particles leave the well at A, and the average distance d traveled before they become trapped again at a new well.

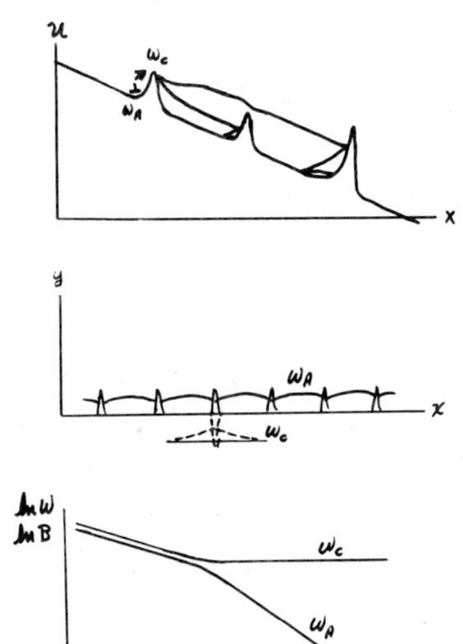

FIG. 2
Schematic sketch of potential energy versus reaction coordinate for repulsive obstacles (a), dislocation displacements for the normal modes ω_A and ω_C (b), and frequency or drag constant versus loop length (c).

In general, both ν and d depend upon B. For $B<2A\omega_A$, overshoot effects occur making d a function of B. The solution for $\rho(x,v,t)$ is centered about the purely mechanical solution for the dislocation displacement. As indicated in Fig. 2a, particles initially localized in well A, diffuse over the barrier and are stopped at the next barrier for large B, but may overshoot this barrier for small B. The distribution function subsequently settles into one approximating a Boltzmann distribution. The result for ν is

$$\nu = \frac{\omega A}{2\pi}\left(\frac{\alpha-\beta}{a}\right)^{\frac{1}{2}}\left[\prod_{j=2}^{N}\omega_j \Big/ \prod_{j=2}^{N}\omega_j'\right] \tag{6}$$

where $a = \beta/2 + (\beta^2/4 + \omega_c^2)^{\frac{1}{2}}$ and $\beta = B/A$ where A is the dislocation mass per unit length. The ratio of products in Eq. (6) is the entropy factor $e^{\Delta S/K}$ considered earlier by Granato et al. (21). For $B < 2A\omega_c$, Eq. (6) becomes the familiar result $\nu = (\omega_A/2\pi)\exp(\Delta S)/k$, while for $B > 2A\omega_c$ this result is decreased by a factor of $2A\omega_c/B$. The possibility of a decreasing ν with increasing B forms the basis of Natsik's theory. However, he compared B with ω_A rather than ω_c. Suzuki and Estrin corrected this and Suzuki suggested that ω_c might be much less than ω_A, making this an important effect.

The equation of motion which determines the normal mode frequencies resembles the Schroedinger equation with a potential which can be represented by delta functions at the obstacle positions with strength f, where $f = -dF/dy$ evaluated at the equilibrium positions and F(y) is the pinning force of the obstacle as a function of the displacement y of the dislocation. This is as indicated in Fig. 2b. The results depend upon the obstacle separation ℓ. For large ℓ, $\omega_A \to \omega_\ell$, where $\omega_\ell = \pi c/\ell$, where c is the shear sound velocity. For small ℓ, $\omega_A \to \omega_p$ where $\omega_p^2 = fc^2/\mu\ell$ and μ is the dislocation line tension. In the former case the dislocation displacement is resisted primarily by the dislocation line tension, while in the latter case it is resisted mainly by the obstacle compliance. In general

$$1/\omega_A 2 = 1/\omega_\ell 2 + 1/\omega_p 2 \tag{7}$$

The (negative) frequency ω_c^2 is found by letting one of the delta functions have negative strength, corresponding to a bound state. The result is independent of ℓ for large ℓ and both ω_A and ω_c are shown in Fig. 2c. One sees that for fixed f, $\omega_c>\omega_A$ for all ℓ, contrary to the supposition of Suzuki. Further, since the critical values of B are given by $2A\omega_c$ for the frequency factor; and $2A\omega_A$ for the displacement d, this curve can also be used to represent the regions of different behavior of ν and d. For phonon damping at $T = 4\theta$, we estimate that $B < 2A\omega_c$. We therefore conclude that the frequency factor will not normally depend on B.

For fixed B, short enough loops will always be underdamped and able to overshoot obstacles. The probability p of overshoot can be obtained from $\rho(x,v,t)$. For overdamped dislocations, overshoot occurs only for $\sigma>\sigma_m$. For underdamped dislocations, overshoot occurs at a lower stress. The probability p is step-like, becoming sharper at lower temperatures. This is the point at which contact is made with the older mechanical inertial model, which simply described this behavior as a step function. Using this result one finds $d = d_o/(1-p)$ for the distance advanced per activation. If a distribution $P_e(\sigma_m)$ of barriers is assumed, then this is replaced by $d = d_o/[\,(1-p)P_e(\sigma_m)d_m]$. Fig. 2a can be used to illustrate how the theory explains the mechanical-thermal dichotomy referred to earlier. For changes of viscosity at constant temperature, the center of the distribution, which follows the purely mechanical path, changes so that the effects appear to have mainly mechanical character. However, for changes with temperature at constant viscosity, the principal effect is a broadening of the distribution, which has a characteristically thermal effect.

Fig. 2c can be used as a phase diagram to separate regions of different behavior according to the magnitude of the viscosity. Below the bottom line for ω_A, inertial effects occur. Between the two curves for ω_A and ω_c, quasistatic effects occur and neither d nor ν are viscosity dependent. Above the curve for ω_c, the frequency factor ν is viscosity dependent.

The theory thus provides a satisfactory foundation for a discussion of thermally activated effects including the effects of viscosity. As a Brownian Motion approach, it is free of objections to previous rate theories based on equilibrium considerations. It has been applied both to macroscopic plastic flow phenomena and to internal friction and gives a good account of the superconducting and anomalous temperature dependence of the flow stress. It delineates the regimes of different types of behavior in terms of ultrasonically measurable dislocation parameters such as the viscosity, dislocation density and segment lengths and brings a certain degree of unification to the two fields of plastic flow and internal friction.

A different approach to some of these questions has been taken by R. B. Schwarz and R. Labusch (22) with computer modelling. Their work may be considered complementary to ours. They deal more easily with statistical aspects of the problem by computer, but it is basically a mechanical approach.

Acknowledgements: *This research was supported by the U.S. Department of Energy Contract No. EY-76-C-02-1198 and the U.S. National Science Foundation Grant DMR-77-10556. †Present address: IBM Thomas J. Watson Research Center, Yorktown Heights, NY 10598.

References

1. H. Kojima and T. Suzuki: Phys. Rev. Lett. 21, 896 (1968).
2. V. V. Pustovalov, V. I. Startsev, D. A. Didenko, and V. S. Fomenko: Fiz. Metal. i Metalloved. 23, 312 (1967).
3. F. Seitz and T. A. Read: J. Appl. Phys. 12, 470 (1941).
4. A. Seeger, in Dislocations and Mechanical Properties of Crystals, ed. by J. C. Fisher, W. G. Johnston, R. Thomson, and T. Vreeland, Jr. (John Wiley and Sons, Inc., New York, 1957), p. 243.
5. A. V. Granato, In Fundamental Aspects of Radiation Damage in Metals, ed. by M. T. Robinson and F. W. Young, Jr. (National Information Services, Springfield, VA, 1976), p. 932.
6. D. Joncich, J. Holder, and A. V. Granato, to be published.
7. A. V. Granato and K. Lucke, to be published.
8. R. Schwarz and A. V. Granato: Phys. Rev. Lett. 34, 1174 (1975).
9. V. S. Natsik: Soviet Physics JETP 34, 1359 (1972).
10. T. Suzuki, in Rate Processes in Plastic Deformation (American Society for Metals, Metals Park, Ohio, 1976).
11. Yu. Z. Estrin, Phys. Stat. Sol. (a) 26, K161 (1974).
12. A. V. Granato: Phys. Rev. Lett. 27, 660 (1971).
13. A. V. Granato: Phys. Rev. B4, 2196 (1971).
14. M. Suenaga and J. M. Galligan: Scripta Met. 5, 63 (1971).
15. K. Kamada and I. Yoshizawa: J. Phys. Soc. Japan 31, 1056 (1971).
16. R. B. Schwarz, R. D. Isaac, and A. V. Granato: Phys. Rev. Lett. 38, 554 (1977).
17. R. B. Schwarz, see paper at this conference.
18. H. A. Kramers: Physica 1, 284 (1940).
19. S. Chandrasekhar: Rev. Mod. Phys. 15, 1 (1943).
20. M. C. Wang and G. E. Uhlenbeck: Rev. Mod. Phys. 17, 323 (1945).
21. A. V. Granato, K. Lücke, J. Schlipf and L. J. Teutonico: J. Appl. Phys. 35, 2732 (1963).
22. R. B. Schwarz and R. Labusch, in Computer Simulation for Materials Applications, Proceedings of International Conference held at NBS, Gaithersburg, MD, edited by R. J. Arsenault, J. R. Beeler, Jr., and J. A. Simmons, Nuclear Metallurgy 20, 650 (1976). See also R. B. Schwarz and R. Labusch, J. Appl. Phys. 49, 5174 (1978).

SOME ASPECTS OF DISLOCATION - POINT DEFECT INTERACTIONS AND THEIR SIGNIFICANCE TO INTERNAL FRICTION

A. D. Brailsford

Engineering and Research Staff-Research
Ford Motor Company
P. O. Box 2053
Dearborn, Michigan 48121

Abstract

A short review of the salient aspects of the Koehler-Granato-Lücke theory of strain-amplitude independent internal friction arising from mobile dislocation segments is given. The subsequent extensions by Thompson and Holmes and Simpson and Sosin to account for point defect pinning and dragging, respectively, are also briefly discussed. These mechanisms invoke the presence of dislocation core-entrapped point defects. An additional damping mechanism arises from diffusive motion of the point defects in solution in response to the time dependent stress field of the vibrating dislocation segments. The magnitude of this effect is estimated. It is possible that it might explain the anomalous low frequency (kHz) dependence of the damping if some trace impurity were to be unusually mobile. The concept of pinning due to irradiation-produced self-defects is also analyzed. The result indicates it to be extremely unlikely that single self-interstitials can act as dislocation pins at low frequencies. Additionally, the topic of Thermodynamic forces upon pinning points is addressed. It is pointed out that besides the entropy force on pinning points, a force originating in the dislocation damping exists. This internal energy force has the opposite effect to that due to vibrational entropy, while the two are of comparable magnitude. As a consequence it is suggested that Thermodynamic forces may be ignored. The paper concludes with a brief discussion of the nature of the point defect capture process by dislocations. It is shown that a previously-ignored interaction arising from induced dislocation glide arises. Some tentative remarks upon the manner in which this may enter into the Internal Friction problem are given.

SOME ASPECTS OF DISLOCATION - POINT DEFECT INTERACTIONS AND THEIR SIGNIFICANCE TO INTERNAL FRICTION

A. D. Brailsford

Engineering and Research Staff-Research
Ford Motor Company
P. O. Box 2053
Dearborn, Michigan 48121

1. Introduction

The past two decades have witnessed an enormous expansion in the study of the Internal Friction of solids. Significant advances have been made in many areas related to dislocations and point defects, while others are currently in an active state of development. Inspection of reviews in standard texts (1-4) and of recent conference reports (5,6) will yield a ready appreciation of the present scope of this burgeoning field.

Because of this activity, it is no longer feasible to give extensive summary reviews of specific defect phenomena associated with the different solid types and at the same time do justice to the many fine experimental and theoretical contributions of individual investigators. This is no less true of internal friction processes attributed to point defect-dislocation interactive effects than it is of those associated with each defect type separately. Thus we shall not even attempt it in exhaustive detail. Rather the approach developed here might in the first instance be viewed as more pedagogic than practical. For our aim will be to take the simplest model of dislocation dynamics, the simplest model of a point defect, and seek the theoretically expected consequences. Experimental results will be referred to only to the extent that they aid the development or suggest unresolved topics.

The above approach is a somewhat blinkered view, admittedly. Yet we believe it is the one which best suits the purposes of this conference. As we shall show, several basic unanswered questions arise even with our rudimentary models of the defects involved. These difficulties, if unexplained, will go hand-in-hand with more sophisticated theoretical developments on other fronts. It seems worthwhile, therefore, to explore afresh what the basic problems are and (in some cases) where the proposed resolution of them might be found.

2. Development of a Perspective

The dislocation dynamics model to be used was first introduced by Koehler (7) and later fully developed by Granato and Lücke (8). In the strain amplitude-independent damping regime, with which we shall be concerned exclusively, this can be described in terms of the motion of independent dislocation segments of length ℓ. One may start on purely phenomenological grounds. For if the dislocation is a linear defect, whose energy increases quadratically with its velocity and linearly with its change in length, then the equation of motion for the displacement, y, due to an applied force (stress), σ, should be of the form

$$A\frac{\partial^2 y}{\partial t^2} + B\frac{\partial y}{\partial t} - C\frac{\partial^2 y}{\partial x^2} = \sigma b \qquad (1)$$

where x is the coordinate along the undisplaced direction of the line and b is the dislocation strength (i.e. Burgers vector). A and C are respectively the effective mass and energy per unit length of the segment and B is a viscous drag coefficient, the latter being introduced to account for energy dissipation.

For σ depending harmonically upon the time, t, a Fourier series solution of Eq. (1) has been given (8) which is valid at all frequencies, Ω. From this the area swept out by the segment (the increment in strain) can be found. It has components both in and out of phase with the applied stress leading, respectively, to a lowering of the modulus and to a contribution to the sound attenuation (or damping) by the solid.

Available estimates derived from elasticity theory led Granato and Lücke to the conclusion that, when applied to the dislocation problem, the segment response was overdamped. That is, that the quantity $\pi(AC)^{\frac{1}{2}}/B\ell$ is much less than unity for expected segment lengths. In this instance, their results for N independent identical segment lengths per unit volume can be cast into the following simple form:

$$\Delta(\ell) = \frac{2\Delta_o \Omega\tau}{[1+(\Omega\tau)^2]} \quad ; \quad \frac{\Delta G(\ell)}{G} = \frac{2\Delta_o}{\pi[1+(\Omega\tau)^2]} \quad . \tag{2}$$

Here Δ is the logarithmic decrement and $(\Delta G/G)$ is the modulus defect, (G being the elastic shear modulus), while the amplitude factor, Δ_o, and the relaxation time, τ, are

$$\Delta_o \simeq \frac{4Gb^2N\ell^3}{\pi^3 C} \quad ; \quad \tau = \frac{B\ell^2}{\pi^2 C} \quad . \tag{3}$$

To compare the above predictions with experiment, one must address the crucial question of the nature of the distribution of segment lengths. It is here that the role of point defects (initially, "misfitting" impurities) first enters the picture. These are conceived to be one entity which can reside on a length of dislocation and serve as discrete anchoring agents, the intervening free portion of line defining the segment of length ℓ, (7,8). Another possible termination of a segment is at a node of the dislocation network (8). Together these define a segment length distribution. Thus if $p(\ell)\,d\ell$ denotes the probability of a segment with length between ℓ and $(\ell + d\ell)$, N must be replaced by $N_T p(\ell)\,d\ell$ (where N_T is the total number density of segments), and a summation (integration) performed over the entire distribution. The observed decrement, for example, then is

$$\Delta = \int_0^\infty \Delta(\ell)p(\ell)d\ell \quad , \tag{4}$$

where it is to be understood that in Δ_o the quantity N is replaced by N_T. The latter may be related to the total line length per unit volume, Λ, via

$$N_T = \Lambda[\int_0^\infty \ell p(\ell)d\ell]^{-1} \quad .$$

Physical assumptions are contained in the commonly quoted examples of $p(\ell)$. Let it be assumed first that the distance between nodes of the dislocation network is random with a mean length, ℓ_N, and, further, that there is <u>no</u> correlation (i.e. interaction) between node and anchoring impurity or between impurity and impurity. Then, if ℓ_i is the mean separation between impurities on the dislocation, the a priori probability of encountering an "anchor" on the line in an element $d\ell$ will be $\left(\ell_N^{-1} + \ell_i^{-1}\right)d\ell$. Thus the probability of meeting an anchor point between ℓ and $(\ell + d\ell)$ given none were encountered prior to this - the same thing as $p(\ell)\,d\ell$ - is $(d\ell/\ell_o)\exp(-\ell/\ell_o)$, where $\ell_o^{-1} = \left(\ell_N^{-1} + \ell_i^{-1}\right)$. Hence, in this case, ℓ_o is the mean length and $N_T = (\Lambda/\ell_o)$. This is the Koehler distribution (7). On the other hand, imagine instead that the interaction between all anchor points were to be long-ranged. It then seems reasonable to suppose that they would be equally spaced, i.e, $p(\ell)\,d\ell = \delta(\ell - \ell_o)\,d\ell$, where $\delta(x)$ is the Dirac delta function. Again we have $N_T = (\Lambda/\ell_o)$.

Clearly, different distributions, that is, different physical assumptions, lead to significantly different predictions. To make the point we use a well-known example. Consider frequencies so low that $\Omega\tau$ is much smaller than unity for the vast majority of segments in the distribution; also adopt a further common assumption - that B and C are independent of segment length. It is then easy to verify that, for the same dislocation density Λ, the Koehler distribution gives a decrement (or modulus defect) a factor (5!) (or (3!)) times the values obtained from the δ function one. This gives clear warning that interpretations of experiment based upon this model must pay heed to the manner in which the length distribution might be modified.

A related question is then raised: how is the predicted damping and associated modulus arising from this dislocation mechanism to be separated from other contributions? Here the field owes much to the pioneering in-pile irradiation studies of Thompson and Holmes (9). They established the following results (for high purity Cu irradiated near room temperature to a fluence of about 4×10^{12} neutrons cm^{-2}): that the modulus defect and decrement (on the whole) decreased with increasing time before eventually reaching saturation values: that Young's modulus for different crystals agreed more closely at the end of the irradiation than at the start: and, finally; that appropriate annealing of irradiated samples and repetition of the irradiation yielded virtually the same results as the earlier experiments. They concluded, then, that such results were consistent with a model in which defects generated by the irradiation migrated to dislocations and there acted in some manner as anchor (or pinning) points in the spirit of the above discussion. And further, that the saturation effects corresponded to complete removal of the dislocation contributions to Δ and $(\Delta G/G)$. Additionally, they presented an anlysis of their experimental results in terms of this pinning mechanism and the damped string model described earlier, using the low frequency form ($\Omega\tau << 1$) believed appropriate to their experimental range ($\sim$ 17 to 50 kHz). It is worth noting, here in passing, that they frequently observed an initial increase in damping and modulus defect prior to the subsequent decline of these quantities with time. These are not consistent with the mechanism as they appear to conceive it.

Nevertheless we shall now explore their development in some detail because it is so fundamental to subsequent work and also because we wish to expose the assumptions involved. First, however, a word about "pinning". Defects are presumed to be created in initially perfect material. They then diffuse (possibly enhanced by an interactive force) to the dislocation, and become anchor points in some manner. Clearly it makes little sense to assert that a defect which is mobile in the perfect lattice is suddenly immobilized when located at the dislocation core. "Pinning" is implicitly a relative term relating to the mobility of a particular line element with and without a defect attached to it. Secondly, to this day, it seems fair to say that there is no consensus as to what constitutes an irradiation-induced pinning point. These factors should be borne in mind as the development proceeds.

For preliminary orientation it is helpful to think of a pin (anchor point) as an isolated defect, no matter whether this be true or not. Let $\dot{q}(t)$ be the capture rate of such objects per unit length of dislocation. It will be assumed that (a) the capture process is equally probable along any element of a line segment, (b) that, once captured, a pin is stuck in position, i.e. its core diffusivity is zero, and (c) that the total dislocation length is fixed. Under these assumptions, the number of segments of length between ℓ and $(\ell + d\ell)$, $n(\ell,t)\,d\ell$, will change through two processes. First, any segment can be disected by the addition of a pin. This will occur at a rate $\dot{q}\ell$ per segment. Also, a segment of length ℓ can be formed by disecting a longer one. This can happen in two ways, because the new pin can be placed a distance ℓ from either end of a longer segment (of length ℓ', say). Thus the number of lengths between ℓ and $(\ell + d\ell)$ increases from this cause at a rate which is the integral of $2(\dot{q}\ell')(d\ell/\ell')n(\ell',t)d\ell'$. (These factors are, successively, the number of ways of disecting a segment, the rate at which ℓ' will capture a pin, the probability it will occur within the interval $d\ell$ on the segment of length ℓ', and the number of segments of length ℓ' that exist in the first place.) It follows that

$$\frac{\partial n(\ell,t)}{\partial t} = \dot{q}(t)\left[-\ell n(\ell,t) + 2\int_{\ell}^{\infty} n(\ell',t)\,d\ell'\right] , \tag{5}$$

which is a trivial generalization of Thompson and Holmes equation to non-uniform capture rates. However, this new feature produces no difficulty as long as $\dot{q}(t)$ remains of one sign. We simply introduce a new "time-like" variable ρ by the definition $\rho = \int_0^t \dot{q}(t')\,dt'$ and define $n(\ell,t)$ as equal to $f(\ell,\rho)$. Thus f satisfies the equation

$$\frac{\partial f(\ell,\rho)}{\partial \rho} = -\ \ell f(\ell,\rho) + 2\int_{\ell}^{\infty} f(\ell,\rho)\,d\ell' . \tag{6}$$

The following sequence of algebraic steps leads to the solution of Eq. (6): (a) Take its Laplace transform with respect to the variable ρ, (b) Differentiate the resulting relation with respect to ℓ to obtain an ordinary first order differential equation with ℓ as the independent variable, (c) Solve this equation, (the constant of integration is zero for fixed total line length), and (d) Invert the Laplace transform. The result of this procedure is given by

$$f(\ell,\rho) = -\left[\int_0^{\infty}\left\{1 + 2(\ell'-\ell)\rho + \tfrac{1}{2}(\ell'-\ell)^2\rho^2\right\}f'(\ell',0)\,d\ell'\right]\exp(-\ell\rho) , \tag{7}$$

where $f' = (\partial f/\partial \ell)$. Thus, once the initial distribution of lengths is specified, its subsequent evolution may be evaluated. For example, the Koehler distribution given earlier changes according to (9)

$$n(\ell,t) = \Lambda\left[(1+n_p)/\ell_o\right]^2 \exp\left[-\ell(1+n_p)/\ell_o\right] , \tag{8}$$

where $$n_p(t) = \ell_o \int_o^t \dot{q}(t')dt' \quad . \tag{9}$$

Consequently the distribution retains its character, the only change being that of a reduction of the mean length in accordance with the number of pins such a length acquires in a given time interval. This particular result, we submit, should be obvious. The initial distribution corresponds to randomly located pins, and there is nothing in the model which is selective in favoring one length class over another. We shall return to the discussion of this topic later in this section.

With the foregoing development of the dislocation dynamics and pinning analysis completed, the groundwork for a major fraction of subsequent investigations is at hand. We will highlight some of these first before attempting a critique.

In their very thorough analysis, Granato and Lücke recognized the importance of the segment length distribution and discussed the ramifications of a Koehler time-independent one. Their conclusion was that the qualitative behavior of Δ and $(\Delta G/G)$ should be as contained in the simple versions embodied in Eq. (2). Namely a maximum in Δ at a frequency $\sim \tau^{-1}$, accompanied by a decrease in the modulus with increasing frequency. This maximum was predicted to be in the MHz, and there it was observed shortly thereafter in high purity Cu by Alers and Thompson (10). These authors were careful to identify the dislocation contribution by measurements of samples both before and after relatively prolonged neutron irradiation, the latter serving to remove the dislocation component of Δ and $(\Delta G/G)$. Moreover, by performing experiments at different temperatures, they were able to confirm the consequences of an anticipated (11) linearly temperature-dependent (increasing) B and assess its magnitude ($B \sim 8 \times 10^{-4}$ dyne sec cm^{-2} at room temperature). Thus the theoretical model was beautifully confirmed at high frequencies, and further experiment has since substantiated its validity in this range.

But a difficulty with the model was noted (10). For, upon using the parameters determined from analysis of their experiments, Alers and Thompson concluded that the predicted decrement for the low kHz range was smaller than the observed values by about two orders of magnitude. However, they failed to consider all their data on a sample-by-sample basis. As emphasized by Granato and Stern (12), it was possible to re-analyze their range of observations so that the predicted decrement at low frequencies was in fact greater than any then observed. Thus the crucial issue became the matter of careful experiments at low frequencies to see if the predicted linear frequency dependence is in fact observed (12). Experiments of this nature have since been performed (13-16) and the same qualitative result is always found. Namely, the decrement is either substantially independent of frequency, or in some cases it even increases (13,16) as the frequency is lowered, in the range at or below $\sim$ 10 kHz in Cu. The theoretical model, for some reason, is there inappropriate.

By and large, the results of irradiation experiments do not shed much light on this very basic problem, in our opinion. We shall simply quote a sampling and let the reader judge the issue personally. First consider the neutron irradiation results of Thompson and Holmes (9). Their frequency range was 17 - 50 kHz, irradiation temperature $\sim 28^{\circ}$C, and defect production rate (very crudely) $\sim 3 \times 10^{8}$ cm^{-3} sec^{-1}. The data were analyzed in terms of the model already presented. In other words the low frequency forms of Eqs. (2) were used as a basis and a seemingly consistent interpretation found. Since it is now known from experiments on samples of comparable purity, in the same temperature range, that the predicted frequency dependence is incorrect (13), a brief re-examination of this early work appears in order.

There is always the possibility that dependences of B and/or C upon frequency conspire, through some unspecified mechanism, to yield a relaxation time which increases with decreasing frequency in the range of measurement, yet leaves $\Omega\tau << 1$. Should this be the case, and also B and C be independent of ℓ, then Eqs. (2), (3) and (8) would still lead to $\Delta \propto (1 + n_p)^{-4}$ and $(\Delta G/G) \propto (1 + n_p)^{-2}$ at fixed frequency. Accordingly, suppose y and z are the ratios, respectively, of the modulus defect and decrement to their unirradiated values (17). It follows that the two relations below should hold:

$$y^{-\frac{1}{2}} - 1 = n_p = z^{-\frac{1}{4}} - 1 \quad . \tag{10}$$

Consequently plots of $y^{-\frac{1}{2}}$ and $z^{-\frac{1}{4}}$ against irradiation time should be coincident. The earliest data of Thompson and Holmes did indeed fulfill this requirement, and, moreover, they were linear in time (9,17). Later experiments with neutron (18), γ-ray (19) and electron (20) irradiation have shown departures from the earliest results in one or both respects. The most one can say, therefore, is that the earliest work indicated that the ℓ_o^2 - ℓ_o^4 dependences were consistent with observation, but the rule is not a universal one. The kinetics problem of obtaining a linear n_p with dose will not be addressed.

Underlying any considerations of the data of Ref. (9) is the question of the (sample-dependent) initial increase of the decrement and modulus defect with irradiation dose. Clearly, it is entirely inconsistent with the pinning picture in its simplest form. We presume it to be an example of a since commonly-observed phenomenon known as the Peaking Effect. After Simpson et al (20) gave particular emphasis to this, it has emerged that the phenomenon is in fact quite a common one. It has been observed in Cu with four types of irradiation, neutron (9), electron (20), γ-ray (21) and proton (22). The characteristics of the effect are somewhat uncertain. Lauzier et al (23) have found it to be a strain-amplitude dependent effect (e^- irradiation at 160°K) which disappears as the strain amplitude is decreased to $\sim 10^{-7}$, though it is clearly present at this level in Fig. 4 of Ref. (9). Further, a recent conclusion of Kerans and Simpson (24), using e^- irradiation at 313°K, is that while the peaking is indeed amplitude-dependent, yet the effect appears to approach a saturation value as the strain amplitude decreases. An additional fascinating data set is provided by the iso-dose irradiation-annealing sequence given in Ref. (20). An as-mounted sample with no prior annealing was found to give no change upon irradiation. It was then annealed at 500°C for 30 minutes prior to the next irradiation, when a peaking effect was observed (maximum decrement about 1.6 times the pre-irradiation value). Repeating the sequence, but now with only a 10 minute anneal, subsequently gave a peak decrement about twice the pre-irradiation value. It was also found that an annealing temperature of 750°C was sufficient to remove the peaking effect entirely. A qualitative inverse correlation between the peak height and pre-irradiation decrement was also noted. We should also mention that the non-dislocation components of Δ and $\Delta G/G$ in these experiments where found by raising the temperature to 500°C and holding there for 10 minutes. This was believed to be sufficient time for irradiation-induced defects to migrate to dislocations and remove their contributions to the measured quantities (20). In each irradiation-anneal sequence, both this "elastic" component and the pre-irradiated decrement were different. In the context of any theoretical model, therefore, it appears to us that this series of measurements is equivalent to a set on different samples even though only one was repeatedly examined.

The observations of this phenomenon led Simpson and Sosin to the development of a somewhat different point of view regarding the dislocation-irradiation-induced defect process as it concerns internal friction. It was their basic contention that instead of immobilizing the segment length upon being trapped at the dislocation, the defect merely introduced a locally enhanced drag force. The potential for this effect is readily apparent by inspection of Eqs. (2) and (3). For suppose one identifies B not with some intrinsic material significance, but rather supposes it due to adding n dragging points on a line. Crudely, then, a maximum in Δ is to be expected as n increases at fixed frequency. On this level, the effect is a strain-amplitude independent one. For further specifics upon the early development of this model the reader is referred to the fine review by Sosin in Ref. (5).

The above admittedly incomplete account of some past highlights sets the stage for what follows. Our approach in preparation for this conference has been to look for hidden assumptions and try and assess if they are valid. The basic problem is obviously the frequency dependence at low frequency. If ever the basic phenomenology should apply, it ought clearly to be in this range rather than any other. Either the entire scheme is inappropriate, which we view as highly unlikely, or some frequency-dependent damping contribution is missing. One possibility is a damping due to dislocation-defect interaction effects. We shall examine this in the section following.

All irradiation experiments are rendered difficult to interpret because of one basic fact. Namely, that they represent the complex convolution of four distinct features - the basic internal friction model, the kinetics of point defect capture by dislocations, the formation of pinning points (in the appropriate model) and the evolution of the line length distribution with irradiation dose. Indeed, the undetached observer might well have voiced some skepticism had all transpired according to the earliest expectations!

These factors are not necessarily independent of each other. If we assume a pin is an isolated defect, as an interpretation of one MHz (γ-irradiation) experiment suggests (25), the length population analysis is applicable as it stands. But if we adopt another interpretation (26) of γ-irradiation experiments in the same range, when only one pin per 10 irradiation defects was formed, we begin to encounter problems. Either the "lost" defects of the variety which cause pinning were trapped in solution (26) or they formed clusters when on the dislocation (intrinsic recombination may be ignored). In the first instance, if the trapping potential has any elastic component we must enquire into what effect such traps have on (a) the damping in the pre-irradiated material and (b) on the kinetics underlying the capture rate at dislocations and traps. Or, in the second instance, by what mechanism do these clusters form? This is not a peripheral matter. For example, suppose just one cluster forms on the dislocation core by homogeneous nucleation of previously trapped defects. Then, if it grows subsequently, this implies that a correlation in position exists between it and all defects arriving at a later time. Thus a random distribution of lengths cannot possibly retain its character unless the isolated captured defect is endowed with zero pinning capability. It appears quite inconsistent to invoke nodal traps and core diffusion to them, for example, while retaining the concept of isolated core-entrapped defects as pinning agents, and to assume that the length distribution remains invariant. Quite clearly, a correlation between pins will develop in such a model of quite different character to any which may or may not have existed initially.

The matter of pin-pin correlations raises another issue. Several years ago Alefeld (27) and Bauer (28) introduced the concept of an entropy force on a defect pin. Its origin lies in the free energy of vibration of two adjacent line segments and has the sense of favoring one long segment (and thus also one of vanishing length) over two shorter lengths of equal magnitude. It is thus a vital matter to consider both in relation to what the length distribution might be in an annealed material and also to how it might evolve under irradiation. This subject is discussed in Section 4.

A final topic to be considered here ultimately relates to what we consider to be a basic omission in all discussions of irradiation experiments in this field. That is, the nature of the defect capture process itself. It is based upon the following observations. First, the number of added pins is inevitably small - adding only one per segment can reduce the decrement and modulus defect by factors of 16 and 4, respectively, in the first model discussed. Whether we equate a pin to one defect or ten the qualitative fact remains; during the course of an experiment one is not averaging over a very large number of capture events per dislocation segment. (We assume, of course, that estimated point defect generation rates are reliable.) Second, the models imply that while the dislocation length responds to the (in-principle) vanishing small applied stress, it is entirely unaffected by the internal stress field of the incoming defect, even though the latter becomes very large as the defect approaches the dislocation. In other words the capture event is assumed to be an entirely passive process as far as the line segment is concerned, the latter simply serving as an inert collector. In many cases we think this may not be correct, and to illustrate the point we have incorporated some calculations of the actual dislocation configuration as a function of the defect position in Section 5. It leads to what we will call a glide-induced dislocation-point defect interaction and has, in the present context, the important feature that the line length is variable during the capture process. We will speculate upon some possible ramifications in the Internal Friction area. However, we should emphasize immediately, such comments are all rather tentative at present.

The general discussion will be returned to briefly in Section 6 after more details of these topics has been given in the intervening sections.

3. Solid Solution Drag and Pinning Drag

One major purpose of this section is to estimate the effect of mobile point defects in solution upon the dislocation drag coefficient, B. As the development rests rather heavily upon past work, we shall begin by summarizing the latter. This will enable us to check one subsidiary but vital assumption alluded to in Section 2 - the assumed independence of B and C upon segment ℓ. It enters, of course, in the interpretation of irradiation experiments in the pinning model, or in discussions of alloying-related problems, with or without irradiation.

For the moment, we regard point defects as immobile objects. They are scattering centers for both electrons and phonons. Since these excitations contribute to the drag force (28, 29), modification of their mean free path could alter B as a result.

Mechanisms of dislocation drag which are not kinematic in origin can be handled conveniently by attributing spatially non-local and temporally retarded viscosities (longitudinal and shear) to the continuum in which the dislocation is moving. Such mechanisms are the thermoelastic effect, electron scattering (30) and drag caused by defects in solution (31). A notable (and so far sole) exception is that due to the flutter mechanism (11, 32, 33), about which we shall here say nothing beyond remarking that we expect little sensitivity of it to defects.

Because of the specific nature of the above viscosities, it is most convenient to work with spatial and temporal Fourier transforms of the displacement field of the continuum. In this formalism, the viscosities (which are really double transforms of response functions) appear as functions of arbitrary frequency, ω, and wave vector $\underline{q}$, viz. $\zeta_\lambda(\underline{q},\omega)$, where λ is a polarization index. These describe the dissipative response of the continuum when a specific plane wave displacement (a deformation wave) exists in the medium. A moving dislocation is a coherent superposition of such waves (34, 35).

The derivation of the electronic and phonon contributions to the frequency and wave vector-dependent viscosities is a rather lengthy affair (29, 30). But in concept, at least, the method is simple. One picks a model for the way in which the phonon or electron energy spectrum is modified by a deformation wave, uses Boltzmann transport theory to compute occupation probabilities for the appropriate "particle" in the steady state, and thereby computes a stress-strain law. From this the frequency and wave vector dependent viscosity can be extracted (29).

Examples for various modes of dislocation motion of fixed configuration are given in Ref. (29). The adaptation to the vibrating segment problem has been provided by Read and Brailsford (36). The basic physical model used by these authors is that of a long straight dislocation lying along the x-axis of the x-y slip plane, with only a segment of length ℓ allowed to move in response to a harmonically time-dependent stress. Such a system consists of a rigid backbone dislocation with a loop of fixed length ℓ but variable extent, $\psi(x,t)$ say, adjacent to it. The continuum displacement field for a dissipative medium with such an internal dislocation set-up can be found. The main approximation used in Ref. (36) was to consider only the quasi-static solution, and some justifications for this were given. Its main effect is to lose the "radiation damping" component of B, which in our case is innocuous in any event (29). The aim was to first obtain expressions for the kinetic energy, T, potential energy, U, and dissipation function Ψ of the system. These were found to be of the following form:

$$\left.\begin{aligned} T &= (\ell/4)\sum_{m,n} A_{mn}\dot{\psi}_m\dot{\psi}_n \quad ; \quad U_L = (\pi^2/4\ell)\sum_{mn} C_{mn}\psi_m\psi_n \\ \text{and} \qquad \Psi &= (\ell/4)\sum_{m,n} B_{mn}\dot{\psi}_m\dot{\psi}_n \quad , \end{aligned}\right\} \qquad (11)$$

with $\psi_n = (2/\ell)\int_0^\ell \psi(x,t)\sin(n\pi x/\ell)\,dx$. Here U_L represents only the self-energy of the loop and its interaction energy with the backbone dislocation. The self energy of the latter is presently of no interest, and the interaction energy with the exciting stress field is obvious. Detailed expressions for the elements A_{mn}, B_{mn} and C_{mn} were given and evaluated in Ref. (36). The essential point to note is that a generalized string model is formally substantiated and a method developed for determining its basic parameters.

Careful examination of the internal friction to be expected from this model showed that the mode coupling contained in Eq. (11) was not likely to be important for length distributions of the Koehler type. The terms $m = n = 1$ in the preceding relations where thereby shown to play the dominant role, thus leading to the string model in its simplest form. Calculated values for A_{11} and C_{11} were also shown to agree quite well with evaluations of

$$A_{11} = (\rho b^2/4\pi)\left[1 + s^4\sin^2\chi\right]\ln(q_D\ell)$$

and

$$C_{11} = (Gb^2/4\pi)\left[3 - 4s^2 + 3(2s^2 - 1)\sin^2\chi\right]\ln(q_D\ell) \quad ,$$

where q_D is a Debye cut-off in wave-vector space and s is the ratio of transverse to longitudinal sound velocities. Isotropic elasticity was used throughout, so that $s^2 = (1 - 2\nu)/2(1-\nu)$, with ν denoting Poisson's ratio. The remaining variable, χ, represents the angle between the Burgers vector and the backbone line direction.

From this one sees that A and C, (with subscripts now discarded), depend very modestly upon segment length. Their ratio, which in part determines the resonant frequency of the segment, is independent of it entirely. We turn next to B, specifically the component associated with longitudinal wave deformations of the continuum. Again, considering only the $m = n = 1$ term, this was found to be (36)

$$B = \frac{4s^4 b^2 \ell}{\pi} \int d\underline{q} \left[\frac{(\hat{\underline{q}} \cdot \underline{n})(\hat{\underline{q}} \cdot \underline{\beta}) \cos (q_x \ell/2)}{\pi^2 - q_x^2 \ell^2} \right]^2 \zeta(\underline{q}) \quad , \tag{12}$$

where $\hat{\underline{q}}$ and $\underline{\beta}$ are unit vectors along $\underline{q}$ and $\underline{b}$ respectively and ζ is the longitudinal viscosity for wave vector $\underline{q}$. It should be clearly understood at this point that the formal development of Ref. (36) relates only to a spatially non-local viscosity, as indicated by the dependence of ζ upon $\underline{q}$ only. However, we believe our later generalization to ω - dependent ζ is warranted.

This issue does not arise in the first instance in examining the ℓ - dependence of B_e and B_s, the electronic and phonon scattering contributions to B, respectively, because the dominant contributions to both viscosities are virtually independent of ω (30). (Relevant frequencies are orders of magnitude less than particle collision frequencies.) The possible effects of immobile point defects centers around two factors in the integrand in Eq. (12). The first is the geometrical quantity in square parentheses, which is associated with how a finite loop might give results differing from an infinitely long straight dislocation. That is, how pinning enters. The second is contained in $\zeta(\underline{q})$, through the dependence of it upon particle mean free path; that is, on one effect of defects in solution. Now ζ in fact depends only upon the magnitude of q in an isotropic solid, so that Eq. (12) may be manipulated into the form

$$B = (s^4 b^2/\, 4\pi) \int_0^{q_D} q\, \zeta(q)\, Y(q\ell)\, dq \quad , \tag{13}$$

where $Y(q\ell)$ is the result of the angular integrations, so-normalized that $Y \to 1$ as $\ell \to \infty$. The present premium on space precludes entering into too much algebraic detail. But the result is simple; $Y(q\ell)$ differs little from $\tanh\,(64\, q\ell/15\pi^3)$, a form suggested by a small $q\ell$ expansion. Thus $Y \sim 1$ for $q\ell \gtrsim 4$. Further, since $q_D \sim 10^8\ \text{cm}^{-1}$, whereas ℓ might be typically of order 10^{-4} cm, the finite loop correction is clearly very small indeed. Moreover, we also know (29, 30) that $\zeta(q)$ is independent of particle mean free path, ℓ_p, as long as $q\,\ell_p > 1$, where ℓ_p is of order 10^{-6} cms for electrons and phonons in Cu at room temperature. Hence we conclude that immobile point defects in solution have negligible effect upon intrinsic drag processes (in normal metals) for segments of typical length.

As a matter of interest, we might mention that we have looked in more detail (37) at the phonon scattering contribution, for which B is linear in temperature, T, at high temperatures. All components are obviously reduced by the presence of $Y(q\ell)$ in Eq. (13). But we have found that for the mechanism under discussion, for example, B assumes its infinite segment length value, in Cu, once ℓ exceeds a modest value $\sim 10^{-6}$ cms. In fact the most significant segment length reduction effect resides (37) in the thermoelastic contributions to B, because these originate in the long wavelength deformation waves (30). In any event, however, such contributions are relatively small; the correction discussed here simply makes then more so.

Any significant contribution of point defects to dislocation damping must therefore reside in their being mobile, the internal stress field of the dislocation inducing defect diffusive-type motion and hence energy dissipation. This will be evidenced by an additional viscosity contribution, and hence, via Eq. (13), by a new drag force component. We shall here simply refer to earlier theoretical work along these lines (31, 38-41) before presenting our own development.

The power dissipation, or two times the dissipative function (42), that is associated with a diffusive defect flux, J, is equal to $(k\,T\,\underline{J}^2/Dc)$ per unit volume (43). Here k is Boltzmann's constant, D the diffusion coefficient and c the local defect volume concentration. We wish to relate this to a dissipative function density which is a generalization of $\tfrac{1}{2}\zeta(\partial\varepsilon/\partial t)^2$, and hence obtain a viscosity contribution arising from this mechanism.

The defect flux is of the usual form, viz $\underline{J} = D(-\text{grad}\, c + (\underline{F}/kT)c)$, where $\underline{F}$ is the force on a defect. We assume that the latter are simple centers of dilatation as used in isotropic elasticity theory. In other words, a defect at $\underline{r}_\alpha$, say, is accounted for by inclusion of a term $-A\,\text{div}\,\underline{u}\,(\underline{r}_\alpha)$ in the potential energy of the system, where $\underline{u}(\underline{r})$ is the lattice displacement field. Thus the displacement induced by this defect at large distances is the usual one (44), i.e., $\kappa(\underline{r}-\underline{r}_\alpha)/|\underline{r}-\underline{r}_\alpha|^3$, where $\kappa = (As^2/4\pi G)$. The force on the defect at r_α is, therefore, $\underline{F} = A\,\text{grad}\,\text{div}\,\underline{u}(\underline{r}_\alpha)$, whence the defect concentration Continuity Equation is given by

$$\frac{\partial c}{\partial t} = -D\,\text{div}\left[-\text{grad}\, c + \frac{Ac}{kT}\,\text{grad}\,\text{div}\,\underline{u}\,(\underline{r},t)\right] \quad . \tag{14}$$

The problem at hand is to solve for c, and hence determine J, for the variation in $\underline{u}(r,t)$ engendered by the dislocation motion. In principle this is possible in the Read-Brailsford set-up defined earlier. If variables with suffix b attached correspond to the presence of the time-independent backbone dislocation alone, and changes due to the loop are assigned the suffix ℓ, then a perturbation solution $\underline{u} = (\underline{u}_b + \underline{u}_\ell)$ and $c = (c_b + c_\ell)$ can be envisaged. Unfortunately, because $\underline{u}_b$ is space-dependent, so is c_b for zero flux: $c_b = c_\infty \exp[A \operatorname{div} \underline{u}_b/kT]$, where c_∞ the point defect concentration at large distances from the dislocation. This, of course, is the usual defect atmosphere distribution of Cottrell-Bilby theory (45). Consequently, when linearized with respect to $\underline{u}_\ell$, Eq. (14) leads to a partial differential equation with variable coefficients and (core) boundary conditions at a time-dependently variable inner surface. Rather than confront this, we shall adopt a much simpler expedient, namely to neglect such atmosphere effects entirely for a start. We suppose in effect that $\underline{u}_b = 0$ and $c_b = c_\infty$ in our reference state, assumptions which reduce the problem to a simple internal Gorsky effect (46) due to the field of the mobile line segment only. We shall attempt to estimate the error involved shortly.

For $c = (c_\infty + c_\ell)$ and $\underline{u} = \underline{u}_\ell$ the solution of Eq. (14) is easily obtainable by Fourier analysis, because all coefficients in the equation become constants. Having solved for a particular $(\underline{q},\omega)$ component of c_ℓ in terms of the corresponding component of $\underline{u}_\ell$, one computes the flux density and hence the dissipation function, Ψ, as a double Fourier sum. A comparison of this with the form involving the viscosity and strain-rate alluded to earlier allows the viscosity contribution, $\zeta_{ss}(q,\omega)$, to be found. One can then insert this in Eq. (12) to find the contribution to the drag coefficient. The pertinent results are

$$\zeta_{ss}(q,\omega) = \frac{D A^2 c_\infty}{k T} \cdot \frac{q^2}{(\omega^2 + D^2 q^4)} \quad ; \quad B_{ss} = \frac{s^4 b^2 A^2 c_\infty}{16\, \pi k T D} \ln\left[1 + \left(\frac{D q_D^2}{\omega}\right)^2\right] \quad , \tag{15}$$

if $Y(q\ell)$ is replaced by unity. The last step is reasonable because ζ_{ss} is small at small q for non-zero ω. The mechanism underlying B_{ss} we shall refer to as solid solution drag.

As far as we can discover, the above result is a new one. It neglects contributions associated with local defect atmosphere formation entirely, we repeat, and it is a moot point as yet as to how good this might be. One test is to compare the result of the above formalism with those for the point defect-induced damping force for an infinitely long straight dislocation moving at constant velocity v. We have carried out this calculation using the viscosity derived above and the formalism developed earlier (29). The result is that a drag coefficient of the form $B_{ss}(v) = (A^2 s^4 b^2 c_\infty/2\pi kTD) I(v/Dq_D)$ is obtained, for an edge dislocation in uniform motion, where

$$I(x) = (4/\pi)\int_0^1 z^2 (1-z^2)^{\frac{1}{2}} \ln\left[1 + (xz)^{-2}\right] dz \quad . \tag{16}$$

Details will be given elsewhere (47), but it is not too hard to see the origin of this result in the following way. First, an inspection of the $q_x\ell$-dependent factor in Eq. (12) shows it to be equal to π^{-4} at vanishing q_x, for any ℓ: it has its first zero at $q_x\ell = 3\pi$. Thus, for integration purposes, it may be estimated by $(3/\pi^3\ell)\,\delta(q_x)$ when ℓ is large. The loop now approximates a uniform advance of the backbone. Setting $\omega = -vq_y$ then simulates uniform translation (29). The evaluation of Eq. (12) by this ruse gives a result differing from that cited above by only the modest factor $(12/\pi^2)$.

A comparison of this estimate of solid solution drag with that found by Eshelby's (31) treatment shows them to be virtually identical except at low velocity (47). The behavior of the drag force in this region is not linear in velocity, but of the form $v \ln(1/v)$. The discrepancy can be traced to the non-local nature of the basic relation of stress to strain, etc. In any event, overall agreement here does not help calibrate the present approximation. What is required is a solution of the Cottrell-Jaswon problem (48), where atmosphere effects are included.

Developments to date on this topic can be gleaned from the recent paper of Takeuchi and Argon (49). The basic shape of the solute drag force versus velocity is well known. It increases initially with velocity at small velocity, attains a maximum value F_{max} at a velocity v_{max}, and subsequently decreases as v^{-1} in the asymptotic region. Values for these quantities are of the form $v_{max} = (pDkT/\beta)$ and $F_{max} = q\,\beta\, c_\infty$, where p and q are numerical coefficients and β, the interaction parameter, is (As^2b/π) in our notation. Values of p range between (49, 50) unity and 4, while q is variously quoted as ∿ unity (49), π(3) and 17 (50). Our own crude model yields $q \simeq 0.47p$ and $p = (\beta q_D/kT)$. Interpreting (β/b) as a defect binding energy (taken here as of order 0.1 eV) we find $p \simeq 8$ at room temperature. In other words, to the extent that drag theories of dislocations and atmospheres in uniform motion yield quantitative values, our simple analysis is not grossly in disagreement with them for the range of parameters indicated.

Vibratory motion might be different; the dislocation cannot then escape from its atmosphere. Only one detailed analysis of defect atmosphere effects (51) is known to us, and this is restricted to high frequencies. We believe also that the ultimate interpretation of the analysis needs modification, for it is tacitly assumed (51) that $\omega\tau$ always tends to infinity at high frequencies. With solid solution drag as the only damping mechanism, inspection of Eq. (15) shows that this may not be true. In fact, Kessler's analysis is found really to relate to the power dissipation per unit length, P, for a dislocation vibrating as a rigid rod. A drag coefficient, B_k, may thus be obtained from $B_k = (2P/\omega^2 a^2)$, where a is the amplitude of vibration. Interpreted in this way, Kessler's analysis leads to $B_k \to (32\beta Dc_\infty/3\omega^2 L^3)\exp(\beta/LkT)$, where L is some inner cut-off radius at the dislocation core. Equation (15), on the other hand, yields $B_{ss} \to (\pi\beta Dq_D^4 c_\infty/64\,kT\omega^2)$ as $\omega \to \infty$. Thus, while the frequency dependence is identical, the coefficients are different. To compare the two, it seems fairest to use Kessler's suggestion that $L \simeq 2b$. Thus, for an f.c.c. lattice, we set $q_D \simeq 4/L$. The ratio of B_k to B_{ss} is then approximately $\gamma^{-1}\exp(\gamma)$, where $\gamma = \beta/LkT$. Hence if now (β/L) is interpreted as a binding energy (∿ 0.1 eV), the above ratio is ∿ 14. Obviously this has the correct sense (i.e. it is larger than unity), because B_{ss} has been derived neglecting atmosphere effects entirely. But the issue is not clear cut, for Ref. (51) ignores the moving boundary condition problem entirely and invokes the cut-off procedure contained in L, a parameter to which B_k is extremely sensitive. A rigorous solution of the vibrating dislocation plus atmosphere problem is thus sorely needed.

We have compared the results of the formalism upon which our calculation of B_{ss} is based with earlier work in as many ways as possible for one purpose. Namely to show that while it is incomplete, the end-result is not outrageously wrong in one parameter domain which is of practical interest. Thus its ramifications for the Internal Friction model are worth a brief mention. Referring first to Eqs. (2), (3) and (15), a novel feature emerges: when ω is equated to the applied frequency Ω one sees that $\Omega\tau_{ss}$ is small at both low <u>and</u> high frequencies. (Here $\tau_{ss} = B_{ss}\,\ell^2/\pi^2 C$). This is a direct consequence of the fact that, once the period of vibration becomes shorter than the time taken by the point defect to diffuse a few lattice spacings, the defect simply sits frozen in place. The transition to this regime comes when $\Omega \sim Dq_D^2$; beyond it $\Omega\tau_{ss} \to 0$, and Δ varies like Ω^{-1} as in the constant B cases, but for an entirely different reason. Or to state matters another way, if $D = D_o \exp(-Q/kT)$, then B_{ss} will be proportional to Ω^{-2} if $Q > kT\,\ell n\,(D_o q_D^2/\Omega)$, i.e. if Q exceeds .6 eV, for room temperature and $\Omega \simeq 15$ kHz. Using this Q value, along with the estimates given earlier, plus $C \sim 10^8$ eV cm^{-1} and $\ell \sim 10^4$ b, leads to $\Omega\tau_{ss} \sim 0.1$ for a fractional defect (impurity) concentration of 0.3 ppm. This is greater than the comparable quantity for anharmonic lattice scattering at this frequency, but of course much less than the latter at the damping maximum in the MHz range, which corresponds to $\Omega\tau = 1$. The end result then is a kHz damping ∿ 10% of that in the MHz range, which appears about correct (13). But it would be premature to attach too much significance to this. Obviously there are a multitude of basically unknown parameters involved. A more satisfactory test would be to investigate the predicted low frequency temperature-dependence of the damping to see if it conforms to the $T^{-1}\exp(-Q/kT)$ expected from solid solution drag in its "high" frequency range. Further, we should add, we have no suggestion to offer on what trace impurity might have such a low motional activation energy as the one used above.

With that we will leave aside solid solution drag and consider approaches other than Kessler's to near-core defect effects. We shall call these pinning drag models because even though the point defects are sometimes not explicitly stated to be on the dislocation line, the end result is that the atmosphere and dislocation are conceived to move as a single entity. For example, we might ascribe a mobility μ to a defect near the dislocation and suppose there to be n_D defects per unit length. Let us ignore intrinsic damping processes for simplicity, and simply glue the defects and dislocation together. We have then $B \to n_D/\mu = (kT\,n_D/D)$. This, when fed into Eqs. (2) and (3), is the basic ingredient of Schoeck's theory (52) of the Snoek-Köster or cold-work relaxation peak attributed to interstitial impurities in b.c.c. metals (1, 2). The assumption is that one counts all defects within some distance r_c of the core and neglects the rest entirely. This seems quite arbitrary (41). Moreover such an hypothesis leads to an obvious problem when applied to long dislocation segments at high frequencies. For on the one hand we have Kessler's result that B varies as $(\text{frequency})^{-2}$, whereas by the above B is independent of frequency. Some kinetic factors evidently are missing (53). Indeed it may be significant that if r_c is adjusted to give agreement with our interpretation of Ref. (51), then $(r_c/L) \sim (\beta/LkT)(\omega\tau_D)^{-1}$, where $\tau_D = (L^2/D)$. Thus parameters measuring the strength of the binding and the time for motion within the potential "well" appear to enter. The whole matter warrants further investigation.

In irradiation damage studies, however, the bulk concentrations of point defects are extremely small. Volume concentrations at most of order 10^{13} cms^{-3} are frequently encountered in such experiments. A generous estimate of the number of defects per atomic plane within one atmosphere radius, $r_o = \beta/kT$, is $\pi r_o^2 b c_\infty \exp(r_o/b)$. This is orders of magnitude less than unity. Thus neglecting point defects except when they reside at the core appears a reasonable physical basis for further discussion. Provided, that is, these irradiation-induced defects are the only ones present (i.e. there are no traps in solution).

The intent of the following analysis is to adopt this view and address the question of what pinning means quantitatively in the Thompson and Holmes (9) picture. It is enough to consider the addition of one "pin", at a point $x = \xi$, on a segment firmly anchored at $x = 0$ and $x = \ell$, $(0 < \xi < \ell)$. Upon application of a stress, the force upon it will be $\sigma b - (\partial y/\partial t)(\mu_c w)^{-1}$ per unit length, where μ_c is its mobility and w its hypothetical width. In the continuum string picture, the preceding drag term of strength $\propto w^{-1}$, and acting over a distance w only, may be replaced by a delta function. The end result for a mobile pin is the replacement of σb in Eq. (1) by $\sigma b - B_c(\partial y/\partial t)\,\delta(x-\xi)$, where $B_c = \mu_c^{-1}$.

For $y(x,t)$ of the form $\phi(x)\exp(i\Omega t)$, an elegant solution of this modified string equation (MSE) is afforded by the Green's function technique (54). Let $\chi(x)$ be the solution for $B_c = 0$ and $G(x,\xi)$ be the solution when $(-i\Omega B_c\phi(x))$ is replaced by unity, both χ and G satisfying the conditions that they be zero at $x = 0$ and $x = \ell$. The solution of the MSE is thus $\phi(x) = \chi(x) - i\Omega B_c\phi(\xi)G(x,\xi)$. Setting $x = \xi$ in the latter allows $\phi(\xi)$ to be found, and hence the general solution can be determined. For our purpose it is convenient to express this in the following way:

$$\phi(x) = \bar{\phi}(x) + \frac{\chi(\xi)\,G(x,\xi)}{g[1 + i\Omega B_c g]} \quad ; \quad \bar{\phi}(x) = \chi(x) - \frac{\chi(\xi)\,G(x,\xi)}{g} , \tag{17}$$

where $g = G(\xi,\xi)$. The point of this mode of expression is clear. As $B_c \to \infty$, when the pin is totally immobile, only $\bar{\phi}(x)$ survives. Thus $\bar{\phi}$ describes the displacement of two adjacent segments of length ξ and $(\ell - \xi)$, each firmly anchored at their ends. The variable $[\phi(x) - \bar{\phi}(x)]$ is thus the incremental associated with the intervening pin being mobile.

The area swept out by the dislocation-mobile pin system follows from Eq. (17). From this the decrement and modulus defect may be obtained. In particular, the relative fractional increment in the modulus defect due to pin mobility, for the length ℓ, is seen to be

$$f = \frac{\chi(\xi)\int_0^\ell G(x,\xi)\,dx}{g[1 + (\Omega B_c g)^2]\int_0^\ell \bar{\phi}(x)\,dx} , \tag{18}$$

upon assuming G and $\bar{\phi}$ are real. (This is true for our subsequent application, but not generally so.) At low frequencies, when intrinsic damping and inertial effects are small, the analysis is particularly simple to complete. Then $G(x,\xi)$ is just the static displacement of a string due to a delta function force of unit strength. (That is, $G = x(\ell - \xi)/C\ell$, $(x < \xi)$; $G = \xi(\ell - x)/C\ell$, $(x > \xi)$, from which g follows.) Similarly, $\chi(x)$ is the displacement of a string of length ℓ due to a stress σ_o, $\chi(x) = (\sigma_o b/2C)\,x\,(\ell - x)$, whence $\bar{\phi}(x)$ is found via Eq. (17). The end result of this analysis is given by

$$f = \frac{3\xi(\ell - \xi)\,\ell}{[\xi^3 + (\ell - \xi)^3]} \cdot \frac{1}{(1 + \Omega^2\tau_c^2)} , \tag{19}$$

where $\tau_c = B_c\,\xi(\ell - \xi)/C\ell$. A quantitative measure of pinning efficiency, η, as counted by its effect upon the modulus defect, must therefore accommodate two factors. Namely where the pin resides relative to anchor points, and what its mobility is. Obviously, if it exists near either anchor point - or anywhere on the line at high enough frequency, $f = 0$; we will call this the Thompson and Holmes limit, $\eta = 1$. At any intervening value of pinning, η may be defined as $\eta = 1 - [I(\phi) - I(\bar{\phi})]/[I(\chi) - I(\phi)]$, where the I's here denote integrals of their arguments over the interval 0 to ℓ. Thus $\eta = 1 - f/(h-1)$, where $h = I(\chi)/I(\bar{\phi})$. Applied to the regime considered, this leads to

$$\eta = (\Omega \tau_c)^2/[1 + (\Omega \tau_c)^2] \quad , \tag{20}$$

which is a useful representation because pinning position and mobility are contained compactly within the one variable τ_c. Clearly, the pinning efficiency is low at low frequencies, when the pin has ample time to move during one period. Asymptotically η approaches unity as the frequency increases.

To attain the Thompson and Holmes limit we require $\Omega \tau_c >> 1$. Thus, if we endow the pin with a mobility $\mu_c = (D_o/kT) \exp(-E_m^c/kT)$, this reads $E_m^c > kT \ln (4D_oC/kT \ell \Omega)$. Or, with $\ell \sim 4 \times 10^{-4}$ cm, $C \sim 10^8$ eV cm^{-1}, at room temperature and a frequency of 15 kHz, $E_m^c \gtrsim .5$ eV or $E_m^c \gtrsim .4$ eV, for $D_o = 1$ or 0.1 cm^2 sec^{-1} respectively. But a self-interstitial-dislocation complex, whatever its detailed character, almost inevitably will have a migrational energy barrier which is lower than that of the isolated defect in the matrix. At these frequencies and temperatures, therefore, a single trapped interstitial cannot "pin".

This is intriguing enough in its own right for us to digress a moment. Instead of thinking of pinning, let us simply load n_D interstitials per unit length on the line. We should eventually attain the limit invoked by Schoeck (52). The appropriate relaxation time becomes, with $n_D\ell = n_I$, $\tau = (n_I\, k\, T\, \ell/\pi^2 D_o C) \exp(E_m^c/k\,T)$, by Eq. (3). Thus to reduce the modulus defect contribution due to this segment by a factor of 2 (see Eq. (2)), we require $E_m^c \sim k\,T \ln(\pi^2 D_o C/n_I\, k\, T\, \ell\, \Omega)$. For the parameters listed above, this does not equal an interstitial motion energy ($\sim$ 0.12 eV) unless $n_I \sim 10^6$, which corresponds to more than one trapped defect per atomic plane. Thus both of these approaches lead to the same conclusion; at low frequencies, trapped self-interstitials cannot "pin" unless they form relatively immobile clusters on the dislocation line. Such a concept has been used by Mercier and Benoit (55) in an interpretation of the internal friction they observed at 1 Hz in n-irradiated Au. The pinning efficiency for MHz frequencies is more complicated to analyze because one can no longer neglect lattice damping. We have as yet not looked into the matter.

The conjecture that an intrinsic point defect-dislocation complex might have a migrational energy different to that of the isolated defect seems initially to have been voiced by Kamel (56). He alluded specifically to a single vacancy when bound to the dislocation and when free, but clearly the question pertains to any point defect. It is, thus a central question in this area of internal friction, one which needs urgent attention. For example, Simpson and Sosin (54) have suggested that, for self-interstitials trapped at dislocations, the motion is not thermally activated at all. The consequences of this hypothesis (54) are clearly quite different from those of conventional pinning, as we have mentioned earlier. Although, the problem looks difficult both analytically and numerically, because of the small energy differences involved if nothing else, yet it would be helpful to have any fundamentally-based handle on this matter. To our knowledge, none at present exists.

With the above remark we must leave this aspect of the general topic. Much more could be said about the relaxation effect associated with pin motion and/or defect dragging (54), as the case may be. The interested reader must be referred to the cited references. We will here next address the issue raised by Eq. (19); that is, where the pin, or pins, might reside relative to the anchor points.

4. Thermodynamic Forces on Pinning Points

In the previous section it was pointed out that as far as the ultimate effect on measurable quantities is concerned, where a pin is situated can be as important as what its mobility might be. Whether or not there are direct elastic interactions between types of pinning points has yet to be considered. What has been suggested is that a force of thermodynamic origin - the entropy force - exists (27, 28). The effect of this on one pin situated between two anchor points is to impel it towards the nearest one. The consequences for interpretations of irradiation studies - or for the impurity component of pinning in unirradiated materials for that matter - would be profound. What we wish to show in this section is that in fact there exists another thermodynamic force with precisely the opposite effect. We shall argue that the net force of thermodynamic origin can be ignored.

It is convenient to begin by summarizing Bauer's treatment (28). This is based upon the observation that the normal modes of vibration of a string such as that described by Eq. (1) are of the form $\omega_p = (C/A)^{\frac{1}{2}}(\pi p/\ell)$, if damping is ignored. Here p is an integer. It runs from unity to N, where $N \simeq (\ell/b)$, this being the member of degrees of freedom the line segment possesses. The Helmholtz free energy contribution, F_V, due to thermal excitation of these vibrations at high temperatures is readily shown to be

$$F_V = kT \sum_{p=1}^{N} \ell n \frac{h\omega_p}{kT} - \frac{1}{2}\frac{h\omega_p}{kT} \quad , \tag{21}$$

where h is Planck's constant divided by 2π. The mode frequencies can be written as $\omega = \omega_{max}(p/N)$ and the sums evaluated to yield

$$F_V \simeq kT\left[N\left\{\ell n\,(\theta_D/T) - 1 - (\theta_D/T)\right\} + \ell n\,(2\pi N)^{\frac{1}{2}}\right] \quad , \tag{22}$$

where $\theta_D = (h\omega_{max}/k)$. Attention is then focussed upon the final term in Eq. (22) because, unlike the remainder, it is non-linear in N. Thus suppose there now existed two such segments side by side, each firmly anchored at the outer extremities of the whole. If there are M atomic planes enclosed in total, and N contained by one segment length alone, then the vibrational free energy is seen to be composed of a term linear in M and independent of N, and another of the form $\delta F_V = kT\,\ell n\{2\pi[N(M-N)]^{\frac{1}{2}}\}$. Hence a thermodynamic force, $F = -b^{-1}d(\delta F_V)/dN$, can be defined. It is entropic in origin and has the form (27)

$$F = (kT/L)(x-L)/x(2-x/L) \quad , \tag{23}$$

upon setting $M = 2L/b$ and $N = x/b$. The sense of this force has already been described.

The following significant aspects of this derivation should be noted. It ignores dislocation drag effects entirely, it utilizes a Debye-type cut-off procedure and, finally, the force originates from a fractional correction of order $N^{-1}\,\ell n N$ to the vibrational free energy. But note that drag is the all-important feature of internal friction. Its effect here will be to endow every mode with a finite lifetime, τ_o say. The latter is evidenced in quantum mechanics as an energy level width. Normally it is of no significance in Statistical Mechanics because ensuing corrections to the free energy are of order $(h/\tau_o kT)$. Yet there are exceptions. One is provided by Dingle's theory of the influence of impurity scattering on the de Haas-van Alphen effect (57). Another is supplied by the present example. For though $h/\tau_o kT$ is indeed small, the fractional correction to the free energy is <u>independent</u> of N and thus cannot be ignored. Indeed, it leads to an effect which, by analogy with the above, we might refer to as an internal energy force. As we shall indicate below, this force has precisely the opposite effect of that due to line entropy in that it favors a pin being positioned centrally rather than at either anchor point.

The calculation is rather lengthy. Consequently, we shall defer all details until a subsequent publication. Here we attempt to convey only the substance of the analysis. This begins with the determination of the <u>complex</u> mode frequencies of Eq. (1) (for $\sigma \to 0$) and an evaluation of the energy-decay of the string following excitation of one mode. The latter is of the type $\exp(-2\omega_I t)$, where $i\omega_I$ is the imaginary part of the frequency. For the vast majority of modes, ω_I is simply (B/2A), whence the lifetime of each quantum mechanical state becomes A/B. Accordingly, if anharmonic scattering dominates the damping, B (and A) are independent of frequency to a first approximation. The lifetime τ_o of every quantum mechanical state is the same. To this degree of approximation, every normal mode frequency is unshifted by damping also. The only change in the Alefeld-Bauer calculation then arises from the contribution of a particular mode level (of energy ε_j) to the density of states at energy ε, $\rho(\varepsilon,\varepsilon_j)$. Their approach is based upon the assumption $\rho(\varepsilon,\varepsilon_j) = \delta(\varepsilon-\varepsilon_j)$. We propose instead that the latter should be (57-59)

$$\rho(\varepsilon,\varepsilon_j) = (h/2\pi\tau_o)\left[(\varepsilon-\varepsilon_j)^2 + (h/2\tau_o)^2\right]^{-1} \quad , \tag{24}$$

to account for damping of the modes. The earlier analysis is appropriate only in the limit of $(h/\tau_o) \to 0$.

The contribution of the vibrational modes to the free energy, F_V, is $-kT\,\ell n Z$, where Z is the partition function. The latter is the product of quantities of the form

$$Z_p = \sum_{n=0}^{\infty}\int_0^{\infty} d\varepsilon\,\rho(\varepsilon, h\omega_p(n+\tfrac{1}{2}))\exp(-\varepsilon/kT) \quad , \tag{25}$$

appropriate to each mode of index p, frequency ω_p. It is in the evaluation of Eq. (25) that the algebraic complexity arises, and it must suffice here to simply state the result we have obtained by a combination of analytical and numerical techniques. Namely, that at high temperatures, Z_p has the form $Z_p \simeq Z_p^o\,(1+g_p)$, where

$$g_p \simeq \frac{h}{4\pi\,\tau_o k T}\,\ell n\left(1+\omega_p^2\,\tau_o^2\right) \quad , \tag{26}$$

and $Z_p^o \simeq [2\sinh(h\,\omega_p/2kT)]^{-1}$, this being the usual result for Z_p when the lifetime is infinite. One must next evaluate the sum of $\ell n\,Z_p$ over all modes to evaluate F_V. To first order in g_p this can be done, with the result that there is a contribution, δF_B, due to the level broadening of the form

$$\delta F_B \simeq -\frac{N\,h}{2\pi\,\tau_o}\left[\ell n\,(\omega_{max}\,\tau_o) - 1\right] \quad , \tag{27}$$

if the cut-off procedure underlying Eq. (22) is used. The form of this expression illustrates the most significant feature of this calculation - δF_B contains the factor N as a multiplier, whereas the entropy correction of Eq. (22) depends upon $\ell n\,(N)$. No matter how small $(h/\tau_o k\,T)$, for large enough segments, δF_B will dominate. In fact the two turn out to be comparable for a realistic range of parameters.

As far as the force on the central pin in the two segment model is concerned, one has to pay heed to the length dependence of the various quantities. We have pointed out earlier (see Sec. 3) that both A and C are proportional to $\ell n\,(q_D \ell)$. Thus, while ω_{max} is independent of ℓ, τ_o is linearly dependent upon the logarithmic factor. The sense of δF_B with variable N is therefore roughly the same as that of $-\,[(N/\ell n\,N) + (M-N)/\ell n\,(M-N)]$, for $N < M$. And this has a minimum at $N = M/2$. Consequently the implied thermodynamic force arising from line damping has precisely the opposite tendency to that resulting from the vibrational entropy.

The basis for the above qualitative conclusion is all we wished to convey. It appears true that in principle, correction terms exist to the lowest order thermodynamic results (i.e. to those terms linear in N as $\tau_o \rightarrow \infty$). But the two considered above have opposite sense. Moreover, the calculation of each depends critically upon fundamentally arbitrary cut-off procedures to the accuracy sought. For example, mode dispersion as a function of wave number is neglected in both. Corrections of this origin will also be of order N. Hence we tend not to attach undue physical significance to either of the mechanisms treated here. We turn instead to a process which we believe may ultimately prove of much greater importance.

5. The Glide-induced Interaction

At this juncture it appeared to us that clustering of trapped defects on a line segment could be a very significant feature of irradiation pinning. Yet there seemed to be no clear idea of how it might come about. Two defects of the type modeled here do not interact at all in perfect material. But in the proximity of a dislocation it appeared that they might. For the shear stress arising from one could induce local glide of the dislocation; the new "loop" so-formed then interacting with the second defect. Our initial concern was whether this could be attractive for certain configurations.

The search for an answer to this, and other questions which have arisen, has taken us further afield than initially envisaged. As a consequence, the work to be described is still incomplete. Nevertheless we believe that some of the results obtained to-date are sufficiently novel to warrant a brief discussion even at this early stage.

The issue has evolved less into a study of dislocation-induced defect-defect interactions than into a question of the capture process of an isolated defect by a dislocation segment. We consider the following simple system: an edge dislocation lying along the y axis of the x - y slip plane with a point defect situated at $(x_o, 0, z_o)$. The segment of dislocation between $y = \xi$ and $y = -\,(\ell - \xi)$ is allowed to glide in response to the internal stress field of the defect. Elsewhere it is considered to be continuously pinned for the moment, as in Ref. (36). When glide occurs there are three internal defects; the backbone dislocation, the adjacent dislocation loop and the point defect itself. The variable energies (for a non-deformable and immobile point defect) are the self-energy of the loop and its interaction with the backbone, which we account for in the line-tension model (36), and the interaction of the loop and point defect. The interaction of the latter with the backbone yields the Cottrell-Bilby interaction. This will be taken as understood.

From the displacement field associated with the defect (see Sec. 3) the shear stress acting in the slip plane in the slip direction, $\sigma(x,y)$, can be found. The energy contributions of interest are therefore the "line" energy of the loop, $\frac{1}{2}C\int_0^{\ell}(d\psi/dy)^2\,dy$, where $\psi(y)$ is the normal displacement of the segment at y, and the interaction term $-\iint\sigma(x,y)\,b\,dx\,dy$ evaluated over the loop area. (The integration over the appropriate remainder of the slip plane gives the Cottrell-Bilby term). Suppose initially we approximate the second of these integrals by using $\sigma(x,y) \simeq \sigma(0,y)$, which amounts to evaluating the stress at the position of the undisplaced line. Minimizing the energy then leads to

$$C\frac{d^2\psi}{dy^2} + \sigma(0,y)\,b = 0 \quad ; \quad \sigma(x,y) = 6G\kappa(x-x_o)z_o/r^5 \quad . \tag{28}$$

(For convenience, we have been appended the general form of σ, for $r = [(x-x_o)^2 + y^2 + z_o^2]^{\frac{1}{2}}$.) The solution for ψ subject to the boundary conditions $\psi = 0$ at $y = \xi$ and $-(\ell-\xi)$ can be found (57). It has the qualitative features one might anticipate as a function of the parameter $(r_\perp = (x_o^2 + z_o^2)^{\frac{1}{2}}$. These are first, that when $(r_\perp/\ell) \gtrsim 10$, the displacement approximates that for a string acted upon by a uniform force. And second, that when this parameter is small ($\lesssim .01$), the shape is little different from the two straight-line segments resulting from application of a point force at the shape's apex, in the attractive case, $z_o < 0$. We can therefore construct the following algorithm to display the dominant dependence of the interaction energy in the present approximation. From $\sigma(0,y)$ construct an equivalent localized stress term of the form $\sigma_o b\,\ell\,\delta(y)$, where $\sigma_o = \ell^{-1}\int_{-\infty}^{\infty}\sigma(0,y)\,dy$. This, as the reader may verify, yields $\sigma_o = -8G\kappa x_o z_o/\ell r_\perp^4$. Next use the Green's function results of Sec. 3 to obtain the amplitude of the line displacement at its maximum, viz $\psi_o = \sigma_o b\,\xi(\ell-\xi)/C$. And, finally, evaluate the interaction energy, E_I, resulting from this mechanism:

$$E_I \sim -\frac{32\eta(1-\eta)}{C}\left(\frac{G\kappa b x_o z_o}{r_\perp^4}\right)^2 \quad , \tag{29}$$

where $\eta = \xi/\ell$. The physical basis of Eq. (29) suggests that the name of glide-induced interaction. Evidently, in the approximate scheme here employed, it is short-ranged, ($E_I \propto r_\perp^{-4}$), and it tends to augment the Cottrell-Bilby interaction.

The above summary is intended for a first appraisal only. It is clear that at sufficiently small $r_\perp$, Eq. (29) cannot be correct. For at such a close approach, $\sigma_o \to \infty$ and so does ψ_o. That is, the loop "overshoots" the source tending to make it glide. It is not hard to appreciate that this deficiency arises from the approximate treatment of $\sigma(x,y)$, since replacing x by zero loses all track of where the segment is relative to the point defect.

In general, the integration of the stress term arising from the defect leads to the expression $2G\kappa z_o b\int[\Phi(\psi-x_o) - \Phi(x_o)]\,dy$, the relevant limits being $-(\ell-\xi)$ and ξ, with Φ the function $\Phi(t) = (t^2+y^2+z_o^2)^{-3/2}$. The prior integral reduces to the form implied by Eq. (28) only if linearized with respect to ψ. We shall choose to disregard the formal limitation of linear elasticity theory to terms of such order, mainly because the deficiency (to this accuracy) has such an obvious physical remedy. It is of interest to see what ensues in such a case, providing one recognizes that the further results may be of more qualitative than quantitative significance.

What we have actually done is a variational calculation. That is, we have taken a trial function

$$\psi(y) = \psi_o(1-y/\xi) \quad ; \quad \psi(y) = \psi_o\left[1 + y/(\ell-\xi)\right] \quad , \tag{30}$$

over the appropriate regions of the segment, and minimized the energy with respect to ψ_o, the maximum displacement. This ultimately leads (60) to the following equation for ψ_o, when $r_\perp/\ell \ll 1$;

$$\frac{\psi_o}{x_o} = \frac{8Gb\kappa(-z_o)\,\xi\,(\ell-\xi)}{C\,\ell\,x_o}\,\frac{(x_o-\psi_o)}{\left\{(x_o-\psi_o)^2 + z_o^2\right\}^2} \quad . \tag{31}$$

The interesting feature of this relation is that its right-hand member changes sign if ψ_o exceeds x_o. This prevents the "overshooting" alluded to earlier, which is now seen to be a consequence of making the (generally) invalid assumption that $x_o \gg \psi_o$. Indeed, a cursory inspection of the orders of magnitude involved shows that when $r_\perp$ is sufficiently small the solution must be $(\psi_o/x_o) \simeq 1$. The dislocation glides in the slip plane until its maximum displacement point is situated directly above the incident defect in this case. It is clear that such occurs because the interaction energy with the defect is much larger than the line energy. The reader can estimate the former by noting that the Φ function introduced above is largest when their argument is small. Hence, upon replacing $\psi(y)$ by ψ_o, $(\simeq x_o)$, in the stress integral, one obtains an energy $\sim -4\,G\,b\,\kappa/(-z_o)$, when $z_o/r_\perp << 1$. This may be contrasted with the Cottrell-Bilby interaction with the backbone dislocation, viz $-4\,G\,b\,\kappa\,(-z_o)/r_\perp^2$, when expressed in the same units.

The results of a numerical analysis of the problem are illustrated in Figure 1. These show the radial dependence of the glide-induced interaction energy when the defect is situated at various angles below the slip plane. The inverse radius and angle dependence is evident at small $r_\perp$. What cannot be discerned in this figure is the transition to the asymptotic form given by Eq. (29). The reason is that with the parameters used it is barely visible on the scale employed. The Cottrell-Bilby and glide-induced interactions found in these calculations are comparable over the range indicated for an angle of $(\pi/4)$. At orientations closer to the slip plane, the glide-induced term is found to be the larger.

We emphasized at the outset the limited quantitative reliability of our model. This is here re-stated. Nevertheless one gets the clear impression that dislocation glide should occur when point defects approach the dislocation. Then, having explored the analytical model to the extent indicated above, we became aware of the numerical analysis by Ingle and Crocker (61) of the vacancy-dislocation interaction in b.c.c. metals. What was most significant to us was that their discrete-atom calculation showed dislocation glide to occur also. It was not as extensive as our model would indicate, nor was it anisotropic, as we find. But, as these authors themselves remark, in their problem the defect (a vacancy) was more properly reflecting the effect of the inhomogeneity interaction (62, 63) that the "misfitting" defect we have discussed. It would be fascinating to learn how any other point defect-dislocation system might deform. Since the inhomogeneity interaction falls of as $r_\perp^{-2}$ rather than as $r_\perp^{-1}$ of Cottrell-Bilby theory, it seems not unreasonable to assume that the induced glide would be somewhat larger than that determined by Ingle and Crocker.

If the general features conjectured upon here are supported by subsequent work, their impact upon the interpretation of irradiation damage studies of internal friction will demand attention. One obviously significant feature is that the force on the defect has a substantial component directed towards the slip plane. Another is that there is also a component tending to center the defect on the line length. This is evidenced analytically by Eq. (29), the numerical result in Figure 1 also shows the same trend. Concomitantly, the defect exerts a force on either pin. Thus if the latter have significant mobility within the core, they will inevitably re-adjust their position with each capture event. Pin-pin correlations will develop. However, we must emphasize that we have no quantitative assessment to offer of the importance of any of these features. The thoughts rest presently upon some highly primitive models. One such is discussed briefly in the Appendix.

6. Discussion

In the foregoing sections we have attempted to acquaint the reader with a few of the problems associated with one internal friction theory and its application to point defect physics. The models considered have been chosen intentionally to be as simple as possible. The experts in the field will have detected numerous omissions. There has been no mention of anisotropic point defect dipoles, Peierls stress, kinks, partial dislocations or Suzuki segregation, for example, and little hint of kinetic theories of irradiation damage. The only excuse the author can offer is one of lack of time and space. There appeared problems enough even at the elementary level attempted. This should not be too surprising. For if the aim is to substantiate the promise of internal friction as a mechanical spectroscopy of the defect solid state, then every step of interpretation demands rigorous inspection. Any approach must be that much more disciplined. Inevitably, the stricter the discipline, the harder each individual step becomes.

Appendix

From the trend depicted in Figure 1 we infer that an appreciable force exists attracting the point defect to the slip plane. Here we shall speculate upon a possible consequence of this, namely that the defect becomes entrapped within the segment core while the latter is still displaced from the backbone. The geometry is that used in Section 5. Let ζ be the distance of the defect from the undisplaced line position. We assume that the line segment deforms very rapidly to the equilibrium configuration dictated by the instantaneous position of its attached defect. Since we assume there are no external forces, this is simply two straight line segments linking the defect to each anchor point. The variable component of the energy of the system is therefore $C\zeta^2\ell/2\xi(\ell-\xi)$.

If the defect has negligible mobility along the segment core, it is easy to show that $\zeta(t)$ will relax to zero exponentially with the characteristic time τ_c defined after Eq. (19). If, however, the core mobility μ_D is significant, we have to solve

$$\dot{\zeta} \simeq -\frac{C\zeta\ell}{\xi(\ell-\xi)}\mu_c \quad ; \quad \dot{\xi} \simeq \frac{C\ell\zeta^2(\ell-2\xi)}{\xi^2(\ell-\xi)^2}\mu_D \quad . \tag{A1}$$

It follows that $\mu_D(\ell-2\xi)\dot{\zeta}^2 = \mu_c^2 C\ell\dot{\xi}$. This may be integrated to yield

$$\xi = \frac{1}{2}\ell - \left(\frac{1}{2}\ell - \xi_o\right)\exp\left[-\left(2\mu_D/\mu_c^2 C\ell\right)\int_o^t \dot{\zeta}^2\,dt\right] \quad , \tag{A2}$$

where ξ_o is the initial distance from the left-hand anchor. Thus while the defect does not move laterally when its core mobility is low, if it is large enough it can center itself upon the length quite rapidly. Suppose we approximate $\dot{\zeta}$ in (A2) by the value appropriate to $t = 0$. Using (A1) we find the characteristic time for this process, τ_D, to be

$$\tau_D \sim (\mu_c/2\mu_D)\left[\xi_o(\ell-\xi_o)/\zeta_o^2\right]\tau_c \quad , \tag{A3}$$

where ζ_o is the initial displacement. The factor in square parentheses is probably large; but the mobility ratio is unknown. No doubt the latter will be significantly different according to whether the defect is considered to be an intrinsic one or an impurity. That is, whether one is concerned about irradiation effects or the establishment of an initial length distribution by annealing. Our intent here is not so much to address such detail as to indicate how the glide-induced interaction introduces new features such as core mobility into the problem. This may present some difficulty with the Thompson and Holmes model. It appears quite conceivable at present that the pinning process is not the simple sequence of chopping up of fixed lengths that these authors envisaged.

References

1. A. S. Nowick and B. S. Berry, Anelastic Relaxation in Crystalline Solids, Academic Press, New York (1972).
2. R. De Batist, Internal Friction of Structural Defects in Crystalline Solids, North Holland Publishing Company, Amsterdam (1972).
3. J. P. Hirth and J. Lothe, Theory of Dislocations, McGraw-Hill Book Company, New York (1968).
4. F. R. N. Nabarro, Theory of Crystal Dislocation, Clarendon Press, Oxford (1967).
5. D. Lenz and K. Lücke (Eds.), Fifth International Conference on Internal Friction and Ultrasonic Attenuation in Crystalline Solids, Vols. I and II, Springer-Verlag, Berlin (1975).
6. P. G. Bordoni, F. M. Mazzolai and M. Nuovo (Eds.), Nuovo Cimento 33B, 1 (1976).
7. J. S. Koehler, Imperfections in Nearly Perfect Crystals, p. 197, John Wiley and Sons, Inc., New York (1952).
8. A. V. Granato and K. Lücke, J. Appl. Phys., 27, 583 and 789 (1956).
9. D. O. Thompson and D. K. Holmes, J. Appl. Phys. 27, 713 (1956).
10. G. A. Alers and D. O. Thompson, J. Appl. Phys. 32, 283 (1961).
11. G. Leibfried, Z. Physik 127, 344 (1950).
12. A. V. Granato and R. Stern, J. Appl. Phys. 33, 2880 (1963).
13. C. R. Heiple and H. K. Birnbaum, J. Appl. Phys. 38, 3294 (1967).
14. R. den Buurman and D. Weiner, Scripta Met. 5, 753 (1971).
15. K. Beissner and E. Biller, Scripta Met. 7, 535 (1973).
16. J. L. Routbart and H. S. Sack, J. Appl. Phys. 37, 4803 (1967).
17. D. O. Thompson and V. K. Paré, Physical Acoustics, Vol. IIIA, p. 293, Academic Press, New York (1966).
18. D. O. Thompson and V. K. Paré, J. Appl. Phys. 31, 528 (1960).

19. D. O. Thompson, O. Buck, R. S. Barnes and H. B. Huntington, J. Appl. Phys. 38, 3051 (1967).
20. H. M. Simpson, A. Sosin and D. F. Johnson, Phys. Rev. B5, 1393 (1972).
21. J. L. Routbort, thesis, Cornell University (1965).
22. R. L. Nielsen, thesis, University of Pittsburgh (1968).
23. J. Lauzier, C. Minier and S. L. Sieffert, Phil. Mag. 31, 893 (1975).
24. R. J. Kerans and H. M. Simpson, J. Appl. Phys. 50, 4739 (1979).
25. P. Winterharger and K. Lücke, J. Appl. Phys. 44, 4855 (1973).
26. R. M. Stern and A. V. Granato, Acta Met. 10, 358 (1962).
27. G. Alefeld, Phil. Mag. 11, 809 (1965).
28. C. L. Bauer, Phil. Mag. 11, 827 (1965).
29. A. D. Brailsford, J. Appl. Phys. 43, 1380 (1972).
30. A. D. Brailsford, see Ref. 5, Vol. II, p. 1.
31. J. D. Eshelby, Phil. Mag. 6, 953 (1961).
32. J. D. Eshelby, Proc. Roy. Soc. A 266, 222 (1962).
33. A. D. Brailsford, J. Appl. Phys. 41, 4439 (1970).
34. A. D. Brailsford, Phys. Rev. 142, 383 (1966).
35. A. D. Brailsford, Phys. Rev. 142, 388 (1966).
36. M. P. Read and A. D. Brailsford, J. Appl. Phys. 45, 3351 (1974).
37. A. D. Brailsford, unpublished.
38. G. Schoeck and A. Seeger, Acta Met. 7, 467 (1959).
39. A. M. Kosevich and V. D. Natsik, Soviet Physics JETP, 24, 810 (1967).
40. A. Seeger and W. Hornung, see Ref. 5, Vol. I, p. 214.
41. A. Seeger, see Ref. 5, Vol. II, p. 232.
42. L. D. Landau and E. M. Lifshitz, Statistical Physics, Addison-Wesley Publishing Company, London (1969), p. 381.
43. K. G. Denbigh, The Thermodynamics of the Steady State, Methuen and Co., Ltd., London p. 53. See also Ref. 3, p. 593.
44. A. D. Brailsford, J. Appl. Phys. 40, 3087 (1969), and references therein.
45. A. H. Cottrell and B. A. Bilby, Proc. Phy. Soc. 62, 49 (1949).
46. G. Alefeld, J. Völkl and G. Schaumann, Phys. Stat. Sol. 37, 337 (1970).
47. A. D. Brailsford, to be published.
48. A. H. Cottrell and M. A. Jaswon, Proc. Roy. Soc. A199, 104 (1949).
49. S. Takeuchi and A. Argon, Phil. Mag. 40, 65 (1979).
50. A. H. Cottrell, Dislocations and Plastic Flow in Crystals, Clarendon Press, Oxford (1961), pp. 136-138.
51. J. O. Kessler, Phys. Rev. 106, 654 (1957).
52. G. Schoeck, Acta Met. 11, 617 (1963).
53. A. Seeger, Phys. Stat. Sol. 55(a), 457 (1979).
54. H. M. Simpson and A. Sosin, Phys. Rev. B5, 1382 (1972).
55. O. Mercier and W. Benoit, Radiation Effects 27, 207 (1976).
56. R. Kamel, Acta Met. 9, 65 (1961).
57. R. B. Dingle, Proc. Roy. Soc. A211, 517 (1952).
58. A. D. Brailsford, Phys. Rev. 149, 456 (1966).
59. G. W. Ford, private communication.
60. A. D. Brailsford, to be published.
61. K. W. Ingle and A. G. Crocker, Acta Met. 26, 1461 (1978).
62. J. D. Eshelby, Proc. Roy. Soc. A241, 376 (1957).
63. R. Bullough and R. C. Newman, Phil. Mag. 7, 529 (1962).

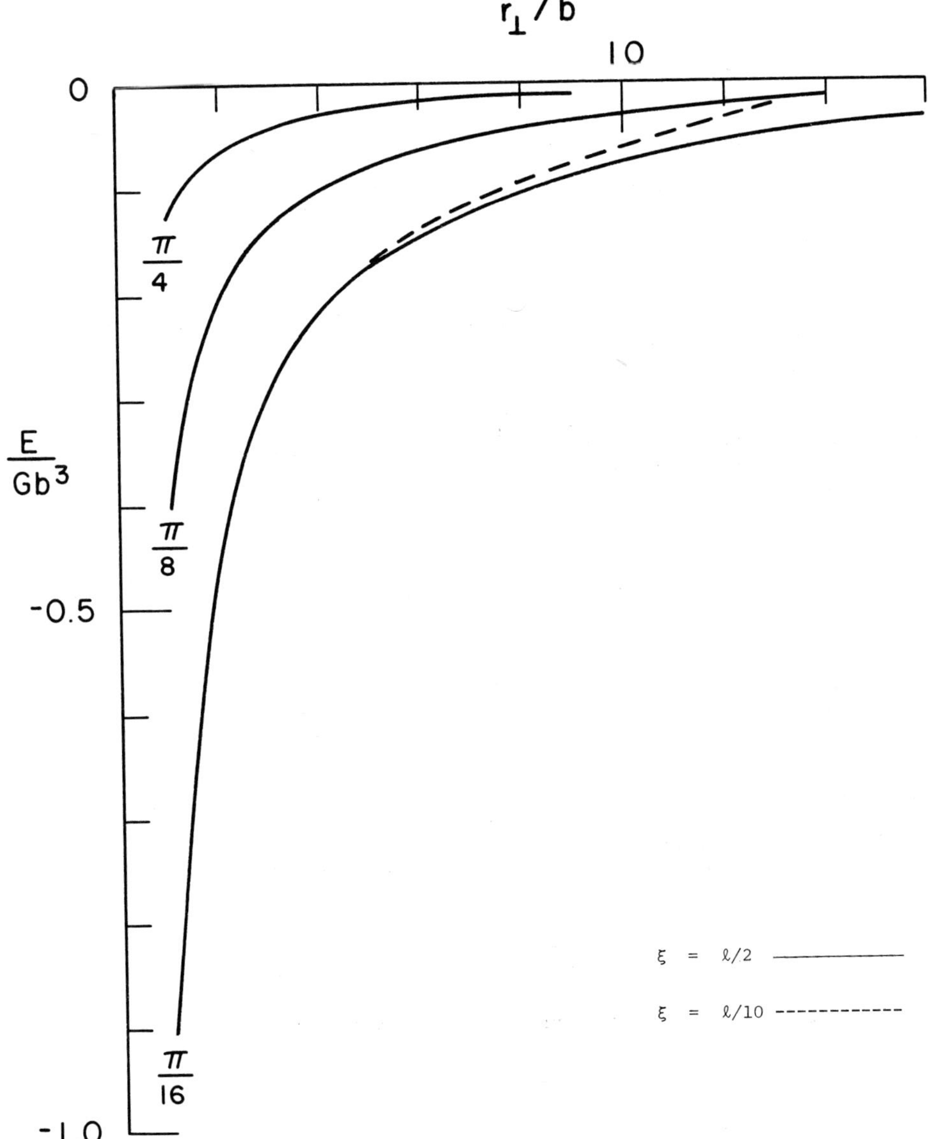

Figure 1. The radial dependence of the glide induced interaction for various orientations below the slip plane of an edge dislocation.

MODE DEPENDENCE OF PHONON SCATTERING BY DISLOCATIONS

Toshiyuki Ninomiya
Department of Physics, University of Tokyo
Tokyo 113, Japan

and

Department of Structural Properties of Materials
The Technical University of Denmark
Building 307, DK-2800 Lyngby, Denmark

Abstract

The cross section of dynamical scattering of thermal phonons by dislocation is calculated for an anisotropic crystal, by including the effect of pinning. Strong dependence of the cross section on the phonon modes is found to come from the Peach-Koehler relation.

1. Introduction

Thermal phonons in a crystal are scattered with dislocations. Recent experiments for LiF by Roth and Anderson (1), together with their previous experiments (2), have revealed that the mechanism of phonon scattering is dynamical ("fluttering"), that slow transverse phonons are scattered much more strongly with dislocations than other modes and that, when the dislocation is pinned, phonons of long wavelength are scattered weakly.

These experimental results make it possible to examine theories of phonon scattering by dislocations in detail. The theories of dynamical scattering have been developed by Garber and Granato (3, 4) and Ninomiya (5-8). By starting with the string model of a dislocation of finite length, Garber and Granato have calculated the scattering strength. On the other hand, Ninomiya has treated the dislocation-phonon interaction in a general form for an infinite dislocation. Although in these treatments equations of motion are given in formula valid for anisotropic case (7-9), explicit calculations of the scattering cross section have been made only for the isotropic case.

The purpose of the present work is to give an explicit form of the scattering cross section for a pinned dislocation in an anisotropic crystal to find what factors are responsible for the above experimental results. We start with treatments given in Refs. 7-9. Although we expect to obtain essentially the same results as those by Garber-Granato's treatments if the latter are modified in accordance with the anisotropy, a general formula will be convenient to use for detailed comparisons with experiments. Detailed calculations and comparisons with experiments will be published elsewhere.

2. Effect of Pinning

We start with treatments given in Refs. 7-9. Since these treatments have been made for an infinite and unpinned dislocation, we have to take the effect of pinning into account additionally.

Let us consider that a dislocation on the z-axis makes a vibration given by the displacement

$$\xi(z) = \sum_{\kappa} \xi_{\kappa}\, e^{i\kappa z} \tag{1}$$

If the dislocation is pinned at the positions z_n, the increase in the pinning energy is given for small displacements as

$$E_P = \frac{P}{2} \sum_n \xi^2 (z_n) = \frac{PL}{2\bar{\ell}} \sum_{\kappa} \sum_{\kappa'} S(\kappa - \kappa') \xi_\kappa \xi^*_{\kappa'} \quad (2)$$

where P stands for the strength of pinning, L is the length of dislocation, and $\bar{\ell}$ is the average separation of pinning points. $S(\kappa)$ is the structure factor for the distribution of pinning points and defined as

$$S(\kappa) = \frac{\bar{\ell}}{L} \sum_n e^{i\kappa z_n} \quad (3)$$

$S(\kappa)$ for simple cases are

(a) Random distribution

$$S(\kappa) = \delta(\kappa)$$

(b) Uniform distribution

$$S(\kappa) = \sum_m \delta(\kappa - \frac{2\pi}{\bar{\ell}} m)$$

where m is an integer.

3. Dislocation-Phonon Interactions

In order to obtain the equations of motion for the pinned dislocation, we have to add the pinning energy, **Eq.** (2), to the previous treatments.

We write the phonon field as

$$\vec{v}(\vec{r}) = \sum_{\vec{k},s} q(\vec{k},s)\, \vec{e}(\vec{k},s)\, e^{i\vec{k}\vec{r}} \quad (4)$$

where $q(\vec{k}, s)$ is the amplitude of the mode $\vec{k}$, s, and $\vec{e}$ is the polarization vector of the mode. The equations of motion for $q(\vec{k}, s)$ and ξ_κ are, then,

$$\omega_o^2(\vec{k},s)\, q(\vec{k},s) + \ddot{q}(\vec{k},s) + \frac{L}{\rho V} \frac{\Phi^*(\vec{k},s)}{\omega_o^2(\vec{k},s)} \ddot{\xi}_\kappa = 0 \quad (5)$$

and

$$\kappa^2 T_\kappa \xi_\kappa + \frac{P}{2\bar{\ell}} \sum_{\kappa'} \left[S(\kappa - \kappa') + S^*(\kappa - \kappa') \right] \xi_{\kappa'} - \sum_{\vec{k},s}^{\kappa} \Phi\, q(\vec{k},s) = 0 \quad (6)$$

where $\omega_o(\vec{k}, s)$ is the eigenfrequency of the mode $\vec{k}$, s in a perfect crystal. The factor Φ is defined as

$$\Phi(\vec{k},s) = - i\, c_{ijk\ell}\, b_i\, n_j\, e_k\, k_\ell \quad (7)$$

where $c_{ijk\ell}$ are the elastic moduli, and $\vec{n}$ is the vector normal to the slip plane. It is easily seen that Φq gives the Peach-Koehler force due to the phonon q. T_κ is the effective line tension of the dislocation and is logarithmically dependent on κ. $\sum^\kappa$ means the summation over $\vec{k}$ with $k_z = \kappa$. The second term of the left-hand side of Equation (6) is the effect of pinning. In this paper we treat the simple case (a). Then, the effect of pinning is to replace $\kappa^2 T_\kappa$ for an unpinned case with $\kappa^2 T_\kappa + P/\bar{\ell}$.

For the incident wave $\overset{o}{q}(\vec{k}_o, s_o)$ with the frequency $\omega = \omega_o(\vec{k}_o, s_o)$, we have the scattered wave $\overset{s}{q}(\vec{k}, s)$ and the dislocation vibration ξ_κ as

$$\overset{s}{q}(\vec{k},s) = \frac{\omega^2}{\omega_o^2(\vec{k},s) - \omega^2 - i\varepsilon} \frac{L}{\rho V} \frac{\Phi^*}{\omega_o^2(\vec{k},s)} \xi_\kappa \tag{8}$$

and

$$\xi_\kappa = \frac{1}{F_p(\kappa^2,\omega^2)} \Phi \overset{o}{q}(\vec{k}_o,s_o) \tag{9}$$

where $k_z = \kappa = k_{oz}$ and

$$F_p = \frac{P}{\ell} + \kappa^2 T_\kappa - \frac{L}{\rho V} \sum_{\vec{k},s}^{\kappa} \frac{\omega^2}{\omega_o^2} \frac{\Phi\Phi^*}{\omega_o^2 - \omega^2 - i\varepsilon} \tag{10}$$

The displacement field due to the scattered wave at large distances from the dislocation is calculated using the method of stationary phase to give

$$\vec{v}(\vec{r}) = \sum_s f(\vec{k},s:\vec{k}_o,s_o) \frac{\vec{e}(\vec{k},s)}{(x^2 + y^2)^{1/4}} e^{i\vec{k}\vec{r}} \tag{11}$$

where

$$f(\vec{k},s:\vec{k}_o,s_o) = \frac{1}{\sqrt{2\pi}\,\rho} \xi_\kappa \frac{\Phi^*(\vec{k},s)\, e^{-i\frac{\pi}{4}\alpha}}{\left\{\left(\frac{\partial\omega_o^2}{\partial k_x}\right)^2 + \left(\frac{\partial\omega_o^2}{\partial k_y}\right)^2\right\}^{1/4} |A|^{1/2}} \tag{12}$$

and

$$A = \sin^2\theta \frac{\partial^2\omega_o^2}{\partial k_x^2} - 2\sin\theta\cos\theta \frac{\partial^2\omega_o^2}{\partial k_x\,\partial k_y} + \cos^2\theta \frac{\partial^2\omega_o^2}{\partial k_y^2}$$

$$\alpha = \mathrm{sgn}A, \quad \theta = \tan^{-1}\frac{y}{x}$$

The differential cross section is, then, given by

$$d\sigma = \frac{\Phi\Phi^*(\vec{k}_o,s_o)}{2\pi\,\rho^2\, F_pF_p^*\, |\mathrm{grad}_{\vec{k}}\, \omega_o^2(\vec{k}_o,s_o)|} \sum_s \frac{d\theta}{|A|} \Phi\Phi^*(\vec{k},s) \tag{13}$$

The total scattering cross section is obtained by integrating Equation (13) over θ. The direct integration over θ is not easy except for the isotropic case and, therefore, the integration is transformed to that in the k_x, k_y space. We finally obtain

$$\sigma_{tot} = \frac{2\Phi\Phi^*\,(\vec{k}_o,s_o)}{\rho\, |\mathrm{grad}_{\vec{k}_o}\, \omega_o^2(\vec{k}_o,s_o)|} \mathrm{Im}\frac{1}{F_p} \tag{14}$$

4. Discussion

As was briefly shown in Section 1, experiments have clarified the strong mode dependence and the pinning effects for the dynamical scattering. We can now see from Equation (14) what factors are responsible for the effects.

1) $\Phi\Phi^*\,(\vec{k}_o,s_o)$:

As was mentioned previously, Φ corresponds to Peach-Koehler relation. Remarkable mode dependence comes from this coefficient. For the vertical incidence of phonon to the (110) slip plane in a cubic crystal, this coefficient is not zero only for the slow transverse mode (1). For the oblique incidence of fast transverse phonon, Φ becomes zero for many symmetric directions of propagation, which means the fast transverse mode is scattered very weakly.

2) $\mathrm{grad}_{\vec{k}}\, \omega_o^2(\vec{k}_o, s_o)$:

This is completely determined with mode of incident phonon. Dependence on the modes comes from the difference in the sound velocities.

3) $\mathrm{Im}\, \frac{1}{F_P}$

F_P has the term where the summation is made over k_x, k_y on the plane vertical to the dislocation, and therefore gives dependence on the type of dislocation. If we write F_P as

$$\mathrm{Re}\, F_P = \frac{P}{\bar{\ell}} + \kappa^2(\text{line tension}) - \omega^2(\text{mass})$$

the dependence comes from the line tension and mass. This term is responsible also for pinning effects. When $\bar{\ell}$ becomes small, the resonance frequency ($\mathrm{Re}\, F_P = 0$) shifts to higher frequency and reduces the cross section for frequencies below the resonance.

We may need comment on the pinning effect given here. If we make the strength of pinning P infinite, the resonance frequency goes to infinity and the whole range of frequency belongs to the low frequency region. This comes from our assumption of random distribution. If the randomness is not complete within the mean free length of phonon, the structure factor $S(\kappa)$ should be that in which the non-randomness is taken into account. If we calculate the case (b) as an extreme case, we can easily see that we have a finite displacement of dislocations even for $P = \infty$. Detailed discussions and comparisons with experiments will be published elsewhere.

Acknowledgments

Part of this work was done while the author stayed at the Technical University of Denmark. The author is grateful for the hospitality. He also thanks Professor Yasushi Wada for general discussions on forward scattering.

References

1) E.P. Roth and A.C. Anderson, Phys. Rev. B20, 768 (1979)

2) A.C. Anderson and M.E. Malinowski, Phys. Rev. B5, 3199 (1972)

3) A.V. Granato, Phys. Rev. 111, 740 (1958)

4) J.A. Garber and A.V. Granato, J. Phys. Chem. Solids 31, 1863 (1970)

5) T. Ninomiya, J. Phys. Soc. Jpn. 25, 830 (1968)

6) T. Ninomiya, Nat. Bur. Stand. (U.S.), Spec. Publ. 317, 315 (1970)

7) T. Ninomiya, J. Phys. Soc. Jpn. 36, 399 (1974)

8) T. Ninomiya, Treatise on Materials Science and Technology, ed. H. Herman (Academic, New York 1975) Vol. 8, p. 1

9) K. Ohashi and Y.H. Ohashi, Philos. Mag. A38, 187 (1978)

THERMAL RESISTANCE ARISING FROM DYNAMIC SCATTERING OF PHONONS BY DISLOCATIONS

G.A. Kneezel * and A.V. Granato
Physics Dept. and Materials Research Laboratory
University of Illinois, Urbana, Illinois
* Present address: Xerox Webster Research Center, Webster, New York

We have calculated the dynamic phonon scattering rates for dislocations and dislocation dipoles and the resulting increase in low temperature thermal resistivity after plastic deformation. We indicate the importance of explicitly including (for both types of defects) the Ninomiya resonance angle effect, phonon focusing, and the resolved shear stress factor for each phonon. We demonstrate that these factors do not conspire together to produce a large effect in [110] LiF, but that for a similar dislocation (or dipole) density the effect would be larger in [110] NaF. Experimental data for deformed and irradiated LiF is fit by assuming a dislocation dipole density of 30 times the observed etch pit density. Finally we predict a negative pinning effect which is still to be confirmed experimentally.

By gamma irradiating plastically deformed rods of LiF and remeasuring the thermal conductivity at low temperature, Anderson and co-workers (1,2) have established that the increase in thermal resistivity after plastic deformation is primarily due to dynamic scattering of phonons by dislocations. Until now however it has not been possible to obtain quantitative agreement between the observed effect and calculations based on either static or dynamic models (2,3). We present the results of a detailed calculation (4) based on the dislocation vibrating string model (5-8) which suggest that the scattering is mainly due to the optical mode of vibration of dislocation dipoles rather than to independently vibrating dislocations.

Background

As originally proposed by Eshelby (9) and amplified by Nabarro (10) if a sound wave or phonon induces a dislocation to vibrate, the incident wave will be dissipated as the dislocation radiates elastic waves (reradiation scattering). Even if the dislocations do not move however, phonons are scattered by the anharmonicity of the dislocation strain field (static strain field scattering) as suggested by Nabarro (10). The first detailed treatment of the effects of dislocations on thermal conductivity was done in 1955 by Klemens (11) who assumed static strain field scattering is dominant and predicted a quadratic temperature dependence of the thermal conductivity. In 1958 Granato (12) used the reradiation scattering cross-section calculated by Nabarro to find that at low temperature reradiation scattering should dominate and result in a temperature exponent of approximately 3.5.

Measurements of the effect of plastic deformation on the low temperature thermal conductivity of alkali halides have been in approximate agreement with the magnitude predicted for the reradiation scattering mechanism but the temperature dependence has not clearly favored one mechanism over the other. As a result there has been considerable controversy as to which mechanism is truly dominant at low temperatures. Of 27 papers published in the past 25 years, 17 have attempted refinements of the static strain field scattering calculation (hoping to increase the predicted magnitude), while the remainder have favored the reradiation mechanism. Of the latter, the most important refinements have been the inclusion of obliquely incident phonons by Ninomiya (13) and the inclusion of pinning by Garber and Granato (14). By including obliquely incident phonons it is found that at a certain critical angle of incidence (the Ninomiya resonance angle) the scattering is much greater than for normal incidence. However, it should be pointed out that due to the complexity of Ninomiya's expressions, the result which has been used in fitting experimental data is that for angles far from the resonance angle--which does not differ significantly from the earlier Nabarro-Granato result (12). The effect of pinning on the reradiation scattering rate (14) is a resonance peak which rises above the asymptotic value of scattering predicted for unpinned dislocations and then a decrease to zero for low frequencies (and hence at low temperatures where most phonons have low frequencies).

By irradiating a plastically deformed specimen the dislocations become pinned. This should have no effect on the static strain field mechanism but should drive up the dislocation resonant frequency so that the reradiation scattering peak moves to higher frequencies (or temperatures) with scattering becoming negligible at lower frequencies (temperatures). Anderson and co-workers (1,2) have performed such measurements on sheared and bent LiF [110] rods and found that for increasing gamma irradiation doses the low temperature thermal conductivity progressively increased to the undeformed sample values. However, as observed by Anderson (3), though the results are completely inconsistent with static strain field scattering, it has not been possible to obtain quantitative agreement with the calculated effects of reradiation scattering either. In particular, the average dislocation lengths deduced from thermal conductivity measurements in freshly deformed LiF (1,2) are much less than has been determined ultrasonically (15). In addition the observed effect (1,2) is broader than predicted, remaining unexpectedly large above 2 K.

However, the qualitative evidence in favor of dominant scattering by vibrating pinned dislocations is so impressive that it raises the question of whether factors previously neglected could account for the discrepancies observed. It is conceivable that factors such as phonon focusing, the Ninomiya resonance angle effect, and the resolved shear stress factors of the phonons mainly responsible for thermal conductivity could conspire together to produce the observed magnitude and temperature dependence of the effect on the thermal conductivity. Our calculation which includes these factors and others previously neglected is outlined in the next two sections.

The Vibrating String Model Including Oblique Incidence of Phonons

In order to incorporate the Ninomiya resonance angle effect into the mathematically simpler vibrating string model formalism (5-7), we use the approach of Nikolayev (8). Although a realistic treatment must consider the case of pinned dislocations it is useful to consider the case of unpinned dislocations first, as the calculation is easier and the results give simple analytic expressions which are the high frequency asymptotes for those of the pinned dislocation case.

A periodic stress wave with wavelength λ_p whose wavefronts are at an angle η to a dislocation along the x direction, hits one end of the dislocation first so that a travelling wave is propagated along the dislocation with wavelength $\lambda = \lambda_p/\sin\eta$. If σ is the shear stress component of σ_p resolved in the x-y glide plane in the direction of the Burgers vector b, the glide force per unit length is $F = b\sigma\cos(kx-\omega t)$ where $k = 2\pi/\lambda$. Treating the dislocation as a damped oscillating string with effective mass A, damping constant B, and line tension C, the equation of motion is

$$A\,\partial^2 y/\partial t^2 + B\,\partial y/\partial t - C\,\partial^2 y/\partial x^2 = b\sigma\cos(kx-\omega t) \qquad [1]$$

The logarithmic decrement describes how quickly the amplitude of the wave decays and is given by $\Delta = \Delta W/2W$ where ΔW is the work done on the dislocation per cycle, $W = \sigma_p^2/2G$ is the energy stored per cycle, and G is the appropriate elastic constant. For unpinned dislocations,

$$\Delta_u = \frac{\pi\Lambda\Omega G b^2 B\omega}{(-A\omega^2 + Ck^2)^2 + B^2\omega^2} \qquad [2]$$

where Λ is the dislocation density, and the resolved shear stress factor is $\Omega = (\sigma/\sigma_p)^2$ and takes account of the fact that only the shear stress component on the slip system moves the dislocation (6,16).

The scattering rate describes the rate at which the incident wave loses its energy to the dislocations and is given by

$$\tau_u^{-1} = \omega\Delta/\pi = \frac{\Omega\Lambda G b^2 s}{\omega A[(R^2-1)^2 + s^2]} \qquad [3]$$

where $s = B/A\omega$ and $R = (C/A)^{1/2}k/\omega = (C/A)^{1/2}\sin\eta/v$. The Ninomiya resonance angle condition corresponds to $R = 1$ while the head-on incidence case treated by Granato (12) corresponds to $R = 0$.

The effect of pinning points at $x = 0$ and $x = L$ is to impose the boundary conditions $y(0) = y(L) = 0$. The Fourier series solution leads to a summation expression for the decrement, the terms of which are

$$\Delta_m = \frac{4\Omega\Lambda G b^2 L^2 \; s u^2 (1-(-1)^m \cos \pi R u)}{\pi^3 C (m(m^2-R^2u^2))^2 [(1-u^2/m^2)^2 + (su^2/m^2)^2]} \qquad [4]$$

where u is the normalized frequency ω/ω_1 and ω_m is the mth normal mode frequency given by

$$\omega_m = m\pi/L \; (C/A)^{1/2} . \qquad [5]$$

Expressions used for the model parameters are those which are appropriate to the case of radiation damping of high frequency phonons at low temperature in the alkali halide structure. Specifically we included (for both edge and screw dislocations) the frequency dependence of the dislocation effective mass (17), the anisotropy (18) and frequency dependence of the dislocation line tension, the wavelength dependence of the radiation damping constant (19), and the resolved shear stress factor (20) for the alkali halide structure. In addition the decrement (or scattering rate) was averaged numerically over the Koehler distribution of loop lengths (5,6).

Figure 1 shows the normalized scattering rate $\tau^{-1}/2\omega_1\Omega\Lambda L^2$ for R = 0, .8, 1 and 1.2 assuming the Debye frequency $\omega_D = 1000\ \omega_1$ and neglecting the wavelength dependence of the radiation damping constant. (Although the wavelength dependence was included subsequently, its effect was found to be not very large.) The pronounced peak which is observed is due to the underdamped character of the resonance. The enhancement which occurs at high frequencies for oblique angles of incidence is illustrated. The greatest enhancement occurs for the Ninomiya resonance angle case (R = 1) but is still considerable for R = .8 and 1.2. At low frequencies the angle of incidence becomes irrelevant as the stress wavelength becomes large enough that the variation in stress along the dislocation is always small.

To obtain the scattering rate from Figure 1 one must multiply by $2\Omega\Lambda\omega_1 L^2$. The effects of Λ and Ω are straightforward: a larger scattering rate comes from a larger density of dislocations and/or a larger resolved shear stress factor. The effect of the average length is shown in Figure 2. The frequency at which maximum scattering occurs is inversely proportional to L while the magnitude of the maximum is proportional to L.

The Effect of Independently Vibrating Dislocations on Thermal Conductivity

At the lowest temperatures the limit to lattice thermal conductivity is due to scattering of phonons by the surface of the sample (21) (boundary scattering) and by defect scattering. In this section we consider boundary scattering and dislocation reradiation scattering, explicitly including phonon focusing, and compare the results to experimental results in LiF.

We used the approach of McCurdy, Maris, and Elbaum (22) (with some modifications) to describe boundary scattering and phonon focusing in a finite anisotropic rod and included reradiation scattering by adding the dislocation and boundary scattering rates for each phonon. The angle of incidence, the resonance angle parameter R, the wavelength dependence of the damping constant and the resolved shear stress factor were calculated for each of the three phonon modes in a octant of spherical angles with $\Delta\theta = \Delta\varphi = 1°$.

Because the dislocation scattering rate changes slowly with frequency we made the dominant phonon approximation (1) which gives the dominant phonon frequency as

$$\omega_{dom} = 7.6\,\pi\,kT/h = 5.0 \times 10^{11}\,T. \qquad [6]$$

Thus if we plot the calculated thermal conductivity versus frequency, the relation $T = 2 \times 10^{-12}\,\omega$ may be used to indicate the corresponding temperature.

In order to determine the importance of the assumptions of a frequency dependent effective mass, the Ninomiya resonance angle effect, and the wavelength dependence of radiation damping, calculations for a [110] LiF rod were made with and without each of these three assumptions. It was found that while the third assumption had only a small effect, the first two broadened the frequency (temperature) dependence of the thermal conductivity and shifted the minimum to higher frequency (temperature). Of these three the most important effect was the Ninomiya resonance angle effect which has been assumed negligible in previous thermal conductivity calculations.

Phonon focusing and the resolved shear stress factor are anisotropic and mode dependent and must be considered together. It is generally true (23) for a [110] rod that fast transverse phonons are focused relatively strongly along the rod, while slow transverse phonons are focused into the sides of the rod and longitudinal phonons are not strongly focused. This suggests that in the absence of dislocation effects the fast transverse mode, having a relatively small velocity and being scattered least by the boundary, provides the major contribution to the thermal conductivity along [110]. A numerical calculation for a [110] LiF rod with length to width ratio of 15 confirms this in that the fast transverse, slow transverse and longitudinal modes contribute 48%, 36% and 16% respectively to the thermal conductivity.

The mode dependence of reradiation scattering arises from the difference in resolved shear stress factors of the three modes. For dislocations on the (110) slip planes perpendicular to the rod axis, C' (and quasi C') phonons are scattered strongly while C_{44} (and quasi C_{44}) phonons are scattered weakly and the scattering of longitudinal phonons is intermediate. The magnitude of the effect of plastic deformation on thermal conductivity depends therefore on whether the fast transverse mode (which makes the dominant contribution to thermal conductivity) is C_{44} or C'. In LiF C_{44} is fast transverse so that phonon focusing and the resolved shear stress factor do not conspire together to produce a large effect, while in NaF and most other alkali halides C' is fast transverse so that for a similar dislocation density the effect on the thermal conductivity will be considerably larger.

An attempt was made to fit the thermal conductivity data of Roth and Anderson (2) for a bent [110] LiF rod with 2.5×10^{11} m^{-2} dislocations as determined by etch pit counts. It was found that this density of dislocations on (110) is not capable of producing an effect as large as was observed and the predicted temperature dependence was too sharp. Moreover in order to position the thermal conductivity minimum at .25 K as was observed, it was necessary to assume an average dislocation length of 6×10^{-8} m (a factor of three less than that determined ultrasonically (15)). Thus, although the inclusion of the Ninomiya resonance angle effect improved the agreement by broadening the predicted temperature dependence and moving the minimum to higher temperature, it was not enough. In addition, the effects of phonon focusing and the resolved shear stress factor did not conspire together to produce the magnitude observed. What this suggested to us is that the defects actually responsible for the effect must be dislocation-like with respect to plastic deformation and reradiation scattering, but must be more numerous, have a higher average resonant frequency and have a broader distribution of resonant frequencies than independently vibrating dislocations. In the next section we show that these conditions are met by optically vibrating edge dislocation dipoles.

The Effect of Vibrating Edge Dislocation Dipoles on Thermal Conductivity

Two dislocations of opposite sign on parallel glide planes a distance d apart have a stable configuration when they are a distance $\sqrt{2}$ d apart (24) and the pair is called a dislocation dipole. Because screw dislocations are able to cross glide and annihilate, most dipoles are of edge character. Dislocation dipoles were first observed and discussed by Johnston and Gilman (25) in their etch pit studies of LiF. They suggested that the approximate range of dipole spacing is 3b to 300b. Because of the difficulty of resolving narrow dipoles, as well as the likelihood that very narrow dipoles do not etch (26) and are invisible under the electron microscope (27), it is difficult to estimate the dipole density and the distribution of spacings. A reasonable dipole density appears to be 10 to 100 times the monopole density, with most having spacings less than a hundred lattice spacings so that they are not easily detected by electron microscopy or etch pitting techniques. A distribution of this character (for d = nb) is the exponential distribution

$$\Lambda(n) = \Lambda_0 \exp(-(n-3)/N) \qquad [7]$$

Long wavelength phonons ($\lambda \gg d$) excite dipoles into the optical mode of vibration (i.e. the dislocations move in opposite directions). For small oscillations a linear term Dy is added to the left side of the equation of motion [1]. The mth normal optical mode frequency is

$$\omega_{op}(m) = (\omega_m^2 + D/A)^{1/2} \qquad [8]$$

where ω_m is given by [4] and D is

$$D = Gb^2/2\pi d^2(1-\nu) \qquad [9]$$

Thus the optical mode resonant frequency is higher than that of independently vibrating dislocations--especially for narrow dipoles. It was found that by assuming an average dipole spacing Nb = 60b the thermal conductivity minimum could be situated at the observed .25 K for the ultrasonically determined average dislocation length (15) of 2×10^{-7} m. Figure 3 shows Roth and Anderson's thermal conductivity data in the deformed sample normalized by the undeformed sample data (2). Also shown are calculated curves assuming dislocation lengths of 2×10^{-7} m, 1×10^{-7} m, 5×10^{-8} m, and 3×10^{-9} m with a dislocation density of 7.5×10^{12} m^{-2} arranged in edge dipoles on the (110) planes. A dipole density 30 times the observed etch pit density is thus necessary to fit the data. The broadening of the temperature dependence has been accomplished by averaging over a distribution of dipole spacings as well as lengths. Discrepancies at low temperature may be attributed to specular rather than diffuse scattering at the surface (2). Discrepancies at temperatures greater than 2 K which become increasingly larger at higher temperatures are due to processes neglected in our treatment--notably point defect scattering, normal and Umklapp processes, and static strain field scattering.

In addition to fitting the experimental data, the model predicts an interesting effect which we call the negative pinning effect. The normal effect of pinning is that after a heavy dose of irradiation the thermal conductivity of a deformed sample progressively recovers toward the undeformed sample data. The negative pinning effect is a small *decrease* in thermal conductivity followed by the usual increase. It is a consequence of the underdamped character of the dislocation or dipole resonance so that the peak scattering rate due to short dislocations exceeds the high frequency asymptotic scattering of longer dislocations as illustrated in Figure 2. The effect is still to be confirmed but should be observable if the thermal conductivity of a deformed sample is continuously monitored during irradiation at constant temperature (e.g. 1 K).

Acknowledgements: This work was funded by NSF grant DMR 77-10556. One of us (G.K.) also gratefully acknowledges the use of the Xerox Rochester Technical Computer Center in completing the calculations begun in his thesis.

References

1. A.C. Anderson and M.E. Malinowski, Phys. Rev. B 5, 3199 (1972).
2. E.P. Roth and A.C. Anderson, Phys. Rev. B 20, 768 (1979).
3. A.C. Anderson (to be published).
4. G.A. Kneezel, Ph.D. Thesis (University of Illinois, 1980).
5. J.S. Koehler, in Imperfections in Nearly Perfect Crystals, edited by W. Shockley, J.H. Hollomon, R. Maurer, and F. Seitz (Wiley, New York, 1952).
6. (a) A. Granato and K. Lücke, J. Appl. Phys. 27, 583 (1956).
   (b) A. Granato and K. Lücke, J. Appl. Phys. 27, 789 (1956).
7. A.V. Granato and K. Lücke, in Physical Acoustics, edited by W.P. Mason (Academic Press, New York, 1966), vol. 4A.
8. V.V. Nikolayev, Fiz. Metal. Metalloved 18, 801 (1964).
9. J. D. Eshelby, Proc. Roy. Soc. Lond. A 197, 396 (1949).
10. F.R.N. Nabarro, Proc. Roy. Soc. Lond. A 209, 179 (1951).
11. P.G. Klemens, Proc. Phys. Soc. Lond. A 68, 1113 (1955).
12. A.V. Granato, Phys. Rev. 111, 740 (1958).
13. (a) T. Ninomiya, J. Phys. Soc. Japan 25, 830 (1968).
    (b) T. Ninomiya, in Fundamental Aspects of Dislocation Theory, edited by A. Simmons, et. al., NBS Special Publication 317, vol. 1 (1970).
14. J.A. Garber and A.V. Granato, J. Phys. Chem. Solids 31, 1863 (1970).
15. G. Wire, Ph.D. Thesis (University of Illinois, 1972).
16. A. Granato and K. Lücke, J. Appl. Phys. 28, 635 (1957).
17. J.D. Eshelby, Proc. Roy. Soc. Lond. 266, 222 (1962).
18. R.M. Stern and A.V. Granato, Acta. Met. 10, 358 (1962).
19. T. Laub and J.D. Eshelby, Phil. Mag. 14, 1285 (1966).
20. R.E. Green and T. Hinton, Trans. Met. Soc. AIME 236, 435 (1966).
21. H.B.G. Casimir, Physica 5, 495 (1938).
22. A.K. McCurdy, H.J. Maris, and C. Elbaum, Phys. Rev. B 2, 4077 (1970).
23. H.J. Maris, J. Acoust. Soc. Am. 50, 812 (1971).
24. A.H. Cottrell, Dislocations and Plastic Flow in Crystals, (Oxford, London, 1953).
25. W.G. Johnston and J.J. Gilman, J. Appl. Phys. 31, 632 (1960).
26. M. Moss, J. Appl. Phys. 36, 3308 (1965).
27. T.R. Cass and J. Washburn, Proc. Br. Ceram. Soc. 6, 295 (1966).

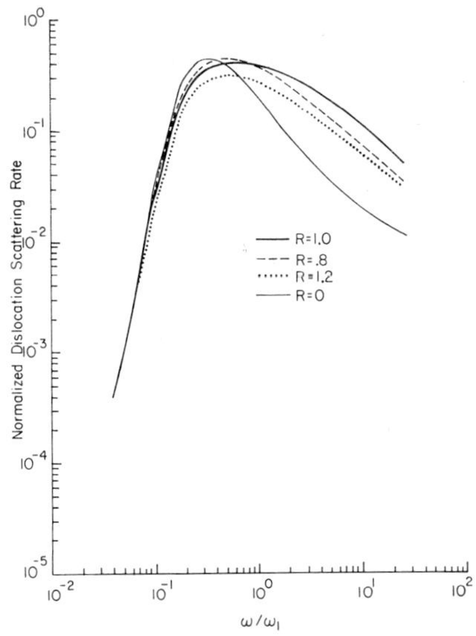

FIG. 1

Normalized scattering rate $\tau^{-1}/2\omega_1\Omega\Lambda L^2$ versus the normalized frequency ω/ω_1, averaged over the Koehler distribution of dislocation lengths for R = 0, .8, 1, and 1.2.

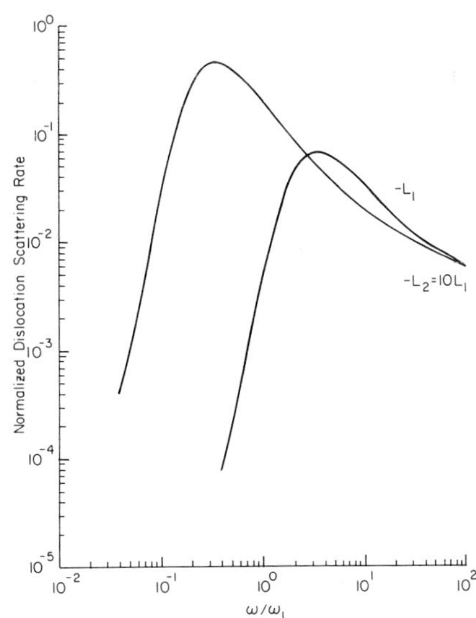

FIG. 2

Normalized scattering rate $\tau^{-1}/2\omega_1(L_2)\Omega\Lambda L_2^2$ versus the normalized frequency $\omega/\omega_1(L_2)$ for average dislocation lengths L_1 and $L_2 = 10L_1$ with R = 0 and $\omega_D = 1000\omega_1(L_2)$.

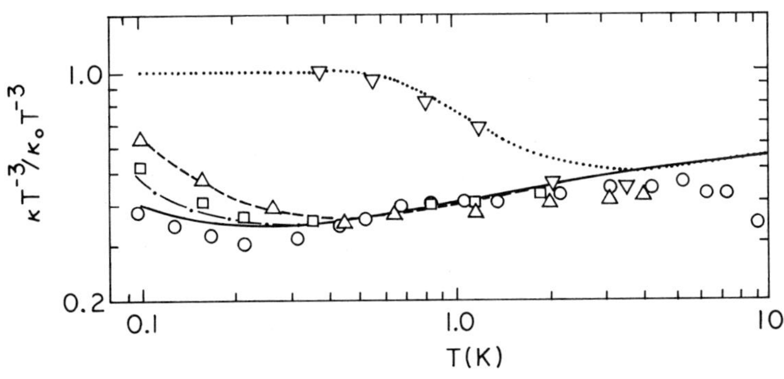

FIG. 3

Measured (2) thermal conductivity ratio of deformed sample (κ_d) to undeformed sample (κ_u) before and after irradiation and calculated curves for edge dislocation dipoles with N = 60 and average lengths as designated.

○ ○ ○ Freshly bent sample;	—— $L = 2\times10^{-7}$m.
□ □ □ After 720 R g irradiation;	– · – $L = 1\times10^{-7}$m.
Δ Δ Δ 2100 R;	– – – $L = 5\times10^{-8}$m.
▽ ▽ ▽ 180,000 R;	· · · $L = 3\times10^{-9}$m.

DISLOCATION DRAG MECHANISMS*

J. M. Galligan

Department of Metallurgy and
Institute of Materials Science
University of Connecticut
Storrs, Connecticut 06268

Abstract

We present some measurements of the influence of electron drag on the mobility of dislocations, in copper-aluminum alloys and zinc crystals. These results show that the electron drag is independent of temperature, and of applied strain rate. In the case of the copper-aluminum crystals, the change in stress with magnetic field varies approximately with the concentration, c, as $c^{\frac{1}{2}}$.

Introduction

At low temperatures the main drag that dislocations feel as they move between barriers arises from the electrons (1). This drag process can be varied, in normal state metals, by switching on or off magnetic fields, which controls the dislocation velocities (2). The variation in the dislocation velocity affects, in turn, the stress for plastic deformation; in copper the effect is quite small - 0.1% - to a readily measurable effect in zinc crystals - 2-5%.

A particular characteristic of the electron drag, measured in copper and aluminum (3), is that the electron drag is independent of the deformation temperature. Such a temperature dependence is that expected according to the basic concepts of the free electron theory of metals (4,5). In the present paper we present some measurements of the temperature dependence of the electron drag in zinc crystal, and in copper-aluminum alloys. In addition we present measurements of the dependence of the change in stress, $\Delta\sigma$, with magnetic field, H, as a function of solute concentration c.

Experimental

The starting materials for the present experiments were spectroscopically pure copper, obtained from ASARCO, with a stated purity of 99.999+%. The aluminum, with a stated purity of 99.9999%, was obtained from Cominco Ltd. The zinc, 99.999+% pure, was obtained from ASARCO. Oriented crystals were produced by the standard Bridgman technique and the crystallographic orientations were checked by standard Laue back reflection techniques. In the case of the copper-aluminum alloys, the crystals were all oriented with the same relative orientation and they showed an extended region of easy glide. The zinc crystals were oriented with their basal planes at 45° to the tensile axis or at 90° to the tensile axis; more about these orientations below.

The deformation, carried out in a laboratory scale tensile machine, was undertaken at 4.2K and below, and, typically, at an applied strain rate of $\sim 10^{-4}$/sec. A superconducting magnet, which could be switched on or off at various precontrolled rates, could be varied between 0 and 0.6 Tesla. The thermometry consisted of a precalibrated Ge-Li crystal.

In order to carry out the present experiments, we deform a crystal at, say, 4.2K, and while we are deforming the crystal, we switch the field on or off. A differential stress measuring technique allows us to measure differences in stresses of the order of 0.3%. An example of the change in stress, associated with a change in field is shown in Fig. 1; in this case the crystal is Cu5%Al. In contrast to the data shown in Fig. 1, when we deform the crystal in an elastic manner, then *no* change in stress is observed, Fig. 2. Further, the change in stress, observed only when the crystal is deforming in a plastic manner, is fully reversible with magnetic field, and the magnitude of the change in stress is *independent* of the rate of change of the magnetic field; the rate of change in field has varied from 6T/sec to 0.06T/sec. These

*Supported by the U.S. Department of Energy

observations rule out any possibility that the change in stress, with change in field, is related to any magnetostriction or heating.

We can now consider the effects of some other, more interesting variables on the drag stress on mobile dislocations. First we show, for zinc crystals, oriented with their basal planes at 45° to the tensile axis that the change in stress, again when the crystal is deforming plastically, is independent of temperature, Fig. 3. These results can be compared with previous observations on the deformation of copper and aluminum crystals where we also observed that $\Delta\sigma$ is independent of temperature. (3)

We now consider the influence of solute concentration - in the case of copper-aluminum alloys - on the electron drag stress. In these experiments, where the crystals are all oriented in a similar manner relative to the tensile axis, we measure the change in stress with solute concentration, Fig. 4. Note that $\Delta\sigma$ increases with solute in a roughly parabolic manner and that $\Delta\sigma$ is easily measured by the present technique in the pure copper. In addition to measuring the change in stress with concentration we have also measured $\Delta\sigma$ as a function of applied strain rate temperature, Fig. 5. Again, the strain rate dependence agrees with the zinc work and with the copper and the aluminum measurements (2).

Discussion

The above results show that we can affect the motion of dislocation by magnetic fields in zinc and in copper aluminum alloys. These results are consistent with a general treatment of dislocation motion at low temperatures, where the dislocations can be considered to move as underdamped oscillators. The temperature dependence of the dislocation drag, which follows directly from considering the dislocation as underdamped oscillators, is consistent with the electron drag mechanisms as discussed by Holstein (4) and by Brailsford (5).

The observed concentration dependence, while varying roughly in a parabolic manner, needs to be checked more carefully at low concentration. This work was supported by the U.S. Department of Energy. I wish to thank my students, L. Motowidlo and Paul Goldman, for allowing me to quote results prior to publication.

References

1. M. Suenaga and J.M. Galligan, Physical Acoustics, 9, 1 (1972). Edited by W. P. Mason and R. Thurston, Academic Press, New York.
2. J. M. Galligan, T. H. Lin and C. S. Pang, Physical Review Letters, 38, 405 (1977).
3. J. M. Galligan and C. S. Pang, Journal of Applied Physics, 50, 6253 (1979).
4. T. Holstein, Physical Review 151, 187 (1966).
5. A. D. Brailsford, Physical Review, 186, 959 (1969).

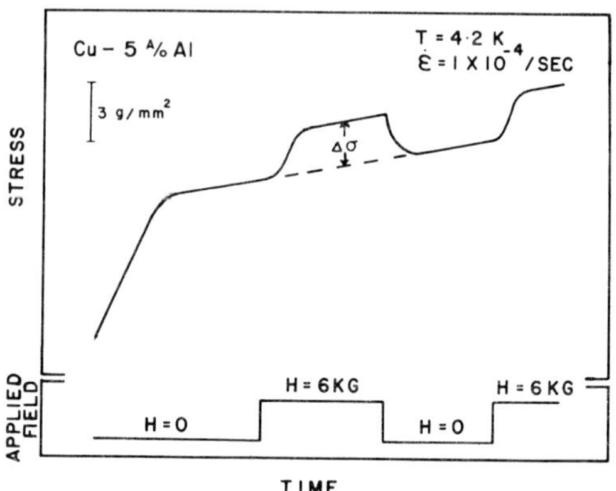

Fig. 1
The change in stress, in a Cu5%Al crystal, deformed at 4.2K while the field is switched from 0.6 Tesla (6 kilogauss) to zero.

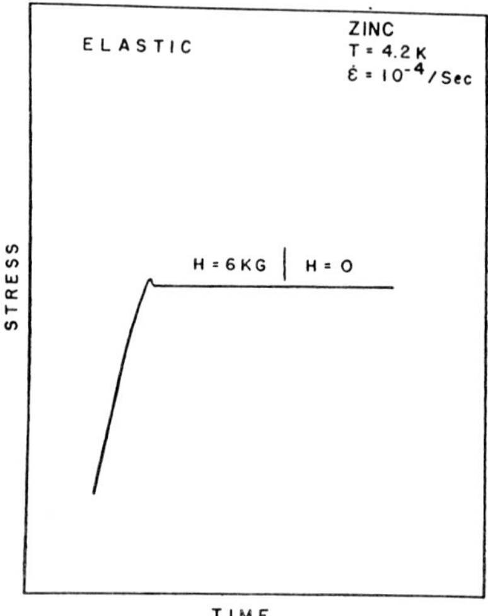

Fig. 2
The change in stress in a zinc crystal - oriented with the basal plane at 45° to the tensile axis. In this experiment we deform the sample elastically, stop the deformation, and then switch the field to on or off.

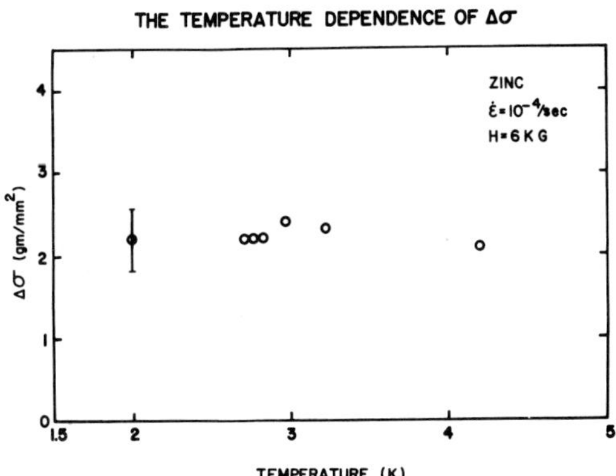

Fig. 3
Some observations on the temperature dependence of the change in stress in a zinc crystal. These measurements are made at constant strain rate and constant change in field.

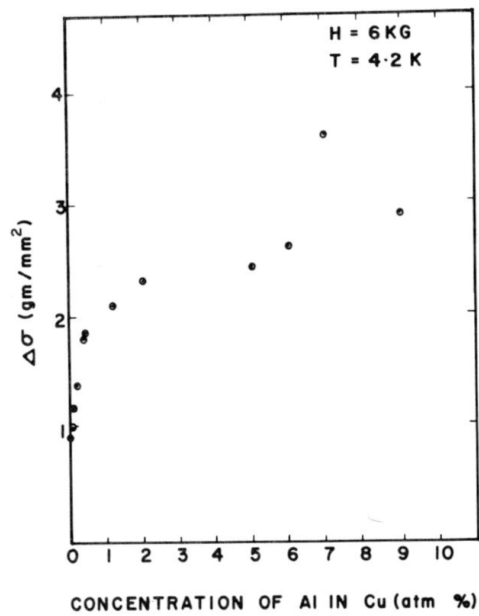

Fig. 4
Concentration dependence of the change in stress in the case of copper aluminum alloys. The reproducibility of the method can be assessed from the measurements of separate specimens of the same composition, i.e. Cu-0.5% Al.

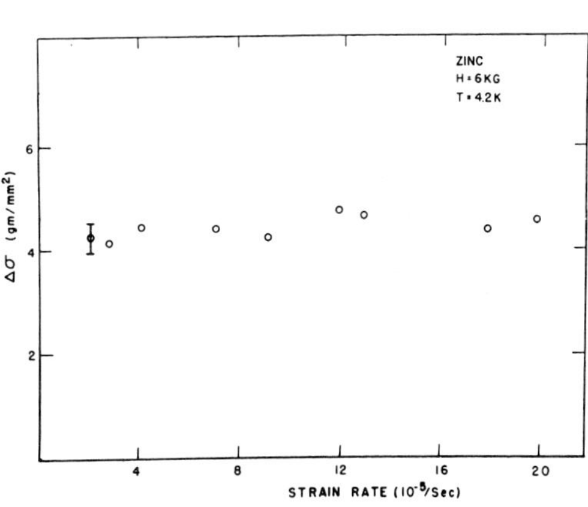

Fig. 5
Some measurements of the strain rate dependence of the change in stress, with magnetic field, in the case of zinc crystals. Deformation carried out at 4.2K.

THEORY OF ACOUSTIC EMISSION

John A. Simmons and Roger B. Clough
Metallurgy Division
National Bureau of Standards
Washington, D.C. 20234

ABSTRACT

A theory of acoustic emission is presented based on a Green's function type of formalism, rather than on the conventional count rate concept. Sources are represented by stress drop rate tensors and conditions are derived from which the source can be considered small in terms of wavelength and distance to the transducer. These "pseudopoint" sources are examined over a restricted frequency bandwidth, called the "informative bandwidth." Such a bandlimited system may be described by a tensor transfer function type of formalism, facilitating the analysis and reducing the inverse problem--where the source is not known *a priori*--to a deconvolution operation. Due to the tensor nature of the source, multiple transducer measurements are necessary to reconstruct the source stress drop. The difficulty of using spectral techniques for data analysis in the presence of multiple sources is discussed. The forward problem for acoustic emission to be expected from plasticity, phase transformations, and cracks is solved within the pseudopoint approximation and the concept of retarded dipole density is introduced to deal with the problem of time delays from propagating inhomogeneity sources. In addition, the strong directionality of the signal with respect to source type and orientation is illustrated by calculating the acoustic emission signals generated by loop expansion of slip and climb (prismatic collapse) type dislocations. This directionality is shown to be the same for isotropic materials as that associated with Mode II/III or Mode I cracks and a comparison is provided between emission from expanding penny-shaped cracks and that from dislocation loops. Finally, the concepts of stress-controlled and materials-controlled acoustic emission are introduced and their significance is discussed in terms of relating AE source descriptions and material mechanisms.

1. Introduction

Acoustic emission (AE) is a term used to describe stress waves emitted by a moving defect or other rapid structural change in a solid body. When an inhomogeneity in a structure under load changes in response to the local stress field, it induces a local stress "drop" in that field. Such a stress drop acts as a center of radiation for elastic waves which propagate throughout the structure to be detected by a transducer and recorded as an acoustic emission signal.

Acoustic emission has important potential as a technique for the study of martensitic transformations [1], twinning [2], crack nucleation and growth [3], stress corrosion cracking [4], and dynamic plastic deformation [5]. In addition it is one of the few passive nondestructive testing methods which can be used for continuous and remote monitoring of structures in difficult environments, such as nuclear reactors [6]. However, in order to realize its potential, quantitative methods must be developed to permit more accurate characterization of AE sources.

The theory presented here formulates the basis for a new type of AE technique in which the source of emission and the AE system are treated as tensor processes so that quantitative information about the source can be reconstructed from AE signals recorded at multiple observatio points. Conventional AE methods, which are extensively catalogued in [7,8], are scalar measures. These techniques utilize the output from a relatively narrow band transducer and measure rms voltage, power, or "AE counts"--counting the number of times a threshold voltage is crossed.

Our first specialization will be to limit the spatial extent of the source. To do this we consider a Taylor's expansion of $\underline{G}^H(\underline{r},\underline{r}',t-t')$ about the point $\underline{r}'_o$:

$$G^H_{ij,k'}(\underline{r},\underline{r}',t-t') = G^H_{ij,k'}(\underline{r},\underline{r}'_o,t-t') + G^H_{ij,k'\ell'}(\underline{r},\underline{r}'_o,t-t')\Delta\underline{r}'_\ell + \ldots, \tag{2.2}$$

where $\Delta\underline{r}'_\ell = \underline{r}'_\ell-\underline{r}'_{o\ell}$. Equation (2.2) leads to a revised form of (2.1)

$$u_i(\underline{r},t) = \int G^H_{ij,k'}(\underline{r},\underline{r}'_o,t-t')\overline{\Delta\dot{\sigma}}_{jk}(t')\,dt' + \int G^H_{ij,k'\ell'}(\underline{r},\underline{r}'_o,t-t')\overline{\Delta\dot{\Gamma}}_{jk\ell}(\underline{r}'_o,t')\,dt' + \ldots \tag{2.3}$$

where the stress drop rate moments are given by

$$\overline{\Delta\dot{\sigma}}_{jk}(t') = \int_V \Delta\dot{\sigma}_{jk}(\underline{r}',t')\,d\underline{r}' \tag{2.4}$$

$$\overline{\Delta\dot{\Gamma}}_{jk\ell}(\underline{r}'_o,t') = \int_V \Delta\dot{\sigma}_{jk}(\underline{r}',t')\,\Delta r'_\ell d\underline{r}', \text{ etc.} \tag{2.5}$$

with $\Delta\underline{r}' \equiv \underline{r}'-\underline{r}'_o$. Equation (2.3) expresses the stress wave signal in terms of a multipolar expansion about the point $\underline{r}'_o$, where the coefficients of the Green's tensor derivatives at time t' are the moments of the stress rate distribution at t'. Equation (2.3) has the advantage of depending on $\underline{r}'$. However, $\underline{r}'_o$ is arbitrary and higher order moments introduce a complexity beyond the capabilities of present day instrumentation. Accordingly, two important approximations for the quantitative description will be developed:

(1) The "pseudopoint" source approximation [42]:

$$u_i(\underline{r},t) = \int G^H_{ij,k'}(\underline{r},\underline{r}'_o,t-t')\,\Delta\dot{\sigma}_{jk}(t')\,dt' \tag{2.6}$$

(2) The Stump-Johnson (S-J) approximation [38]:

$$u_i(\underline{r},t) = \iint G^H_{ij,k'}(\underline{r},\underline{r}',t-t')\,\dot{m}(\underline{r}',t')\,M_{jk}d\underline{r}'dt' \tag{2.7}^*$$

where $\underline{M}$ is a constant matrix.

The pseudopoint approximation is considered first. This concept was presented by the present authors in 1975 [49]. It has also been used in seismological applications (c.f., pp. 801-825 [25]). In this approximation, the multipolar expansion (2.3) is truncated at the dipole term which "lumps" all of the distributed stress drop at one point $\underline{r}'_o$, where it is expressed by its zeroth moment as in (2.4). This will provide a simplification in the formalism but will be shown later to require a band limiting operation to avoid "white noise" errors.

A heuristic analysis for the error associated to assuming a fixed point source is as follows: figure 2a shows a simple configuration in which the pseudopoint source is assumed to remain within the domain R during the emission event. There is an approximation error introduced by placing the source at $\underline{r}'_o$, chosen at a "central" location in R, instead of at $\underline{r}'$ at t'. The first part of the error consists in missing the "central" location in R. This "error" produces a change in the Taylor's expansion of $\overset{4}{\nabla}\underline{G}^H$ due to a shift in the origin with a consequent change in the dipole density. For each of the moment terms $\overline{\Delta\dot{\sigma}}_{jk}$, it is possible to choose a

*Equation (2.7) is a slight generalization of the approximation proposed by Stump and Johnson [38] which is the concatenation of the two approximations (2.6) and (2.7).

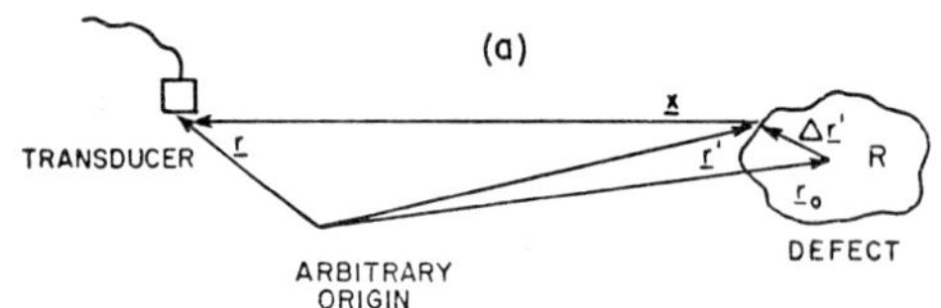

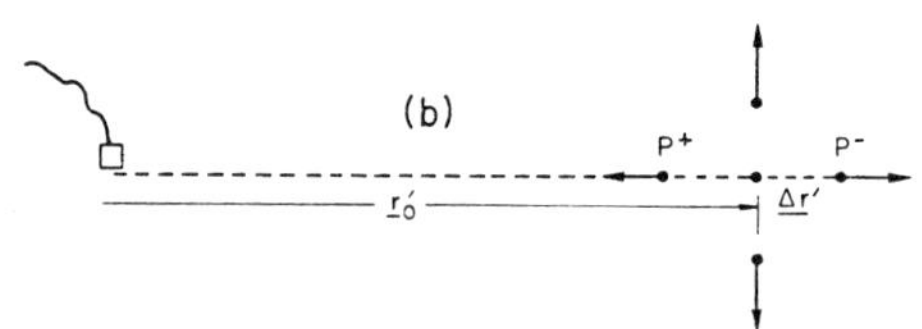

FIG. 2
Pseudopoint approximation of emitter: a) single source; b) multiple sources.

"centroid" position for $\underline{r}'_o$ so that the first moment vanishes, but this position will vary in general from component to component unless the S-J hypothesis is valid, as discussed later. Such a variation in central position can occur, for instance, when several different emission processes occur very close together almost simultaneously. For such situations the best one can do is to choose a value for $\underline{r}'_o$ which is, in some sense, the "centroid" of the centroids. We shall assume that such an optional position for $\underline{r}'_o$ has been found and will estimate errors with respect to that $\underline{r}'_o$. Although errors in choosing $\underline{r}'_o$ are of significance in dealing with the inverse problem and have been discussed by Backus and Mulcahy [36], they need not concern us in dealing with the forward problem. Suppose next that $\underline{G}^H$ is thought of in an approximate way as a superposition of self-similar wave components, each traveling with a characteristic frequency and velocity, $v(\omega)$:

$$G^H_{ij,\hat{k}'}(\underline{r},\underline{r}',t-t') \sim \int D_{ij\hat{k}}(\omega,\Theta,\phi,|\underline{r}-\underline{r}'|)\exp[i\omega((t-t') - \frac{|\underline{r}-\underline{r}'|}{v(\omega)})]\, d\omega \tag{2.8}$$

where Θ and ϕ are spherical angles associated to $(\underline{r}-\underline{r}')/|\underline{r}-\underline{r}'|$. This is an optimistic approximation since, in general, there is a range of v's for each ω that must be taken into account, and this will cause partial interference effects. For each ω the error in $\underline{G}^H$ caused by using $\underline{r}'_o$ contains two components:

(a) the error caused by change in the coefficient $D_{ij\hat{k}}(\omega,\Theta,\phi,|\underline{r}-\underline{r}'|)$; and

(b) the phase error associated with the transit time $\Delta\underline{r}'\cdot\underline{x}/v(\omega)$, where $\Delta\underline{r}' \equiv \underline{r}'-\underline{r}'_o$ and $\underline{x} \equiv \underline{r}-\underline{r}'$.

Unless there is extreme Θ or ϕ dependence (which could happen near a surface), the worst errors in this approximation will occur when $\underline{r}$, $\underline{r}'_o$ and $\underline{r}'$ are in a straight line so we set $\Delta r' = \Delta\underline{r}'\cdot\underline{x}$. If it is then assumed that

$$D_{ij\hat{k}}(\omega,\Theta,\phi, |\underline{r}-\underline{r}'|) \sim k(\omega)\cdot|\underline{r}-\underline{r}'|^{\alpha} \tag{2.9}$$

for some α, the relative error introduced by the pre-exponential coefficient $\underline{D}$ is

$$E_D \sim \frac{\alpha\,\Delta r'}{|\underline{r}-\underline{r}'_o|}\,. \tag{2.10}$$

The phase error introduces an error term of the same frequency as the carrier wave, and we use as an error measure $(\text{energy})^{1/2}$, or the ratio of the maximum amplitudes:

$$E_\Theta \sim 1-\exp[\omega\Delta r'/v(\omega)] \sim \frac{2\ f\Delta r'}{v(\omega)} \tag{2.11}$$

where $f \equiv \omega/2\pi$. The magnitude of the error is then bounded by

$$E \underset{<}{\sim} E_D + E_\Theta = \left(\frac{\alpha}{|\underline{r} - \underline{r}_o'|} + \frac{2\ f}{v(\omega)}\right)\Delta r'. \tag{2.12}$$

This equation shows two major factors contributing to the pseudopoint source approximation. One is inversely related to the distance from the source to the transducer (usually $|\alpha| \underset{<}{\sim} 4$) and the other contains a term linear with frequency (with possible dispersion modifications).

The magnitude of the error depends on the source mechanism and the bound given in (2.12) is a "worst case" type bound in the sense that it assumes that $\underline{r}_o'$ is not at the centroid of the emission pattern. As an example consider emission in steel from an expanding semicircular type source. For this source, the centroid moves with time and for simplicity, we choose $\underline{r}_o'$ to be d/2 units from the center, where d is the ultimate radius of the semicircle. In addition, the exponent $\alpha \underset{<}{\sim} 4$ and $v(\omega) \underset{>}{\sim} c$, the shear velocity of 3.14 mm/μs in steel. Then at a distance of 10 cm, according to (2.11) one can expect an error of

$$E \sim (0.02 + f)d \tag{2.13}$$

where d, the diameter of the semicircle, is measured in mm, and f is measured in MHz. In the case of a source with $\underline{r}_o'$ at the centroid, the error is of the second order since the first moment term contributes no error. A similar analysis in this case leads to

$$E \underset{<}{\sim} \frac{\alpha(\alpha-1)}{8}\, d^2 + \left|\, 1-\cos\frac{fd}{v(\omega)}\right| \tag{2.14}$$

where d is the diameter (mm) of the source. For a 0.5 mm source, (2.13) leads to a maximum frequency of 180 Hz to keep the error under ten percent while (2.14) leads to a maximum frequency of about 900 KHz. It should be emphasized that .5 mm in this example refers to the area of active emission, not the possible overall flaw size.

The crude error approximations given by (2.12) and (2.14) leads to the conclusion that at distances from the source, large compared to source size, the critical assumption needed to approximate the source as a fixed pseudopoint source is given by

$$f \underset{\sim}{<} f_I(d) \tag{2.15}$$

where $f_I(d)$ defines the upper limit to what we shall call the "informative bandwidth." Below this frequency, equation (2.6) may be approximated using a fixed point source. Above this frequency the information provided by the signal is unavailable unless a more complex approximation or a specific dynamic defect model is used.

We turn next to eq. (2.7), the Stump-Johnson (S-J) approximation [38]. This is a physical assumption which insists that the event can be described by a single parameter, or single mechanism, type emission source. In this case the mechanism is characterized by the constant matrix $\underline{M}$, while its space-time history is given by the scalar function $\dot{m}(\underline{r}',t')$. Even equation (2.7) offers great experimental difficulty and we reduce it to its multipole form, truncating at the dipole term:

$$\begin{aligned} u_i(\underline{r},t) &= \int G^H_{ij,k'}(\underline{r},\underline{r}_o',t-t')\ \bar{\dot{m}}(t')\ M_{jk}dt' \\ &+ \int G^H_{ij,k'\ell}(\underline{r},\underline{r}_o',t-t')\ \bar{\dot{\Gamma}}_\ell(t')\ M_{jk}dt' \\ &= \int G^H_{ij,k'}(\underline{r},\underline{r}_o'(t'),t-t')\ \bar{\dot{m}}(t')\ M_{jk}dt'. \end{aligned} \tag{2.16}$$

Here $\bar{\dot{m}}$ and $\bar{\dot{\Gamma}}$ are derived from $\dot{m}$ in an analogous fashion to (2.4).

The first form of (2.16) is the dipole form of the S-J approximation and can be used in the transfer function formalism. The second form of (2.16) places $\underline{r}_o'$ at $\underline{r}_o'(t')$, the centroid of the source distribution at time t', a concept which can only be rigorously defined for a single parameter process. This approximation goes one step beyond the pseudopoint approximation (while restricting the types of processes permitted) and offers a way of characterizing source motion other than that inferred using the pseudopoint approximation. It should be particularly applicable for relatively small sources moving at moderate velocities over distances large compared to the source size.

In order to complete the derivation of the transfer function formalism, it is necessary to model the response of the transducer and instrumentation. The approximation made is that of a "non-disturbing" transducer monitoring a given area or volume of the body, which we denote by S_T. By "non-disturbing" we mean that the change in the waveform caused by the presence of the transducer can be neglected compared to the magnitude of the waveform itself. For such a situation, consider the transducer to be of a kind sensitive to displacements, although other linear elastic field measures could be used. The point-by-point response function of the transducer will be denoted by $TP_i(\underline{r},t)$, $\underline{r}\epsilon S_T$ and refers to the voltage at time t produced by a δ-function displacement in the i-direction at the point $\underline{r}$ at time zero. With this definition, the voltage at time t due to the emitting source will be

$$V(t) = \iint_{S_T} TP_i(\underline{r},t-t')G^H_{ij,k''}(\underline{r},\underline{r}_o'',t'-t'')\,\overline{\Delta\dot{\sigma}}_{jk}(t'')\,d\underline{S}dt'' \tag{2.17}$$

In frequency form (2.17) becomes the <u>transfer function formalism</u>:

$$V(\omega) = \int TP_i(\underline{r},\omega)G^H_{ij,k''}(\underline{r},\underline{r}_o'',\omega)\overline{\Delta\dot{\sigma}}_{jk}(\omega)\;d\underline{S}$$

$$= T_{jk}(\omega)\overline{\Delta\dot{\sigma}}_{jk}(\omega) \tag{2.18*}$$

valid for ω within the informative bandwidth. Here

$$T_{jk}(\omega) = \int_{S_T} TP_i(\underline{r},\omega)G^H_{ij,k''}(\underline{r},\underline{r}_o',\omega)\;d\underline{S} \tag{2.19}$$

defines the transfer tensor of the structure and instrumentation while $\overline{\Delta\dot{\sigma}}(\omega)$ is the stress drop tensor of the source. We note that the integration over the transducer face given by (2.19) has a tendency to cancel out "high" frequency components (those with wavelength greater than the transducer diameter) producing a <u>de facto</u> low pass filter.

3. The Forward Problem: Consequences of the Transfer Function Formalism

Equation (2.18), the transfer function formalism, has a number of important consequences:

(1) By far the most important consequence of the transfer function formalism is that information about the source is passed frequency by frequency to the output through the transfer tensor. For acoustic emission to be detectable, then, the transfer tensor must have significant components in that frequency band where the signal-to-background noise ratio for the source history is highest. One can then only expect to gain that information about the source history as is contained in that frequency band. By maximizing this useful band, one maximizes the possibility of finding enough information about the source to detect its signature. On the other hand, in certain situations no significant information about the source may pass to the output.

---

*Whenever possible in this paper we shall denote the Fourier transform of a function by a change of variable notation: $V(\omega) = \int V(t)e^{i\omega t}dt$, noting as well that this definition is the same as that used in mathematics applications while most electrical engineering applications use $e^{-i\omega t}$. The definition using $e^{+i\omega t}$ is preferred here, since in certain signal processing applications the sense of the time direction is maintained.

For instance, in a ductile versus brittle material comparable stress drops occur (although usually a substantial amount of potential energy is retained in the zero frequency internal stress fields), but since the time scale of the ductile fracture process will be longer, most information will be sent at lower frequencies--often below those that can be practically measured due to noise or instrumentation--so that one finds "no acoustic emission" [43,44].

The frequency-by-frequency transfer of information allows powerful signal processing and inversion techniques to be employed. These will be discussed in the context of the inverse problem.

(2) The need to band limit the signals in order to remain inside the informative bandwidth has the effect of filtering the output while lowering both spatial and temporal resolutions to permit the pseudopoint approximation to be used for moving sources of finite extent. Another way to look at sources of error outside the informative bandwidth is illustrated in figure 2b, where we imagine the source decomposed into four approximate point sources of the same type starting at time zero and moving rapidly outward for a fixed period of time. From this simple figure one can see several of the difficulties associated with detailed acoustic emission analysis and an indication of why we feel that it is important to employ the pseudopoint approximation within the informative bandwidth as an initial approach. If the velocities of the four point sources differ, the centroid, $\underline{r}'_o$, of the pseudopoint source will move and be detectable through the first moment tensor. In the examples cited in (2.12) and (2.14) this forced a substantial drop in the upper frequency of the informative bandwidth. Even if the four velocities are identical, and equal to V, say, there are still two contributing factors giving rise to error in a point approximation. The first is the spatial extent of the source. The second is the Doppler shift associated to the component of source velocity in the direction of the transducer. The Doppler shift is contained within the error analysis for the approximation drawn from (2.8). The phase error

$$\frac{\Delta r'}{v(\omega)} \pm \frac{Vt}{v(\omega)}$$

occurs in estimating the transit time of the two points P^+ and P^-. Ignoring the fixed time shift $\Delta r'/v(\omega)$ which corresponds to the spatial extent of the source, the relative time shift will cause a time scale error of $1 \pm V/v(\omega)$ which will give rise to an apparent shift in the frequency content of the source for frequencies outside of the informative bandwidth; within the informative bandwidth a rapid source will appear as a "delta function" by decreasing the effective duration and increasing the effective magnitude.

By using information from within the informative bandwidth we can obtain the source moment tensor as a function of frequency if we know the approximate location of the source (also a factor in determining the informative bandwidth). In real time this band-limiting gives rise to what we call convolution smoothing [42]. Such convolution smoothing, illustrated in figure 3, has the effect of smoothing out high frequency wiggles and dynamic stochastic effects which might occur with microscopic high velocity starts and stops or with spatial variations in the exact defect growth mechanism during the emission process. To extend frequency information beyond the informative bandwidth, one must deal with specific dynamic models describing both spatial extent and source type as possibly multi-parameter functions of time. One example of this is Wadley, et al. [44]. Finally, we mention that if the source type or the velocities of the four point sources in figure 2b interact with each other (as is the case for modeling inhomogeneities) then another, different type of error is introduced associated to the delay in receipt of information at a given point of what other points are doing. This type of error is taken up in section four.

(3) Because the complex voltage $V(\omega)$ consists of the sum of a series of terms, phase interference effects are important. Because of this, simple amplitude spectrum analysis will be invalid except for single parameter sources which can be described by the S-J approximation. Most non-stationary time series signal analysis methods are applicable to S-J sources.

(4) Because of the number of stress rate components that must be separated from each other, multichannel measurements are, in principle, essential for inversion to find an unknown source. This can mean from six channels if the pseudopoint approximation is used with no momentum term sources (due to density changes) up to 15 channels if the dipole form of the S-J approximation and density changes and surface monopole sources are all included. A conceptual example of this multichannel system is shown in figure 4.

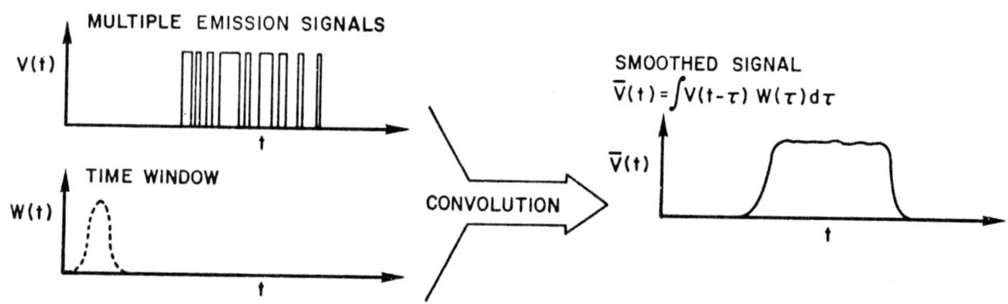

FIG. 3
Convolution smoothing effect on stochastic emission process.

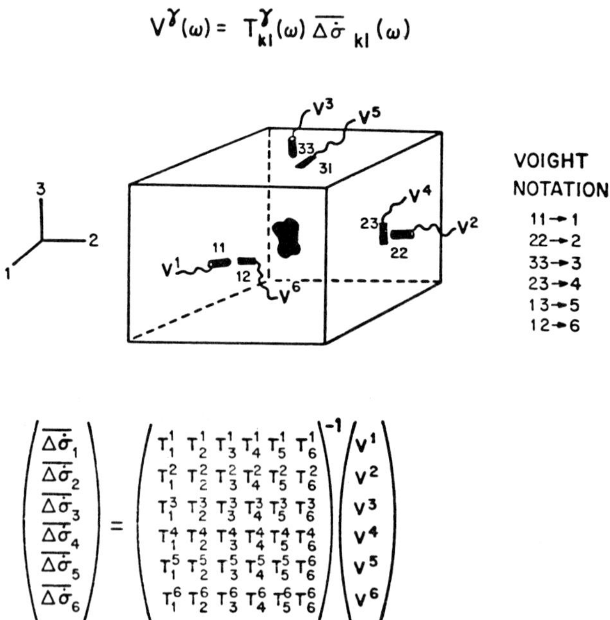

FIG. 4
Schematic of multichannel system for complete measurement of source stress drop rate tensor. Cylinders represent displacement measurements normal to surface; rectangles represent in-plane surface displacement measurements. The inverse matrix operation is shown below.

4. The Forward Problem: Emission From Cracks, Phase Transformations, and Plasticity

As an example of the above theory, the acoustic emission from a small region undergoing plastic deformation, cracking, or other inhomogeneity propagation will be given. We shall solve the integral equation (A32) of Appendix A in the far field for an elliptical crack or an ellipsoidal inhomogeneity undergoing an arbitrary self-similar change of shape. Plasticity will automatically be included. The near field problem has been treated for self-similar cracks expanding at constant rate by Burridge and Willis [17].

We describe the moving region by $\Sigma_{a(t)}$ and we denote by $\Sigma_{a(t)}(\underline{r})$ the Heaviside function for the region $\Sigma_{a(t)}$. Using the fact that the difference in moduli, $\underline{\Delta C}$, between the inhomogeneity and the surrounding matrix is constant over $\Sigma_{a(t)}$, we write (A32) as

$$
\begin{aligned}
u_m^T(\underline{r},t) = & \iint G_{mi,\hat{j}'\ell'}(\underline{r},\underline{r}',t-t')\Delta C_{i\hat{j}k\ell}u_k^T(\underline{r},'t')\Sigma_{a(t')}(\underline{r}')d\underline{r}'dt' \\
& + \iint G_{mi,\hat{j}'}(\underline{r},\underline{r}',t-t')\Delta C_{i\hat{j}k\ell}u_k^T(\underline{r}',t')\overset{4}{\nabla}\cdot\Sigma_{a(t')}(\underline{r}')d\underline{r}'dt' \\
& + \iint G_{mi,\hat{j}'}(\underline{r},\underline{r}',t-t')(C+\Delta C)_{i\hat{j}k\ell}\beta_{k\ell}^P(\underline{r}',t')\Sigma_{a(t')}(\underline{r}')d\underline{r}'dt' \\
& - \iint G_{mi,\hat{j}'}(\underline{r},\underline{r}',t-t')\Delta C_{i\hat{j}k\ell}\beta_{k\ell}^o(\underline{r}',t')\Sigma_{a(t')}(\underline{r}')d\underline{r}'dt \quad . \qquad (4.1)
\end{aligned}
$$

Now we can expect that for small a(t) the first term in (4.1) will become negligible compared to the second term for two reasons: the greater drop-off in $|\underline{r}|$ with increasing $|\underline{r}-\underline{r}'|$ (even when $\hat{j} = \ell = 4$) and the increase in surface to volume ratio for small inhomogeneities. We further use the informative bandwidth approximation to assume that $\underline{G}(\underline{r},\underline{r}',t-t')$ does not depend on $\underline{r}'$ and we choose t_o so that Σ_o is null at t_o. Then

$$
\begin{aligned}
u_m^T(\underline{r},t) = & - \Delta C_{ijk\ell}\int_{t_0}^{t_1} G_{mi,j'}(\underline{r},\underline{r}_o',t-t')\left(\int_{\partial\Sigma_{a(t')}} u_k^T(\underline{r}',t')dS_\ell'\right)dt' \\
& + (C+\Delta C)_{i\hat{j}k\ell}\int_{t_0}^{t_1} G_{mi,\hat{j}'}(\underline{r},\underline{r}_o',t-t')\left(\int_{\Sigma_{a(t')}} \beta_{k\ell}^P(\underline{r}',t')d\underline{r}'\right)dt' \\
& - \Delta C_{i\hat{j}k\ell}\int_{t_o}^{t_1} G_{mi,\hat{j}'}(\underline{r},\underline{r}_o',t-t')\left(\int_{\Sigma_{a(t')}} \beta_{k\ell}^o(\underline{r}',t')d\underline{r}'\right)dt' \qquad (4.2)
\end{aligned}
$$

or, in terms of the Heaviside Green's tensor, making use of the fact that t_o lies before any activity and t_1 lies after it:

$$u_m^T(\underline{r},t) = -\Delta C_{ijk\ell} \int_{t_0}^{t_1} G_{mi,j'}(\underline{r},\underline{r}'_o,t-t') \left[\frac{d}{dt'} \int_{\partial\Sigma_{a(t')}} u_k^T(\underline{r}',t')dS'_\ell \right] dt'$$

$$+(C+\Delta C)_{i\hat{j}k\ell} \int_{t_0}^{t_1} G_{mi,\hat{j}'}(\underline{r},\underline{r}'_o,t-t') \left[\frac{d}{dt'} \int_{\Sigma_{a(t')}} \beta_{k\ell}^P(\underline{r}',t')d\underline{r} \right] dt'$$

$$-\Delta C_{i\hat{j}k\ell} \int_{t_0}^{t_1} G_{mi,\hat{j}'}(\underline{r},\underline{r}'_o,t-t') \left[d/dt' \int_{\Sigma_{a(t')}} \beta_{k\ell}^o(\underline{r}',t')d\underline{r}' \right] dt' \quad . \tag{4.3}$$

Equations (4.2) and (4.3) are the basic integral equations for either far field scattering or acoustic emission involving small defects. To solve (4.2) or (4.3) we first observe that they are of such a form that the apparent source is a time dependent dipole located at $\underline{r}'_o$. If we stop the growth at some intermediate time t*, place ourselves just outside $\Sigma_{a(t^*)}$ and wait a short time, the final disturbance associated to the defect will pass and we will be in the static lacuna before the arrival of secondary boundary reflections whose interactions with the defect we will ignore. Therefore, the dipole density describing the dynamic motion must be consistent with that for the static problem, which can be written $\underline{D}(t^*)$. If the dynamic dipole density is given by $\underline{D}(t,t^*)$, then

$$\lim_{t \to t_1} \underline{D}(t,t^*) = \underline{D}(t^*) \tag{4.4a}$$

or, in terms of dipole rates

$$\int_{t_0}^{t_1} \dot{\underline{D}}(t,t^*)dt = \underline{D}(t^*) \tag{4.4b}$$

with $\underline{D}(t_o) = 0$. Here it has been assumed that t_1 is sufficiently greater than t* that the process can "come to rest." The simplest solution to (4.4) is to assume

$$\underline{D}(t,t^*) = \underline{D}(t^*), \tag{4.5}$$

which we call the slow defect approximation.

It is possible to improve slightly on (4.5) even within the context of the pseudopoint approximation by using the concept of retarded dipole density. Our purpose for presenting the retarded dipole approximation will not be to provide detailed quantitative improvement in the slow defect approximation, since within much of the informative bandwidth the slow defect approximation should be adequate, but rather to indicate new features which will occur for rapidly moving defects and to permit estimates of the frequencies at which these features become significant.

Examination of (4.2) shows that while the second and third terms on the right hand side act purely as driving terms, the first term on the right hand side involves the instantaneous surface displacements at the surface of the inhomogeneity. We shall assume that the contributions associated to this term "switch on" to their static configuration in ellipsoidal (or elliptical) rings at a slightly delayed time. Figure 5 schematically shows an expanding elliptical defect. The retarded dipole approximation is equivalent to assuming that the current dynamic configuration is made up of two regions, an inner retarded volume on which the slow defect approximation is used together with an "instantaneous" outer ring where no inhomogeneity corrections are applied to the driving term dipole sources. This leads to the modified form for (4.3):

$$u^T_m(\underline{r},t) = -\Delta C_{ijk\ell} \int_{t_0}^{t} G^H_{mi,j'}(\underline{r},\underline{r}'_o,t-t') \left[\frac{d}{dt'} \int_{\partial\Sigma_{a(t_R(t'))}} u^T_k(\underline{r}',t')\, dS_\ell \right] dt'$$

$$+(C+\Delta C)_{i\hat{j}k\ell} \int_{t_0}^{t} G^H_{mi,\hat{j}'}(\underline{r},\underline{r}'_o,t-t') \left[\frac{d}{dt'} \int_{\Sigma_{a(t')}} \beta^P_{k\ell}(\underline{r}',t')d\underline{r}' \right] dt'$$

$$-\Delta C_{i\hat{j}k\ell} \int_{t_0}^{t} G^H_{mi,\hat{j}'}(\underline{r},\underline{r}'_o,t-t') \left[\frac{d}{dt'} \int_{\Sigma_{a(t')}} \beta^o_{k\ell}(\underline{r}',t')d\underline{r}' \right] dt' \ . \qquad (4.6)$$

That is, at any time t the magnitude of the inhomogeneity correction term--the first term in (4.2) or (4.3)--is obtained by solving the static dipole problem over $\Sigma_{a(t_R)}$ to which are added the current driving terms.

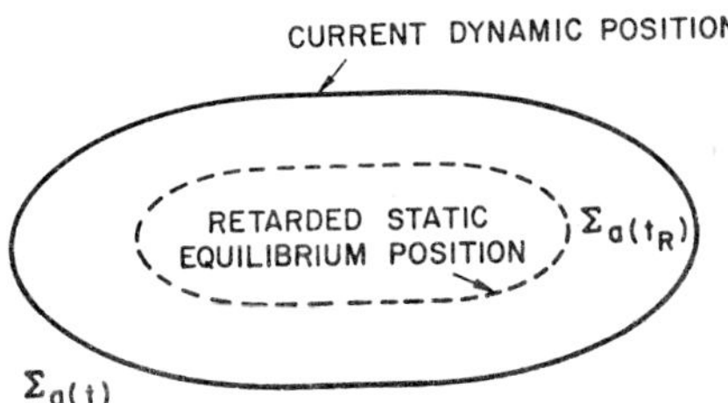

FIG. 5

Determination of the retarded dipole density for an expanding elliptical defect.

To complete this approximation we need the time at which the contribution of the ring associated to $a(t^*)$ should be "switched on." At time t^*, elastic waves begin to radiate out from the dotted ring in figure 5 "communicating" the disturbance associated to $\partial\Sigma_{a(t^*)}$. When the last of these disturbances has covered the dotted ring we neglect any secondary effects and "switch on" the ring, adding it to the "slow defect" core in quasi-static equilibrium. Except for corrections due to strong directional anisotropy of the slowest "shear" wave vis-a-vis the orientation of the ellipsoid this is the time for that slowest wave to travel backward from $a_1(t^*)$ to $-a_1(t^*)$ on the other side of the ring where $a_1(t)$ is the largest semi-axis of the ellipsoid (ellipse). If we use v_c for the appropriate minimum elastic wave speed, this condition leads to a delay time of $2a_1(t^*)/v_c$. Thus, at any time t the retarded time, t_R, is defined by

$$t = t_R + 2a_1(t)/v_c \quad . \tag{4.7}$$

By comparison the slow defect approximation is equivalent to

$$t = t_R . \tag{4.8}$$

We now conclude the solution using equation (4.6). If we freeze the configuration on the retarded volume at $t_R(t^*)$ and assume quasi-static equilibrium, so that $u(\underline{r},t) = u(\underline{r},t^*)$ for $\underline{r}$ near $\partial\Sigma_{a(t_R(t^*))}$, $\beta^P(\underline{r},t) = \beta^P(\underline{r},t^*)\Sigma_{a(t_R(t^*))}(\underline{r})$, and $\beta^o(\underline{r},t) = \beta^o(\underline{r},t^*)\Sigma_{a(t_R(t^*))}(\underline{r})$, we may then replace $G^H_{mi,j}(\underline{r},\underline{r}'_o,t-t')$ by $G^\infty_{mi,j}(\underline{r}-\underline{r}'_o)$, the static Green's tensor of the infinite body--which is independent of t' and is the value taken by G^H in the static lacuna containing $\Sigma_{a(t_R(t'))}$.

Then for $\underline{r}$ near $\partial\Sigma_{a(t_R(t^*))}$, (4.6) becomes

$$\begin{aligned} u^T_m(\underline{r},t^*) &= \Delta C_{ijk\ell} G^\infty_{mi,j}(\underline{r}-\underline{r}'_o) \int_{\partial\Sigma_{a(t_R(t^*))}} u^T_k(\underline{r}',t^*)\, dS'_\ell \\ &\quad -(C+\Delta C)_{ijk\ell} G^\infty_{mi,j}(\underline{r}-\underline{r}'_o) \int_{\Sigma_{a(t_R(t^*))}} \beta^P_{k\ell}(\underline{r}',t^*) d\underline{r}' \\ &\quad +\Delta C_{ijk\ell} G^\infty_{mi,j}(\underline{r}-\underline{r}'_o) \int_{\Sigma_{a(t_R(t^*))}} \beta^o_{k\ell}(\underline{r}',t^*) d\underline{r}' \end{aligned} \tag{4.9}$$

which is a direct generalization of equation (3.20) in Simmons and Bullough [47] where the solution for the sphere was given. The solution to (4.9) for an ellipsoidal inhomogeneity is given, for instance, by Kinoshita and Mura [48], where it is shown that

$$\begin{aligned} \bar{G}_{m\ell ik} &\equiv -\frac{1}{2} \int_{\partial\Sigma_a} (G_{mi,k}(\underline{r}-\underline{r}'_o) dS_\ell + G_{\ell i,k}(\underline{r}-\underline{r}'_o) dS_m) \\ &= (a_1 a_2 a_3/8\pi) \int_{\tilde{\Omega}} (\tilde{G}_{mi}(\underline{\xi})\bar{\xi}_k\bar{\xi}_\ell + \tilde{G}_{\ell i}(\underline{\xi})\bar{\xi}_k\bar{\xi}_m)/(a_n^2\bar{\xi}_n^2)^{3/2}\, d\Omega \quad . \end{aligned} \tag{4.10}$$

Here a_1, a_2, a_3 are the lengths of the semi-axes of the ellipsoid, $\underline{\xi}$, is a vector on the reciprocal unit sphere, and $\tilde{G}_{mi}(\underline{\xi})$ is the Fourier transform of $\underline{G}^\infty$ evaluated on the unit reciprocal sphere:

$$\tilde{G}_{ik}(\xi) = (C_{ijk\ell}\xi_j\xi_\ell)^{-1}, \quad \xi_j\xi_j = 1 \quad . \tag{4.11}$$

The value of $\underline{G}$ for the isotropic spherical inclusion was given in Simmons and Bullough [47]:

$$\bar{G}_{m\ell ik} = (1/6\mu)\ [\delta_{mi}\delta_{\ell k} + \delta_{mk}\delta_{\ell i} - (1/(5(1-\nu)))(\delta_{m\ell}\delta_{ik} + \delta_{mi}\delta_{\ell k} + \delta_{mk}\delta_{i\ell})]\ 1 \tag{4.12}$$

Using a six-dimensional vector terminology where vectors are symmetric 3 x 3 matrices following, for instance, the Voight convention, and defining:

$$\underline{\upsilon}(t^*):\ \upsilon_{k\ell} = (1/2)\int_{\partial\Sigma_a(t_R(t'))} (u_k^T(\underline{r},t^*)dS_\ell + u_\ell^T(\underline{r},t^*)dS_k) \tag{4.13a}$$

$$\bar{\underline{\beta}}^P(t^*):\ \bar{\beta}^P_{k\ell}(t^*) = \int_{\Sigma_a(t_R(t^*))} \beta^P_{(k\ell)}(\underline{r},t^*)d\underline{r} \tag{4.13b}$$

$$\bar{\underline{\beta}}^o(t^*):\ \bar{\beta}^o_{(k\ell)}(t^*) = \int_{\Sigma_a(t_R(t^*))} \beta^o_{(k\ell)}(\underline{r},t^*)d\underline{r} \tag{4.13c}$$

where "( )" around subscripts indicates symmetrization, we can integrate equation (4.8) around $\partial\Sigma_{a(t_R(t^*))}$ to obtain

$$\underline{\upsilon}(t^*) = \bar{\underline{\underline{G}}}(\underline{\underline{\Delta C}}\underline{\upsilon}(t^*) - (\underline{\underline{C}}+\underline{\underline{\Delta C}})\bar{\underline{\beta}}^P(t^*) + \underline{\underline{\Delta C}}\bar{\underline{\beta}}^o(t^*)) \tag{4.14}$$

from which

$$\underline{\upsilon}(t^*) = (\underline{\underline{I}}-\bar{\underline{\underline{G}}}\underline{\underline{\Delta C}})^{-1}\bar{\underline{\underline{G}}}(\underline{\underline{\Delta C}}\bar{\underline{\beta}}^o(t^*) - (\underline{\underline{C}}+\underline{\underline{\Delta C}})\bar{\underline{\beta}}^o(t^*)) \quad . \tag{4.15}$$

Comparison of equation (4.6) and (2.6) together with (4.15) gives:

$$\overline{\Delta\dot{\underline{\sigma}}}(t) = -\underline{\underline{\Delta C}}\dot{\underline{\upsilon}}(t_R(t)) + (\underline{\underline{C}}+\underline{\underline{\Delta C}})\bar{\underline{\beta}}^P - \underline{\underline{\Delta C}}\bar{\underline{\beta}}^o \tag{4.16a}$$

for the spatial components and

$$\overline{\Delta\dot{\sigma}}_{i4}(t) = -\rho\dot{\bar{\beta}}^P_{i4} + \Delta\rho\dot{\bar{\beta}}^o_{i4} \tag{4.16b}$$

for the time component of the stress drop rate tensor. The superscript "•" in (4.16a) refers to time differentiation; this can be related to t_R in the first term of the RHS of (4.16a) using (4.7). One of the key features apparent in (4.16) is the time delay required for rapidly moving inhomogeneities to reach the quasi-static emission condition which is characteristic of the slowly moving defect. For plasticity there is no inhomogeneity correction and the key modifications associated to dynamic plasticity arise from interference due to spatial extent and to Doppler effects. But inhomogeneities require time to reach local quasi-static equilibrium; and this gives rise to another source of high frequency error. From (4.4) we see that any dynamic inhomogeneity growth must be consistent with the slow defect approximation at sufficiently low frequencies. The retarded dipole approximation gives an indication of the frequency range at which the slow defect approximation begins to break down and what the initial corrections should look like, but the factor $2/v_c$ in (4.7) is obviously only an estimate. The expression given in (4.16) gives the stress drop rates for inhomogeneity motion and/or plasticity. However, it is not applicable to microcracking, since the driving terms in (4.2), for instance, vanish if the volume of the ellipsoid goes to zero. In the case of a crack, a different driving term is required--the value of the local tractions across the crack surface just before cracking. When the ellipsoid is reduced to an ellipse with $\underline{\underline{\Delta C}} = \underline{\underline{-C}}$, (4.6) takes the form

$$u^T_m(\underline{r},t) = C_{ijk\ell} \int_{t_0}^{t} G^H_{mi,j'}(\underline{r},\underline{r}'_o,t-t') \left[\frac{d}{dt'} \int_{\Sigma_{a(t_R(t'))}} \Delta u^T_k(\underline{r}',t)dS_\ell \right] dt' \tag{4.17}$$

where Δu^T_K is the discontinuity in displacement across the crack surface in the direction of the normal n_ℓ.

The retarded dipole formulation remains essentially intact when the driving terms are dropped. For the crack this model has the interpretation that within the retarded volume the crack has its equilibrium configuration; while outside, the crack surface has been "cut" up to the instantaneous crack tip position, but not yet begun to move. The solution to (4.9) is now that obtained for a static crack. The elliptic crack problem has been solved in an elegant treatment by Willis [45] for a surface traction vector $\underline{P}(x)$ with polynomial coefficients in which he postulated and then proved that $\underline{P}(x)$ gives rise to a displacement discontinuity $\underline{Q}(\underline{x})(1-x_1^2/a_1^2-x_2^2/a_2^2)^{1/2}$ where $\underline{Q}(\underline{x})$ is a vector with polynomial coefficients of the same type as $\underline{P}$. Within the pseudopoint formulation it will suffice to consider $\underline{P}$ to be of degree zero; i.e., constant. In this case we have for a circular crack with the normal in the $\underline{x}_3$ direction

$$\underline{Q} = (8a(1-\nu)/\pi\mu) \begin{vmatrix} 1/(2-\nu) & 0 & 0 \\ 0 & 1/(2-\nu) & 0 \\ 0 & 0 & 1/2 \end{vmatrix} \underline{P}, \tag{4.18a}$$

and for elliptic cracks

$$\underline{Q} = (2a_2(1-\nu)/\pi\mu) \begin{vmatrix} E(k)-(\nu/k^2)(E(k)-k'^2K(k)) & 0 & 0 \\ 0 & E(k)-(\nu k'^2/k^2)(K(k)-E(k)) & 0 \\ 0 & 0 & E(k) \end{vmatrix} \underline{P} \tag{4.18b}$$

with $a_2 < a_1$, $(k')^2 = 1-k^2 = a_2^2/a_1^2$ and ν = Poisson's ratio. In the general anisotropic case

$$\underline{Q} = \underline{\underline{W}} a_2 \underline{P} \tag{4.18c}$$

where $\underline{\underline{W}}$ is a constant generally non-diagonal matrix obtained using equation (6.8) in Willis [45]. Defining

$$\bar{P}^o_i(t) = \int_{\Sigma_{a(t_R(t))}} \sigma^o_{ij}(\underline{r},t_R(t)dS_j \tag{4.19}$$

we have for a crack with unit normal $\underline{\nu}$:

$$\overline{\Delta\dot{\sigma}}_{im} = (2\pi/3)C_{ijk\ell}\nu_j\nu_\ell\nu_m W_{kn} \frac{d}{dt} (a_2(t_R(t))\bar{P}^o_n(t_R(t))). \tag{4.20}$$

$\overline{\underline{\Delta\dot{\sigma}}}$ has no time components since cracking is not accompanied by a bulk density change in the initial state. In both (4.16) and (4.20) it should be noted that if the respective driving terms, $\underline{\beta}^P, \underline{\beta}^o, \underline{P}^o$ are independent of time, the rate of change of defect volume determines the acoustic emission magnitude at any instant.

To illustrate the above results we give a comparison for emission from dislocation loops and cracks in an infinite isotropic body. The dislocation problem was studied by Nabarro [9] and more recently by Mura [10] and by Malén and Bolin [11]. A special case for modeling the crack has recently been independently proposed by Wadley, et al. [44].

Figure 6 shows a dislocation loop on a plane in an infinite body. We take the origin at the "centroid" of the loop and place the "detector" at $\underline{r}$. The loop has a plastic distortion rate related by the elastic constants to the stress drop rate

$$\Delta\dot{\sigma}_{jk}(\underline{r}',t) = -C_{jk\ell m}\dot{\beta}^P_{\ell m}(\underline{r}',t) \tag{4.21}$$

where

$$C_{jk\ell m} = \lambda\delta_{jk}\delta_{\ell m} + \mu(\delta_{j\ell}\delta_{km} + \delta_{jm}\delta_{k\ell}) \tag{4.22}$$

and λ, μ are the Lamé constants. The plastic distortion rate itself is derivable from the mechanics of loop motion as follows. Let $\underline{b}$ be the Burgers vector with magnitude b and $\underline{v}_N$ be the velocity of the loop in the normal direction in the plane which we take as the 3-direction to fix the notation. Let s be the line length along the loop. For $0 \le s \le L(t)$ the position at any point along the loop in the (x_1', x_2') plane is $x_1' = x_1'(s,t)$, $x_2' = x_2'(s,t)$, and $x_3' = 0$. Then the plastic distortion rate is

$$\dot{\beta}^P_{\ell m}(\underline{r}',t) = \int_0^{L(t)} b_\ell\delta_{m3}\delta(x_1'-x_1'(s,t))\ \delta(x_2'-x_2'(s,t))\ \delta(x_3')v_N(s,t)\ ds \tag{4.23}$$

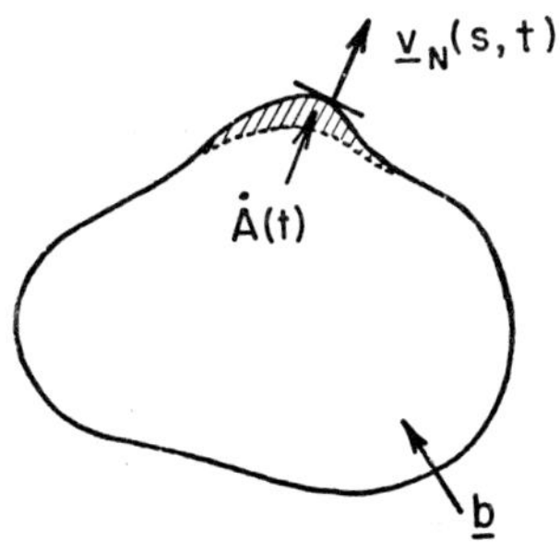

FIG. 6

Schematic of dislocation loop for infinite body calculation; the plane is normal to the $\underline{x}_3 = \underline{\nu}$ direction, $\underline{b}$ is the Burgers vector, $\underline{v}_N$ is the velocity normal to the dislocation, and A(t) is the rate of area slipped.

The zeroth moment for the loop is then

$$\overline{\Delta\dot{\sigma}}_{jk}(t) = -C_{jk\ell 3}b_\ell \int_0^{L(t)} v_N(s,t)\ ds \tag{4.24}$$

Using $G^H_{ij,k'} = -G^H_{ij,k}$ together with formula (B2) from Appendix B and (4.24), equation (2.6) yields

$$
\begin{aligned}
u_i(\underline{r},t) = (C_{jk\ell 3}b_\ell/8\pi\rho r c^3) \Big\{ & 2\alpha^3 \Delta_1\dot{A}(t-r/a) - 2\Delta_o\dot{A}(t-r/c) + \\
& + (\Delta_2 + 2\Delta_1)(\alpha^2/r)A_o(t-r/a) - (1/r)(\Delta_2 + 2\Delta_o)A_o(t-r/c) \\
& + \Delta_2(3/r^3)[c^2t^2(A_o(t-r/c) - A_o(t-r/a)) - 2ct\,(A_1(t-r/c) \\
& - A_1(t-r/a) + (A_2(t-r/c) - A_2(t-r/a))]\Big\},
\end{aligned}
\tag{4.25}
$$

where $\alpha = \frac{c}{a}$, $\underline{n} = \underline{r}/r$,

$$\dot{A}(\tau) \equiv \int_0^{L(\tau)} v_N(s,\tau)\,ds, \tag{4.26}$$

is the area swept out per unit time with time moments

$$A_n(\tau) \equiv c^{n+1}\int_{-\infty}^{\tau} (\tau')^n\,\dot{A}(\tau')\,d\tau' = \int_{-\infty}^{c\tau} (\tau'')^n\,\dot{A}(\tau''/c)\,d\tau'', \tag{4.27}$$

and where

$$
\begin{aligned}
\Delta_o &\equiv \delta_{ij}n_k - n_in_jn_k \\
\Delta_1 &\equiv -n_in_jn_k \\
\Delta_2 &\equiv 5n_in_jn_k - (\delta_{jk}n_i + \delta_{ik}n_j + \delta_{ij}n_k).
\end{aligned}
\tag{4.28}
$$

Equation (4.25) gives the pseudopoint approximation for the emission from a general planar dislocation loop. Since this is a single parameter process, equation (4.25) also satisfies the S-J condition. When two or more dislocations with different Burgers vectors or different slip planes are activated almost simultaneously, but follow different time histories, the S-J approximation will not be valid. This will be discussed in more detail in connection with the inverse problem (§5).

In frequency space, (4.25) clearly exhibits the form of a S-J process:

$$u_i(\underline{r},\omega) = (C_{jk\ell 3}b_\ell/4\pi\rho r c^3)(F_o\Delta_o + F_1\Delta_1 + F_2\Delta_2)\,\dot{A}(\omega) \tag{4.29}$$

where we have used (B-3), (B-4), and (B-5) from Appendix B. Equation (4.29) fits directly into the transfer function formalism (2.18) by including the transducer response. For the purposes of connecting (4.29) to an acoustic emission signal, we could think of point transducers sensitive to only one direction and flat in frequency. Then, of course, (4.29) remains unchanged, but now u_i is proportional to the voltage associated to the transducer sensitive to displacements in the i direction.

Finally to be able to relate (4.25) and (4.29) directly to the dislocation density one has the identities

$$
\begin{aligned}
C_{jk\ell m}\nu_m b_\ell\Delta_o/\rho c^2 &= b_i(\underline{n}\cdot\underline{\nu}) + \nu_i(\underline{n}\cdot\underline{b}) - 2n_i(\underline{n}\cdot\underline{b})(\underline{n}\cdot\underline{\nu}) \\
C_{jk\ell m}\nu_m b_\ell\Delta_1/\rho c^2 &= -(\lambda/\mu)n_i(\underline{\nu}\cdot\underline{b}) - 2n_i(\underline{n}\cdot\underline{b})(\underline{n}\cdot\underline{\nu}) \\
&= (2-\alpha^{-2})\,n_i(\underline{\nu}\cdot\underline{b}) - 2n_i(\underline{n}\cdot\underline{b})(\underline{n}\cdot\underline{\nu}) \\
C_{jk\ell m}\nu_m b_\ell\Delta_2/\rho c^2 &= 10n_i(\underline{n}\cdot\underline{b})\,(\underline{n}\cdot\underline{\nu}) - 2n_i(\underline{b}\cdot\underline{\nu}) - 2b_i(\underline{n}\cdot\underline{\nu}) - 2\nu_i(\underline{n}\cdot\underline{b})
\end{aligned}
\tag{4.30}
$$

TABLE 1

Displacement Matrix for Expanding Dislocation Loop (Infinite Body)

$Q_o \equiv (b\dot{A}(\omega)/4\pi rc)(F_o - 2F_2)$

$Q_1 \equiv (b\dot{A}(\omega)/4\pi rc)(-F_1\alpha^{-2} + 4F_2)$

$Q_2 \equiv (b\dot{A}(\omega)/4\pi rc)((2-\alpha^{-2})\, F_1 - 2F_2)$

Nonzero Components	$\dot{\bar{\beta}}^P_{\ell m}$ →					
	ℓm=11	22	33	23	13	12
$u_1(1,0,0,\omega)$ =	Q_1	Q_2	Q_2	0	0	0
$u_2(0,1,0,\omega)$ =	Q_2	Q_1	Q_2	0	0	0
$u_3(0,0,1,\omega)$ =	Q_2	Q_2	Q_1	0	0	0
$u_3(0,1,0,\omega)$ =	0	0	0	Q_o	0	0
$u_1(0,0,1,\omega)$ =	0	0	0	0	Q_o	0
$u_2(0,0,1,\omega)$ =	0	0	0	0	0	Q_o
$u_2(0,0,1,\omega)$ =	0	0	0	Q_o	0	0
$u_3(1,0,0,\omega)$ =	0	0	0	0	Q_o	0
$u_1(0,1,0,\omega)$ =	0	0	0	0	0	Q_o

TABLE 2

Directionality Factors for δ Function Components of the Two Basic Types of Dislocation Source

Source Type	Relative Displacement Magnitudes at Spherical Wavefront	
	Longitudinal Wavefront (Radial Displacement)	Shear Wavefront (Tangential Displacement)
Prismatic Dislocation (Climb) or Mode I Microcrack	$\alpha(1-\alpha^2) + \frac{\cos 2\theta_p}{\alpha}$	$\sin 2\theta_p$
Slip Dislocation or Mode II/III Microcrack	$\frac{\sin 2\theta_s}{\alpha}$	$\cos 2\theta_s$

$\alpha = \frac{c}{a}$ = ratio of shear to longitudinal wave velocities

θ_p = angle between radial observation direction, $\underline{r}$, and dislocation plane normal, $\underline{\nu} = \underline{b}$.

θ_s = angle between radial observation direction, $\underline{r}$, and the $\underline{\nu}\mathrm{x}\underline{b}$ direction.

where we have replaced the 3-direction for the normal direction of the loop plane by the unit vector $\underline{\nu}$.

The acoustic emission response for the planar dislocation loop can then be obtained by combining (4.25-4.28) for a time domain representation or (4.29) and (B-3 - B-5) for a frequency domain representation. This is illustrated for real time in table 1 for the three coordinate Burgers vector directions and for the three coordinate directions in space. A specific example of the acoustic emission signal for axial locations associated to emission from the two basic types of circular dislocation loop expanding at constant radial velocity is shown in figure 7. The directionality of emission in the far field from the sources in figure 7 is summarized in table 2. Far field emission is dominated by the δ function terms so that the emission is broken down into two spherical wavefronts travelling at the longitudinal and shear wave velocities. As would be expected, the longitudinal wave has only radial and no tangential components, while the shear wave has only tangential and no radial components. The amplitudes of the waves are modulated by a directionality factor (table 2), through which the source types may be distinguished.

Emissions from individual dislocations are orders of magnitude too small to be detected ($\sim 10^{-9}$ µm in the above example). Thus, emission from plastic deformation must be thought of as the superposition of the emissions of the individual dislocations making up the slip process. This can vary from localized rapid multiplication of dislocations which appears on a very short time scale to bulk plasticity which can be expected to be continuous in character. If the slip mechanisms have identical time histories the S-J approximation should be valid, but if multiple slip systems operate, generally with different time histories, the pseudopoint approximation will be required.

Emission from an expanding elliptical crack is summarized by (4.20). More specifically, using (4.18) for a crack in the $\underline{x}_3 = 0$ plane of an isotropic body, one can see that the directional character of emission from cracks will be identical with that of emission from dislocations and the emission will be of Mode I or Mode II/III type depending on the character of the initial tractions σ_i^o existing at the time of cracking. The differences between dislocations and cracks then are: (i) crack emission is stress controlled--as discussed in §5* and (ii) the defect volume, and, therefore, the emission rate for cracks is generally an order of magnitude greater than for individual dislocations.

To get a concrete feeling for the comparison of crack and dislocation emission, we consider a circular crack expanding under the same conditions as those given for the dislocation loop in figure 7. The effective $\underline{\beta}^P$ for such a crack is given by (4.20) with the elastic constant factor removed:

$$\bar{\underline{\beta}}^P_{crack} = (16a^3(1-\nu)\tau^0/3\mu)(1/(2-\nu),1/2) = (2.12,\ 1.76)\ (v^3\tau^0/\mu)\ t^3 \tag{4.31}$$

for Mode II/III and Mode I, respectively, where τ^0 is the magnitude of the surface traction, while for a dislocation loop

$$\bar{\underline{\beta}}^P_{dis} = \pi a^2 b \cong 9 \times 10^{-10} v^2 t^2 \text{ meters}^{-1} \ . \tag{4.32}$$

The ratio of the emission rates is then

$$\dot{\bar{\underline{\beta}}}^P_{crack}/\dot{\bar{\underline{\beta}}}^P_{dis} \cong (3.5 \times 10^9,\ 2.9 \times 10^9)(v\tau^0/\mu)t \quad \text{meters}^{-1} \tag{4.32}$$

$$\cong (3.5 \times 10^{11}, 2.9 \times 10^{11})(\tau^0/\mu)t \quad s^{-1}$$

*Actual dislocation motion which can be detected by acoustic emission may well be associated with dislocation pileups which are partially stress controlled. Also, high velocity dislocation drag exerts a fractional backstress so that high speed deformation and associated dislocation multiplication processes are very likely to be, in part, also stress controlled.

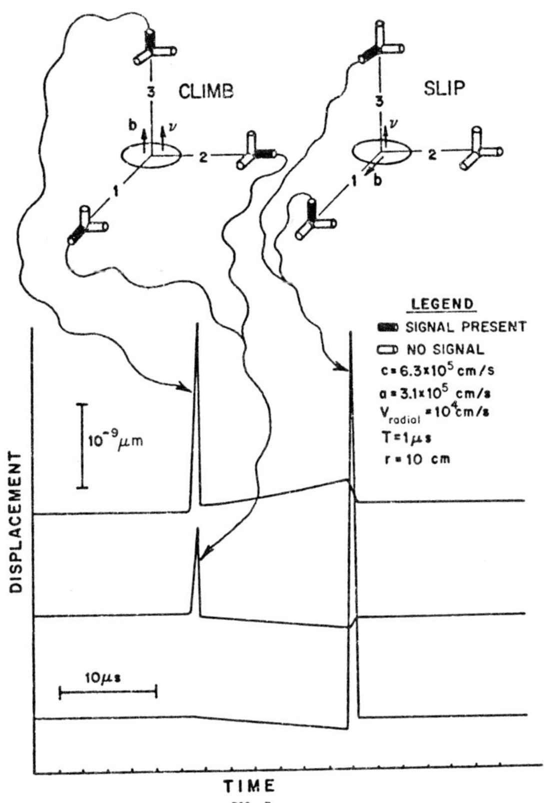

FIG. 7

Multichannel measurement of displacement produced by expansion of dislocation loops of prismatic (climb) and slip type in aluminum. The source expands for 1 μs at a radial velocity of 10^4 cm/s and then stops. Cylinder axis represents direction of measurement of displacement at that location. These are darkened if signal (shown below) is present and left blank if no signal is produced.

For a typical value of $\tau^o/\mu \sim 0.01$ and in the microsecond time range this means emission rates of $(3.5 \times 10^3, 2.9 \times 10^3)$ t μs^{-1} with a final static value given for the ratios of $\dot{\bar{\beta}}^P_{crack}$ to $\dot{\bar{\beta}}^P_{dis}$ at 10^{-6} s of $(2.4 \times 10^3, 2.0 \times 10^3)$. Thus we see that for uniform expansion, crack emission has a characteristic parabolic growth rate. It, therefore, tends to have a lower frequency content relative to single dislocation emission which has a linear growth rate. This parabolic growth rate may be a characteristic signature for crack emission in some materials which do not exhibit slow crack acceleration or dislocation multiplication processes which simulate the same AE buildup. In any event the orders of magnitude greater emission from (non-atomic size) cracks than from single dislocations, due to differences in defect volume, is apparent.

Finally, we observe that for cracks of the speed in this example, the retarded dipole density approximation has some significance. From (4.7), $t_R \cong 0.935$ t so that the crack velocity will appear to be ~ 93.5 m/s and the growth $\sim 1.07 \times 10^{-6}$ s. For a crack of this size whose centroid does not move with time, the informative bandwidth should be in the neighborhood of 2.25 MHz so that this apparent velocity shift might be detectable. In fact for this bandwidth the apparent crack velocity should be measurable if the apparent event time $\gtrsim 4.5 \times 10^{-7}$ s, i.e., the apparent crack velocity $\sim v_{shear}/14$. This means a possible correction of ~ 17 percent due to retarded dipole considerations.

TABLE 3.

STRESS CONTROLLED EMISSION

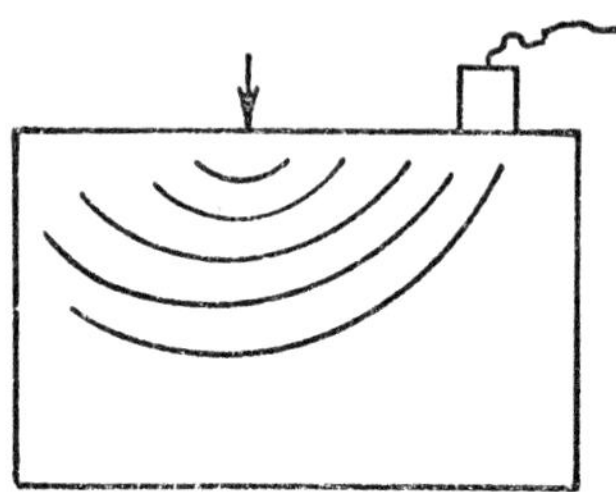

Characteristics:

1. Numerous randomly oriented defect sources and wavelengths >> defect size.

2. $\bar{\dot{\sigma}} \sim \sigma\, \bar{\dot{m}}(t)$ orientation
   $\bar{\dot{\sigma}} \sim M\, \dot{\sigma}(t)$ time

MATERIAL CONTROLLED EMISSION

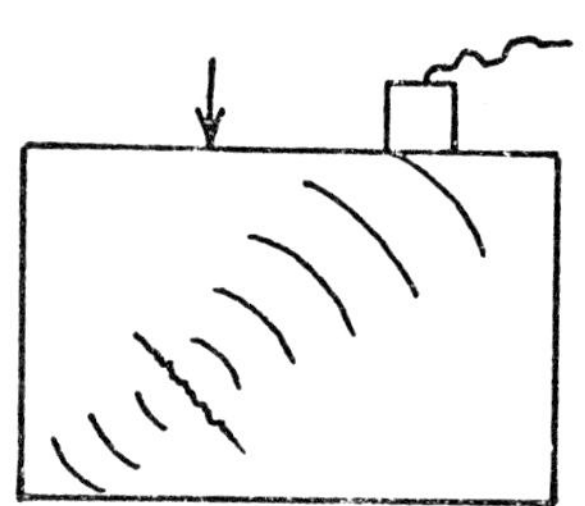

Characteristics:

1. Directional Sources (orientation)
   Source nucleation barriers (time)

EXAMPLES

STRESS CONTROLLED ORIENTATION STRESS CONTROLLED TIME	STRESS CONTROLLED ORIENTATION MATERIAL CONTROLLED TIME
Plasticine	Plumb bob indentation on glass Stress induced martensitic transformation--small platelets
MATERIAL CONTROLLED ORIENTATION STRESS CONTROLLED TIME	MATERIAL CONTROLLED ORIENTATION MATERIAL CONTROLLED TIME
Laminate Composite flow	Grain boundary cracking Stress induced transformation Large grains Oriented slip avalanche

5. The Inverse Problem

While the determination of the acoustic emission to be expected from a given source is an essential component in understanding AE, it is finding solutions to the inverse problem--in which one must characterize the source from recorded emission signals--that is needed to make AE useful as a technique in materials science and nondestructive evaluation. Specific solutions involve the development of new algorithms in signal processing and remote sensing. Our efforts in this direction will be presented elsewhere; here we present a few general considerations and difficulties which must be kept in mind in dealing with the inverse problem.

The goal, of course, is to identify the AE source from the AE signals. This identification can take a number of forms: (1) location or spatial identification, (2) event recognition or time identification, and (3) orientation or mechanism identification. This refers to distinguishing differences in stress drop histories due to source orientation or source mechanism effects.

Acoustic emission has the potential for carrying out any of the three above identifications, but not unequivocally. Both the pseudopoint and S-J approximations require source location and band limiting in order to be valid. Sophisticated algorithms for source location have been extensively studied in the geophysical literature [25,35,38]. In AE even greater relative precision in source location is possible by using non-AE techniques as well as by calibration methods to develop location signatures. The band limited process automatically introduces both time and space resolution limits, as was apparent from the discussion of the pseudopoint approximation. Events occurring within these limits must be lumped together as one event. Furthermore, many different source mechanisms might occur within such a space-time region giving rise to an acoustic emission event whose apparent source can in no way be linked with the individual events, thus making interpretation very difficult.

An example of this occurs with the breaking of a glass capillary [46], which is now a standard way of producing a step unloading force in a structure. One can produce a similar step unloading force in a block of glass by indenting with a sharp conical indenter, for example, a plumb bob. When the tiny point of glass under the indenter shatters, the typical characteristic step force unloading signal is produced. And yet the actual source mechanism was brittle fracture with the production of dozens or hundreds of cracks over a period of the order of a nanosecond. The problem of unravelling this incredibly complex history associated to the source in such a situation is analogous to trying to ascertain the paths of individual molecules in a gas. In many cases of acoustic emission one can expect that the emission source will reflect not the physical micromechanisms involved, but rather the local internal stress state being relieved. We refer to this effect as stress-controlled emission as opposed to material-controlled emission (table 3). Almost all acoustic emission is to some extent stress controlled as has been noted in §4 for both inhomogeneities and cracks.

To find the stress drop history associated with a given event one must solve for the six to fifteen unknown components in the stress drop tensor of equation (2.16). To find these components, at least as many observations are necessary. Thus, finding the source stress drop history requires at least six multiple channels in order to perform an inversion (unless one knows the time history in advance, in which case it is conceivable that multiple time observations on one channel could be used instead). Figure 4 shows a conceptual example of the system of equations governing multichannel observations. In frequency space these are algebraic equations; in real time they are convolution equations.

The measurement that is illustrated in figure 4 shows the spectral matrix which represents the transfer function for such a multi-dimensional measurement, as well as the output voltages and the unknown stress drops at the source. The simplest representation of equation (2.18) is in a type of Voight notation, where the subscripts "jk" are replaced with "α", where $\alpha = 1,2,\ldots,6$. Thus

$$V^{\gamma}(\omega) = T^{\gamma}_{\alpha}(\omega)\overline{\Delta\dot{\sigma}}_{\alpha} \tag{5.1}$$

Here each "γ" represents a particular transducer location as well as the orientation at that location. This is shown in figure 4, which suggests a conceptual experimental configuration for measuring acoustic emission which will be algebraically determinate so that inversion is possible.

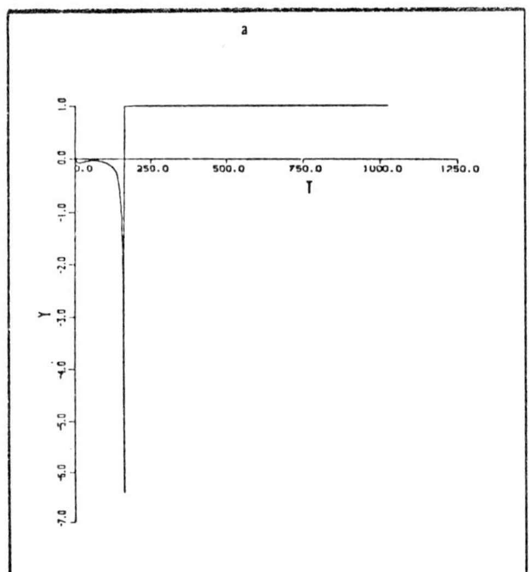

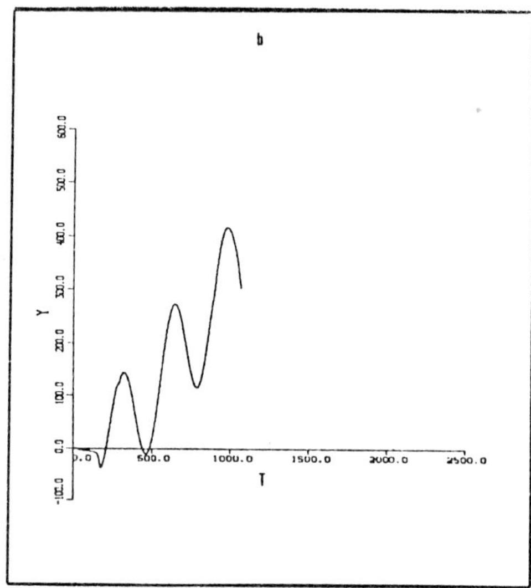

FIG. 8

a) Vertical displacement surface to surface Green's function for vertical point force on a half-space. This is used as a kernel convolved with "unknown" function shown in Figure 9a. The number of time increments (1024 points total) are displayed on the T axis. Arbitrary units are used on the Y axis.

b) Convolution of functions in Figures 8a and 9a. This models an "output" function from which the goal is to deconvolve the Green's function shown in Figure 8a.

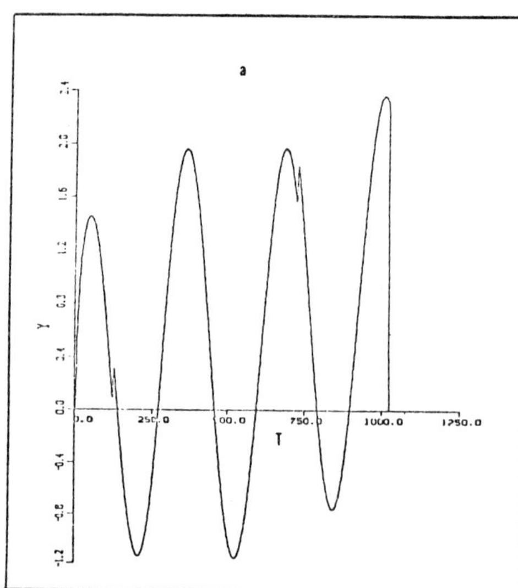

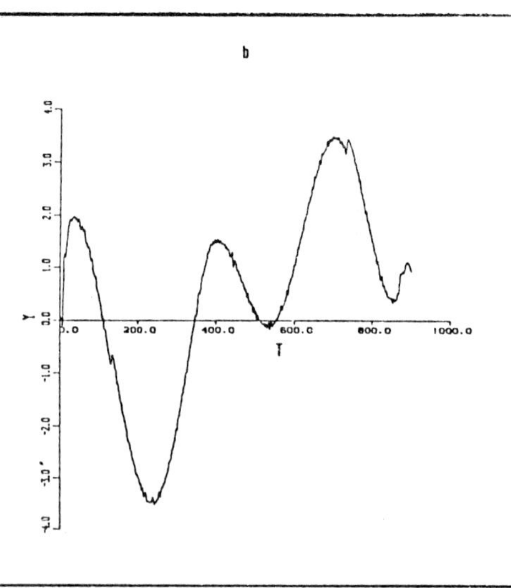

FIG. 9

a) "Unknown" function on convolution example.

b) Result of deconvolving function in Figure 8a from that in Figure 8b to recover "unknown" function using division of FFT's of the individual functions. Extensive error due to time aliasing is apparent.

An important simplification can be made for single source reconstruction by use of the S-J approximation. Let there be two sources operating within a small volume of material. They have orientation matrices $\underline{M}_a$ and $\underline{M}_b$ and time histories $\dot{\bar{m}}_a(t)$ and $\dot{\bar{m}}_b(t)$. Their combined or equivalent emission then in frequency representation is

$$\underline{M}_{eq}\dot{\bar{m}}_{eq}(\omega) \neq \underline{M}_a\dot{\bar{m}}_a(\omega) + \underline{M}_b\dot{\bar{m}}_b(\omega) \ . \tag{5.2}$$

But if the S-J approximation holds, then $\dot{\bar{m}}_a(\omega) = \dot{\bar{m}}_b(\omega)$ and

$$\underline{M}_{eq}\dot{\bar{m}}_{eq}(\omega) = (\underline{M}_a + \underline{M}_b)\dot{\bar{m}}_a(\omega) \tag{5.3}$$

so that all channels have frequency spectra which are proportional.

It might appear that one can merely solve these equations algebraically in frequency space to carry out the inversion process, but this is unfortunately not always true. The recorded signals, V^γ, have to be truncated at some fixed time unless the V^γ have decayed to zero before this time. If the output signals must be truncated, then the artificial truncation renders the solution of the convolution incomplete. Such solutions demand signal processing to avoid excessive errors. Figures 8 and 9 show examples of such a situation for a one parameter deconvolution problem where merely dividing FFT's (fast Fourier Transforms) leads to great distortion, partly due to this truncation effect and partly due to the fact that FFT's are actually operations on a circle rather than a line so that the time axis wraps around on itself allowing time aliasing to occur during inversion.

The inverse problem also has another difficulty common to many remote sensing problems, noise amplification. Very often almost undetectable amounts of noise on the output channels will be amplified by the inversion algorithm and completely swamp out the actual solution. Again advanced filtering and numerical algorithms are often required.

Knowing the stress drop history is an important aid to providing orientation and mechanism identification. However, even knowing the stress drop history, one still must perform a signature analysis of some sort to correlate the orientation and mechanism with the local stress drop produced. The advantage of the inversion process is that it permits removal of AE signal coloration due to the structure and instrumentation. However, in many cases complete or precise inversion may be impossible due to the difficulties cited above or due to the difficulty of obtaining the transfer tensor for the structure. In addition to dealing with general inversion methods, then, there is also a need to develop partial inversion, or signature analysis, methods which could be combined with empirical data to correlate the source mechanism with AE output in a more direct manner.

Summary

A conceptual framework has been developed for quantitative defect characterization using acoustic emission techniques. In this theory the acoustic emission source has been characterized by the changes in internal stresses it induces when it responds inelastically to the structural loads. Two distinct types of problems relevant to AE have been discussed: The forward problem--to calculate the AE signal from a known source; and the inverse problem--to determine the source characteristics from detected AE signals. The representation of an AE source event using multipolar expansions appears to be too complex for the state of today's technology, so two simplifying representations have been introduced, the pseudopoint representation and the Stump-Johnson (S-J) representation. The pseudopoint approximation allows multiparameter processes while the (S-J) approximation is a single parameter approximation allowing higher source moments to be included. Aside from rather weak conditions on the ratio of source size to distance from transducer, the key restriction to apply these approximations is a band limiting process to limit the frequencies to within what we call the informative bandwidth. Outside the informative bandwidth specific models of the AE mechanism are needed to analyze the AE signals. Estimates on the magnitude of the cutoff frequency as a function of defect size were given and shown to be also dependent on defect shape. Both of these approximations permit the development of the transfer function formalism (2.18) which allows one to consider transformation of information frequency by frequency from the source to the signal. Another important consequence of this formalism is the fact that unless single parameter processes are involved, conventional spectral analysis methods cannot be used.

As an application of the above formalism the forward problem has been solved for AE from plastic deformation, inhomogeneities, and cracks within the pseudopoint approximation. The slow defect and retarded dipole density approximations were introduced--the latter to describe effects associated with the finite amount of time needed for inhomogeneities to come to local quasi-static equilibrium and detailed solutions given for spherical inhomogeneities and elliptic cracks in isotropic media. The AE signal magnitude was shown to be directly related to defect volume in all three cases--to the change in local prestress state for cracks and inhomogeneities with an additional density change effect for inhomogeneities. The strong directionality and Burgers vector/mode dependence was explicitly shown for circular dislocation loops and penny-shaped cracks and a detailed numerical comparison of the emission signal strength from cracks and dislocations was presented.

The inverse problem was also touched upon and some of the difficulties associated to the necessity of multichannel inversion to reconstruct the source tensor have been discussed. The necessity of band limiting to obtain a usable transfer function formalism automatically reduces spatial and temporal resolution and with it the possibility of distinguishing fine structure within the controlling mechanism. In the extreme case there will be no orientation effects other than those dictated by the local internal stress field being relieved. We refer to this effect as stress controlled AE. In other cases where the material imposes substantial constraints on the release mechanism, distinct orientation and/or time history effects will be evident; this is referred to as mechanism-controlled AE. All AE is stress controlled to some degree.

Detailed discussions on particular algorithms for both partial and complete AE signal inversion will be presented elsewhere.

Acknowledgements

We wish to thank J. R. Willis, H. N. G. Wadley, and J. R. Rice for their valuable comments. The major portion of this work was performed under a joint program of NBS with the Electric Power Research Institute on Acoustic Emission Spectral Measurement of Defect Motion.

Appendix A

Dynamic Formalism for AE from Defect Motion

The derivation we present here is a simple extension of the static formulation given in Simmons and Bullough [47] in the proceedings of the precursor conference to this one. The generalization to dynamics is very simple for this case since space-time tensors of degree two do not appear, rendering the exterior calculus unnecessary. We shall use ordinary lower case Roman indices to refer to space variables (i,j, etc. = 1,2,3) and capped indices to refer to space-time variables ($\hat{i},\hat{j}$, etc. = 1,2,3,4).

We take

$$\sigma_{i4} \equiv -\rho v_i \tag{A1}$$

to extend the stress tensor to a 3 x 4 tensor $\sigma_{i\hat{j}}$ and

$$\beta^T_{m\hat{j}} \equiv u_{m,\hat{j}}$$

or (A2)

$$\underline{\beta}^T \equiv \overset{4}{\nabla}\underline{u}$$

to extend the distortion tensor. Here $\overset{4}{\nabla}$ is the four dimensional gradient operator.

The dynamic equations of equilibrium then take the form

$$\overset{4}{\nabla}\underline{\sigma} + \underline{f} = 0 \tag{A3}$$

exactly analogous to the static form. We also extend the elastic constants $C_{ijk\ell}$ by defining

$$C_{i\hat{j}k4} \equiv C_{i4k\hat{j}} \equiv -\rho\delta_{4\hat{j}}\delta_{ik} \,. \tag{A4}$$

The equation for the dynamic Green's tensor is

$$C_{i\hat{j}k\hat{\ell}}\, G_{im,\hat{j}\hat{\ell}}(\underline{X},\underline{X}') + \delta_{km}\delta(\underline{X}-\underline{X}') = 0 \tag{A5}$$

where $\underline{x}$ is the space-time point (x_1,x_2,x_3,t). $\underline{G}$ satisfies the reciprocity relation

$$G_{mn}(\underline{x}',\underline{x},t'-t) = G_{nm}(\underline{x},\underline{x}',t'-t), \tag{A6}$$

so that we can write down the displacement at $\underline{x}',t'$ due to a distribution of forces, surface tractions and initial conditions.

$$\begin{aligned} u_m(\underline{X}') = &\int_{VT} G_{mi}(\underline{X}',\underline{X})f_i(\underline{X})dVT + \int_{(\partial V)\times T} G_{mi}(\underline{X}',\underline{X})t_i(\underline{X})dST \\ &+ \rho\int_V \left(G_{mk}(\underline{X}',\underline{X}_o)\dot{u}_k(\underline{X}_o) - \dot{G}_{mk}(\underline{X}',\underline{X}_o)u_k(\underline{X}_o)\right) dV \end{aligned} \tag{A7}$$

*Note the primed and unprimed variables have been interchanged in these expressions compared to those in Simmons and Bullough [47].

where $VT = V\times(t_o,t_1)$ is the space-time cylinder associated to the volume V and time interval $t_o < t < t_1$; ∂V is the boundary of V, and $X_o = (\underline{x},t_o)$.

For acoustic emission purposes we are interested only in changes in an existing field, so we can write an exact analogy to the static situation:

$$\Delta u_m(\underline{X}') = \int_{VT} G_{mi,\hat{j}'}(\underline{X}',\underline{X})\ \Delta\sigma_{i\hat{j}}(\underline{X})\ dVT + \int_{(\partial V)xT} G_{mi}(\underline{X}',\underline{X})\ \Delta\tau_i(\underline{X})\ dST \tag{A8}$$

where we have assumed that the initial conditions for the changes are zero. In equation (A8) $\Delta\underline{\sigma}$ and $\Delta\underline{\tau}$ are characterized as sources. The signals arising from these sources must be distinguishable from the sources themselves to avoid double counting. This is usually not a problem if one has reasonably localized sources and a detector at a position without sources, although secondary errors can occur if one mislocates the source (Backus and Mulcahy [36]).

If we now introduce the Heaviside Green's tensor associated to a point load with a step function time history:

$$G^H_{mi}(\underline{X}',\underline{X}) = G^H_{mi}(\underline{x}',\underline{x},t'-t) \equiv \int_o^{t'-t} G_{mi}(\underline{x}',\underline{x},t'-t-\tau)\ d\tau, \tag{A9a}$$

so that

$$\frac{\partial}{\partial t} G^H_{mi}(\underline{x}',\underline{x},t) = G(\underline{x}',\underline{x},t), \tag{A9b}$$

equation **(A8) becomes**

$$\Delta u_m(\underline{x}',t') = \int_{VT} G^H_{mi,\hat{j}}(\underline{x}',\underline{x},t'-t)\ \Delta\dot{\sigma}_{i\hat{j}}(\underline{x},t)\ dVT + \int_{(\partial V)\times T} G^H_{mi}(\underline{x}',\underline{x},t'-t)\ \Delta\dot{\tau}_i(\underline{x},t)\ dST \tag{A10}$$

where we avoid the point $\underline{x}' = \underline{x}$, $t' = t$; that is, we keep the detector away from the sources.

Following the development in Simmons and Bullough, we define the operator F by

$$\beta^F_{m\hat{n}}(\underline{X}') = \int_{VT} G_{mi,\hat{k}\hat{n}'}(\underline{x}',\underline{x},t'-t)\ C_{i\hat{k}j\ell}\beta_{j\ell}(X)\ dVX = \int_{VT} G_{mi,\hat{k}\hat{n}'}(\underline{x}',\underline{x},t'-t)\ \sigma_{i\hat{k}}(X)\ dVX \tag{A11}$$

which gives the dynamic elastic distortion fields induced either by force fields or surface tractions. The second form of (A11) defines the operation form $\underset{\sim}{F}^{\sigma}$ which operates on the stress field rather than the distortion field. We define the residual operator R

$$R = I-F \tag{A12}$$

so that the residual distortion field

$$\underline{\beta}^R = \underline{\beta} - \underline{\beta}^F. \tag{A13}$$

Of course, all fields here are thought of as generalized function fields [50,51] to account for singularities.

The formalism to solve the external dipole density problem is then exactly analogous to the static case. We once again consider an externally applied dynamic stress dipole distribution $\underline{\sigma}^A$, where σ^A_{i4} represents a negative applied momentum density. $\underline{\beta}^P$ is again plastic distortion with the term β^P_{i4} corresponding to a plastic velocity field where the material flows plastically but incorporates no elastic stresses. Summarizing the derivation in [47]

$$\underline{\sigma}^A + \underline{\sigma} = \underline{\sigma}^T \tag{A14}$$

$$\underline{\beta}^P + \underline{\beta} = \underline{\beta}^T \tag{A15}$$

$$\beta^T_{i\alpha} = u^T_{i,\alpha} \tag{A16}$$

$$[\sigma]_{mn} = \sigma_{[mn]} = -\sigma^A_{[mn]} \tag{A17}$$

$$\overset{\sigma}{F}(\underline{\sigma}^T) = 0 \tag{A18}$$

$$\overset{\sigma}{F}(\underline{\sigma}) \equiv \underline{\sigma}^F = -\overset{\sigma}{\underline{F}}(\underline{\sigma}^A) \tag{A19}$$

$$\left.\begin{array}{l} F(u_{i,\hat{j}}) = u_{i,\hat{j}} \\ \\ R(u_{i,\hat{j}}) = 0 \end{array}\right\} \text{ for any } \underline{u}\,. \tag{A20}$$

Therefore,

$$R(\underline{\beta}^T) = 0\,, \tag{A21}$$

$$R(\underline{\beta}) \equiv \underline{\beta}^R = -R(\underline{\beta}^P), \tag{A22}$$

$$\underline{\beta} = \underline{\beta}^F + \underline{\beta}^R = F(-\underline{\sigma}^A) + R(-\underline{\beta}^P)\,, \tag{A23}$$

and

$$\underline{\beta}^T = \underline{\beta}^P + \underline{\beta} = \overset{\sigma}{F}(-\underline{\sigma}^A) + F(\underline{\beta}^P) = \overset{\sigma}{F}(\underline{\underline{C}}\underline{\beta}^P - \underline{\sigma}^A)\,. \tag{A24}$$

Finally,

$$\underline{\sigma} = \underline{\underline{C}}\underline{\beta} - [\underline{\sigma}^A] \text{ and} \tag{A25}$$

$$\underline{\sigma}^T = \underline{\underline{C}}\underline{\beta} + (\underline{\sigma}^A), \tag{A26}$$

with "[ ]" denoting antisymmetrization and "( )" denoting symmetrization, complete the dipole source solution.

We next extend the analyses in Simmons and Bullough to include dynamic effects for inclusions and cracks under prestress. We assume initial elastic fields β^o and σ^o in a homogeneous medium and we imagine that a small volume element Σ undergoes a change in space-time elastic modulus $\underline{\underline{\Delta C}}$ (which includes density) and simultaneously undergoes a stress free strain $\underline{\beta}^P$ to give rise to a new distortion field denoted by $\underline{\beta}^S$. The change in elastic distortion, which we denote by $\underline{\beta}$ is then

$$\underline{\beta}^S - \underline{\beta}^o \equiv \underline{\beta} = - \underline{\beta}^P + \underline{\beta}^T \quad (A27)$$

where

$$\beta^T_{i\hat{j}} = u^T_{i,\hat{j}} \quad (A28)$$

The change in stress, $\underline{\sigma}^T$ is then given by

$$\begin{aligned} \underline{\sigma}^T &= (\underline{\underline{C}} + \underline{\underline{\Delta C}}(\underline{x}))\underline{\beta}^S - \underline{\underline{C}}\underline{\beta}^o \\ &= (\underline{\underline{C}} + \underline{\underline{\Delta C}})\underline{\beta} + \underline{\underline{\Delta C}}\ \underline{\beta}^o. \end{aligned} \quad (A29)$$

$\underline{\sigma}^T$ must, of course, satisfy the dynamic elastic equilibrium equations as does $\underline{\sigma}^o$.

We define

$$\underline{\sigma} \equiv \underline{\underline{C}}\underline{\beta} \quad (A30)$$

and

$$\underline{\sigma}^A \equiv \underline{\underline{\Delta C}}(\underline{x})(\underline{\beta} + \underline{\beta}^o) \quad (A31)$$

so that (A14) is satisfied.

The conditions of the dipole problem have been achieved with the inhomogeneity effects incorporated as an applied dynamic "force dipole" distribution which is here symmetric, hence a "double couple" distribution because of the elastic constants. Assuming no surface sources, the solution is

$$\begin{aligned} \underline{u}^T(\underline{X}') &= \int_{VT} \overset{4}{\nabla}\underline{G}(\underline{X}',\underline{X})(\underline{\underline{C}}\underline{\beta}^P(\underline{X}) - \underline{\sigma}^A(\underline{X}))\ dVT \\ &= \int_{VT} \overset{4}{\nabla}(\overset{4}{\nabla}\underline{G}(\underline{X}',\underline{X})\underline{\underline{\Delta C}}(\underline{X}))\ \underline{u}^T(\underline{X})\ dVT \\ &+ \int_{VT} \overset{4}{\nabla}\underline{G}(\underline{X}',\underline{X})(\underline{\underline{C}} + \underline{\underline{\Delta C}}(\underline{X})\underline{\beta}^P(\underline{X})\ dVT \\ &- \int_{VT} \overset{4}{\nabla}\underline{G}(\underline{X}',\underline{X})\underline{\underline{\Delta C}}(\underline{X})\underline{\beta}^o(\underline{X})\ dVT \end{aligned} \quad (A32)$$

where we have used (A15), (A16), and integration by parts. This equation describes both dynamic plasticity and inhomogeneity effects.

*A similar concept has recently been independently proposed by Willis [53].

Appendix B

Dynamic Heaviside Green's Tensor for an Infinite Isotropic Elastic Body

Consider an infinite body which is elastically isotropic. $\underline{G}^H$ is then a function of $\underline{r}-\underline{r}'$ and t-t'. At the origin at time t = 0 there is a step function of force unloaded in the jth direction. The dynamic Heaviside Green's function gives the displacement in the ith direction at point $\underline{r}$, time t due to that unloading, and is given by [52]:

$$8\pi\rho G^H_{ij}(\underline{r},t) = H(t-r/a)((\delta_{ij}-n_in_j)/a^2r + (t^2/r^3)(3n_in_j - \delta_{ij}))$$
$$+ H(t-r/c)((\delta_{ij}+n_in_j)/c^2r + (t^2/r^3)(\delta_{ij}-3n_in_j)) \quad \text{(B1)}$$

where ρ is the material density, H( ) is the Heaviside step function, a is the longitudinal velocity, c is the shear wave velocity, $r \equiv |\underline{r}|$, δ_{ij} is the Kronecker delta and $n_i = \frac{r_i}{r}$.

An examination of the space-time geometry of the function shows that it is possible to differentiate twice with respect to the t^2/r^3 terms without introducing Dirac delta function terms, but that one must introduce these functions in the derivatives of 1/r. The first spatial derivative of the Heaviside Green's function then is

$$8\pi\rho G^H_{ij,k}(\underline{r},t) = \delta(t-r/a)(-2n_in_jn_k/a^3r) + \delta(t-r/c)((2n_in_jn_k-2\delta_{ij}n_k)/c^3r)$$
$$+ H(t-r/a)[(3n_in_jn_k - \delta_{jk}n_i + \delta_{ik}n_j + \delta_{ij}n_k)/a^2r^2$$
$$+ (3t^2/r^4)(\delta_{jk}n_i + \delta_{ik}n_j + \delta_{ij}n_k - 5n_in_jn_k)] \quad \text{(B2)}$$
$$- H(t-r/c)[(3n_in_jn_k - \delta_{jk}n_i + \delta_{ik}n_j - \delta_{ij}n_k)/c^2r^2$$
$$+ (3t^2/r^4)(\delta_{jk}n_i + \delta_{ik}n_j + \delta_{ij}n_k - 5n_in_jn_k)]$$

In frequency space (B2) takes the form

$$4\pi\rho c^3 r G^H_{ij,k}(\underline{r},\omega) = F_o\Delta_o + F_1\Delta_1 + F_2\Delta_2 \quad \text{(B3)}$$

where $\alpha \equiv c/a$ and

$$\Delta_o \equiv \delta_{ij}n_k - n_in_jn_k$$
$$\Delta_1 \equiv -n_in_jn_k \quad \text{(B4)}$$
$$\Delta_2 \equiv 5n_in_jn_k - (\delta_{jk}n_i + \delta_{ik}n_j + \delta_{ij}n_k)$$

$$F_o \equiv - [(\pi c/r)\delta(\omega) + e^{-i\omega r/c} - (ic/\omega r)e^{-i\omega r/c}]$$

$$F_1 \equiv \alpha^2[(\pi c/r)\delta(\omega) + \alpha e^{-i\omega r/a} - (ic/\omega r)e^{-i\omega r/a}]$$

$$F_2 \equiv (\pi c/2r)(\alpha^2-1)\delta(\omega) + (ic/\omega r)(\alpha^2 e^{-i\omega r/a} - e^{-i\omega r/c}) \quad \text{(B5)}$$
$$+(3c^2/\omega^2r^2)(\alpha e^{-i\omega r/a} - e^{-i\omega r/c}) + (3ic^3/\omega^3r^3)(e^{-i\omega r/c} - e^{-i\omega r/a})$$

From (B2) also follows the second spatial derivatives:

$$
\begin{aligned}
8\pi\rho G^{H}_{ij,k\ell}(\underline{r},t) = {} & \delta'(t-\tfrac{r}{a})\ \left[(2/a^{4}r)(n_i n_j n_k n_\ell)\right] + \\
& \delta'(t-\tfrac{r}{c})\ \left[(2/c^{4}r)(\delta_{ij}n_k n_\ell - n_i n_j n_k n_\ell)\right] \\
& \delta(t-\tfrac{r}{a})\ \left[-2(\delta_{ij}n_k n_\ell + \delta_{jk}n_i n_\ell + \delta_{ik}n_j n_\ell + \delta_{i\ell}n_j n_k \right.\\
& \left. +\delta_{j\ell}n_i n_k + \delta_{k\ell}n_i n_j - 10\, n_i n_j n_k n_\ell)\right] + \\
& \delta(t-\tfrac{r}{c})\ \left[2(4\delta_{ij}n_k n_\ell + \delta_{jk}n_i n_\ell + \delta_{ik}n_j n_\ell + \delta_{i\ell}n_j n_k \right.\\
& \left. + \delta_{j\ell}n_i n_k + \delta_{k\ell}n_i n_j - 10\, n_i n_j n_k n_\ell - \delta_{ij}\delta_{k\ell})\right] \\
& + H(t-r/a)\Big\{-(8\pi/15a^2)\left[\delta_{ik}\delta_{j\ell}+\delta_{jk}\delta_{i\ell}+\delta_{ij}\delta_{k\ell}\right]\delta(\underline{r}) \\
& \qquad -(1/a^2r^3)\left[(\delta_{ik}\delta_{j\ell}+\delta_{jk}\delta_{i\ell}+\delta_{ij}\delta_{k\ell}) \right.\\
& \left. - 3(\delta_{ik}n_j n_\ell+\delta_{jk}n_i n_\ell+\delta_{i\ell}n_j n_k+\delta_{j\ell}n_i n_k+\delta_{k\ell}n_i n_j-\delta_{ij}n_k n_\ell-5n_i n_j n_k n_\ell)\right] \\
& + (3t^2/r^5)\left[(\delta_{ij}\delta_{k\ell}+\delta_{i\ell}\delta_{jk}+\delta_{ik}\delta_{j\ell}) - 5(\delta_{ij}n_k n_\ell+\delta_{jk}n_i n_\ell+\delta_{i\ell}n_j n_k \right.\\
& \left. + \delta_{ik}n_j n_\ell + \delta_{j\ell}n_i n_k+\delta_{k\ell}n_i n_j+35n_i n_j n_k n_\ell)\right]\Big\} \\
& + H(t-r/c)\Big\{(8\pi/15c^2)\left[(\delta_{ik}\delta_{j\ell}+\delta_{jk}\delta_{i\ell}) - 4\,\delta_{ij}\delta_{k\ell}\right]\delta(\underline{r}) \\
& \qquad + (1/c^2r^3)\ (\delta_{ik}\delta_{j\ell}+\delta_{jk}\delta_{i\ell}-\delta_{ij}\delta_{k\ell}) \\
& - 3(\delta_{ik}n_j n_\ell+\delta_{jk}n_i n_\ell+\delta_{i\ell}n_j n_k+\delta_{j\ell}n_i n_k+\delta_{k\ell}n_i n_j-\delta_{ij}n_k n_\ell-5n_i n_j n_k n_\ell) \\
& - (3t^2/r^5)\ \left[(\delta_{ij}\delta_{jk}+\delta_{i\ell}\delta_{jk}+\delta_{ik}\delta_{j\ell})-5(\delta_{ij}n_k n_\ell+\delta_{jk}n_i n_\ell+\delta_{i\ell}n_j n_k \right.\\
& \left. +\delta_{ik}n_j n_\ell+\delta_{j\ell}n_i n_k+\delta_{k\ell}n_i n_j+35n_i n_j n_k n_\ell)\right]\Big\}
\end{aligned}
\tag{B6}
$$

References

1. J. Baram and M. Rosen, Scripta Met., Vol. 13, pp. 565-568, 1979.

2. J. P. Toronchuk, Materials Evaluation (Oct. 1977) pp. 51-53.

3. C. B. Scruby, H. N. G. Wadley, and J. E. Sinclair, "The Relationship Between Acoustic Emission and Fracture Mechanisms," in Proc. of Int. Conf. on Pressure Vessels, Inst. of Mech. Eng., London, 1979.

4. P. McIntyre and G. Green, Brit. J. Nondestructive Testing, (May 1978) pp. 135-140.

5. S. H. Carpenter and F. P. Higgins, Metall. Trans. A, Vol. 8A pp. 1629-1632, 1977.

6. K. E. Stahlkopf and G. Dau, Nuclear Safety, Vol. 17, pp. 33-42, 1976.

7. For a recent bibliography of these methods, see T. F. Drouillard, "Acoustic Emission, A Bibliography with Abstracts," Plenum Press, N.Y., 1979.

8. A. G. Evans and M. Linzer, Ann. Rev. Mater. Sci., Vol. 7, pp. 179-208, 1977.

9. F. R. N. Nabarro, Philos. Mag., Vol. 42, pp. 1224-1231, 1951.

10. T. Mura, Bull. Amer. Phys. Soc., Vol. 6, p. 521, 1961; see also T. Mura., Philos. Mag., Vol. 8, pp. 843-857, 1963.

11. K. Malén and L. Bolin, Phys. Stat. Sol. (b) Vol. 61, pp. 637-645, 1974.

12. L. B. Freund, "The Analysis of Elastodynamic Crack Tip Stress Fields," in Mechanics Today, Vol. 3, ed. by S. Namat-Nasser, Pergamon Press, pp. 55-91, 1976.

13. B. R. Baker, J. Appl. Mech., Vol. 29, pp. 449-458, 1962.

14. K. B. Broberg, Arkiv für Fysik, Vol. 18, pp. 159-192, 1960.

15. J. D. Achenbach and R. Nuismer, Int. J. Frac. Mech., Vol. 7, pp. 77-88, 1971.

16. L. B. Freund, J. Mech. Phys. Solids, Vol. 20, pp. 129-140, 1972.

17. R. Burridge and J. R. Willis, Proc. Camb. Phil. Soc., Vol. 66, pp. 443-468, 1969.

18. A. Niazy, "Elastic Displacement due to a Propagating Crack in an Infinite Medium: An Exact Solution," Proc. International Conf. on Dynamic Crack Propagation, ed. G. Sih, Noordhoff, pp. 501-511, 1972.

19. J. D. Eshelby, J. Mech. Phys. Solids, Vol. 17, pp. 177-199, 1969.

20. B. V. Kostrov, Int. J. Fract., Vol. 11, pp. 47-56, 1975.

21. Y. I. Bolotin, V. A. Greshnikov, Y. B. Drobot, and L. A. Maslov, Measurement Technology, Vol. 17, pp. 64-66 (in translation), Dec. 1974.

22. L. A. Maslov and B. Shchigrin, Sov. Jour. of NDT, Vol. 13, pp. 83-90 (in translation), 1977.

23. L. B. Freund, J. Appl. Mech., Vol. 39, pp. 601-602, 1972.

24. J. D. Achenbach and J. G. Harris, J. Appl. Mech., Vol. 46, pp. 107-112, 1979.

25. K. Aki and P. G. Richards, "Quantitative Seismology," W. H. Freeman, San Francisco, 1980.

26. J. R. Rice, "The Mechanics of Earthquake Rupture," Proc. Int. School of Physics "Enrico Fermi," Italian Physical Society, Course LXXVIII (1979, Varenna on Lake Como, Italy) on Physics of the Earth's Interior, ed. E. Boschi, North-Holland Publ. Co. (publication expected 1980-81).

27. J. D. Eshelby, "The Continuum Theory of Lattice Defects," in Solid State Physics, F. Seitz and D. Turnbull, eds., Vol. 3, pp. 79-144, Academic Press, N.Y., 1956.

28. E. Kröner, Erg. Angew. Math., Vol. 5, pp. 1-179, 1958.

29. N. Haskell, Bull. Seis. Soc. Amer., Vol. 54, pp. 1811-1842, 1964.

30. R. Burridge, and L. Knopoff, Bull. Seis. Soc. Amer., Vol. 54, pp. 1875-1888, 1964.

31. A. T. deHoop, "Representation Theorems for the Displacement in an Elastic Solid and Their Application to Elastodynamic Diffracton Theory," D.Sc. Thesis, Technische Hogeschool, Delft, 1958.

32. K. Aki , Bull. Earthquake Res. Inst., Tokyo Univ., Vol. 44, p. 73, 1966.

33. J. Brune, J. Geophys. Res., Vol. 75, pp. 4997-5009, 1970.

34. M. Randall, Bull. Seis. Soc. Amer., Vol. 61, pp. 1321-1326, 1971.

35. V. Kostrov, Physics of the Solid Earth, pp. 13-21 (in translation), No. 1, 1974.

36. G. E. Backus and M. Mulcahy, Geophys. J. Roy. Astr. Soc., Vol. 46, pp. 341-361, 1976; Vol. 47, pp. 301-329, 1976.

37. B. W. Stump., "Investigation of Seismic Sources by the Linear Inversion of Seismograms," Ph.D. Thesis, Univ. of California, Berkeley, 1979.

38. B. W. Stump and L. R. Johnson, Bull. Seis. Soc. Amer., Vol. 67, pp. 1489-1502, 1977.

39. F. Gilbert, J. Res. Astr. Soc., Vol. 22, pp. 223-226, 1970.

40. R. W. B. Stephens and A. A. Pollock, J. Acoustic. Soc. Amer., Vol. 50, pp. 904-910, 1971.

41. C. A. Tatro, "Design Criteria for Acoustic Emission Experimentation," in Acoustic Emission, STP 505, ASTM, Philadelphia, Pa. (1972).

42. J. A. Simmons and R. B. Clough, "Theoretical Aspects of Acoustic Emission Spectra," Proc. Eighth World Conference on Nondestructive Testing, Cannes, France (Sept. 1976).

43. R. B. Clough and J. A. Simmons, "Reproducible Acoustic Emission Signatures by Indentation in Steels," submitted to Materials Evaluation, ASNT.

44. H. N. G. Wadley, C. B. Scruby, and G. Shrimpton, "Quantitative Acoustic Emission Source Characterization," to be published in Acta Metallurgica.

45. J. R. Willis, Int. J. Eng. Sci., 6, pp. 253-263, 1968.

46. F. R. Breckenridge, C. E. Tschiegg, and M. Greenspan, J. Acoust. Soc. Amer., Vol. 57, pp. 626-629, 1975.

47. J. A. Simmons and R. Bullough, "Internal Stress and the Incompatibility Problem in Infinite Anisotropic Elasticity," in Fundamental Aspects of Dislocation Motion, NBS Spec. Pub. 317, edited by J. A. Simmons, R. deWit, and R. Bullough, Vol. I, pp. 89-124, 1970.

48. N. Kinoshita and T. Mura, Phys. Stat. Sol. (a), Vol. 5, pp. 759-768, 1971.

49. R. B. Clough and J. A. Simmons, "Fundamental Aspects of Acoustic Emission," (First) Int. Conf. on NDE in the Nuclear Industry, Denver, December, 1975.

50. I. M. Gelfand and G. E. Shilov, Generalized Functions, Vol. 1, Academic Press, NY, 1964.

51. G. DeRham, Varíetés Differéntiables, Hermann and Cie, Paris, 1955.

52. T. Mura, "The Continuum Theory of Dislocations," Advances in Materials Research, H. Herman, ed., Vol. 3, pp 1-108, Wiley, NY, 1968.

53. J. R. Willis, "A Polarization Approach to the Scattering of Elastic Waves", to appear in Jour. Mech. Phys. Solids.

DISLOCATION SOURCES

J.C.M. Li

Department of Mechanical and Aerospace Sciences
University of Rochester
Rochester, N.Y. 14627, U.S.A.

Abstract

Some important sources of dislocations are reviewed. These include Frank-Read source; condensation of point defects in quenching, electron irradiation, neutron irradiation, and ion bombardment; prismatic punching of dislocation loops from particles, particle sources activated by external stresses and the concept of a critical misfit for dislocation generation; grain boundary sources from grain boundary dislocations or ledges; surface sources and the effect of a thin hard layer; and dislocation generation from cracks. The interaction between a crack and many dislocations is studied in some detail and the recent observation of a dislocation-free region in front of a crack is interpretted as due to the difficulty of nucleation of dislocations from crack tips. This interpretation and its associated effect on the shape and the extent of the plastic zone in front of a crack should find application also in amorphous metals and polymers.

1. Introduction

Early in the development of dislocation theory, it is recognized (1) that homogeneous nucleation of dislocation loops by external stress is unlikely. As shown by Hirth (2), for a nucleation rate of one loop per cc per sec, the critical stress needed approaches that of theoretical strength or the strength of whiskers. Even in the case of supersaturated vacancies, Hirth and Lothe (3) showed that, for Al quenched from 873°K, the maximum nucleation rate of dislocation loops as predicted by homogeneous nucleation theory is negligible. In fact, recent experimental evidences do indicate heterogeneous nucleation of dislocation loops by the condensation of point defects. Thus dislocations are nucleated heterogeneously by either chemical or mechanical driving forces. This review attempts to identify as much as possible the heterogeneous sites or sources of dislocation nucleation.

2. Frank-Read Source

The first mechanism proposed to understand the origin of the large number of dislocations needed for plastic flow was due to F.C. Frank and W.T. Read (4,5). Their discovery was a milestone in the field of dislocations. Various mechanisms of work hardening, yielding, fracture, creep, etc. start with a postulation of Frank-Read sources.

A few years after their proposal, Dash (6) seemed to be the first to supply an experimental evidence for such a source to operate in Si. Later Ruff and Wilsdorf (7) observed the operation of a single-ended source in a thin film of 302 stainless steel inside an electron microscope. More evidences were reported since then. An example is due to Milevskii and Chuvilin (8) who showed in an Si single crystal doped with Ga and saturated with Au, many coplanar and concentric loops were generated by deformation at 700°C and $16MN/m^2$ shear stress. With the high voltage electron microscope, T. Imura (Metal Physics Department, Nagoya University, Furocho, Chikusaku, Nagoya, Japan) was able to observe directly the operation of Frank-Read sources in Al. In fact, he showed a movie at this conference. Other in situ experiments were reported by Kubin, et al (158).

Since the Frank-Read mechanism needs a dislocation segment to start with, it is really a multiplication mechanism rather than a nucleation mechanism. An important way of producing a Frank-Read segment is by partial cross slip of a screw dislocation. In materials where cross slip is easy, a complete loop may not appear if cross slip of its screw portion takes place before the completion of the loop. In such a case the original Frank-Read mechanism becomes a cross slip or double cross slip mechanism of multiplication proposed by Koehler (9). The cause for cross slip is most likely the interaction with other dislocations (10). A recent x-ray topographic study of bent Cu plates reported by Minari, et al. (143) showed cross slip of screw segments caused by repulsion between half loops. While the Frank-Read multiplication mechanism may be important when the dislocation density is very low, the mechanism involving multiple cross slip seems to be the way in which a shear band loaded with dislocations develops in the material, as demonstrated by Johnston and Gilman (11). See also Gilman (12) for some later references on the experimental results of multiplication studies.

The operation of the Frank-Read source by climb instead of slip is known as the Bardeen-Herring source (13). A recent example is shown by Edelin and Levy (14) in quenched Mg. Instead of cross slip, dislocation multiplication by climb can be achieved by a combined slip and climb process as illustrated by Westmacott et al. (15) and by Embury and Nicholson (16) in quenched Al-Mg alloys, and by Edington and West (17) in quenched Al-Ag alloys. Dislocation climb was observed directly during electron irradiation of Cu-Ni alloys by Barlow and Leffers (18) and during oxidation of zinc by Roessler and Burns (19,20).

Instead of normal dislocations, Frank-Read sources can be operated by partial dislocations or superdislocations. The effect of atomic order on the operation of Frank-Read sources was discussed by Moine, et al. (21). They found that the variation of the flow stress of an Fe-Co-2a/oV alloy with temperature exhibited a maximum near the critical temperature of ordering. This maximum was explained by assuming that, at high values of the long range order parameter, the dislocation sources were operated by paired dislocations while at low values of the parameter they were operated by single dislocations.

Instead of in the lattice, a Frank-Read source was seen by Buzzichelli and Mascanzoni (117) in a grain boundary of a Nb-steel (0.2w/oC, 0.15w/oNb, 1.2w/oMn, 0.3w/oSi) deformed at 250°C. The source was surrounded by loops of grain boundary dislocations.

3. Condensation of Point Defects

As mentioned in the Introduction, nucleation of dislocation loops from supersaturated solutions of point defects is likely to be heterogeneous. The nature of sources is not clear in most instances. However, there are now considerable evidences for heterogeneous nucleation and these evidences are collected below. Some possible sources are suggested.

3.1. Quenching

Nucleation of dislocation loops in Al-base alloys supersaturated with vacancies by quench aging was reviewed by Smallman and Eikum (22) in 1965. Except for existing dislocations which could climb or form helices by absorbing vacancies, the nature of other kind of heterogeneities suitable for vacancy condensation was not clear. However, a particle was frequently observed to be at the center of dislocation loops. Examples were Westmacott, Barnes and Smallman (15) in Al-3.5a/oMg and Al-7.3a/oMg alloys quenched into silicon oil from 550°C and Embury and Nicholson (16) in Al-7w/oMg and Al-9.3w/oMg alloys, quenched and then aged at 160-250°C. That these particles might be due to impurities was suggested by the observation (22) that loops were rarely found in highly purified alloys.

The impurity effect was reported also by Westmacott et al. (23). When pure Al, an Al-1.2a/oMg alloy, and an Al-1.2a/oSi alloy were all heated to 550°C and quenched into 20°C silicon oil, loops less in number but large in size were observed in pure Al, more but smaller in the Al-Mg alloy and even more and yet smaller in the Al-Si alloy. Later DeMelfi and Siegel (24) showed a steady decrease of loop density ($5.3x10^{13}$/cc to $1.7x10^{13}$/cc) and a corresponding increase in loop size (1200 to 1800Å) in Al quenched from 580°C into 40°C distilled water as a function of increasing time (0 to 48 hours) of a pre-quench annealing treatment in air at 600°C. This treatment was known to purify Al (removing H_2?). Similar effects were observed by Kritani, et al. (25) when a high purity Al was zone refined. Shimomura and Yoshida (26) used Al in three purity levels heated to 620°C for 10 hours in different atmospheres (H_2, wet air, CO, and dry air) and quenched into 10°C paraffin oil. For the same atmosphere they found the usual effect of impurities. For the same impurity level, defects were preferentially formed by quenching in H_2 or wet air. As pointed out by Lally and Partridge (27) such atmosphers might increase the concentration of H_2 in the specimens and form bubbles on cooling. These bubbles may be the sites for loop nucleation.

In Au, Segall and Clareborough (28) showed the effect of quenching atmosphere on the nucleation of loops and tetrahedra. Samples were heated at 1030°C for 3 hours in O_2 or CO and quenched into water at 0°C. They were then stored in liquid N_2 and later heated in a bath of some aging temperature (40-200°C). In all cases less but larger loops or tetrahedra were found in specimens annealed in O_2 than in CO. More loops than tetrahedra were found at higher aging temperatures or if annealed in O_2. Later Siegel (29) showed that, with more dissolved impurities, a higher density of tetrahedra was observed in Au quenched from 700-900°C into iced brine and aged at 40-60°C. Similarly Johnston, et al. (142) showed that an addition of 200ppm Mg in Au increased the density of tetrahedra by a factor of 4 ($4x10^{14}$/cc instead of $1.2x10^{14}$/cc, both quenched from 930°C into iced brine, preaged 5 min. at 0°C and aged 1 hour at 100°C) and that an addition of 500ppm Mg caused dislocation loops to nucleate instead of tetrahedra. Prequench annealing for one hour at 950°C in 10% H_2 produced the highest density of defects ($4x10^{15}$/cc). The effect of reducing atmosphere was considered to be caused by hydrogen. Without annealing in H_2, a specimen aged at 55°C in aqua regia instead of O_2 or air produced five times more defects. Among others, heterogeneous nucleation of tetrahedra and dislocation loops was emphasized by Cotterill and Segall (30), Quader and Dodd (31), and Hussein and Dodd (32).

In other metals such as Ag, Cu and Pt, Westmacott (33,34) showed the effect of residual gasses in inhibiting the growth of dislocation loops. When too many nucleating sites were available, small vacancy clusters were formed instead of loops. In Pt, instead of black dots, Westmacott (34) observed square faulted loops in {100} planes with <100> edges when the specimen was degassed in 10^{-4} torr vacuum at 1250°C, quenched in He, and immersed into a salt bath at 430°C. Here again, the suggestion of Lally and Partridge (27) on the possibility of hydrogen bubbles as nucleation sites should be investigated.

3.2 Electron Irradiation

Some evidence of heterogeneous nucleation of dislocation loops caused by electron irradiation was obtained by the observation of saturation of loop density with dose and the effect of impurity on loop density. For example, Bourret (35) irradiated Ni with 1 Mev electrons at a dose rate of $4.2x10^{17}$ e/cm^2/s inside the microscope and observed the formation of Frank (interstitial) loops of (a/3)<111> Burgers vectors. A critical dose for the formation of visible clusters was $3x10^{19}$ e/cm^2. The loop density increased with dose but saturated at about $2x10^{16}$/cc after $1.5x10^{20}$ e/cm^2 for Ni of 6"9" purity. Many more loops of smaller size were seen in Ni of 4"9" purity. Loop growth was observed after density saturation with a 1/3 power law between diameter and time reaching a diameter of about 43Å. Some in situ experiments were also reported by Urban (36) using 650 kev electrons. Bourret (35) also irradiated Fe with 1 Mev electrons at a dose rate of $1.7x10^{18}$ e/cm^2/sec inside the microscope. The loop density reached a limit of about $2x10^{15}$/cc after a dose of about 10^{21} e/cm^2. Loop growth was also observed after density saturations and the diameter reached 170Å.

As another example Shimomura (37) irradiated Au by 3 Mev electrons to doses of $6x10^{18}$ e/cm^2 at 130°K. Interstitial loops of (a/3)<111> Burgers vectors were observed. The critical dose for visible defects was about 10^{18} for 6"9" Au and $3x10^{18}$ for 5"9" Au. While the density of loops saturated at 10^{15}/cc (loop size about 50Å, increased to 70Å upon further irradiation) for the 6"9" Au, it did at about $3x10^{16}$/cc for the 5"9" Au.

More direct evidences for heterogeneous nucleation were supplied by Barlow and Leffers (18,38) who studied in detail the nucleation of dislocation loops in Cu(-20w/o)Ni alloys under 1 Mev electron irradiation. The material was rolled into 0.25mm plates and vacuum annealed for four hours at 800°C. The specimen had a strong texture with {100} parallel to the plate. These specimens after electropolishing (to ~5000Å) were electron irradiated at 250°-600°C under a dose rate of ~10^{-2}dpa/sec or $1.6x10^{24}$ electrons/m^2/sec in a vacuum of $5x10^{-7}$ torr. When a new area was irradiated, there was an incubation period of about 10 sec. (at 400°C) before the appearance of the first dislocation loop. During this period a number of "black dots" was formed. When a black dot developed into a loop, the black dot itself disappeared. Then a new black dot appeared at the same location and later transformed into a loop again. Three types of loops were generated, rectangular loops of a<100> Burgers vector,<011> edges and {100} habit planes perpendicular to the Burgers vector, normal prismatic loops of (a/2)<110> Burgers vector and no habit plane and single- and double-faulted Frank loops with (a/3)<111> Burgers vector and {111} habit planes. All loops were of interstitial type. Most Frank loops appeared to nucleate at random and they became unfaulted when touched by other dislocations. Most prismatic loops and all rectangular loops were nucleated at a source or a black dot. The source density was low in 1w/oNi and increased with Ni content up to 5w/o. Source density and source activity in alloys of 5-20w/oNi were similar. Prismatic loops seemed to nucleate randomly at 250°C. But random emission decreased with increasing temperature and at above 450°C all loops appeared to originate from sources. Above 500°C the source activity began to decline and at 600°C no more loops were nucleated. Preirradiation at a lower temperature (e.g. 250°C) produced a high source density at a higher temperature (e.g. 450°C). The rate of emission from sources increased with increasing temperature up to 500°C and so was the proportion of rectangular loops. In thick foils with high Ni content, continuous emission of many loops produced a dense netwrok reaching 10^{14}/m^2 dislocation density. The sources were considered as platelets of Ni atoms on {100} planes.

If Ni platelets are the sources of interstitial loops in Cu, the situation should be different without Ni. Kenik and Mitchell (39) showed in pure Cu the formation of Frank loops at the beginning of irradiation ($1.4-3.1x10^{-4}$dpa/sec at 200-300°C). These loops grew to a maximum size of 3000Å and gradually became unfaulted and slipped out of the foil. No further loops were generated apparently due to a lack of sources. A similar behavior was reported by Makin in pure Ni (40). However, for Ni preirradiated by $3.1x10^{17}-6.9x10^{18}$neutrons/cm^2 (>1 Mev) at 50°C, Harbottle (41) observed a homogeneous distribution of black dots which grew into a network of interstitial loops upon electron irradiation at 400°C ($1.6x10^{23}$e/m^2/sec or 10^{-3}dpa/sec). Most of these loops were faulted Frank loops lying on {111} planes. No continuous emission of loops was observed. The dislocation density increased to a maximum and then decreased.

In a Cu-1.1w/oBe alloy (18) heat-treated to produce a supersaturated solid solution (no visible precipitate) electron irradiation at 400-450°C caused rectangular loops to appear before precipitates in one to two minutes. In this alloy, {100} GP zones were known to form which could serve as sources for dislocation loop nucleation.

3.3 Neutron Irradiation

The problem of nucleation of dislocation loops was addressed by Makin, et al. (42) who counted the number of defects in various diameter ranges in Cu as a function of dose between $(0.61-3.79)x10^{18}n/cm^2$ and concluded that homogeneous nucleation by chance collision of vacancies (or interstitials) did not fit the experimental results. They proposed a heterogeneous nucleation theory in which the nuclei were formed with the heated spikes produced by knock-on atoms.

Experimental evidences for heterogeneous nucleation were supplied by the effect of impurities studied by the Harwell group in Mo. See a review by Eyre (43) and a series of five papers published in Phil. Mag. by them (Eyre, et al. (44,45) Maher, et al. (46,47) and Eyre and Maher (48)). Two impurity levels were examined, one commercial purity with resistivity ratios of 140-200 and a zone refined material with ratios of 7000-8000. After irradiation at 77°K to three doses in the range $5x10^{17}$ to 10^{20} n/cm^2, the defects were generally finer but more numerous (about a factor of 10 at low doses) in the less pure material. Furthermore, the density of defects seemed to saturate with increasing dose at about 10^{17}/cc in the less pure material. The larger defects were in the form of loops having Burgers vectors in the <110> and <111> directions. About 96% of them were interstitial loops in the less pure material and 87% in the purer material. Materials with three purity levels (resistivity ratios, $R_{293}/R_{4.2}$, of 200, 2000, and 8000) were irradiated at 473°K at a dose of $6x10^{19}$n/cm^2. The density of defects was again very much higher in the impurer material than in the purer ones.

3.4. Ion Bombardment

Dislocation loops were observed by Brandon and Bowden (49) in gold after two hours ion bombardment by 75ev A^+ at a total dose of $3.5x10^{17}$ions/cm^2. Similarly square loops on {100} planes with <100> edges were observed by Masters (50) in iron after irradiation by protons or 150 kev Fe^+ ions. They were interstitial loops. Diepers (51,52) found that loop (interstitial) density saturated in Cu irradiated at low temperatures (<130°K) by 5 kev A^+ ions and then only loop growth was observed, indicating heterogeneous nucleation of these loops. Thomas and Balluffi (53) found also that impurities (by irradiating with electrons or A^+ ions) were required to nucleate interstitial loops in Au when bombarded by 270ev gold ions.

4. Particles and Inclusions

As mentioned earlier, direct observation of dislocation loop nucleation at small particles was reported by Barlow and Leffers (18,38) during electron irradiation in Cu-Ni alloys. In quenched Al-Mg alloys, particles were frequently seen at the center of dislocation loops (15,16). Evidently the misfit between the particle and the hole it occupied caused sufficient strain energy to assist nucleation of dislocation loops. Sometimes the misfit strain energy was large enough to cause dislocation generation without the condensation of point defects. Such situations are reviewed below.

4.1. Prismatic Punching of Loop Rows from Particles

Prismatic punching was suggested by Seitz (85) as a mechanism for relieving punching stresses by the generation of dislocation loops whose Burgers vectors were parallel to the punching directions. Some early observations of prismatic dislocation loops generated from particles or cavities were from J.W. Mitchell's group. For example, Barber, Harvey and Mitchell(54) heated NaCl with $HAuCl_4$ at 650°C for four hours after slight bending observed rows of loops (decorated by Au) which seemed to have emitted from a cavity or a particle. The rows were in <110> directions, sometimes in the form of a helix. The loops were usually of uniform size but sometimes larger for the ones farther away from the cavity. They were probably vacancy loops. Later Bartlett and Mitchell (55) observed loop rows which seemed to have generated from Au particles in AgCl. The rows again were in <110> directions. The Au particles probably grew too fast to have the stress relaxed by vacancy absorption. These loops were very likely interstitial loops. Similar observations were made by Jones and Mitchell (56) and Parasnis and Mitchell (57) when they saw loops of uniform size which seemed to be generated from a Hysil glass sphere in AgCl on cooling. Often the loops were in the form of a figure 8 or other closed helices whose axis-directions were always <110>. The spacings between these loops were analyzed by Bullough and Newman (58) in terms of elastic repulsion between loops and a lattice friction stress.

Barnes and Mazey (59) observed rows of loops generated from precipitate particles (unknown composition) in high purity copper foils (.50μm) heated to 1055°C for five minutes in an argon atmosphere and quenched in water at 21°C. The rows were in <110> directions and the spacings between the loops were consistent with the interactions formulated by Bullough and Newman (58). Based on such formulation, an estimate of lattice friction stress was $2x10^{-4}$ of shear modulus.

A similar lattice friction stress was calculated by Eikum and Thomas (60) in an Al-6.5w/oMg alloy quenched from 550°C into iced brine and aged at 100°C for 100 hours. From the loop spacings, a lattice friction of $3.4x10^{-4}$ of shear modulus was estimated. They also observed loops which seemed to have punched out from a particle at a grain boundary. These loops had the same shape as the particle.

In zone-melted Mo, Lawley and Gaigher (61) also observed loop rows parallel to <111> which seemed to have emitted from a particle probably Mo_2C. From the loop spacings, they estimated using Bullough and Newman's (58) method a lattice friction of 10^{-4} shear modulus (averaged from a few measurements ranging from $7x10^{-5}$ to $1.3x10^{-4}$). Since the lattice friction at room temperature was about $9x10^{-4}$ shear modulus, they deduced that the loops were formed above 900°C during cooling. Later Meakin and Lawley (62) confirmed that the loops were interstitial loops created due to differential thermal expansion between Mo and the particle. Also in Mo, Votava (84) observed loop rows generated from hard particles both during cooling and after stretching.

Carrington, Hale and McLean (63) heated pure iron to 950°C for half hour and air cooled. They found loop rows in <111> directions which seemed to have emitted from particles. Similarly Humphreys and Martin (64) observed loop rows in <101> directions in Cu-1.1w/oCo alloys solution treated at 950°C and aged at 700°C for 150 hours to produce a noncoherent precipitate of nearly pure fcc Co. The loops had the same size as the particle which was about 800Å.

Loop rows in <11$\bar{2}$0> directions were observed by Lally and Partridge (27) in Mg quenched from moist air at 625°C into water. The loops were in the size range of 300 to 5000Å. The nucleation site was considered to be high pressure H_2 bubbles caused by the reduced solubility during quenching. Heating a quenched foil on a hot stage revealed a small hole after all the dislocations were annealed out. Since these loops could be interstitial loops, condensation of vacancies might shrink their sizes while being pushed out from the bubble. Such size distribution was quite apparent in the micrographs.

Wagenblast and Glenn (157) observed interstitial dislocation loops generated from metastable carbide particles in Fe by rapid heating. Fe-C solid solutions were aged at 150-200°C for 4-6 hours to produce large (~0.5μm) metastable carbide, Fe_xC particles. These particles were disks lying on {100} planes. Inserting a sample strip (0.1mm thick) into 450°C helium for eight min. and quenched in water produced 4-6 interstitial loops per particle. These loops had dimensions approximately the same as the particle and lay on {100} planes parallel to the particle. They appeared to have been emitted from both faces of the particle. A change of heating time from two to 165 minutes produced no effect on the loops but the Fe_xC particles changed their appearance to several small globular particles after 10 minutes heating time. For faster heating rate, the specimen was immersed directly into a lead bath. There the dislocations were less numerous and more localized around the particle. Some of the dislocations had (a/2)<111> Burgers vectors. For even faster heating rate, a specimen surface was exposed to a laser beam. A chromel-alumel thermocouple attached to the opposite surface of the specimen showed that the temperature peaked at 600°C within 10 msec, dropped off to 500°C in about 350 msec and to 300°C in 1 sec. No dislocations were produced after this treatment implying that longer time was needed for dislocation generation.

4.2. Particle Sources Activated by External Stresses

So far the misfit between the particle and the hole it occupies is introduced or enhanced by temperature changes. Of course there are other ways of creating misfits. An important one is the external stress. There are numerous observations of dislocation generations from particles by external stresses. Some examples are given by Gilman (65) in LiF, Stein and Low (66) in silicon iron, Leslie (67) in an Fe-0.45w/oMn-0.017w/oC alloy aged 24 hours at 100°C and cold rolled 3%, Hornbogen (68) in an Fe-0.9w/oCu alloy aged 100 hours at 600°C and strained 1%, Damiano and Brown (69) in Zn, Webb and Hayes (70) in ice, Lindroos and Miekk-oja (71) in Al-Mg alloys, Ravi (72) in silicon steam oxidized at 900-1250°C and Gulden and Nix (73) in an Al-4w/oCu-3w/oSi alloy. An example of pressure produced misfit is that of Radcliffe and Warlimont (74) who found dislocation generation by pressurizing Fe-C alloys. The dislocations appeared more frequently at larger or more asymmetric particles. This effect of particle size on dislocation generation was shown also by Young (86) in Cu oxidized at 900°C and furnace cooled. More dislocations were found around larger particles.

An example showing the importance of dislocation sources from particles in affecting mechanical properties was given by Patel and Chandhuri (75). When "dislocation-free" Si was grown with 10^{18} oxygen/cc, it had a yield stress of 93 MN/m^2 at 800°C. However, after the sample was heat treated at 1000°C for 63 hours to form Si-O clusters, the yield stress dropped to 48 MN/m^2 almost one-half of the as-grown value. Further heat treatment of 134 hours reduced the yield stress again to 32 MN/m^2. Since this treatment also reduced the yield drop, the yielding was associated with the dynamics of dislocations. The reduction of yield stress was due to profuse generation of dislocations at particles as indeed observed by these authors using etch pits.

Even without experimental observation of dislocation generation from particles, particle sources are useful concepts in understanding mechanical properties. For example, Edwards, et al. (76) proposed a model for the creep of zinc containing a large volume fraction (0.05-0.3) of small (<1μm) Al_2O_3 or W particles by assuming that dislocations were generated from these particles above a threshold stress taken as the plateau stress in a flow stress-temperature plot. These dislocations were piled up against grain boundaries and the rate controlling step was the climb of these dislocations. The proposed equation was in general agreement with the experimental creep rate-stress relations for the case of 36v/o 0.6μm Al_2O_3. The predicted creep rate was only about a factor of five smaller than the experimental value. However, later McNelley et al. (77) found that, for the case of 28v/o 0.6μm W particles, the predicted creep rate was a factor of 40 to small. Nevertheless, any revised model still should take into consideration the generation of dislocations from particles.

4.3. Critical Misfits Needed for Dislocation Generation

As mentioned before, the misfit between a particle and the hole it occupies can have many causes such as transformation, pressurization, heating and cooling elastic and plastic deformation, etc. When the misfit reaches a certain value, dislocations are generated. This problem was analyzed by Ashby and Johnson (68) For a given misfit between a spherical particle and the spherical hole it occupied, the energy of formation of a dislocation half loop at the particle-matrix interface first increased with the size of the loop, reaching a maximum which represented the activation energy, and then decreased to a minimum before it increased again. The critical misfit needed to reduce the activation energy to zero represented the upper limit for spontaneous dislocation nucleation. Such critical misfit (about 5% difference between the radii of the particle and the hole) was independent of particle size and was the same as the critical stress criterion of Weatherly (69). On the other hand, the misfit which gave a zero minimum energy of formation represented the lower limit since the driving force then is zero. This lower limit decreased with increasing particle size. While quantitatively the model was not very accurate since it involved several simplifying assumptions, all the experimental data on incoherent particles were found to agree fairly well with the lower limit estimates.

For example, Ashby, Gelles and Tanner (80) examined dislocation generation by pressurization (5 to 25 Kbar at room temperature) in Cu-0.1w/oSi, Cu-0.18w/oA and Cu-.08w/oAl alloys (4mm diameter, 0.5mm thick) internally oxidized (by packing the sample disks in a mixture of Cu_2O, Cu, and Al_2O_3 powder at 1000°C for 24 hours) to form spherical amorphous SiO_2 or crystalline polyhedral Al_2O_3. The pressure was the cause for misfit due to the difference in compressibility between the particle and the matrix. They found that at a given pressure, particles larger than a critical size were surrounded by dislocations while the small ones were not. These dislocations formed a plastic zone which extended one or two particle diameters out from the particle. The critical size was found to be smaller for larger pressures consistent with the lower limit predictions of the theory. Since the Al_2O_3 particles were not spherical, they needed a much smaller misfit (or pressure) for dislocation generation.

Generation of dislocations (or loss of coherency) from a coherent particle is more difficult than an incoherent particle. Phillips (81) reported for a Cu-3.36a/oCo alloy aged at 600-700°C, the initial Co rich particles were fully coherent. Since Co has a 2% smaller atom spacing than Cu, the first dislocation may be generated when the particle diameter is 50 Burgers vectors or 130Å. The observed diameter was about 500Å. However, the loss of coherency could be assisted by plastic deformation and then it occurred at a diameter between 180 and 250Å. These are close to the lower limit estimate of Ashby and Johnson (68). For the same alloy, Brown et al. (82) observed loss of coherency by neutron irradiation (a fast neutron dose of $2x10^{17}/cm^2$ plus a thermal neutron dose of $3x10^{17}/cm^2$) when the diameter reached 300Å.

Weatherly and Nicholson (83) studied several semi-coherent precipitates, θ' in Al-Cu, δ in Al-Cu-Mg, β' in Al-Mg-Si, and γ' in Ni-Cr-Ti-Al. While θ' lost coherency by nucleating dislocation loops within the precipitate, others lost coherency by attracting lattice dislocations. For θ', the loss of coherency occurred at particle thicknesses between 250 and 500Å. A balance between the elastic energy of misfit and the interfacial energy of the incoherent particle showed that the critical thickness should be about 550Å, consistent with observations.

5. Grain Boundary Sources

For commercial alloys which are polycrystalline, the most important dislocation sources are probably grain boundaries even though the multiplication mechanism still may be the multiple cross slip inside the grains. Because of its importance, many observations have been reported on it. They are reviewed below:

5.1. Experimental Evidences for Dislocations Emitted from Grain Boundaries

Some early observations were collected in three recent reviews (87-89). By etch pitting Fe-3%Si, Worthington and Smith (90) showed preyielding dislocation activity at grain boundaries at a stress level independent of grain size but much below the yield stress. Similarly Carrington and McLean (91) found that 95% of slip lines contact grain boundaries at one end during early yielding indicating that slip was originated at grain boundaries. Brentnall and Rostoker (92) also observed dislocation activity near grain boundaries and triple junctions during microyielding in silicon iron.

With electron microscopy, Bonfield and Li (93) showed high dislocation density near grain boundaries in Be, Hornbogen (94) showed dislocation loops emitted from grain boundaries in an Fe-3.17a/oP alloy, Orlov (95) observed dislocations generated from grain boundaries at a stress which was only 10% of yield stress in commercial pure iron, Murr (96) found rows of dislocations originated from grain boundaries in a 5% cold rolled 304 stainless steel, and Odegard and Thompson (97) saw dislocation loops attached to grain boundaries after 1000 hours at 295°K and 90% of yield stress in a Ti-6Al-4V alloy.

Partial dislocations were observed to originate from grain boundaries in Ni by Singh and Tangri (98) and Malis, et al. (99) and in stainless steel by Warlimont (100) and Murr (96). Malis and Tangri (101) reported in high purity Cu that dislocation generation from grain boundaries started at about $1x10^{-4}$ tensile plastic strain; both perfect and Shockley partial dislocations were produced in roughly a 2:1 ratio. Some details of dislocation emission from grain boundaries were reported by Mascanzoni and Buzzichelli (102) who showed in pure α-Fe a segment of dislocation in a grain boundary producing a dislocation loop in the lattice. Part of the operation was activated in the electron microscope during observation.

The effect of surfaces on grain boundary sources in Fe-3%Si was studied by Tandon and Tangri (103). It was found that grain boundaries near the surfaces were easier to emit dislocations than those inside. With increasing depth from the surface, both the number of grains showing slip activity and the length of slip lines extending from a grain boundary decreased. As pointed out by Barrett (104), the surface grains were easier to deform because they were less constrained than the interior grains. It follows that grain boundaries near surfaces have a larger capacity for dislocation generation than those inside because it is less likely for the surface boundaries to stop generation due to back stresses caused by the dislocations they emit.

More macroscopic but indirect evidence of grain boundary dislocation sources was shown by Tandon and Tangri (105) in their acoustic emission measurements during deformation of silicon iron in the preyield and yield regions. The emission counts were lower for larger grain sizes in the Lüders band region. Another indirect evidence was supplied by Mastel et al. (106) who observed dislocation channeling after deforming a neutron-irradiated Mo. Two levels of C were used, 10ppm and 150ppm. Both showed channels originated at grain boundaries For the high C alloys, channels were also originated at carbide particles.

Some quantitative measurements on dislocation density were reported by Murr and Hecker (107) who showed in 304 stainless steel that dislocation density near the grain boundaries was markedly higher than that in the interior. Typical numbers at 2.8% strain in tension of a 0.23mm sheet specimen were $10^{11}/cm^2$ at the boundary and $10^9/cm^2$ inside the grain. Another micrograph showed 10^{12} and 10^{10} respectively. However, this ratio of 100 decreased with increasing strain approaching unity at about 8.7% strain. Boundaries of annealing twins were also effective dislocation sources especially at non-coherent steps.

5.2. Mechanisms for Grain Boundary Sources

The mechanism of grain boundary sources was conceived as an adsorbed dislocation (108) stabilized by interacting with the grain boundary structure. The Burgers vector is of course arbitrary, including all lattice dislocations and also partial dislocations. Price and Hirth (109) (see also Hirth (110)) proposed a screw dislocation segment residing in the grain boundary and capable of continuous generation of dislocations. As proposed, grain boundary sliding is a necessary ingredient of the process. The pinning mechanism as well as the origin for such segment were not made clear. However, there is no question about possible "sites" or low energy positions for the adsorbed dislocations (111) in the grain boundary and the stress needed to desorb them certainly could be lower than the yield stress.

These adsorbed dislocations are known also as ledges in the grain boundary. Although a ledge may appear as a step in the boundary if the Burgers vector of the adsorbed dislocation has a component perpendicular to the grain boundary, it is not a stress-free step. Nucleation of dislocations from a stress-free step would be more difficult than that from a step on the free surface since a negative dislocation is introduced simultaneously at the step of the grain boundary. Nucleation of dislocations from steps on the free surface was analyzed by Hirth (2) who showed that it requires about one-third of the theoretical strength to nucleate a dislocation loop at a surface step at room temperature. Thus the mechanism as detailed by Orlov (112) is probably not likely. However, Orlov's general concept of external and internal surfaces as dislocation sources is consistent with observations.

Another mechanism which also requires grain boundary sliding is that of Gleiter, et al. (113) and Bäro, et al. (114). They proposed that grain boundary dislocations whose Burgers vectors may not be those of lattice dislocation could glide in the grain boundary and at grain boundary kinks or bends emit lattice dislocations.

Bäro, et al. (114) observed also dislocations on two slip planes in Ni-13.5a/oAl emitted from the same ledge in a grain boundary. This led them to suggest a mechanism in which a grain boundary dislocation could dissociate into two dislocations each of lattice Burgers vectors and emit them on two slip planes whose intersection was at the grain boundary dislocation. If the grain boundary dislocation had a large Burgers vector, it could emit these two lattice dislocations alternately. No grain boundary sliding was needed for this mechanism to operate.

5.3. The Nature of Grain Boundary Dislocations or Ledges

The fact that a lattice dislocation can be adsorbed onto a grain boundary was shown by Isida and Henderson Brown (115) who crept pure iron, an Fe-0.75%Mn alloy, and an Fe-0.75%Mn-0.01%N alloy, at 500°C and 40-80MN/m^2. The grain boundaries in the annealed and undeformed samples were featureless. After creep, dislocations were seen to join grain boundaries, some partly. Dislocations in the grain boundary could change their contrast with diffraction conditions but the lattice dislocations could not. Burgers vector determinations showed that lattice dislocations retained their Burgers vectors after they joined the grain boundary and there was no evidence of dissociation. They proposed that grain boundary sliding was due to both slip and climb motion of the lattice dislocation in the grain boundary.

Later Isida, et al. (116) studied dislocation images in the grain boundary of an Fe-0.75%Mn alloy. Slip traces were seen attached to grain boundary dislocations. Specimens rolled and annealed at 850°C for 30 min. (0.014mm grain size) were strained 0.3-1% at 300°C. While clear dislocation images in the grain boundaries were seen after about 0.3% strain, the density of images increased so much at about 1% strain that discrimination between individual dislocations was difficult. Dislocation images disappeared after 1% strain leaving serrated diffraction fringes in the boundary. They found also that while there was no change of dislocation images if heated below 300°C, the image started to fade away if heated above 300°C leaving also the serrated diffraction fringes in the boundary.

A direct observation of lattice dislocations joining grain boundaries was reported by Buzzichelli and Masconzoni (117) who made an in situ observation by deforming pure iron in the electron microscope and showed some dislocations already in the grain boundary and others about to join it.

Quantitative measurements of ledge density were reported by Venkatesh and Murr (118) who showed that ledge density in the grain boundary of Ni increased with cold reduction and for the same cold reduction it was higher in samples of larger grain size. Earlier, Murr (96) found more ledges in high angle grain boundaries than in low angle ones. Later Murr and Venkatesh (119) correlated ledge density with grain boundary hardness and matrix hardness. More recently, Malis and Tangri (101) found in high purity Cu that ledge density increased rapidly between 0 and 3×10^{-4} tensile plastic strain and then more gradually. As just mentioned, Isida et al. (116) found in an Fe-Mn alloy that ledge density first increased and then decreased upon plastic straining.

The geometry and crystallography of grain boundary ledges were studied by Murr, Horylev and Lin (120). They found that the ledges were frequently at the intersection of {111} planes of one of the grains. In one figure, they showed a sample of Ni rolled and annealed at 1060°C for 15 min. All the ledges were associated with {111} planes. Most of them were single dislocation ledges. But a range of b to 8b of ledge heights were observed in a coherent twin boundary in 304 stainless steel. Large ledge steps were associated with coherent and incoherent twin boundaries in both Cu-5a/oAl and 304 stainless steel. For the three materials studied, ledges were most abundant in Ni, less in Cu-5a/oAl, and only frequently in 304 stainless steel.

The annealing of grain boundary dislocations (ledges) can be done inside the electron microscope and the temperature at which the image starts to fade away (spreading temperature) can be recorded. Pumphrey and Gleiter (121) did such experiments and found that the spreading temperatures were 140°C for Al-0.9%Mg alloy, 220°C for Ni, 300°C for Ni-5%Al alloy and 500°C for anstenitic stainless steel. Later, they (122) found that the spreading temperature could be lowered by 50-100°C if the grain boundary was at a non-equilibrium state (loaded with vacancies or dislocations or having a small radius of curvature). More recently Lojkowski and Grabski (123) showed the effect of purity on the spreading temperature. While such temperature was 30-40°C for Al containing 1.7%Mg_2Si, it was 0-10°C for .995 pure Al and -100°C to -75°C for 6"9" Al.

5.4. The Hall-Petch Relation and the Grain Boundary Sources

Grain boundary sources were used to understand the Hall-Petch relation (87,108). The number of dislocations generated at the time of yielding from grain boundary sources should be proportional to the grain boundary area and hence the average density of dislocations is inversely proportional to grain size. By using the linear relationship between flow stress and the square root of dislocation density, a Hall-Petch relation results.

The foregoing model assumes that the ledge density in the grain boundary or the grain boundary source density is the same for all grain sizes. Since this may not be ture, Venkatesh and Murr (118) measured the ledge density in cold rolled (10-70%) and annealed (800-1325°K) Ni samples with different grain sizes. They showed Hall-Petch relations for all samples with the same ledge density and found that the Hall-Petch slope increased with the ledge density in the grain boundary.

Bailon, et al. (124) determined the Hall-Petch relation in Armco iron and also the flow stress-dislocation density relationship and concluded that the dislocation density was proportional to the grain boundary area per unit volume. Thus, they confirmed the correlation between grain size effects and grain boundary sources.

6. Surface Sources

The importance of dislocation sources at surfaces can be demonstrated by the observation of Laister and Jenkins (125) on single crystals of gallium arsenide. With appropriate choices of strain rate and temperature, chemically polished specimens can be made to fracture in a brittle manner. But the same specimens with the surfaces lapped with a #600 alumina powder can be deformed plastically at the same strain rate and temperature. Similarly for dislocation-free germanium, Patel and Chandhuri (126) found a 25% reduction in yield stress during tensile deformation when the specimen had a #600 SiC grit finish. However, the surface finish had little effect if the dislocation density in the crystal was already high such as found by Bell and Bonfield (127).

Dislocation generation at surfaces was demonstrated by Gilman (65) in LiF. He found that heterogeneities such as cleavage steps, glide bands, dislocation loops, dirt particles, as-grown precipitates and radiation-induced precipitates were all sources of dislocations. A simple way of introducing dislocations was to scratch the surface and then apply stress to the crystal. As did Gilman, many others used such a method to determine dislocation velocities as affected by stress, temperature, and composition. Examples are Gilman and Johnston (128) in LiF, Stein and Low (66) in solicon iron, Chandhuri, Patel and Lubin (129) in Ge, S. Schäfer (130) in Ge and Patel and Freeland (131) in Silicon. Scratches were also used to study photoplasticity in Ge (132). Nixon et al. (133) bent a Cu-1.05a/oAl single crystal and found that dislocations generated from the surface can propagate several mm into the bulk.

The reasons for easier generation of dislocations at or near surfaces than in the bulk are lower strain energy for dislocations and less extent of internal stress buildup at the source by the dislocations generated. The stress needed to operate a segment source is also smaller at the surface (134) than in the bulk. Some recent efforts using x-ray topography to reveal surface sources were not successful. Examples are Nøst and Nes (135) on Al, Young and Sherrill (136) on Cu, Minari, et al. (137) on Cu, and Lohne and Rustad (138) on Al. However, Gerward made scratches on the surface of Si single crystal slices and found that dislocations were nucleated by quenching above 900°C. A calculation of thermal stresses due to quenching showed that they were below the yield stress at 1000°C. Hence, the dislocations were generated from the stresses of the scratch. This notion was confirmed by the fact that in repeated quenchings from 1000 and 1200°C, no new dislocation loops were generated. The dislocations only moved a little further after the latter quenches. Nittano (140) also used x-ray topography on Cu whickers and found that concaves on crystal surfaces were effective sites for dislocation generation.

Kramer and Balasubramanian (141) found in Fe-3%Si polycrystals after deformation the dislocation density was the highest at the surface ($10^8/cm^2$ at 1.5% strain) and decreased with increasing depth from the surface reaching a constant value after about 100μm (grain size 85μm). This gradient of dislocation density decreased with increasing strain (A ratio of about 30 at 1% strain and 5 at 1.5% strain). However, this phenomenon may not be the result of surface sources since Tandon and Tangri (103) found that grain boundaries near the surfaces were easier to emit dislocations than those inside the specimen.

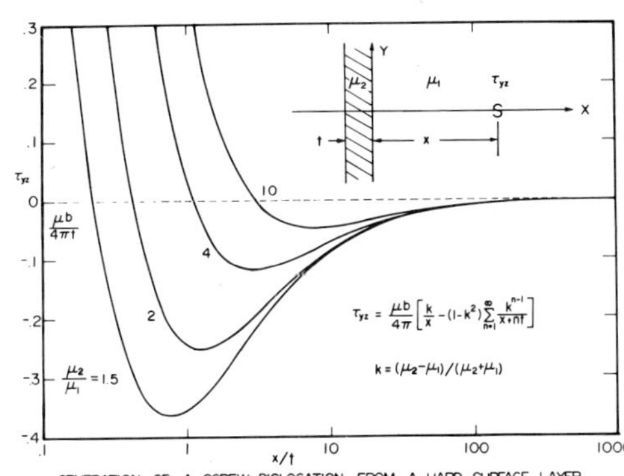

Fig. 1

The effect of a thin hard layer over the surface on the dislocation interaction was shown by Head (144). Because of the attraction due to the free surface combined with the repulsion due to the hard layer, stable positions are created for dislocations near the surface. These are shown in Fig. 1 where the stress or the force on a screw dislocation is plotted as a function of distance. As in the case of grain boundaries, these stable positions enable the surface to adsorb dislocations which can act later as dislocation sources.

7. Dislocation Generation from Cracks

A special surface source of dislocations is the crack tip. The difference between brittle and ductile fracture depends on the extent of plastic deformation associated with crack propagation. Experiments on plastic zones are numerous but direct observations of dislocation emissions from crack tips are rare. Some recent examples are Kobayashi and Ohr (145) who did in-situ shear cracking (Mode III) experiments on thin foils of Mo and W, Ohr and Narayan (146) on similar experiments in stainless steel, and Majumdar and Burns (147) who showed dislocations at crack tips in bulk MgO revealed by TEM after thinning. The dislocation distribution near the crack tip revealed a region void of dislocations next to the crack. The attraction between the crack and its emitted dislocations was suspected to be the cause (147).

Theoretical analysis of crack-dislocation interactions was reported by Louat (148) who used a dislocation model for a finite, double-ended two-dimensional crack interacting with a parallel screw dislocation outside the crack in an isotropic medium. The same problem for the semi-infinite crack was solved by Asaro (149) and more recently by Majumdar and Burns (150) by conformal mapping. Also in isotropic medium, Rice and Thomson (151) formulated the interaction between a dislocation of general Burgers vector and a crack by a thermodynamic technique. In anisotropic media, Atkinson (152) and later Barnett and Arsaro (153) presented procedures for solving the elastic problem involving a crack and a dislocation. Using this method, Asaro (149) formulated a general image force theorem for a single dislocation located near the tip of a semi-infinite crack. However, this theorem cannot be applied to the case of many dislocations since a second dislocation changes the interaction between the first dislocation and the crack.

In the following, the interaction between a semi-infinite crack and many dislocations of arbitrary Burgers vector will be formulated to determine the effect of dislocation nucleation at the crack tip on the dislocation distribution in front of it. A dislocation model for the crack will be used following the method of Louat (148). For simplicity, an isotropic medium will be assumed.

7.1. Dislocation Model of a Crack

As is well known, the elastic properties of a crack can be satisfactorily described by a continuous distribution of infinitesimal dislocations. This is possible since dislocations can produce stresses on the crack plane to cancel the applied stresses so that a stress-free surface is created in an infinite medium. For example, a semi-infinite crack on the xz plane between $x=-\ell$ and $x=0$ can be modelled by a pileup of dislocations parallel to $-z$. The Burgers vector of these dislocations may orient in x,y or z directions depending on the mode of the crack. Although the Burgers vector is defined by a RHFS (right-handed Burgers circuit and finish-to-start closure vector) convention, no crystal structure is really needed. The dislocastions can be the Volterra types. Let $f(\alpha)d\alpha$ be the total Burgers vector of dislocations between $x=\alpha$ and $x=\alpha+d\alpha$, the stress produced at $x=\alpha'$ in the xz plane by these dislocations is given by $Af(\alpha)d\alpha/(\alpha'-\alpha)$ where A is given in Table 1 where μ is shear modulus and ν is Poisson's ratio. The condition for a stress-free surface is then

Mode	Applied Stress, σ	A	Direction of Burgers Vector
I	σ_{yy}	$\mu/2\pi(1-\nu)$	y
II	σ_{xy}	$\mu/2\pi(1-\nu)$	x
III	σ_{yz}	$\mu/2\pi$	z

Table 1.

$$A \int_{-\ell}^{0} \frac{f(\alpha)d\alpha}{\alpha-\alpha'} = \sigma \tag{1}$$

which should be satisfied for any α' between $-\ell$ and 0.

The solution for this equation in the case of constant σ is given by Leifried (154) and Head and Louat (155) as follows

$$f(\alpha) = \frac{\sigma}{\pi A}\sqrt{\frac{\ell+\alpha}{-\alpha}} \tag{2}$$

which is plotted in Fig. 2. The unit in the plot is $2\sigma(1-\nu)/\mu$ for Mode I (Burgers vector in the y direction) or Mode II (Burgers vector in the x direction) cracks and is $2\sigma/\mu$ for a Mode III crack. The dislocations are of the edge type in the first two modes and the screw type in the third mode. To differentiate between these cases, $f_x(\alpha)$ can be used for σ_{xy}, $f_y(\alpha)$ for σ_{yy} and $f_z(\alpha)$ for σ_{yz}. For a mixed-mode crack, dislocations of all three Burgers vectors may be needed, each type is proportional to the corresponding stress as given by Eq. (2). When all the applied stress components are constant along the crack plane, the distribution function is the same for each type and can be combined to be considered as a distribution of mixed dislocations. It is noted that applied stress components σ_{xx}, σ_{zz}, and σ_{xz} do not produce stresses on the crack surface and therefore will not affect the distribution of dislocations.

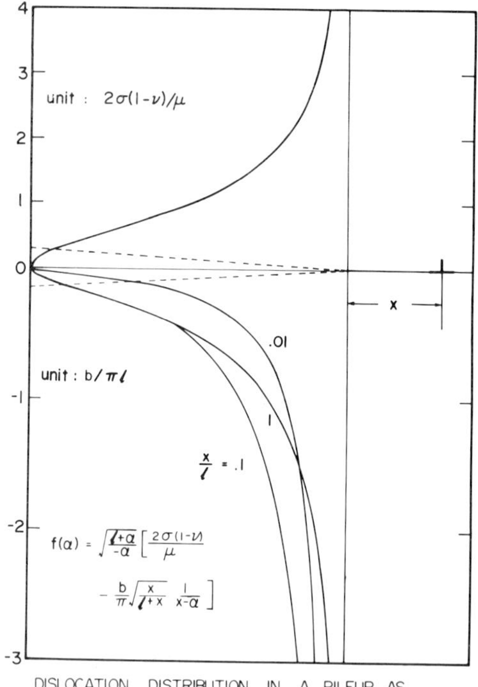

DISLOCATION DISTRIBUTION IN A PILEUP AS AFFECTED BY ANOTHER DISLOCATION

Fig. 2

The crack opening displacement at $x=\alpha$ is given by

$$u(\alpha) = \int_{\alpha}^{0} f(\alpha)d\alpha = \frac{\sigma}{\pi A}\left(\sqrt{-\alpha(\ell+\alpha)} + \ell \sin^{-1}\sqrt{-\alpha/\ell}\right). \tag{3}$$

Here again, the displacement is in the y direction for Mode I, x direction for

Mode II and z direction for Mode III. The displacement is in the direction of Burgers vector for the mixed mode. Eq. (3) shows that the displacement is zero at the crack tip (x=0) and is a maximum at $\alpha=-\ell$ with a maximum value of $\sigma\ell/2A$. Hence the problem is exact for a dislocation pileup in an infinite medium or for a crack whose opening displacement is constant between $x=-\ell$ and $x=-\infty$. It is only approximate for a semi-infinite crack with a free surface at $x=-\ell$. However, the situation near the crack tip is sufficiently accurate for the present purpose.

For example, the stress (σ_{yy} for Mode I, σ_{xy} for Mode II and σ_{yz} for Mode III) at x=x in front of the crack is given by

$$A\int_{-\ell}^{0} \frac{f(\alpha)\,d\alpha}{x-\alpha} + \sigma = \sigma\sqrt{\frac{\ell+x}{x}} \tag{4}$$

or a stress intensity factor of $\sigma\sqrt{2\pi\ell}$ caused by the applied stress. The interaction of the crack with any other defects of short range stresses such as small particles, dislocation loops or dilute solute distributions can be studied through the stress field of the dislocation model. However, for defects of long range stresses such as a dislocation which produces stresses on the crack surface, the dislocation distribution modelling the crack has to be modified.

7.2. Dislocation Crack Interactions

The additional distribution which is needed to maintain the stress-free surface of the crack due to an outside dislocation such as the one in front of the crack at x=x in Fig. 2 can be obtained from the following condition:

$$\int_{-\ell}^{0} \frac{f(\alpha)\,d\alpha}{\alpha'-\alpha} = \frac{b}{x-\alpha'} \tag{5}$$

where the Burgers vector of the dislocation modelling the crack is in the same direction as the Burgers vector of the dislocation outside the crack ($f_x(\alpha)$ for b_x, $f_y(\alpha)$ for b_y and $f_z(\alpha)$ for b_z) but of opposite sign. The solution for Eq. (5) is

$$f(\alpha) = -\frac{b}{\pi(x-\alpha)}\left(\sqrt{\frac{x(\ell+\alpha)}{-\alpha(\ell+x)}}\right) \tag{6}$$

which is plotted in Fig. 2 with a unit of $b/\pi\ell$ for $x/\ell=0.01$, 0.1 and 1.0. Such a distribution can be considered as image dislocations for the outside dislocation, although the total Burgers vector is

$$\int_{-\ell}^{0} f(\alpha)\,d\alpha = b\left(\sqrt{\frac{x}{\ell+x}} - 1\right) \tag{7}$$

which is less than b in magnitude and approaches b only when x=0. Nevertheless, the distribution of Eq. (6) represents a model for the crack in the absence of an applied stress but in the presence of a dislocation. With the applied stress, the model of the crack is described by both distributions of Eqs. (2) and (6). Note that the Burgers vector used may be different for the two distributions.

Thus, without the applied stress the outisde dislocation is attracted to the crack by its image dislocations. The attractive force is given by

$$Ab\int_{-\ell}^{0} \frac{f(\alpha)\,d\alpha}{x-\alpha} = -\frac{Ab^2\ell}{2x(x+\ell)} \tag{8}$$

where $f(\alpha)$ is given by Eq. (6) and A is $\mu/2\pi(1-\nu)$ for $b=b_x$ or b_y and $\mu/2\pi$ for $b=b_z$. It is interesting to note that for $x\ll\ell$ this attraction is the same as that due to a single parallel image dislocation at a distance 2x away. This result agrees with those already published (147,149,151).

With the applied stress, additional interaction due to the crack stress field as given by Eq. (4) takes place also. For example, for the case of a

Mode II or Mode III crack, the force exerted on an edge (Mode II) or screw (Mode III) dislocation emitted from the crack tip is, in terms of τ (τ_{xy} for edge and τ_{yz} for screw):

$$\tau = \sigma\sqrt{\frac{\ell+x}{x}} - \frac{Ab}{2x} \tag{9}$$

This relation is plotted in Fig. 3 for the case of Mode III crack and a small x/ℓ ratio. As pointed out by Majumdar and Burns (147), the dislocation is attracted toward the crack tip at small distances and gets repelled only after a critical distance ($x \ll \ell$):

$$x_0 = A^2b^2/4\sigma^2\ell \tag{10}$$

at which $\tau=0$. This critical distance represents also an unstable equilibrium position and is smaller for larger σ. Thus contrary to usual expectation, the stress singularity at the tip of the crack cannot generate a dislocation at any σ. The dislocation needs to be created by nucleation to appear at or beyond x_0. The formation energy of the dislocation can be obtained by making a cut in the xz plane between the crack and the dislocation and reversibly removing the dislocation by displacing the two cut surfaces relative to each other by a distance b in the direction of the Burgers vector. The stress involved for the reversible work is ($f(\alpha)$ is given by Eq. (6)):

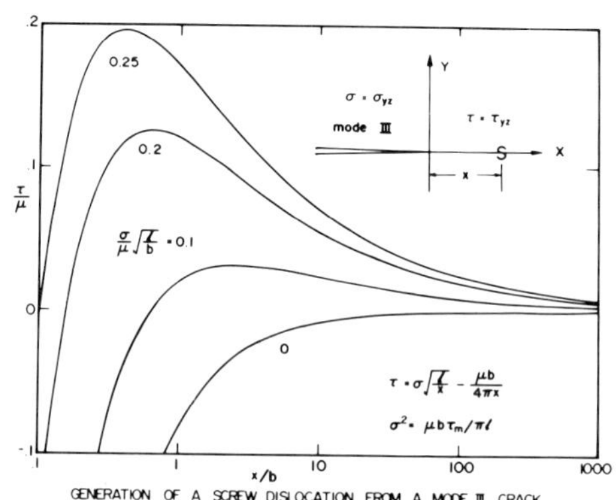

Fig. 3.

$$\tau(x') = \sigma\sqrt{\frac{\ell}{x'}} + A\int_{-\ell}^{0}\frac{f(\alpha)d\alpha}{x'-\alpha} + \frac{Ab}{x'-x} = \sigma\sqrt{\frac{\ell}{x'}} + \frac{Ab\sqrt{x}}{\sqrt{x'}(x'-x)} \tag{11}$$

so that the formation energy is

$$E = -\int_0^{x-r_0}\int_0^b \tau(x')dbdx' = \frac{Ab^2}{2}\ln\frac{4x}{r_0} - 2\sigma b\sqrt{\ell x} \tag{12}$$

per unit length with r_0 being the core radius of the dislocation. It has a maximum value at $x=x_0$ and is equal to

$$E_m = \frac{Ab^2}{2}\left(\ln\frac{4x_0}{r_0} - 2\right) \tag{13}$$

per unit length. This maximum energy is that required for nucleation and is obviously very large for small σ but decreases for increasing σ.

With lattice friction there are in general two positions of the same τ for each value of the applied stress as given by Eq.(9) and shown in Fig. 3. When a dislocation appears anywhere between the two positions, the applied stress can pull it to the farthest one from the crack. The minimum σ needed is ($x \ll \ell$)

$$\sigma^2 = 2\tau_m Ab/\ell \tag{14}$$

where τ_m is the maximum τ for a given σ. The maximum τ is at $x_m=4x_0$:

$$x_m = A^2b^2/\sigma^2\ell. \tag{15}$$

Thus with lattice friction the nucleation of dislocation is more difficult since it has to reach $4x_0$ to overcome both the attraction of the crack and the lattice friction.

The situation is similar for the generation of an edge dislocation from the tip of a Mode I crack as shown in Fig. 4. The shear stress is the largest in the plane oriented at $\theta=70.5°$ as shown. The image force is given by Rice and Thomson (151) and by Asaro (149) and is the same as that due to a negative dislocation at 2r away.

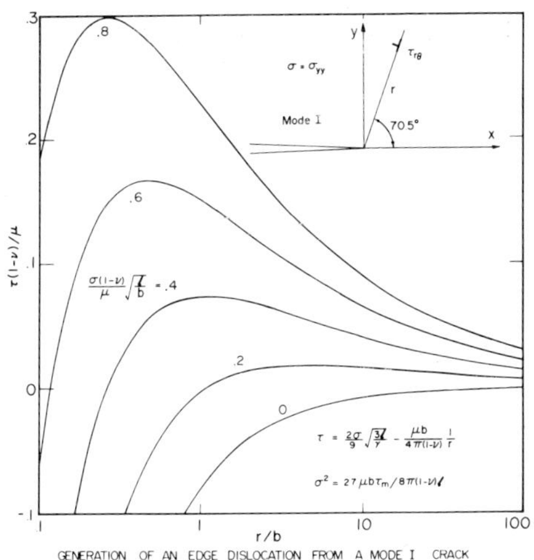

Fig. 4

7.3. Generation of Two Dislocations From a Crack

The dislocation model for the crack in the presence of two dislocations will contain three terms, one given by Eq. (2) and two given by Eq. (6). Hence the interaction between one of the dislocations and the crack contains also three terms. Let the dislocations be located at x_1 and x_2 as shown in Fig. 5. They are edge dislocations from a Mode II crack or screw dislocations from a Mode III crack. The force exerted on dislocation 1 in terms of the shear stress (τ_{xy} for edge and τ_{yz} for screw) is

$$\tau_1 = \sigma\sqrt{\frac{\ell}{x_1}} - Ab\left(\frac{1}{2x_1} + \frac{1}{x_1+\sqrt{x_1x_2}} + \frac{1}{x_2-x_1}\right) \qquad (16)$$

for large ℓ. The first term is due to the applied stress σ (σ_{xy} for Mode II and σ_{yz} for Mode III): the second term (or the first term in the parentheses) is its own image force; the third term is due to the image distribution of the second dislocation; and the fourth term is due to the stress field of the second dislocation. The stress at the second dislocation is the same as Eq.(16) except that the subscripts 1 and 2 are interchanged.

Let $\tau_1=\tau_2=\tau$ be the frictional stress. Then the two equations can be combined to give

$$\tau = Ab\left(\frac{1}{2\sqrt{x_1x_2}} + \frac{1}{x_1+x_2-2\sqrt{x_1x_2}}\right). \qquad (17)$$

Thus for a given τ and x_1x_2, x_1+x_2 and hence both x_1 and x_2 can be calculated from this equation. These values can be substituted into Eq. (16) to obtain σ. The results are shown in Fig. 5. As in the case of a single dislocation emitted from the crack, there are two configurations in general for the same σ and τ. The applied stress is sufficient to pull the dislocations from any configuration between the two to the farthest one from the crack. The minimum stress required to do this is shown also for comparison. It is seen that, for the same frictional stress, the external stress required is about 40% more than that for a single dislocation.

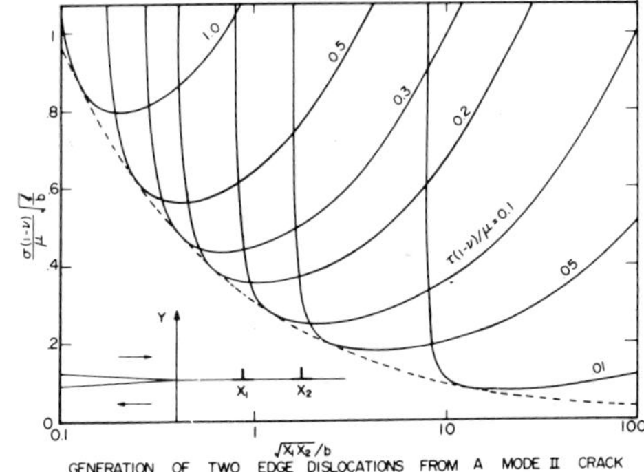

Fig. 5.

To obtain analytically the stress required for the case of two dislocations the two equations of the type of Eq. (16) (τ_1 and τ_2) can be added to give:

$$2\tau = \sigma\sqrt{\ell}p - \frac{Ab}{2} p^2 \tag{18}$$

where

$$p = \frac{1}{\sqrt{x_1}} + \frac{1}{\sqrt{x_2}} \tag{19}$$

The value of p for which τ is a maximum at constant σ or σ is a minimum at constant τ is given by:

$$1/p_m^2 = A^2b^2/\sigma^2\ell . \tag{20}$$

Thus the σ needed is

$$\sigma^2 = 4\tau_m Ab/\ell \tag{21}$$

which can be compared with Eq. (14). It shows that the σ needed is $\sqrt{2}$ times that for the case of a single dislocation in agreemwnt with the numerical result of Fig. 6 just discussed.

The positions of the two dislocations, x_1 and x_2, at the situation of minimum σ or maximum τ can be obtained by expressing p in terms of τ_m and substituting into Eq. (17) to obtain a relation between x_1 and x_2 (no σ or τ). Express x_1+x_2 and hence x_1 and x_2 in terms of x_1x_2 and then express σ and τ_m also in terms of x_1x_2. It is found that the farthest position, x_2 is

$$x_m = 15.13\ A^2b^2/\sigma^2\ell \tag{22}$$

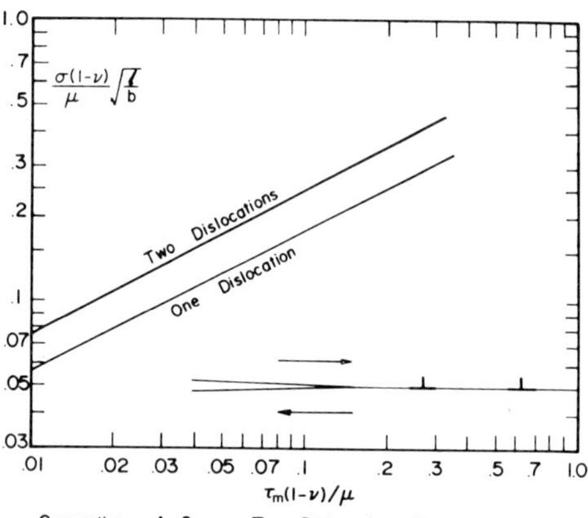

Generation of One or Two Dislocations from a Mode II or III Crack

Fig. 6

which is about 15 times larger than that given by Eq. (15). The other dislocation is at x_1=0.1197 x_2. Burns/Majumdar have similar results to this problem.

7.4. Generation of Many Dislocations From a Crack

The foregoing discussion of the case of two dislocations can be generalized. Let the dislocations be located in the xz plane at $x_1, x_2, \ldots, x_n$ in front of the crack. Then the stress at the ith dislocation is ($x \ll \ell$).

$$\tau_i = \sigma\sqrt{\frac{\ell}{x_i}} - Ab\left[\frac{1}{2x_i} + \sum_{j\neq i}\left(\frac{1}{x_i+\sqrt{x_ix_j}} + \frac{1}{x_j-x_i}\right)\right]. \tag{23}$$

As before, let $\tau_i = \tau$ for all i and add up all the equations. The result is:

$$n\tau = \sigma\sqrt{\ell}p - \frac{Ab}{2}p^2 \tag{24}$$

where

$$p = \sum_i \frac{1}{\sqrt{x_i}} \tag{25}$$

which has a critical value for maximum τ or minimum σ given by Eq. (20). The minimum σ required is thus

$$\sigma^2 = 2n\tau_m Ab/\ell \tag{26}$$

which shows that for a given σ and τ_m, a crack has a limited capacity of generating dislocations, namely n of them as given by Eq. (26). It turns out that Eq. (26) agrees with the limiting form for large lattice friction of the results of Bilby, Cottrell and Swinden (156).

However, the situation of minimum σ or maximum τ for n dislocations indicates that a further increase in σ or a decrease in τ will cause another dislocation to be emitted. In other words the analysis assumes no nucleation barrier for dislocation generation at the crack tip or the stress intensity factor is zero. The same assumption was used in the analysis of Bilby, Cottrell and Swinden (156). In fact, numerical results for large n in the situation of minimum σ or maximum τ approaches their results in the limit of a semi-infinite crack. As shown by Bilby, et al. (156) the dislocation distribution in front of the crack increases without limit near the crack tip. In the numerical results using discrete dislocations, even though there is a space between the crack tip and the first dislocation, it is very small and is certainly not of the magnitude observed by Ohr, et al. relative to the other dislocations.

It seems obvious that a dislocation-free zone near the tip of the crack has its origin at the nucleation barrier for dislocation generation at the tip of the crack. By assuming a critical stress intensity factor needed for dislocation emission, the minimum σ for maintaining n dislocations before the appearance of the (n+1)th one is larger than the case without the nucleation barrier. Furthermore, a dislocation-free zone appears whose size increases with the difficulty (critical stress intensity factor) for dislocation emission at the tip of the crack. For a blunted crack, see (160).

ACKNOWLEDGEMENTS

Financial supports from NSF through contract DMR78-12807, DOE through contract DE-AC02-76ER02296.A003, and ARO through contract DAAG29-80-C-0109 are gratefully acknowledged. Fruitful discussions with S.J. Burns, B.S. Majumdar, S.M. Ohr and S.L. Chang on crack dislocation interactions are appreciated.

REFERENCES

1. F.C. Frank, Symposium on Plastic Deformation of Crystalline Solids (Carnegie Inst. of Tech. and Office of Naval Research, Pittsburgh, 1950) p. 89.
2. J.P. Hirth, The Relation Between the Structure and Mechanical Properties of Metals (H.M. Stationary Office, London, 1963) pp. 218-28.
3. J.P. Hirth and J. Lothe, Theory of Dislocations (McGraw-Hill 1968) p. 559.
4. F.C. Frank and W.T. Read, Symposium on Plastic Deformation of Crystalline Solids (Carnegie Inst. of Tech. and Office of Naval Research, Pittsburgh, 1950) p. 44.
5. F.C. Frank and W.T. Read, Phys. Rev. 79, 722-3 (1950).
6. W.C. Dash, J. Appl. Phys. 27, 1193-5 (1956).
7. A.W. Ruff, Jr. and H.G.F. Wilsdorf, Acta Met. 13, 694-5 (1965).
8. L.S. Milevskii and Yu.N. Chuvilin, Sov. Phys. Solid State 20(11), 1883-6 (1978).
9. J.S. Koehler, Phys. Rev. 86, 52-9 (1952).
10. J.C.M. Li, J. Appl. Phys. 32, 593-9 (1961).
11. W.G. Johnston and J.J. Gilman, J. Appl. Phys. 30, 129-44 (1959).
12. J.J. Gilman, Micromechanics of Flow in Solids (McGraw-Hill, 1969) p. 190.
13. J. Bardeen and C. Herring, Imperfections in Nearly Perfect Crystals (John Wiley & Sons,1952) p. 261.
14. G. Edelin and V. Levy, Phil. Mag. 27, 487-97 (1973).
15. K.H. Westmacott, R.S. Barnes and R.E. Smallman, Phil. Mag. 7, 1585-96 (1962).
16. J.D. Embury and R.B. Nicholson, Acta Met. 11, 347-54 (1963).
17. J.W. Edington and D.R. West, J. Appl. Phys. 37, 3904-5 (1966).
18. P. Barlow and T. Leffers, Phil. Mag. 36, 565-583 (1977).
19. S.J. Burns and B. Roessler, Phys. Stat. Sol. (a) 13, K91-2 (1972).
20. B. Roessler and S.J. Burns, Phys. Stat. Sol. (a) 24, 285-98 (1974).
21. P. Moine, J.P. Eymery and P. Grosbras, Phys. Stat. Sol. (b) 46, 177-185 (1971).
22. R.E. Smallman and A. Eikum, "Lattice Defects in Quenched Metals" (Ed. by R.M.J. Cotterill et al. Academic Press 1965) pp. 591-637.

23. K.H. Westmacott, R.S. Barnes, D. Hull and R.E. Smallman, Phil. Mag. 6, 929-35 (1961).
24. R.J. DiMelfi and R.W. Siegel, Phil. Mag. 24, 279-94 (1971).
25. M. Kritani, T. Nishikawa and Sh. Yoshida, J. Phys. Soc. Japan 27, 67-73 (1969).
26. Y. Shimomura and Sho. Yoshida, J. Phys. Soc. Japan 22, 319-31 (1967).
27. J.S. Lally and P.G. Partridge, Phil. Mag. 13, 9-30 (1966).
28. R.L. Segall and L.M. Clareborough, Phil. Mag. 9, 865-77 (1964).
29. R.W. Seigel, Phil. Mag. 13, 337-58 (1966).
30. R.M.J. Cotterill and R.L. Segall, Phil. Mag. 8, 1105-25 (1963).
31. M.A. Quader and R.A. Dodd, Phil. Mag. 17, 575-93 (1968).
32. A.A. Hussein and R.A. Dodd, Phil. Mag. 24, 1441-58 (1971).
33. K.H. Westmacott, Phil. Mag. 24, 475-80 (1971).
34. K.H. Westmacott, J. Phys. F, 3, L217-9 (1973).
35. A. Bourret, Phys. Stat. Sol. (a) 4, 813-25 (1971).
36. K. Urban, Phys. Stat. Sol. (a) 4, 761-72 (1971).
37. Y. Shimomura, Phil. Mag. 19, 773-94 (1969).
38. T. Leffers and P. Barlow, Phil. Mag. 32, 491-6 (1975).
39. E. Kenik and T.E. Mitchell, Radia. Effects 24, 155-160 (1975).
40. M.J. Makin, Phil. Mag. 25, 761-7 (1972).
41. J.E. Harbottle, Phil. Mag. 27, 147-57 (1973).
42. M.J. Makin, A.J. Whapham and F.J. Minter, Phil. Mag. 7, 285-299 (1962).
43. B.L. Eyre, J. Phys. F. 3, 422-70 (1973).
44. B.L. Eyre, D.M. Maher and A.F. Bartlett, Phil. Mag. 23, 409-38 (1971).
45. B.L. Eyre, D.M. Maher and A.F. Bartlett, Phil. Mag. 23, 439-65 (1971).
46. D.M. Maher, M.H. Loretto and A.F. Bartlett, Phil. Mag. 24, 181 (1971).
47. D.M. Maher, B.L. Eyre and A.F. Bartlett, Phil. Mag. 24, 745-65 (1971).
48. B.L. Eyre and D.M. Maher, Phil. Mag. 24, 767-97 (1971).
49. D.G. Brandon and P. Bowden, Phil. Mag. 6, 707-10 (1961).
50. B.C. Masters, Phil. Mag. 11, 881-93 (1965).
51. H. Diepers, Phys. Stat. Sol. 24, 235-44 (1967).
52. H. Diepers, Phys. Stat. Sol. 24, 623-33 (1967).
53. L.E. Thomas and R.W. Balluffi, Phil. Mag. 15, 1117-35 (1967).
54. D.J. Barber, K.B. Harvey and J.W. Mitchell, Phil. Mag. 2, 704-8 (1957).
55. J.T. Bartlett and J.W. Mitchell, Phil. Mag. 3, 334-41 (1958).
56. D.A. Jones and J.W. Mitchell, Phil. Mag. 3, 1-7 (1958).
57. A.S. Parasnis and J.W. Mitchell, Phil. Mag. 4, 171-9 (1959).
58. R. Bullough and R.C. Newman, Phil. Mag. 5, 921-6 (1960).
59. R.S. Barnes and D.J. Mazey, Acta Met. 11, 281-6 (1963).
60. A. Eikum and G. Thomas, Acta Met. 12, 537-45 (1964).
61. A. Lawley and H.L. Gaigher, Phil. Mag. 8, 1713-23 (1963).
62. J.D. Meakin and A. Lawley, Phil. Mag. 10, 737-43 (1964).
63. W. Carrington, K.F. Hale and D. McLean, Proc. Roy. Soc. A259, 203-227 (1960).
64. F.J. Humphreys and J.W. Martin, Phil. Mag. 16, 927-57 (1967).
65. J.J. Gilman, J. Appl. Phys. 30, 1584-94 (1959).
66. D.F. Stein and J.R. Low, J. Appl. Phys. 31, 362-9 (1960).
67. W.C. Leslie, Acta Met. 9, 1004-22 (1961).
68. E. Hornbogen, Trans. ASM 57, 120-32 (1964).
69. V.V. Damiano and N. Brown, Trans. AIME 230, 926-8 (1964).
70. W.W. Webb and C.E. Hayes, Phil. Mag. 16, 909-25 (1967).
71. V.K. Lindroos and H.M. Miekk-oja, Phil. Mag. 16, 593-610 (1967).
72. K.V. Ravi, Met. Trans. 4, 681-9 (1973).
73. M.E. Gulden and W.D. Nix, Phil. Mag. 18, 217-28 (1968).
74. S.V. Radcliffe and H. Warlimont, Phys. Stat. Sol. 7, K67-9 (1964).
75. J.R. Patel and A.R. Chandhuri, J. Appl. Phys. 33, 2223-4 (1962).
76. G.R. Edwards, T.R. McNelley and O.D. Sherby, Phil. Mag. 32, 1245-64 (1975).
77. T.R. McNelley, G.R. Edwards and O.D. Sherby, Acta Met. 25, 117-124 (1977).
78. M.F. Ashby and L. Johnson, Phil. Mag. 20, 1009-22 (1970).
79. G.C. Weatherly, Phil. Mag. 17, 791-9 (1968).
80. M.F. Ashby, S.H. Gelles and L.E. Tanner, Phil. Mag. 19, 757-71 (1969).
81. V.A. Philips, Trans. AIME 230, 967-76 (1964).
82. L.M. Brown, G.R. Woolhouse and U. Valdre, Phil. Mag. 17, 781-9 (1968).
83. G.C. Weatherly and R.B. Nicholson, Phil. Mag. 17, 801-31 (1968).
84. E. Votava, Acta Met. 11, 1105-6 (1963).
85. F. Seitz, Phys. Rev. 79, 723-4 (1950).

86. F.W. Young, Jr., Direct Observation of Imperfections in Crystals (Ed. by J.B. Newkirk and J.H. Wernick, Interscience, 1962) pp. 103-113.
87. J.C.M. Li and Y.T. Chou, Met. Trans. 1, 1145-59 (1970).
88. H. Gleiter, J. Less Common Metals 28, 297-323 (1972).
89. L.E. Murr, "Interfacial Phenomena in Metals and Alloys" (Addision-Wesley, Reading, Mass. 1975).
90. P.J. Worthington and E. Smith, Acta.Met. 12, 1277-81 (1964).
91. W.E. Carrington and D. McLean, Acta Met. 13, 493-9 (1965).
92. W.D. Brentnall and W. Rostoker, Acta Met. 13, 187-98 (1965).
93. W. Bonfield and C.H. Li, Acta Met. 12, 577-83 (1964).
94. E. Hornbogen, Trans. ASM 56, 16 (1963).
95. L.G. Orlov, Soviet Phys. Solid St. 9, 1828-35 (1968).
96. L.E. Murr, Met. Trans. 6A, 505-13 (1975).
97. Ben C. Odegard and A.W. Thompson, Met. Trans. 5, 1207-13 (1974).
98. R.N. Singh and K. Tangri, Met. Trans. 1, 3151-7 (1970).
99. T. Malis, D.J. Lloyd and K. Tangri, Phys. Stat. Sol. (a) 11, 275-86 (1972).
100. H. Warlimont, Iron Steel Inst. Spec. Rept. 93, London, 1965, p. 73.
101. T. Malis and K. Tangri, Acta Met. 27, 25-32 (1979).
102. A. Mascanzoni and G. Buzzichelli, Phil. Mag. 22, 857-60 (1970).
103. K.N. Tandon and K. Tangri, Met. Trans. 6A, 809-813 (1975).
104. C.S. Barrett, Structure of Metals (McGraw-Hill, 1952), p. 334.
105. K.N. Tandon and K. Tangri, Materials Science and Engineering, 20, 47-54 (1975).
106. B. Mastel, H.E. Kissinger, J.J. Laidler and T.K. Beirlein, J. Appl. Phys. 34, 3637-8 (1963).
107. L.E. Murr and S.S. Hecker, Script. Met. 13, 167-171 (1979).
108. J.C.M. Li, Trans. AIME 227, 239-47 (1963).
109. C.W. Price and J.P. Hirth, Mat. Sci. Eng. 9, 15-18 (1973).
110. J.P. Hirth, Met. Trans. 3, 3047-67 (1972).
111. J.C.M. Li, Acta Met. 8, 296-311 (1960).
112. L.G. Orlov, Soviet Phys. Solid St. 9, 1836-9 (1968).
113. H. Gleiter, E. Hornbogen and G. Bäro, Acta Met. 16, 1053-67 (1968).
114. G. Bäro, H. Gleiter and E. Hornbogen, Mat. Sci. Eng. 3, 92-104 (1968-9).
115. Y. Ishida and M. Henderson Brown, Acta Met. 15, 857-60 (1967).
116. Y. Ishida, T. Hasegawa and F. Nagata, Trans. Japan Inst. Metals 9, Suppl. 504-8 (1968).
117. G. Buzzichelli and A. Masconzoni, Phil. Mag. 24, 497-508 (1971).
118. E.S. Venkatesh and L.E. Murr, Mat. Sci. Eng. 33, 69-80 (1978).
119. L.E. Murr and E. Venkatesh, Metallography 11, 61-79 (1978).
120. L.E. Murr, R.J. Horylev and W.N. Lin, Phil. Mag. 22, 515-42 (1970).
121. P.H. Pumphrey and H. Gleiter, Phil. Mag. 30, 593-602 (1974).
122. P.H. Pumphrey and H. Gleiter, Phil. Mat. 32, 881-5 (1975).
123. W. Lojkowski and M.W. Grabski, Scripta Met. 13, 511-4 (1979).
124. J.-P. Bailon, A. Loyer and J.-M. Dorlot, Mat. Sci. Eng. 8, 288-98 (1971).
125. D. Laister and G.M. Jenkins, J. Materials Science 8, 1218-1232 (1973).
126. J.R. Patel and A.R. Chandhuri, J. Appl. Phys. 34, 2788 (1963).
127. R.L. Bell and W. Bonfield, Phil. Mag. 9, 9-36 (1964).
128. J.J. Gilman and W.G. Johnston, "Dislocations and Mechanical Properties of Crystals" (John Wiley & Sons, New York, 1957) p. 116.
129. A.R. Chaudhuri, J.R. Patel and L.G. Rubin, J. Appl. Phys. 33, 2736-46 (1962).
130. S. Schäfer, Phys. Stat. Sol. 19, 297-312 (1967).
131. J.R. Patel and P.E. Freeland, Phys. Rev. Letters 18, 833-5 (1967).
132. A. Cavallini, P. Gondi and A. Castaldina, Phys. Stat. Sol. (a) 32, K55 (1975).
133. W.E. Nixon, M.H. Massey and J.W. Mitchell, Acta Met. 27, 943-50 (1979).
134. B. Pichaud, F. Minari and J. Kellerhals, Phil. Mag. A38, 593-602 (1978).
135. B. Nøst and E. Nes, Acta Met. 17, 13-20 (1969).
136. F.W. Young and F.A. Sherrill, J. Appl. Phys. 42, 230-7 (1971).
137. F. Minari, B. Pichaud and L. Capella, Phil. Mag. 31, 275-84 (1975).
138. O. Lohne and O. Rustad, Phil. Mag. 25, 529-36 (1972).
139. L. Gerward, Phys. Stat. Sol (a) 2, 797-802 (1970).
140. O. Nittono, Japan. Journal of Applied Phys. 10, 188-196 (1971).
141. I.R. Kramer and N. Balasubramanian, Acta Met. 21, 695-9 (1973).
142. I.A. Johnston, P.S. Dobson and R.E. Smallman, Proc. Roy. Soc. A315, 231-8 (1970).
143. F. Minari, B. Pichaud and L. Capella, Phil. Mag. A 41, 281-9 (1980).
144. A.K. Head, Phil. Mag. 44, 92-4 (1953).
145. S. Kobayashi and S.M. Ohr, Proc. 37th Ann. Mtg. EMSA (Ed. by G.W. Bailey, 1979) p. 424; also private communication.

146. S.M. Ohr and J. Narayan, Phil. Mag. A 41, 81-89 (1980).
147. B.S. Majumdar and S.J. Burns, Scripta Met. 14, 653-6 (1980).
148. N.P. Louat, Proc. First Int. Conf. on Fracture (Sendai, Japan, 1965) pp. 117-32.
149. R.J. Asaro, J. Phys. F: 5, 2249-55 (1975).
150. B.S. Majumdar and S.J. Burns, to be published.
151. J.R. Rice and R. Thomson, Phil. Mag. 29, 73-97 (1973).
152. C. Atkinson, Int. J. Fracture Mech. 2, 567-74 (1966).
153. D.M. Barnett and R.J. Asaro, J. Mech. Phys. Solids 20, 353 (1972).
154. G. Leibfried, Z. Phys. 130, 214 (1951).
155. A.K. Head and N. Louat, Austr. J. Phys. 8, 1 (1955).
156. B.A. Bilby, A.H. Cottrell and K.H. Swinden, Proc. Roy. Soc. 272, 304-14 (1963).
157. H. Wagenblast and R.C. Glenn, Met. Trans. 1, 2299-2304 (1970).
158. L.P. Kubin, F. Louchet, D. Caillard and J.L. Martin, The Sixth International Conference on High Voltage Electron Microscopy, 1980.
159. J.E. Sinclair, Nucl. Metall. 20, 288-99 (1976).
160. Sinclair (159) made some numerical calculations of dislocation distributions in front of a blunted crack without a nucleation barrier and obtained a maximum in dislocation density. However, the maximum is located at a distance comparable to the crack tip radius. If the crack tip blunting is ignored, the dislocation density tends to infinity at the tip in agreement with BCS (156). It seems obvious that if a nucleation barrier for dislocations is introduced, the maximum of dislocation density will be located farther away from the crack tip even without blunting.

DEFECTS IN REGULAR CELLULAR STRUCTURES AND THEIR ROLE IN GRAIN GROWTH MECHANISMS

A.G. Crocker*, P.J. Rinous* and M.O. Tucker**

* Department of Physics, University of Surrey, Guildford, England, GU2 5XH
** Central Electricity Generating Board, Berkeley Nuclear Laboratories
Gloucestershire, England GL13 9PB

Abstract

The effects of allowing a single rogue grain in an otherwise uniform packing of cells to either shrink or grow is investigated. It is shown using a two-dimensional model, that in both cases the structure remains in equilibrium and the energy is constant until grain corners meet and become unstable. Irreversible grain growth then occurs and the migrating interface between the recrystallised region and the original structure is characterised in terms of climbing dislocations. A similar mechanism arises in the corresponding three-dimensional model.

Introduction

In order to obtain a better understanding of grain growth processes in crystalline materials, possible mechanisms, based on the behaviour of a single rogue grain (or point defect) in an otherwise regular array of grains, have been examined. In particular, the way in which the total grain boundary area, and hence the energy of the system, changes as a single grain either shrinks or grows, until unstable nodes are generated at its corners, has been considered in detail. These nodes then dissociate so that a nucleus of new larger grains is created. The interface between this region and the original array of grains may be described in terms of characteristic dislocations, and the nucleus is able to grow by means of the climb of these defects. In the present paper this mechanism of grain growth is first illustrated using a two-dimensional model of regular hexagons. The corresponding three-dimensional case of an array of regular tetrakaidecahedra is then considered. The mechanism is distinct from an earlier model due to Hillert (1-3) which is based not on a point defect but on a pre-existing dislocation.

The Two-Dimensional Model

Assuming that grain boundary energy is isotropic, the only regular two-dimensional polycrystal which can be in equilibrium is the array of hexagonal grains shown in Fig. 1(a). In this case the grain boundaries, or rather grain edges, all meet symmetrically at 3-fold nodes at angles of $2\pi/3$. Consider now the uniform contraction of one of these grains as shown in Fig. 1(b). This process clearly leaves unchanged the geometry of the grain corners and also, rather surprisingly, the total length of the grain edges. Therefore as the area of the rogue grain changes the system remains in equilibrium and the total energy of the grain edges is constant. Thus it is possible that this process will occur in practice.

Consider now the situation which arises when the rogue grain of Fig. 1(b) shrinks to a point as shown in Fig. 1(c). There are now six pentagonal grains meeting at an unstable 6-fold node. This may dissociate symmetrically into four 3-fold nodes in one of two equivalent ways, one of which is shown in Fig. 1(d). In this diagram the extent of the dissociation has been chosen so that the central node and its three nearest neighbours are in equilibrium, the angles between the grain edges being $2\pi/3$. However the surrounding ring of six nodes is not in equilibrium, the angles between the triplets of edges being $\pi/2$, $2\pi/3$ and $5\pi/6$ in each case. In practice the polycrystal will attempt to restore equilibrium in two inter-related ways. Grain edges will become curved and nodes will change their locations. However in the present analysis only the latter mechanism will be considered. A possible sequence of events is then shown schematically in Figs. 1(e)-(h). First the three pentagonal grains of Fig. 1(d) shrink to points to produce the configuration of Fig. 1(e) consisting of a central 3-fold node with three 5-fold

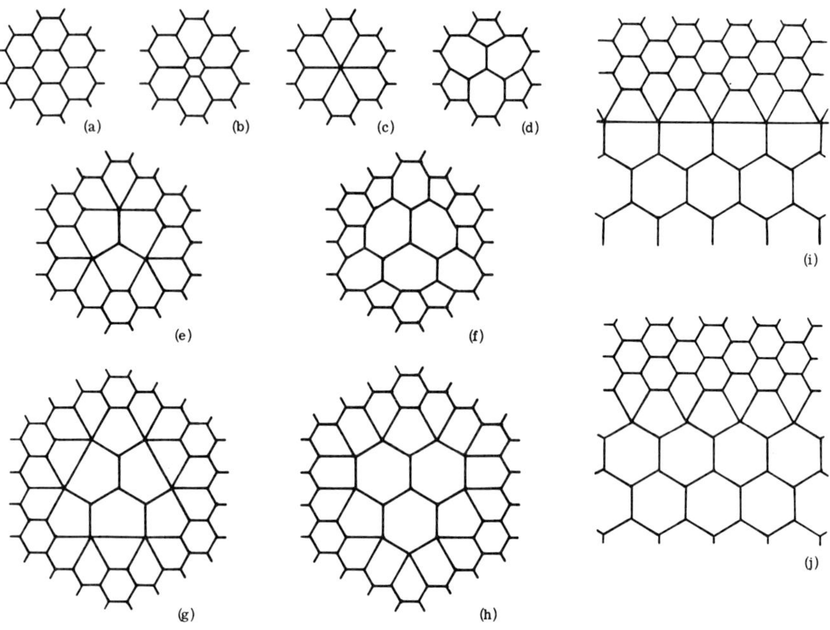

FIG. 1

Grain growth in a two-dimensional model polycrystal arising from the contraction of single grain (a)-(h), and two representations (i,j) of the resulting interfacial structure.

nearest neighbours. These 5-fold nodes then dissociate symmetrically into three 3-fold nodes, which for the configuration shown in Fig. 1(f) are in equilibrium. However twelve unstable 3-fold nodes have now been generated, which cause the six small pentagonal grains to shrink until the configuration of Fig. 1(g) is reached. This involves six 5-fold nodes of the same type as those occurring in Fig. 1(e). These in turn are eliminated by symmetric dissociation followed by grain shrinkage until the structure of Fig. 1(h) is obtained.

At this stage of the process the true nature of the recrystallised structure is revealed for the first time. It consists of three hexagonal grains, each three times larger than the original cells and oriented at right angles to them. This nucleus is seen to be surrounded by a ring of 18 pentagonal cells, which forms an interface between the parent and product hexagonal structures. These cells have three distinct shapes but all of the unstable nodes have identical geometry consisting of four edges meeting at angles of $\pi/3$, $\pi/2$, $\pi/2$ and $2\pi/3$. These nodes now dissociate into pairs of 3-fold nodes and the resulting small pentagonal cells shrink to give a configuration which has nine unstable 5-fold nodes, identical to those occurring in Fig. 1(e) and 1(g). The same mechanism of dissociation followed by grain shrinkage may therefore occur to give a larger nucleus of twelve product hexagonal cells surrounded by a ring of 30 pentagonal interfacial cells. These cells are of exactly the same type as those occurring in Fig. 1(h). Recrystallisation therefore proceeds by repeating the mechanisms which lead from Fig. 1(e) to 1(h). Indeed this process is irreversible as at every nodal dissociation the total length of grain edge and hence the energy of the system is reduced.

During the recrystallisation process the interface of the parent and product structures alternates between configurations involving 5-fold and 4-fold nodes. Figures 1(g) and 1(h) illustrate the rather special cases of these two structures which arise from very small triangular nuclei of the product structure. In order to characterise the interface further it is convenient to consider extended straight sections of boundary and Figs. 1(i) and (j) represent the resulting regular configurations. The mechanism of migration of the interface into the parent phase can now be followed more closely. The 5-fold nodes of Fig. 1(i) dissociate into three 3-fold nodes in such a way that the quadrilateral grains shrink to give the structure of Fig. 1(j). The 4-fold nodes of Fig. 1(j) then dissociate into two 3-fold nodes in such a way that the small pentagonal grains shrink to reproduce the structure of Fig. 1(i). This process may be described as climb of the epitaxial dislocations forming the interface. Defects such as steps and corners may of course arise in these linear boundaries and their structures delineated. In particular the structure of $4\pi/3$ corners in the two limiting boundary structures are contained in Fig. 1(g) and (h), and corresponding $2\pi/3$ corners arise in subsequent stages of the recrystallisation process.

A complementary mechanism of two-dimensional grain growth arises if the rogue grain of Fig. 1(b) is considered to grow rather than to shrink. Again at first there is no change in the geometry of grain corners, or of total edge length, so that energetically the mechanism is acceptable. However when the rogue grain becomes four times its original size it meets its second nearest neighbours and six unstable 4-fold nodes are formed. Each of these dissociates into two 3-fold nodes resulting in the large grain shrinking until it is only three times its original size. At this stage it is surrounded by six 4-fold nodes such as those shown in the interface of Fig. 1(j). Further dissociation and hence recrystallisation therefore occurs as in the model of Fig. 1, and in particular the resulting interface alternates in structure between Fig. 1(i) and (j). The contraction and growth mechanisms are nevertheless distinct the former being centred on a node and having $3(n + 1)^2$ large hexagonal cells while the latter is centred on a cell and has $6n + 1$ large cells ($n = 0, 1, 2....$).

The Three-Dimensional Model

If a three-dimensional polycrystal is to be in equilibrium, three grain boundaries should meet at equal angles of $2\pi/3$ along each grain edge and four grain edges should meet at equal angles of $\phi = \cos^{-1}(-1/3) = 109.47°$ at each grain corner. No regular array of polyhedral grains can satisfy these conditions but the body centred cubic packing of regular tetrakaidecahedra is the best approximation (4). This polyhedron has six square and eight hexagonal faces as shown in Fig. 2(a) and when packed in a b.c.c. array the hexagons separate first neighbours and the squares second neighbours as indicated in Fig. 2(b). In this structure the edge angles

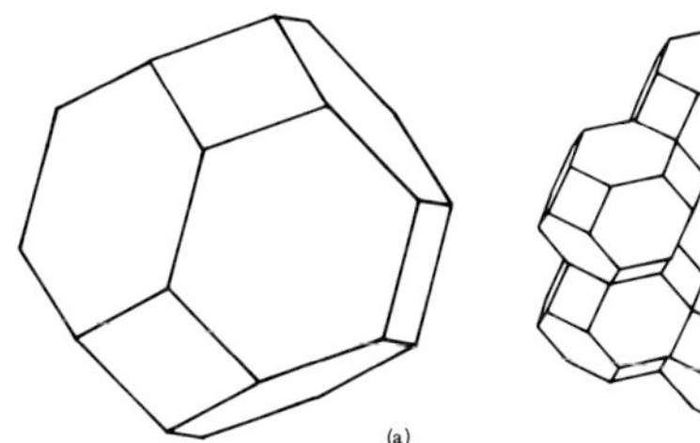

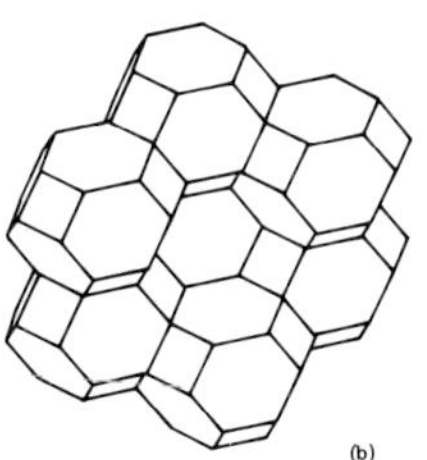

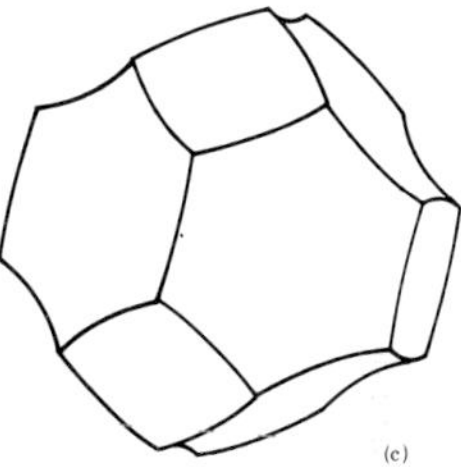

FIG. 2

A plane-faced tetrakaidecahedron (a), six of these polyhedra packed to form part of a b.c.c. array (b), and an equilibrium curvilinear tetrakaidecahedron (c).

are ϕ(x1) and $\psi = \cos^{-1}(-1/\sqrt{3}) = 125.27°$(x2) and the corner angles are $2\pi/3$ (x4) and $\pi/2$ (x2). The exact equilibrium conditions can in fact be met, as illustrated in Fig. 2(c), if the hexagonal faces take up a convoluted form with 3-fold symmetry and zero net curvature whilst the square faces remain flat (4, 5). However in the present paper polycrystals based on regular polyhedra will be considered.

As in the two-dimensional model, allow one of the tetrakaidecahedra to change its volume in a uniform manner either by shrinking or by growing until unstable nodes are generated. During this process the angles at grain edges and grain corners remain constant. Thus the square faces remain square but the length L of their edges changes. Thus L provides a convenient measure of the progress of the mechanism. Figure 3 summarises schematically the cases of L = 2, 3/2, 1, 1/2 and 0, where L = 1 corresponds to the original regular packing. The figure shows the shapes and multiplicities of the grain faces for the rogue grain and its eight first and six second nearest neighbours.

First consider the rogue grain to shrink so that its six square faces become smaller and its eight hexagonal faces become truncated equilateral triangles. A critical configuration is reached at L = 0 when the squares and hence the truncations become points, so that the grain has the form of an octahedron. Meanwhile as shown in Fig. 3, seven faces of the first nearest

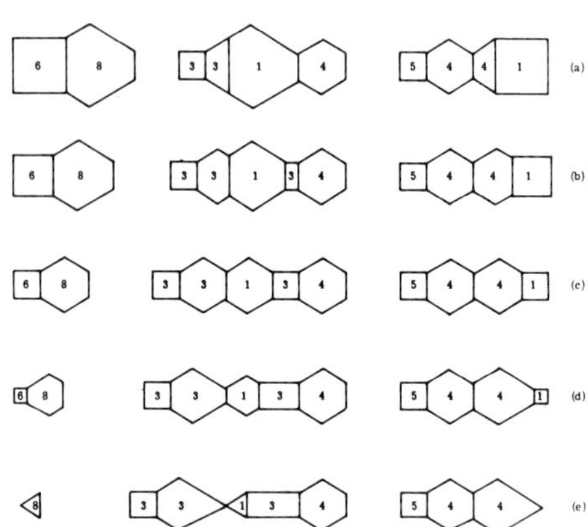

FIG. 3

Shapes and multiplicities of faces of polyhedra representing a rogue grain (1st column) its eight nearest neighbours (2nd column), and its six second neighbours (3rd column) in a three-dimensional polycrystal. Five sizes of rogue grain are shown, the lengths L of the edges of the square faces being 2, 3/2, 1, 1/2, and 0 in (a)-(e) respectively.

neighbours have been changing their shapes together with five faces of the second neighbours, one of which disappears at L = 0. At this stage eight grain edges meet at each of the six corners of the octahedron. These corners are therefore unstable and dissociate into two 5-fold nodes enabling the octahedron to shrink further until it disappears. The six 5-fold nodes associated with the corners of the octahedron then combine to form a single 6-fold node which is itself unstable so that further dissociation will occur. Similarly the remaining 5-fold nodes created at L = 0 and located at the junctions of four first and one second nearest neighbour grains are unstable and dissociate.

Consider now the growth of the rogue grain, its square faces increase in size and its hexagonal faces again become truncated triangles. The process continues until, at L = 2, 6-fold nodes are formed at its corners. The situation corresponds, as shown in Fig. 3, to the elimination of three square faces on each of the first nearest neighbours, a further four faces changing their shape. Again five faces change on the second nearest neighbours but none are lost. The structure is now unstable so that dissociations of grain corners are expected to occur.

Whether or not the growth and contraction mechanisms illustrated in Fig. 3 will occur in practice will depend on the changes in the total grain boundary area, and hence in energy, which are involved. In units of the energy of a perfect grain these are found to be - 0.017 for L = 3/2 and 1/2, and - 0.070 for L = 2 and 0. Thus, unlike the two-dimensional case, where the corresponding energies are constant, changes do occur in this three-dimensional model. In particular there is a small reduction in energy as the rogue grain changes its volume in either sense from that of the perfect structure, which therefore appears to be marginally unstable. The fact that the energy changes are symmetric about the initial uniform shape is also remarkable as the structures involved are quite distinct. The dissociations which follow the formation of unstable nodes at L = 2 and L = 0 lead to further substantial energy reductions and hence irreversible grain growth, as in the two-dimensional case. Thus a nucleus of large product cells is again generated and the interface between these and the small parent cells can again be described in terms of arrays of dislocations.

Discussion

The models developed in this paper are clearly capable of improvement and extension. For example curved cell faces and edges could be introduced. In particular in the two-dimensional model, the cell edges at the straight interfaces of Figs. 1(i) and 1(j), and the intermediate structures between these two limiting cases might be allowed to take up curved forms which are in equilibrium, subject to conservation of area of the individual cells. In this way migration of the interface by means of the climb of the interphase boundary dislocations could be followed. Indeed an analysis of this type has recently been carried out on the development of regular bimodal and trimodal distributions of two-dimensional cell structures (6). Unfortunately the situation is more complex for grain growth nucleating from a point defect as illustrated for example in Fig. 1(d) to (h). The corners and steps in the interfaces then have long range strain fields so that the curved cell edges are not localised at the defect. However by restricting the range of the strain field, information could still be obtained about this dynamic process. The steps in the interphase boundaries are effectively transformation dislocations with properties corresponding for example to twinning dislocations in crystals (7). Determination of the equilibrium curvature of cell edges near steps would thus be of interest. Other types of interface corresponding to anti-phase boundaries may be generated if grain growth is nucleated at two or more centres. For example the large cells of Fig. 1(h) occupy the sites of only one-third of the small cells of Fig. 1(a). Two other origins for the product structure are therefore possible and different origins result in boundaries when recrystallised regions meet. Note however that none of these mechanisms rely on the pre-existence of dislocations in the cell structure as assumed by Hillert in his model of grain growth (1,2). The present mechanism is simply based on the remarkable fact that energetically there is nothing to prevent an individual grain, or any number of isolated grains, from growing or contracting until unstable nodes are generated.

Only the early stages of the corresponding three-dimensional model have been examined. The most striking fact in this case is that the energy of the system actually decreases slightly as the rogue grain either grows or shrinks. The effect could be an artefact arising from the use of polyhedra with plane faces. If cell shapes based on the model with curved faces shown in Fig. 2(c) had been adopted it is possible that constant energy might have been obtained as in the two dimensional case. However preliminary calculations suggest that allowing the faces to take on an equilibrium curvature would result in the energy changes being increased rather than being eliminated. Clearly the structure of the parent-product interface needs to be examined in detail for this three-dimensional model. This will involve a study of interfacial dislocations, including steps and corners, and of the climb mechanisms which occur. Anti-phase boundaries will again arise and the effects of variations in grain size needs to be considered. In this connection it is worth noting that a successful model of inter-linked grain edge porosity using a bimodal distribution of grain sizes has recently been developed (8). This is based on equal numbers of large and small irregular tetrakaidecahedra packed in a caesium chloride structure.

In conclusion it is necessary to emphasise the need for a powerful analytic procedure for investigating the properties of defects in cellular structures. Theories of defects in crystals provide a natural basis for such an analysis but the additional requirement of incorporating the sizes and shapes of the cells and not simply their locations is a demanding challenge.

This paper is published by permission of the Central Electricity Generating Board.

References

1. M. Hillert, Acta Met. 13, 227 (1965).
2. J.W. Cahn and G.E. Padawer, Acta Met. 13, 1091 (1965).
3. J.E. Morral and M.F. Ashby, Acta Met. 22, 567 (1974).
4. C.S. Smith, Metal Interfaces, p.65, ASM, Cleveland (1952).
5. W. Thompson, Phil. Mag. 24, 503 (1887).
6. P. Hartland and A.G. Crocker, to be published.
7. J.W. Christian and A.G. Crocker, in "Dislocations in Solids, Vol. 3" Nabarro ed. (North Holland, 1980), p.165.
8. P.J. Rinous, M.O. Tucker and A.G. Crocker, to be published.

OBSERVATIONS OF GRAIN BOUNDARY STRUCTURE

R.C.POND

Department of Metallurgy and Materials Science
The University of Liverpool
P.O. Box 147. Liverpool L69 3BX

Abstract

The dislocation model of grain boundary structure is first summarised, and its main advantages and limitations considered. The notions of reference structure and reference lattice are outlined in order to illustrate the characterisation of primary and secondary dislocations. The idea that primary dislocations are perturbations of the fundamental periodicity in single crystals, and that secondary dislocations are perturbations of periodic arrays of primary dislocations is developed.

Most of the experimental observations presented are transmission electron micrographs. This technique has been used to study grain boundary steps and dislocation displacement fields and, recently, to resolve directly the atomic structure of a boundary in Ge. In the discussion of the observations emphasis is placed on their interpretation in terms of the dislocation model of grain boundaries rather than on experimental procedures.

Introduction

The purpose of this paper is to describe experimental observations of grain boundaries and to discuss the interpretation of these in terms of the dislocation model of boundary structure. Burgers (1) and Bragg (2) originally proposed the modelling of grain boundaries as arrays of dislocations and since that time the model has become increasingly sophisticated. Frank (3) demonstrated how the Burgers vector content of a low angle boundary can be predicted using his now famous formula. Read and Shockley (4) derived an expression relating the energy of low-angle boundaries to their dislocation content and also provided insight into the nature of strain fields resulting from periodic arrays of dislocations and the consequences of perturbations in such arrays. An expression relating the dislocation content of high-angle boundaries to the relative orientation of the adjacent crystals was derived by Bilby, Bullough and Smith (5). Later, Bollmann (6) showed that an equivalent expression can be obtained by considering the registry between two misorientated crystal lattices (the 'O' lattice theory). Bollmann also introduced the concept of the d.s.c. lattice which has been very important in clarifying the nature of dislocations in high-angle boundaries. Recently, Pond and Bollmann (7) have shown how the symmetry of grain boundaries can be classified, and various researchers have considered the topological properties of grain boundary dislocations (e.g. 8) and their role in mechanisms such as deformation and boundary migration (e.g. 9). The main achievements and limitations of the dislocation model of grain boundaries are summarised in the next section.

Over the past thirty years a wide range of experimental techniques has been used to investigate grain boundaries and the resolving power of these methods has improved dramatically. At the beginning of this period the highest resolution observations were of etch pits on the surfaces of specimens at the points of emergence of dislocations (e.g. 10). Present day electron microscopes are capable of directly resolving the atomic structure of boundaries in materials such as oxides and semi-conductors where the lattice parameters are not too small. In addition to atomic resolution, transmission electron microscopy is an excellent technique for studying dislocation displacement fields (10) and for establishing the orientations of abutting crystals. Consequently, this technique has been used extensively in the study of grain boundary structure and is the main technique considered in this article.

The Dislocation Model of Grain Boundary Structure

We now consider the modelling of a bicrystal as an array of dislocations. To introduce the discussion an isolated dislocation in a single crystal is considered first. The structure of a perfect single crystal is most concisely described by specifying its lattice and basis. The lattice is a purely geometrical concept in which each lattice point has identical surroundings. The crystal structure can be imagined to be created by locating one basis (i.e. an atom or group of atoms) at each lattice point. The lattice and crystal so defined are the *reference lattice* and *reference* (or undislocated) *structure*. A dislocation in such a structure introduces a perturbation of the translational symmetry (periodicity) of the crystal and has a displacement field associated with it. The displacement field can be described adequately by linear elasticity theory except close to the core of the dislocation. The geometrical character of the dislocation is specified by its line direction and Burgers vector $\underline{b}$. The latter is the closure failure (following some convention such as RH/FS) of a circuit constructed in the reference lattice corresponding to a closed circuit constructed in "good" material in the dislocated structure.

We note that the basic requirement of a reference lattice is that it should bear some correspondence to the reference structure. In principle it is possible to use a reference lattice other than the crystal lattice, for example one which has identical orientation to the crystal lattice but different lattice parameters. A correspondence would still exist between this alternative reference lattice and the reference structure, and the definition of $\underline{b}$ would be unaffected. However, the crystal lattice is the most convenient choice not only because of the simple correspondence between it and the reference structure but also because if the closure failure is a translation vector it follows that the associated dislocation is *perfect*.

The dislocation model of low-angle grain boundaries is well understood (e.g. 6), and the displacement field of the boundary can be formulated elegantly using continuum elasticity theory (13). The Burgers vector of individual dislocations can still be defined as long as circuits can be constructed around the dislocations and the distortions are not so great that the correspondence to the reference lattice cannot be recognised. However, as the dislocations become more closely spaced the displacement field surrounding a particular dislocation becomes progressively less like the displacement field of an isolated dislocation with the same $\underline{b}$ due to interaction with the other dislocations in the boundary.

Read and Shockley (4) showed that the interaction of dislocation displacement fields leads to a local minimum in the energy versus misorientation relationship when the dislocations are equi-spaced (i.e. when they are separated by an equal (integral) number of slip planes). Any additional angular deviation across such a special boundary requires the periodicity of the dislocations to be perturbed, and, as far as the long range displacement field is concerned, this is formally equivalent to introducing into the periodic array a partial dislocation at the location of the perturbation. The Burgers vector of the partial dislocation is inversely proportional to the period of the unperturbed array. This process can continue indefinitely in the sense that a periodic arrangement of perturbations introduced into a periodic array of dislocations can be regarded as a new array with a different fundamental periodicity and this, in turn, may be perturbed.

If a high-angle boundary is modelled as an array of dislocations and a single crystal is taken as the reference structure the spacing of the dislocations becomes so small that the physical aspects of these dislocations bear little resemblance to isolated dislocations with the same $\underline{b}$. Nevertheless, geometrically speaking the dislocations can still be identified and Bullough (14) has shown how the $\underline{b}$ content of such a boundary can be expressed. Bollmann (6) has presented an equivalent treatment ('O' lattice theory) using one of the crystal lattices as reference lattice (Bullough has considered both this and the median lattice as reference lattices). As an illustration consider fig 1(a) which shows a periodic boundary structure between f.c.c. crystals with misorientation [001] 36.9° and boundary plane $(3\bar{1}0)$. The bicrystal can be imagined to have been created by the introduction of $\frac{1}{2}[\bar{1}10]$ edge dislocations into a single crystal until the final spacing is $\frac{1}{4}[130]$ (i.e. one half of the boundary period). Note that the extra half planes indicated in the figure are $(\bar{2}20)$ so that each dislocation comprises two such half planes, and note also that each successive dislocation can be regarded as having been introduced between the extra half planes of the preceeding dislocation. This boundary can also be regarded as an array of [100] edge dislocations, each comprising two (200) extra half planes, spaced $\frac{1}{2}[130]$ apart.

"DISLOCATION MODELLING"

It is important to recall that on the basis of a geometrical theory such as described above (6,14) it is only possible to predict the total Burgers vector content of a boundary. Thus, the Burgers vectors of individual dislocations and to what extent they may be dissociated cannot be predicted by geometrical arguments. In order to describe the relative position of the crystals in fig. 1(a) for example, it would be necessary to consider the dissociation of the crystal dislocations. It is clear that it is only possible to predict the details of the structure in the immediate vicinity of a boundary by the introduction of some physical hypothesis into an otherwise geometrical theory. The physical hypotheses suggested so far, for example that the energy of grain boundaries is reduced if some of the atoms occupy coincidence sites (15), have not gained general support from experimental observations and atomistic calculations (12).

Now there is not a unique dislocation description for a particular boundary since it is always possible to make alternative choices of reference structures (and reference lattices). Let us consider an alternative choice for the boundary in fig. 1(a). First, consider the choice of reference lattice. Fig. 1(c) shows the pattern created by the coincidence related crystal lattices; one set of crystal lattice sites has been coloured black for clarity and hence the pattern is referred to as a dichromatic pattern. Using coincidence site lattice (c.s.l.) terminology the reciprocal density of coinciding lattice sites, Σ, is 5. In addition to the black and white crystal lattices three other possible candidates for reference lattice are depicted in this figure. On the left is one unit cell of the c.s.l. (body centred tetragonal), on the right are several unit cells of the d.s.c. lattice (body centred tetragonal), and in the centre are two unit cells of the median lattice (face-centred cubic). (The primitive translation vectors of the d.s.c. lattice are the three shortest independent vectors which join black to white crystal lattice sites). The c.s.l., median and d.s.c. lattices all have identical orientation but different lattice parameters. Bollmann (6) has shown that the d.s.c. lattice is a particularly convenient choice because the translation vectors of this lattice are the Burgers vectors of *perfect* interfacial dislocations. The notion of perfection in these cases is somewhat different from that for crystal dislocations because the $\underline{b}$ do not necessarily correspond to translation vectors of the reference structure. However, as described further below, it can be shown that the structure of the interface on either side of a perfect grain boundary dislocation is identical.

We are now able to describe fig. 1(a) as a dislocation array with respect to alternative reference structures. For example, let the reference structure be the symmetrical tilt boundary constructed by locating atoms at white lattice sites on one side of the $(2\bar{1}0)$ plane in fig. 1(c) and at black lattice sites on the other, and let the reference lattice be the d.s.c. lattice. Fig. 1(a) can then be obtained by introducing an array of $^1/5\,[\bar{2}10]$ edge dislocations into the reference structure with a spacing of $\frac{1}{2}[130]$ as depicted in fig. 1(b) (note that each $^1/5\,[\bar{2}10]$ dislocation is perfect and comprises two extra $(\bar{4}20)$ half planes). It is also permissible to choose the structure of fig. 1(a) itself as the reference structure (i.e. as the dislocation free state). This is a particularly convenient choice for considering structures which have a small angular deviation from that in fig. 1(a). This is illustrated in fig. 1(d) which shows a perfect d.s.c. dislocation, with $\underline{b} = {}^1/10\,[3\bar{1}0]$, introducing a small angular deviation from the exactly periodic structure. Note that $\underline{b}$ is perpendicular to the interface and its magnitude is a rational fraction of the boundary period. It is possible to see that this dislocation corresponds to a perturbation of the arrangement of crystal dislocations in the description referred to a single crystal (fig. 1(a)) in accordance with Read and Shockley's (4) proposals. In fact, as far as elastic displacement fields are concerned, networks of d.s.c. dislocations in high-angle grain boundaries are completely analogous to low-angle boundaries (16). The extra half-planes and lattice plane distortions around the sessile edge dislocation have been indicated in fig. 1(d).

Bollmann (6) has proposed that crystal dislocations in a boundary referred to a single crystal as reference be termed *primary* dislocations, and d.s.c. dislocations be called *secondary* dislocations. The idea of secondary dislocations has received extensive experimental support and the model is useful not only for analysis of dislocations which contribute to the relative orientation of adjacent crystals but also for dislocations whose motion causes, for example, the shear of the adjacent grains past each other. This is illustrated in fig. 1(e) which shows schematically a perfect $^1/10\,[130]$ edge dislocation in the $(3\bar{1}0)$ boundary. Note that this secondary dislocation can also be regarded as a perturbation of the primary dislocations. Also, note that there is a step associated with the core of this dislocation; step heights can be predicted by geometrical means (8,17).

In the foregoing discussion perfect primary and secondary dislocations have been considered. Primary dislocations are always perfect non-primitive dislocations when referred to a d.s.c. lattice. Partial primary dislocations are known to exist in single crystals with low stacking-fault energy, and it is possible that partial secondary dislocations can also exist. Partial secondary dislocations would separate regions of boundary with different structures, but the $\underline{b}$ of such defects cannot be predicted by geometrical means.

An interesting possibility is the existence of partial secondary dislocations which separate domains of interface which have structures related by symmetry. These dislocations can exist in boundaries where the holosymmetric form includes both translation and point symmetry elements. For example, if an equilibrium boundary structure is characterised by a particular relative position of the adjacent grains (relative position is specified by the displacement, $\underline{t}$, of the black crystal relative to some arbitrarily chosen reference position) and this structure has lower symmetry than the holosymmetric form, then energetically degenerate equilibrium structures with symmetry related $\underline{t}$ can exist. Interfacial domains of the symmetry related structures would be separated by partial secondary dislocations with $\underline{b}$ equal to the difference of relative displacement between the adjacent structures. The multiplicity of related domains can be predicted if the point symmetry of the holosymmetric and equilibrium structures are known (11). This will be illustrated further by experimental observations in the next section.

In this brief summary it has been shown in outline that all grain boundaries can be modelled as arrays of dislocations, but that there is no unique dislocation description for a particular boundary because the choice of reference structure is arbitrary. It has also been mentioned that dislocations are perturbations to the translation symmetry of a periodic object and have displacement fields associated with them. Primary dislocations are perturbations of the periodicity of single crystals, and secondary dislocations can be regarded either as perturbations in periodic bicrystal structures or as perturbations in periodic arrays of primary dislocations. The displacement fields of primary and secondary dislocations become progressively less like the fields of isolated dislocations with the same $\underline{b}$ as the spacing between the dislocations decreases. Thus, a description of a grain boundary in terms of dislocations has greatest physical significance when, by choice of suitable reference structure, the component dislocations (primary or secondary) are reasonably well separated in the boundary. The two cases where this is so and which have been confirmed by experimental observations are low-angle boundaries (i.e. arrays of primary dislocations) and boundaries where the misorientation includes a relatively small deviation from an exactly periodic boundary structure (i.e. arrays of secondary dislocations). In these cases the longer range components of the displacement fields can be predicted using the dislocation model in conjunction with continuum elasticity theory. On the other hand, dislocation modelling does not appear to be useful for predicting the structural details in the immediate vicinity of a boundary, i.e. the material corresponding to the dislocation cores, which is hereafter referred to as the boundary core material.

Experimental Observations

The object of this section is to describe some experimental observations of boundary structure. In the accompanying discussion emphasis is placed on the interpretation of the observations in terms of the dislocation model rather than on experimental techniques (these latter have been reviewed elsewhere recently (11,18)). The observations presented concern the preferred orientations of grain boundary planes, primary dislocations, perfect and partial secondary dislocations and the core structure of boundaries. A discussion of plane matching defects has been omitted on the grounds that these features are not currently understood on the basis of a physical theory, and because they have been discussed elsewhere recently (11,20).

Preferred boundary plane orientations

A considerable number of experiments have been carried out to discover the orientations of favoured boundary planes between c.s.l. related grains. High resolution observations are particularly valuable because it is known that boundaries which appear planar macroscopically can be faceted on a microscopic scale (e.g. 21). Fig. 2 is an electron micrograph of two c.s.l. related gold crystals ($\Sigma = 11$, [110] 50.5°) (22). The common [110] axis is perpendicular to the foil and the boundaries are seen to be faceted and to be orientated perpendicular to the foil surface. Careful analysis of the boundary planes showed that some facets corresponded to low index planes of the c.s.l., but this was not generally the case for <110>tilt boundaries. In similar experiments on [100] gold foils Goodhew et al. (22) found no tendency for coincidence boundaries to facet. Other observations of preferred boundary orientations have been reviewed (22) and it is found in general that for high-angle periodic boundaries, with certain exceptions such as the coherent twin in f.c.c. materials, observed boundary planes cannot be predicted by arguments based on geometrical aspects of c.s.l.s. or 'O' lattices.

According to the dislocation model of grain boundaries it might be expected that preferred planes are those with minimum dislocation content. However, as explained in the previous section, primary dislocations in high-angle boundaries bear little resemblence to isolated dislocations with the same $\underline{b}$. It is therefore questionable whether a criterion suitable for low-angle boundaries, where the elastic energy of the dislocation strain fields is an important contribution to the total energy, is also appropriate for high-angle boundaries. In the high-angle case the dominant contribution to the total boundary energy generally arises from the interaction of all atoms in the core of a boundary, and it seems unlikely that this contribution can be predicted except by atomistic calculations.

Fig. 3(a) is an electron micrograph of a grain boundary between twin related ($\Sigma = 3$) crystals of aluminium. The boundary is faceted in a complex manner and it was established experimentally (23) that the facet indices are (using the coordinate system of the upper crystal) (111), the coherent boundary, and $(\bar{2}11)$, $(\bar{1}55)$, $(1\bar{3}\bar{1})$, $(2\bar{2}0)$, and the crystallographic variants of these in the $\Sigma = 3$ system. Atomistic calculations indicated that these boundary structures had lower energies than other geometrically possible but experimentally unobserved boundaries (23). The features a, b, c and d indicated in fig. 3(a) are $(1\bar{3}\bar{1})$ and $(2\bar{2}0)$ steps on the main $(1\bar{2}1)$ facet. This is confirmed in fig. 3(b) which was recorded with the electron beam parallel to $[11\bar{2}]$ so that the steps are seen edge on (although the specimen has been reloaded upside down between the two observations). These features did not possess long range displacement fields since they did not give rise to strong strain contrast using any diffraction condition. Judging from the offset of the fringes across the steps (the interpretation of the fringe contrast will be discussed later) the steps are approximately 30 nm high. The observations were carried out using a heating stage in order to observe grain boundary migration and it was found that migration involved the movement of the steps along the $(1\bar{2}1)$ facet. The origin of the steps, the nature of the driving force, and the mechanism of migration are not known. However, the observation that $(1\bar{3}\bar{1})$ and $(2\bar{2}0)$ type facets are mobile while $(1\bar{2}1)$ facets are not is interesting because the former types have structures which are independent of the location of the boundary plane whereas the latter does not (17).

Low-angle boundaries

We now consider recent developments in the study of low-angle boundaries. The simple model of primary dislocation walls has been supported by many electron microscope investigations (e.g. 24). However, as mentioned in the previous section, using geometrical theories it is only possible to predict the Burgers vector content of a given boundary,and recent investigations (25, 26) have explored other aspects such as the stability of grain boundary planes, the choice of different possible Burgers vectors and the effect of dislocation dissociation in the grain boundary.

Fig. 4 is from the work of Bourret and Desseaux (25) and shows two [011] tilt boundaries in Ge recorded in bright field multi-beam diffraction conditions with the electron beam nearly parallel to $[01\bar{1}]$ so that all defects are seen end on. Fig. 4(a) shows a boundary with a misorientation of 1.64° and comprising two types of dislocations, A and B. Type A dislocations are perfect edges with $\underline{b} = \frac{1}{2}[0\bar{1}1]$ forming a simple tilt boundary on $(0\bar{2}2)$. Type B are 60° dissociated dislocations with total Burgers vector ½<101> (note that the stacking-faults can be seen on the image as short line segments perpendicular to the $(2\bar{1}1)$ boundary plane). Fig. 4(b) shows a boundary with misorientation 3.8° comprised mainly of type C dislocations which are triplets of identical Frank partials. By using a 500 KeV microscope it was possible to identify the nature of the stacking-faults and Burgers vectors of these defects directly from high-resolution images. Fig. 5 is a high-resolution filtered image showing three $^{1}/_{3}[\bar{1}1\bar{1}]$ partials separated by stacking-fault on the $(\bar{1}1\bar{1})$ plane (note the three extra half planes adding successively along the defect). In the boundaries described above the Burgers vector content was found to be in complete agreement with Frank's formula but it is clear that the detailed structure of low-angle boundaries can be complex, especially in materials with low stacking-fault energy.

Medium-angle boundaries

Before considering observations of high-angle boundaries the transition from the low to high-angle regime, which is generally taken to be the misorientation range $\sim 10^{\circ}-20^{\circ}$, is discussed briefly. In the low-angle case both the long-range displacement field and the core structure of a boundary are described most advantageously in terms of primary dislocations. In the high-angle case long-range displacement fields, if present, are better described in terms of secondary dislocations, and the individual character of primary dislocations in a boundary core no longer has physical significance. In the medium-angle range it is reasonable to expect a gradual transition between these two cases. To illustrate the case of medium-angle boundaries we consider first a boundary in Al with misorientation 10.8°; it will be seen that although the primary dislocations constituting the boundary are spaced about 6|b| apart (i.e. sufficiently far apart for the description of the core in terms of individual dislocations to be physically meaningful) the long-range displacement field is described most advantageously in terms of secondary dislocations.

Fig. 6 is an electron micrograph showing the 10.8° boundary at a position where it is joined by a very low-angle boundary, G, in which the individual primary dislocations are clearly resolved (27). The misorientation across this medium-angle boundary is close to that for the Σ = 41 periodic boundary ([100] 12.67°) and the features, H, were interpreted as one component of a network of secondary dislocations, spaced about 20 nm apart with b = 1/82[41,5,$\bar{4}$], accommodating the angular deviation. The faint features J are thought to be secondary dislocations with b = 1/82[0$\bar{1}$9] forming another component of the network. Note that although |b| for these latter dislocations is only 16% of that for ½<110> dislocations these defects can still be detected experimentally when sufficiently well separated. Further interesting features are seen at K,L and M; at these points primary dislocations from one of the crystals impinge on the boundary and dissociate into secondary dislocations. For example, at K a primary dislocation can be regarded as dissociating into one H type and four J type dislocations which can just be detected in the image (28).

A second example of a medium-angle boundary is shown in fig.7. This micrograph shows an array of secondary dislocations with b = 1/50[7$\bar{1}$0] accommodating a tilt deviation of 1.08° from the Σ = 25 periodic boundary ([001] 16.26°) in Ge (29). These dislocations are spaced 255 nm apart and at Q a primary dislocation impinges on the boundary and interacts with the secondary network. Bollmann et al. (29) have suggested that the dissociation components in this case include partial secondary dislocations (see later section). The core structure of this boundary has also been studied by viewing the boundary end on (30). It is possible to obtain contrast in regions of dilatation, such as at dislocation cores, by observation in bright-field multibeam diffraction conditions at the Scherzer defocussing distance. Bourret and D'Anterroches (30) found that the primary dislocation structure of this tilt boundary is an array of edge dislocations of type <100> spaced 2 nm apart and their observations indicate that the dislocation cores are separately distinguishable.

High-angle grain boundaries: secondary dislocations

The structure of high-angle boundaries is discussed here by first considering observations of secondary dislocations and subsequently investigations of boundary core structures. Many observations of secondary dislocation networks accommodating deviations from high-angle periodic structures have been reported in the literature (e.g. 31,32). In many instances the bicrystal specimens have been specially manufactured, as in the extensive work on gold and MgO by Balluffi and his co-workers (e.g. 33,34). Burgers vector determination for secondary dislocations by contrast experiments is considerably more difficult than for isolated primary dislocations (11). However, substantial evidence supporting the d.s.c. model of perfect secondary dislocations has been obtained by other methods, for example by correlation of accurate determinations of angular deviations with measurements of secondary dislocation network spacings. Observations of the dissociation of primary dislocations impinging on grain boundaries, as in fig. 6, are also important in this respect (35).

Fig.8(a) is an electron micrograph of a boundary in gold between crystals close to the Σ = 5 orientation relation ([001] 36.9°) taken from the work of Cosandey and Bauer (36). The boundary plane is (3$\bar{1}$0) and is seen end on along [001], i.e. corresponding to the schematic diagrams in fig. 1. The exact misorientation was found to be [001] 38.5 ∓ 0.5° and hence there is an angular deviation of [001] 1.6 ± 0.5° which could be accommodated by an array of secondary edge dislocations with b = 1/10[3$\bar{1}$0] (see fig. 1(d)) spaced 5.1∓1.6 nm. The image contrast in this case arises from the boundary's displacement field and relaxations where the boundary intersects the free surface. Two periodic strain contrast effects can be seen in the image; one corresponds to a long-range displacement field and has a periodicity of 7 ∓ 1 nm, and the second corresponds to a short range field with spacing 0.62 ∓ 0.05 nm. The former spacing agrees well with the anticipated separation of the 1/10[3$\bar{1}$0] secondary dislocations, and the latter corresponds to a boundary core displacement field since the spacing is, within experimental error, that of the exactly periodic boundary, i.e. $\frac{a}{2}$[130] where a is the lattice parameter. Thus, although it would be difficult to interpret the contrast quantitatively, this observation reveals simultaneously the existence of both the secondary dislocation array and a core structure. Fig. 8(b) is a lattice-fringe image of a $(3\bar{2}0)$ $\Sigma = 13$ boundary, but it is not straightforward to interpret the contrast of such images (36).

It has been demonstrated by observation of boundaries in specimens prepared from deformed or irradiated material that secondary dislocations are involved in mechanical processes. For example, irregular arrangements of secondary dislocations have been observed in materials where grain boundary sliding has occurred (37,38,39). In general, secondary dislocations must move by a combination of glide and climb since their $\underline{b}$ are only parallel to a boundary in special instances. It is to be expected, therefore, that grain boundary dislocations can act as sources or sinks of point defects under the action of appropriate chemical potentials, and this has been observed recently (39).

It has been frequently observed that the images of primary dislocations impinging on boundaries gradually disappear, especially if the temperature of the specimen is raised (e.g.40,41). This type of observation has been interpreted in two different ways. The first suggestion is that the dislocation core gradually becomes delocalised in the boundary so that the dislocation's strain field is eventually insufficient to give rise to detectable contrast (40). The alternative explanation is that the incoming dislocation dissociates into secondary dislocations which either have $|\underline{b}|$ so small that they are not detectable or which interact with a pre-existing network of secondary dislocations. In the latter case the pre-existing network may or may not be resolved experimentally but the important point is that interaction causes a perturbation of the network. This perturbation may not be localised after relaxation but the process does not involve any delocalisation of individual defects (28,42).

High-angle boundaries: core structures

Using high resolution transmission electron microscopy it is now possible to resolve directly the atomic structure of materials with relatively large unit cell parameters. However, the image contrast only corresponds to a projection of the charge density in a specimen when certain demanding experimental requirements, such as very thin specimens, are fulfilled (18). Studies of $\Sigma = 3$ twins (30) and a $\Sigma = 9$ boundary in Ge (43) have been published. The lattice parameters of metals are too small for this technique to be used to resolve atomic positions although lattice plane images of grain boundaries have been obtained (e.g.fig.8b).The core structure of metallic boundaries have been studied using other techniques such as X-ray diffraction and a method for establishing the relative position of coincidence related crystals as is described below.

Fig. 9(a) is a high resolution electron micrograph of a $\Sigma = 9$ $(2\bar{2}1)$ boundary in Ge (43) obtained using the 500 KeV Kyoto instrument which has a point to point resolution of 0.2 nm. The boundary is viewed end on and the electron beam is parallel to the common $[110]$ direction. Each dark spot in the image corresponds to pairs of atomic columns spaced 0.14 nm apart. Fig.9(b) is a photographically blurred image of a wire model of the tetrahedral bonding in the region of the boundary and the close correspondence to the electron micrograph is evident. The structure of this boundary is somewhat complicated by the occurrence of twinning, indicated by arrows in fig. 9(a). However, the region to the right of AA' in fig. 9(a) is free of these features. The boundary structure has been sketched schematically in fig. 9(c) to illustrate the positions of the atoms in the columns and to show that there are no unsatisfied covalent bonds at the interface. Note that the relative position of the two crystals is such that there is mirror glide symmetry across the interface; in fact the bicrystal structure is holosymmetric with layer space group $p2_1'ma'$ (7) where primed symbols refer to symmetry elements which relate black to white atoms and unprimed symbols relate atoms of the same colour (fig. 9(c)).

Grain boundary core structure has also been investigated using X-ray diffraction. In the case of periodic boundaries the interfacial displacement field acts as a diffraction grating and gives rise to diffraction spots in addition to those caused by crystal planes (44). The "thickness" of a boundary core (or short range displacement field) can be established by studying the profiles of grain boundary relrods (45). Thus, it is possible, in principle, to establish a core structure by measuring the intensities of extra reflections. However, due to experimental limitations, it has been possible to collect only a relatively small number of relative intensity measurements. The approach used, therefore, has been to use computer simulation methods to predict possible boundary structures using a variety of inter-atomic potentials and to look for agreement between observed intensities and calculated intensities for the simulated structures.

Good agreement has been found for a $\Sigma = 13$ (001) twist boundary in Au and the calculated structure of the four (002) planes adjacent to the boundary is shown in fig. 10 (44,45). The atoms are shown as points and neighbouring atoms in each plane have been joined by lines to indicate the puckered nature of the planes. For clarity the interplanar spacing has been increased by about four times. A particularly noteworthy feature of this boundary is the displacement towards the boundary of the atoms at the apices of the shaded pyramids. We note that the layer space group of this bicrystal is $p42_1'2$ (as opposed to the other possible holosymmetric form $p42'2$ (7)) and that systematic absences in the array of extra reflections supports the diagnosis of a $2_1'$ axis (the $2_1'$ axes are located halfway between the first black and white planes and are parallel to $[5\bar{1}0]$ and $[150]$, the 2' axes are parallel to $[32\bar{0}]$ and $[2\bar{3}0]$ at the same level, and the 4 fold axis is parallel to $[001]$).

It is possible to establish the relative position of two crystals when they have a simple coincidence relation such as $\Sigma = 3$ or 9 (46). This technique employs transmission electron microscopy and depends on the presence of low index atomic planes which have the same orientation and spacing in the adjacent crystals. If a boundary is observed so that only one set of such 'common' planes is diffracting the resulting contrast depends on whether the planes are aligned across the interface or offset. No contrast is observed at the interface if the planes are aligned but stacking-fault-like fringes result if the planes are offset. It is possible to establish the extent of the offset by measuring the intensity of the fringes. Thus, by repeating this procedure for different sets of common planes the relative position of the two crystals can be found. Coherent and incoherent twin boundaries in Al, stainless steel and Fe3%Si have been investigated in detail using this method (11).

An example of fringe contrast revealing an offset of common diffracting planes is shown in fig. 11. This fig. shows another section of the $\Sigma = 3$ boundary depicted in fig. 3; the facets A,B,C,D are $(\bar{1}55)$ types, K $(1\bar{2}1)$, E,F,G,H $(\bar{2}11)$, and J is a $(13\bar{1})$ type. Fig. 11(a) and (b) have been obtained by diffraction from different sets of common planes, and we note, for example, that whereas the fringe contrast has similar intensity for the $(\bar{1}55)$ and $(\bar{2}11)$ facets in (a), it is quite different in (b) where the $(\bar{1}55)$ facets are virtually invisible. The details of the relative position of these two crystals have been published elsewhere (23), and it is noted that for the $(\bar{2}11)$ and $(1\bar{2}1)$ facets the relative displacement of the crystals includes a component with magnitude 0.02 nm perpendicular to the boundary thereby reducing the density of the boundary material (46).

These boundaries have also been investigated using computer simulation techniques and excellent agreement was found between the predicted and observed relative positions. The simulations for $(\bar{1}55)$ and $(1\bar{2}1)$ type facets are shown in fig. 12; the upper structure is $(\bar{1}55)$ and the lower two are symmetry related forms of the $(1\bar{2}1)$ types all viewed along common $<20\bar{2}>$ directions. The traces of the common (111) planes (parallel to the coherent boundary plane) have been indicated in the figure in order to show the correspondence of these structures to the image contrast in fig. 11(b), i.e. the (111) planes are virtually aligned across the $(\bar{1}55)$ boundary but are offset by about one third of the inter-planar spacing across the $(1\bar{2}1)$ boundary. The black or white nature of the extreme fringes indicates the sense of the relative displacement of the two crystals, and we note that this contrast is opposite for facets F and G, for example, indicating that these facets have the two symmetry related forms of $(1\bar{2}1)$ structure shown in fig. 12. A recurring feature of the computed Al structures was the occurrence of close-packed groups of atoms in the boundaries such as the trigonal prismatic and tetrahedral groups circled in fig. 12. In addition, it was found that the displacement of atoms from their parent crystal sites was often small in low energy structures, considerably less, for example, than the displacements found in the Au twist boundary (fig.10).

Partial secondary dislocations

Partial secondary dislocations exist either at the junction of facets across which the relative displacements of the adjacent crystals are different, or separating domains on a planar interface with different relative displacements. The former type must exist at the junctions of all facets in figs. 3 and 11, and examples, with $|\underline{b}| = 0.18a$, can be seen bounding facet G in fig. 11(a). The second type of partials have been observed in secondary dislocation networks in $\Sigma = 25$ (29) and $\Sigma = 5$ (31) boundaries in Ge.

It was mentioned earlier that partial dislocations also separate interfacial domains with structures related by symmetry such as those shown for the $(1\bar{2}1)$ $\Sigma = 3$ boundary in Al, fig. 12. The holosymmetric symmetry in this case is p 2'mm', where 2' is parallel to $[111]$, m is parallel to $(20\bar{2})$ and m' is parallel to the boundary plane $(1\bar{2}1)$. The equilibrium structures, designated

α_1 and α_2, have the space symmetry plml, and they are related by the elements 2' and m' destroyed by the relaxation from the holosymmetric reference structure. The defect M in fig. 11(b) is a dislocation separating α_1 and α_2 domains (note that the fringe contrast is complementary across the defect) (17).

Perfect secondary dislocations can dissociate into partials separating symmetry domains, and an experiment illustrating this is shown in fig. 14. Fig. 14 (a) shows a region of $(1\bar{2}1)$ $\Sigma = 3$ boundary shown previously in fig. 3(b) (note the steps in bottom right-hand corner). A primary dislocation with known $\underline{b}$ in the upper crystal was forced, by specimen heating, into the boundary where its dissociation was observed. The anticipated dissociation into perfect secondary dislocations was

$$\tfrac{1}{2}[10\bar{1}] \rightarrow {}^{1}/_{3}[11\bar{1}] + {}^{1}/_{6}[1\bar{2}\bar{1}]$$

but the ${}^{1}/_{3}[11\bar{1}]$ dislocation dissociated into partials as follows (17)

$$^{1}/_{3}[11\bar{1}] \rightarrow [0.22,0.22,0.2\bar{2}] + [0.11,0.11,0.1\bar{1}]$$

the two partial secondary dislocations, TQ and SQ in fig. 14(b), bound an α_1 type domain, and the reversal of the fringe contrast inside and outside this domain can be seen. This latter dissociation is shown schematically in fig. 13. Fig. 13 (a) shows a perfect ${}^{1}/_{3}[11\bar{1}]$ secondary dislocation in an α_1 boundary; the defect is a pure edge dislocation and has a step associated with its core. Fig. 13 (b) shows the situation after dissociation and we note that the step is now associated with one of the partials and that, unlike the perfect dislocation, the partials do not have extra half plane description even though they are pure edge dislocations. Finally we note that the form of networks of partial dislocations which accommodate angular deviations from exactly periodic structures depends on the symmetry of the domain structures present (11).

Conclusions

It is clear that the combination of skillful experimental work and improvements of instrumental capabilities, especially over the last decade, has contributed to a significant advance in our understanding of grain boundary structure. The progress made in studies of perfect and partial secondary dislocations and the preliminary observations of grain boundary core structures are particularly noteworthy achievements.

The observations described in this article indicate the probable complexity of a general understanding of grain boundary structure in all materials. It is evident that boundary structure depends not only on geometrical considerations but also on the detailed nature of the chemical bonding. For example, in the case of low angle boundaries in Ge the extent of primary dislocation dissociation was found to depend on both the stacking-fault energy and the misorientation (26). It is also clear that there is a need for further work in areas where our understanding is limited. One such area is the structure of boundaries where the adjacent crystals are not so misoriented that there is a small angular deviation from a periodic structure. The observations of "plane matching" defects are interesting in this context and further work to elucidate the nature of these features would be valuable. This type of investigation is also especially relevant to boundaries in non-cubic materials where comparatively few relative orientations lead to reference lattices with three dimensional translation symmetry.

Acknowledgement

The author wishes to thank the organisers and sponsors for their invitation to this stimulating and enjoyable meeting.

References

1. J.M.Burgers, Proc. Phys. Soc. 52, 23 (1940).
2. W.L.Bragg, Proc. Phys. Soc. 52, 54 (1940).
3. F.C.Frank, see e.g. W.T.Read, Dislocations in Crystals, McGraw-Hill, 1953.
4. W.T.Read and W.Shockley, Phys. Rev. 78, 275 (1950).
5. B.A.Bilby, R.Bullough, and E.Smith, Proc. Roy. Soc. A 231, 263 (1955).
6. W.Bollmann, Crystal Defects and Crystalline Interfaces, Springer-Verlag (1970).
7. R.C.Pond and W.Bollmann, Phil. Trans. Roy. Soc. Lond. 292, 449 (1979).
8. A.King and D.A.Smith, this issue.
9. D.A.Smith, this issue.
10. R.Castaing and A.Guinier, Comples Renders 228, 2033 (1949).
11. R.C.Pond, Grain Boundary Structure and Kinetics, A.S.M. Seminar Series, Milwaukee (1979).
12. A.P.Sutton and V.Vitek, this issue.

13. G.Saada, Acta Metall. 27, 921 (1979).
14. R.Bullough, Phil. Mag. 12, 1139 (1965).
15. M.L.Kronberg and F.H.Wilson, Trans. A.I.M.E. 185, 501 (1949).
16. D.H.Warrington and W.Bollmann, Phil. Mag. 25, 1192 (1972).
17. R.C.Pond, Proc. Roy. Soc. Lond. A 357, 471 (1977).
18. R.Gronsky, as ref. 11.
19. D.H.Warrington and M.Boon, Acta Metall. 23, 599 (1975).
20. R.Schindler, J.E.Clemans and R.W.Balluffi, Phys. Stat. Sol. (a), 56, 749 (1979).
21. D.Vaughan, Phil. Mag. 22, 1003 (1970).
22. P.J.Goodhew, T.Y.Tan and R.W.Balluffi, Acta Metall. 26, 557 (1978).
23. R.C.Pond and V.Vitek, Proc. Roy. Soc. A 357, 453 (1977).
24. J.M.Penisson and A.Bourret, Phil. Mag. A 40, 811 (1979).
25. A.Bourret and J.Desseaux, Phil. Mag. A 39, 405 (1979).
26. A.Bourret and J. Desseaux, Phil. Mag. A 39, 419 (1979).
27. D.J.Dingley and R.C.Pond, Acta Metall. 27, 667 (1979).
28. R.C.Pond and D.A.Smith, Phil. Mag. A 36, 353 (1977).
29. W.Bollmann, G.Silvestre and J.J.Bacmann, Phil. Mag. in press.
30. A.Bourret and C.D'Anterroches, 37th Ann. Proc. E.M.S.A., 388 (1979).
31. J.J.Bacmann, G.Silvestre, W.Bollmann and M.Petit, Phil. Mag. in press.
32. W.A.T.Clark and D.A.Smith, Phil. Mag. A 38, 367 (1978).
33. T.P.Darby and R.W.Balluffi, Phil. Mag. 36, 153 (1977).
34. C.P.Sun and R.W.Balluffi, Scripta Metall. 13, 757 (1979).
35. W.Bollmann, B.Michaut and G.Sainfort, Phys. Stat. Sol. (a) 13, 637, (1972).
36. F.Cosandey and C.L.Bauer, Phil. Mag. in press.
37. R.C.Pond, D.A.Smith and P.W.J.Southerden, Phil. Mag., 37, 27 (1978).
38. G.R.Kegg, C.A.P.Horton and J.M.Silcock, Phil. Mag. 27, 1041 (1974).
39. R.W.Balluffi, as ref. 11.
40. P.H.Pumphrey and H.Gleiter, Phil. Mag. 30, 593 (1974).
41. R.A.Varin and K.Tangri, Scripta Metall. 14, 337 (1980).
42. C.A.P.Horton, J.M.Silcock and G.R.Kegg, Phys. Stat. Sol. (a) 26, 215 (1974).
43. O.L.Krivanek, S.Isoda and K.Kobayashi, Phil. Mag. 36, 431 (1977).
44. P.D.Bristowe and S.L.Sass, Acta Metall. 28, 575 (1980).
45. J.Budai, W.Gaudig and S.L.Sass, Phil. Mag., A 40, 757 (1979).
46. R.C.Pond, J. of Micros. 116, 105 (1979).

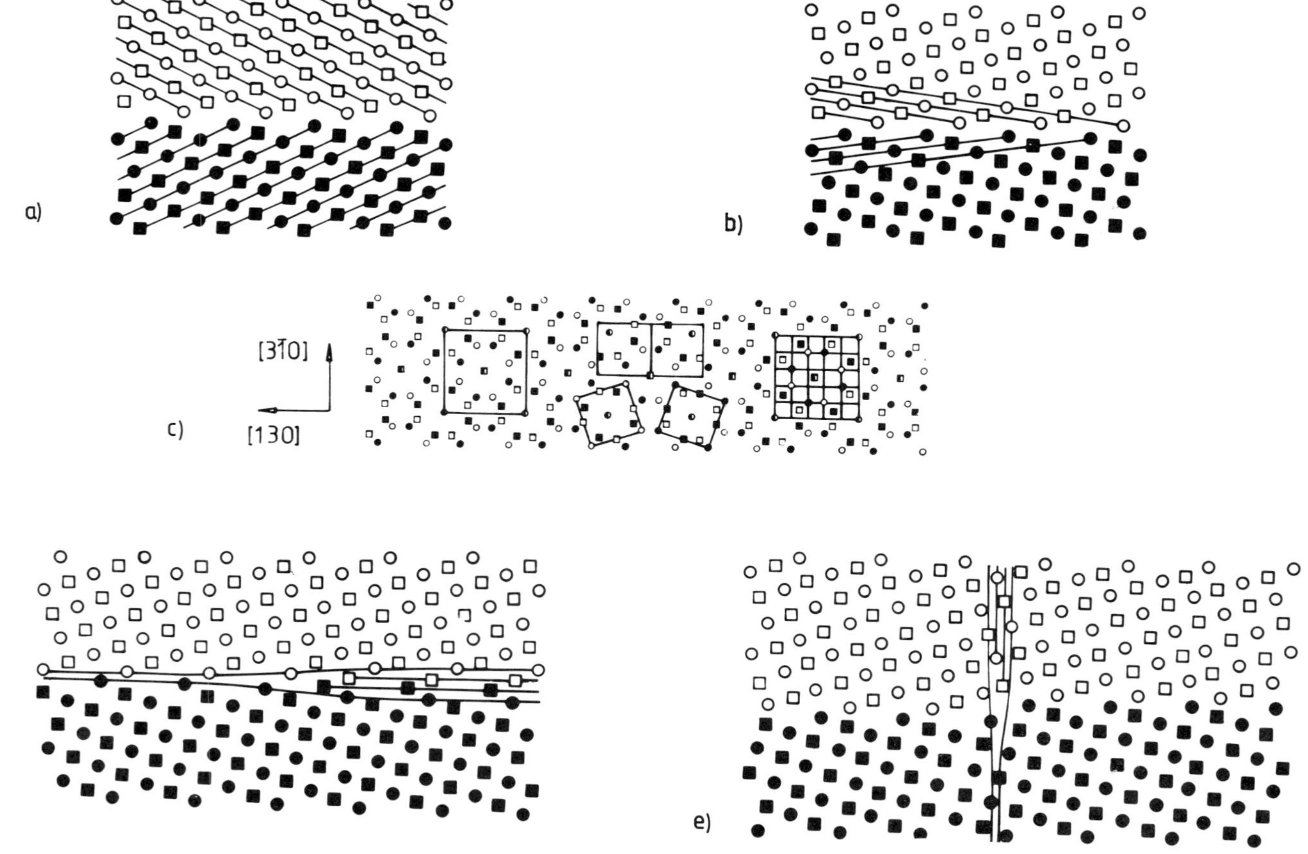

FIG. 1

Schematic illustration of $(3\bar{1}0)$, $\Sigma = 5$ boundary between f.c.c. crystals. Circular and square symbols represent the .....ababab..... stacking along [001], and indices refer to the white crystal (open symbols). The boundary is modelled as an array of $\frac{1}{2}[\bar{1}10]$ primary dislocations in (a) and as $^1/_5[\bar{2}10]$ secondary dislocations in (b). The coincidence site, median and d.s.c. lattices are shown in the dichromatic pattern (c). (d) and (e) show $^1/_{10}[3\bar{1}0]$ and $^1/_{10}[130]$ secondary edge dislocations respectively.

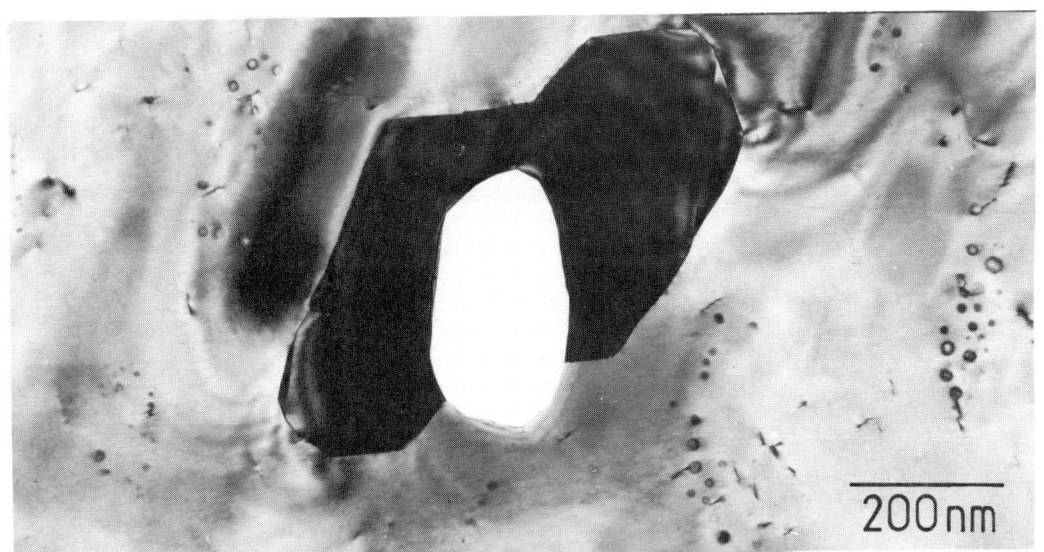

FIG. 2

Faceted Σ = 11 boundaries in gold (Courtesy P.Goodhew)

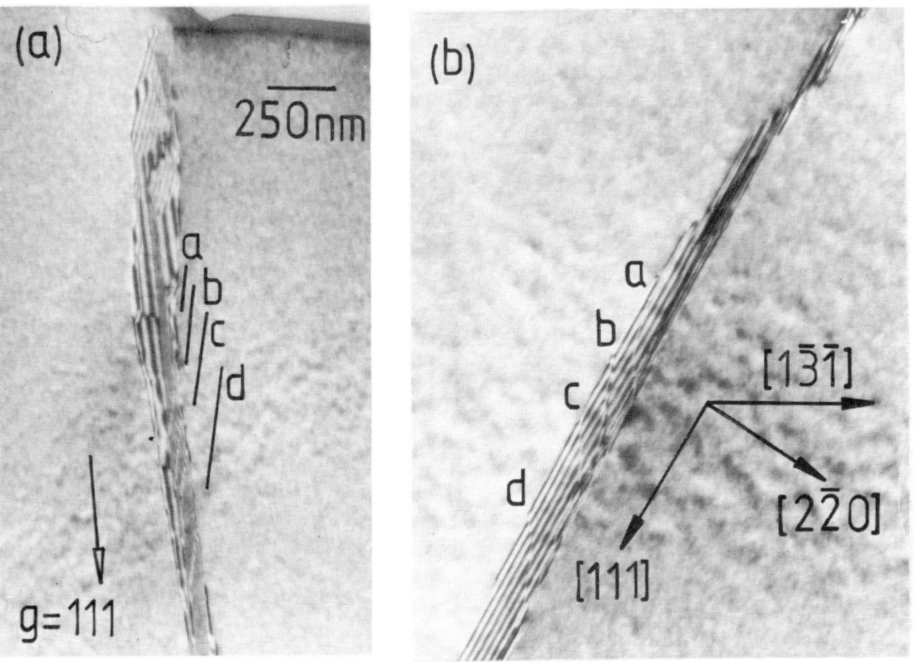

FIG. 3

Faceted Σ = 3 boundary in aluminium

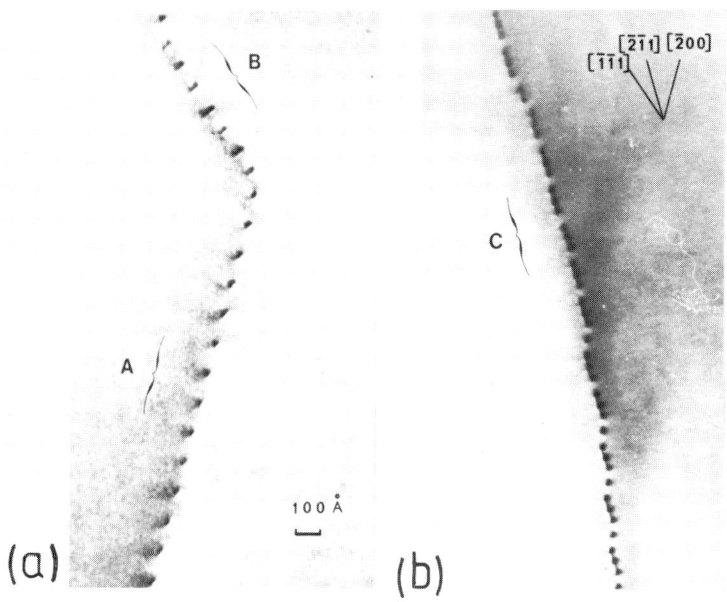

FIG. 4

Low-angle [011] tilt boundaries in germanium (Courtesy of A.Bourret and Phil.Mag.)

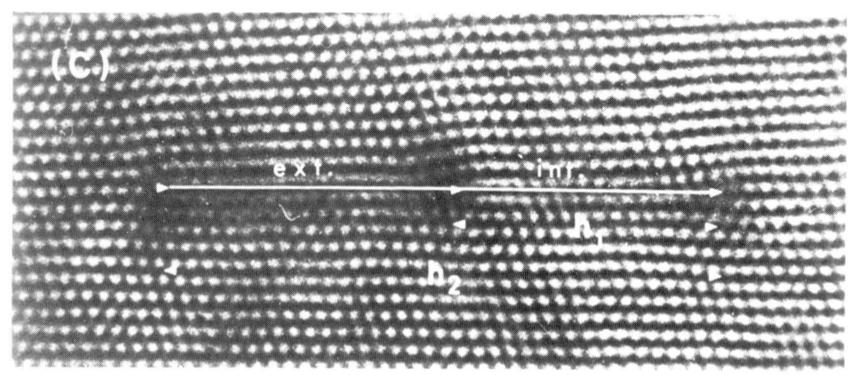

FIG. 5

High-resolution filtered image of type-C dislocation (Courtesy of A.Bourret and Phil. Mag.)

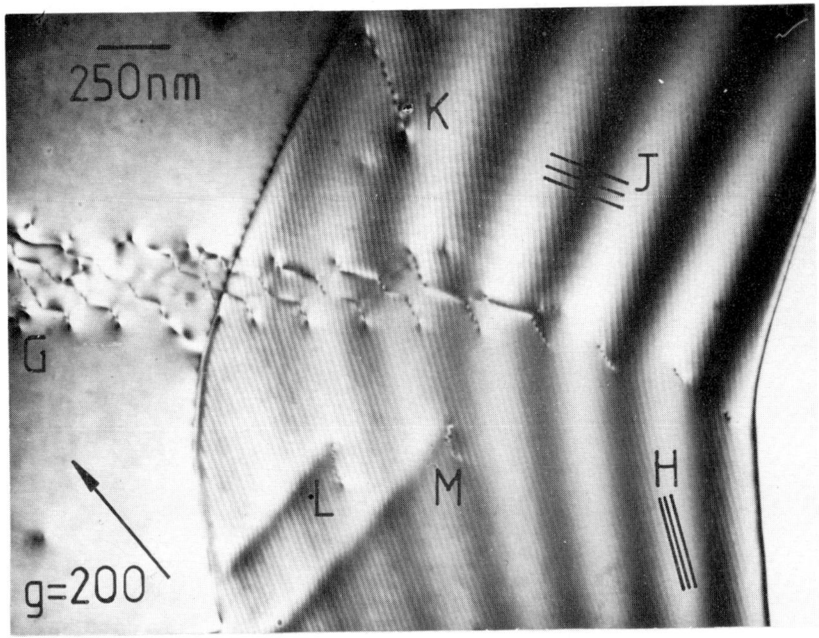

FIG. 6
Medium-angle boundary in aluminium close to the Σ = 41 orientation

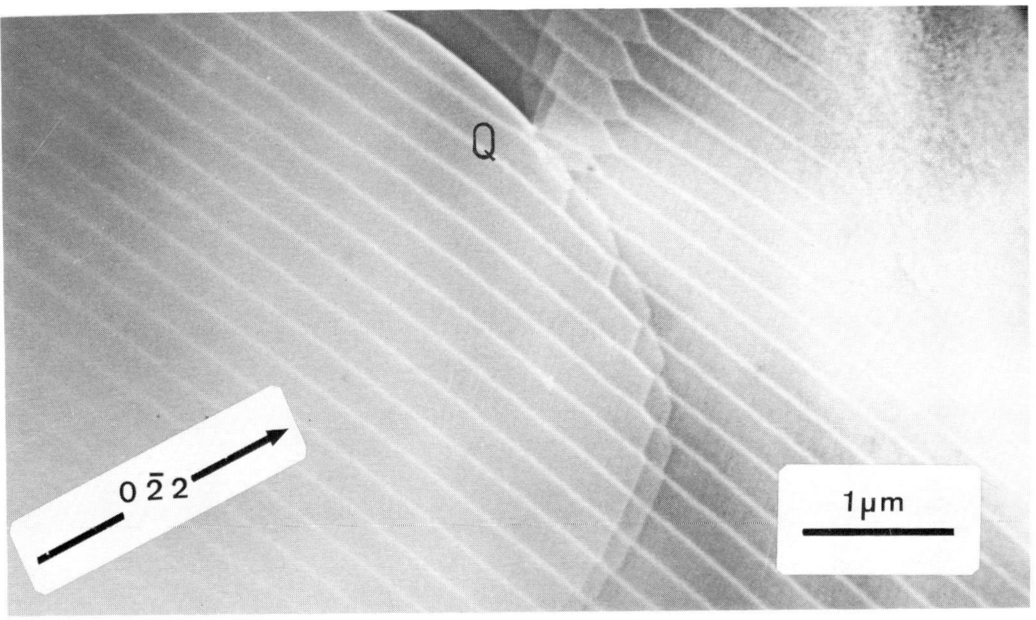

FIG. 7
Medium-angle boundary in germanium close to the Σ = 25 orientation (Courtesy J.J.Bacmann)

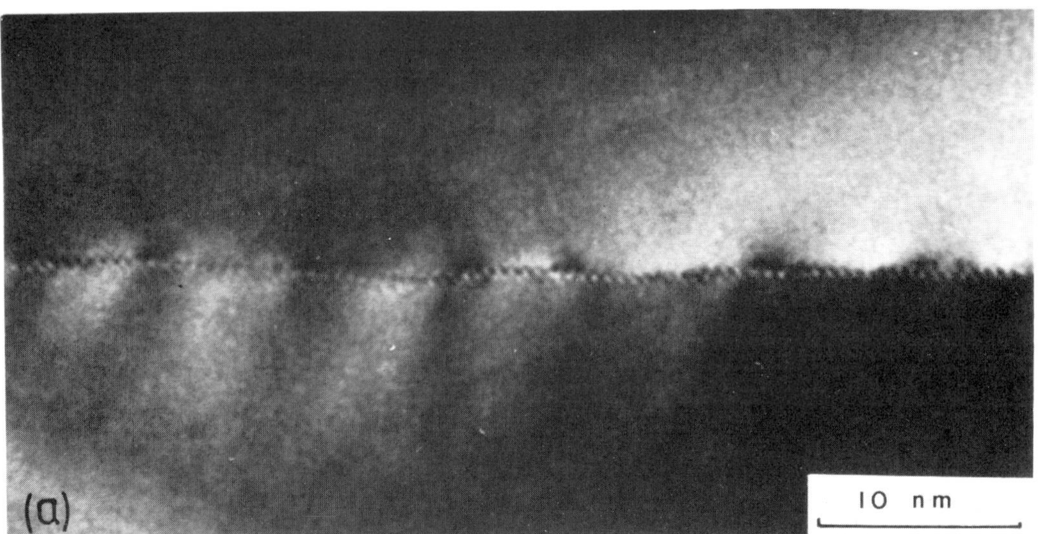

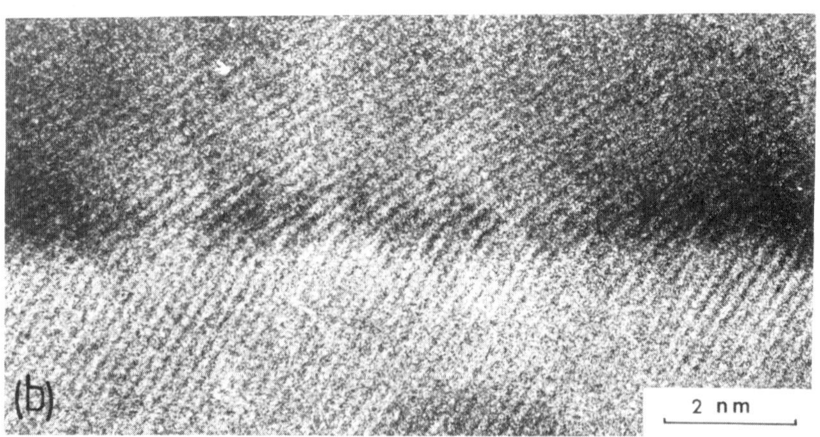

FIG. 8

(a) strain contrast image of a $(3\bar{1}0)$ boundary in gold close to the $\Sigma = 5$ orientation viewed along the common [001] direction. (b) lattice-fringe image for (200) lattice planes at a $(3\bar{2}0)$ tilt boundary in gold close to the $\Sigma = 13$ orientation. (Courtesy of C.L.Bauer).

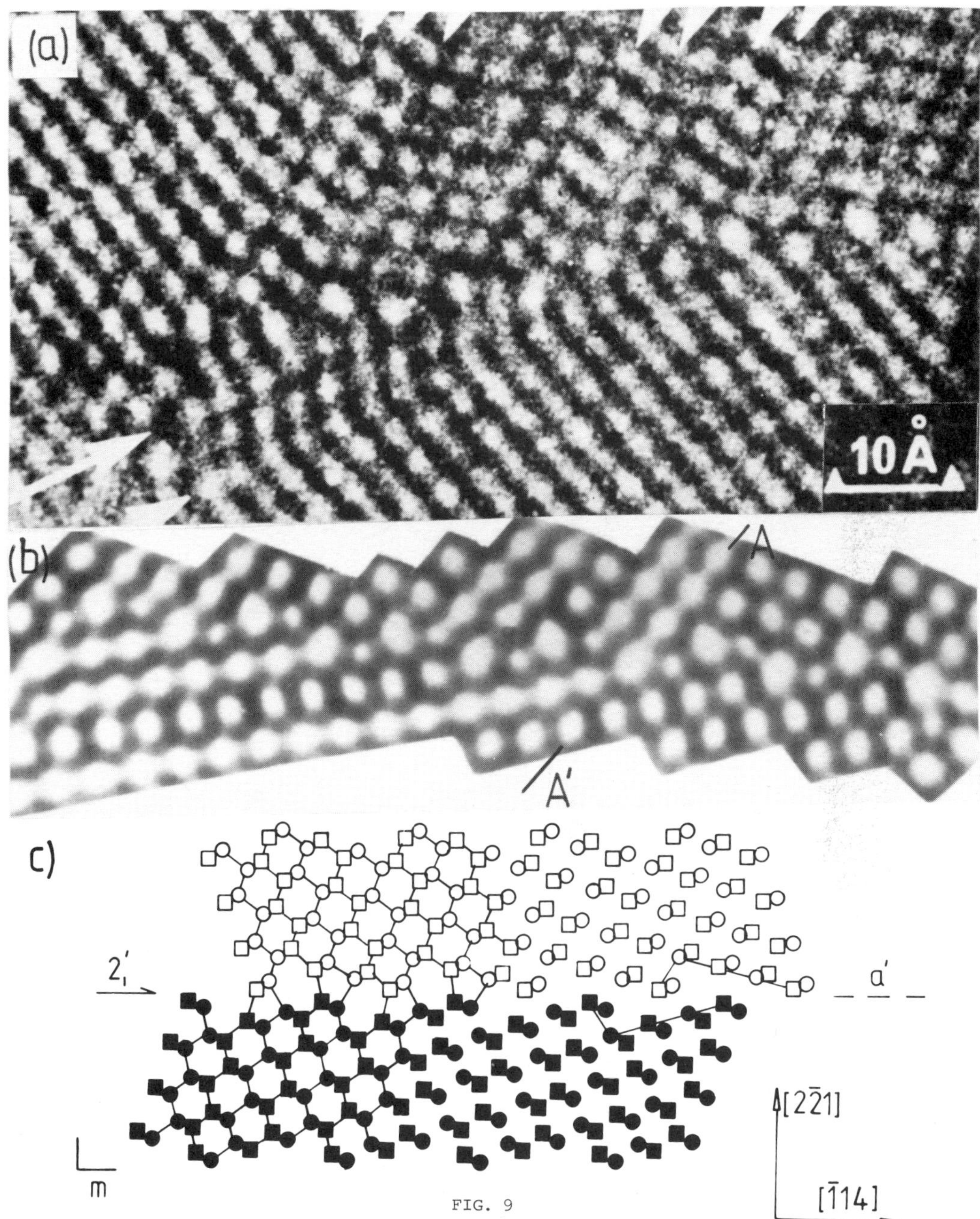

FIG. 9

(a) High-resolution image of a grain boundary in Ge viewed along the common [110] direction. The arrows indicate twinning on (111) planes. (b) Blurred photographic image of atomic model of the boundary. (c) Schematic illustration showing the atomic positions and symmetry in the $\Sigma = 9$ section of the boundary to the right of AA'. (Courtesy of O.Krivanek and Phil. Mag.)

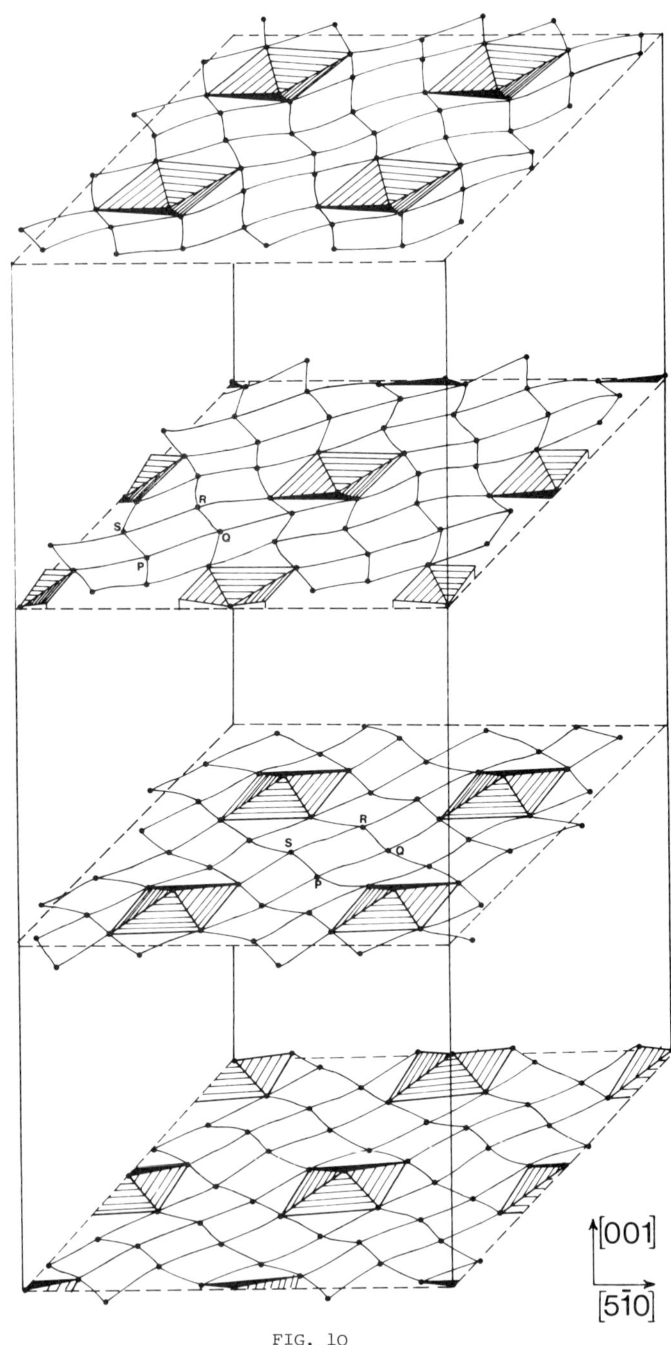

FIG. 10

Calculated structure of [001] twist boundary (Σ = 13) in gold. (Courtesy of S.Sass)

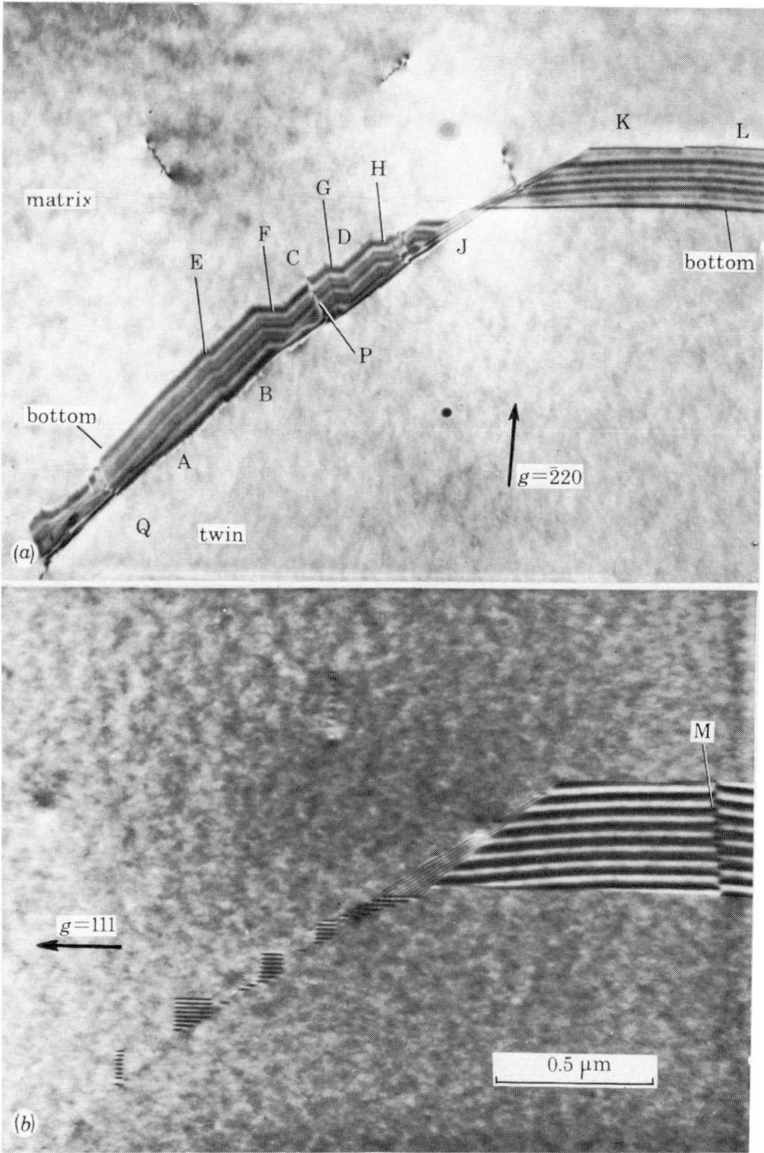

FIG. 11

Faceted twin (Σ = 3) boundary in aluminium. The adjacent crystals are orientated for diffraction by the same set of atomic planes, and the stacking-fault-like contrast reveals that the diffracting planes are not aligned across the boundary (Courtesy of Royal Society).

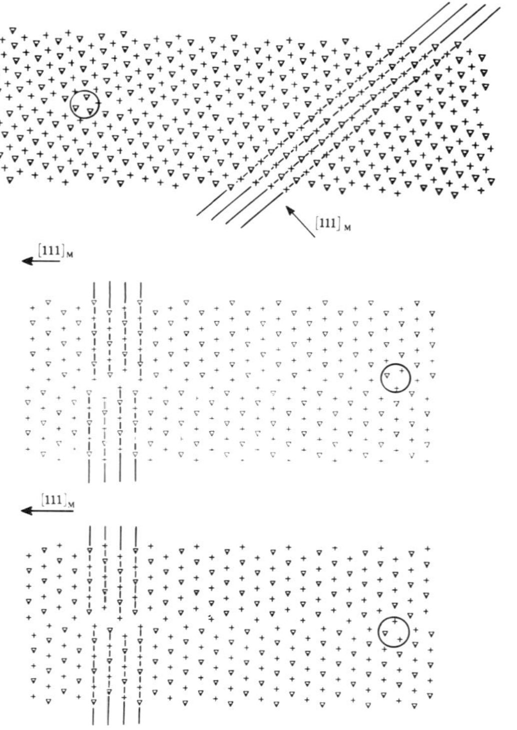

FIG. 12

Calculated structures of the $(\bar{1}55)$ and symmetry related forms of $(1\bar{2}1)$ boundaries in the $\Sigma = 3$ system for aluminium. Triangles and crosses represent the ...ababab... stacking along the common <011> directions. (Courtesy of Royal Society).

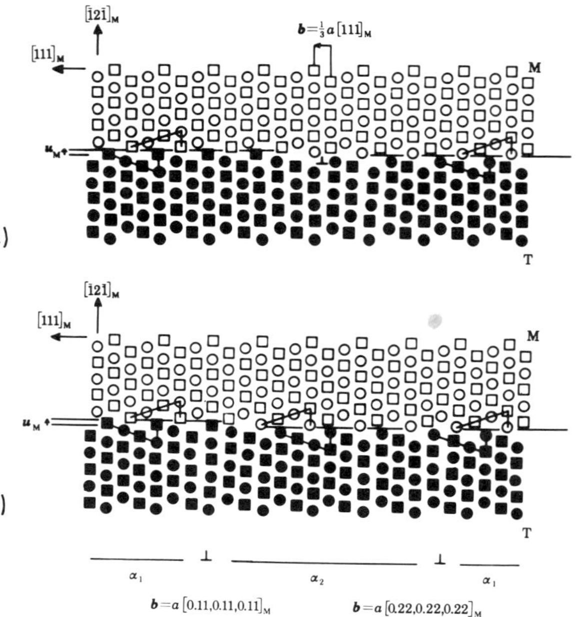

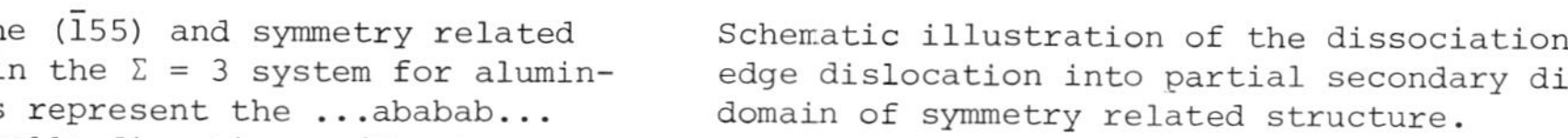

FIG. 13

Schematic illustration of the dissociation of a $^{1}/3\,[\bar{1}\bar{1}\bar{1}]$ secondary edge dislocation into partial secondary dislocations bounding a domain of symmetry related structure.

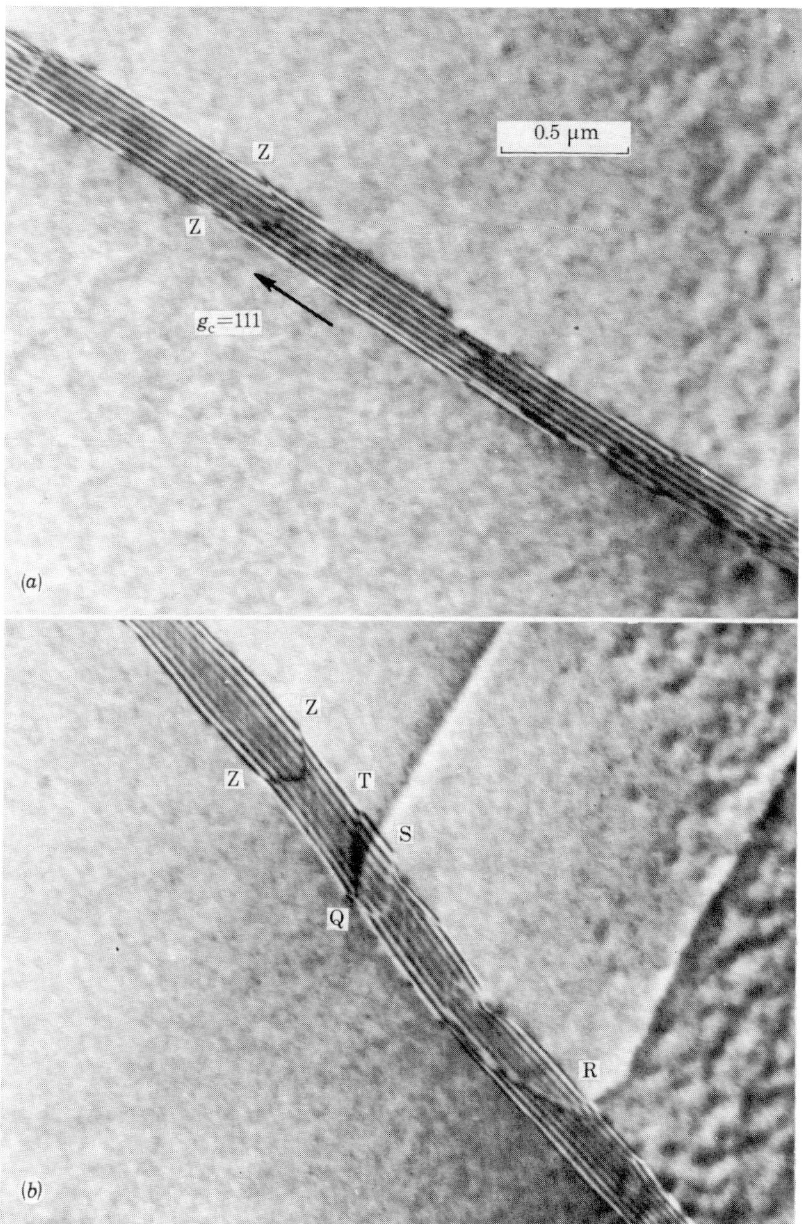

FIG. 14
Observation of the dissociation of a primary dislocation into perfect and partial secondary dislocations in a $(1\bar{2}1)$ boundary between twin ($\Sigma = 3$) related grains of aluminium. In (a) the boundary is shown before the specimen was heated. The slip traces of a primary dislocation and the dissociation are shown in (b). (Courtesy of Royal Society.)

Grain Boundary Dislocations and the Topography of Grain Boundary Planes

A.H. King

Department of Materials Science and Engineering,
Massachusetts Institute of Technology,
Cambridge, MA 02139

D.A. Smith
IBM Thomas J. Watson Research Center,
Yorktown Heights, NY 10598

Abstract

Some topographic properties of grain boundary dislocations are demonstrated, and their effects on the properties of the dislocations are discussed. It is shown that conventional rules concerning dislocation reactions may be inadequate for many grain boundary dislocations.

Introduction

Dislocations which exist in the structures of ordered grain boundaries or interphase boundaries share the properties of crystal lattice dislocations: they have Burgers vectors which are lattice translations in the reference lattice in which they are defined (usually the crystal lattice for a crystal lattice dislocation and the DSC lattice for a grain boundary dislocation). The line sense of the dislocation determines, through the use of the appropriate Burgers circuit and convention, the direction of the Burgers vector. Burgers vector is conserved at dislocation nodes. Grain and some interphase boundary dislocations have an additional set of properties which relate to steps in the interface plane: in general, a grain boundary will be stepped at the core of a grain boundary dislocation, fig. 1. Such steps must, for obvious topographical reasons, be conserved at all dislocation nodes. A similar behavior occurs at the core of a 1/6 <112> transformation dislocation in the fcc-hexagonal phase boundary in cobalt fig. 2.

The reason for the existence of steps at the cores of perfect grain boundary dislocations (gbds) is that, in addition to shifting the DSC lattice by a whole DSC vector, the gbd shifts the coincidence-site lattice (CSL) by an amount which is not, in general, a CSL vector. This can be seen in fig. 1, where coincident sites are shown with filled symbols. It has been demonstrated theoretically and experimentally [1,2] that the grain boundary plane has for a given orientation, a favoured path through the CSL: translations of the boundary plane parallel to its normal, away from its favoured path, result in higher energy structures (unless the translation is equal to a CSL vector and an identical path at a different level is followed). Since a gbd shifts the CSL, it shifts the favoured path of the grain boundary plane which then steps at the gbd core to maintain its minimum energy structure.

In figure 1 and all following examples, the grain boundary plane is taken to pass through coincident lattice sites: this is a graphical and formal convenience which is not designed to represent the atomic structure of any real boundary: any structure of the boundary plane, provided it is always the same for a given boundary orientation, would obey the same rules and lead to the same results.

Formal Geometrical Aspects

The height of the step associated with a gbd is equal to the displacement of the CSL parallel to the boundary normal, or if we define a step vector, **s**, connecting coincident sites on either side of the dislocation (or before and after its passage) then the step height is given by $\mathbf{s} \cdot \mathbf{n}$ where $\mathbf{n}$ is the boundary normal. Some care must be taken in the definition of $\mathbf{s}$ and $\mathbf{n}$ so that all step heights derived from this analysis have self consistent units and signs. We adopt the following conventions:

1. Crystal 1 and crystal 2 are labelled according to the disorientation convention: crystal 2 is found from crystal 1 by means of the smallest possible right-hand rotation angle about an axis **a**.
2. The reference lattice for all vectors is the crystal lattice of crystal 1.
3. The grain boundary plane is defined by **n**, which points from grain 1 into grain 2 and is the reciprocal lattice vector corresponding to the grain boundary plane in the reference crystal.
4. The Burgers vector of the dislocation is defined by the FS/RH (perfect crystal) convention and is the shift of crystal 2 with respect to crystal 1, which is held stationary.
5. The step vector **s** joins an "old" coincident site to a "new" one; i.e., it runs from a site which was coincident before translation of lattice 2 by **b** to one which is coincident after the operation. (See fig. 3.)
6. Following the above conventions, the step height h is given by $\mathbf{s} \cdot \mathbf{n}$, and is the number of crystallographic planes parallel to the boundary plane which the step comprises in crystal 1. The step is said to be positive or negative if h is a positive number or a negative number respectively.

A positive step may be visualized simply, by viewing the dislocation along its positive line direction: if the grain boundary plane lies horizontally across the field of view, then a positive step implies that the boundary plane is higher on the right-hand side of the dislocation than on the left-hand side. For a negative step, the opposite is true. (Fig. 4.) Note that a reversal of any one of the conventions 3 to 6 changes the sign of the step: reversal of any two of the conventions leaves the sign unchanged. The step height defined by h is in effect the height of the stress-free step which would be left on crystal one at the position of the gbd core after separation of the two crystals along the boundary plane.

The step vector, defined using crystal 1 as the reference lattice, is always a vector of lattice 1. It is not, however a unique vector, since any vector from an "old" coincident site to any "new" one is a valid step vector, and gives a geometrically allowed step height for the dislocation in question. For example, in fig.1, the step vector is $1/2[01\bar{1}]$; an equally valid step vector giving the same step height is $[01\bar{2}]$ another valid step vector is $1/2[10\bar{3}]$, which gives a negative step of 3 (062) layers, corresponding to the boundary passing through the layer of coincident sites immediately below the current boundary plane in the right-hand side of fig. 1. Any pair of step vectors **s** and **s**′, referred to the same lattice as reference, is related by

$$\mathbf{s} = \mathbf{s}' + \mathbf{c} \tag{1}$$

where **c** is a CSL vector.

The step vector is reversed if the Burgers vector is reversed, and it is thus a pseudo-vector, like the Burgers vector since it also reverses with the line direction of the dislocation. This also reverses the sign of the step, since the direction of viewing, convention 6, is reversed. If two Burgers vectors are added to form a third, then addition of the step vectors associated with the Burgers vectors gives a step vector appropriate to the product dislocation, although addition of two minimum step vectors need not, yield the minimum step vector for the product. This behavior occurs with an hexagonal network of Shockley dislocations in a {111} oriented twin boundary where topographic considerations require the step heights indicated in fig. 5 i.e., within each mesh of dislocations the boundary is at one of three different levels.

When the step height is measured using crystal 2 as the stationary reference frame, but maintaining all other conventions including the direction of **n**, the step height determined differs from that determined in the reference frame of crystal 1 by $\mathbf{b} \cdot \mathbf{n}$ [3]. This difference represents the material removed parallel to the boundary plane in the Volterra operation to form the component of **b** parallel to **n**. Thus only dislocations with their Burgers vectors parallel to the boundary plane may have equal step heights (including sign) on both crystals, creating restrictions on the step forms which may be associated with different dislocation characters, as summarized in fig. 6.

A pure step, which has no net dislocation character may also be considered in terms of this analysis: for a zero Burgers vector, the step vector is simply (zero plus) a CSL vector. A pure step always, therefore, has height equal (in both crystals) to an integer multiple of the CSL periodicity parallel to **n**.

Step Conservation

It is clear for reasons of topography and topology respectively, that in any dislocation reaction, the step height, as well as the Burgers vector must be conserved. This means that the component of **s** parallel to **n** must be conserved, although in cases such as that illustrated in fig. 1, where several step vectors give the same step height, the in-plane component need not be conserved for a dislocation reaction in a planar boundary. Where the grain boundary plane changes, however, **s** rather than h must be conserved since it has components $\mathbf{s} \cdot \mathbf{n}_1$ and $\mathbf{s} \cdot \mathbf{n}_2$ perpendicular to two boundary planes, both of which must be conserved. The conservation of **s** may have consequences where gbd reactions take place at facet junctions [4] or for dislocations moving in curved boundaries.

Energy Associated with a Step

The total energy for a crystal lattice dislocation is the sum of the long range elastic, core and interaction energy terms. For gbds the core term includes a contribution from the step in the boundary plane: this may be crudely estimated as an increase in boundary area multiplied by grain boundary free energy, γ [5] and refined to take account of anisotropy of γ, the angle of the step and interaction terms in arrays [3]. Both estimates, however, predict step energies proportional to step height. To predict the direction of dislocation reactions, it is usual to assume that the elastic energy terms dominate the total dislocation energy, and to apply the b^2 criterion. For grain boundary dislocations this approach may be inadequate, if the step energy term is comparable with the elastic energy term, and the progress of a reaction produces opposing changes in total step and elastic energies, especially if the dislocations are part of a closely spaced array. The elastic energy per unit length is approximately μb^2 and the corresponding step energy is approximately γh. When these are of the same order of magnitude, the b^2 criterion becomes unreliable, so to derive a "safety condition" we can write

$$\gamma h << \mu b^2 \tag{2}$$

giving

$$\frac{\gamma}{\mu} << \frac{b^2}{h} \tag{3}$$

For a typical metal, μ has a value of approximately 5×10^{10} N m^{-2} and γ has a value of approximately 1 joule m^{-2} so γ/μ is typically 2×10^{-11} m. A reasonable condition for the safe use of the b^2 criterion is then that b^2/h exceeds 10^{-10} m. Since $|b|$ decreases, on average, as Σ, the number of lattice sites of each crystal per coincident lattice site increases and s_{min} is found not to vary so strongly, in most metals the b^2 criterion may be unreliable when Σ exceeds approximately 9. Where the step energy exceeds the elastic energy, we expect gbd reactions to proceed in the direction of minimization of total step height: unfortunately there appears to be no regime where such a criterion is unequivocally superior to the b^2 criterion. Computer simulation is perhaps the best theoretical tool for establishing the range of validity of the b^2 criterion. We note that recent work [6] has predicted the existence of doubly primitive DSC dislocations in the $\Sigma=11$ and $\Sigma=27a$ systems in aluminum and copper where the b^2 criterion would predict splitting into the primitive Burgers vectors, but where this would lead to large steps in the grain boundary plane.

Conclusion

Step heights at grain boundary dislocation cores can be determined graphically in the context of a self consistent set of conventions linking the Burgers vector and core step height of a dislocation. For dislocations with small Burgers vectors, especially in closely spaced arrays the step properties can be dominant. Conservation of step height at nodes is the topographical counterpart of the familiar topological requirement that Burgers vector is conserved at dislocation nodes.

Acknowledgements

This work was supported partially by the NSF MRL program under contract number DMR 78-24185.

References

1. R.C. Pond and V. Vitek, Proc. R. Soc. Lond., A 357 (1977) 453.
2. R.C. Pond, Proc. R. Soc. Lond., A 357 (1977) 471.
3. A.H. King and D.A. Smith, Acta. Crystallogr. (1980) In Press.
4. W. Bollmann, B. Michaut and G. Sainfort, Phys. Stat. Sol. (a) 13 (1972) 637.
5. R.C. Pond and D.A. Smith, Phil. Mag. 36 (1977) 353.
6. A.P. Sutton and V. Vitek, Scripta Met. 14 (1980) 129.

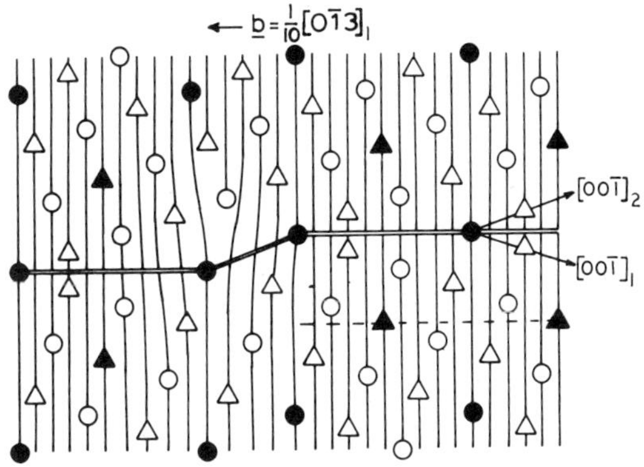

Fig. 1. Diagram showing the geometrical features of a 1/10 [013] dislocation in a (031) boundary in an fcc metal. The Δ and 0 symbols show the ABA stacking sequence of the {200} planes. Coincidence sites are shown solid.

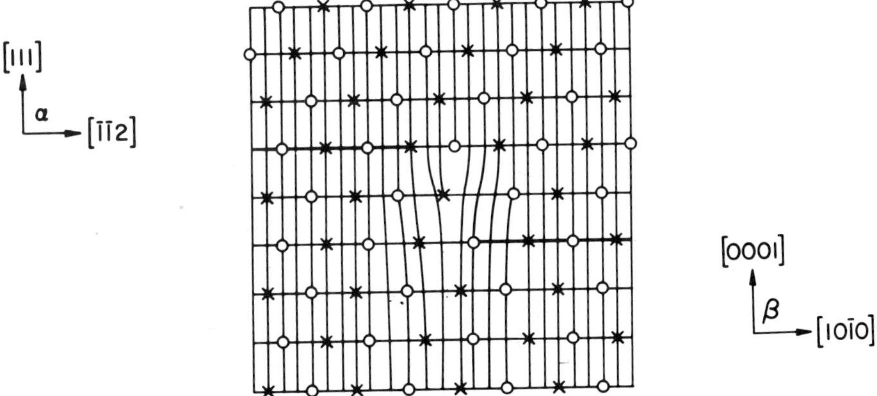

Fig. 2. Diagram showing the geometrical features of a 1/6 <112> transformation dislocation in an fcc-hcp interface the $\{111\}_{fcc}$ $\{0001\}_{hcp}$ plane is perpendicular to the diagram which is a projection on a $\{110\}_{fcc}$ $\{11\bar{2}0\}_{hcp}$ plane.

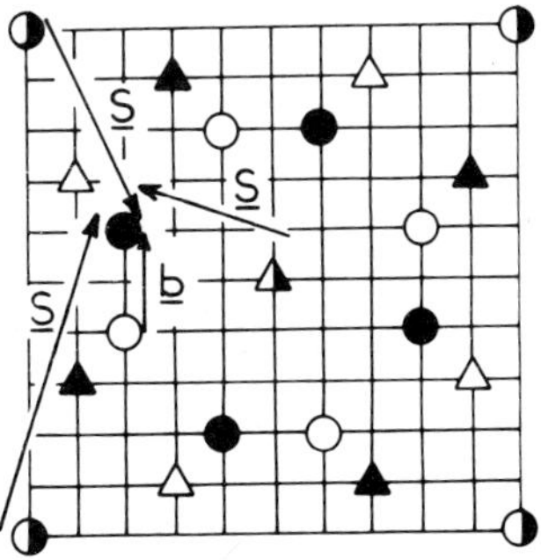

Fig. 3. Diagram showing the unit cell of the $\Sigma=5$ coincidence site lattice for the fcc case and indicating several possible step vectors, filled and open symbols represent lattices 1 and 2 respectively.

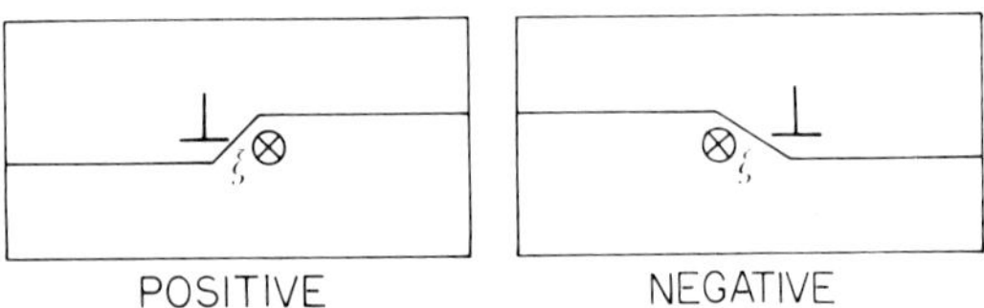

Fig. 4. Diagrams distinguishing positive and negative steps. The symbol ξ represents the line direction of the dislocations and is positive into the page.

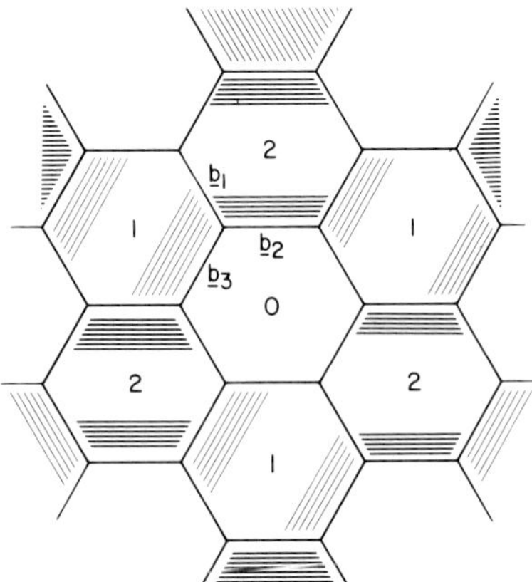

Fig. 5. Diagram indicating the variations in boundary level which are geometrically necessary between different meshes of a hexagonal grid of 1/6 <112> dislocations in a {111} twin in an fcc metal. Note the necessity for some steps of height twice the {111} interplanar spacing, d_{111}.

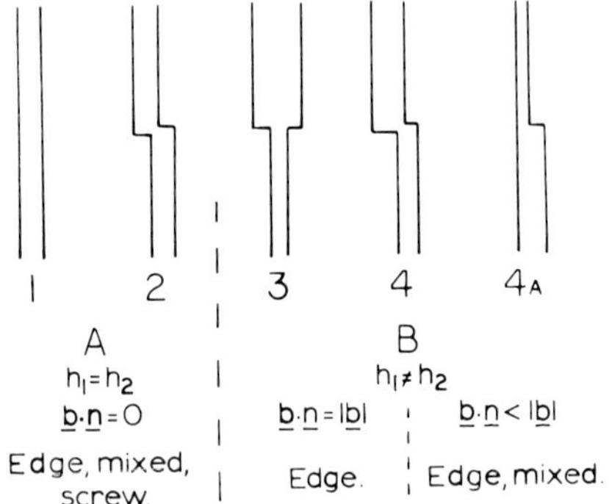

Fig. 6. Diagrams illustrating step forms possible at grain boundaries by showing the configurations that would result if the grains were separated along the boundary.

THE DISLOCATION STRUCTURE OF HIGH ANGLE GRAIN BOUNDARIES

A. P. Sutton and V. Vitek

Department of Materials Science and Engineering and the
Laboratory for Research on the Structure of Matter
University of Pennsylvania
Philadelphia, PA 19104 U.S.A.

ABSTRACT

The fundamental problems of describing unambiguously the dislocation structure of high angle grain boundaries are discussed. This is done by extending recent work on symmetrical tilt boundaries which addressed these problems to more general boundaries. Some experimental evidence for the proposed scheme is mentioned together with a new interpretation of the phenomenon of core spreading.

1. Introduction

In a purely formal sense, a dislocation content can be ascribed to any grain boundary. To do this a reference structure has to be chosen. The corresponding network of dislocations preserves that structure and the Burgers vector content of the boundary accomodates the required misorientation of the grains; the plane of the network is identified with the boundary plane. A natural reference structure is the ideal crystal lattice. Corresponding dislocations are then lattice dislocations and for any grain boundary the Burgers vector content is given by Frank's formula (1). Thus, in the case of low angle boundaries we obtain the Read-Schockley model (2). In this case, individual dislocations are well separated with clearly localized strain and stress fields. Hence, these dislocations can be observed, for example, in the electron microscope. On the other hand, in the case of high angle boundaries the separation between corresponding lattice dislocations may be smaller than the lattice parameters. No localization of the stress and strain can then occur and, furthermore, the elasticity theory becomes inapplicable. Therefore, this dislocation description, although geometrically valid, is physically entirely formal.

However, a dislocation description of high angle grain boundaries which involves localized dislocation has been proposed (3). In this case, the reference structure is not the ideal lattice but an ideal, dislocation-free, coincidence boundary. The dislocations are then appropriate d.s.c. dislocations whose Burgers vectors are translation vectors which preserve the corresponding coincidence site lattice (c.s.l.). A fundamental problem in this description of high angle grain boundaries is deciding which is the correct reference structure appropriate to any particular case, since, at least in principle, an infinite number of different descriptions exist, although only boundaries with relatively low reciprocal density of coincidence sites, Σ, are appropriate. The geometrical theories (i.e. the c.s.l. and O-lattice theories) cannot address this problem because the description as to what is the appropriate reference structure is energetically based,(4). This problem led many workers to resort to ad hoc criteria for choosing the reference structures, such as that $\Sigma \leq n$ where n is a small integer. For example, in ref.(5) n=19 was suggested.

Another question which arises in connection with the above description of high angle boundaries is the transition between two adjacent reference structures. Given two adjacent reference structures in a misorientation range it is not clear how the transition from the first dislocation description to the second occurs as the misorientation changes from the first reference structure to the second.

These questions were studied recently for symmetrical, tilt boundaries with <110> and <100> rotation axis in f.c.c. materials (6). In the present paper we first summarise the principal conclusions reached in ref. (6) and then show their validity for symmetrical grain boundaries with higher index tilt axes, specifically <111>. We further demonstrate that these conclusions can be applied to asymmetrical tilt boundaries, i.e., boundaries for which the crystallographic planes parallel to the boundary are different in each grain. The latter boundaries are likely to be much more common in polycrystalline materials than the symmetrical ones and, therefore, this step represents an extension of the discussed scheme to a more general class of boundaries.

2. Symmetrical <110> and <100> Tilt Boundaries

The calculations in ref. (6) have been made using the pair potentials for aluminium (7) and copper (8) and the following are the principal conclusions reached:

(i) There are two general classes of high angle grain boundaries: "favoured" and "unfavoured." A favoured boundary contains only one type of unit of structure and the boundary consists of an infinite sequence of these identical units. The stress field of a favoured boundary shows no variation parallel to the boundary plane, i.e., no localization of stress in certain parts of the boundary exists. An unfavoured boundary consists of a mixture of units of the two nearest favoured boundaries in the misorientation range. The corresponding boundary stress field shows line sources of stress centred at sites where the two species of units meet. These lines are intrinsic dislocations of the unfavoured boundary and the map of boundary stresses uniquely reveals the dislocation content of the boundary. Thus, the favoured boundaries are the reference structures for defining the dislocations which are an integral part of the structure of the unfavoured boundaries. These dislocations possess similarly localized stress and strain fields as lattice dislocations in low angle boundaries, and can, at least in principle, be directly observed.

(ii) The transition in the dislocation description of unfavoured boundaries lying between A and B favoured boundaries occurs at the boundary corresponding to a 1:1 mixing of the units: ...ABAB... Between ...AA... and ...ABAB... there are more A than B units and the boundaries contain d.s.c. dislocations (with Burgers vector $+\underline{b}_A$ say), corresponding to the A favoured boundary coincidence system, with their cores being the B units. Between ...ABAB... and ...BB... the boundaries consist of opposite sign d.s.c. dislocations (with Burgers vector $-\underline{b}_B$) corresponding to the B favoured boundary coincidence system, with their cores being the A units. At the ... ABAB ... boundary either description can be applied.

(iii) Which boundaries are favoured is not governed by the value of Σ but by energetic considerations and will, in general, vary with the material.

3. Symmetrical Tilt Boundaries with [11$\bar{1}$] Tilt Axis

As an example of symmetrical boundaries with higher index tilt axes we present the results of atomistic calculations of the structure of some symmetrical tilt boundaries with [11$\bar{1}$] tilt axis carried out using the potential for aluminium (7). They are Σ = 3(1$\bar{2}\bar{1}$), 60° [11$\bar{1}$], Σ = 7(2$\bar{3}\bar{1}$), 38.21° [11$\bar{1}$] and Σ = 19(3$\bar{5}\bar{2}$), 46.83° [11$\bar{1}$]. Unlike the <110> boundaries all three boundaries have translations along the tilt axis: 0.235a, 0.283a, and 0.277a for the (1$\bar{2}\bar{1}$), (2$\bar{3}\bar{1}$) and (3$\bar{5}\bar{2}$) boundaries respectively (a=4.033Å is the lattice parameter). The (1$\bar{2}\bar{1}$) boundary is favoured and contains a sequence of identical, connected and interlocked trigonal prisms as seen in Fig. 1 (max. distortion of 1st nearest neighbour distance < 2%). The unit of the boundary is characterised by the vector ½ [101], as shown by the arrow in Fig. 1. The (2$\bar{3}\bar{1}$) boundary (Fig. 2) is also favoured and also consists of identical interlocked trigonal prisms (max. distortion of 1st nearest neighbour distance < 4%). This boundary has three c.s.l. sites per period of the boundary and the vector characterising the unit of the boundary is 1/6 [415], as shown by the arrow in Fig. 2. Fig. 3 shows the structure of the (3$\bar{5}\bar{2}$) boundary. It also contains three c.s.l. sites per period of the boundary and the vector characterising its unit is 1/6 [718] (shown by the arrow in Fig. 3). This boundary is not favoured and each of its units has decomposed into one unit of each of the (1$\bar{2}\bar{1}$) and (2$\bar{3}\bar{1}$) boundaries according to the equation:

$$1/6\ [718] \rightarrow 1/2\ [101] + 1/6\ [415]$$

Thus, we see that the (3$\bar{5}\bar{2}$) boundary is the 1:1 boundary separating the dislocation description based on the (1$\bar{2}\bar{1}$) and (2$\bar{3}\bar{1}$) favoured boundary reference structures. To reveal the intrinsic dislocation content of this boundary the hydrostatic stress field map of this boundary is presented in Fig. 4 which clearly shows line sources of stress.

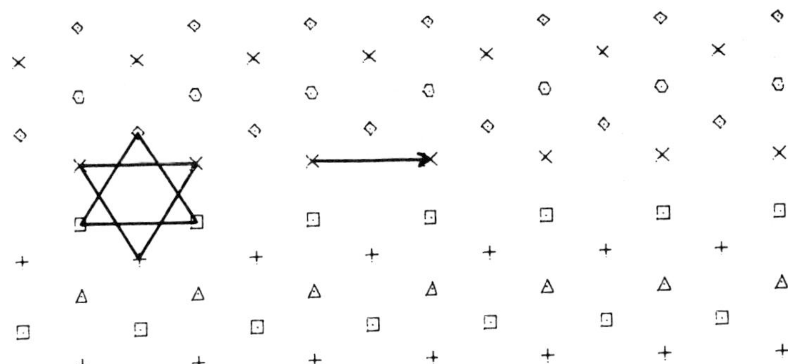

Fig. 1. Fully relaxed structure of $\Sigma=3(1\bar{2}\bar{1})$, 60°$[11\bar{1}]$ boundary in aluminium

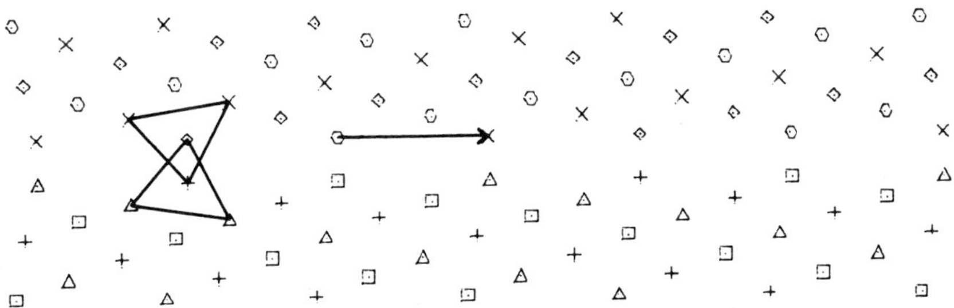

Fig. 2. Fully relaxed structure of $\Sigma=7(2\bar{3}\bar{1})$, 38.21°$[11\bar{1}]$ boundary in aluminium

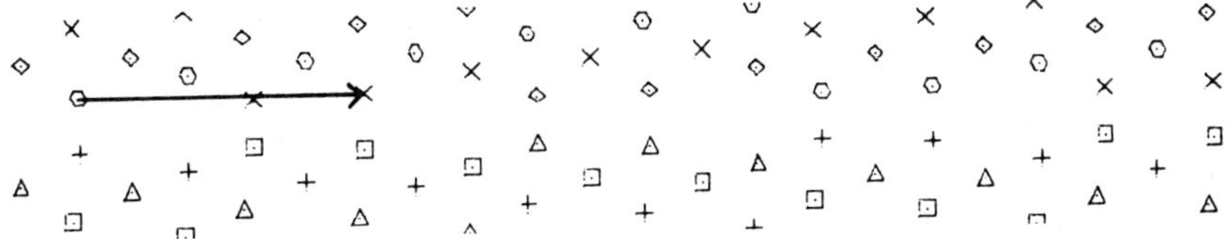

Fig. 3. Fully relaxed structure of $\Sigma=19(3\bar{5}\bar{2})$, 46.83°$[11\bar{1}]$ boundary in aluminium

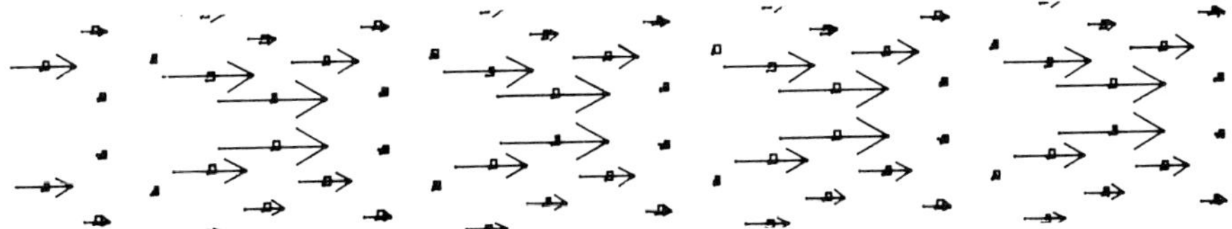

Fig. 4. The hydrostatic stress field map of the same boundary as Fig. 3. Arrows pointing to the right denote hydrostatic compression and to the left tension. The arrow length is proportional to the magnitude of the hydrostatic stress.

4. Asymmetrical Tilt Boundaries

In this section we will show how the above scheme may be extended to asymmetrical tilt boundaries by presenting the results of atomistic calculations of the structure of the $\Sigma=17$ $(\bar{1}\bar{1}11)_U/(77\bar{5})_L$, 124.12° $[1\bar{1}0]$ in aluminium. (Subscripts U and L denote the upper and lower grain, respectively; the planes are indexed with respect to the upper grain.) This boundary, shown in Fig. 5, was found to be unfavoured and decomposed into one unit of each of the nearest favoured boundaries, i.e., the $\Sigma=3(001)_U/(22\bar{1})_L$, 109.47° $[1\bar{1}0]$ and the $\Sigma=9(\bar{1}\bar{1}5)_U/(11\bar{1})_L$, 141.06° $[1\bar{1}0]$ boundaries.

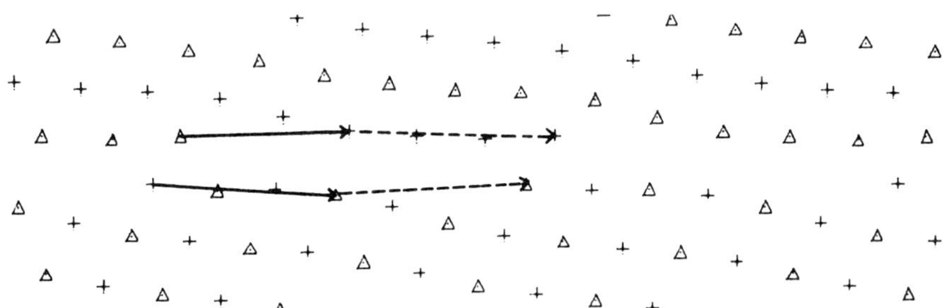

Fig. 5. Fully relaxed structure of $\Sigma=17(\bar{1}\bar{1}11)_U/(77\bar{5})_L$, 124.12° $[1\bar{1}0]$ in aluminium; an asymmetrical tilt boundary

With respect to the upper grain the decomposition of the units may be written as

$$\frac{1}{4}\,[11\ 11\ 2] \rightarrow \frac{3}{2}\,[110]_U + \frac{1}{4}\,[552]_U$$

or equivalently, with respect to the lower grain the decomposition may be expressed as

$$\frac{1}{4}\,[5\ 5\ 14]_L \rightarrow \frac{1}{2}\,[114] + \frac{3}{4}\,[112]$$

where $1/4\ [11\ 11\ 2]_U/1/4\ [5\ 5\ 14]_L$ is one unit of the $\Sigma=17$ boundary,

$\frac{3}{2}\,[110]_U/1/2\ [114]_L$ is one unit of the $\Sigma=3$ boundary shown by the dashed arrows

and $1/4\ [552]_U/3/4\ [112]_L$ is one unit of the $\Sigma=9$ boundary shown by the full arrows.

Thus this $\Sigma=17$ boundary is the 1:1 boundary separating the dislocation description based on the above $\Sigma=3$ and $\Sigma=9$ favoured reference structures.

5. Discussion

After carrying out an extensive number of calculations for symmetrical and asymmetrical tilt boundaries for different tilt axes and with different interatomic potentials we are led to the general conclusion that all tilt boundaries are related to some favoured boundary reference structures by appropriate arrays of d.s.c. dislocation networks. The reference structures (favoured boundaries) may be quite far apart (say, 20-30°) but nevertheless the dislocations always remain localized and distinct since the highest density of dislocations corresponds to the 1:1 boundaries where the dislocations are separated by one unit of either species. When all the favoured boundaries are known for a given tilt axis, and there is always only a small number of them, the structure of all the other intervening boundaries can be predicted using the above decomposition scheme. The spirit of this description of high-angle grain boundary structure in terms of structured units exists in the work of Bishop and Chalmers (9). While they were able to show, geometrically, how units could be combined to construct other boundaries the fundamental problem of which structural units to choose remained; there also seems to be no distinction made between decompositions producing faceting and decomposition resulting in intrinsic dislocation formation.

An experimental investigation which strongly supports our general conclusions is the studies of the rotation of a large number of copper single crystal balls which were originally randomly spread on a planar single crystalline substrate (10). On the basis of these observations it has been concluded in ref. (10) that "the observed rotations of all balls, irrespective of their angular deviation from a low energy relationship, suggest that all grain boundaries are structurally related to low energy boundaries."

This description of high angle grain boundary structure also leads to a simple interpretation of experiments which report the so-called "core-spreading" of grain boundary dislocations (11). When an extrinsic lattice dislocation enters an unfavoured boundary which may be far from the nearest favoured boundaries (e.g., the $\Sigma=19$ $(3\bar{5}\bar{2})$, 46.83° $[11\bar{1}]$ shown above) it may or may not be incorporated into the existing dislocation array. If it is incorporated it will dissociate into the d.s.c. dislocations of the nearest favoured boundary. However, these dislocations may be very closely spaced and thus although localized, they will be irresolvable in the T.E.M. This process will have the appearance of dislocation core spreading.

A complete geometrical analysis of high angle grain boundary structure based on the calculations of the type presented in this paper, is presently being developed for tilt, twist and mixture of tilt and twist and will be published later (12). This analysis, for example, also provides a clear physical explanation of the existence of 'plane matching' dislocations (13,14) but a more detailed discussion of this phenomenon is outside the scope of this paper.

Acknowledgements

This research was supported by the National Science Foundation, the MRL Program under Grant DMR79-23647 and by the Department of Energy Grant No. DE-AC02-79ER10429.

References

1. F. C. Frank, "Report of the Symposium on the Plastic Deformation of Crystalline Solids," Carnegie Institute of Technology, Pittsburgh, 1950, p. 150.
2. W. T. Read and W. Shockley in "Imperfections in Nearly Perfect Crystals," Wiley, New York, 1952, p. 352.
3. W. Bollmann, "Crystal Defects and Crystalline Interfaces," Berlin: Springer-Verlag (1970).
4. P. J. Goodhew and D. A. Smith, Scripta Metall. 14, 59, 1980.
5. P. H. Pumphrey in "Grain Boundary Structure and Properties," Ed. G. A. Chadwick and D. A. Smith, Academic Press, London, pp. 139-200 (1976).
6. A. P. Sutton and V. Vitek, Scripta Metall., 14, 129 (1980).
7. L. Dagens, M. Rasolt and R. Taylor, Phys. Rev.(B) 11, 2726 (1975).
8. K. M. Miller and P. D. Bristowe, Phys. Stat. Sol.(b) 86, 93 (1978).
9. G. H. Bishop and B. Chalmers, Scripta Metall. 2, 133 (1968).
10. H. Sauter, H. Gleiter and G. Bäro, Acta Met., 25, 467, 1976.
11. P. H. Pumphrey and H. Gleiter, Phil. Mag., 30, 593, 1974.
12. A. P. Sutton and V. Vitek, to be published.
13. P. H. Pumphrey, Scripta Metall., 6, 107 (1972).
14. R. Schindler, J. E. Clemans and R. W. Balluffi, Phys. Stat. Sol.(a), 56, 749 (1979).

DISSOCIATED DISLOCATIONS IN (111) TWIST BOUNDARIES

C. B. Carter* and H. Foll**

*Department of Materials Science and Engineering, Cornell University, Ithaca, NY 14853

**I.B.M. Research Center, P. O. Box 218, Yorktown Heights, NY 10598

Abstract

In face-centered cubic and diamond cubic materials small-angle, pure-twist grain boundaries on a 111 plane are found to consist of alternating K_S and P_S dislocation nodes. When the stacking fault energy is sufficiently low both types of node can have a dissociated form: in this paper, the geometry and structure of extended K_S nodes is discussed. In particular, it is shown that at least one jog must be present at the node, and for dislocation symmetry to be maintained two or three jogs are required. Observations showing the variation in extension of K_S nodes which may be explained by this analysis are given and the results are related to the structure of dislocations in $\Sigma 1$ and $\Sigma 3$ {111} grain boundaries.

1. Introduction

In the small-angle regime, twist boundaries on {111} planes in materials with the face-centered cubic or diamond-cubic structures are found experimentally to take the form of hexagonal networks of perfect screw dislocations. Dislocations with three different Burgers vectors are present and meet at two types of alternating nodes which are referred to as P_S and K_S. (The subscripts refer to the nodes being symmetric such that the three dislocations can all be of pure screw character and make angles of 120° with one another.) If the stacking-fault energy is sufficiently low the perfect dislocations can dissociate. As a result it is found experimentally that the P_S node is usually widely dissociated (1-3) with the Shockley partial dislocations enclosing an intrinsic stacking-fault. However the K_S nodes are found to vary greatly in size being completely constricted in some cases and widely extended in others (1,3-5). When a K_S node is extended it is assumed to contain an extrinsic stacking-fault and this along with the possible difficulty of moving the partial dislocations at these nodes and the potential of stronger solute pinning, has been proposed as an explanation for the variability in the observed extension of the K_S nodes. Observations on double ribbons in the same materials show that the extrinsic stacking-fault energy is invariably similar to the value for the intrinsic stacking-fault (5-7). Furthermore, detailed studies of double ribbons also suggest that the partial dislocations enclosing extrinsic stacking-faults behave in a manner similar to the familiar Shockley partial dislocations which enclose intrinsic stacking faults (6).

In this paper it will be shown that the extension of K_S nodes involves the movement of jogs and that it is the presence or absence of these jogs that will control the ease with which the K_S nodes can dissociate. A recent observation which is closely related to this is that the presence of jogs can also cause both K_S and P_S nodes to be constricted (3).

2. Extended Nodes in $\Sigma 1$, {111} boundaries.

The pure twist boundary is found to consist of repeated node-pairs of the type shown in Fig. 1a - the notation of Frank (8) and Thompson (9) is used throughout this paper (see (10) for a summary). In a material with a sufficiently low intrinsic stacking-fault energy, the P_S node will usually be dissociated as shown in Fig. 1b. Since it is generally found that, in such materials, the extrinsic stacking-fault energy is within ∿20% of the value for the intrinsic fault, the K_S node can also dissociate with four possible resulting configurations shown in Fig. 2. For energetic reasons, the configuration shown in Fig. 2a is unlikely to occur in practice, but it is of interest because it shows a general feature of the dissociation: the number of dislocation junctions increases but there are then only three dislocations involved at each junction. In this case the three 'new' perfect dislocations in the node now form a P_s node. However, the new P_s node cannot dissociate because to do so would involve the creation of the forbidden A-upon-A stacking sequence; if instead the P_s node were on the plane adjacent to the stacking-fault, dissociation could occur with the resulting configuration shown in Fig. 2b. Thus the so-called cross-tie partial dislocations in Fig. 2b must lie on a plane adjacent to the original, intrinsic stacking-fault. In the third possible configuration, shown in Fig. 2c, the Shockley partial dislocations appear to have crossed through the node; the situation is actually more complex since the cross-over would again produce the forbidden A-upon-A stacking: each of the partial dislocations enclosing the triangle of extrinsic stacking-fault is actually composed of 2 (or 3) Shockley partial dislocations on adjacent planes. Although both the configurations in Figs. 2b and 2c have been observed, the configuration shown in Fig. 2d is actually more common and can form by simple rearrangement of the previous forms. This configuration minimizes the interaction between different partial dislocations by maximizing the angles between them (5).

The mechanism by which the K_s node dissociates is actually complex and may not be unique. In a small angle $\Sigma 1$, 111 pure-twist boundary the dislocations meeting at the K_s node are assumed to be coplanar. Symmetric dissociation of the node involves the nucleation of three jogs whose Burgers vectors sum to

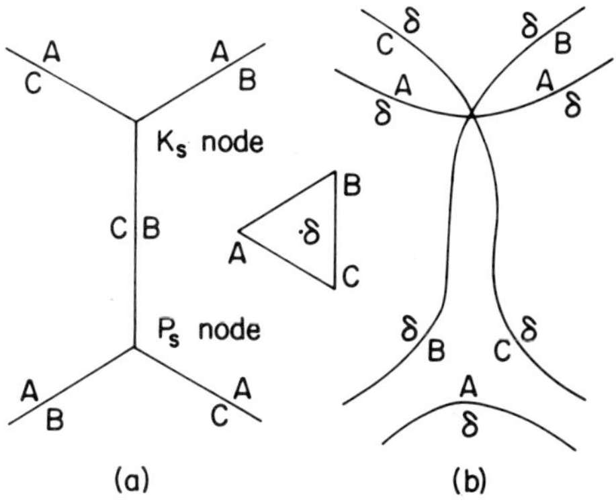

Fig. 1 K_s and P_s node pair: a) without dissociation b) with dissociation.

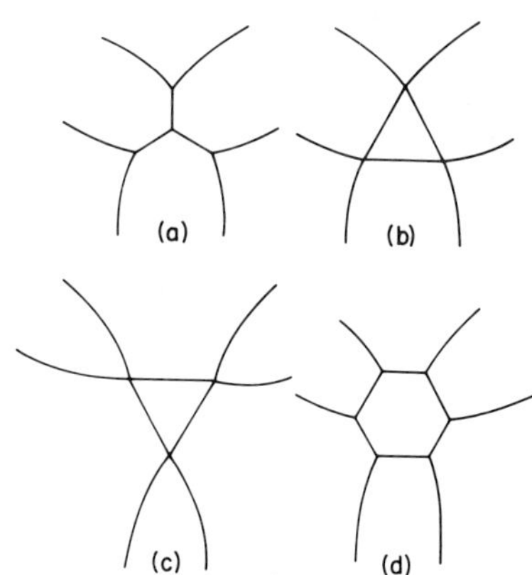

Fig. 2 Four different possible configurations of the extended K_s node.

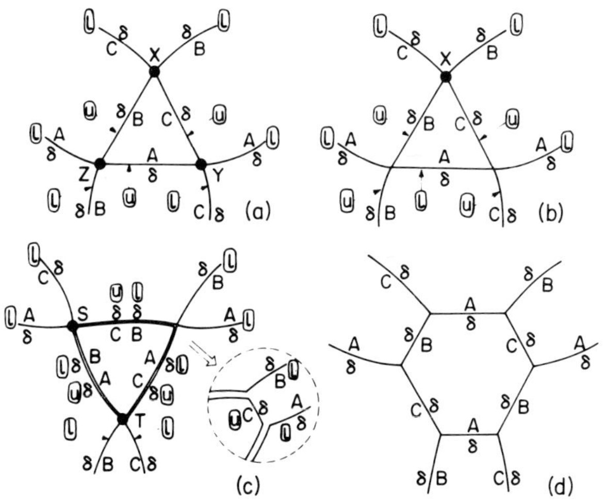

Fig. 3 Different arrangements of jogs involved in the extension of K_s nodes; u and l refer to the upper and lower of two adjacent {111} planes. In (c) ST is parallel to edge AC of the Thompson Tetrahedron. (d) shows schematically the effective Burgers vectors in the final configuration.

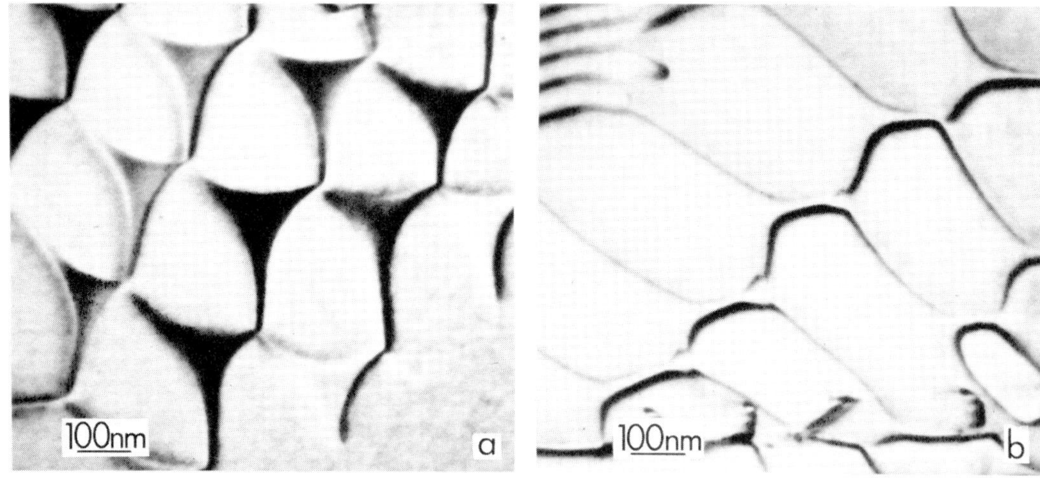

Fig. 4 (a) Constricted K_s nodes in Cu-16at%Al (b) K_s nodes in Cu-16at%Al which are extended by different amounts.

zero as shown in Fig. 3a for the geometry of Fig. 2b. In order to arrive at the geometry shown in Fig. 3a, at least one of the jogs must climb. If a single jog is present, ab initio, at the K_s node, then the same node can dissociate by glide - in this case the Shockley partial dislocation Aδ can cross the constricted node and convert part of the intrinsic stacking fault to extrinsic as shown in Fig. 3b.

The node extension discussed for Figs. 3a and b thus requires either the presence of a single pre-existing jog or the nucleation and climb of new jogs. An alternative mechanism is shown in Fig. 3c where the node has the shape discussed for Fig. 2c. In this case the K_s node can become extended by the nucleation of two jogs whose Burgers vectors have the same magnitude but opposite sign. The jogs can glide apart by moving in the <110> direction parallel to their Burgers vector. In this example the dissociation can be imagined as occurring by a short length of dislocation with Burgers vector CA cross-slipping and then dissociating on the plane adjacent to the original fault plane; one partial dislocation, δA(u) (u and l signify whether a partial dislocation lies on the upper or lower of two adjacent {111} planes) then remains along the line joining the two jogs and interacts with partial δB(l) to give a complex partial dislocation with Burgers vector Cδ(l)-the other partial dislocation δC(u) bows out to react in a similar manner with partial dislocations δB(l) and δA(l). There is, of course, no reason then for the node to be constricted at Z and the configuration shown in the inset is then expected to occur. As was the case for the node in Fig. 3a, the node in Fig. 3c can subsequently rearrange to form the configuration with effective Burgers vectors shown in Fig. 3d. Before considering the implications of this discussion for understanding the geometry of {111} twist boundaries experimental observations which support the proposed model will be discussed.

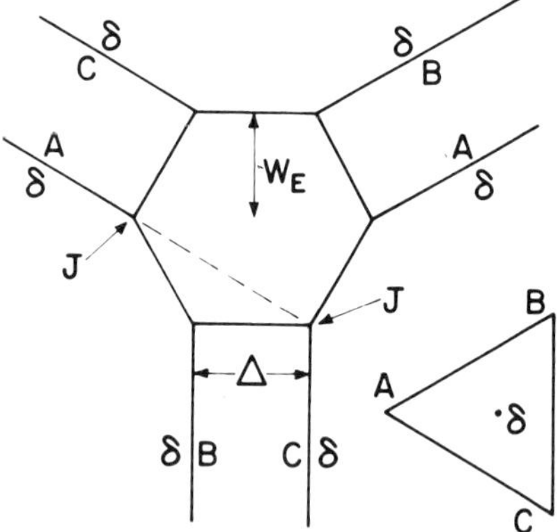

Fig. 5 The extension of a symmetric, pure screw K_s node by glide alone. The two jogs, J, have Burgers vectors $\pm\underline{AC}$ and W_E is the radius of the inscribed circle.

3. Experimental Results and Discussion

Experimentally it is found that K_s nodes are extended by varying amounts and can be totally constricted. An example of the latter is shown in Fig. 4a where it can be seen that the contrast from the intrinsic stacking-faults contained in the P_s node changes abruptly from node to node suggesting a relatively large step, between the nodes; such large jogs would cause the

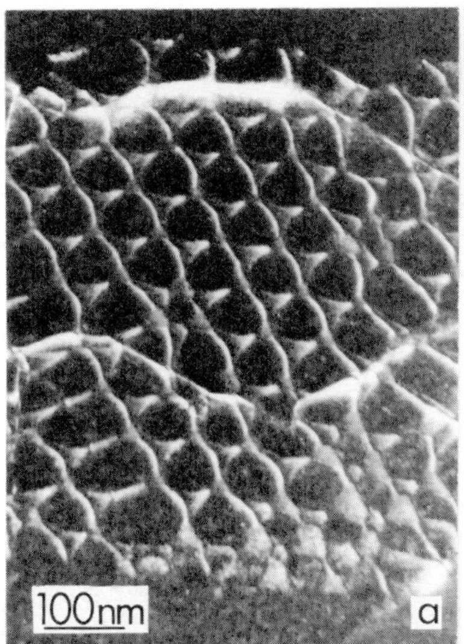

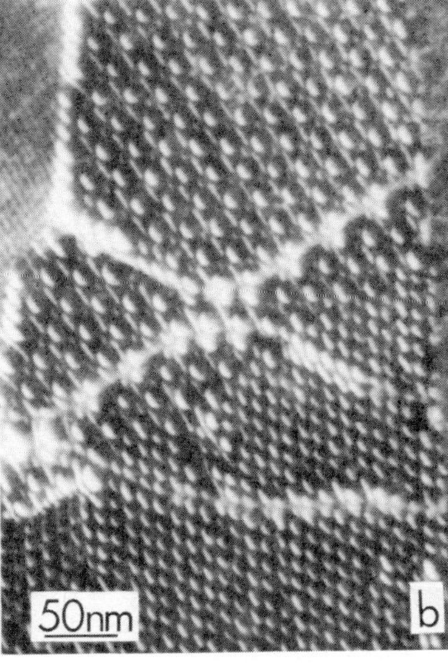

Fig. 6 (a) A small angle $\Sigma 1\{111\}$ grain boundary in Si. The star pattern may be in effect a small nucleus of 3 like boundary. (b) A small angle boundary showing in one region extended P_s and K_s nodes and in an adjacent region the hexagonal pattern of a $\Sigma 3$ boundary.

K_s nodes to be totally constricted. A row of K_s nodes extended by differing amounts can be seen in Fig. 4b; in this case, the specimen has been annealed after deformation and this would have allowed climb to occur - the amount of climb and, thus dissociation, could vary locally. Widely extended K_s nodes have been observed in as-deformed material (3). By the arguments given above, such nodes could occur if a single jog were present at the node - such jogs were identified at P_s nodes where they cause the node to be at least partly constricted (3).

By considering for example the case of a K_s node in the screw orientation where all three perfect dislocations are initially coplanar, an estimate can be made of the maximum extension possible in the absence of climb. The situation is shown schematically in Fig. 5 where the assumption is made that the length of the cross-tie partial dislocations will not exceed Δ, the dissociation width of the screw perfect dislocations. Simple geometric considerations lead to the result

$$W_E = \frac{3^{1/2}}{2} \Delta$$

where W_E is the radius of the inscribed circle of the K node. Early calculations (11) showed that the corresponding value for the P_s node could, to a good approximation, be expressed as

$$W_I = 2.4\Delta$$

which thus leads to the relation

$$W_E \simeq \frac{1}{3} W_I$$

This relation is precisely that found experimentally by many authors and which earlier led to the conclusion (12) that $\gamma_E \sim 3\gamma_I$ (γ is a stacking-fault energy). If jog climb is allowed and the configuration is the symmetric one shown in Fig. 3b the node may be able to more closely approach elastic equilibrium. Hence in Silicon where the intrinsic and extrinsic stacking-fault energies only differ by $\sim$14% (7) and deformation is carried out at elevated

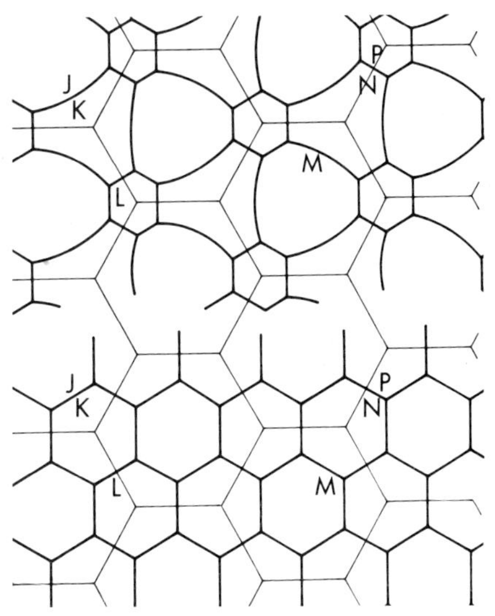

Fig. 7 Schematic diagrams of two [111] grain boundaries showing the same misorientation from exact $\Sigma 1$ and $\Sigma 3$ coincidence.

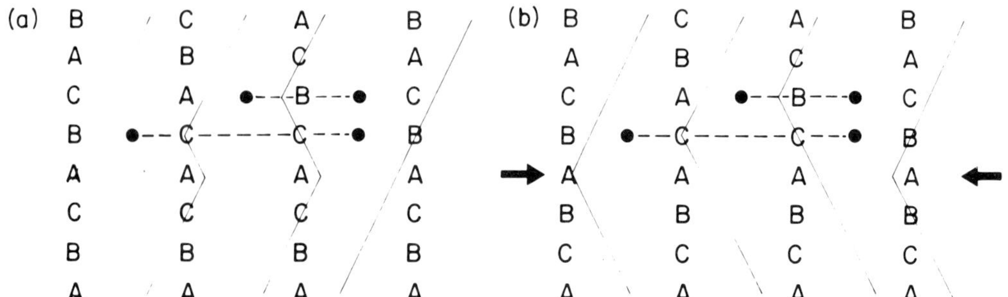

Fig. 8 The stacking sequence for the Σ1 and Σ3 boundaries shown in Fig. 7: e.g. along JKL and MNP.

temperatures, the node sizes (as determined by W) are very similar (7,13,14) as would have been expected on the grounds of simple elasticity theory (12).

4. The Relation to Grain Boundaries

The 1/2°Σ1,{111}Si grain boundary in Fig. 6 consists of alternating extended P_s and K_s nodes. By the analysis given above, either there is a single jog present at each K_s node, in which case the boundary would not be of the pure twist type, or climb has occurred. If climb has occurred then the boundary will no longer be glissile.

The question then arises as to whether other boundaries also contain jogs; this can be answered by considering Fig. 7 which shows the relation between a Σ1 and Σ3 {111} boundary with the same deviation from exact coincidence. The network of perfect dislocations dissociates as shown in Fig. 7a where it is assumed that the K_s nodes are all of the symmetric type shown in Fig. 3b, and contain jogs for example at R,S and T (e.g. along both JKL and MNP). The stacking sequence in the different parts of this boundary are shown in Fig. 8a; if a perfect Σ3 boundary is now added to the Σ1 boundary the stacking sequence shown in Fig. 8b results, i.e. all stacking faults are removed, and the dislocation network is expected to look like that shown in Fig. 7b. The steps seen in Fig. 8b (which again occur along both JKL and MNP) are associated with the screw secondary dislocations of the Σ3 boundary which have Burgers vector a/6<112>. Since the dislocations have the same Burgers vectors as the original Shockley partial dislocations and they lie in the same glide plane as before, it must be concluded that any jogs present in the Σ1 boundary will be present in the Σ3 boundary. The physical explanation for this is that since the secondary dislocations are associated with steps in the boundary, they produce steps/jogs on dislocations which cross them. The implications of this conclusion for other boundaries e.g. Σ5, where steps are also associated with secondary dislocations is being examined.

This work was supported in part by the NSF through the Materials Science Center at Cornell University.

References

1. M.J. Whelan, Proc. Roy. Soc. (1959).
2. A.W. Ruff, Met. Trans. 1 (1970).
3. C.B. Carter, Phil. Mag. in the press (1980).
4. M.H. Loretto, Phil. Mag. 10, 467 (1964).
5. P.C.J. Gallagher, Phys. stat. sol. 16, 95 (1966).
6. C.B. Carter, submitted for publication (1980).
7. H. Foll and C.B. Carter, Phil. Mag. A40, 497 (1979).
8. F.C. Frank, Report of the Conference on Defects in Crystalline Solids (London:Phys. Soc.) p.159 (1954).
9. N. Thompson, Proc. Phys. Soc. 66B, 481 (1953).
10 J.P. Hirth and J. Lothe, Theory of Dislocations (New York:McGraw-Hill).
11. L.M. Brown, Phil. Mag. 10, 441 (1964).
12. W.F. Hartl, R. DeWit and R.E. Howard, J. Appl. Phys. 38, 1 (1967).
13. E. Aerts, P Delavignette, R. Siems, and S. Amelinckx J Appl. Phys. 33, 3078 (1962).
14. I.L.F. Ray and D.J.H. Cockayne, Proc. Roy. Soc. A 325, 543 (1971).

DISLOCATIONS AND FREE SURFACES IN THE MICROPLASTIC DEFORMATION OF F.C.C. METALS : AN X-RAY TOPOGRAPHIC STUDY

F. Minari and B. Pichaud
Equipe de Recherche Associée au C.N.R.S. n° 545
Université Aix-Marseille III, Centre de Saint-Jérôme
13397 MARSEILLE Cedex 4, France

Abstract

At the very beginning of the deformation of a crystal, surface sources are activated before internal sources of same length. X-ray topographs of slightly deformed f.c.c. metal crystals exhibit particular configurations of dislocations induced by the presence of the external boundary. The experimental facts can be explained by a simple model of surface source and its different mechanisms of expansion under an applied stress.

Introduction

In all the microdeformation tests performed on f.c.c. metals in conjunction with direct imaging of single dislocations by X-ray topography, the external boundaries of the crystal are found to play an important role at two different levels :

- the activation of the sources giving rise to the first fresh dislocations, from the double point of view of the critical shear stress and of the selection among the twelve available slip systems,
- the further development of these dislocations.

The aim of this paper is to review the main results we have obtained in this field in order to show how the concept of surface source can explain the experimental facts. We present first some theoretical considerations about a very simple model of surface source. For a given slip system, one can expect from these considerations a behaviour different from the bulk during the activation and the expansion of such a source. This behaviour is in agreement with our experiments on Cu crystals. In addition, important experimental facts have been observed, which had not been immediately deduced from the model : these facts concern a multiplication mechanism and the failure of Schmid's Law. We show that they are quite consistent with the consequences of the model of surface source, and we propose a criterion for the selection of the observed slip systems.

The simplest model of surface source

In the unstressed state the source is considered as a straight segment of dislocation of length L anchored at some point A and perpendicular to the free surface (Fig. 1). When submitted to an external stress, this segment can expand by motion in its slip plane .

Activation : If the applied stress-field is known and sufficiently weak, an equilibrium configuration of the source can be computed by balancing the different forces (applied force, self-stress, image force) acting on the dislocation in a series of points distributed along the line (1,2). Friction forces are neglected. A series of configurations can be determined by increasing the stress step by step until the equilibrium can no longer be obtained. The last equilibrium corresponds to the critical shear stress (c.r.s.s.). An example of results is given in Fig. 2 for a pure screw and a pure edge source 2.55 µm long (10^4 b in Cu) in a homogeneous stress-field.

The c.r.s.s. varies with the character of the source i.e. the angle ϕ between the source in the unstressed state and $\vec{b}$. If the c.r.s.s. is put equal to

$\sigma_c = B \frac{\mu b}{2\pi L} \mathrm{Log} \frac{L}{r_o}$, the same calculation performed for increasing values of ϕ allows the variations of B versus ϕ to be determined as shown in Fig. 3 (curve a). They can be compared with the variations of the same factor for Frank-Read sources within the bulk (curve b) : in the last case, sources with Burgers vectors symmetric about the pure edge orientation have symmetrical equilibrium configurations for the same value of the c.r.s.s. This does not hold for surface sources, because whatever the Burgers vector, the source and its image are always symmetric about the free surface. The ratio of the c.r.s.s. relative to bulk and surface sources varies from 1.3 (ϕ = 150°) to 2.6 (ϕ = 60°).

Expansion : The motion of the source around the anchoring point results in the formation of a half-loop.

We consider now the further expansion of this half-loop. The most characteristic features are obtained in the case of a stress field which is maximum on the free surface (pure bending of a flat sample). According to the orientation of the crystal and the obstacles encountered by the dislocation, different behaviours can be expected :

- Free expansion

If no obstacle impedes its motion the half-loop expands over long distances by its two emerging segments because of the maximum stress at the free surface. Pure bending providing a zero stress surface at half the thickness of the plate, the major part of the half-loop stabilizes in this surface, as a straight line. In a sample whose thickness is much smaller than the lateral dimensions, this results in the formation of an elongated half-loop.

- Cross-slip of the leading segments of the half-loop

If an obstacle lies on the path of the mobile emergence of the half-loop, the last can cross-slip along its screw part to avoid the obstacle (Fig. 4). This cross-slip can be facilitated by the presence of a Hirsch jog along the dislocation (3). If the resolved shear stresses in the primary and the cross-slip planes are nearly equal, a weak obstacle is sufficient to induce this deviation. In this case the repetition of the process during the elongation of the half-loop can lead to a zig-zag shaped dislocation (4) (Fig. 5).

Experimental evidences

Measurement of the c.r.s.s. for single dislocations :

Pure tension is difficult to achieve experimentally on thin (∿ 100 µm) metal crystals and lead to randomly dispersed values, so we have performed pure bending tests on copper crystals on the X-ray topographic camera. The applied stress was deduced from the curvature directly measured by the misorientation between two points of the crystal. The results are in fairly good agreement with the model of surface source (5).

Configurations of expanded half-loops :

The first dislocations activated by the bending stress are always elongated half-loops. They are straight if no obstacle opposes their expansion. On the contrary, Fig. 6 is an example of easy cross-slip observed in a sample with such an orientation that the Schmid factors (related to the tension-compression axis perpendicular to the bending axis) of the two glide systems $[1\bar{1}0]$ (111) and $[1\bar{1}0]$ $(11\bar{1})$ were nearly equal. Successive cross-slip provide zig-zag dislocations. It was not possible to determine the nature of the obstacles inducing these deviations because they were too small to be visible on the topographs. However the obstacle is clearly identified when it consists of a repulsive dislocation in the same plane. This is the case of Fig. 7 where the repulsive emergences of two half-loops moving in opposite senses in $(\bar{1}11)$ have repelled one another and cross-slipped into $(1\bar{1}1)$. This particular "X configuration" which can be observed only in the case of a deformation by bending, allows the primary plane to be determined unambiguously since in this plane only two branches of the "X" are exactly aligned. This fact has been very useful to confirm the hypothesis reported below about the selection of the activated systems.

Unexpected observations

Three unpredicted facts clearly related to the presence of an external boundary have been observed.

Multiplication :

Fig. 8 shows two steps of the progressive deformation of a copper sample oriented for single slip. Several long half-loops with Burgers vectors $[0\bar{1}1]$ have been nucleated in the same plane (111). This plane being nearly perpendicular to the photographic plate, the existence of several loops is attested on the topograph by the enhancement of their image contrast. The repulsive interaction between these dislocations causes the bowing of portions of the loops out of their plane (the phenomenon is clearly visible because of the small angle between the cross-slip plane (111) and the photographic plate). The expansion of these portions by motion of one extremity in the cross-slip plane results in their emergence on the free surface, where they unlink from the original dislocation. By this mechanism, a new dislocation is created and the original one is regenerated (6). The process is depicted in Fig. 9.

<110> {110} glide :

It has been observed for crystal orientations near {001} (Fig. 10) and develops long dislocations similar to <110> {111}. In macrodeformation tests, where internal sources provide nearly all the deformation, this glide is not observed at room temperature because even if a source lies in a {110} plane, the screw part which always forms before the source reach the critical shape, immediately cross-slips in {111} where it has a lower energy. On the contrary, the computation of the equilibrium configurations of surface sources shows that for $0 < \phi < \pi/2$ these sources exhibit no screw part up to their critical shape : such a source initially in {110} can be activated in this plane.

Failure of Schmid Law :

All the microdeformation tests published in the literature including ours, show that the first activated surfaces sources always have a Burgers vector making a small angle α (<30°) with the trace of the glide plane in the free surface. In addition the Schmid factor is reasonably high (∿ 0.25). A criterion had been proposed (7) to explain a few cases of selection among {111} planes only in Al. We have extended this criterion by addition of the possible glide on {110} in Cu and by examining systematically all the orientations of Burgers vector and crystal faces. The principle of the criterion is to compare for a given orientation the energies of the emergent segment of the source up to a constant depth when it is supposed in equilibrium without any applied stress, either in a {111} plane or the other, or in the {110} containing $\vec{b}$. Details of the calculations will be published elsewhere. If one value of this energy is less than 0.5 to 0.7 times the others, the emergence will stay in the corresponding plane and the dislocation will be activated (if the remaining part of the source is in the same plane). If the two lowest values are nearly equal, the emergence can utilize alternately and repeatedly the corresponding planes to reach the surface : the source will be sessile.

This criterion applied by numerical computation to all possible orientations, in addition to a Schmid factor higher than 0.25 predicts that no system can be activated with α greater than about 30°.

R e m a r k :

This criterion may seem inconsistent with the observation of zig-zag dislocations. In fact it must be considered that in this case the cross-slip is accidental and not spontaneous, and that at the beginning of its developpement the source was in the primary plane selected by the criterion. This point of view is strongly supported by a carefull examination of all the "X configurations" for the orientation of Fig. 7 : it can be seen that two couples of planes can provide X configurations : $(\bar{1}\bar{1}1)$, (111) and $(1\bar{1}1)$ $(\bar{1}11)$, but in each couple, only one plane is selected by the criterion : this means that for a given couple, the primary plane is always the same, so on the topograph the aligned dislocation segments of the configuration must also be always the same : this is effectively verified.

Conclusion

Our results describe the basic events which can occur in the immediate vicinity of the free surface during the early stages of the deformation of a f.c.c. metal like copper. These elementary mechanisms which result from a relatively easy cross-slip can be somewhat altered in materials exhibiting different physical characteristics, especially different stacking fault energies. This point is being checked at present by experiments on silver base alloys.

References

1 - D.J. Bacon, Phys. Stat. Sol. 23, 527 (1967)
2 - B. Pichaud, F. Minari and J. Kellerhals, Phil. Mag. A 38, 593 (1978)
3 - P.B. Hirsch, Phil. Mag. 7, 67 (1962)
4 - B. Pichaud and F. Minari, Phil. Mag. 34, 1128 (1976)
5 - J. Kellerhals, F. Minari and B. Pichaud, Phil. Mag. A 39, 341 (1979)
6 - F. Minari, B. Pichaud and L. Capella, Phil. Mag. 31, 275 (1975)
7 - O. Lohne, Phys. Stat. Sol. a, 23, 709 (1974).

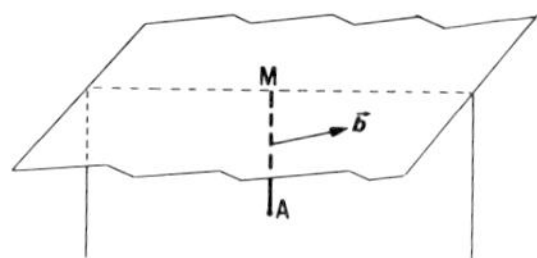

FIG. 1
Surface source.

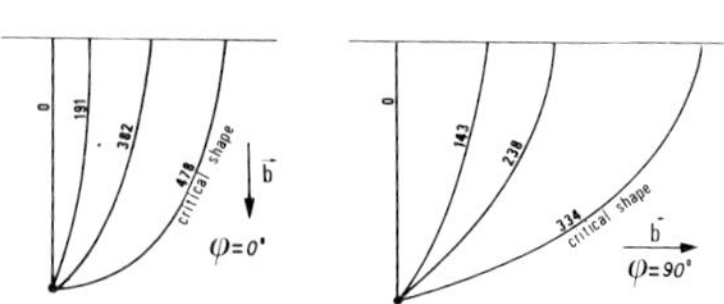

FIG. 2
Successive equilibrium shapes for increasing stresses given in 10^4Pa (Source length 2.55μm=10^4b)

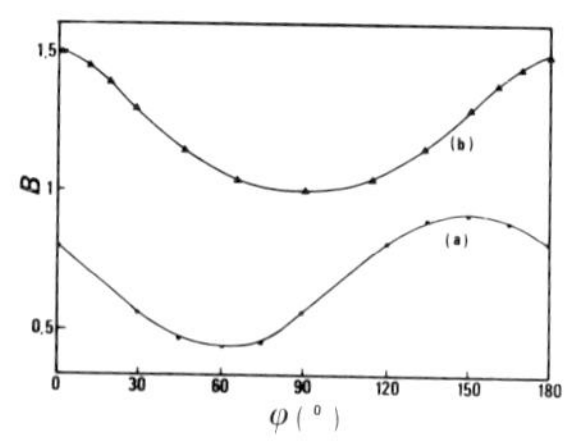

FIG. 3
Variations of B versus ϕ for surface (a) and internal (b) sources.

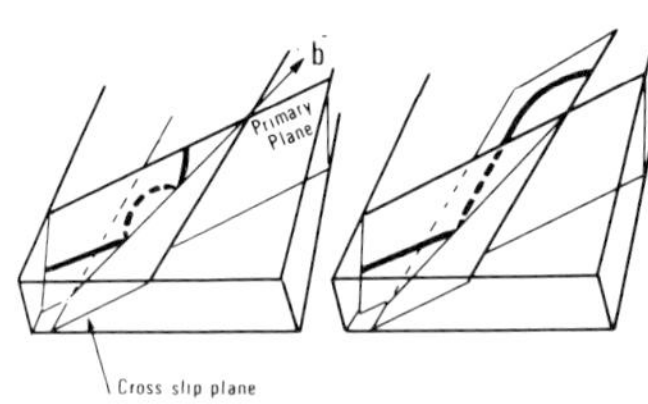

FIG. 4
Cross slip of the leading segment.

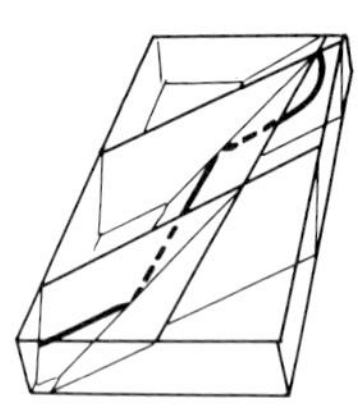

FIG. 5
Zig-zag dislocation.

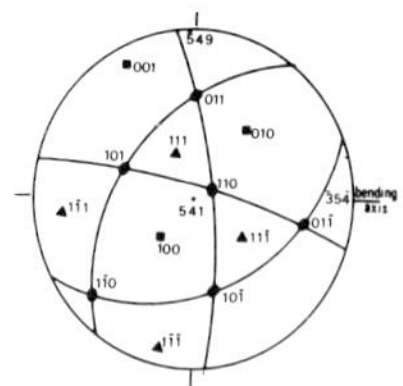

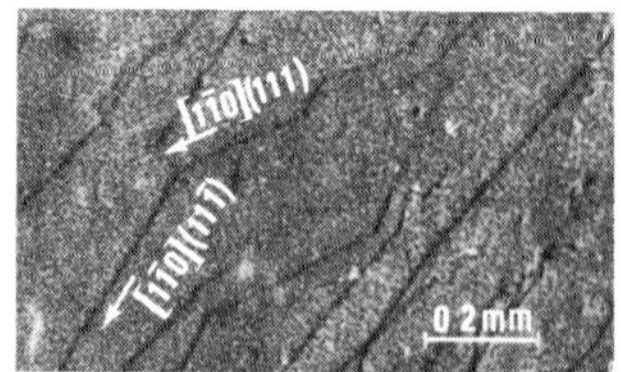

FIG. 6
X-Ray topograph of zig-zag dislocations in a crystal with surface orientation near (110). The two observed systems are [$1\bar{1}0$](111) and [$1\bar{1}0$]($11\bar{1}$) (Schmid factors 0.29 and 0.28).

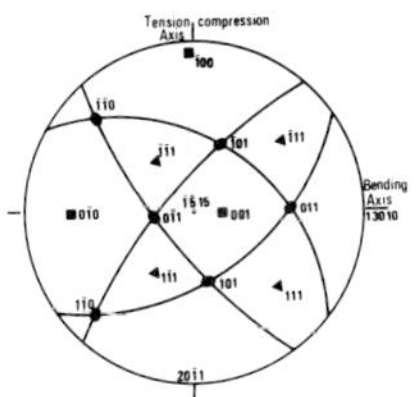

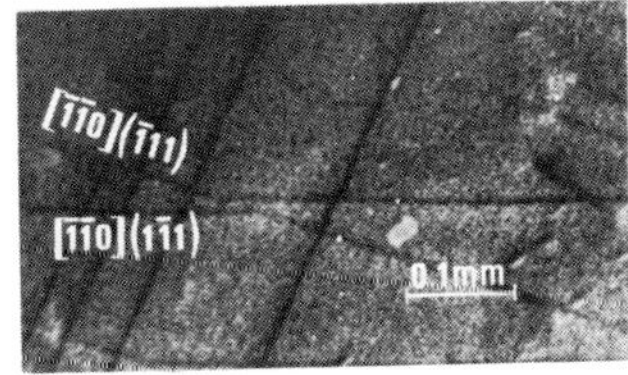

FIG. 7
X-Ray topograph of an "X configuration" in a copper crystal resulting from the repulsive interaction of two half loops in the same primary plane (111)

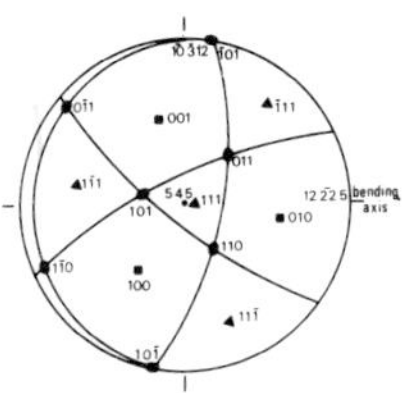

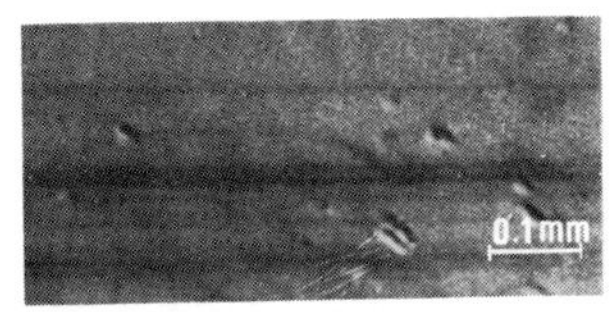

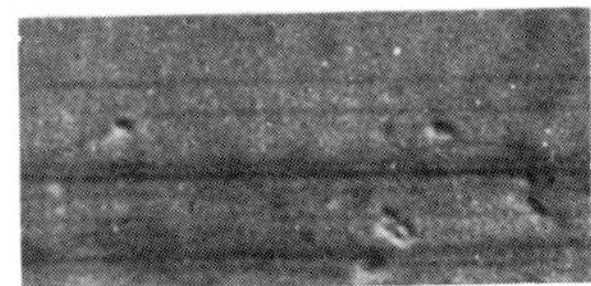

FIG. 8
Multiplication induced by cross-slip.

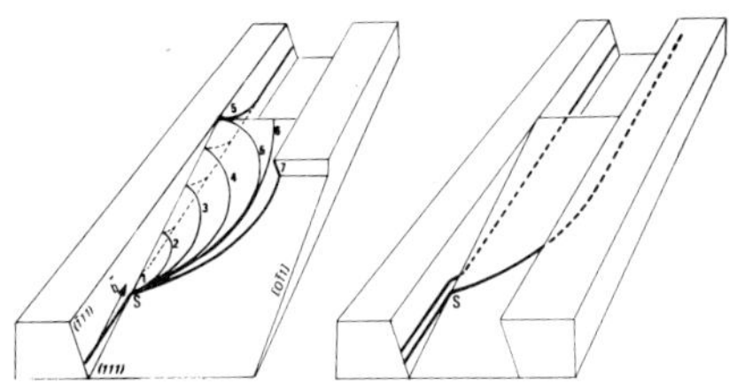

FIG. 9
Mechanism of the multiplication by emergence of the deviated segment on the external boundary.

FIG. 10
X-Ray topograph of a copper crystal showing evidences of <110> {110} glide in addition to usual systems <110> {111}.
Orientation of the sample same as FIG. 7.

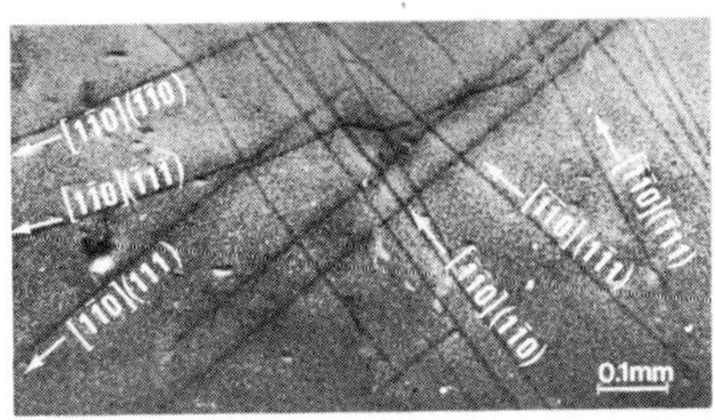

THE CLIMB OF DISSOCIATED DISLOCATIONS

D. Cherns, G.J. Hardy, P.B. Hirsch,
A. Ourmazd and H. Saka
Department of Metallurgy and Science of Materials,
University of Oxford, Parks Road, Oxford OX1 3PH

ABSTRACT: The weak-beam technique of electron microscopy has been used to directly study the climb of dissociated dislocations under conditions of high supersaturation of point defects introduced by electron irradiation. In a Cu-13%Al alloy irradiated at room temperature, prismatic loops (assumed to be of interstitial type because of the dislocation bias) are nucleated on the individual partials. The loop Burgers vectors are a function of partial orientation and are such as to minimize the elastic energy of the resultant configuration of partial plus loop, and to maximize the edge component of the loops. In Cu-13%Al irradiated at temperatures in excess of 200°C complex dislocation configurations are observed which are quite different from those observed at room temperature. These results may be understood in terms of the nucleation of prismatic loops, followed by their growth in directions controlled by the climb force on the parts of the prismatic loops not lying in the glide plane of the dislocations and their interaction with the non-parent partial. In this paper the mechanisms of climb of dissociated dislocations in Cu-13%Al are explained. The results are compared with preliminary observations carried out on a Cu-10%Al alloy and on pure Si and the effect of partial separation on the climb mechanisms is discussed.

Introduction

The climb of dissociated dislocations in f.c.c. metals and alloys under conditions of a high supersaturation of point defects introduced, for example, by irradiation or quenching is not well understood. Various models have been proposed for double jog nucleation. The early model of Stroh(1) requires a prior constriction of the partials before jog nucleation can occur; later models (2-4) deal with the formation of extended double jogs without requiring prior constriction of the partials. In Thomson and Balluffi's model (4), illustrated in fig. 1, a prismatic loop with the same Burgers vector as that of the total dislocation is nucleated on one of the partials. After interaction of the loop and partial, a dipole of partials is generated which glides across the original glide plane and an adjacent plane accompanied by dissociation of dislocation segments out of the glide plane. Subsequent interaction with the non-parent partial creates the extended double jog.

Recent experimental observations have been made of dislocation climb in a Cu-13%Al alloy electron-irradiated at room temperature in the high voltage electron microscope(5-7) Prismatic loops were observed to nucleate on the individual partial dislocations whose separations varied in the range 20-40nm for screw-edge orientations of the total dislocation. The prismatic loops, assumed to be of interstitial type owing to the dislocation bias, had Burgers vectors which depended on partial orientation and were such as to minimize the interaction energy with the parent partial and to maximize the climb force. Thus, for some orientations of partial dislocations, the Burgers vector of the nucleated loop differs from that of the total dislocation (c.f. fig. 1). In this paper we first compare the climb of dislocations in Cu-13%Al under electron irradiation at room and at elevated temperatures and show that they lead to markedly different dislocation configurations. It is then shown that the results can be explained fully by a single model of total dislocation climb. The climb mechanism is seen to be more complex than that illustrated in fig 1. Results are then presented which show that the climb model here extends to Cu-Al alloys of lower aluminium content (and smaller partial separations) and to silicon. The general significance of these results is discussed.

Experimental

(111) specimens were prepared with dislocations introduced by bending either at room temperature (Cu-Al alloys) or at 700-900°C in an argon or nitrogen atmosphere (silicon). Dislocations lying in the (111) foil plane were examined using the weak-beam technique(8) in a JEM 100B electron microscope both before and after electron irradiation in the Oxford AEI EM7 electron microscope operating in the range 0.3-1.0MeV. Specimens were irradiated both at room temperature and at elevated temperatures achieved in a standard heating stage.

Fig. 2 and 3 compare segments of 60° dislocations in Cu-13%Al irradiated at room temperature and 200°C respectively. In fig. 2 a dislocation of total Burgers vector **AB** (using Thompson's notation) is seen before irradiation (fig. 2(a)) and after irradiation (figs 2(b) and 2(c)). In figs. 2(a) and (b) the dislocation is seen in $|\mathbf{g.b}| = 2$ so that both partials $\mathbf{b}_1 = \mathbf{A\delta}$ and $\mathbf{b}_2 = \mathbf{\delta B}$ are in contrast. In fig. 2(c) **Aδ** and **δB** both have $|\mathbf{g.b}| = \frac{1}{3}$ and show little contrast although the stacking fault contrast of the original dislocation is clearly visible. In both figs 2(b) and (c) new dislocations are seen as a result of irradiation labelled W, X, Y and Z and shown schematically. A detailed analysis of these dislocations is given elsewhere(6); we note the result, that W,X,Y and Z are all dislocation dipoles of Shockley type **Bδ/δB** (W) **Aδ/δA** (Y) and **Cδ/δC** (X and Z). The dipoles are either imaged in relatively strong "inside" contrast (where the images of the two segments lie inside the dipole) or weaker "outside" contrast (images outside the dipole); for example compare Y and W in fig. 2(b) which show inside and outside contrast respectively and both have $|\mathbf{g.b}| = 1$. Fig. 3 shows a 60° dislocation irradiated at 200°C both before irradiation (fig 3(a)) and after irradiation (figs 3(b) and (c). In figs 3(a) and (b) the dislocation is seen in $|\mathbf{g.b}| = 2$ so that both partials $\mathbf{b}_1 = \mathbf{A\delta}$ and $\mathbf{b}_2 = \mathbf{\delta B}$ are in contrast. In fig. 3(c) the top partial $\mathbf{b}_2 = \mathbf{\delta B}$ has $\mathbf{g.b} = 0$ and is out of contrast but new dipoles, R, introduced by irradiation are visible and appear to be on the line of **δB** at points where this partial is distorted (fig. 3(b)). An extended analysis showed the Burgers vectors of the new dipoles to be **Cδ/δC**, so that $|\mathbf{g.b}| = 0, 1$ in $\mathbf{g} = 2\bar{2}0, 02\bar{2}$ respectively.

In comparing the results in fig. 2 and 3 it appears that new loops are visible on both partials after the room temperature irradiations and only on one partial **δB** after the high temperature irradiations. Moreover the loops on the 30° partial **δB** are predominantly **Aδ/δA** and **Cδ/δC** in the room and high temperature irradiation cases respectively. High temperature irradiations showed similar results to those in fig. 3 for most 30-90° orientations of dislocation **AB**, namely new dipoles **Cδ/δC** on the partial of greater screw character. For dislocations **AB** in exact edge orientation a different result was obtained; dipoles were observed on both partials of type **Aδ/δA** on **δB** and **Bδ/δB** on **Aδ**. For dislocations **AB** in screw to 30° orientations dipoles **Cδ/δC** were observed; in this case the final configurations were similar to those observed in room temperature irradiations.

Theoretical Models

In order to explain the high temperature results in Cu-13%Al we take the model of loop nucleation introduced in previous papers (5-7) and consider the interaction of loops nucleated on one partial(parent) with the other partial (non-parent). The case of the 60° dislocation in fig. 3 is shown in fig. 4; for simplicity, only the nucleation on the partial of greater edge character **Aδ**, ie. the partial with greater preference for interstitials, is illustrated. Prismatic interstitial loops **AB** and **AC** are both equally favoured on the pure edge partial **Aδ** by the criteria of low interaction energy and a large climb force. In figs. 4(b) and (c) the interactions of these loops with the parent partial are shown. The loop **AB** interacts with **Aδ** to form a partial **δB** on the original glide plane and a segment **AB** below the original glide plane dissociates to form **Aδ** and **δB**. The resulting image contrast is that of a Shockley dipole **Bδ/δB**. Similarly the loop **AC** gives rise to a dipole **Cδ/δC**. As a result of the climb force on the segments **AB** and **AC** out of the glide plane the new loops tend to rotate in the directions shown. The arrangement in fig 4(c) shows the dipoles joined together whereupon they can become detached from the original partial **Aδ** to form a zig-zag chain. This agrees well with the experimental arrangement in fig. 2 (note, however, that for climb to take place by an interstitial mechanism in this case, loops **AB** and **AC** should lie above the plane of the paper; otherwise the mechanism and final dislocation arrangements are equivalent). Eventual interaction of the dipole chain with the non-parent partial will occur as continued operation of the climb force causes the new dislocation segments to grow towards the opposite partial and/or as secondary nucleation causes the segments of the dipole to grow further apart perpendicular to the glide plane favouring the elastic interaction. There is some evidence for the latter possibility from in situ weak-beam observations of the growth process at elevated temperatures in the HVEM (Cherns, unpublished). The partial **B**δ and δ**B** on the original glide plane can then interact and annihilate causing segments of the total dislocation to climb completely. However, where dipoles **C**δ/δ**C** exist, complete interaction with the original partial δ**B** will not occur and dipoles **C**δ/δ**C** will remain now on the non-parent partial. The interaction is shown in fig. 4(d) and it is seen that both the final dislocation type and configuration (twist due to the climb force) are in agreement with the results in fig. 3. For dislocations of 30-90° orientations the twist of nucleated loops on the nearer edge partial, say **A**δ, favours the formation of both **AB** and **AC**. We believe the model in fig. 4 accounts adequately for the general observations in this orientation range. For pure edge dislocations **AB** the most favoured dislocations are **AB** on both **A**δ and δ**B** and should nucleate on both partials giving **A**δ/δ**A** on δ**B** and **B**δ/δ**B** on **A**δ (with no tendency to twist). Provide both dipoles grow at the same rate no interaction can occur with the non-parent partial and both dipoles will be observed. This is in agreement with the experimental observations noted earlier. For screw to 30° dislocations the similar observations at room and elevated temperatures have been satisfactorily explained(6,7).

Discussion

It has thus been shown that dislocation climb in Cu-13%Al under conditions of high supersaturation of interstitials produced by electron irradiation can be explained by prismatic loop nucleation on the partial with greater edge component followed by an interaction with the non-parent partial. The resulting climb is more complex than that envisaged by Thomson and Balluffi (4) in that, for most dislocation orientations, loops with Burgers vectors differing from that of the original dislocation are generated. Consider now the effect of partial separation on the climb mechanism. We expect that, as the stacking fault width is reduced, interaction with the non-parent partial will eventually affect the loop nucleation event favouring the generation of **AB** rather than **AC** on a partial **A**δ of dislocation **AB**. As **AC** is suppressed the operation of the Thomson and Balluffi mechanism becomes more likely. A second point is that the reduced partial separation will favour an earlier interaction of the nucleated loops with the non-parent partial making the observation of the nucleation stage of loops on their parent partial less likely.

Preliminary observations of dislocation climb have been made on a Cu-10%Al alloy where partial separations are in the range 10-20nm, and on high purity silicon (n-type, resistivity 500 ohm cm) where partial separations are in the range 5-8nm. In Cu-10%Al irradiated at room temperature dislocation dipoles similar to those in Cu-13%Al irradiated at room temperature were observed. However, some additional dipoles were found of the type seen in fig 3 i.e. consistent with the irradiation of Cu-13%Al at elevated temperatures. These results are consistent with the greater elastic interaction between nucleated loops and the (closer) non-parent partial but suggest that the nucleation event is unaffected. In silicon irradiated at 600°C new loops were also observed as shown in fig. 5. Here an approximately 40° dislocation is imaged before and after irradiation in $|\mathbf{g.b}| = 2$ (figs. 5(a) and (b)) with both original partials $\mathbf{b}_1 = \delta\mathbf{B}$ and $\mathbf{b}_2 = \mathbf{A}\delta$ in contrast. Fig. 5(c) is taken in $|\mathbf{g.b}| = 1$; the top near screw partial **A**δ has $\mathbf{g.b} = 0$ and is out of contrast. However, new dipoles are visible along the line of **A**δ whose Burgers vectors were found to be **C**δ/δ**C**. These observations can be explained in detail by applying the climb model in fig. 4. The following tentative conclusions may be drawn from these further experiments. The climb model described here may apply to materials of medium stacking fault energy including, for example, austenitic stainless steels. The effect of complex dislocation configurations produced by climb on irradiation creep and void formation should, therefore, be considered. Another question of interest is the possible role of loop nucleation in the formation of helices on screw dislocations in quenched or irradiated material. In both cases one question of interest is whether the new dipoles can be removed by suitable dislocation interactions allowing the dislocation to assume its original form. Preliminary theoretical work suggest that such mechanisms are possible (7) but no detailed experimental evidence is yet available.

REFERENCES

1. A.N. Stroh Proc. Phys. Soc. *67B* (1954) 427.
2. B. Escaig, Acta. Met. *11*, (1963) 595.
3. H. Kimura, R. Maddin and D. Kuhlmann-Wilsdorf Acta Met. *7*, (1957) 145, 154.
4. R.M. Thomson and R.W. Balluffi, J. Appl Phys. *33*, (1962) 803.
5. C.B. Carter, D. Cherns, P.B. Hirsch and H. Saka, Proc. of the 9th Int. Congress on Electron Microscopy, Toronto (Microscopical Soc. of Canada) (1978) Vol I p.324.
6. D. Cherns, P.B. Hirsch and H. Saka, Proc. 5th Int. Conf. on the Strength of Metals and Alloys, Aachen (1979) p.295.
7. D. Cherns, P.B. Hirsch and H. Saka, Proc. Roy. Soc. London in press.
8. D.J.H. Cockayne, I.L.F. Ray and M.J. Whelan, Phil Mag. *20*, (1969) 1265.

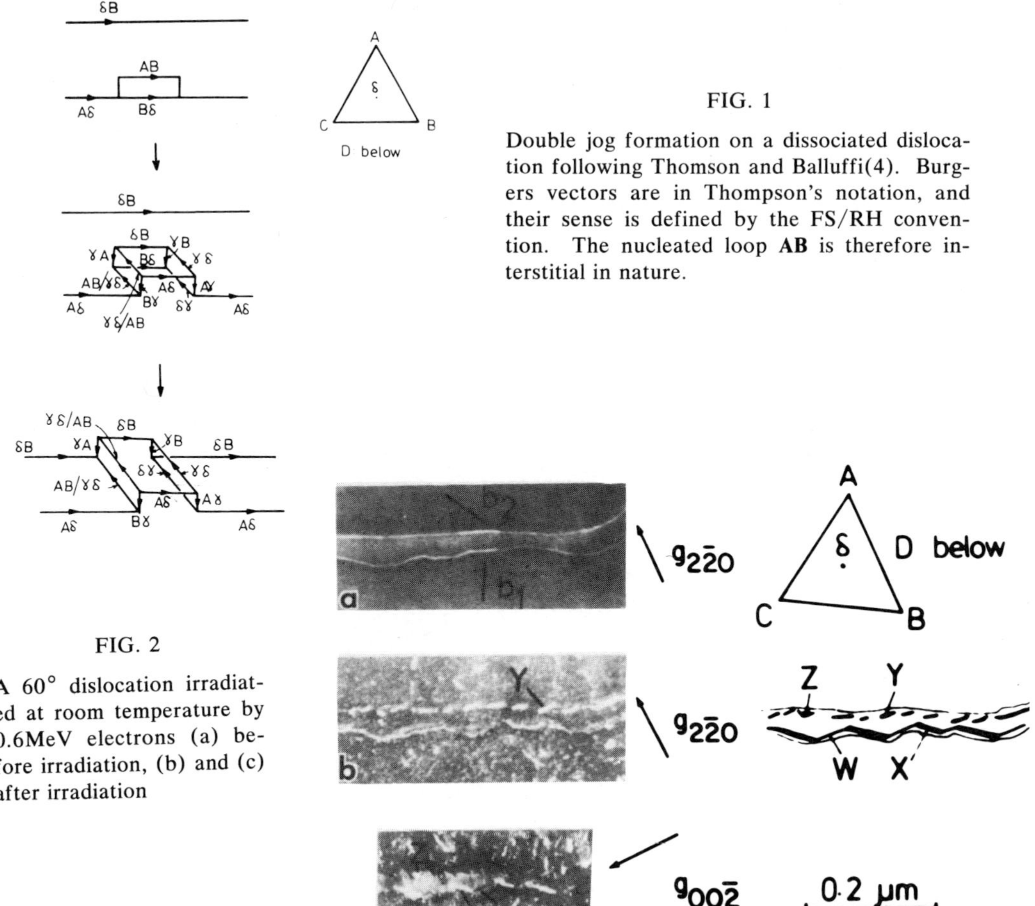

FIG. 1

Double jog formation on a dissociated dislocation following Thomson and Balluffi(4). Burgers vectors are in Thompson's notation, and their sense is defined by the FS/RH convention. The nucleated loop **AB** is therefore interstitial in nature.

FIG. 2

A 60° dislocation irradiated at room temperature by 0.6MeV electrons (a) before irradiation, (b) and (c) after irradiation

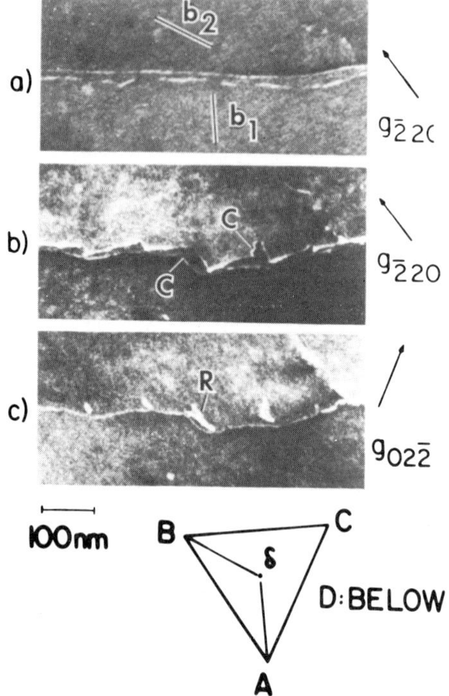

FIG. 3

A 60° dislocation irradiated at 200°C (a) before irradiation, (b) and (c) after irradiation.

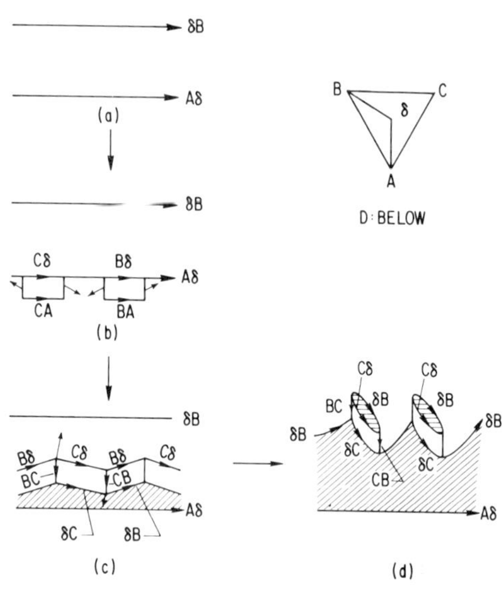

FIG. 4

Theoretical model for climb of a 60° dislocation.

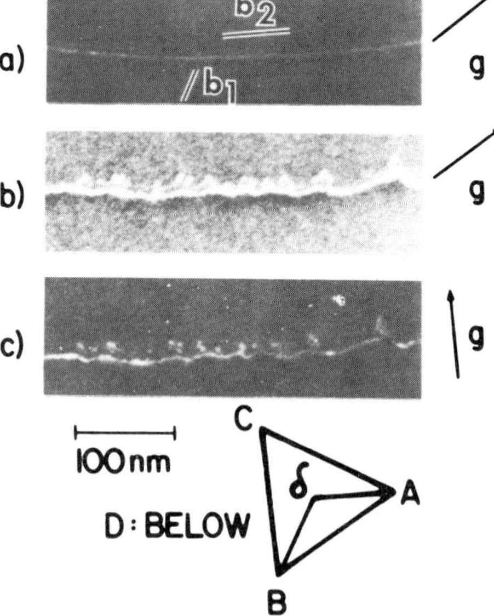

FIG. 5

A 40° dislocation in silicon irradiated at 600°C by 0.8MeV electrons (a) before irradiation, (b) and (c) after irradiation.

On the Contribution of Grain Boundary Dislocations to Grain Boundary Sliding, Migration and Point Defect Emission

David A. Smith
IBM T.J. Watson Research Center
P.O. Box 218
Yorktown Heights, N.Y. 10598

Abstract: An attempt is made to account for the crystallographic dependence of grain boundary sliding migration and point defect emission in terms of the diffusion limited movement of intrinsic and extrinsic grain boundary dislocations.

Introduction

All experimental measurements on bicrystals and textured polycrystals suggest that grain boundary properties depend on the crystallography and chemical composition of the boundaries.(1) Observations and calculations imply that many, and possibly all, grain boundaries in cubic metals may contain dislocations.(2-4) Their Burgers vectors depend on the crystallographic relationship of the grains whilst their configuration is governed by the orientation of the boundary plane and the treatment of the specimen. Transmission electron microscopy (TEM) has shown that the arrangement of grain boundary dislocations (gbds) is regular in annealed material; (5) however after deformation, point defect absorption or grain boundary migration has occurred the observation of distorted networks, spiral configurations and interaction with crystal lattice dislocations shows that gbds can move and multiply. (6-10) In general the movement of gbds is diffusion limited since their Burgers vectors are usually not parallel to the grain boundary plane. In the isotropic approximation the elastic interaction of gbds and solutes is similar to that of crystal dislocations. Most gbds observed in cubic metals are thought to have Burgers vectors governed by the appropriate DSC lattice. (11)

Movement of Grain Boundary Dislocations

Movement of gbds results in relative translation of the grain centers, usually requires a point defect flux and in many cases is accompanied by boundary migration, fig. 1. Since sliding, migration and point defect emission can all be consequences of the movement of gbds all three processes can occur together in such a way that any one is accompanied by the other two as part of the accommodation process. The nature of the accommodation required depends on the constraints to which the specimen is subject. For example in migration it is expected that the overall dislocation movement is such that, except at external surfaces, their shears cancel whilst in sliding the contributions to migration might cancel. However the responses, via its dislocation structure, of a grain boundary interacting with a shear stress, (grain boundary sliding) a tensile stress, (Coble or Herring-Nabarro creep) or a chemical stress, (recrystallization, diffusion induced boundary migration) will depend in related ways on the structural characteristics of the boundary S (dislocation density, Burgers vectors, inclination of the Burgers vectors to the grain boundary plane), the driving force P, frictional resistances P_i and absolute temperature T so that expressions of the type:

$$R \propto S \frac{[\Omega(P - P_i)]^n}{kT} f \exp -(\frac{\Delta g^*}{kT}) \quad (1)$$

can be written for any behavior of a grain boundary which is the consequence of gbd movement; Δg^* is an activation energy related to the movement of gbds which is a function of grain boundary structure and composition, f is an attack frequency, Ω is the atomic volume and n is usually predicted to be 1 or 2.

Two essential aspects of the movement of gbds are that (a) movement is in general by glide and climb and that (b) just as in twinning some atoms may be translated to their correct sites but others shuffle. Consequently a single diffusive atomic jump may trigger the movement of a group of N atoms which results in more sliding or migration than otherwise expected per thermally activated event. This

behavior will appear as an orientation dependent pre-exponential term in eq. (1) For a dislocation with **b** inclined at an angle ϕ to the grain boundary plane and where the core step height is h, N is $h/b \sin \phi$. N can take any value between one and infinity for h = **b** and $\phi = 10°$, N ~ 6. The density, ρ, of moving intrinsic or structural dislocations is a linear term in S and $R \propto (P-P_i)$ but, for spiral sources of extrinsic dislocations $\rho \propto P - P_i$ and consequently $R \propto (P-P_i)^2$.
The relative importance of intrinsic and extrinsic dislocations in grain boundary processes is still being investigated. The movement of intrinsic gbds by glide-climb in kinematically unconstrained bicrystals provides a mechanism for grain rotation. (12) However an intrinsic array may not necessarily contain the dislocations appropriate for a particular process; eg for sliding, two sets of dislocations with independent components of their Burgers vectors parallel to the grain boundary plane are required and this configuration does not occur in boundaries which are tilt deviations from coincidence site lattice orientations. Similarly migration by a gbd mechanism requires h to be non-zero but $\frac{1}{11}$<113> dislocations with h large in $\Sigma = 11$ (13) related boundaries seem not to occur; instead dislocations with **b** = $\frac{2}{11}$<113> apparently and no core step are observed. (14)

The table summarizes the nature of the dislocations which might account for various grain boundary processes for tilt & twist deviation from csl misorientations

Table

	Sliding	Migration	Coble Creep	Point Defect Emission
Tilt Deviation	E	E(I?)	I	I
Twist Deviation	I	E	E	E

E extrinsic I intrinsic

The classification I does not exclude the possibility of an extrinsic contribution.

An important feature of the motion of gbds is that it involves the movement of (I) or in (E) a dislocation network. This raises the question of the nature of gbd sources the mechanism of escape from the potential well occupied by a dislocation in a network and how gbds can be created or destroyed at the intersection of a grain boundary and the external surfaces. However in a polycrystal in which the majority of of grains are enclosed (note not in thin films), crystal and grain boundary dislocations constitute a continuous interconnected network. The TEM reveals that there are numerous imperfections in this network with the topological properties of dislocations in an otherwise regular array. Such network defects have also been observed in arrays of misfit dislocations in epitaxial systems. (15) Whilst in principle a dislocation network might move as a whole, motion by the propagation of network defects, fig. 2, is much more plausible.

The interactions of gbds have much in common with those of crystal dislocations; gbds can bow between pinning points, interact with solutes, bypass grain boundary precipitates, dissociate and recombine before movement can occur on an alternate plane, pile up against obstacles etc. Since gbds are generally confined to the grain boundary plane the possibilities for forming jogs and kinks are limited and in general where dislocations intersect, or a point defect is added to a gbd a perturbation with combined jog and kink character is expected.

The diffusional fluxes when gbds move may be over distances comparable with the dislocation spacing or the grain diameter according to the type of process involved; eg growth of the spiral shown in fig. 3, which is a possible mechanism for grain boundary migration, requires a flux between the different leaves of the spiral and since $\mathbf{b}_1 + \mathbf{b}_2 + \mathbf{b}_3 = 0$ for no deformation overall, the process requires zero net flux. In contrast Coble creep (irrespective of the details of the atomic mechanism) requires the transfer of atoms from boundaries under compression to those under tension.

Experiments with bicrystals are frequently used to investigate properties of a specific grain boundary. To relate such results to the behavior of polycrystalline materials requires a suitable averaging procedure, for the variations in S, P and Δg^* and account to be taken of the influence of surrounding grains on the behavior of particular pairs of grains. Propagation of sliding past grain triple junctions requires dislocation reactions(8) analogous to, but more complicated than, those which occur in cross slip of an extended crystal dislocation. In a bicrystal, boundary migration may occur without the need for zero strain whilst in a polycrystal additional sources of dislocations must operate to minimize the overall deformation. Any variations of the energy required to transport atoms across a grain boundary appear in the activation energy Δg^*. Prediction of Δg^* and its dependence on solute segregation and grain boundary crystallography are only possible at the crudest level at present. However many aspects of the variation of grain properties such as sliding and mobility can be correlated at least qualitatively with the corresponding changes in defect structure.

Grain Boundary Sliding

The gbd model of grain boundary sliding is consistent with phenomenological knowledge of the sliding process. These general observations and their explanation in terms of the gbd model are given below. (8)

Sliding is intermittent and variable in a particular boundary. Dislocations travel different distances thus giving variable strains and are subsequently released by a thermally activated process, generally diffusion.

The activation energy observed for sliding sometimes appears to be that for grain boundary diffusion and in some other measurements corresponds to bulk diffusion. In the grain boundary dislocation description of sliding, the rate-determining process is the diffusion necessary to permit the climb of gbds. The diffusion path may be in the boundary or through the bulk, depending on the nature of the dislocations responsible for the boundary sliding.

Solutes and precipitates impede sliding but crystal deformation is impeded by solutes more than the sliding process. The gbds interact with solutes and precipitates in a way similar to that of crystal dislocations since, in the isotropic approximation, the elastic fields of gbds and climbing and, consequently, for a particular solute concentration, solutes would be expected to impede the glide of dislocations *relatively* more than climb.

The grain strain is often a function of the sliding strain. Crystal slip generates both gbd sources and obstacles to gbd movement.

Sliding is often accompanied by migration. This may be a consequence of the steps which exist at the cores of some gbds.

Low angle grain grain boundaries do not slide extensively. The constituent crystal lattice dislocations are free, between nodes, to respond separately to an external stress and move in their glide planes rather than the grain boundary plane.

Measurements on bicrystals suggest that there is a threshold stress for sliding and that slide hardening, analogous to work hardening, occurs. Possible origins for the threshold stress are breakaway of gbds from solute atmospheres or an increase in the density of gbd sources after the impingement of crystal lattice dislocations on grain boundaries.

Sliding rates calculated according to the gbd model are in agreement with those observed but the dependence of sliding rate on stress has not been fully explained.

Grain Boundary Migration

The observed variation of migration rates and driving force dependences are also within the ranges predicted by the gbd model. In addition many of the other experimental data can be rationalized. (16)

Impurities may impede grain boundary migration by pinning grain boundary dislocations and limiting the freedom of the boundary structural features to rearrange. diffusive transport may also be altered as a consequence of segregation.

A flux of vacancies is sometimes required for migration this flux feeds the dislocation climb process. A vacancy flux is not required for a step mechanism of migration although a high vacancy concentration would enhance diffusive transport in the boundary, where necessary.

High-angle grain boundaries generally migrate faster than low-angle boundaries and with a lower activation energy (grain boundary diffusion rather than bulk diffusion); both may migrate by a glide-climb process but diffusion is faster in high-angle grain boundaries.

$\Sigma = 7$ related tilt boundaries migrate more rapidly than corresponding twist boundaries since DSC dislocations give a collective process in the former but not the latter case so that the pre-exponential in the tilt case is larger. Diffusion induced grain boundary migration can be accounted for in terms of substitutional solution and grain boundary diffusion by a vacancy mechanism of solute on the hypothesis that the sources of vacancies required are gbds. Breakaway from the solute can occur since the driving force is high.

Point Defect Absorption

Limited observations of *variations* in width of void, dislocation loop and precipitate free zones adjacent to grain boundaries of different characters suggest that grain boundaries cannot all be regarded as perfect, neutral and unsaturable sinks as is usual. (13) A gbd mechanism for point defect absorption implies a bias and a variation of properties with either dislocation spacing or Burgers vector. Whilst the number of sites for point defect absorption increases as the dislocation density increases, the range of the elastic field of each dislocation generally decreases so that the observation of a maximum sink strength at a certain dislocation (17) density is not unexpected. The bias is also calculated to decrease as the dislocation density increases.

Conclusions

It is clear that many questions concerning dislocation mechanisms for grain boundary processes remain unanswered. Nevertheless, the dislocation description seems potentially to offer a satisfactory way to account systematically for the orientation dependence of many grain boundary properties.

A much deeper understanding of grain boundary properties can be expected as an increased knowledge of grain boundary structure is related to further precise data.

ACKNOWLEDGEMENTS

I am indebted to Drs. W.A.T. Clark, CRM Grovenor, A.H. King, R.C. Pond, K. Reading, A.P. Sutton, V. Vitek and Miss CMF Rae for the benefit of their insights in relating grain boundary structure to properties, through and after our interactions in Oxford and to Dr. K.N. Tu of IBM Yorktown for his unstinting help and encouragement since my arrival at IBM.

REFERENCES

1. B. Chalmers and H. Gleiter, Prog. Matls. Science, *16*, (1972) Pergamon, Oxford.
2. R.W. Balluffi, Y. Komen and T. Schober, Surf. Sci. *31*, 68. (1972).
3. A.P. Sutton and V. Vitek, Scripta Met. *14*, 129 (1980).
4. D.Y. Guan and S.L. Sass, Phil. Mag. A *39*, 293 (1979).
5. W.A.T. Clark and D.A. Smith, Phil. Mag. A, *38* 367 (1978).
6. G.R. Kegg, C.A.P. Horton and J.M. Silcock, Phil Mag. *27*, 1041 (1973).
7. A.H. King and D.A. Smith, Phil Mag A in press (1980).
8. R.C. Pond, D.A. Smith and P.W.J. Southerden, Phil. Mag. A *37*,27, (1978).
9. W. Bollmann, B. Michaut and G. Sainfort, Phys. Stat. Sol.(a), *13*, 637, (1972).
10. T. Schober and R.W. Balluffi, Phil. Mag. *24*, 165, (1971).
11. H. Grimmer, W. Bollmann and D.H. Warrington, Acta Crystallogr. *A30*, 197, (1974).
12. G. Herrmann, H. Gleiter and G. Bäro, Acta Metall. *24*, 353 (1976).
13. A.H. King, D. Phil thesis, Oxford University, England (1979).
14. H. Ichinose and Y. Ishida, Private communication.
15. D. Cherns and M.J. Stowell, Thin Sol. Films *29*, 127 (1975).
16. C.M.F. Rae and D.A. Smith, Phil. Mag. A *41*, 477 (1980).
17. A.H. King and D.A. Smith, Radiation Effects, in press.

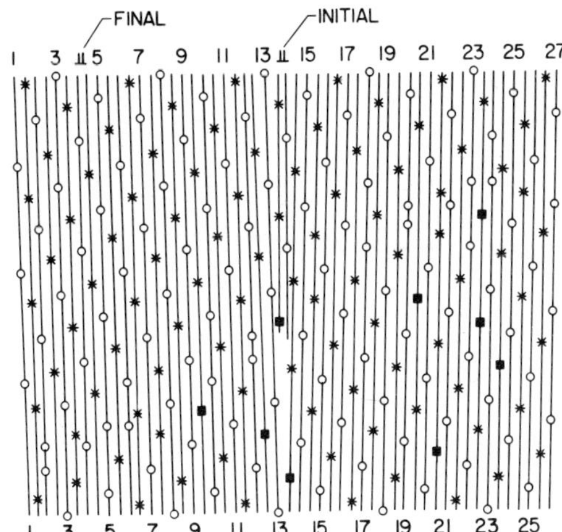

Fig. 1. Schematic diagram of a $\frac{1}{5}$<210> dislocation in a Σ = 5 related asymmetrical tilt boundary in a bcc metal. The ABAB-stacking is represented by the two kinds of symbols; the solid squares indicate atoms that might have been introduced by diffusion induced grain boundary migration.

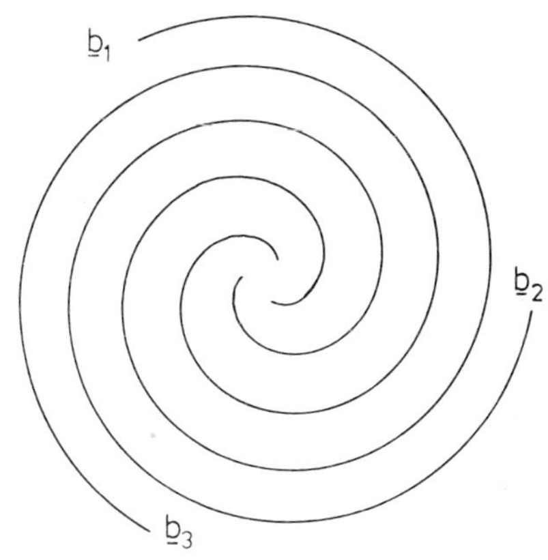

Fig. 3. Schematic diagram suggesting a spiral dislocation mechanism for migration such that there is no overall deformation .

(Courtesy CMF Rae and Philosophical Magazine)

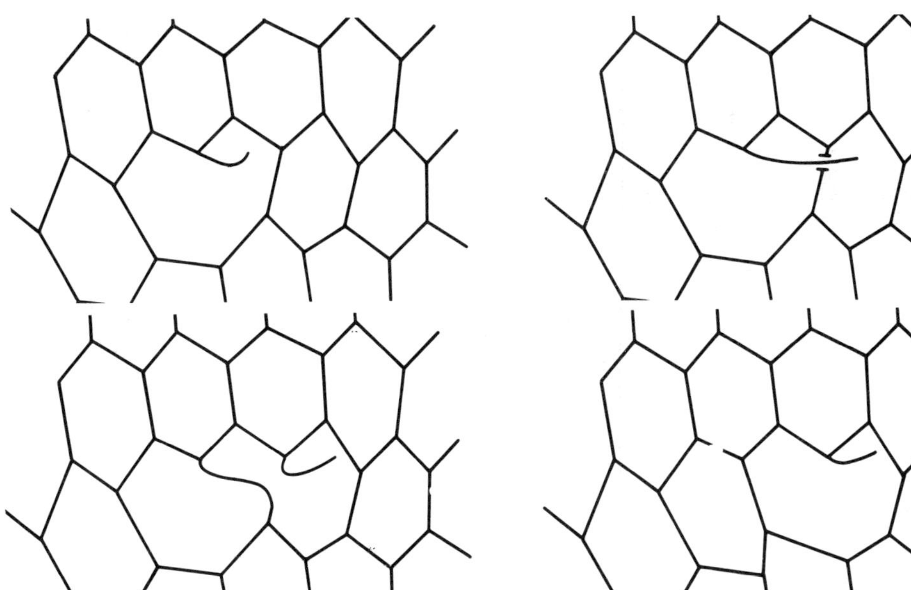

Fig. 2. Schematic diagrams suggesting how a dislocation may propagate through a network of dislocations.

THE ROLE OF VACANCIES IN THE KINETICS OF GRAIN BOUNDARY MOTION

K.LÜCKE, Y.ESTRIN, and G.GOTTSTEIN+)

Institut für Allgemeine Metallkunde und Metallphysik, RWTH Aachen, D-51 Aachen, West Germany

+) Materials Science Division, Argonne National Laboratory, Argonne, Illinois, USA

Abstract

The effect of vacancy production and annihilation on the kinetics of the grain boundary motion (GBM) has been considered for two different mechanisms of boundary displacement: (i) GBM resulting from direct atomic jumps across the boundary, without participation of vacancies; (ii) GBM associated with material transport across the boundary via vacancy diffusion. The vacancy drag experienced by a moving tilt boundary due to an elastic interaction with its vacancy atmosphere has been estimated. A Kirkendall-type phenomenon accompanying the GBM has been predicted for mechanism (ii).

Introduction

Two basically different mechanisms of the grain boundary motion (GBM) should be considered when studying the microscopic kinetics of boundary displacement:
(i) The GBM occurs by direct jumps of atoms across the boundary, without participation of vacancies. The vacancies can be considered here as "foreign atoms" (except that their total number may change due to sinks and sources in the boundary and in the crystal bulk).
(ii) The atom transport from one phase to the other across the grain boundary (GB), required for the GBM, can occur only via vacancy jumps.

In both cases one of the most important features of the GBM is the vacancy generation accompanying it. Vacancies can be produced or annihilated at sources and sinks within a GB, the GB structure being reproduced and thus remaining unchanged(1). The rate of the vacancy production within a GB is determined by the difference of the thermal equilibrium, c_{Bo}, and the actual, c_B, vacancy concentration in the GB and by the average life time τ of a vacancy within the GB. This mechanism of vacancy generation can be operative in a stationary as well as in a moving GB. Another one is of purely dynamic origin and is connected with the consumption of dislocations of the cold worked phase by a moving GB. The surplus volume of this phase associated with dislocations is divided up into individual vacancies (2,3).

In the present report we will illustrate the role of the vacancy generation for mechanism (i) of the GBM by considering the elastic interaction between vacancies and a moving GB. The analysis will follow our treatment as suggested in Ref. 4, based on a continuous diffusion equation for vacancies and providing a means of determining self-consistently both the steady state GB velocity and the vacancy concentration profile around the moving GB. A quite new approach to the investigation of mechanism (ii) which is based on a one-dimensional discrete analogue of the diffusion equation will be also presented.

A Self-Consistent Method for Determining the Steady State GB Velocity and the Vacancy Concentration Profile Around a Moving GB in Case (i)

As indicated in the Introduction, mechanism (i) will be treated along the lines of Ref. 4. We confine ourselves to a one-dimensional system: a planar GB moving uniformly from left to right in the positive direction ξ normal to the GB plane. A balance between the production and the removal of vacancies via the diffusion and annihilation at some sinks in the bulk leads to establishing a steady state regime of the GBM. Denoting the velocity of this steady state motion as v, we can represent the source term in the diffusion equation for vacancies,

$$\dot{c} = D\partial^2 c/\partial\xi^2 - \beta(c-c_o) + \dot{q}, \quad (1)$$

as

$$\dot{q} = \{\frac{b}{\tau}(c_{Bo}-c_B)+Qv\}\delta(\xi-vt). \quad (2)$$

Here c is the volume concentration of vacancies, c_o its thermal equilibrium value, D the diffusion coefficient, Q the vacancy production capacity of the GB (i.e. the vacancy concentration that would be left behind the GB in the absence of vacancy removal), β a phenomenological parameter determining the annihilation rate of vacancies, b the lattice constant. As discussed above, $\dot{q}$ is made up of two production terms. The δ-function indicates that the vacancy generation is confined to a narrow region where the GB is located at a given time t. The boundary conditions are chosen in the form

$$c = c_o \text{ at } \xi\to\pm\infty, \text{ assuming}$$

thermal equilibrium far from the GB which is maintained due to bulk sinks incorporated in the second term on the right-hand side of Eq. (1).

Going over to the moving refernce frame connected with the GB, $x=\xi-vt$, and looking for a steady state vacancy concentration distribution in the new coordinate system, we obtain for the surplus vacancy concentration $C=c-c_o$:

$$DC'' + vC' - \beta C + \{\frac{b}{\tau}(c_{Bo}-c_o)-\frac{b}{\tau}C_B+Qv\}\delta(x) = 0 \quad (3)$$

where a prime indicates differentiation with respect to x and $C_B=c_B-c_o$. Eq. (3) can be easily solved for x<0 and x>0. With the boundary condition C=0 at $x\to\pm\infty$ and the condition of continuity at x=0, one has

$$C = C_B\exp(k_1x) \quad \text{for } x<0, \quad (4)$$

$$C = C_B\exp(-k_2x) \quad \text{for } x>0 \quad (5)$$

where $k_1=\{(v^2+4\beta D)^{1/2}-v\}/2D$; $k_2=\{(v^2+4\beta D)^{1/2}+v\}/2D$. The integration constant C_B can be found from the condition that in the steady state, the sum of the diffusional fluxes from the GB into the bulk (in both directions), $DC_B(k_1+k_2)_{x=0}$, be equal to the number of vacancies produced by a unit area of the GB per unit time, $\frac{b}{\tau}(c_{Bo}-c_o)-\frac{b}{\tau}C_B+Qv$. If v were determined by some independent mechanism indifferent to the presence of vacancies (no vacancy-GB interaction), this relationship between the GB velocity and the vacancy concentration in the GB whould readily give the value of C_B (and the whole vacancy distribution profile, Eqs. (4), (5)):

$$C_B = \{Qv+\frac{b}{\tau}(c_{Bo}-c_o)\}/\{(v^2+4\beta D)^{1/2}+\frac{b}{\tau}\}. \quad (6)$$

The above-mentioned condition provides a convenient means of determining steady state regimes of the GBM also in a general case where the GB velocity v may depend on the vacancy concentration. Such a dependence gives an additional relationship between v and C_B which enables one to determine self-consistently both the GB velocity and the vacancy concentration profile.

To simplify the consideration, we confine ourselves to the case of a large life time in the GB, omitting the terms containing b/τ in Eq. (6). The latter can be rewritten in a convenient form

$$(1/f^2-1)^{1/2}s(f)=\nu \qquad (7)$$

Here the following notation has been introduced: $f=C_B/Q$, $v=v_o s(f)$ where $s(f)$ represents the concentration dependence of the GB velocity, v_o being a concentration-independent factor (bearing the dimensionality of v); the dimensionless vacancy removal rate ν is equal to $2(\beta D)^{1/2}/v_o$.

It is easily seen that the functional dependence of the left-hand side of Eq. (7) on f determines whether a steady state can exist for any values of the parameter ν. If this function monotonically decreases from infinity to zero as f varies from zero to unity, there exists one, and only one, solution of Eq. (7) determined by the intercept of its plot with the horizontal line corresponding to ν. A rapidly increasing function s(f) may result in a non-monotonic f-dependence of the left-hand side of Eq. (7) which, under certain conditions, can lead to the existence of more than one steady state regime of the GBM (4,5). In what follows, the case of a monotonically decreasing left-hand side of Eq. (7) is realized.

Vacancy Drag of a Tilt Boundary

Due to the asymmetric vacancy distribution around a moving GB, a net force on the GB is exerted by the ensemble of vacancies, provided a force interaction between a vacancy and a GB exists. This force, superimposed on the driving force of the GBM, can retard the moving GB (attractive interaction) or accelerate it (repulsive interaction) (4).

Here we consider the retardation of a tilt boundary by the vacancy atmosphere produced due to its motion (the vacancy drag). The tilt boundary is constructed of equidistant infinite edge dislocations lying along the z-direction in the yz-plane, with a spacing h. As is well-known, the interaction potential $U=U(x,y)$ between such a boundary and a vacancy with coordinates (x,y) has the form (6,7)

$$U = U_o \sin 2\pi\frac{y}{h}/(\cosh 2\pi\frac{x}{h}-\cos 2\pi\frac{y}{h}) \qquad (8)$$

where U_o has the order of $Gb^3(b/h)$ (G is the shear modulus). This potential, short-range with respect to x, is periodic with respect to y. It is reasonable to assume that within the action radius of the potential (8) (of order h) the vacancies rearrange in the y-direction lying in the yz-planes corresponding to the minima of the potential U with respect to y. In these planes, the x-dependence of their interaction potential with the GB is described as

$$U_m=-U_o/\sinh(2\pi|x|/h) \qquad (9)$$

indicating the repulsive nature of the interaction.

With this interaction potential, the average "vacancy force " p_{vac}, acting on a unit area of the moving GB from the vacancy atmosphere, can be estimated. This vacancy force can be represented in the form

$$p_{vac} = \int_{-\infty}^{\infty} C(x)\frac{dU_m}{dx}dx = C_B\int_0^{\infty}\{\exp(-k_2x) - \exp(-k_1x)\}\frac{dU_m}{dx}dx \qquad (10)$$

To estimate this integral, the characteristic distances of the decay of the vacancy concentration, k_1^{-1} and k_2^{-1}, should be considered. Representing the annihilation rate constant β as $D\rho$, we see that at sufficiently low GB velocities $k_{1,2}^{-1}$ have roughly the order of magnitude of $\rho_{1,2}^{-1/2}$ ($\rho_{1,2}$ being the dislocation densities at the left and the right side of the GB, respectively), $k_{1,2}h \ll 1$. Since the characteristic decay distance of the vacancy-GB interaction force has the order of h, a series expansion of the exponentials in Eq. (10) is possible. Retaining the linear terms of this expansion and truncating the integral at the lower limit of $x \cong b$, we get an estimate for p_{vac}:

$$p_{vac} \cong C_B\rho_2^{1/2}hU_o \cong (\rho_2 b^2)^{1/2}(Gb^3)C_B. \qquad (11)$$

Denoting the unperturbed driving force for the boundary motion as p and the boundary mobility as m, we have for the boundary velocity:

$$v = v_o(1 - \chi f) \qquad (12)$$

where $v_o = mp$ and $\chi = (b^2\rho_2)^{1/2}(Gb^3)Q/p$; $s(f) = 1-\chi f$.

Due to the difference in the vacancy removal rates ν_1 and ν_2 in the recrystallized and the cold work phases, respectively, and because of our assumption of sufficiently low boundary velocities, another form of Eq. (7) should be used:

$$2(1/f - \chi) = \nu_2. \tag{13}$$

f can be readily determined from Eq. (13). Substitution of this value into the expression for the retarding vacancy force $p_{vac} = \chi f p$ gives

$$p_{vac}/p = 2\chi/(2\chi+\nu_2) \cong 2\chi/\nu_2 \tag{14}$$

Eq. (14) shows that p_{vac} can constitute a substantial part of the driving force ($Q \cong \rho/b$ may be taken for an estimate). Thus the elastic interaction between a moving tilt boundary and its vacancy atmosphere may result in a noticeable dragging effect (up to several percent of the p-values for recrystallization).

Kirkendall-Type Effects Accompanying the GBM in Case (ii)

For the mechanism (ii) of the GBM, we neglect the force interaction between a GB and its vacancy atmosphere to concentrate on the diffusion kinetics of the GBM. We introduce a discrete model that can be illustrated as follows:

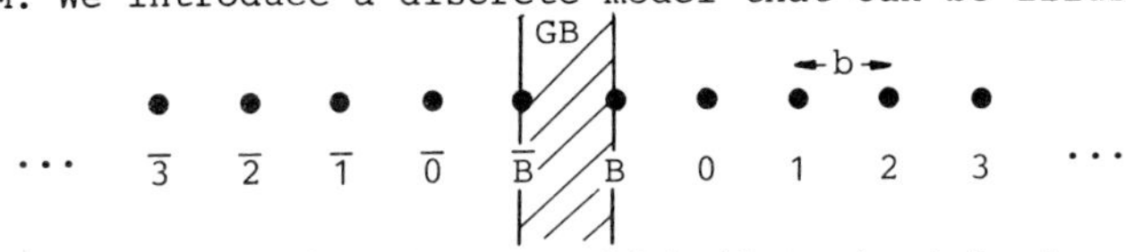

GB ← b →

... $\bar{3}$ $\bar{2}$ $\bar{1}$ $\bar{0}$ $\bar{B}$ B 0 1 2 3 ...

The dots mark here the sites at which the material planes are located. As will be shown below, the right half-crystal is shortening whereas the left one is lengthening. It is convenient to consider the right (B) and the left ($\bar{B}$) GB planes at rest, the material planes moving from right to left with some velocity v_2 in the right half-crystal (crystal II) and v_1 in the left half-crystal (crystal I). The material arriving at plane B disappears at some rate annihilating vacancies at the same rate (it is the assumed property of GB's that their structure remains unchanged in this process). Analogously, material emerges from plane $\bar{B}$, and vacancies are produced with the same rate. The vacancy migration will be described by a discrete chain of rate equations representing a discrete analogue of the diffusion equation. A "convection flux" of vacancies associated with the motion of materials planes from right to left will be taken into account.

It is beneficial to consider the both half-crystals separately. Starting from crystal II, we introduce the vacancy concentration at site n, c_n, i.e. the vacancy concentration in the material plane located at a given time at site n (see the Fig.), and consider a steady state (c_n independent of time). All concentrations are defined per atomic site.

The chain of equations for n=0,1,2,... reads as follows

$$\nu_o\{c_{n-1} + c_{n+1} - 2c_n + (\psi-1)c_B\delta(n)\} + \frac{v_2}{b}(c_{n+1} - c_n) = 0 \tag{15}$$

with the boundary conditions that the vacancy concentration in the B-plane is given as c_B and that far from the GB ($n\to\infty$) is given as c_i. Here ν_o denotes the unperturbed jump frequency of vacancies between the neighbouring sites, the factor ψ indicates that the jumps from the plane B into plane 0 occur at a reduced frequency owing to the existence of a potential well for the vacancies in the GB. $\delta(n)$ is equal to unity if n=0 and is equal to zero otherwise. The solution of Eq. (15) with these boundary conditions reads (8):

$$c_n = c_i + (\psi c_B - c_i)/(1+v_2/b\nu_o)^{n+1} \tag{16}$$

For crystal I, the solution of the analogous chain of equations leads to the result

$$c_{\bar{n}} = \text{const} = c_{\bar{B}}(\psi+v_1/b\nu_o)/(1+v_1/b\nu_o) \tag{17}$$

where $c_{\bar{B}}$ is the vacancy concentration in the $\bar{B}$-plane.

To determine the "integration constants" c_B and $c_{\bar{B}}$, we consider the detailed balance in plane B and plane $\bar{B}$. For our model consideration it is assumed that the thermal equilibrium vacancy concentration c_o is equal for crystals I and II. Accordingly, the equilibrium concentrations within planes B and $\bar{B}$ are $\psi^{-1}c_o$. The rate of vacancy annihilation in plane B (each vacancy carrying away a lattice site resulting in the shortenung of crystal II), $(c_B-\psi^{-1}c_o)/\tau\nu_o$, is equal to the dimensionless rate of shortening of crystal II:

$$(c_B-\psi^{-1}c_o)/\tau\nu_o = v_2/b\nu_o. \qquad (18)$$

A similar relation between the rate of lengthening of crystal I and the rate of vacancy generation in plane $\bar{B}$ reads:

$$(\psi^{-1}c_o-c_{\bar{B}})/\tau\nu_o = v_1/b\nu_o. \qquad (19)$$

Following the original theory of Lücke et al. (9,10), we assume that the jump frequency for vacancies from $\bar{B}$-plane into B-plane is by a factor of g larger than ν_o. Here $g=\exp(pb^3/2kT)$, kT has the usual meaning. The frequency of backward jumps from plane B into plane $\bar{B}$ is, inversely, smaller than ν_o by a factor g^{-1}. This asymmetry incorporates the effect of the driving force p.

With these model assumptions, the equations of detailed balance for the two GB planes,

$$c_{\bar{B}}g + c_F - c_B g^{-1} - c_B\psi + (v_2/b\nu_o)c_F + \gamma(\psi^{-1}c_o-c_B) = 0 \qquad (20)$$

$$c_{\bar{F}} + c_B g^{-1} - c_{\bar{B}}\psi - c_{\bar{B}}g - (v_1/b\nu_o)c_{\bar{B}} + \gamma(\psi^{-1}c_o-c_{\bar{B}}) = 0, \qquad (21)$$

combined with Eqs. (18), (19), determine self-consistently the boundary concentrations c_B and $c_{\bar{B}}$ (together with the whole vacancy concentration profile, Eqs. (16), (17)) and the velocities v_1 and v_2. In Eqs. (20), (21) the abbreviation $\gamma=(\tau\nu_o)^{-1}$ has been used. To avoid confusion, we denoted the concentrations in planes 0 and $\bar{0}$ as c_F and $c_{\bar{F}}$, respectively. Substituting these quantities from Eqs. (16), (17) and excluding from the obtained set of equations all unknowns but v_1, we get a third-order equation whose coefficients show that it has only one positive root. For small velocities, $v_1/b\nu_o \ll \psi$, this root can be represented as

$$v_1=\psi^{-1}b\nu_o\{\gamma c_o(pb^3/kT)(1-c_i)\}/\{g(1-c_i)+g^{-1}(1-c_o)+\gamma(1-c_i)(1-c_o)\}. \qquad (22)$$

The equation for v_2 has a similar form:

$$v_2=\psi^{-1}b\nu_o\{\gamma c_o(pb^3/kT)(1-c_o)\}/\{g(1-c_i)+g^{-1}(1-c_o)+\gamma(1-c_i)(1-c_o)\}. \qquad (23)$$

It is interesting to note that generally $v_1 \neq v_2$, their ratio depending on the relation between c_o and c_i: $v_1/v_2 = (1-c_i)/(1-c_o)$. Since after cold work c_i is typically larger than c_o, v_1 is smaller than v_2. It means that the rate of shortening of crystal II is higher than the rate of lengthening of crystal I. We term this phenomenon "Kirkendall-type GBM". A detailed treatment of it will be given elsewhere (8).

References

1. K.Lücke and G.Gottstein, Acta Met. (in the press)
2. C.S.Smith, cited by W.W.Mullins, Acta Met. 4, 421 (1956)
3. K.Lücke and H.P.Stüwe, In: Recovery and Recrystallization of Metals, ed. by L.Himmel, Interscience, New York (1963), p. 171
4. Y.Estrin and K.Lücke, Acta Met. (in the press)
5. B.V.Petukhov and Y.Estrin, J Phys. F: Metal Physics 10, L223 (1980)
6. S.Amelinckx and W.Dekeyser, Solid State Physics 8, 325 (1959)
7. J.P.Hirth and J.Lothe, Theory of Dislocations, p. 670, McGraw Hill (1968)
8. Y.Estrin, K.Lücke, and G.Gottstein (to be published)
9. K.Lücke and K.Detert, Acta Met. 5, 628 (1957)
10. K.Lücke and H.P.Stüwe, Acta Met. 19, 1087 (1971)

A PRELIMINARY STUDY OF PERSISTENT SLIP BANDS IN AgCl

S. L. Ogin and L. M. Brown
Cavendish Laboratory, Madingley Road, Cambridge CB3 OHE

Abstract

Optically transparent and stress birefringent AgCl single crystals have been subjected to cyclic deformation in an attempt to study persistent slip and crack formation. Preliminary results indicate that persistent slip bands behave rather like those in metals, and that they are accompanied by internal stresses whose principal axes are parallel and perpendicular to the plane of the band.

Introduction

Recent work on the fatigue of metals at low plastic strain amplitudes has centred on the nucleation and properties of persistent slip bands - those regions of the crystal which remain active after cyclic hardening has produced the state of cyclic saturation. The two-phase system of persistent slip bands and relatively inactive matrix has been observed in a wide variety of crystal structures and temperatures (1,2,3,4,5,6). The PSBs are thought to act as sites for crack nucleation; recent observations strongly support this idea (7).

Transmission electron microscopy has revealed that the persistent slip bands in copper are composed of walls of dislocation dipoles of predominantly vacancy type (8). This observation suggests that the PSBs may have a macroscopic internal stress associated with them, caused by the greater concentration of vacancy dipoles in the PSB than in the matrix, and that this internal stress would encourage crack nucleation at the interface between the persistent slip band and the matrix structures (8).

Silver chloride is a ductile ionic crystal which has been used many times in deformation studies and has been called a 'model' metal. It possesses the useful property of stress birefringence which enables internal stresses to be directly observed.

Several workers have used the material in fatigue studies (9,10,11) where they have noted properties similar to those in fatigued metals: the formation of extrusions and the appearance of PSB-type structures. It is therefore hoped that a more comprehensive study of fatigue in silver chloride will help to clarify current ideas about the role of persistent slip bands and internal stresses in the fatigue of metals.

Experimental Methods

Single crystals of silver chloride were grown by the Bridgman method from 99.9% nominal purity material (supplied by Johnson, Matthey & Co. Ltd.). The crystals were grown in a split boron nitride mould in a vertical temperature gradient in a vacuum of 10^{-2} - 10^{-3} torr. All the crystals had their long axes within 5° of [100] and had faces within 5° of {100} except crystal S10 which had faces 16° from {100}. All the crystals had small angle boundaries which gave them a striated appearance, but the misorientations were not greater than 4°. The crystals were mechanically polished with emery paper and then chemically polished in molar potassium cyanide solution. The etchant for the etch pit studies on the {100} faces was 3N sodium thiosulphate solution. After polishing, the gauge sections of the crystal were all about 6 mm x 4 mm, with a length of 18 mm. The crystals were not annealed after growth.

Silver chloride slips by pencil glide at room temperature (12,13) so any plane containing a <110> direction is a possible glide plane. However, the [100] - axis orientation encourages slip on the two orthogonal pairs of 100 planes which lie at 45° to the tensile axis (see figure 2).

The crystals were fatigued at constant plastic strain amplitudes in a modified Instron 1253 dynamic testing machine. Surface and birefringence observations were made with an MeF Reichert polarising microscope in transmission and reflection.

Results

§ (1) Cyclic hardening and saturation

i. Hardening curve

The general features of a typical cyclic hardening curve at low plastic strain amplitude ($0.1\% < \varepsilon_p < 0.5\%$) are shown in figure 1. The precise shape of the hardening curves depends on both the magnitude of ε_p and the strain rate of the test. The curve shows the mean value of the compressive and tensile stress (the tensile peak stresses in the tests were usually about 4% higher than the peak stresses in compression).

After an initial period of rapid hardening, the crystals soften, reaching a minimum resolved stress of 2.2N mm^{-2}. The crystal then hardens, but not smoothly and at a much lower rate, until a saturated and constant stress of 2.8N mm^{-2} is reached after about 10^3 cycles. The hysteresis loops are square shaped for the first 300 cycles, but then become pointed during the slow hardening phase. Figure 1 shows inset a hysteresis loop at saturation.

ii. Appearance of slip traces

Coarse slip traces are first noticed at the edges of the crystal at about 10^3 cycles. These traces multiply and spread across the gauge face of the upper part of the crystal over the next 2500 cycles. By cycle 3500, the slip traces have stopped spreading. The progress of these coarser slip traces is very similar to the appearance and propagation of persistent slip bands in copper (14).

iii. X-ray asterism

Back reflection Laue photographs before and after fatiguing do not show any increase in the asterism of the spots. In fact, the asterism if anything is reduced by fatigue.

iv. Saturation stress and strain rate dependence

The crystals show a strong rate dependence of the saturation stress, very reminiscent of body-centred cubic metals (16). The saturation stress depends logarithmically on strain rate, a factor of 10 in strain rate producing a 10% increase in saturation stress.

§ (2) Microstructure

i. Division of crystal volume into blocks

An examination of the crystals fatigued into saturation at various plastic strain amplitudes indicates that each crystal was divided by the deformation into several blocks. The division of the crystal into blocks was confirmed by surface observations both before and after etching and by birefringence observations. The blocks extend between the parallel surfaces of the crystal and have the following general features (see figure 3):

(a) In each block most slip occurs on only one pair of orthogonal glide systems and an adjacent block glides mostly on the conjugate glide systems. In effect, the deformation structure of each block is rotated 90° with respect to the other block, and hence the surface for which screw dislocations emerge for one block is the surface for which edge dislocations emerge for the other block and vice-versa.

(b) The blocks interpenetrate by extending finger-like projections into each other and the boundaries between blocks are often sharp but not crystallographic.

(c) Slip traces on the surface of a block tend to be interrupted by the fingers from a neighbouring block.

Figure 4 shows the surface features associated with the block structure. Figure 6(a) shows one of the most characteristic etching structures at high magnification. On the crystal face which contains the operative slip vectors, a 'shredded wheat' or 'maze-like' structure appears, well-organised and associated with a well-defined spacing. The structure consists of walls on cube planes, as demonstrated by etching figures on the orthogonal face, i.e. the face with which the slip directions make angles of approximately 45°. This is shown schematically in fig. 3. As fig. 6(a) shows, this structure is interrupted by strong bands parallel to a single slip vector.

ii) Persistently slipping bands and extrusions

Within each block of the crystal, features can be recognised which are characteristic of persistent slip bands. If a crystal is polished and etched to reveal the structures just described, and then subjected to a few hundred further cycles of deformation, coarse wavy extrusions and possibly slip steps reform where the maze-like structure is interrupted. Figure 5 shows the correlation between the etching structures and the slip markings. Another characteristic feature of persistent slip is the increasing coverage of the crystal surface by persistent slip bands as the plastic strain amplitude is increased (1); this feature is also observed in AgCl.

iii) Birefringence experiments

The crystals were examined by transmitted polarised light and with a sensitive tint plate. The tint plate converts stress birefringence normally seen as black and white, to colour, with the colouration depending on whether the stresses in the crystals are tensile or compressive.

It has been difficult to obtain convincing results on the birefringence of the fatigued crystals because of the wavy nature of the glide planes at room temperature, which means that usually the birefringence is sampled not from a single glide plane, but from a series of corrugated planes. However, the following basic results seem to be fairly well established:

(i) A birefringence is associated with the persistent slip bands, that is, disorganised etch pit bands which run in <110> directions through the maze-like structure; fig. 6(b) shows this.

(ii) The birefringence has maximum brightness when the bands are oriented at 45° to the crossed polarisers. Almost complete extinction occurs when the bands are parallel or perpendicular to the polarisers, indicating that the principal stresses in the bands are parallel or perpendicular to the bands.

(iii) The tint plate indicates that the symmetry of the birefringence in the bands is probably different from that in a simple glide plane. In a persistent slip band, the colouration is relatively uniform across the band, whereas it changes discontinuously from red to blue across a glide band. Often the bands and the matrix appear with different colourations; the bands, for example, appearing blue against a matrix of yellow, implying that the bands and the matrix are in opposite states of stress.

(iv) The colouration of the bands and the matrix seems to be dependent upon the part of the cycle in which the test has been stopped. For example, blue bands against yellow matrix compressive strain (zero load), become yellow bands against blue matrix at tensile strain (zero load) suggesting a reversal in the signs of the internal stresses around the hysteresis loop.

§ (3) Wall Spacing and Saturation Stress

Wall spacings (d) within the regularly spaced ladder structure of PSBs in other fatigued materials (e.g. Cu, Mg) have been found to be related to the saturation stress (σ_{sat}) by the expression: $\sigma_{sat} = (3 \pm 1)\ \mu b/d$ (5). (μ is the shear modulus, taken as $1.2 \times 10^4 Nmm^{-2}$, and b is the Burgers vector equal to 3.9 Å for AgCl.)

This relationship is also applicable to fatigued silver chloride if the wall spacings between the etch pits in what appears to be the 'matrix' region are used; the 'matrix' region being the maze-like structure of fig. 6 (a).

Table 1

Crystal	$\dot{\varepsilon}_p$ (s^{-1})	ε_p	σ_{sat} (N/mm^2)	d (μm)	$\frac{\sigma_{sat} d}{\mu b}$
S4	.024	0.3%	2.71 ± .05	4.8 ± .2	2.8
S7	.016	0.2%	2.85	4.7	2.9
S2	.016	0.2%	2.46	5.6	2.9
S7	.0016	0.2%	2.30	5.6	2.8

Table 1 shows some results from a variety of crystals. The product σ_{sat} d is much more constant than either σ_{sat} or d taken alone, and it seems probable that the spacing of the walls in the maze-like structure plays the same role as the wall spacing in the ladder structures observed in metals. As is common with these observations, there is some unexplained variability which may perhaps be attributed to variations in the impurity content of the crystals. Further data will be worthy of more careful analysis, following Basinski et al. (6).

Discussion

The cyclic deformation of silver chloride appears to have properties which are similar to both f.c.c. and b.c.c. materials. The strong dependence on strain rate is characteristic of b.c.c. metals, for example Fe, Mo, Nb (16). The formation and propagation of persistent slip bands is similar to f.c.c. copper. Most remarkable, however, is the observation revealed by etching that the relatively inactive matrix has an ordered structure, and active PSBs have disordered structure. This is the reverse of all observations hitherto. It is possible, of course, that the etch pit technique is not sensitive enough to detect structure in the PSBs.

The division of the crystal into blocks is probably due to grip effects - slight misalignment of the grips or grip constraints. The fact that the boundaries between the blocks are non-crystallographic suggests that slip in cycled silver chloride is fine and homogeneous, at least before cyclic saturation is reached. There must be very little incompatability between blocks. The well-ordered maze-like structure or 'matrix' structure appears to account for the cyclic hardening and in an as yet ill-understood way plays the role of wall structure in metals.

The birefringence results are as yet incomplete and difficult to interpret because of the wavy glide. However, they demonstrate the existence of macroscopic internal stresses associated with persistent slip.

Acknowledgments

SLO is grateful to the S.R.C. for financial support. It is a pleasure to thank Dr. A. T. Winter and Mr. A. Robinson for their help.

1. A. T. Winter, Phil. Mag. 30, 719 (1974).
2. P. J. Woods, Phil. Mag. 28, 155 (1973).
3. H. Mughrabi, Mat. Sci. Eng. 33, 207 (1978).
4. A. Robinson in L. M. Brown, Metals Science, p 315 Aug/Sep (1977).
5. R. Kwadjo and L. M. Brown, Acta Metall. 26, 1117 (1978)
6. Z. S. Basinski, A. S. Korbel and S. J. Basinski, Acta Metall. 27, 191 (1979).
7. K. Katagiri, A. Omura, K. Koyanagi, J. Awatani, T. Shiraishi and H. Kaneshiro, Met. Trans. 8A, 1769 (1977).
8. J. G. Antonopoulos, L. M. Brown and A. T. Winter, Phil. Mag. 33, 87 ((1976).
9. B. A. Kuznetsov, Sh. Tek. Fiz. 26, 5, 1034 (1959).
10. P. J. E. Forsyth, J. Inst. Met. 82, 449 (1953).
11. R. W. K. Honeycombe, Proc. Roy. Soc. A242, 213 (1957).
12. J. F. Nye, Proc. Roy. Soc. A198, 190 (1949).
13. R. P. Carnahan, T. L. Johnston, P. J. Stokes and C. H. Li, Trans. Met. Soc. AIME 221, 1, 45 (1961).
14. A. T. Winter, Ph.D. Thesis, Cambridge (1974)
15. P. J. E. Forsyth, The Physical Basis of Metal Fatigue, (Blackie) p 49 (1969).
16. A. Robinson, Private communication.

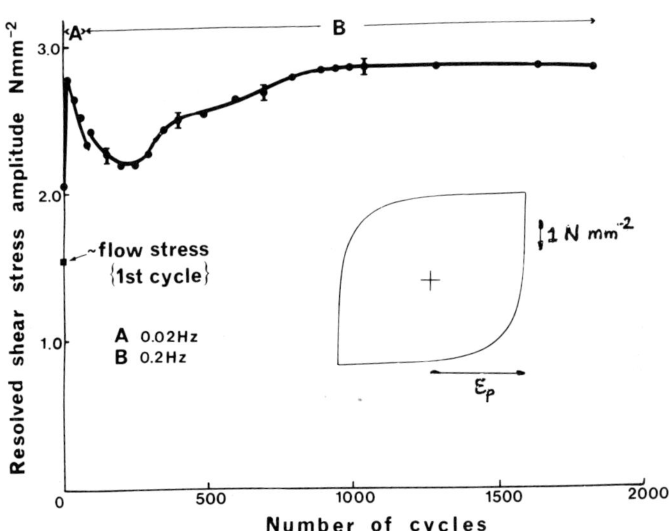

FIG. 1

Hardening curve of crystal S10. Inset shows a hysteresis loop taken at cycle 1500.

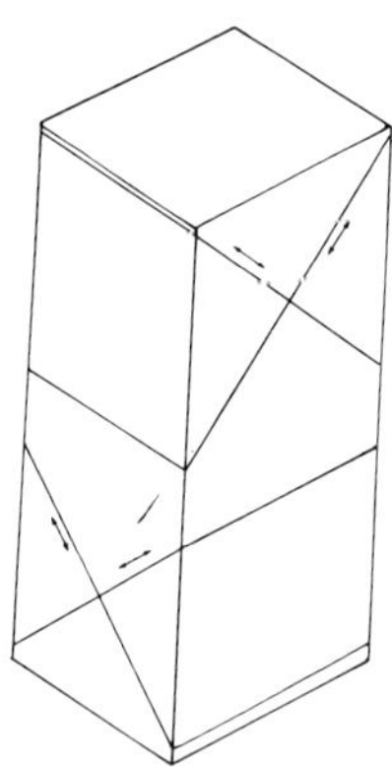

FIG. 2

The slip geometry of AgCℓ crystals: Equally strained <1$\bar{1}$0>{110} glide systems.

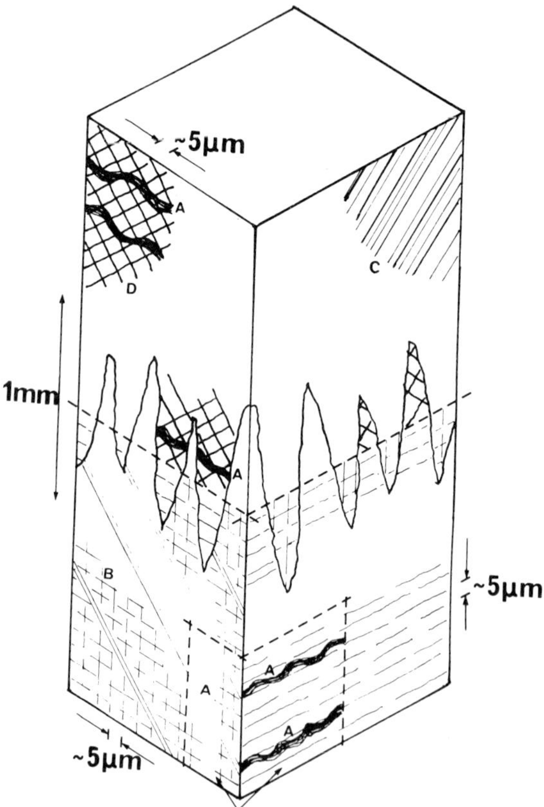

FIG. 3

Schematic diagram of two interpenetrating blocks showing finger-like projections. Areas above the central dashed lines show the microstructural details after fatiguing to saturation. Areas below the dashed lines show the etching microstructure.

A. Straight bands represent the disorganised etch pit structures which have been identified as the persistent slip bands (PSBs).Wavy slip markings on adjacent faces are also shown.
B. Well organised maze-like structure which is interrupted by the persistent slip bands.
C. Fine slip markings parallel to the operative slip vector. When etched, some of these markings are revealed as PSBs.
D. Slip markings appearing on same surfaces as wavy PSBs; these seem less imprtant and are not described in the text. Their interpretation is obscure at present.
E. Adjacent faces of the crystal showing the reappearance of slip markings coinciding with the disorganised etch pit bands after refatiguing. See also Fig. 5.

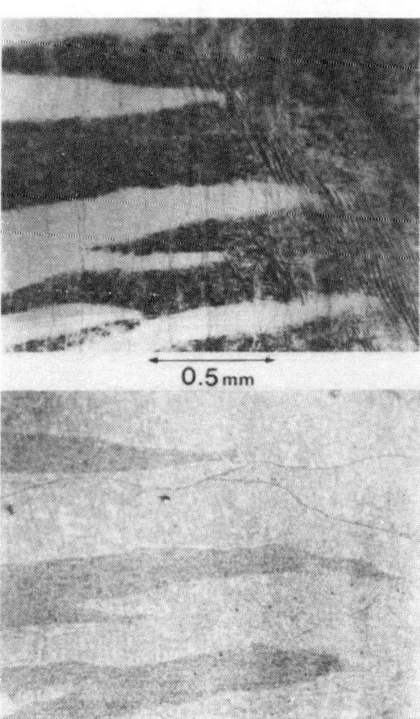

FIG. 4

Finger-like interpenetrating blocks, as illustrated in fig. 3. Surface features before etching (above) and after etching (below) are shown.

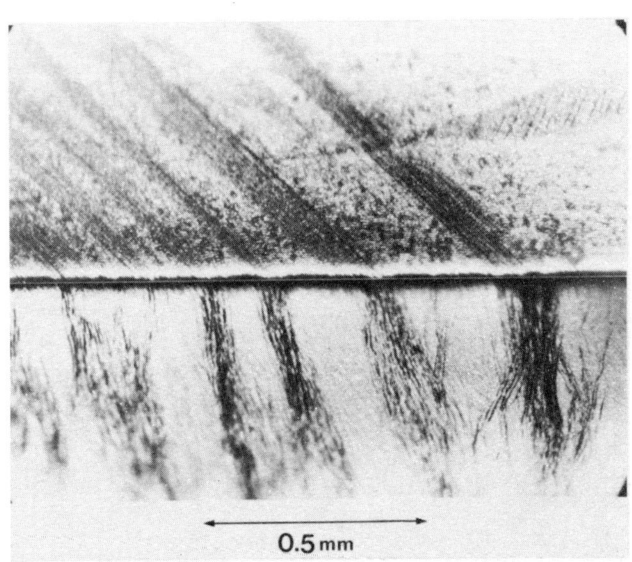

FIG. 5

Two adjacent surfaces of a crystal which has been etched and then refatigued. Disorganised etch pit bands correspond with wavy slip markings on an adjacent face, see Fig. 3, region E.

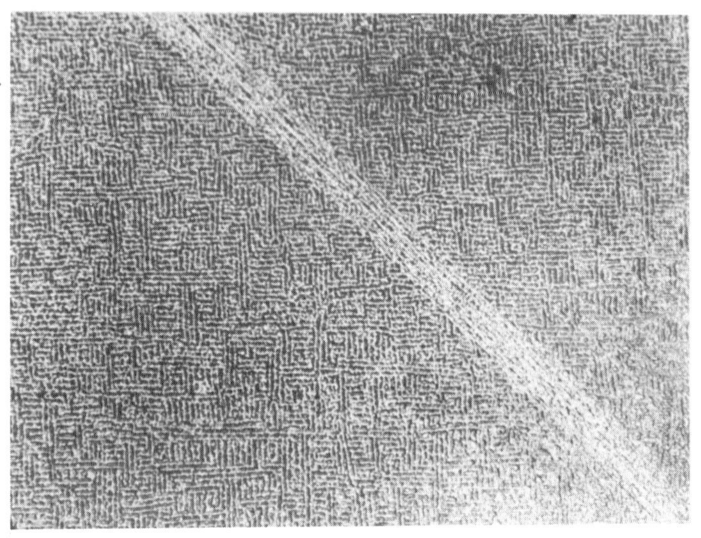

0.2 mm

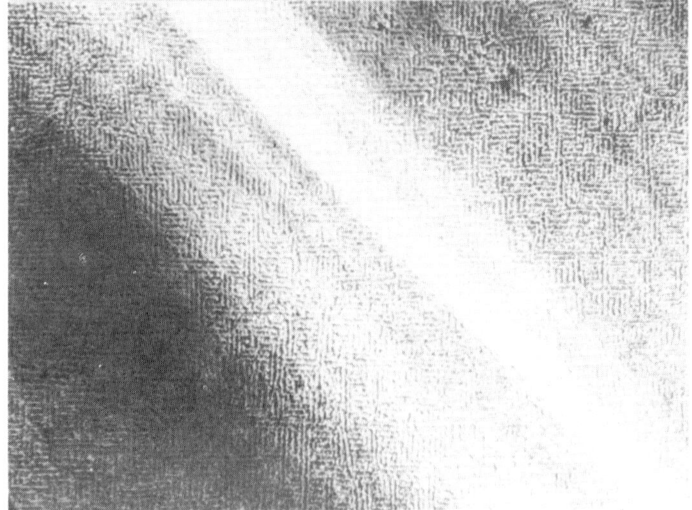

FIG. 6

a) Characteristic etching (maze-like) structure on surface containing operative slip vector. Structure is interrupted by a strong band of disorganised etch pits parallel to a single slip vector (persistent slip bands).

b) Birefringence associated with the persistent slip bands. The crossed polarisers are parallel to the edges of the photograph.

INDEX